The Anthropocene: Politik—Economics—Society—Science

Volume 30

Series Editor

Hans Günter Brauch, Peace Research and European Security Studies (AFES-PRESS), Mosbach, Baden-Württemberg, Germany

More information about this series at http://www.springer.com/series/15232
http://www.afes-press-books.de/html/APESS.htm
http://www.afes-press-books.de/html/APESS_30.htm

Úrsula Oswald Spring · Hans Günter Brauch
Editors

Decolonising Conflicts, Security, Peace, Gender, Environment and Development in the Anthropocene

With a Foreword by Johan Galtung and a Preface by Betty Reardon

Springer

Editors
Úrsula Oswald Spring
Centre for Regional Multidisciplinary Research (CRIM)
National Autonomous University of Mexico (UNAM)
Cuernavaca, Morelos, Mexico

Hans Günter Brauch
Peace Research and European Security Studies (AFES-PRESS)
Mosbach, Baden-Württemberg, Germany

More on this book is at: http://www.afes-press-books.de/html/APESS_30.htm.

ISSN 2367-4024 ISSN 2367-4032 (electronic)
The Anthropocene: Politik—Economics—Society—Science
ISBN 978-3-030-62315-9 ISBN 978-3-030-62316-6 (eBook)
https://doi.org/10.1007/978-3-030-62316-6

Copy-editing: PD Dr. Hans Günter Brauch, AFES-PRESS e.V., Mosbach, Germany.
English Language Editor: Dr. Vanessa Greatorex, Chester, England.

The cover photo was taken by Hans Günter Brauch at the Gandhi Ashram in Ahmedabad on 28 November 2018 who granted permission for it to be used here.

This Springer imprint is published by the registered company Springer Nature Switzerland AG
The registered company address is: Gewerbestrasse 11, 6330 Cham, Switzerland

This book is dedicated especially to
the people of the world who have experienced colonialism,

global and social injustice

multiple forms of violent conflicts,

lack of security, peace, and gender equality.

It is written especially for the young generation who during this century will face the

projected challenges of global environmental change and climate change

in the Anthropocene that have been addressed in this book

by peace researchers from all parts of the world

and from the Global South in particular

Foreword

Dear Reader,

Please have a look at the Table of Contents of the book in your hands. There are 25 chapters, spanning fairly well the field of what today is called peace studies, peace research, or something similar. Each chapter has a specialist guiding the reader in an aspect of that field. And then try to imagine that some decades ago, the field did not exist at all.

When I became the world's first professor of peace research at the major university in a country, the University of Oslo in Norway, it was greeted by a key conservative politician/spokesman (Carl Joachim Hambro) with the words "professorat i vroevi"—professorship in nonsense.

I am happy to say that the "nonsense" has spread. There are at least half a thousand peace researchers spanning the world, and no doubt many more to come. An idea whose time had come? No. Very much overdue, I would say. Two world wars overdue. Or massive inequality overdue. Or massive suffering and death by exploitation overdue. Today, all these issues are key topics of peace studies. You will find something on all of this, and much more, in this book, so pluralistic, so rich in a multiplicity of ideas developed by excellent authors, and so competently put together by the editors. A deep thank you to the authors and to the editors who designed it all.

Happy reading!

Johan Galtung, dr. dr. h.c. mult.
Director
Transcend International
for Peace and Development

Johan Galtung, dr. dr. h.c. mult. a professor of peace studies, was born in 1930 in Oslo, Norway. Email: galtung@transcend.org. He is a mathematician, a sociologist, a political scientist, and a pioneer of peace studies. He founded the International Peace Research Institute, Oslo (1959), the world's first academic research centre on peace studies, and the *Journal of Peace Research* (1964). He has helped to establish dozens of other peace centres around the world. As a professor of peace studies, he has taught at universities all over the world, including Columbia (New York), Oslo, Berlin, Belgrade, Paris, Santiago de Chile, Buenos Aires, Cairo, Sichuan, Ritsumeikan (Japan), Princeton, Hawaii, Tromsø, Bern, Alicante (Spain), and many others on all continents. He has motivated thousands of individuals to dedicate their lives to the promotion of peace and the satisfaction of basic human needs. He has mediated in over a hundred conflicts between states, nations, religions, civilisations, communities, and persons since 1957. His contributions to peace theory and practice include the conceptualisation of peace-building, conflict mediation, reconciliation, non-violence, theory of structural violence, theorising about negative vs. positive peace, peace education, and peace journalism. His unique imprint on the study of conflict and peace stems from a combination of systematic scientific inquiry and the Gandhian ethics of peaceful means and harmony. He has conducted research in many fields with his original contributions to peace studies, human rights, basic needs, development strategies, a world economy that sustains life, macro-history, the theory of civilisations, federalism, globalisation, the theory of discourse, social pathologies, deep culture, peace and religions, social science methodology, sociology, ecology, and future studies. In 2008, he founded the TRANSCEND University Press (https://www.transcend.org/tup/). Thirty-six of his books have been translated into thirty-three languages, for a total of 134 book translations. He is founder (in 2000) and rector of the TRANSCEND Peace University (https://www.transcend.org/tpu/), the world's first online Peace Studies University. He is also the founder and director of TRANSCEND International (https://www.transcend.org/), a global non-profit network for peace, development, and the environment, founded in 1993, with over 500 members in more than seventy countries around the world. As a testimony to his legacy, peace studies are now taught and researched at universities across the globe and contribute to peacemaking efforts in conflicts around the world. He was jailed in Norway for six months at the age of twenty-four as a conscientious objector to serving in the military. In jail, he wrote his first book, *Gandhi's Political Ethics*, together with his mentor, Arne Naess. As a recipient of over a dozen honourary doctorates and professorships and many other distinctions, including a Right Livelihood Award (also known as the Alternative Nobel Peace Prize), Johan Galtung remains committed to the study and promotion of peace.

Preface

As a feminist peace educator, I find this volume to be uniquely attuned to the peace problematic of a complex, rapidly changing world. The complexity and speed of change has increased exponentially since the papers published here were presented at the 2018 General Conference of the *International Peace Research Association* (IPRA). Yet, the editors have framed the volume in a manner sharply relevant to the 2020 challenges confronting all members of the peace knowledge community, researchers, educators, and activists. A global pandemic, a resurgence of the nuclear threat, intensification of authoritarianism, severe weather events, undeniable revelations of human inequality, systematic deprivation and oppression, protested throughout the world, now comprise the problematic of peace. Never have we been in so great a need of such frames as those presented by the editors, Hans Günter Brauch and Úrsula Oswald Spring.

Framed within an exposition of the origins and characteristics of the Anthropocene geological epoch, this work integrates multiple and varied issues, ranging from the quotidian to the cosmic, from the intimate and personal violence of child abuse to the global violence of the systemic, continued rape of Planet Earth. In this frame, the various issues can be seen as interrelated elements of the larger challenge of comprehensive peace. The editors enable us to view peace issues in the integrated, holistic,

Earth-centred manner, so urgently needed by the whole field of peace knowledge. Each editor delineates fundamental elements of their frame. Oswald Spring presents the current problematic in a fresh perspective on the evolution of the present problematic within the Anthropocene—the geological age in which human intervention in our planet's living systems has brought us to this existential crisis. Brauch, reviewing the stages of conceptualisation of the Anthropocene, demonstrates a "rethinking of the evolutions of peace research" in the convergence of what I would term the ecological imperative—the need to place the fate of Earth at the centre of all economic and political decisions. This frame provides a set of invaluable conceptual tools for learning to make the crucial changes essential to the survival of humanity and our planet.

The editors have, as well, shown some of the strides in defining the peace problematic that I have observed over the many decades of my association with IPRA. The IPRA of 1972, when I first attended a general conference in Gyor, Hungary, was markedly different from the association of 2018 that is represented in this volume. I see a deepening of the ever-broadening field, comprised of diverse practitioners. IPRA, half a century ago, was celebrating the establishment of relationships between European researchers from both sides of a melting Iron Curtain. Very few women, only two peace educators and virtually no researchers from the Global South were in attendance in Gyor in 1972. The gathering was far from the global association that convened in Ahmedabad in 2018. As noted in Oswald Spring's chapter, its organisational culture was formed by European men. And, I would note, focused primarily on problems of war and weapons as research topics with little attention to educating the public about the problems.

How different from the 1972 gathering was that in 2018! The papers the editors have selected show researchers from the world over, men and women discussing a far wider variety of forms of violence as defined over the last five decades, including gender violence. For years, gender was ignored, and then resisted as irrelevant to the research on arms and conflict that dominated the programmes of the general conferences. But in the 1980s, the Peace Education Commission embraced gender issues as integral to the education realm of peace knowledge. Not until a paper presented by a Finnish peace educator was environmental violence considered within the purview of the field. Peace educators will celebrate the way that this volume places both gender and ecology at the very centre of the peace knowledge project. We see now a diversity of substance beyond even the frames of structural and cultural violence, for several decades standard features of the conceptual map of peace research.

Also to be celebrated is the posing of the peace problematic as a decolonisation process that exposes epistemological imperialism within the larger historical reality of multiple injustices of Western colonialism. Westernisation as "progress", still a common view among Northern policy-makers, is revealed as a force that oppressed masses of the human family as it accelerated the consumption of the planet. This process perspective is projected on the large screen of history and forms the backdrop for multiple issues, illuminated in small frame focus so as to highlight the interrelationships among effects of extraction, oppression, and the cannibalisation of Earth on human lives.

The effects on human lives are the realm of the problematic that preoccupies peace educators who deal directly with those most vulnerable to these effects. Peace education strives to educate for critical capacities to prepare the deprived for resistance and liberation and aims to prepare the privileged to understand the consequences of the effects to our whole species, to develop a capacity for empathy with suffering humanity, and to take responsibility for a humanly abused planet. We seek ways to present the specificities of the human within the holism of the planetary. Hans Günter Brauch's introduction of peace ecology, calling upon Oswald Spring's delineation of five pillars of peace, offers us what we seek. The concept of peace ecology is a fine heuristic tool for educating the suffering for resistance and liberation, and developing in the privileged capacities for empathy and responsibility. It is a key conceptual convergence in the overall convergences revealed through the lens of the Anthropocene Epoch. I see it as an example of a holistic, comprehensive mode of thinking, an epistemology and a mindset essential for the transformative learning upon which our survival depends. Such a reconceptualisation of the peace problematic can free the field of peace knowledge from the limits of the linear, reductionist world-views of its Western origins, just as authentic and comprehensive political decolonisation could free the human family from the sexism, racism, exploitation of the vulnerable, and despoliation of the planet imposed by Euro-American, patriarchal imperialism. Such a conceptualisation is, as well, an intellectual frame for life affirming norms that could impede such reckless individualist behaviours as those that prevent containment of the COVID-19 pandemic, manifest the extractive greed that proposes to mine and further pollute the seas, and enacts the patriarchal hubris of nuclear "advancement".

As I write this in August of 2020, I speculate on the themes and issues that will comprise an IPRA general conference a decade hence. What might the cogent research questions posed in this book have produced in data and knowledge that could enable us to transcend the planetary damage of the anthropocentric and patriarchal thinking evolved through the Anthropocene Epoch? Will papers recount how strategies were researched and enacted to resuscitate the possibility of democracy, to eliminate nuclear weapons, to establish firm benchmarks to mitigate climate change, to design non-violent security systems and just conflict resolution procedures, to move towards social equity and gender equality? Will discussions be framed so as to indicate the integral interrelationships among all these realms of peace design? Will it reflect Brauch's call to rethink the evolution of peace research?

Oswald Spring and Brauch have given us a foundation for the research and learning that might make such a conference programme possible. As peace researchers, educators, and activists, we can build upon what they provide towards a body of peace knowledge that could contribute to the survival of the planet and the lives it sustains.

New York, USA
August 2020

Betty A. Reardon

Betty A. Reardon is a feminist peace and human rights educator-activist who has spent six decades in the development and dissemination of work in this field. The founder of *the International Institute on Peace Education* (IIPE) and the original peace education graduate specialisation at Teachers College Columbia University, and one of the civil society originators of UN Security Council Resolution 1325 on Women, Peace, and Security, she has worked in all regions of the world towards international cooperation between peace educators. Her writings on peace education and gender issues have been widely published and include *Comprehensive Peace Education* (Teachers College Press, 1985); *Educating for Human Dignity* (University of Pennsylvania Press, 1995); *Tolerance: The Threshold of Peace* (UNESCO, 1997); and *Education for a Culture of Peace in a Gender Perspective* (UNESCO, 2001). Most recently, she co-edited with Dale T. Snauwaert *Betty A. Reardon: A Pioneer in Education for Peace and Human Rights* and *Betty A. Reardon: Key Texts in Gender and Peace* (Springer Briefs on Pioneers in Science and Practice, PSP Vols. 26 and 27, 2015 and 2015). On the occasion of her ninetieth birthday in 2018, Dale T. Snauwaert published an edited book on: *Exploring Betty A. Reardon's Perspective on Peace Education: Looking Back, Looking Forward.* Her publications are archived in the Ward M. Canaday Centre for Special Collections at the University of Toledo.

Acknowledgements

This volume emerged from papers that were orally presented during the 27th Conference of the *International Peace Research Association* (IPRA) in Ahmedabad, Gujarat, India from 27 November to 1 December 2018. We thank all sponsors—especially the IPRA Foundation—who supported the participation of selected colleagues with submitted written papers from developing and low-income countries. The co-editors of this book are grateful to all the authors who passed the double-blind anonymous peer-review process and subsequently revised their papers, taking the many critical comments and suggestions of these reviewers into account. Each chapter was assessed by at least three reviewers from different countries and continents.

We would like to thank all the reviewers who spent much time reading and commenting on the submitted texts. They made detailed perceptive and critical remarks and suggestions for improvements—even for texts that could not be included here—that will help the young authors to improve their scientific writings. The goal of both editors has been to enhance the quality of all submitted texts.

The following colleagues (in alphabetical order) contributed anonymous reviews:

- Adebayo, Joseph Olusegun, Ph.D., Durban University, NRF Innovation Postdoctoral Fellow in Africa Studies, University of Cape Town, South Africa.
- Akude, John Emeka, independent specialist on Africa, former Lecturer and Researcher, Institute of African Studies, University of Cologne; former adviser, German Federal Ministry for Economic Cooperation and Development (BMZ), Senior Researcher at the German Development Institute (DIE/GDI) in Bonn, Germany.
- Al, Serhun, Ph.D. from Univ. of Utah, Asst. Prof., Faculty of Business, Political Science and International Relations, Izmir University of Economics, Izmir, Turkey.
- Ameglio Patella, Pietro, Prof. at UNAM, Facultad de Filosofía y Letras, UNAM, Mexico City, Mexico.

- Amisi, Baruti Bahati, Dr. (Ph.D.), Independent Researcher, Durban, University of Technology's International Centre of Nonviolence (ICON), Durban, South Africa.
- Angom, Dr Sidonia, Ph.D., Chairperson, Taskforce of the Government of Uganda to start a Constituent College of Agriculture at Gulu University, Karamoja, Uganda; author of *Women in Peacemaking and Peacebuilding in Northern Uganda,* Kampala, Uganda.
- Arach, Omar, Dr., Universidad Nacional de la Patagonia Austral, Santa Cruz, Argentina.
- Badmus, Isiaka Alani, Assoc. Professor, Department of Peace and Conflict Studies, Afe Babaola University, Ado Ekiti, Nigeria.
- Bal, Bhargava Mansee, Entrepreneur, Researcher, Educator at Environmental Design Consultants and Professor at School of Architecture and Planning, Amity University Chhattisgarh, Raipur, India.
- Baser, Bahar, Dr., Associate Professor, Centre for Trust, Peace and Social Relations, Coventry, England, UK.
- Benitez Manaut, Raul, Prof. Dr., Faculty of Political Science, National Autonomous University of Mexico (UNAM), Mexico City, Mexico.
- Braga, Camila, Ph.D., Post-Doctoral Fellow, International Relations Institute, University of São Paulo (USP/FAPESP), São Paulo, Brazil.
- Brauch, Hans Günter, Adj. Prof., PD, Dr., Free University of Berlin (ret.) and AFES-PRESS chairman, Mosbach, Germany.
- Caballero-Anthony, Mely, Professor of International Relations, S. Rajaratnam School of International Studies (RSIS), Nanyang Technological University, Singapore.
- Cabot, Charlène, M.A., Independent Reviewer, Research in climate-conflict nexus, France; current position, Manager of the Response Innovation Lab in Kampala, Uganda.
- Chatterjee, Monalisa, Ph.D. from Rutgers University, Assist. Prof. (Teaching) of Environmental Studies, University of Southern California, Dornsife, California, USA.
- Chivasa Norman, Dr., Post-doctoral Research Fellow, International Centre of Nonviolence, Peacebuilding Programme, Durban University of Technology, Durban, South Africa.
- Collin Harguindeguy, Laura; Prof. Dr., Directora del Centro de Investigaciones Políticas y Sociales, El Colegio de Tlaxcala, A.C. Mexico.
- Collins, Andrew E., Prof. Dr., Director, Disaster and Development Network, Northumbria University, Norththumbria University, Newcastle on Tyne, UK.
- Curran, David, Ph.D., Assist. Prof., Centre for Trust, Peace and Social Relations, Coventry University, Coventry, England, UK.
- Dalby, Simon, Prof. Dr., CIGI Chair in the Political Economy of Climate Change, Balsillie School of International Affairs, Wilfrid Laurier University, Waterloo, Canada.
- Galaviz Armenta, Tania, Prof. Dr., UAEM, CICSER, Cuernavaca, México.

- Gómez Alvarado, Luis Alberto, Dr., Universidad Nacional Autónoma de México, Facultad de Contaduría y Administración (UNAM, FCA), Mexico City, Mexico.
- Harris, Geoffrey, Prof. Dr., International Centre of Nonviolence (ICON), Technical University of Durban, South Africa.
- Henfrey, Tom, European Network of Community-led Initiatives on Climate Change and Sustainability (Ecolise), Brussels, Belgium from the UK.
- Holsti, Kal, Prof. Dr., Prof. Emeritus, University of British Columbia, Canada, now a Research Associate with the Centre for International Relations in the Liu Institute, Canada.
- Hove, Mediel, Ph.D., Assoc. Prof., University of Zimbabwe, History Department, War, Peace and Strategic Studies Section, Harare, Zimbabwe and International Centre of Nonviolence (ICON), Technical University of Durban, South Africa.
- Irene, Oseremen Felix, Dr., National Open University of Nigeria (NOUN), Abuja, Nigeria
- Jahn, Egbert, Prof. Dr., Emeritus, University of Mannheim, Germany.
- Jiminez Guzman, Maria Lucero, Prof. Dr., UNAM, CRIM, Cuernavaca, Mexico.
- Kakakhel, Shafqat, Ambassador (1994–1998), former UN Assistant General and Deputy Executive Director, United Nations Environment Programme (1998–2007); Chair, Board of Governors, Sustainable Development Policy Institute since 2010; Chair, Board of Directors, Civil Society Coalition for Climate Change, Islamabad, Pakistan (2019–2020).
- Kiyala, Jean Chrysostome Kimbuku, Ph.D., Honourary Research Associate, International Centre of Nonviolence, Durban University of Technology, South Africa.
- Krell, Gert , Prof. Dr., Em. Johann Wolfgang Goethe Universität Frankfurt, Germany
- Kriesberg, Louis, Ph.D., Prof. Emeritus of Sociology, Syracuse University, Syracuse, NY, USA.
- Kurebwa, Jeffrey, Dr., Lecturer, Bindura University of Science Education, Bindura, Zimbabwe.
- Lantos, Nóra Anna, Ph.D., Department of Social Psychology, Eötvös Loránd University, Budapest, Hungary.
- Lundy, Patricia, Prof. Dr., Ulster University, School of Applied Social and Policy Science and in the Faculty of Arts, Humanities & Social Sciences, Belfast, UK.
- McKerrow, Raymie E., Professor Emeritus, Ohio University, Athens, OH, USA.
- Merwe, Hugo van der, Prof. Dr., Director of Research at Centre for the Study of Violence and Reconciliation (CSVR), Cape Town, South Africa.
- Mesjasz, Czeslaw, Assoc. Prof. Dr., Cracow Economic University, Cracow, Poland.
- Messerschmid, Clemens, Dr., independent water expert, consultant, Ramallah, Palestine from Germany.

- Misati, Joseph Akuma, Dr., Lecturer, Cooperative University of Kenya, Nairobi, Kenya.
- Nadiruzzaman, Md., Dr., Research Fellow (CLICCS-B3), Centre for Earth System Research and Sustainability (CEN), Climate Change and Security (CLISEC) Research Group, Institute of Geography, University of Hamburg, Hamburg, Germany.
- Nayak, Amar, Prof. Dr., Xavier Institute of Management, Bhubaneswar, India.
- Ngwenya Maqeda, Dumisani, Ph.D., Post-doctoral Research Fellow, International Centre of Nonviolence, Peacebuilding Programme, Durban University of Technology, Durban, South Africa, GTH, Bulawayo, Zimbabwe.
- Olokesusi, Femi, Prof. Dr., College of Social Management Sciences, Afe Babalola University, Ado-Ekiti, Ekiti State, Nigeria.
- Olorunfemi, Felix Bayode, Ph.D., Assoc. Prof., Nigerian Institute of Social and Economic Research, Ibadan, Nigeria.
- Ombati, Mokua, Ph.D. candidate, Anthropology and Human Ecology Department, Moi University Kenya, Eldoret, Kenya.
- Orihuela Gallardo, Maria del Carmen, Dr., postdoctoral fellow, UNAM, CRIM, Cuernavaca, Mexico.
- Oswald Spring, Úrsula, Prof. Dr., UNAM, CRIM, Cuernavaca, Mexico.
- Padilla Menendez, Luis Alberto, Amb., Prof. Dr., Director, International Relations and Peace Research Institute (IRIPAZ), Guatemala City, Guatemala.
- Patel, Mukundbhai A., Prof. Dr., Foundation for Peaceful Change, FPC, Ahmedabad, Gujarat, India.
- Pelling, Mark, Ph.D., Professor of Geography, King's Centre for Integrated Research on Risk and Resilience—An IRDR International Centre of Excellence and Challenge Leader, Resilience to Environmental Shocks and Change Portfolio Global Challenges Research Fund (GCRF), London, UK.
- Pócsik, Andrea, Ph.D., Film Culture and Media Theory, senior lecturer, Pázmány Péter Catholic University, Budapest, Hungary.
- Qarmout, Tamer, Ph.D. from University of Saskatchewan, Canada, Assist. Prof. in Public Policy and Public Administration, Doha Institute for Graduate Studies, Qatar.
- Ramos Muslera, Esteban A., Dr., Institute for Democracy, Peace, and Security (IUDPAS), National Autonomous University of Honduras (UNAH), Honduras.
- Reychler, Luc, Prof. Dr. Emeritus, Catholic University Leuven, Leuven, Belgium.
- Richards, Prof. Dr. Howard, Universidad de Santiago de Chile (USACH), Research Professsor of Philosophy (Ret.), Earlham College, USA.
- Roy, Sajal, Ph.D. Candidate, MPhil (Bergen), Institute for Culture and Society, Western Sydney University, Penrith, New South Wales, Australia.
- Selim, Gamal, Assoc. Prof. Dr., British University, Cairo and University of Port Said, Egypt.

- Singer-Browdoski, Mandy, Dr., Research Associate, UNESCO World Action Programme, Education for Sustainable Development, Free University of Berlin, Faculty of Education and Psychology, Institut Futur, Berlin, Germany.
- Snodgrass, Lyn, Prof. Dr., Head of the Department of Political and Conflict Studies at Nelson Mandela University in Port Elizabeth, South Africa.
- Steinweg, Reiner, Dr., Peace and Conflict Researcher, Educator on Theatre and Specialist on Gandhi, former editor of the Journal Friedensanalysen (Analyses on peace), from Germany living in Linz, Austria.
- Stensvold, Anne, Prof. Dr., Religious Studies, University of Oslo, Department of Humanities, Department of Culture Studies and Oriental Languages, Oslo, Norway.
- Suarez, Marcial A. G., Ph.D., Assist. Prof., International Relations and Strategic Studies, Fluminense Federal University, Rio de Janeiro, Brazil.
- Subert, Maria, Assistant Prof. Dr., City University of New York, Hostos C. C., NewYork, USA.
- Tadjoeddin, Zulfan, Associate Prof. Dr. and Director of Academic Programme, Humanitarian and Development Studies, School of Social Sciences, Western Sydney University, Australia.
- Tapia Uribe, Medardo, Prof. Dr., Centro Regional de Investigaciones Multidisciplinarias (CRIM), Universidad Nacional Autónoma de México (UNAM), Cuernavaca, Mexico.
- Uwimbabazi, Penine, Prof. Dr., Deputy Vice Chancellor for Academics, Protestant Institute of Arts and Social Sciences, Ruanda.
- Vesa, Unto, Emeritus Research Fellow, Tampere Peace Research Institute (TAPRI), Tampere University, Finland.
- Ware, Helen, Ambassador, Prof. Dr., Faculty of Humanities, Arts, Social Sciences and Education, School of Humanities, Arts, and Social Sciences, University of New England, Armidale, New South Wales, Australia.
- Weiffen, Brigitte, Visiting Prof. Dr., Department of Political Science, University of São Paulo, Brazil.
- Weiss, Matthew, Assist. Prof. Dr., Department of Public Affairs and Security Studies, University of Texas Rio Grande Valley, USA.
- Wielanga, Cori, Ph.D., from the University of KwaZulu Natal, senior lecturer in the Department of Political Sciences and the Deputy Director of the Centre for the Study of Governance Innovation, University of Pretoria, South Africa.

This book is the result of international teamwork between *Prof. Dr. Úrsula Oswald Spring* as Co-Secretary General of IPRA (2016–2018), who was the organiser of the 27th IPRA Conference in Ahmedabad, Gujarat, India, with the local support of Dr. Mukund Patel; and *PD Dr. Hans Günter Brauch* as IPRA's cashier (2017–2018), who, with *Úrsula Oswald Spring* and *Unto Vesa*, also managed the IPRA travel grants for young participants from Africa, Asia, and Latin America that were made available by the IPRA Foundation, through the fundraising of Prof. Dr. Matt Meyer (New York) and Prof. Dr. Polly Walker (USA). The co-editors are grateful to Rajib Timalsina, Assistant Professor, and Roshan Pokharel, Assistant Professor,

Department of Conflict, Peace and Development Studies, Tribhuvan University, Nepal, and to Dr. Senthan Selvarajah, the then editor of the IPRA Newsletter, from the UK and Sri Lanka for their superb local organisational support in Ahmedabad.

The editors want to thank all the financial supporters, the local organiser, and the above anonymous reviewers who made the 27th IPRA Conference in Ahmedabad a success on the occasion of the 150th birthday of Mahatma Gandhi. They are also grateful to Prof. Dr. Johan Galtung, one of the co-founders of IPRA in 1964, for writing the Foreword and to Prof. Dr. Betty Reardon of Columbia University in New York, USA, for writing the Preface.

Last but not least, we thank our colleagues at Springer International Publishing in Heidelberg (Dr Christian Witschel, editorial director for Earth Sciences, Environment and Geography, Ms Marion Schneider and Ms Birke Dalia) and in Chennai, Tamil Nadu, in India (Mr Arulmurugan Venkatasalam, and the team of typesetters and graphic designers). The co-editors are grateful to Dr. Vanessa Greatorex (United Kingdom) for her language editing and to IPRA for funding her work. As IPRA's Co-Secretary General 2017–2018) *Prof. Dr. Úrsula Oswald Spring* (Mexico) organised the IPRA Conference 2018 in Ahmedabad, Gujarat (India) while *PD Dr. Hans Günter Brauch* (Germany) was responsible for the finances of this IPRA Conference, managed the peer-review process and did the copy editing.

Cuernavaca, Mexico
Mosbach, Germany
September 2020

Úrsula Oswald Spring
Hans Günter Brauch

Contents

Abbreviations

AA	German Federal Foreign Office
ACCORD	African Centre for the Constructive Resolution of Disputes, Durban, South Africa
ACLEDP	Armed Conflict Location and Event Data Project
ADA	Amigos dos Amigos [Friends of Friends]
AFD	Alternative für Deutschland
AFES-PRESS	Arbeitsgemeinschaft Friedensforschung und Europäische Sicherheitspolitik—Study Group Peace Research and European Security Studies
AFK	Arbeitsgemeinschaft für Friedens- und Konfliktforschung e.V. (German Association for Peace and Conflict Studies)
AFPRA	African Peace Research Association (founded in 1985)
AFPREA	Africa Peace Research and Education Association
AHRC	Asian Human Rights Commission
AI	Artificial Intelligence
AKF	Aga Khan Foundation
AMAN	The Asian Muslim Action Network
Amul-GCMMF	Gujurat Cooperative Milk Marketing Cooperative Ltd.
ANC	African National Congress
ANZUS	Treaty consisting of the USA, Australia, New Zealand, in 1951
APP	Agriculture Promotion Policy
APPRA	Asia-Pacific Peace Research Association
AQIS	Al Qaeda of the Indian Subcontinent
ARSA	Arakan Rohingya Salvation Army
ASER	Annual Status of Education Report
ASPR	Austrian Study Centre for Peace and Conflict Resolution
AU	African Union
AWG	Anthropocene Working Group
BAU	Business As Usual
BBC	British Broadcasting Corporation

BCE	Before Common Era
BFVP	Bódvalenke Fresco Village Project
BMBF	German Federal Ministry for Education and Research
BMU	German Federal Ministry for the Environment, Nature Conservation, and Nuclear Safety
BNP	Bangladesh Nationalist Party
BOA	Bank of Agriculture of Nigeria
BP	Before Present
CAPAZ	German-Colombian Peace Institute
CAVR	The Timor Leste Commission for Reception [of Timorese who had supported Indonesia], Truth and Reconciliation
CBD	Convention on Biological Diversity
CC	Climate Change
CCGPS	Civil Society Groups for Peace and Security
CCMS	The Committee on the Challenges of Modern Society of NATO
CDC	Centre for Disease Control
CES	Community Enterprise System
CGT syndrome	Chosenness-Glory-Trauma syndrome (by Johan Galtung)
CILSS	Permanent Interstate Committee for Drought Control in the Sahel
CLAES	Centro Latino Americano de Ecología Social (Latin American Centre for Social Ecology)
CLAIP	Consejo Latinoamericano de Investigación para la Paz (Latin American Peace Research Association)
CLAs	Citizen-Led Assessments
CMPS	Conflict Management and Peace Science
CNTE	National Coordination of Education Workers
CO_2	Carbon Dioxide
CoE	Council of Europe
CoHR	Commission on Human Rights in South Sudan
COP 21	Conference of Parties (UNFCCC) in Paris in 2015
COVID-19	Global pandemic in 2020
CPA	Comprehensive Peace Agreement
CRECO	Constitution and Reform Education Consortium
CRISPR	Clustered Regularly Interspaced Short Palindromic Repeats
CTAP	Conflict, Terrorism, and Peace group, University of Auckland
CV	Comando Vermelho [Red Command]
CVEX	Truth Commission for El Salvador (Comision de la Verdad Para El Salvador)
D3E	Centre for Development, Economy, Ecology, and Equity
DDR	Demobilisation, Disarmament, and Reintegration
DGFK	German Association on Peace and Conflict Research
DMA syndrome	Dualism-Manicheism-Armageddon syndrome (by Johan Galtung)
DoD	Department of Defense (USA)

DPA	Dialogic Performance Analysis
DPC	District Peace Committees
DSF	Deutsche Stiftung für Friedensforschung (German Foundation for Peace Research)
DvPCs	Divisional Peace Committees
ECLAC	Economic Commission for Latin America and the Caribbean
ECOWAS	Economic Commission of West African States
EGRA	Early Grade Reading Assessment
EIA	Energy Information Agency
ELI	Environmental Law Institute
ENCOP	Environment and Conflicts Project
ENVSEC	Environmental Security Initiative
EPU	European Peace University
ERO	Education Review Office
ESA	Earth Systems Analyses
ESRL	NOAA Earth System Research Laboratories
ESS	Earth Systems Sciences
EU	European Union
EUPRA	European Peace Research Association
F.D. Roosevelt	Franklin Delano Roosevelt [US President, 1933–1945]
FAO	UN Food and Agriculture Organisation
FDMN	Forcibly Displaced Myanmar Nationals
FDN	Família do Norte [Northern Family Criminal Organisation]
FGD	Focus Group Discussions
FPO	Farmer Producer Organisation
GATT	General Agreements on Tariffs and Trade
GDP	Gross Domestic Product
GEC	Global Environmental Change
GECHS	Global Environmental Change and Human Security Project
GHG	Greenhouse Gases
GKST	Gereja Kristen Sulawesi Tengah
GP	Gram Panchayat or Ward
GSSP	Global Boundary Stratotype Section and Point
GSU	General Service Unit
GWSP	Global Water Security & Sanitation Partnership
HCNM	High Commissioner on National Minorities
HDI	Human Development Index
HIV/AIDS	Human Immunodeficiency Virus Infection and Acquired Immune Deficiency Syndrome
HuJI	Harkat ul Jihad Islami
HUJI-A	Harkat ul Jihad Islami-Arakan
IBRD	International Bank for Reconstruction and Development
ICBL	International Campaign to Ban Landmines
ICG	International Crisis Group
ICP	International Coalition for West Papua

ICRC	International Committee of the Red Cross
ICS	International Council of Science
ICS	International Commission on Stratigraphy
ICTJ	International Centre for Transitional Justice
ICTR	International Criminal Tribunal for Rwanda
ICUPRI	International Christian University Peace Research Institute
IDPs	Internally Displaced Persons
IDR	Indonesian Rupiah
IEP	Institute for Economics and Peace, Sydney, Australia
IFAD	International Fund for Agricultural Development
IGAD	Intergovernmental Authority on Development
IGBP	International Geophysical and Biological Programme
IGOs	Inter-Governmental Organisations
IISS	International Institute for Strategic Studies
IMF	International Monetary Fund
INC	Indian National Congress
INEGI	National Institute of Statistics and Geography [Mexico]
INR	Indian Rupee
IO	Industrial Organisations
IOM	International Organisation on Migration
IPCC	Intergovernmental Panel on Climate Change
IPRA	International Peace Research Association
IR	International Relations
IRIPAZ	Instituto de Relaciones Internacionales e Investigación para la Paz (Guatemala)
IRO	International Refugee Organisation
ISA-	International Studies Association, Peace Studies Section
ISE	Institute of Social Ecology
ISEE	International Society for Environmental Economics
ISI	Pakistan Inter-Services Intelligence
ISÖ	Institut für Sozialökologie, Siegburg, Germany
ISOE	Institut für Sozialökologische Forschung, Frankfurt
ISSC	International Social Science Council
IT	Information technology
ITD	Inter- and Transdisciplinarity
IUCN	International Union for Conservation of Nature
IUDPAS	Instituto Universitario en Democracia Paz y Seguridad [Honduras]
IUGS	International Union of Geological Sciences
IWT	Indus Water Treaty
KMMTTP	Kaladan Multi-Modal Transit Transport Project
KNBS	Kenya National Bureau of Statistics
KNCHR	Kenya National Commission on Human Rights

KSI	Dutch Knowledge Network on Systems Innovation and Transition
LGA	Local Government Association
LPCs	Location Peace Committees
LVC	La Via Campesina
MAD	Mutual Assured Destruction
MATCC	Myanmar's Anti-Terrorism Central Committee
MEA	Millennium Ecosystem Assessment
MFF	Multiannual Financial Framework
MIBUC	Military-Industrial-Bureaucratic-University-Congress Complex [in USA]
MOOC	Massive Open Online Course on Environmental Security and Sustaining Peace
MSG	Melanesian Spearhead Group
NASA	National Assessment of Student Achievement
NASPA	National Adaptation Strategy and Plan of Action on Climate Change
NATO	North Atlantic Treaty Organisation
NATO CCMS	NATO's Committee on the Challenges of Modern Society
NCIC	National Cohesion and Integration Commission
NCPACS	The National Centre for Peace and Conflict Studies, New Zealand
NDCs	National Determined Commitments (Paris Agreement, 2015)
NEMA	National Emergency Management Agency
NFD	Northern Frontier District
NGEU	Next Generation EU
NGN	Nigerian Naira
NGO	Non-governmental Organisation
NIRSA	Nigeria Incentive-Based Risk Sharing System for Agricultural Lending
NOAA	National Oceanic and Atmospheric Administration
NOAA/ESRL	NOAA Earth System Research Laboratories
NRC	National Research Council
NS	Natural Sciences
NSC	National Steering Committee
NSC-68	National Security Council Paper No. 68 [1950, USA]
NSPCC	National Society for the Prevention of Cruelty to Children
OAS	Organisation of American States
OCHA	United Nations Office for the Coordination of Humanitarian Affairs
ODIHR	Office for Democratic Institutions and Human Rights
OECD	Organisation of Economic Cooperation and Development
OHCHR	United Nations Office of the High Commissioner on Human Rights

OSCE	Organisation of Security and Cooperation in Europe
OSI	Open Society Institute
PA	Painter A
PAR	Participatory Action Research
PB	Painter B
PCC	Primeiro Comando da Capital [First Capital Command, criminal organisation]
PCDMB	Post-Conflict and Disaster Management Branch (UNEP)
PDHRE	People's Decade for Human Rights Education
PG	Painter G
PGP	International Priority Graduate Programme, Japanese Government
PIC	Person in Charge
PIF	Pacific Islands Forum
PIK	Potsdam Institute for Climate Impact Research
PJSA	Peace and Justice Studies Association
PNG	Papua New Guinea
PO	Producer Organisations
PPF	Provincial Peace Forum
Ppm	Parts per million
PRIF	Peace Research Institute Frankfurt
PRIO	Peace Research Institute Oslo
PSAJ	Peace Studies Association of Japan
PSS(I)	Peace Science Society International
R2P	Responsibility to protect
RPF	Rwandan Patriotic Front
RSO	Rohingya Solidarity Organisation
SAARC	South Asian Association for Regional Cooperation
SCN	Social Cohesion Network
SDG	Sustainable Development Goal
SEATO	South East Asia Treaty Organisation
SECASA	South Eastern Centre against Sexual Assault and Family Violence
SGBV	Sexual and Gender-based Violence
SIPRI	Stockholm International Peace Research Institute
SJPG	Social Justice Policy Group
SPLA-IO	Sudan People's Liberation Army-In Opposition
SPLM	Sudan People's Liberation Movement
SPLM/A	Sudan People's Liberation Movement/Army
SS	Social Sciences
STRN	Sustainability Transition Research Network
TAPRI	Tampere Peace Research Institute
TC	Terceiro Comando [Third Command, criminal organisation, Brazil]
TCC's	Troop-contributing Countries

TRC	Truth and reconciliation commissions
UCI	University of California in Irvine
UDHR	United Nations Universal Declaration of Human Rights
UK	United Kingdom
ULMWP	United Liberation Movement of West Papua
UN	United Nations
UNAH	Universidad Nacional Autónoma de Honduras
UNAM	National Autonomous University of Mexico
UNCCD	United Nations Convention to Combat Desertification (1994)
UNCh	United Nations Charter
UNDP	United Nations Development Programme
UNECE	United Nations Economic Commission for Europe
UNEP	United Nations Environment Programme
UNESCO	United Nations Environment, Science and Cultural Organisation
UNFCCC	United Nations Framework Convention on Climate Change
UNGA	United Nations General Assembly
UNHCR	United Nations High Commissioner for Refugees
UNHRC	UN Human Rights Council
UNICEF	United Nations Children's Fund
UNIDIR	United Nations Institute on Disarmament Research
UNMISS	The United Nations Mission in South Sudan
UNODC	The United Nations Office on Drugs and Crime
UNSC	United Nations Security Council
UNSC	United Nations Statistical Commission
UNSG	Secretary General of the United Nations
UNTEA	United Nations Temporary Executive Authority
UNU	United Nations University
UPEACE	The United Nations-mandated University for Peace
US(A)	United States of America
USA	United States of America
USD	US dollar
USIP	US Institute of Peace
VDC	Village District Council
VHP	Vishwa Hindu Parishad
VNSA	Violent Non-state Actors in Brazil
WBGU	Wissenschaftlicher Beirat Globale Umweltfragen (German Advisory Council on Global Change)
WCED	World Commission on Environment and Development [Report]
WEF	World Economic Forum in Davos
WHO	World Health Organisation
WMR	World Migration Report
WP	Warsaw Pact
WPNCL	West Papua National Coalition Liberation Front

WPNLA	West Papuan National Liberation Army
WTO	Warsaw Treaty Organisation
WTO	World Trade Organisation
WUSC	World University Service of Canada
WWII	World War II

Chapter 1
Decolonising Peace in the Anthropocene: Introduction Towards an Alternative Understanding of Peace and Security

Úrsula Oswald Spring

Abstract The 27th General Conference of the *International Peace Research Association* (IPRA) was held in Ahmedabad in November 2018. Mahatma Gandhi's political truth, sacrifice, nonviolent resistance, selfless service and cooperation were linked to the present peace and security concerns, especially on behalf of researchers from Asia, Africa and Latin America. The colonial burdens imposed on the Global South have increased through neoliberal globalisation of over-indebtedness to financial capitalists, the IMF and the World Bank. Multiple internal conflicts have emerged, often aggravated by promoted proxy wars, the illegal arms trade and violent armed struggles between neighbouring countries, often in the interests of industrialised nations. These complex processes have been aggravated during the new phase of Earth's history, the Anthropocene, with catastrophic climate extremes, where again the Global South is highly unprotected and has limited resources to adapt and protect its population. Therefore, security and peace research require deep decolonisation processes, starting from bottom-up and including the most vulnerable – generally women, girls and ethnic minorities – who are victims of disasters, war crimes and intrafamilial violence. Large global military budgets of almost two trillion USD have in many countries further frozen the public resources for health, food and education that are crucial for sustainable human development. This deeply embedded patriarchal and dominant behaviour reinforces the internal colonialism, and also the mechanisms of power and control exerted by external forces, post-colonial countries and neoliberal corporations. To overcome this perverse circle of multiples crises – health, human, social, economic, political, environmental, military, cultural and civilizational risks – in which the world has become entangled, a decolonised, engendered, sustainable and participative security and peace analysis may contribute to a different way of thinking. This includes gender equity and an alternative socio-green economy that may promote transition processes to decarbonised and dematerialised consumption, where everybody has the right to live in a culturally and naturally diverse society that may restore the exploited and destroyed Mother Earth and the societies in the Global South.

Prof. Dr. Úrsula Oswald Spring, research professor, CRIM-UNAM, Cuernavaca, Morelos, Mexico.
Email: uoswald@gmail.com.

Ú. Oswald Spring and H. G. Brauch (eds.), *Decolonising Conflicts, Security, Peace, Gender, Environment and Development in the Anthropocene*,
The Anthropocene: Politik—Economics—Society—Science 30,
https://doi.org/10.1007/978-3-030-62316-6_1

Keywords Anthropocene · Cosmopolitical security and peace · Decolonised · Dependency · Engendered security and peace · Globalisation · Global south · Internal colonialism · Neoliberal postcolonialism · Nexus of food, water and energy security · Patriarchy

1.1 Introduction

1.1.1 Gandhi's Socio-Political Imaginary for Decolonisation

The 27th General Conference of the *International Peace Research Organisation* (IPRA) was held in Ahmedabad, in Gujarat in India, from the 24–28 November 2018, in the town where Mohandas Karamchand Gandhi established his Ashram for self-reliance and from where he led the process of India's independence from the British Empire. During a dynamic cycle between pressures and negotiations, Gandhi radically transformed the Indian Nationalist Movements though nonviolent actions and the promotion of religious pluralism. His strategy was a combination of truth, sacrifice, nonviolent resistance, selfless service and cooperation, whereby every Indian should be brave but not a coward. He argued that this independence war should be fought with actions without violence, but with conviction and unity. He insisted that "one should fight a war with the weapons of truth and nonviolence". He advocated for the Indians to boycott British institutions and products in a non-violent way. From South Africa he brought back the idea of *Satyagraha* (Sanskrit: सत्याग्रह satyāgraha), civil disobedience or nonviolent resistance and fighting with peace (see Chaps. 5 and 6 below). The non-cooperation with British colonial orders was intended to debilitate the economic, ideological and legal colonial power system. Especially important was Gandhi's protest march of 241 miles to the west coast of Gujarat, where all participants harvested salt on the shores of the Arabian Sea to evade the salt tax imposed by Britain. In response, the colonial power imprisoned over 60,000 peaceful protesters, which reinforced the support for Gandhi's non-violent strategy. Further, in his quest to reject British domination and encourage Indian self-sufficiency, he produced his own woven cotton cloths and promoted a subsistence vegetarian food culture. In political terms and internally in the country, he protested without success against the local custom of discrimination against the lowest cast, the 'untouchables', insisting that all human beings are equal. Finally, following the debilitation of the British Empire after WWII and the increasing non-violent opposition to imperialism within India, in August 1947 the country achieved independence, but India and Pakistan were divided into two separate states, and deadly riots between Hindus and Muslims emerged and continue today. Gandhi was assassinated by a Hindu extremist on 30 January 1948 and more than 1.5 million people marched in his funeral procession. He said, "After I am gone, no single person will be able completely to represent me. … But a little bit of me will live in many of you. If each puts the cause first and himself last, the vacuum will to a large extent be filled."

During the IPRA General Conference in Ahmedabad, multiple interventions discussed the evolution of the *Satyagraha* principle and its influence in the present security and peace conceptualisation in the Global South and the Global North.

1.1.2 Occidental Peace as a Measure of Colonial Interests

Peace research and security studies were developed by scholars and policy advisers as two different academic research programmes after the two World Wars, primarily in Europe in the aftermath of the Versailles Peace Treaty (1919). The first two chairs of peace research were set up in Geneva and at Aberystwyth University (Wales) primarily in the idealist-liberal tradition. Security studies emerged after World War II in the United States and were influenced by political refugees from Europe (Morgenthau, Herz, Spykman) who were promoting a realist tradition among thinkers of war and military strategy. Initially, nearly all pioneers of both opposite and competing academic research programmes – security and peace – were white men, while many leaders of the pacifist peace movement in Europe and North America were female (Bertha von Suttner, Jane Adams, Elise Boulding, Betty Reardon). After the Cold War, East and West continued to adapt the conceptual framework, and researchers from the South brought in the decolonising thinking of security and peace to deal with the post-colonial ideological and financial control from the Global North and their former colonisers. The United States expanded its influence after 1989 as the sole superpower, and Francis Fukuyama (1992) interpreted the *End of the History* as a success of the Western liberal capitalist model. This author insisted that since the French Revolution, and especially after the disintegration of the Soviet Union, capitalist democracy has proven to be a better political, ethical and economic system that was able to end the Cold War and bring peace to the world. However, there exists no globally accepted democratic model, but different historically developed liberal democratic states. Perry Anderson (1991) reminded peace researchers that all the present liberal democracies had consolidated on conquest, slavery, racism, sexism, colonialism, raw material extraction and female exploitation. These colonial burdens were increased with a neoliberal model of over-indebtedness of the southern countries during the postcolonial globalisation, whereby multiple internal conflicts emerged, often aggravated by proxy wars, the illegal arms trade and armed struggles between neighbouring countries. Within this complex world situation, new communication systems (IT) and artificial intelligence (AI) emerged, together with the progressive destruction of the natural resources (air, soil, biodiversity and water), emerging climate extreme events, and a brutal concentration of wealth controlled by a few multinational corporate enterprises. Both processes created a break or caesura in the evolution of Earth and humanity and were called 'Anthropocene' and 'Globalisation'.

1.1.3 Anthropocene and Globalisation: Start and Impacts

There has been a controversial discussion among natural scientists as to when the Anthropocene began. Some researchers (Ruddiman 2001) have argued that the Anthropocene began with the agricultural revolution (12,000 to 7,000 years ago). Crutzen and Stoermer (2000) initially referred to the Industrial Revolution and especially to Watt's invention of the steam engine in 1782 as its starting point. The *Anthropocene Working Group* (AWG) referred to the first secret testing of an atomic bomb on 16 July 1945 in Alamogordo as the major marker for the start of this new geological epoch, while Steffen et al. (2018) pointed to socio-economic and ecological trends they had observed since 1950, which they called the 'Great Acceleration', due to the major increase in the burning of cheap fossil fuels (coal, oil and natural gas) and the concentration of CO_2 in the atmosphere.

Scientists also called this phase 'the atomic age' or the 'nuclear era', but important scientific discussions are still going on within the geological community and look set to continue before the term Anthropocene is officially accepted as the new ecological epoch of Earth. At the time of writing (2020) neither the *International Commission on Stratigraphy* (ICS)[1] nor the *International Union of Geological Sciences* (IUGS)[2] has officially accepted the term as a global subdivision (epoch) of geological times. This decision by several responsible scientific bodies of geologists is still pending. Only when it has been ratified can the mid-twentieth century (1950) officially be recognised as the starting point of this new common era (Dvorsky 2016), with multiple global changes in the physical–chemical composition of the Earth, the atomic age, massive population growth, increasing urbanisation and the globalisation of trade, finances, chains of productions, IT and AI. Thus, this phase may be called the Anthropocene Epoch of Earth's history that was brought about by the 'Great Acceleration' after WWII (Ellis 2018).

Vaclav Smil (2011) claimed that 10,000 years ago, the composition of humankind and animals on Earth was 1% human beings and 99% wild animals. Today, humankind has increased its share to 32%, wildlife has collapsed to 1%, and livestock has increased to 37%. Two hundred years ago there existed fewer than a billion people, but in 2020 we were more than 7.9 billion people. Species are disappearing a thousand times faster than they would without the human impacts, since *homo sapiens* has existed and had impacted for the last 200,000 years compared with the 4.5 billion years of Earth history. Beyond doubt, since 1950 anthropogenic changes have accelerated the changes on Earth. Since 1970 the population has doubled and the number of wild animals has more than halved. Only five times before has Earth

[1]The International Commission on Stratigraphy (ICS) is the largest and oldest constituent scientific body in the International Union of Geological Sciences (IUGS); at: https://stratigraphy.org/. The ICS has seventeen subcommissions, among them the Subcommission on Quaternary Stratigraphy, which is chaired by Emeritus Professor Jan Zalasiewicz: at: https://stratigraphy.org/subcommissions#quaternary. This subcommission http://quaternary.stratigraphy.org/ set up the Anthropocene Working Group (AWG); at: http://quaternary.stratigraphy.org/working-groups/anthropocene/ that was headed since 2008 by Prof. em. Jan Zalasiewicz.

[2]See the International Union of Geological Sciences (IUGS): at: https://www.iugs.org/.

lost so many species due to natural events in such short time spans, but this sixth massive extinction has been described by Smil (2011) as the result of biological annihilation.

Brett Milligan (2015: 1) argued that Earth has been in constant movement, where "biomes grow and shrink. The oceans expand. The Pacific Plate grinds and slips against the North American Plate, traveling northwest a few inches every year" and millions of salmon fight their way upstream twice a year. Thus migration not only refers to birds, fishes and humans, but also to plates, ecosystems and rivers, often called landscape migrations (Milligan 2015). Furthermore, there is the change produced by humans through infrastructural buildings, such as dams, industrialisation, urbanisation, tourism, and beach development. The shrinking of the Aral Sea and the Chad Sea are among the most extreme examples of anthropogenic alteration of a wide natural resource. But there are also more subtle changes produced by humans, especially with modern biotechnology and particularly with CRISPR,[3] which may alter the evolution of species, including human beings. This process was understood by Beck (2009, 2011) as part of the 'global risk society'. These environmental impacts were a result of a deeply consolidated global power system, called patriarchy, understood as a social system in which a few men maintain primary power and predominate in roles of political and economic leadership through violence, discrimination, exploitation and slavery. All these processes have limited and ideologically transformed reflections on security and peace, and have impeded investigations into the origin of violence and the exercise of despotic power, which drastically discriminates against women and girls in particular, but also southern countries and poor people.

1.1.4 Patriarchy, Colonialism and Neoliberal Postcolonialism

Historically there has been a conceptual development in the gendered impacts of wars and the impact of wars on women and girls. The analysis has focused on times when women were affected in war situations and their bodies were used as a battlefield or on the impacts of new wars (Kaldor 2012) or organised crime (Oswald Spring 2020). Females' bodies are still being used as sexual objects in conflict situations, which has increased the traditional intrafamilial violence and discrimination at home. Women and girls are also involved in relief and recovery actions. Petö (2016: 5) wrote that men continue to be pressured to protect the 'motherland', while women and children are in charge of the protection of the nation. Faced with increasing gender violence during wars and during the so-called non-war phases, in 2000 the *UN Security Council* (UNSC) launched Resolution 1325, in order to improve activities to prevent gender violence prevention. This was an initiative by the Namibian Minister Netumbo Nandi-Ndaitwah (as a temporary member of the UNSC), who, with the support of multiple feminist organisations, succeeded in getting this UNSC Resolution adopted.

[3]CRISPR is an abbreviation for *Clustered Regularly Interspaced Short Palindromic Repeats*. It represents the next-generation Cas9 protein engineered to deliver maximum editing efficiency.

However, with the 'new wars' and organised crime, violence is becoming simultaneously local and transnational, intrafamilial and global. Mary Kaldor (2012) has argued that the logic has changed and old conflicts organised by states, rebel groups and elites are now changing their objective by integrating human trafficking, pornography, child abuse, terrorism, drug and arms trafficking and organised crime. Therefore the UNSC approved Resolution 2467 in 2019 to protect children, women and the vulnerable from these emerging new threats. This new violence has changed socially constructed identities and roles for women and men, which include different cultural meanings, but also new social attributes. The hypermasculine risk and violence, often related to the pressure on men to provide and protect (Giménez 2020), often induces rebel groups and gangs to get involved in illegal economic activities, such as drug, organ and human trafficking, including female and male rape together with torture. All these new upcoming and interrelated processes as a result of these new wars were addressed in UNSC Resolution 2467, but only marginally discussed by the security and peace community, basically by women.

The environmental destruction and the loss of biodiversity by this patriarchal destructive behaviour have also created new threats to humanity, which go further than conflicts or a war, involving also global health, well-being and the survival of the human species. The present epidemic of COVID-19 indicates further that several leading countries in the world and permanent members of the UNSC have based their security and peace policies on outdated theories of negative, radical and liberal peace, linked to their androgenic nationalist considerations that are based on armed terror and nuclear threats, instead of consolidating a solid public and universal health and educational system that is able to prevent and deal with the new upcoming risks. The incapacity, the suffering and the avoidable deaths of their citizens during the outbreak of COVID-19 has proven that their security policies have wasted trillions of USD on military expenditure and arms (SIPRI 2019). Security and peace researchers have been unable to propose an alternative thinking to care for and protect their own society. This militarisation has produced incredible numbers of dead people, socio-environmental destruction, and additional insecurity risks, as this new pandemic and future ones, but also climate change, global environmental change, hunger, air pollution related to unsustainable fossil fuel burning, and emerging chronic diseases as a result of an industrialised fast-food culture with increasing garbage, are unaddressed dangers, and a growing population in the Global South means that even more people are at risk. This wasted investment in military 'insecurity', creating policies in financial and scientific terms together with armed conflicts, has proven the failure of the dominant neoliberal model and is morally and intellectually obliging the researchers of security and peace to replace the present dominant armed security and peace paradigms with a decolonised, humanised (Adger et al. 2014) and equal understanding of security and peace for the future survival of humanity and nature.

As a result of this dominant radical and liberal peace, which was reinforced by the military-technological boom after WWII, human health and also ecosystem services have dramatically deteriorated, leaving water, air, soil, biodiversity and oceans deeply polluted and physically and chemically altered. Furthermore, the UN World Population programme (2019) has projected unprecedented population growth in the twenty-first century:

> The world's population is projected to grow from 7.7 billion in 2019 to 8.5 billion in 2030 (10% increase), and further to 9.7 billion in 2050 (26%) and to 10.9 billion in 2100 (42%). The population of sub-Saharan Africa is projected to double by 2050 (99%). Other regions will see varying rates of increase between 2019 and 2050: Oceania excluding Australia/New Zealand (56%), Northern Africa and Western Asia (46%), Australia/New Zealand (28%), Central and Southern Asia (25%), Latin America and the Caribbean (18%), Eastern and South-Eastern Asia (3%), and Europe and Northern America (2%).[4]

More food will be required to feed all these people; however, corporate agribusiness has transformed the traditional food production of thousands of years into an industrial business, controlled by global multinational enterprises, toxic agrochemicals and genetically modified seeds. Overweight has nearly tripled since 1975, and in 2017 4.7 million people died prematurely due to excess of weight and its consequences because 13% of adults are obese and 39% of the population is overweight, including one fifth of children.[5]

Colonial and post-colonial power and violence are always related to profits, exploitation and slavery, today in modern sweatshops. This socio-economic inequality has had destructive impacts on the Global South and has reinforced the former colonial and present neocolonial power structures. At the time of writing in 2020, one per cent of the world's population owns twice the wealth of 6.2 billion people, and only 4 cents in each dollar of tax revenue is collected from taxes on wealth and can be redistributed to relieve poverty, whether in the Global South or the wealthiest countries (Oxfam 2020). Speculation on stock markets and with foreign currency is mostly tax-free and promotes money-laundering, fiscal paradises, poverty rise, violence, the devaluation of local currencies, and the consolidation of social inequality in the Global South, and also within all other countries. This unequal access to wealth is responsible for 10,000 people dying prematurely every day because of a lack of affordable health care, and for millions of children (25% more girls) being left without schools. Across the globe, men own double the wealth of women. The annual unpaid female care work has been estimated by UNICEF at 10.8 trillion USD, three times the value of the existing tech industry.[6] Security and peace researchers generally do not think in systemic terms and rarely include in their analyses environmental, health, educational, military, economic and social factors due to the present neo-colonial model of production, consumption and unequal access to wealth. The COVID-19 pandemic has opened up some new ways of understanding the complex linkages of the global risk society (Beck 2009), because so many people – including those in the Global North – have been affected by the situation with insufficient warning or governmental action. But there are more complex interrelations

---

[4]See UN World Population Prospects 2019: Highlights (June 2019); at: https://population.un.org/wpp/Publications/Files/WPP2019_10KeyFindings.pdf (10 August 2020).

[5]See the World Health Organisation; at: https://www.who.int/features/factfiles/obesity/en/.

[6]See at: https://www.oxfam.org/en/5-shocking-facts-about-extreme-global-inequality-and-how-even-it.

with climate change impacts which got completely forgotten during the health emergency and which will also affect all countries, unless we overcome the neoliberal consumerism and seriously decarbonise and dematerialise our way of living together and live with real sustainable alternatives.

Looking beyond these issues to environmental deterioration, including the depletion and pollution of resources such as air, water, soil and biodiversity, the present globalisation process has produced the destruction of ecosystems and their services, habitat devastation, and the extinction of wildlife, plants and microorganisms, and global environmental change and climate change are deteriorating both land and oceans and leading to the loss of glaciers and permafrost (IPCC 2013, 2014, 2019). Both hemispheres are and will be further affected by climate change impacts, but in the Global South, more than half of the people still earn their living from agriculture and depend on their natural resources. Land-grabbing, promoted by corporate enterprises, is expelling people from their traditional livelihoods. Additionally, the undesirable disturbance of the environment, understood as a global planetary process that has produced a new phase of Earth history, is reducing well-being and the potential to survive.

The outcomes of the Anthropocene, as a new historical phase of Earth that is superseding the former Holocene that has worked for the past 12,000 years, will produce new risks and threats. Given these complex and interrelated uncertainties, the question emerges: how do we restore the socio-economic, political, environmental and cultural equilibrium between humankind and Earth to reduce the military-technological boom according to the dominant model of globalisation during the Anthropocene epoch?

1.1.5 Decolonising Science, Thinking and New Social Representations

Without doubt, it is not only security and peace research which requires a deep decolonisation process, but also all the different scientific disciplines (archaeology, biology), their imperialist collections (museums, plant and animal collections), and their rewards for the androgenic reproduction of colonised thinking (jobs, citations, publications, prices).

Colonialism can be understood as the establishment of foreign rule over a distant territory and the control of its people. Generally associated with European imperial powers, colonialism and the colonial project include political and legal domination over a subordinate people, the exploitation of human and natural resources and the redistribution of those resources to benefit imperial interests, and the construction of racial and cultural difference that privileged the colonial ruler over the populations they ruled. Colonialism, which started in the late fifteenth century, is one of the fundamental social, cultural, and political forces that shaped our contemporary world. It is one of the phenomena that have structured modernity with regard to racial and economic hierarchies, which continues to have profound effects on communities worldwide" (Kroll-Zeldin 2016: 2).

Anthropology[7] (Assad 1973; Lewis 1973) began in the UK as a colonial approach to better understand and more speedily conquer the southern countries by manipulating their native leaderships and opposition. Theology (Heaney 2016) helped to consolidate the social representations of dominated people and encouraged them to accept these rules of discrimination and exploitation as the 'Law of God'. The conquest of Latin America was achieved with firearms and the Catholic creed. Furthermore, communication sciences and psychology (Bingham 1947) have similarly reinforced the manipulation of the present neoliberal model by inducing global consumerism, backwards-orientated reactionary nationalisms, war propaganda, and violence within households and society. The dominant colonised school system is based on competition and has reinforced the system in each student through individual qualifications, access to jobs, leadership, personal and university ratings, Nobel Price distributions, publication rates and other recognition for research, whereby the mechanisms of evaluation are deeply embedded in competitive patriarchal thinking and thus orientated towards male white scientists (Blazquez 2014), who reproduce the colonial ideology of occidental superiority. Roan (2018) argued that the success of British colonialism and imperialism was linked to medical science in the conquest of Africa, while the Latin American colonisation decimated the native population with the new illnesses introduced by the Spaniards and Portuguese, because the immune system of the native people had no defence against these new viruses and bacteria imported by the Europeans. A few centuries later health improvements brought by the colonisers undoubtedly reduced local diseases, and the vaccination campaigns increased the life expectancy of indigenous people by several decades, although typical tropical illnesses such as malaria were only marginally studied, generally when they affected the white colonisers and their armed forces.

How is security and peace research influenced by this complex patriarchal mindset? Advocating a type of irenic security theory in which critical reflections on ontological claims about security are sought, Richmond (2014) asserted that in the name of security and peace the decolonisation process should take into account feminist approaches and increase the stance on positive peace. Psychologists have begun this process with personal self-training or systemic emotion-focused management. Meanwhile, conscious disengagement with oppression in social movements in the Global South is empowering people and freeing them from colonised mentalities (Phillips et al. 2015). Freire (1968, 1992) favoured bottom-up education over the dominant position of standpoint feminism (Harding 1988), thus putting the ball in the court of the colonised people and enabling them to take charge of changing their lives. As Gandhi maintained, they should improve their personal health and livelihood, but also understand the enveloping imperialist conditions that limit the safe development

---

[7]"The historical origins of anthropology are rooted in the colonial enterprise, thus forever linking colonialism and anthropology. As such, colonialism is one of the most widely explored and written about subjects in the history of anthropology" (Kroll-Zeldin 2016: 2).

of humankind and nature through the predatory neoliberalism controlled by small corporate occidental multinational enterprises.

Thus, Fontan (2012) stated that decolonising peace must start from bottom-up, and take especial care of the most vulnerable, generally women and girls, who were victims of war crimes or intrafamilial violence. However, this process involves different racial groups, social classes and castes, gender relations, religious creeds, economic imperialism and authoritarian political systems, which are all related to an old but diverse embedded colonial patriarchal representation system in which discrimination, exploitation and violence have been integrated into the social imaginary for thousands of years. These social representations are so deeply assimilated in daily life and mostly unquestioned and invisible that most societies regard the complex and culturally diverse patriarchal mindset and the present system of hierarchical monopolisation of power as normal (Mies 1985).

In synthesis, a decolonised epistemological foundation of global power advocates for the liberation of the oppressed from bottom-up. This scientific foundation includes the decolonisation of the present forms of knowledge assimilation, the shift from an atomic understanding of well-being that was lost in the global consumerism, the environmental restoration of lost and deteriorated ecosystems and their services, and an outspoken rejection of the present structural forces of oppression, exploitation and violence. This deconstruction of the dominant imperialist and patriarchal mindset includes structural and personal changes in actions, thinking, consumption and care about others and ourselves. Only with a decolonised understanding of the present threats and challenges may we be able to live in a more sustainable and peaceful way, in which the decolonised response to the system of imperialist oppression is promoting a step-by-step transformation of the present dominant and unjust socio-political system into equality, sustainability and an alternative process of learning and teaching from infancy onward:

> Decolonisation promises to make science more appealing by integrating its findings more firmly with questions of justice, ethics and democracy. Perhaps, in the coming century, success with the microscope will depend on success in tackling the lingering effects of imperialism (Roan 2018: 16).

1.1.6 Research Questions

This chapter asks why the multiple theoretical and conceptual approaches of security and peace (negative, radical, positive, liberal, structural, environmental, cultural, cosmopolitan) have been unable to explain the present destruction of the Earth and the dramatic increase in gender and social inequality between North and South and also within each country between a small elite and a mass of impoverished people living in miserable conditions. All these accepted security and peace approaches were developed primarily by Western male researchers, thus they implicitly include the dominant legal, economic, social, political and cultural superiority of the Occident within a patriarchal, colonial and imperialist world-view. How can we move toward a true security and peace concept and practice when the contemporary militarism, globalisation and disasters have killed millions of people – mostly innocent women and children – since 1945? Why did the present security and peace policy in the East and West never produce stability for a global world order and why did the dominant military doctrine, based on physical strength and nuclear-armed threats, never achieve its proposed preventive goals? Why did this militarised model of development and resource extraction increase inequality between the Global North and South, but also within each nation and between genders? One possible hypothesis is that all these androgenic Western approaches always worked ex post through the sole method of calibrating threats by promoting more arms and higher military expenditures, e.g. through radical militarisation in order to maintain the dominant patriarchal system of power.

Confronted with these facts, this chapter – and, indeed, the whole book – explores new ways to understand security and peace from a bottom-up perspective. It emphasises a 'decolonised' approach to security and peace in the Global South, where countries, economies and cultures were historically and militarily subordinated by patriarchy, colonialism, slavery, militarism, mercantilism, capitalism and neoliberalism. Women and men dominated by these forces are reflecting on their historical and present reality of neo-colonialism, imperialism and exploitation in Latin America, Asia, Africa, and also industrialised countries. There are specific empirical proposals for overcoming the existing violent structures historically embedded in the post-colonial period, where local elites and corrupt governments often support, from a subordinate position, the imperialistic behaviour of the dominant elites. Additionally, there are also some researchers in the Global North who express their solidarity with the conditions in the Global South and are joining their reflections on decolonising. They have understood that neoliberalism has dismantled most of the welfare state and increased economic inequality, social vulnerability[8] and poverty in their industrialised countries too.

1.1.7 Structure and Organisation of This Chapter

This chapter first addresses the necessity of decolonising the peace concept (Sect. 1.1.5) and in Sect. 1.2 deepens the dominant security and peace paradigms. It starts by asking why decolonised peace is necessary. Later it introduces the concept of internal colonialism (Sect. 1.2.1). From the perspective of violence as the basic system of colonial domination and exploitation, the text critically reviews the realist approach to security and peace resulting from the Westphalian Peace (1648) and the consolidation in Central Europe of nation states with hierarchical power structures based on military technologies that are taking hostage billions of innocent human

[8]See for instance in 2020 the very high death rates of Covid-19 and poverty levels in the most industrialised countires such as USA, U.K., Italy, Spain, France, but also the emerging nations such as Brazil, Mexico and Iran.

beings and Mother Earth (Sect. 1.2.2). The liberal approach to peace is linked to universal ethical principles (Kant 1795), which have benefited small elites and some industrialised countries thanks to colonialism, mercantilism, industrialisation, two World Wars, and global institutions such as the United Nations. The UN Security Council, with the veto rights of its five permanent members and a geopolitical thrust in countervailing power through military security, was able to consolidate the imperialist structure through global economic institutions, represented by the Washington Consensus (the World Bank, the International Monetary Fund, later GATT, and today the World Trade Organisation) (Sect. 1.2.3). Several authors in this book (e.g. Wasike, Bakari, Esuruku, Oluwaleye) are exploring decolonising approaches, and ask what the decolonisation of a political science curriculum would entail (see the chapters by Cocodia, Ferreira, Ameglio, Hernández, Ariza, Ali and Bal, Momen, Nayak, Timalsina and Pamungkas in this book).

The next subchapter synthesises some Westernised understandings of realist and liberal peace (Falk 2019), which have also led to the overexploitation of natural resources and a dysfunctional global response to global environmental change and climate change (Sect. 1.2.4), in which stable peace, sustainable development (Brundtland 1987), Millennium Development Goals (UN 2000), Sustainable Development Goals (UN 2015) and the Paris Agreement on climate change (2015) are timid efforts to save the planet and the survival of humankind. But none of these efforts are touching the deep-grounded cultural and military imperialist structures of patriarchy and dominance. Confronted with these global institutions, ideologies, colonialism and exploitation of natural resources, Southern independence movements have struggled since 1806 (Venezuela) in Latin America, and later from 1824–1947 in India and after 1950 in almost all African countries to overcome the colonial yoke. However, multiple proxy and new wars during the last and the present centuries have made it clear to southern researchers that security and peace are deeply interlaced with cultural behaviours and habits based on military violence and patriarchal warfare (Reardon 1996). These profound links between violence, peace, security and gender violence have motivated UNESCO and the wider international community to promote the concept of a culture of peace. The United Nations General Assembly (UNGA) designated 1990 the International Year of the Culture of Peace.[9] However, neither the origin nor the execution of that year, nor later the International Decade for a Culture of Peace and Non-Violence for the Children of the World (2001–2010), questioned the traditional violent understanding, in which the Occident is the dominant producer of cultural and economic goods. UNESCO tried through these global efforts to include some subordinate social groups who had rebelled and reinforced their own cultural background (Black Africans, Latino culture). The imperialist hegemony of the Anglo-Saxon white male dominance in the United States was also challenged by the empowerment of Afro-Americans and the massive new immigrations from

[9]The promotion of a culture of peace was part of the central concerns of UNESCO, and the concept of a 'Culture of Peace' was born in Africa, and developed for the first time on a global scale by UNESCO during the International Congress on "Peace in the Minds of Men", organized in Yamoussoukro, Côte d'Ivoire, in 1989.

Latin America and Asia, which created a highly culturally diverse landscape, and Huntington (1993) developed the discussion about a clash of civilisation (Sect. 1.2.5) that might globally affect the dominant colonised mindset. As the origin of violence goes back to the consolidation of patriarchy, Sect. 1.2.6 proposes an engendered and diverse understanding of security and peace.

The text continues with a review of peace in the Anthropocene (Sect. 1.3), where proxy wars, resource depletion and the armed nexus of water, food and energy security (Sect. 1.3.1) are still producing militarised borders with an increasing number of refugees and migrants, who struggle to survive in a world against them (Sect. 1.3.2). The chapter continues to question the unipolar global imperialist system and explores a diverse decolonised paradigm (Sect. 1.4) in which the Global South has realised that it will continue to be exploited, indebted and without any opportunity to achieve a similar level of development as industrialised countries (Sect. 1.4.1), which is unlikely to be in its best interests anyway. Finally, in Sect. 1.5, there is a short presentation on the twenty-five chapters of this book, which add depth to the proposal for a decolonised security and peace for all humankind to enable us to survive in the era of greater global disturbance in the Anthropocene.

1.2 Why Decolonising Peace?

Undoubtedly the origin of violence and the dominant development of the concepts of security and peace have been deeply anchored in the occidental patriarchal and violent world-view, which still dominates in the realist military approach. The increasing globalisation has had an impact on the theory and global promotion of liberal peace (Falk 2019), affecting the ideological manipulation of global institutions, which have generated growing inequality, violence and exploitation of subordinated humans and natural resources. Stiglitz (2007) noted the growing inequality between countries and also within industrialised nations, while at the World Economic Forum in Davos, Oxfam (2020) declared that there had never previously existed such disparity between a small rich elite and the majority of poor people globally, where the richest 1% own more than twice as much wealth as 6.9 billion people and where poverty is especially dramatic among women and girls. Increasing social unrest as a result of the growing inequality, hunger and lack of basic social services has induced – often 'forced' – governments in the Global South to invest in military tools and, with the support of the global elite and their military advisers, repress the uprisings of non-conformists. Therefore, the realist approach to security and peace is still one of the dominant processes that prevails in countries where inequalities and poverty are dramatic, but also in emerging countries, where the new wars (Kaldor 2012) are related to drug and human trafficking, money-laundering in tax havens, and the export of arms and military tools from imperialist countries.

1.2.1 Internal Colonialism

Pablo González Casanova (1963) introduced the concept of internal colonialism and referred to the mechanisms of control within a country, which are always related to external forces, former colonial powers and neoliberal corporations. After the global triumph of capitalism over communism, social democracy and national liberation projects, the globalising and neoliberal policy of corporate business, supported by large political-military complexes, tended towards an integration of inter-, intra- and transnational colonisation. This neoliberal global domination of markets and workers, and increasingly also of global consumption, allowed the distribution of the surplus within each country to be appropriated. This process is supported by large transnational companies and the financial-speculative capital that has allowed a massive concentration of wealth by consolidating intra-national and transnational networks between the local bourgeoisie and the global elite. In addition to providing financial capital, internal colonialism produces a de-industrialisation process in most emerging countries, often in the form of local cheap assembly complexes (*maquila*) which use the cheap labour force of Southern workers without traditional labour rights or social protection. The system additionally features the primacy of extractive industries (mining) over industries of transformation (manufacture); land-grabbing; and the eviction of peasants and indigenous people from their historical agricultural land, subsistence food facilities and cultural heritage sites in favour of transnational development projects promoted by global tourist resorts, productive enclaves and mining.

Better global health conditions related to vaccination have produced demographic crises in the rural areas of most Southern countries. This situation is worsened by the expropriation and deterioration of local land, which has led to the loss of rural livelihoods. Lack of land, better education and the expectation of new labour opportunities have induced mass migration to urban areas. Since the 1950s rural-urban movement to national and regional capitals within migrants' own countries, and to other countries as cheap labour in agricultural and low-status service activities, has characterized the social movements in the Global South. The mass migration of cheap labour forces to industrialised countries has also exerted pressure on the labour conditions and salaries of non-immigrant workers, led to protest marches, and been a pretext for governments to control their borders via legislation and later militarily close them to illegal and unskilled immigrants from developed countries.

This dual dependency on the international and local bourgeoisie was also called dependency theory by researchers at the *Economic Commission for Latin America and the Caribbean* (ECLAC). Ruy Mauro Marini (1973), among others, has explained the dialectic of the dual dependency, which referred to the resource flows (financial, natural and human resources) from a 'periphery' of poor and underdeveloped states to 'core or central' wealthy states, producing poverty in the Global South and the concentration of wealth in industrialised countries. Thus, underdevelopment is a result of dominant institutions of the world economy (corporate enterprises supported by the World Bank, the IMF and the WTO) through the low price of raw materials from the South, high interest rates for poor countries, technological dependency, and structural adjustment programmes to pay for the debt service. In poor countries, the crises have been aggravated by the corruption of governments and the local bourgeoisie, and the dual exploitation of the people by the local-national and the international elite has aggravated the existing colonial structural dependency.

Dependency theory and internal colonialism were analysed first by Latin American researchers, and later by Asian (Nu 1975) and African scientists (Amin 1977). Samir Amin asserted that the law of value creation and accumulation is crucial to understand the present inequality. Within the present system it is impossible to catch up. Consequently, delinking has been a prerequisite for the development of productive forces in the Global South, such as China, but some industrialised countries have also experimented. To overcome global inequality there has been no lack of theory or ways to deal with inequality, but in practice there are often complex intrinsic interrelations between the local – often rural – bourgeoisie and global corporate capitalism. Most independence movements have learned that these mechanisms of local and international exploitation were traditionally linked to complex processes of control, which worked simultaneously through armed repression, socio-economic exploitation, ideological justification, religious domination, cultural alienation and systematic discrimination, whereby the militarist power dominated over the ideological manipulation. Thus, to understand both internal colonialism and external dependency, it is crucial to explore the evolution of the realist approach, which not only consolidated the unequal system of wealth, but also produced two World Wars and multiple regional and local confrontations in order to consolidate the present corporate neoliberal military-socio-economic and cultural domination of humankind and Earth.

1.2.2 The Realist Approach to Peace

The Thirty Year War (1618–1648) in Central Europe was one of the most deadly, due not only to military losses, but also to hunger and plagues, which killed soldiers and civilians directly, caused famines, destroyed livelihoods, disrupted the commerce of food and forced large numbers of people to relocate elsewhere as refugees. The war was initially about the Catholic hegemony (the Holy Roman Empire). The Habsburg Emperor wanted to impose the Catholic creed in all states, while the Protestant princes in the North and East fought for religious freedom and independence. Gradually the war developed into a general conflict which involved most European powers. In 1644 peace negotiations began in Westphalia and a peace agreement was signed in 1648 in the towns of Münster (Roman Catholic) and Osnabrück (Protestant). The Peace of Westphalia brought an end to the Eighty Year's War between Spain and the Netherlands and the Thirty Year's War in Germany and Central Europe. This peace agreement had a long-standing influence on international relations, due to the fact that besides religious freedom, the sovereignty of fully autonomous states (*rex est imperator in regno suo*) and the absolute power of rulers was granted. The balance of power between states generated a relatively stable distribution of power in Europe, which later produced a kind of power equilibrium under the assumption that an unbalanced power arena is dangerous for regional and global security. However, it could not prevent two World Wars.

This peace is still reflected in the Charter of the United Nations. However, some of the five veto powers in the UNSC have promoted military interventions and have not respected the sovereignty of fully autonomous states (Libya). Therefore, Ivan Illich (1973) criticised this dominant realist approach to power and control based on the military control exerted by these powerful nations. He called for a change in the realist mindset in order to overcome the prevailing military understanding of power and control in emerging nations. Based on Gandhi's self-reliance, he asserted that a radical monopoly occurs when people give up their native ability to do what they can do by themselves for a so-called major tool, named 'the better'. He understood that a radical monopoly exists in the industrial institutionalisation of values, whereby people need protection and are paid for this monopoly by accepting military, colonial and ideological control mechanisms. Schools are critical in promoting an ideological hegemony and control to children and parents. Thus, Illich (1971) proposed 'de-schooling' – abolishing the present system of education – insisting that there is no future for school reform within the present patriarchal mindset and realist world system.[10]

Illich (1975) deepened the analysis of all types of colonial imposed structures and argued that the present health system is not preventing diseases or curing people. It is only benefitting some pharmaceutical corporations and private corporate hospital chains, where chemotherapy, radiations, hospitalisation and expensive medicaments for ill persons who are able to pay for them, is increasing their suffering and creating dependency on expensive treatments and medicaments. For poor people there is no health support in this profit-interested medicine and they often return totally impoverished to traditional medical treatments, which are increasingly also controlled by corporate multinational enterprises. In any case, health is not provided in most cases and in the industrialised countries the so-called health care is often prolonging the length of suffering. Illich opposed this realist approach towards livelihood by pharmaceutical companies and allied governments and called for less medicine and more health, less school and more education. His radical understanding of existing monopolies in education and health brought him to an alternative socialisation and health care system for poor people. He stated that people and southern countries must

[10]"It would not be more feasible if we tried to create alternative institutions by building in the style of today's schools. Neither new teachers' attitudes towards their students, nor the proliferation of new physical or mental tools and methods (in the classroom or in the bedroom) nor, finally, the attempt to extend the responsibility of the pedagogue until it encompasses the entire lives of your students will result in universal education. The current search for new institutional funnels must be reversed towards the search for its institutional antipode: educational plots that increase so that everyone transforms each moment of their life into a moment of learning, of sharing, of caring. We trust that we are providing necessary concepts for those who carry out such broad research on education – and also for those seeking alternatives to other established service industries" (translated from Spanish, Illich 1971, Introduction).

understand the postcolonial yoke, which only produce dependency on finances and technology from outside. Changes of mindset and, later, personal and social liberation movements first begin ideologically and culturally. During the 500 years of colonialism indigenous and poor people have developed alternative ways to survive. In most colonialised countries social movements have developed first resistance and later resilience to protect their proper cultural goods against the conquerors and today's post-colonial enterprises.

Many so-called democratic countries and authoritarian states (Russia, China, North Korea, etc.) maintain their militaristic origins, reflected in their cultural expressions such as parades, uniforms, music, flags and carnivals with armed groups and military institutions that coexist with civil onesThis militarisation.[11] The growing insecurity in all aspects of daily life is also increasingly militarised, and in the USA in summer 2020 social protests against racism were contained by the National Guard. Globally, military budgets have been growing since the mid-1990s, while the support for development, health, schools and food has declined.

National security and defence policy are dominant in the debate about state budgets, and, with the support of lobbyists, the 'Military-Industrial-Media-Academia Alliance' has convinced the Congress to increase the military budget on the grounds that people need protection and security, which can allegedly only be provided by force in the present post-colonial global system. As an example, in 2018 the global defence expenditure was 1.822 trillion USD, 2.6% up on 2017 (SIPRI 2019). The IISS data indicated that in 2019 the United States, China, Saudi Arabia, Russia and India were the world's top military spenders. In 2018 the US defence budget was 684.6 billion USD; that of China was 181.1 billion USD; Saudi Arabia spent 78.4 billion; Russia 61.6 billion, India 60.5 billion, the UK 54.8 billion and France 52.3 billion USD.[12] Governments never explain against whom their people need protection. Without doubt today people need protection against corporate multinational conglomerates, banks, financial speculation and the privatisation of public services, which are destroying their livelihoods and threatening their survival.[13]

Partly due to this militarisation of life, the excess of military expenditure, and the neoliberal economic model, in 2020 the US, the UK, Italy and France failed to grant preventive health security to their people or protect them adequately from the COVID-19 pandemic that was fairly predictable after the epidemic broke out in China in late 2019 and early 2020.

Will this global pandemic bring a change to the prevailing business-as-usual understanding of a realist security and peace by proposing a budget for coping with the 'Militarism virus'? The predictable health, economic and food threats were also ignored or severely underfunded in most industrialised countries. While the US had

[11] This militarisation is also visible during the actual epidemic of COVID-19 in most southern countries, where the military is taking over civilian functions of public security and greater control of society (through mandatory curfews), when all the people should voluntarily stay at home to avoid the propagation of the pandemic.

[12] See at: https://www.iiss.org/blogs/military-balance/2020/02/global-defence-spending.

[13] But none of these items are on the socio-economic and development agenda, or addressed within military expenditures.

sophisticated programmes to counter biological warfare threats, the world's only superpower lacked the face masks and protective equipment for medical personnel and its people. This was a result of a massive build-up of military expenditure, a drastic cutback in public health expenditure and reduction and privatisation of hospitals.

Since the Peace of Westphalia (1648) the realist understanding of peace has prevailed in the military and economic sphere. Many specialists in International Relations have supported a more liberal approach to peace – one of new ideological controls for normal citizens and trained people who are analysing their political arena nationally and globally.

1.2.3 The Liberal Approach to Peace

A central problem related to liberal peace is the increase in the dilemma produced by the capitalist and especially the neoliberal ideology, among multiple other contradictions: growth vs. sustainable environment; public goods vs. private profits; governmental guidelines vs. private profits; collective well-being vs. private accumulation; democracy vs. soft power; food sovereignty vs. corporate fast food imposition, etc. These dilemmas hide the real threats of monopolised corporate profits and the dramatic global increase in poverty (Stiglitz 2007; Piketty 2013).

During the Cold War, the American trade union leader Charles Levinson (1978) claimed that corporate interests are so deeply intertwined with the military complex and financial speculation in the stock market in the East and West that at that time a war between the two superpowers was impossible. Knowing about these global interlinks, however, both superpowers were continuing to increase the amounts they invested in military research, star wars and modern arms of destruction. After the end of the Cold War, the situation has not changed significantly. Since 1996 military expenditure has increased in the US and by 2020 China and Russia have emerged as geopolitical forces that challenge the sole remaining superpower, but not the dominant neoliberal system. Therefore the liberal security and peace system has been protecting the present system of global accumulation by corporate capitalism not only in the USA and Europe, but also increasingly in China, South Korea, Japan and in other Asian countries.

As Naomi Klein (2007) explained, this liberal understanding of the economy based on militarism is creating new threats and lacks preventive actions to mitigate and adapt to these unknown risks. She claimed that due to climate change impacts and all other health and economic threats we are living in a permanent shock situation. Climate change management is highly complex. The impacts of climate change are mostly affecting poor countries, which are not responsible for the massive emissions of greenhouse gases (GHG). The liberal understanding of business-as-usual (BAU) is prioritising the status quo of the present accumulation model. Klein argued that in this liberal approach any emergency today will be resolved by a capitalist mercantilisation process (e.g. insurance, hospitalisation) or when conflicts emerge among the poor and discriminated in the US and by the exterminator arm of military-police repression and tear gas in the Global South.

Thus, the central problems of liberal security and peace are related to the dilemmas outlined above. Mansoob Murshed (2019) stated that the policy of liberal peace assumes that the democratic state will negotiate any dispute and will not engage in new armed disputes, due to economic interdependence (see Levinson 1978). Instead of liberal peace, Gartzke (2007) proposed that we should speak about capitalist peace, which is not peaceful in the Global South. It has imposed proxy wars, military putsches, warlords, debt burdens, economic crises, extractivism, land-grabbing, and financial pressure to get the required resources in the most advantageous conditions for the corporate economy. The institutions of the Washington Consensus are protecting these interests, and Latin America is the region saddled with the highest demand for billions of dollars, after destructive mining companies were expropriated by national governments due to wide-ranging damage on people and natural resources. Most of these settled disputes increase the compensation sums charged to Latin American countries, due to the industrialised corporation also adding the potential profits calculated during various decades of the obtained concession. Panels of the *World Trade Organization* (WTO) generally give corporate enterprises the right to charge poor Global South governments these exorbitant sums without remediating any of the long-term destruction of natural resources, pollution or poverty resulting from the excessive depleted natural resources. This unequal treatment in the global liberal political-economic arena has benefited corporate multinationals and for decades indebted the Southern countries, taking away crucial financial resources for basic services and disaster management to deal with climate change impacts.

Further, this type of global liberal security and peace also took away money for human and socio-economic development, measured by GDP, technological modernisation, chemical fertilizers in agriculture, industrial modernisation and other technical tools (IT, artificial intelligence), mostly provided by industrialised countries. In cultural terms, this prevents the development of proper science and technology innovations in the Global South and increases the dependence on industrialised countries. Further, there is also an ethical contradiction in this liberal security and peace approach, related to the rising social and environmental losses in the Global South in the form of the consequences of the imported technologies that are polluting the air, water and land; destroying the associated biological diversity; damaging ecosystem services; and causing the loss of food sovereignty for poor rural people (Oswald Spring 2020). Without any doubt, all these phenomena are exacerbated by climate change impacts (IPCC 2012, 2013, 2019), which are greatly affecting the poor countries, even though they have played little part in causing GHG emissions.[14] The industrialised countries are providing only limited support to deal with the growing impacts of climate disasters, forced migration and emerging conflicts, where again the liberal security and peace approach mitigates first with foreign aid or debts and

[14]The current members of the G20 have accepted responsibility for 78% of GHG emissions, but have not contributed sufficiently to the reduction of these gases, which have affected above all the Global South.

later with arms and military advice when the situation is getting out of control or the interests of global corporations are threatened.

Without doubt the present COVID-19 pandemic is expressing the failures of this liberal security and peace approach. The WHO recommended that 7.6 billion people globally should stay at home, because of the danger of infection at gatherings, sweatshops of mass production, large chain stores, sports events and entertainment in big arenas. Also the global casino of financial speculation momentarily collapsed, and in emerging countries activities shifted back from the financial sector to credit unions, public banks, cooperative sectors and other solidarity arrangements, especially among women. Manufacturers of goods and the service-based economies revived traditional barter systems, and informal voluntary sectors in local currencies and numerous non-monetary transactions helped to mitigate the height of the pandemic, the survival of poor people and the economic costs. As people are confined to their houses, widespread decentralisation and an impressive growth in self-reliant community activities have emerged. In the times of pandemic the economy has become regenerative rather than extractive, and the poverty gaps and inequality of the money-obsessed, exploitative models are slowly reducing. Instead of liberal security and peace, a new type of everyday peace has emerged, where commerce is replaced by barter and common interchange, and where survival, solidarity, care and well-being are prioritised, especially in the poor Southern countries. Social transformations with justice are also reducing the existing gender, class, race and ethnic gaps, and the care activities of female unpaid work are increasingly recognised.

The pandemic also had a great impact on the environmental sector when air and water pollution declined. The air quality in Wuhan and other cities improved, and safe water in the canals in Venice brought back dolphins and other marine creatures. This demonstrates the potential for an alternative decolonised peace, where the Global South in particular is offering diverse cultural approaches to a decolonised peace, adapted to each region and concrete necessities.

1.2.4 Natural Resources and Ecological Peace

The apocalyptic threat of nuclear war is still one of the greatest dangers of ambitious androgenic leaders who are not conscious of the real global destruction that would be wrought by a nuclear war. But the Partition of India and Pakistan during independence negotiated by the British Empire, radical groups such as the All-Indian Muslim League, the Indian National Congress and the Sikhs in Punjab led to 15 million uprooted people and to a permanent conflicts between both nations. Both countries became involved in a nuclear arms race. It is not only these two nations who are responsible for the increasing global insecurity, but also the other nuclear powers whose nuclear tests have seriously affected the environment. However, there are other phenomena that are also disturbing the ecological peace. The present industrialised and factory-raising of pigs, chickens, cows and other ruminants, such as sheep and goats, instead of traditional grazing, has in some cases converted these animals into carriers of diseases. Even if they often do not show any symptoms, they can transmit diseases to humans (e.g. AH1N1 or swine flu). Consequently, around the world the slaughter and consumption of animals dropped dramatically during the COVID-19 pandemic not only as a result of fear of disease, but also due to lack of money or meat supply. Further, the industrial production of animals has added almost 15% to the annual growth of global greenhouse gases (GHG). Big meat producers, and especially fossil fuel multinational corporations, may become the next 'stranded assets', due to a lack of investors and the massive development of renewable energy to achieve, by 2050, governmental commitments to keep the increase in the global average temperature to below 2 °C (if possible no more than 1.5 °C) above the global average temperature that existed during the industrial revolution. With the present changes in food habits, increasing population groups are switching entirely to plant-based foods, while others have reduced their consumption of meat, fish, and dairy products.

Charles Darwin (1859), other biologists and geneticists have claimed that the basic DNA of the human species varies little from that of plants and wild animals. Humankind evolved with other species in the biosphere of the planet by natural selection, often responding to stresses in habitats and environments. Therefore we all come from the African continent and the success of the twenty-first century was related to our abilities to cooperate, share and evolve in increasingly larger populations and complex organisations. Today more than half of the world's population lives in urban settlements and, based on current trends, extreme population growth will produce more than 10 billion people by 2100. Humankind grew from roving bands of nomads to live in settled agricultural villages, then towns, and then the megacities of the twentieth century, but the intensive use of fossil fuels, massive deforestation and land use changes, together with the destruction of crucial ecosystem services, has brought us to a new era of Earth's history, the Anthropocene. Several thousand years ago these social changes in human history in irrigated small cities also allowed the consolidation of the power system of patriarchy (Oswald Spring 2019), which represents the basis of the present violent and discriminatory global system.

In order to restore ecosystems worldwide and peace with nature, governments, enterprises and organised society need to promote a global shift from the industrialised towards an organic and regenerative agriculture, along with plant-based foods, beverages and more saltwater-grown foodstuff. Only by planting billions of trees around the world, along with climate-smart agriculture (FAO 2013, 2016a), may the damaged ecosystems be restored and the polluted air, soil (FAO 2016b) and water (Garatuza et al. 2011) recover. As a consequence of these deep environmental changes, by 2100 the global temperature may have reduced and the concentration of CO_2 in the atmosphere may have stabilised at the safe level of 350 parts per million. However, higher sea levels will remain for several centuries, and many people in coastal cities may be obliged to evacuate toward higher ground. All these changes towards environmental peace will require deep cultural changes in behaviour, substitution of business-as-usual, solidarity, support, a new understanding of the intrinsic links between humankind and nature that involves a more holistic view, including cultural elements, and also a new security and peace understanding far from the dominant realist and liberal paradigm of patriarchal violence and control.

1.2.5 Cultural Diversity of Peace

Some mass media and politicians are using social terrorisation to create fear, more shocks and a feeling of impotence in the face of new, powerful, unknown and lethal processes. The collective imagination, the media and the official discourses often produce confusion between the threat and the reality of death and destruction. COVID-19, a nano-virus impossible to see, but with a dramatic capacity to infect, allows pharmaceutical corporations to control the collective imaginary by offering the sole solution: medicines and vaccines to restore a peaceful cultural alternative. However, there may be soon new viruses, which may require still more transnational pharmaceuticals to deal with more dangerous threats. So far we have not worked on systemic answers and we have taken away the peace from nature by massively deforesting Earth and, through instigating loss of habitat, causing wild animals to live near our cities.

This one-dimensional thinking led Huntington (1993) to develop the thesis of a 'clash of civilisation', where he postulated that cultural and religious identities will be converted into the primary source of conflicts and new wars would be fought not between countries, but between cultures. In this approach Huntington refuted Fukuyama's (1992) thesis of the 'end of history' after the end of the Cold War, when human rights, liberal democracy and free market economy would become the sole remaining socio-economic and ideological alternative for all nations. In his thesis Huntington analysed inter-civilisational wars in Yugoslavia, Chechnya and India-Pakistan. He claimed that the belief of the West in the universality of its values, of democracy and of the free market was naïve and that no 'universal norms' exist that may prevent conflicts in the framework of the liberal peace postulated by Fukuyama. With regard to both the COVID-19 pandemic and climate change impacts, neither theory applies. Only through global efforts from bottom-up in a diverse cultural settlement that is environmentally sustainable may history have more chapters and culture produce not only clashes, but also solidarity and co-operation.

In the global discussion on cultural peace, UNESCO (2002) adopted the proposal which the Gambian government had submitted to the UNGA in 1999, and after ten months of negotiation the participating nations launched an International Year for the Culture of Peace. The eight action areas included a general programme comprising peace through education; sustainable socio-economic development; respect for all human rights; equity between women and men; democratic participation; understanding, tolerance and solidarity; participatory communication and a free flow of information and knowledge; and international security and peace. This general agreement did not touch on the delicate and conflicting themes of cultural peace, the realist and liberal understanding of peace, or the militarisation of almost all countries, hence this global effort failed to improve the security and peace agenda globally.

Without any doubt, modern conflicts will change in the future and will increasingly involve internet propaganda and cyber warfare, where a widespread misunderstanding is confusing money with wealth and well-being. The cyber-revolts exhibit all the fault lines in the present human societies through 'fake news': from racism and ignorance, conspiracy theories, xenophobia and scapegoating of 'the other' to the various cognitive biases, such as the technological determinism (geoengineering), theory-induced blindness, and the fatal, widespread media implosion and belief in alternative media.

These top-down efforts from UNESCO have had poor impacts in traditional, ethnic, peasant and indigenous societies, where multiple cultural changes and transformations are going on, based on their own deep cultural groundwork activities and 500 years of resilience. Indigenous people such as the Aymara-Quechua coalition in Bolivia have fought against the dominant repressive white political elite that embarked on a dramatic exploitation of their country's natural and human resources. The indigenous Aymara promoted among their ethnic groups a different lifestyle called 'living-well', based on an alternative paradigm to the generalised corporate neoliberalism, exclusive globalisation, and accumulation and growth philosophy of accumulation ad infinitum. Their traditional behaviour starts with caring about their community and the nature that is sustaining their communal life. Nature and humankind are complementary, thus harmony with nature, community life, cosmic rights, reciprocity and social control are crucial. The elderly have extensive life experience and can teach the rest of the community how to drink, eat, dance, work and enjoy without excess or destruction, in order to maintain the cosmic unity. These indigenous societies have understood that there exists only one good and not a better life, thus accumulation and hoarding are not the keys to an integrated livelihood in harmony with nature and happiness. However, corporate interests linked to a Protestant-inspired belief system and voracious economic interests in abundant mine resources organised a military coup and dismissed the indigenous government. Nevertheless, they could not defy their cosmovision of unity between humankind and nature.

There are multiple other efforts going on globally, such as the Zapatistas' autonomous shell political organisation in the indigenous region of Chiapas. During a low-intensity ongoing war with the Mexican government until 2018, these indigenous people established a totally independent indigenous government. After two decades of internal socialisation and training, without any governmental support they achieved a socio-environmental improvement in all their communities, based on their cultural beliefs. They were able to eradicate the endemic corruption and eliminated all femicides by overcoming the patriarchal world-view through equity between women and men. They based their socio-cultural representations on traditional storytelling and established a new equilibrium of internal powers, where women have the same access to temporary limited political leadership as men. Systematically, they deconstructed (Chomsky 1988), re-interpreted, argued, criticised and confronted existing social habits and are now producing democratic and inclusive alternatives, including care for the most vulnerable. Nobody is suffering from extreme poverty and everybody has enough healthy food.

However, this deconstruction of cultural, ethnic and racial discrimination through a new form of cultural peace was not achieved all over the world. Recent racist remarks, based on a 'colonial hangover', that the vaccines against COVID-19 should only be tested on African humans obliged the WHO to postulate that we "will follow all the rules for testing any vaccines or therapeutics in exactly the same route – whether it is Europe, Africa… We will use the same protocol and if there is a need to be tested elsewhere, to treat human beings the same way – equally the same way – as we are doing under the Solidarity trials" (Dr. Tedros Adhanom Ghebreyesus, 8 April 2020). Among other problems, President Trump withdrew financial support for the WHO. These racist comments show that there is still a profound sense of superiority in Western societies and that white male Eurocentrism and the denigration of 'the other' continue to belong to the cultural imperialism that brought colonialism, exploitation and discrimination to the Global South.

Jean Piaget (1950) argued that most nationalist and selfish actions are employed in the second phase of human development, but are overcome when the third stage of social co-operation (equalisation and mutual respect) is achieved. Ahead of UNESCO and other multilateral organisations, the WHO, with an Ethiopian Secretary-General, has tried to implement global actions during the present second stage of cultural evolution (egoism and unilateral respect), while the West and the corporate world are not yet prepared for the third stage of human evolution. The reasons are historically diverse, but understandable within the model of neoliberal egoist behaviour, where more than half of the population is living in misery and only a small elite is able to abuse human and natural resources. Thus, promoting social cohesion for each socio-cultural reality in a divided world is historically only possible with an Aymara or Zapatista approach. However, the present world model is still locked in the second phase of evolution, based on a realist and liberal peace and military security paradigms. For instance, nowadays governments and multilateral organisations are heavily pressurising their citizens to stay at home, mostly without any compensation, due to the emergency of the COVID-19 pandemic. In the Global South, where most workers and especially marginalised women have no social support system, they need to work daily to get the income to feed their families. Further, these regions are often also associated with terrorism and organised crime, and actions to avoid contagion by this lethal virus may be a second choice. Thus it is impossible to grant to every human being survival, security, health and solidarity in normal times and even harder during this exceptional pandemic.

The dominance of racism, patriarchal political control and unilateral nationalist behaviour in the remaining superpower also indicates that the second phase of a cultural evolution is deeply influenced by reactionary nationalism and that deep changes in occidental society are required to achieve cultural peace. The present pandemic could be an opportunity to move towards the third stage of social coexistence, where mutual respect, equality, solidarity and care about the most vulnerable predominate, often exercised by women and girls.

1.2.6 Engendered Diverse Peace: A Cosmopolitical Approach to Peace

By 2020, over several thousand years, patriarchy has emerged in all societies. Urbanisation and female education have slowly eroded the total control over the female population. Women are empowering themselves globally. When they took control of their own bodies, fertility rates tumbled in many countries even before the turn of the twenty-first century. Widespread revolts against the top-down narrow economic model of globalisation and its male-dominated elites, intrafamilial violence, femicides and discrimination led to disruption of the unsustainable paths of development driven by fossil fuels, nuclear power, militarism, profit, greed, androgenic research and egocentric leadership, in synthesis by the realist mindset.

In many countries high global military budgets of almost two trillion USD have frozen the public resources for health, food and education that are crucial for human development. To meet human needs, the world must gradually shift from tanks and battleships to less expensive, less violent forms of security that may grant the necessary well-being to all citizens. By the twenty-first century, international competition for power also focused on social propaganda, persuasion technologies, infiltration and ideological control of the global internet, besides heavy weapons and nuclear arms. There is no doubt that the best managed countries in the present pandemic were Germany, Taiwan, New Zealand, Iceland, Finland, Norway and Denmark. What do they have in common? All of them are led by women and these women are stepping up to show how to manage a messy patch for our human family, when an unexpected pandemic[15] emerged. More gender equality in political and economic leadership may be able to turn away the threats of future pandemics, nuclear war, global environmental change and climate change and give humankind the potential to survive in the more complex epoch of the Anthropocene. This gender behaviour disagrees with the often populist demagoguery of Presidents Bolsonaro and Trump and Prime Minister Johnson, who think that they can resolve with rhetoric the complex problems of this new phase of the Anthropocene, in which COVID-19 is just one of the multiple threats that are coming.

1.3 The Anthropocene May Pose Severe Survival Threats

The Anthropocene Epoch of Earth and human history may generate global threats for the survival of people that could also affect the conditions of security and peace. But people who face global risks might find potential alternatives through post-neoliberal diverse and sustainable forms of co-operation, including the active participation of women, children and senior citizens to address new risks. However, in most poor

[15]See at https://www.forbes.com/sites/avivahwittenbergcox/2020/04/13/what-do-countries-with-the-best-coronavirus-reponses-have-in-common-women-leaders/amp/?__twitter_impression=true.

countries there still exist multiple problems and conflicts that challenge the survival of the people locally and sometimes also regionally.

1.3.1 Proxy Wars and Resource Depletion: Conflicts on the Nexus of Water, Soil, Food and Energy

Proxy wars are armed conflicts between states or non-state actors that occur on behalf of other parties or countries that are not directly involved in the hostilities. Interested countries are supplying arms, technology and money for wartime activities. They often start to support a weaker party which would be unable to begin or continue a conflict alone and thus relies on interested allied nations or non-state actors to continue the fight. By winning or supporting the conflict these nations acquire some advantages in the form of resources or global influence. Proxy wars were very frequent during the Cold War and, including the 'secret war' in Laos, were fought in Vietnam between the US and the Soviet Union and between guerrillas and several Central American governments on behalf of the US to avoid 'a domino effect' due to a spread of communism in the region after the Cuban Revolution. Another example occurred in the Middle East between Saudi Arabia and Iran. Lengths, scales and often intensities of proxy wars increase with greater external support and when the belligerent parties are not interested in peace talks or diplomatic negotiations, an example being the longstanding war that went on in Colombia and was directly related to the illegal drug trafficking of cocaine, the arms trade and money laundering in the US. These proxy wars have had long-standing impacts on people and infrastructure, and Afghanistan is an example of how the mujahedeen were supported by US intelligence agencies during the Soviet-Afghan conflict in the 1980s, but later, Al-Qaeda, which emerged from these guerrilla groups, was accused by the US of being responsible for the terrorist attack on 11 September 2001.

Modern legal systems and the consolidation of global international organisations (WTO, IMF) based on international law will need fewer proxy wars to gain access to strategic resources, because private concessions for mining extraction are a new way of resource depletion in the Global South. Mining activities began during colonial times and have continued and intensified today through new concessions that were granted to multinational corporations. Further, supported by the illegal arms trade from industrialised countries and illegal mining groups, warlords with private armies are increasing the mining conflicts, especially in Africa. But so-called legal mining extraction faces increasing opposition, due to the environmental destruction resulting from opencast mines and the pollution of strategic resources such as water, air and land in southern countries. Mining has also exacerbated multiple existing internal conflicts between ethnic groups and local landlords (Arach). In Latin America Canadian mining companies are generating the highest conflict potential and most of their concessions lack the legal consent of the indigenous people with regard to the right to land and water use, as stipulated in Art. 169 of the ILO Convention.[16] In 2019, in Latin America 245 mining conflicts existed, 43 of which occurred in Mexico, 41 in Chile, 39 in Peru, 29 in Argentina, 26 in Brazil, 14 in Colombia and 10 in Guatemala, among others.[17] All these conflicts are limiting the safe access of native people to water, food, energy and their cultural security.

During the *World Economic Forum* (WEF 2011), the global elite was concerned about the potential risks related to the nexus between food, water and energy security. Food security was introduced early by the FAO (1983) with the aim of eliminating hunger and malnutrition. It refers to "a situation that exists when all people, at all times, have physical, social and economic access to sufficient, safe and nutritious food that meets their dietary needs and food preferences for an active and healthy life" (FAO 1996). However, multinational food enterprises are more interested in selling industrialised low quality transformed food at high prices and controlling the supply of basic grains to the Global South with genetically modified organisms.

Facing the cruel reality of long-standing hunger, La Via Campesina (LVC 2018) developed an alternative concept of food sovereignty, which stresses that women are key food producers globally, but lack official and private support. Food sovereignty includes the basic right to consume safe, sufficient, and culturally accepted food produced locally with native seeds; access to land, credit and the basic means of production, especially for women, girls, indigenous people and peasants; access to land, water, native seeds, credit, technical support and financial facilities for all participants; and local production and the trade of agricultural products in local markets (Oswald Spring 2019, 2020).

There is no food security or life without safe water. Water maintains all types of ecosystem, but is also related to economic, societal and health security. About 70% of water is still used in agriculture, and climate change is increasing the water stress in the highly populated drylands of the Global South.

> …water represents one of the great diplomatic and development opportunities of our time. It's not every day you find an issue where effective diplomacy and development will allow you to save millions of lives, feed the hungry, empower women, advance our national security interests, protect the environment, and demonstrate to billions of people that the United States cares, cares about you and your welfare …[18]

Energy security is closely linked to geopolitical events and armed conflicts in the world, especially in the Middle East, but also to oil prices, oil storage capacity, petrochemistry and flexibility in the daily production and refinement capacity. The US *Energy Information Agency* (EIA 2015) stated that stable oil prices and a continuous supply of hydrocarbons are a prerequisite for energy security. Therefore, not only the military dimension, but also political, economic, societal and environmental considerations are relevant for a long-term global energy security.

---

[16]C169 – Indigenous and Tribal Peoples Convention, 1989; see, at: https://www.ilo.org/dyn/normlex/en/f?p=NORMLEXPUB:12100:0::NO::P12100_ILO_CODE:C169.

[17]See at. https://www.statista.com/statistics/824501/latin-america-mining-conflicts-by-country/.

[18]Speech of Secretary of State Hillary Rodham Clinton, at the National Geographic building in Washington DC on 22 March 2010.

Further, the WEF stated that human well-being, human development and peaceful coexistence require healthy ecosystems and ecosystem services with safe water, energy and food supply, the regulation of climate change, and support for the nutrient cycle, carbon sequestration and all the cultural services related to a healthy environment. In its 11th Global Risk Report (WEF 2016), the World Economic Forum assessed the economic, environmental, geopolitical, societal and technological risks. The report noted as key failures the lack of climate change mitigation and adaptation, thus producing scenarios of profound social instability with large-scale migration, state collapse and crises. Climate change impacts triggered by extreme weather events may increase food crises in different parts of the world, along with biodiversity loss, whereby ecosystems may collapse. As a result, new financial crises may increase unemployment and social unrest, confronting governments and infrastructure with unknown challenges and unprecedented environmental migrations. In 2020, the COVID-19 pandemic affected billions of people, caused an economic downturn with millions of unemployed people and personal losses of humans, well-being, unemployment and a severe recession globally. How will we deal with these complex security challenges within a dominant realist or liberal military understanding of peace?

1.3.2 Militarised Borders, Refugees and Migrants

The number of forced displacements due to wars, conflicts, disasters or public insecurity has been growing since the mid-1990s. The UNHCR has identified three core reasons: first, civil, international and internationalised wars have created a large number of refugees; second, the present model of realist security and peace has also produced new conflicts and re-ignited existing ones, e.g. in Syria, Iraq, South Sudan, and at lower level in different parts elsewhere; third, the rate at which solutions are being found for refugees and internally displaced people has been on a falling trend since the end of the Cold War. In 2020 UNHCR estimated that 65.6 million people were forced from their homes, 22.5 million of whom were refugees from war situations. The UN institution estimates further that one in every 113 people globally is now an asylum-seeker, an internally displaced person, or a refugee. This has never existed before and will increase with new climate disasters and global instability.

Further, socio-economic and environmental migration add an additional flow of moving people. Migration has always been a complex phenomenon and is based on the theories on *push factors* (e.g. poverty, lack of jobs, erosion, landless people, and public insecurity), pull factors (better socio-economic, educational, health and environmental conditions) and *mediation factors*, such as trajectories, diaspora and known people in the immigration place. But drought, land-grabbing and disasters with a lack of government support or private investment in drylands have destroyed multiple productive conditions and crop yields. Further, seriously deteriorated soils have increased internal and international migration, due to the loss of soil fertility, harvests, food, income and livelihoods. The international collapse of agricultural commodity prices is also increasing migration in Guatemala, as a result of the collapse of the coffee price in 2019. Rural people in developing countries depend on natural resources for their food and survival and when these are lost due to climate change or mismanagement, these people search for a better livelihood outside to grant their survival.

UN data estimates that in 2019 there were 272 million international migrants, which represents an unprecedented increase from 51 million in 2010. Globally, 3.5% of the world population is migrating and furthermore a total lack of estimates of the number of internal migrants exists. Their number can only be estimated indirectly from the growing area of shanty towns, where people are living in very reduced and often highly risky spaces.

International migrants confront increasingly militarised borders and the SDG 2030 are intended to facilitate orderly, safe and responsible migration. In 2019, Europe hosted 83 million international migrants, the US 59 million and Northern Africa and Western Asia 49 million.[19] Based on their birthplace India has a diaspora of 17.5 million; Mexico of 11.8 million, China of 10.7 million, the Russian Federation of 10 million and the Syrian Arab Republic of 8 million migrants (WMR 2020: 3).

Rising nationalism in the US and Europe, xenophobic nationalist aggression and short-term electoral interests are limiting the prospects of engendered, sustainable and culturally diverse security and peace for international migrants and war refugees in these industrialised nations. Families, women and unaccompanied children are the most vulnerable migrants, highly exposed to all types of physical, sexual and psychological violence, due to illegal or undocumented migration. They will be confronted with capture, rape, family separation, jail, deportation and lengthy asylum processes. Over the last two years all socio-economic causes were eliminated as a justification for asylum in the US. Thus, there is growing concern over how future security and peace will grant everyone, including the most vulnerable people, a liveable world with human rights, respect, equity, equality and dignity.

1.4 From an Unipolar World Towards Diverse Decolonised Systems

The present crisis of the COVID-19 pandemic, the lockdown of billions of people, massive unemployment and the rising deaths with no possibility of receiving a dignified burial indicate that neoliberal capitalism is reaching its limit. The exhaustion of capitalism does not prove the benefits of socialism, and the past experiences of socialism indicate that it is not the way to a desirable future either. Both systems are based on patriarchal violence, militarism, discrimination, authoritarianism and

[19] In 2019 about ten countries hosted most of the migrants: US 51 million or 10 per cent; Germany and Saudi Arabia 13 million each, Russia 12 million, UK 10 million, United Arab Emirates 9 million, France, Canada and Australia around 8 million each and Italy 6 million international migrants.

resource depletion, in contexts which also apply to the present century. What alternatives are there to overcome this perverse circle of multiples crises – health, human, social, economic, political, military, cultural and civilizational risks – in which the world has become entangled? How do we finance a socio-environmental agenda with a gender perspective without restricting social and economic freedoms, especially of the corporate elite?

After WWII winners and losers together agreed on a Marshall Plan, which helped Europe to recover fast from the destruction and to establish a solid economic bloc against the consolidation of the Soviet Union. Now, isn't it time for an economic summit to focus on socio-environmental concerns with gender equity at the centre, linking up public and private efforts towards the promotion of a sustainable and equitable global system? The present COVID-19 pandemic should not end with charity, guilt or fear, but with a sustainable engendered security and peace approach that allows the planet to adopt a solidary scheme of wealth redistribution with the consolidation of institutions that guarantee social rights to everybody (third phase with solidarity); otherwise, the political arena will lose further governance, while violence and selfish behaviour will increase and provoke crises (second phase of Piaget 1950, 1972). The forced eruption of the pandemic is not to save the critical stage of neoliberal capitalism, but to grant a new order that assures the future, in which a decolonised Global South is treated equally, climate change and global environmental change are efficiently combatted, and the harmony between humankind and nature is re-established.

1.4.1 The Global South is Left Alone, Indebted and Poor

Like a lot of idealists, we can find meaning in our lives if we fight for a higher cause: saving children today and tomorrow, protecting the environment and biodiversity, equalising women's rights and respecting cultural diversity. Let's join Greta Thunberg and all other progressive people to save children, women, the poor, nature, humanity and all the other creatures that live with us on Planet Earth. By working in this way, we will make sense of our lives, thus the concrete question is how to save Mother Earth and also humankind.

First, the empowerment of women will give them freedom to choose their family size, thereby controlling population growth, and through education women will promote gift or solidarity economies (Collin 2020; Vaughan 2007) that foster economic development from bottom-up by integrating constructive social changes in complex social settlements.

Second, living well, promoted by indigenous Aymara, is a philosophy of people lifting themselves out of poverty via a subsistence economy within their own cultural and socio-economic context. There is only one good life and there is not a better one.

Third, the education of women and girls, together with their active involvement in political and economic decisions, will allow them to choose modern family-planning, smaller families and limit the patriarchal religious control over their female bodies, the economy and nature.

Fourth, sustainable agriculture, reductions in consumerism and waste generation, healthy organic food, less meat, the recovery of deteriorated soils, massive reforestation, and the reuse and recycling of water will improve the health of people, boost immune systems and reduce the conditions for the emergence of new viruses, due to animals recovering their natural habitats.

Fifth, sustainable urbanisation with renewable energy, public transportation, roof gardens, parks with fruit trees, orchards and vertical agriculture in balconies, together with water and waste recycling, ecosystem restoration and food sovereignty, will enable even the poorest people to enjoy a dignified quality of life without slum development.

Sixth, climate-induced extreme events result in disasters because of the lack of early warning, disaster risk reduction preparation, mitigation, adaptation and resilience-building in vulnerable regions. Ecosystem restoration has always been a powerful ally which can reduce extreme events, such as flash floods. For instance, massive reforestation helps to retain soils and water in mountainous areas; mangrove protection reduces the strong winds and waves of hurricanes; and integrated basin management mitigates the impacts of drought and increases the infiltration of rain into aquifers. All these efforts will create local jobs, reduce disasters, loss and damage, increase the quality of life, and improve livelihoods.

1.5 Content of the Book

This book is divided into four parts, starting with a theoretical understanding of peace research in the Anthropocene. This introductory chapter has questioned the dominant realist and liberal understanding of security and peace, developed primarily by white male researchers. The author has proposed a decolonised security and peace concept, with gender equality and preventive behaviour in a world affected by stronger and more frequent climate disasters, economic crises and a pandemic.

Part I: Peace Research Epistemology for the Anthropocene

In this part five chapters address epistemological problems of peace research approaches for the Anthropocene Epoch of Earth and human history. This part starts with a research proposal for a new peace ecology in this new epoch, which is followed by a suggested transformative and participative peace with a theoretical and methodological proposal for an epistemology for peace and conflict studies, while the third text explores peaceful societies through the perspective of social cohesion and discusses the power of paradigms for normative and interdisciplinary research. The last two texts discuss the national and universal importance of the nonviolent policy of Mohandas K. Gandhi and a conceptualisation of a disobedient peace based on non-cooperation with inhuman orders (Sects. 1.5.1–1.5.5).

1.5.1 Peace Ecology in the Anthropocene

Hans Günter Brauch, the editor of this book series, outlines 'peace ecology' as a new scientific approach that aims at to build bridges between peace research and ecology and discusses peace research within the context of the Anthropocene. In 2000, the Dutch Nobel Laureate Paul J. Crutzen introduced the 'Anthropocene' as a new epoch of Earth's history. Direct human interventions into the Earth System through the accumulation of greenhouse gases and *carbon dioxide* (CO_2) in the atmosphere have caused multiple societal impacts resulting in rapid increases in production, consumption, urbanisation, pollution, migration, crises and conflicts. This chapter develops 'peace ecology' in the context of the 'Anthropocene'.

After a detailed conceptual introduction, the chapter discusses five alternative starting points of the Anthropocene: the Agricultural and Industrial Revolutions, the Columbian Exchange, the nuclear era and the 'Great Acceleration'. The chapter offers a conceptual mapping of the Anthropocene and interprets the Anthropocene as a turning point, context and challenge for science and politics. For the Anthropocene it suggests a rethinking of peace and the evolution of peace research since the end of World War I, World War II and the Cold War in selected countries and the development of three international peace research organisations. A reconceptualisation of peace in the Post-Cold War Era (1990-2020) and in the Anthropocene has also taken place. The text reviews the evolution and rethinking of several ecology concepts and (political) geoecological approaches in the Anthropocene.

A major part reviews past bridge-building initiatives between peace research and ecology by various scholars (Kenneth and Elise Boulding, Arthur H. Westing et al.) and approaches that were suggested in studies on environmental security and in case studies by Günter Bächler and Thomas Homer Dixon on environmental degradation, scarcity and stress as causes of and as conflictive outcomes. Since 1990 debates have evolved on environmental peacemaking and post-conflict peacebuilding and on climate change, security and conflict linkages. While older bridge-building efforts have addressed issues related to violence, recent discourses have addressed issues of sustainability transition and their impact on sustainable peace initiatives in the Anthropocene.

The suggested peace ecology approach and research programme is conceived as a holistic scientific project for the Anthropocene that requires the fragmentation of scientific and political knowledge to be overcome and combined with holistic perspectives and transformative approaches that facilitate the move from knowledge to action. The author proposes an ecological peace policy in the Anthropocene by developing strategies and policies for addressing the challenges in the Anthropocene. The author concludes by proposing a peace ecology research programme and an ecological peace policy in the Anthropocene.

1.5.2 *Transformative and Participative Peace: A Theoretical and Methodological Proposal of an Epistemology for Peace and Conflict Studies*

Esteban Ramos, a peace researcher at the National University of Honduras, is looking at vulnerable people and proposes a theoretical approach for decolonising conflicts in the most violent country in Latin America: Honduras. His approach starts from bottom up and links it to social participation in order to achieve a 'transformative peace' that allows people to overcome the present violence and insecurity. Epistemologically he justifies the need to work with the whole population, not just from top-down with governmental functionaries, in order to overcome the increasing violence. Corruption, gang-organisation, poverty, migrants and family disintegration are all factors of this new war (Kaldor 2012) that obliges peace researchers to promote different peace education and actions for peacebuilding with the subjects and the objects of conflict. He further asserts that his new episteme of peace and conflict studies is based on Freire's action-reflection approach, inserted into social-praxical daily life. He uses medical science symbols to cure the social and violence illnesses, and his theoretical development of peace and conflict studies allows a different epistemological perspective in Honduras. Thus, the conception of a transformative peace puts the human being at the centre as his practical builder of social reality that makes it possible to configure an epistemological framework in the discipline of peace and conflict studies, which requires the full involvement of the protagonists of peace, violence and conflict.

1.5.3 *Peaceful Societies Through Social Cohesion? the Power of Paradigms for Normative and Interdisciplinary Research*

Katarina Marej, a German Social Anthropologist at Münster University, addresses the functions and constructions of collective identities and memories and their transformation modes and conditions in micro–macro processes. The focus is centred on the manifold connections between culture, politics, education and the role of academia for socio-cultural changes. Her research is orientated towards qualitative approaches, including interdisciplinary transformative research, semiotics, and deep-hermeneutics, as well as (inter)textuality, (de-)construction, and postcolonial analyses. She asks how the *Sustainable Development Goals* (SDG) 2030 can be achieved when positive peace is rarely put on the political agenda. She argues that social cohesion and the reconstruction of social tissue is an issue of growing interest because with nine billion humans soon likely to be on Earth we need new ways to get food and live peacefully together within a context of diverse societies and cultural contexts. This approach challenges the existing paradigmatic positions and business-as-usual and she analyses ontological and axiological aspects for finding alternative methodologies and epistemologies to increase interdisciplinary communication. This may be a tool for shaping peaceful social cohesion in a global society and a way to achieve decolonised security and peace management.

1.5.4 The National and Universal Importance of the Non-violent Policy of Mohandas K. Gandhi

Egbert Jahn, an Emeritus Professor at Mannheim University in Germany, explores India's attainment of national independence, primarily as the result of non-violent actions – or was it the moderate British colonial policies in a complex war situation? His chapter analyses numerous domestic political factors in India and Britain related to the weakening of the British Empire. This geopolitical conjuncture paved the way for the success of the largely non-violent independence movement led by Mohandas Karamchand Gandhi. His philosophy and actions were not orientated towards obtaining power in an independent India and he only rarely took up official positions in the Indian National Congress. Widely respected by the Indian masses and elsewhere, he stressed the importance of taking responsibility for one's own actions, but also noted the challenges that can limit the resistance against injustice in one's own surroundings. His tools were non-cooperation and civil disobedience, carefully prepared to escalate non-violent policies. Gandhi's basic principles of non-violent social behaviour and policies have universal significance, and neoliberal democracies and dictatorships are progressively losing their legitimacy among citizens who are suffering from injustice, poverty, exploitation and discrimination.

1.5.5 Disobedient Peace: Non-cooperation with Inhuman Orders

Pietro Ameglio, from the National Autonomous University of Mexico (UNAM), continues with Gandhi's non-cooperation and his chapter develops a practical theory of formal and informal education that justifies direct action. His conceptual approach is based on three decades of experience of non-violent civil resistance in Mexico and Latin America. His learning is conceptualised and systematised in this chapter, focusing on the reflection-action processes which enable us to oppose 'inhuman orders'. He considers this action to be a key concept and practice for building 'disobedient peace', which he uses to conceptualise and enrich studies on peace. However, the complexity of 'disobedience' requires us to have the ability to analyse and understand the social order in order to work out how to act in solidarity and develop a comparable epistemology. In this case it is a key factor for a decolonised peace process able to question top-down inhuman orders and promote humanised bottom-up actions.

Part II: Conflicts, Peace, Gender, Families and Vulnerable People

In Part II researchers from the Global South and some from the Global North analyse the emerging problems related to chaotic urbanisation, land and herder conflicts in the Sahel, ethnic clashes and poverty that have changed the traditional livelihoods in patriarchal rural societies. Sexual abuse, human rights, the imposition of traditional male values on land, women and especially the youth have produced new responses whereby the Roma, who are living without territories, are adapting culturally and socially to the changing conditions of a globalised world (Sects. 1.5.6–1.5.14).

1.5.6 Risks, Mitigation and Adaptation to Urban Climate Change Impacts in the Global South from a Gender Perspective

Úrsula Oswald Spring from the National Autonomous University of Mexico reviews the Talanoa Dialogue proposed by the Bali President to reduce impacts and increase peacebuilding in regions where climate change effects will increase the number and intensity of disasters in urban areas and often force people to leave high-risk areas and settle down elsewhere. She further examines how governments and urban dwellers are promoting multilevel governance actions to prevent conflicts and increase disaster risk reduction, mitigation and adaptation, especially among extremely exposed social groups. She analyses the dual vulnerability of women, children and the elderly and the potential for empowerment from bottom-up, and reviews the interchange of experiences between cities in the Global South. Without any doubt, cities are the most significant producers of GHG emissions and therefore key to reducing climate change impacts and also air pollution by promoting clean energy, sustainable transportation and a radical carbon sink of greenhouse gases in highly agglomerated regions. Urban agriculture on roofs, balconies and parks, the use of bicycles, more local activities and greater integration of urban colonies are citizen efforts to improve air quality. All these activities may help to prevent conflicts and disasters, especially when stakeholders are trained in participative governance and conflict resolution within a framework of cultural and socio-environmental diversity.

1.5.7 Conflicts in Kenya: Drivers of Conflicts and Assessing Mitigation Measures

Charles Wasike from the University of Nairobi in Kenya argues that it is necessary to understand the origins, trends and evolution of conflicts within the context of the political history of a country in order to deal with upcoming threats to security, which differ in both form and dynamics. Kenya suffers from widespread violence, exemplified by high levels of sexual and gender-based violence, intercommunal confrontations, political conflicts during electoral cycles and an increasing number of terrorist attacks. The author asserts that the incidence, gravity and intensity of the phenomena have increased over the years. Empirically, he compares the multiple and overlapping confrontations in the Rift Valley, the peripheral pastoralist drylands nearby Nairobi, and the coastal region. Ethnic intolerance, border conflicts, political party influence, internal colonialism and competition over land and other crucial resources such as water emerge in any conflict region. The proliferation of small arms, weak public security and the high poverty rate are factors that have increased local and regional risks of violence in Kenya, where the corruption of the elite and the manipulation of local communities has opened space for violent Islamist activities in the north-eastern part, which has borders with Somalia and Ethiopia. The official politicisation of the counterterrorism response and the scapegoating of certain ethnic and religious groups have served to consolidate Al-Shabaab's propaganda. The author's proposal for reducing conflicts is similar to that needed in other emerging countries: support for sustainable economic development, the reduction of inequalities, provision for basic needs, gender equality, transparency and equal power relations between women and men.

1.5.8 Human Rights and Sexual Abuse of the Girl-Child in Nigeria: Implications for Development

Janet Monisola Oluwaleye from the Ekiti State University in Nigeria has studied the sexual abuse of girl-children, using secondary sources. She has written a qualitative analysis of the nature and scope, causes, and consequences of the sexual abuse of girl-children, including poverty, hawking, lack of values and parental control, pornographic pictures and abuse in social media. She scrutinises the get-rich-quick mentality, unemployment, broken homes, single parenting, health hazards, population growth, bad governance, weak law enforcement and the culture of backwardness and cheating. All these factors are reinforcing abuse inside the household and in local and national society. The author cannot find any differences in illegal behaviour between rich and poor people, teachers, religious leaders or laymen, especially when family members and neighbours are involved in this crime. She finishes by recommending realistic and effective government policies, job promotion, interventions by community and religious leaders, gender empowerment and high penalties for offenders to curb these criminal acts.

1.5.9 *The Farmer-Herder Conflicts in Nigeria's Open Space: Taming the Tide*

Aminu Bakari Buba from the Federal University of Kashere, Gombe State in Nigeria, has studied the nomadic Fulbe group in Nigeria and its conflicts with farmers over shared natural resources (pasture, water and productive land for livelihoods). Environmental changes in the northern part of Nigeria have produced more irregular precipitation and rising temperatures, which are aggravated by land degradation, droughts, overgrazing, declining food production and less water availability. Thus, the farmer-herder conflicts might increase in the coming years. To find alternatives, the author proposes articulating to the government the historical antecedent of the Fulbes' mobility and migration as a lifestyle, and recommends engaging in strategies to effectively combat the desertification process in conditions of growing climate change.

1.5.10 *Climate Rituals: Cultural Response for Climate Change Adaptations in Africa*

Mokua Ombati from the Anthropology and Human Ecology Department of Moi University in Kenya reviews the threat to Africa's development and growth posed by climate change, and explores climate-resilient development pathways which are capable of reducing risks and creating opportunities for development. He analyses the interactions between climate and culture, traditional knowledge and technologies. He frames the weather and climate-related modification rituals and asserts that over the centuries cultures across Africa have adapted to climate variation. He has selected representative cases within different socio-cultural, economic, political, geographical and anthropological settings across Africa that demonstrate how climate variations are successfully managed within their civilisations and historical backgrounds, where 'climate rituals' are a prototype of cultural responses to a changing climate that are able to reduce upcoming conflicts.

1.5.11 *Ethnically-Charged Wartime Sexual Violence: The Agony of the South Sudanese Refugees in Uganda*

Robert Senath Esuruku from Makarere University in Kampala, Uganda explores sexual violence, including rape, gang rape, sexual mutilation, castration, and forced nudity since the outbreak of civil war in South Sudan in December 2013. He relays the narratives of some of the survivors, who are largely living in refugee settlements in the West Nile region of Uganda, where they are trying to recover from the physical, psychological and social impacts of the ethnically-charged sexual attacks. He also examines how sexual violence was part of a deliberate strategy to terrorise, degrade, shame and humiliate both the victims and their ethnic and political groups.

1.5.12 *Traditional Conflict and Peacemaking Processes: The Case of Kurdish Tribes in Mardin, Turkey*

Safiye Ateş Burç from Artuklu University in Turkey explores a traditional approach to peacebuilding based on local loyalties in everyday life among Kurdish tribes in the south-eastern part of Turkey. The existence of these tribes was historically characterised by conflicts and reconciliations, where *Xwin* (blood feud) was and continues to be the unique traditional conflict management of this community. *Bêş/bêj* (blood money) is the most common form of traditional peacemaking and *nanê aşitiyê* (a dinner of peace) is accepted as a traditional peacebuilding ritual. The author states that among Kurdish tribes' traditional practices plays an effective role in peacemaking and conflict resolution.

1.5.13 *Family Shame and Eloping Couples: A Hindustani Warp in Time – Steps in Progress Towards Non-violence*

David Evans, an Australian researcher, analyses the Hindustani Warp as an approach to non-violence, analysing its social emergence in relation to a forbidden love marriage and its consequences, predominantly in India and Nepal, but also in Pakistan. Patriarchal social rules and taboos around love marriage have led to couples who transgress them being punished, persecuted and murdered. The author's research examines media reports, personal communications and Hindustani love legends to scrutinise local and national cultural obstacles to positive acceptance of love marriages by families and communities. A shift in socio-cultural thinking and behaviour rather than unquestioning adherence to androcentric centuries-old marriage rules is needed to overcome patriarchal customs and promote greater gender equity in this traditional mindset.

1.5.14 *We Are not Victims: The Roma, an Outdoor Art Gallery and the Same Old Story – Critical Thinking in Communication for Development*

Maria Subert from the City University of New York analyses a historically disadvantaged ethnic group, the Roma, and peacebuilding attempts to 'help' which embrace co-operation, dialogue, collaborative problem-solving, social justice and communal well-being to prevent conflict and different forms of oppression in their community. The author scrutinises the truth behind occidental helpers, who have historically lacked the skills and insight to provide the right kind of empowering support to communities which primarily just want the opportunity to direct their own lives and provide for themselves. The inhabitants of Roma or Gypsy villages in Hungary find 'help' unacceptable because it restricts villagers to limited external perceptions of their cultural identity, builds a need industry around them, and portrays Roma as passive, incapable and dependent on non-Roma surroundings and support. This paternalistic help overlooks the ethical norms of the locals. Consequently, the author highlights the need for "(1) major concepts of identity in the helper's narrative that reveal the helper's position and that of those helped; (2) narratives of material and non-material needs; and (3) narratives of pride and need as dialectical counterparts".

Part III: Peacebuilding in the Anthropocene

Part III analyses processes, norms and ideologies of peacekeeping, governance concerns in non-state actors, grass-root post-conflict peacebuilding, hydrodiplomacy, refugee crises and transitional justice which takes into account internal colonialism (Sects. 1.5.16–1.5.21).

1.5.15 A Discourse on the Norms and Ideologies of Peacekeeping

Jude Cocodia from Niger Delta University in Nigeria explores the international community's need to prevent human catastrophes. He scrutinises liberal values in projects that provide security and prevent conflict areas from becoming terrorist enclaves. His chapter adopts a traditional approach to the challenges of peacekeeping, its neo-realist manipulations and the implications for troop-contributing countries. He critically revisits the ideological foundations of peace studies and peacekeeping theories. Robust missions are usually large and have generated unintended effects, such as sex trafficking, an increase in dual economies, and manipulation to sustain autocratic regimes. The author concludes that the legitimacy and authority of international peacekeeping organisations derives from its ability to harmonise these different motives and aims into cohesive action.

1.5.16 Governance by Violent Non-state Actors as a Challenge to Sustainable Peace in Brazil

Marcos Alan S.V. Ferreira from the Universidade Federal da Paraíba in Brazil examines the role of *violent non-state actors* (VNSA) and their capacity to offer governance in poor urban communities in Brazil, where state security forces fail to provide their citizens with protection, human dignity and social security. The author examines two criminal organisations – *Primeiro Comando da Capital* [PCC] (First Capital Command) and *Família do Norte* [FDN] (Northern Family). The governance by VNSA has been transformative, first in the prison system, and later reaching poor neighbourhoods and creating new practices and norms to regulate social spaces. By promoting education and income, open spaces and legitimacy for these VNSAs, rules of conviviality, conflict resolution processes and peaceful ways to discourage the theft of material goods, many young people have combatted cultural, structural and direct violence inside their favelas. However, organised crime and the lack of a clear public policy is undermining the power of VNSA nowadays, and hindering the efforts of peace researchers and social scientists who are aiming for a more peaceful Brazilian society.

1.6 Art of Peace: Cultural Practices and Peacebuilding in Mexico

Alfonso Hernández Gómez from the Autonomous Metropolitan University in Mexico reviews the poor results of the strategy used by the military and the police to control drug cartels and local gangs. Using ethnographic research and direct fieldwork in some of the most violent neighbourhoods in Mexico, he has contributed to innovative strategies which favour cultural activities for peacebuilding as an alternative to militarised violence prevention. Cultural actions may engage citizens and make them resilient, although the use of arts for non-violence needs a multilayered intercultural approach that reveals the subtle forms of change within the individuals. Cultural work cannot erase violence by itself and peacebuilding is not that homogeneous, but depends on the specific conditions of violence in each cultural background. Cultural practices are a means of healing violence, since they promote mutual respect, joy and solidarity, help to restore a sense of community and rebuild social tissue – positive outcomes which were all achieved when adopted by those who have suffered violence in Mexico.

1.6.1 Grass-Roots Post-Conflict Peacebuilding: A Case Study of Mosintuwu Women's School in the Poso District in Central Sulawesi, Indonesia

Putri Ariza Kristimanta from the Centre for Political Studies at the Indonesian Institute of Sciences focuses on the grass-roots peacebuilding movement in a post-conflict society in the Poso District of Indonesia, where a communal conflict occurred (1998–2001). Based on a case study with a qualitative approach, her chapter explores the contribution of women to peacebuilding and peace education. The author has found that grass-roots movements are essential in peacebuilding efforts, because they contribute to the creation of sustainable peace by restoring the social conditions that facilitate the implementation of peace treaties, the dissemination of peace issues, the establishment of a culture of peace, and the teaching and understanding of conflicts and peace, and contribute towards the independence of society. Women who attended the Mosintuwu Women's School played a critical role in peacebuilding because they prioritised transformation, empowerment, healing and forgiveness, together with gender justice and equality. Further, the peace process highlighted in the media tended to be patriarchal and elitist, while the voices and peace stories that came from women tended to be calm. Love increased in peaceful society and women represented a role model for their environment by advocating community rights, sharing opinions in male-dominated discussion forums, and advocating in cases of violence against women and children at the village level, thus questioning the dominant patriarchal paradigm of peace management by the governments.

1.6.2 Hydro-Diplomacy Towards Peace Ecology: The Case of the Indus Water Treaty Between India and Pakistan

Saima Sabit Ali from Pakistan and Bhargava Mansee Bal from India analyse the Indus Water Treaty (1960) between India and Pakistan as a successful case of hydro-diplomacy. The Treaty covers six rivers of the Indus River Basin located in the two countries, namely the Indus, Jhelum, Chenab, Ravi, Beas and Sutlej, where water matters have snowballed into the geopolitics of the two countries. The Treaty has survived frequent tensions, including conflicts, and has provided a framework for irrigation and hydropower development in the region for six decades. The impacts of climate change now require the close monitoring and sharing of information to safeguard the ecological systems in both countries. Water has a connective capacity to encourage people and politicians to unite in dialogues, especially when billions of people on both sides depend on a collective culture of care for the environment. The impacts of climate change are exacerbated by population growth, and affect most water-stressed regions, thus the complications that lie deeper are the conflicting geopolitics that dominate water management. Hydro-diplomacy coupled with water co-operation offers hope in complex transboundary water situations, especially where one riparian is stronger politically, militarily, economically, or technically, and also within regions that encompass estranged neighbours and unstable border settings. The Treaty has also built bridges over the common concern to tackle the challenge of climate change, which in turn has helped to enforce compliance with the Treaty, built confidence, and reduced the distrust between two nations with border conflicts in Kashmir.

1.6.3 *Rohingya Refugee Crisis: Implications for Regional Security*

Nurul Momem from Bangladesh analyses the situation of more than one million stateless Rohingya people who have fled to Bangladesh from Myanmar since August 2017. This stateless population is considered to be a threat to regional security and peace in South Asia. The author explores the security challenges to local livelihoods; the political competition between different factions; and the future risks of radicalisation among the statelessness of the ethnic minority Rohingya in a very poor host country that is not fully capable of satisfying the basic needs of all refugees. The current Rohingya refugee situation is characterised by vulnerabilities which the *United Nations* and regional governments have largely failed to treat – not solely the humanitarian crisis, but also the activities of the Rohingya within and outside the refugee camps, which may become a threat to the security and peace of the host country.

1.6.4 *An Unsustainable Price: The Opportunity Costs of Transitional Justice*

Helen Ware from the University of New England in Australia reviews the role of transitional justice, not only as a post-conflict measure, but also as a practical action plan. Citizens in post-conflict areas experience inequities on a daily basis. Such situations require actions – not just statistical reports – but economic policy-makers and service delivery departments are slow to transform existing inequities. Without any doubt, transitional justice is crucial in violent conflict situations. The author argues that during the transition period, in addition to the justice issues, greater priority should be given to economic policies so that governments will be able to achieve economic progress and well-being for their citizens. This may improve their daily lives, which cannot be accomplished by the delivery of truth and forensic justice alone.

Part IV: Peace, Development and Education

Part IV reviews the design of the Sustainable Development Goals 2030 (SDG 2015) at local level and with regard to peacebuilding. Without doubt, citizen-led peace education from bottom-up and orientated towards action is a primary tool for increasing participatory governance and resolving conflicts in communities and at national and regional level. Development infrastructure is crucial for improving livelihoods, but maladapted projects frequently cause more damage and expel people from their homeland, whereas integrated public collaboration over water, sewage and roads, which includes the advice of local women and men, may improve the livelihoods and health of everybody (1.5.22–1.5.24).

1.6.5 Simultaneous Intervention Strategies at Local Ecosystems for Sustainable Development Goals and Peace: Design and Systems Perspectives

Amar Nayak from the Xavier University Bhubaneswar in India is re-visualising the *Sustainable Development Goals* (SDG) from a design and systems science perspective to provide a better framework for sustainable interventions at a lower cost. The author argues that simultaneous interventions rather than sequential interventions on each of the six dimensions in an ecosystem could help achieve the seventeen SDGs without damaging the environment. He criticises the reductionist approach of science with a limited perspective of hybrid competition and co-operation instead of exploring sustainability from a co-operative perspective. His methodologies explore alternatives for sustainability that go beyond enquiries, since empiricism is rarely consistent with inter-disciplinary system thinking, action research and holistic processes for exploring new alternatives towards successful transition processes.

1.6.6 Citizen-Led Assessment and the Participatory Approach to Peace Education in Nepal

Rajib Timalsina from Tribhuvan University in Nepal has studied the long-time instability when power was transferred to 753 new local governments in the education and health care sectors. His chapter focuses on two aspects of current education governance. The first is the problematic nature of global and national agendas in learning assessment, where traditional 'pen and paper' based assessments which do not match the needs of local communities were enforced. In this regard, the author discusses an exemplary project on participatory pedagogic methods that has mobilised hundreds of young undergraduate students to assess the reading and mathematical skills of children aged between five and sixteen years. The second focus is the problem of teachers' sometimes violent disciplinary techniques of instruction. Participatory exercises trained teachers to use non-violent communication and solve classroom challenges peacefully. This participatory peace education approach used two innovative intervention models: 1) a citizen-led learning assessment model and 2) non-violent communication tools to promote a culture of peace in schools. This research project concluded that citizen-led assessment methods on measuring the learning outcomes of the children became a very useful tool for new local bodies to practise evidence-based decision-making in Nepal. Engaged citizens can hold local governments accountable for ensuring better learning and also serve as resources that can contribute to improved learning processes. This citizen-led assessment made decisions through peaceful means but also trained a pool of youth that may be able to improve the quality of education locally. This method further encouraged children to stay in school because it is child-friendly and peaceful, therefore it simultaneously addressed the school dropout problem.

1.6.7 Can Infrastructure-Based Development Bring Peace to West Papua?

Cahyo Pamungkas from East Jakarta in Indonesia discusses positive impacts on peace-building arising from road construction in West Papua between 2014 and 2018. Since 2014 the Indonesian government has given more attention to coping with separatism in the study region. One of the national government efforts was promoting several road construction projects to increase the connectivity between districts in West Papua. The author argues that infrastructure-based development has had only a minor impact on improving the livelihoods of the indigenous West Papuans, so this policy is not entirely successful from a peacebuilding perspective. His research suggests that the national government should conduct a mediated political settlement via dialogues with the separatist group, because the implemented development did not benefit the indigenous population, who suffered from the violent conflict. The Trans-Papua highway has not yet improved the livelihoods of the indigenous population, and though job diversification has occurred, during the study period it was unable to shift most people from subsistence agriculture towards other economic activities. Conversely, the road is still seen as an immense threat because it facilitates the entry of drugs and alcohol, while imported pigs have endangered indigenous pig-raising activities. The hybrid peacebuilding approach from the Indonesian government, combining both liberal and illiberal peace, has not eradicated the internal colonialism between Jakarta and West Papua, because it was not rooted in cultural values, and the imposed development ideology, complemented with a military security approach, has not even granted negative peace.[20]

References

Adger, W.N.; Pulhin, J.M.; Barnett, J.; Dabelko, G.D.; Hovelsrud, G.K.; Levy, M.; Oswald Spring, Ú.; Vogel, C.H., 2014: "Human Security", in Field, C.B.; Barros, V.R.; Dokken, D.J.; Mach, K.J.; Mastrandrea, M.D.; Bilir, T.E.; Chatterjee, M.; Ebi, K.L.; Estrada, Y.O.; Genova, R.C.; Girma, B.; Kissel, E.S.; Levy, A.N.; MacCracken, S.; Mastrandrea, P.R.; White, L.L. (Eds.), 2014: *Climate Change 2014: Impacts, Adaptation, and Vulnerability, Part A: Global and Sectoral Aspects:*

[20] *Nonviolence* is American English and *non-violence* is British English. As this book is in British English based on *The New Oxford Style Manual* (Oxford - New York: Oxford University Press 2016): 61 the recommended rule on the hyphenation of prefixes is to "use a hyphen to avoid confusion of mispronunciation". This rule also applies for compound words beginning with non- (p. 413). These rules were implemented in this book and in the index. This author preferred the American spelling of 'nonviolence'.

Contribution of Working Group II to the Fifth Assessment Report of the Intergovernmental Panel on Climate Change (Cambridge – New York: Cambridge University Press): 755–791.

Amin, Samir (1977): *Imperialism and Unequal Development*, at: https://maoistrebelnews.files.wordpress.com/2015/04/imperialism-unequal-development.pdf.

Anderson, Perry, 1991: *Imagined Communities: Reflections on the Origin and Spread of Nationalism* (London: Verso).

Arach, Omar, 2018: "'Como un Ejército en Territorio Enemigo': Acerca de la Violencia Epistémica en la Expansión Megaextractiva", in: Oswald Spring, Úrsula; Serrano Oswald, Serena Erendira (Eds.): *Riesgos Socioambientales: Paz y Seguridad en América Latina* (Cuernavaca: CRIM-UNAM): 123–138.

Asad, Talal (ed. 1973): *Anthropology and the Colonial Encounter* (Atlantic Highlands:, Humanities Press).

Beck, Ulrich, 2009: *World at Risk* (Cambridge: Polity Press).

Beck, Ulrich, 2011: "Living in and Coping with World Risk Society", in: Brauch, Hans Günter; Oswald Spring, Úrsula; Mesjasz, Czeslaw; Grin, John; Kameri-Mbote, Patricia; Chourou, Béchir; Dunay, Pal; Birkmann, Jörn (Eds.): *Coping with Global Environmental Change, Disasters and Security: Threats, Challenges, Vulnerabilities and Risks* (Berlin – Heidelberg – New York: Springer-Verlag, 2011): 11–16.

Bingham, Water, 1947: "Military Science in War and Peace", in: *Science*, (22 August): 155–160.

Blazquez Graff, Norma (Ed.), 2014: *Evaluación Académica: Sesgos de Género* (Mexico City: CEIICH-UNAM).

Braudel, Fernand, 1950: *A History of Civilizations* (London: Penguin Books).

Braudel, Fernand, 1993: *A History of Civilizations*, translated by Richard Mayne (New York: Penguin Books).

Brundtland, Gro Harlem, 1987: *Report of the World Commission on Environment and Development: Our Common Future.*

Chomsky, Noam, 1988: *El Lenguaje y los Problemas del Conocimiento: Conferencias de Managua I* (Madrid: Visor).

Crutzen, Paul J., 2002: "Geology of Mankind", in: *Nature*, 415, 6867: 23.

Darwin, Charles, 1859: *The Origin of Species* (London: John Murray).

Dvorsky, George, 2016: "New Evidence Suggests Human Beings are a Geological Force of Nature"; at: https://gizmodo.com/new-evidence-suggests-human-beings-are-a-geological-for-1751429480.

EIA, 2015: *Annual Energy Outlook* (Washington: EIA).

Ellis, Erle, 2018: *Anthropocene: A Very Short Introduction* (Oxford, Oxford University Press); at: doi: https://doi.org/10.1093/actrade/9780198792987.001.0001.

Falk, Richard, 2019: "Cosmopolitan Path to Peace", in Kulnazarova, Aigul; Popovski, Vesselin (Eds.): *The Palgrave Handbook of Global Approaches to Peace* (Cham: Palgrave Macmillan): 29–46.

FAO [Food and Agriculture Organization of the UN], 1983: *The State of Food and Agriculture* (Rome: FAO).

FAO, 1996: "Rome Declaration on World Food Security and World Food Summit Plan of Action", *World Food Summit, 13–17 November* (Rome: FAO).

FAO [Food and Agriculture Organization of the UN], 2013: *Climate-Smart Agriculture: Sourcebook*; at: https://www.fao.org/3/a-i3325e.pdf.

FAO, 2016a: *Climate Change and Food Security: Risks and Responses* (Rome: FAO).

FAO, 2016b: *Status of World's Soil Resources* (Rome: FAO).

Fontan, Victoria, 2012: *Decolonizing Peace* (Lake Oswego, Dignity Press).

Freire, Paulo, 1968: *Pedagogia do Oprimido* (Rio de Janeiro: Paz et Terra).

Freire, Paulo, 1992: *Pedagogía de la Esperanza*; at: www.cronicon.net/paginas/Documentos/paq2/No.11.pdf.

Fukuyama, Francis, 1992: *The End of the History and the Last Men* (New York: Free Press).

Garatuza, Jaime; Rodríguez, Julio César; Watts, Christopher, 2011: "Environmental Monitoring and Crop Water Demand", in: Oswald Spring, Úrsula (Ed.) *Water Resources in Mexico: Scarcity, Degradation, Stress, Conflicts, Management and Policy* (Berlin Heidelberg: Springer): 101–110.
Gartzke, Erik, 2007: "The Capitalist Peace", in: *American Journal of Political Science*, 51,1: 166-191.
Giménez Guzmán, M. Lucero, 2020: "Importancia del Estudio de las Masculinidades: Mandatos de Género y Algunas de sus Consecuencias", in: Oswald Spring, Úrsula; Hernández Pozo, M.R.; Velázquez Gutiérrez, M. (Eds.): *Transformando al Mundo y a México: Objetivos de Desarrollo Sostenible 2030: Justicia, Bienestar, Igualdad y Paz con Perspectiva de Género* (Cuernavaca: CRIM-UNAM [in print]).
González, Casanova, Pablo, 1963: *Sociedad Plural, Colonialismo Interno y Desarrollo* (París: UNESCO).
Harding, Sandra, 1988: *Is Science Multicultural? Postcolonialism, Feminism, and Epistemologies* (Bloomington: Indiana University Press).
Heaney, Robert (Ed.) 2016; *From Historical to Critical Post-Colonial Theology: African Christian Series Studies* (San Jose, Pickwick Publications).
Huntington, Samuel, 1993: "The Clash of Civilization", in: *Foreign Affairs*, 72,3 (Summer): 22–49.
Illich, Ivan, 1971: *Deschooling Society* (New York: Harper & Row).
Illich, Ivan, 1973: "Inverting Politics, Retooling Society", in: *The American Poetry Review,* 2,3 (May/June): 51–53.
Illich, Ivan, 1975: *Némesis Médica: La Expropiación de la Salud* (Barcelona, Barral Editores).
IPCC [Intergovernmental Panel on Climate Change], 2013: *Climate Change 2013: The Physical Science Basis: Working Group I Contribution to the IPCC Fifth Assessment Report*, (Cambridge: Cambridge University Press).
IPCC [Intergovernmental Panel on Climate Change], 2014: *Climate Change 2014: Impacts, Adaptation, and Vulnerability: Working Group II Contribution to the IPCC Fifth Assessment Report* (Cambridge: Cambridge University Press).
IPCC [Intergovernmental Panel on Climate Change], 2019: *The Ocean and the Cryosphere in a Changing Climate* (Cambridge, Cambridge University Press).
Kaldor, Mary, 2012: *New and Old Wars: Organised Violence in a Global Era* (Cambridge, Polity Press).
Kant, Immanuel, 1795: *Zum ewigen Frieden: Ein philosophischer Entwurf* (Leipzig: Insel Verlag).
Klein, Naomi, 2007: *The Shock Doctrine: The Rise of Disaster Capitalism* (New York: Knopf Canada).
Kroll-Zeldin, Oren, 2016: *Colonialism* (Oxford: Oxford University Press).
Levinson, Charles, 1978: *Vodka Cola* (London – New York: Gordon & Cremonesi).
Lewis, Diane, 1973: "Anthropology and Colonialism", in: *Current Anthropology*, 14,5: 581-602.
LVC, 2018: "Food Sovereignty Now"; at: https://viacampesina.org/en/wp-content/uploads/sites/2/2018/02/Food-Sovereignty-A-guide-Low-Res-Vresion.pdf.
Mansoob Murshed, Syed, 2019: "The Liberal Peace: Challenges to Development, Democracy, and Soft Power", in: Kulnazarova, Aigul; Popovski, Vesselin (Eds.): *The Palgrave Handbook of Global Approaches to Peace* (Cham: Palgrave Macmillan): 109–128.
Marini, Ruy Mauro, 1973: *Dialéctica de la Dependencia* (Mexico City: Era).
Mies, Maria, 1985: *Patriarchy & Accumulation on a World Scale: Women in the International Division of Labour* (London: Zed Books).
Milligan, Brett, 2015: "Landscape Migration: Environmental Design in the Anthropocene", in: *Places Journal* (June); at: https://placesjournal.org/article/landscape-migration/.
Nu, U., 1975: *Saturday's Son* (New Haven: Yale University Press).
Oswald Spring, Úrsula, 2019: "Toward Engendered-Sustainable Peace to End Patriarchal Violence, in: Kulnazarova, Aigul; Popovski, Vesselin (Eds.): *The Palgrave Handbook of Global Approaches to Peace* (Cham: Palgrave, Macmillan): 601–624.
Oswald Spring, Úrsula, 2020: *Earth at Risk: Rethinking Peace, Environment, Gender, and Human, Water, Health, Food, Energy Security, and Migration* (Cham: Springer International Publishing).

Oxfam, 2020: *Time to Care Report* (London: Oxfam).
Petö, Andrea, 2016: "How are Anti-Gender Movements Changing Gender Studies as a Profession?", in: *Religion and Gender*, 6,2: 297-299.
Phillips, Nia L.; Adams, Glenn; Salte, Phia S., 2015: "Beyond Adaptation: Decolonising Approaches to Coping with Oppression", in: *Journal of Social and Political Psychology*, 3,1: 365-387.
Piaget, Jean, 1950: *Introduction à l'Épistémologie Génétique* (Paris: PUF).
Piaget, Jean, 1972: *Psychology and Epistemology: Towards a Theory of Knowledge*, (Harmondsworth: Penguin Press).
Piketty, Thomas (2013). *Le Capital au XXI Siècle*, Editions du Seuil, Paris.
Reardon B.A., 1996: *Sexism and the War System* (New York: Syracuse University Press).
Richmond, Oliver, 2014: "Decolonizing Security and Peace: Mono-Epistemology versus Peace Formation", in: *Conflict Resolution and Ontological Security, Peace Anxieties* (Boca Raton): Taylor and Francis): 172–192.
Roan, Deb Roy, 2018: "Decolonise Science: Time to End Another Imperial Era"; at: https://theconversation.com/decolonise-science-time-to-end-another-imperial-era-89189.
Ruddiman, William, 2001: "Earth Climate. Past and Future", in: *Global and Planetary Change*, 30, 3–4: 329–330.
SIPRI, 2019: *SIPRI Yearbook 2019* (Oxford: Oxford University Press).
Smil, Vaclav, 2011: "Harvesting the Biosphere: The Human Impact", in: *Population and Development Review*; at: https://doi.org/https://doi.org/10.1111/j.1728-4457.2011.00450.x.
Steffen, Will et al., 2018: "Trajectories of the Earth System in the Anthropocene", in: *PNAS*; at: https://www.pnas.org/content/115/33/8252.
Stiglitz, Joseph, 2007: *Globalisation and its Discontent* (New York: W.W. Norton).
UNESCO, 2002: *Universal Declaration on Cultural Diversity* (Paris: UNESCO).
UNGA, 2015: *Sustainable Development Goals* (New York: UN General Assembly).
Vaughan, Genevieve, 2007: *Women and the Gift Economy* (Toronto: Inanna Publications and Education).
Waters, Colin N.; Zalasiewicz, Jan; Summerhayes, Colin, et al., 2016: "The Anthropocene is Functionally and Stratigraphically Distinct from the Holocene", in: *Science*, 351, 6269 (8 January): 2,622.
WEF [World Economic Forum], 2011: *Annual Report 2010-2011* (Davos: WEF).
WEF [World Economic Forum], 2016: *Annual Report 2015–2016* (Davos: WEF).
WMR, 2020: *World Migration Report,* (Geneva: ILO).

Part I
Peace Research Epistemology for the Anthropocene

Chapter 2
Peace Ecology in the Anthropocene

Hans Günter Brauch

Abstract 'Peace ecology' is a scientific approach that aims to build bridges between peace research and environmental studies. In 2000, Paul J. Crutzen introduced the Anthropocene as a new epoch of Earth's history. Geologists still need to identify evidence in the sediments, e.g. from nuclear explosions and the testing of nuclear and hydrogen weapons in the atmosphere, that such a transition has actually occurred. Direct human interventions into the Earth System through the accumulation of greenhouse gases and *carbon dioxide* (CO_2) in the atmosphere have caused multiple societal impacts, resulting in rapid increases in production, consumption, urbanisation, pollution, migration, crises and conflicts. Peace ecology in the Anthropocene era of Earth and human history can be conceptualised on the basis of five conceptual pillars: peace, security, equity, sustainability and gender. This chapter develops 'peace ecology' in the context of the Anthropocene in ten sections. After a detailed conceptual introduction (2.1), the second Sect. (2.2) discusses five alternative starting points of the Anthropocene: the Agricultural or the Industrial Revolutions, the Columbian Exchange, the Nuclear era and the 'Great Acceleration', while the third Sect. (2.3)

PD Dr. Hans Günter Brauch, chairman of AFES-PRESS, chairman of the board of the *Hans Günter Brauch Foundation on Peace and Ecology in the Anthropocene* (HGBS), and editor of this Anthropocene (APESS) book series; Email: brauch@onlinehome.de.

This text addresses two themes the author has been working on for some time and thus builds on several previous texts that were authored and co-authored on 'peace ecology' and the 'Anthropocene' (Brauch 2012, 2014, 2016, 2016a, 2017); Brauch et al. (2011, 2015); Brauch et al. (2016); Brauch et al. (2017); Crutzen/Brauch (2016); Oswald Spring et al. (2009); Oswald Spring et al. (2014; 2014a). Additional new texts by this author are in preparation and will be published in the years to come.

The author appreciates the critical and very constructive comments he received on a first rough draft of 31 January 2020 from Prof Dr. Holsti, Vancouver (Canada); Prof Dr. Simon Dalby, Waterloo, Canada; Prof Dr. Úrsula Oswald Spring, CRIM, UNAM, Cuernavaca (Mexico); Ambassador Dr. Luis Alberto Padilla Menendez, Guatemala City (Guatemala); and Prof Dr. Czeslaw Mesjasz, Cracow (Poland). Brief comments were received from Prof Dr. Geoff Harris, Durban (South Africa) and Prof Dr. Janos Bogardi, the former Vice Rector of UNU and founding director of UNU-EHS in Bonn (Germany). For information on ongoing work by peace researchers on environment issues the author is grateful to Prof Dr. Peter Wallensteen (Sweden), Prof Nils-Petter Gleditsch (Norway) and Mr Unto Vesa (Finland).

Ú. Oswald Spring and H. G. Brauch (eds.), *Decolonising Conflicts, Security, Peace, Gender, Environment and Development in the Anthropocene*,
The Anthropocene: Politik—Economics—Society—Science 30,
https://doi.org/10.1007/978-3-030-62316-6_2

offers a conceptual mapping of the Anthropocene and the fourth Sect. (2.4) interprets the Anthropocene as a turning point, context and challenge for science and politics. For the new context of the Anthropocene Sect. 2.5 offers a rethinking of peace and the evolution of peace research since the end of World War I, World War II and the Cold War in selected countries and the development of three international peace research organisations: IPRA, PSS(I) and ISA-PEACE. A reconceptualisation of peace in the Post-Cold War Era (1990–2020) and in the Anthropocene is also taking place. Section 5.6 reviews the evolution and rethinking of several ecology concepts (human, political, social) and (political) geo-ecological approaches during the Anthropocene. Section 2.7 reviews several bridge-building initiatives between peace research and ecology that were previously developed by scholars (e.g. Kenneth and Elise Boulding, Arthur H. Westing et al.) and were suggested during the conceptual debate on environmental and ecological security and in the empirical case studies by Günter Bächler (Switzerland) and Thomas Homer Dixon (Canada) on environmental degradation, scarcity and stress as causes and on conflictive outcomes. Since the end of the Cold War, from a policy perspective, debates have evolved on environmental peacemaking and post-conflict peacebuilding and on climate change, security and conflict linkages. While older bridge-building efforts stemming from peace research have addressed issues related to violence, more recent discourses emerging from environment and sustainability studies have addressed issues of sustainability transition and their impact on sustainable peace (Peck 1998) initiatives in the Anthropocene Sect. 2.7. Section 2.8 focuses on the suggested peace ecology approach and research programme as a holistic, enlightening and critical scientific project for the Anthropocene. For this it is necessary to overcome the fragmentation of scientific and political knowledge to incorporate holistic perspectives and transformative approaches that facilitate the move from knowledge to action. In Sect. 2.9 the author addresses the need to develop an ecological peace policy for the second phase of the Anthropocene (2020–2100) by developing strategies and policies to surmount the challenges in the Anthropocene. In Sect. 2.10 the author concludes by proposing a peace ecology research programme and an ecological peace policy in the Anthropocene (2.10).

Keywords Action · Conceptualisation · Earth and human history · International order · Mindset · Peace concept · Peace ecology · Peace research · Security concept · Sustainable development · Sustainable peace · Sustainability transition · Transformative science · World-view

2.1 Introduction

This chapter proposes a new research and study programme in the social and political sciences that links two established programmes on peace research and environmental studies and several ecological perspectives (social, political, human ecology) as a new 'peace ecology approach' which will be contextualised as a new holistic approach in the Anthropocene, which will be introduced in this text as a *turning point* in Earth, political, economic, strategic and human history, as a *context* for this new programme and as a major *challenge* posed by the consequences of *global environmental change* (GEC) and *climate change* (CC) that necessitate major policy strategies and decisions for adaptation, mitigation and resilience.

The concept of the Anthropocene was first coined by the Nobel Laureate in Chemistry (1995), Paul J. Crutzen (The Netherlands) on 23 February 2000 during a conference of the *International Geophysical and Biological Programme* (IGBP) in Cuernavaca in Mexico, where he claimed that the Anthropocene has replaced the 'Holocene' as a 'New Epoch of Earth History' (2.2). Crutzen's claim (Crutzen/Stoermer 2000; Crutzen 2002) triggered both a global scientific discourse across many scientific disciplines and an awareness of emerging societal and political problems.

A few research questions will be discussed in Sect. 2.1.1, a conceptual theoretical framework will be outlined and programmatic policy-orientated ideas will be developed in Sect. 2.1.2, and the structure of this chapter will be sketched in Sect. 2.1.3.

2.1.1 Research Questions

This chapter will address the following research questions:

- When did the Anthropocene as a new era of 'Earth history' begin? (2.2)
- How has the Anthropocene been defined and interpreted in the natural and social sciences as well as in the humanities and the arts? (2.3)
- Why does the Anthropocene as a new epoch of Earth's history matter for the study of peace and ecology and for peace research and environmental studies? (2.4, 2.5, 2.6)

2.1.2 Theoretical Framework and Research Methods

This chapter addresses the emergence of the Anthropocene concept since Crutzen coined it in February 2000 as a *new epoch of earth history* in the natural sciences and as a *context*, *turning point* and *challenge* for political and human history since 1945 and for policies aiming at a sustainability transition (Brauch et al. 2016) extending until the end of the twenty-first century. The following theoretical frameworks and research methods are appropriate:

- *Conceptual theory*: In other words, addressing the Greek origins (‘*anthropos*’ and ‘*cene*’)[1] of the Anthropocene concept, and reviewing similar concepts that have been suggested by scientists before the year 2000 (Crutzen/Stoermer 2000) and alternative concepts that have been proposed by other scientists (e.g. the ‘Capitalocene’ (Moore 2016; Altvater 2016).
- *Conceptual history*: Compared with the conceptual history (Koselleck 2006; Müller/Schmieder 2016) of major concepts like ‘peace’ (Janssen 1975: 543–591; Biser 1972, II: 1113–1115; Kimmenich 1972, II: 1115–1119; Huber 1972, II: 1119–1122) and ‘security’ (Brauch et al. 2008, 2009, 2011) in the social sciences, the history of the Anthropocene concept started in the year 2000 and is limited to only two decades.
- *Conceptual mapping*: Since Crutzen coined the concept at a meeting of the IGBP in Cuernavaca in February 2000, the Anthropocene concept has rapidly spread beyond geology, global environmental change and climate change in the natural sciences to many disciplines in the social sciences, the humanities and the arts (Brauch 2021).
- *Programmatic approach*: The major goal of this chapter is programmatic, and suggests linking recent research programmes in the social sciences – ‘peace research’, ‘environmental studies’ and ‘ecology’ from distinct ecological perspectives – to create a new ‘peace ecology research programme’ whose object of study is limited to research themes since 1945 up to the present (August 2020), reviewing the past seventy-five years since the end of WWII.
- *Holistic policy-relevant transformative research approach*: The aim of this approach is to contribute to policies of ‘sustainability transition’ in order to avoid catastrophic climate change (Schellnhuber et al. 2006), resource wars, and climate-change-induced crises, violent conflicts and wars (Brauch 2002; WBGU 2008, 2011; Brauch/Oswald Spring 2016). The time perspective covers the next eighty years up to the year 2100 – i.e. the same time perspective addressed in the first five assessment reports of the IPCC (1990, 1996, 2001, 2007, 2013), in which the approaches and proposals differ due to divergent policy mindsets and scientific world-views.

[1]On the Greek origins and meanings of the term Anthropos, access: Frederik Arends’ (1943–2020) oral contribution on “Conceptual Origins: ‘Anthropos’ and ‘Politik’ as seen by the Greeks”; at: http://www.afes-press-books.de/html/events.htm [use the password: ‘anthropocene 2017’ to listen]. An obituary of Dr. Jacob Frederik Martinus Arends, who passed away on 3 June 2020, is at: http://www.afes-press-books.de/html/PDFs/2020/Obituary_Jacob_Frederik_Martinus_Arends.pdf. See also: https://www.afes-press-books.de/html/PDFs/2020/Brauch_Personal_Memories_on_Frederik_Arends.pdf

2.1.3 Structure of the Chapter

This chapter first reviews the emergence of the Anthropocene concept in the natural sciences and the discussion among geologists on when this new epoch of Earth's history began – was it during the Neolithic or Agricultural Revolution, the Columbian Exchange (1492–1600), the Industrial Revolution (1750–1850), the first nuclear weapons test on 16 July 1945, or since the 'Great Acceleration' that started in approximately 1950 (2.2)?

It then tries to conceptually map how the *Anthropocene* concept has spread across all disciplines within only two decades and resulted in about a thousand printed books and more than five thousand peer-reviewed articles in scientific journals (2.3). In the social and political sciences the Anthropocene has been discussed as a *turning point*, *context* and *challenge* (2.4).

The argument then turns to the limited impact the *Anthropocene* concept has had since 2000 on two scientific research programmes and the inadequate bridge-building between them, including the limited rethinking of peace and peace research during this new era, and the extent to which it has had an impact on the rethinking of ecology concepts and approaches.

2.2 "We Are Now in the Anthropocene"

During a meeting of the IGBP in Cuernavaca in late February 2000, Paul Crutzen disagreed with a colleague who referred to the Holocene when discussing recent factors in global change often dominated by human actions, and exclaimed, "We are now in the … the Anthropocene!" He recalled this debate when speaking to a BBC Journalist (Falcon-Lang 2011), and this exchange was later confirmed by Will Steffen (2013), who had witnessed it.

> I was at a conference where someone said something about the Holocene. I suddenly thought this was wrong. The world has changed too much. No, we are in the Anthropocene. I just made up the word on the spur of the moment. Everyone was shocked (cited by Lewis/Maslin 2018: 21).

Tritschler (2014), who curated the first exhibition at the German Technological Museum in Munich with the Rachel Carson Centre for Environment and the Society, wrote:

> Crutzen has repeated this story of a sudden flash of insight, a 'eureka' moment of our day, on multiple occasions (Schwägerl 2015), and Will Steffen has confirmed it, thus codifying an attractive founding myth about the origins of the term (Steffen 2014: 486). Crutzen tells an additional story as part of this founding myth: after colleagues suggested that he should claim ownership of the term, he discovered that Stoermer had come up with the term independently. The joint authorship of the article in the *IGBP Newsletter* that initiated the entire current debate was a reflection of this co-creation. For the atmospheric chemist Crutzen, the neologism was meant to firmly emphasise the anthropogenic causes of climate change, which at the beginning of the twenty-first century were still heavily contested. To

him, the new term offers a powerful tool to validate the ongoing process of anthropogenic climate change, which has impacted the Earth in such an enduring manner that it needs a new geological era to depict it properly (Tritschler 2016).

2.2.1 The Industrial Revolution as the Start of the Anthropocene?

Crutzen and Stoermer (2000) initially suggested that the Anthropocene began with the Industrial Revolution, and especially with Watt's invention of the steam engine in 1782. Zalasiewicz (2008), who has chaired the *Anthropocene Working Group* (AWG) since 2008, shared Crutzen's proposal in an article in the *Observer*, in which he wrote: "The Anthropocene does seem geologically real and could be judged to have begun in 1800." In a scientific article – written by twenty-one geologists – Zalasiewicz (2008a) repeated that "a point can be made for [the Anthropocene's] consideration as a formal epoch in that, since the start of the Industrial Revolution, Earth has endured changes sufficient to leave a global stratigraphic signature distinct from that of the Holocene".

Steffen et al. (2011) argued a decade later that "the term Anthropocene … suggests that the Earth has left its natural geological epoch … called the Holocene. Human activities have become so pervasive and profound that they rival the great forces of Nature and are pushing the Earth into planetary *terra incognita*" (cited by Angus 2016).

Influenced by the discussions during the Dahlem workshop (Steffen et al. 2004), the results of the *Millennium Ecosystem Assessment* (MEA 2005), and the projects of the IGBP, Steffen et al. (2005) later agreed that "the twentieth century can be characterised by global change processes of a magnitude which have never occurred in human history" (Steffen et al. 2015). They further agreed that human-environment interactions increased after 1950 due to the "amplified human impact on the environment" in the period they termed the 'Great Acceleration' (Angus 2016: 42).

While there was much support in the natural science community for Crutzen's thesis that humankind had triggered a transition from the 'Holocene' (the period of Earth's history from the end of the glacial period some 11,700 years BCE to the present), the point at which the Anthropocene actually started has been contested.

Instead of the Industrial Revolution, the following major alternative starting points for the Anthropocene Epoch of Earth's history have been suggested:

- The Neolithic or Agricultural Revolution (12,000 BP);
- The Impact of the Columbian Exchange (1492–1600);
- The beginning of the Nuclear Era and the Nuclear fallout from nuclear testing (1945) that coincided with the end of World War II and the gradual evolution of a new political, economic, security and ideological order;
- The Great Acceleration (since 1950).

2.2.2 The Neolithic or Agricultural Revolution as the Start of the 'Thin' Anthropocene

A minority of natural scientists, especially Ruddiman (2003, 2005, 2016), argued that humans have impacted on the climate since the Neolithic or Agricultural Revolution (Lewis/Maslin 2015, 2018: 141–142; Renn 2020: 348–354).

> Ruddiman (2003, 2005, 2016) claims that the Anthropocene has had a significant human impact on greenhouse gas emissions, which began not in the industrial era, but rather 8000 years ago, as ancient farmers cleared forests to grow crops. Ruddiman's work has, in turn, been challenged with data from an earlier interglaciation. … Furthermore, the argument that 'something' is needed to explain the differences in the Holocene is challenged by more recent research showing that all interglacials differ. Although 8000 years ago the planet sustained a few million people, it was still fundamentally pristine. This claim is the basis for an assertion that an early date for the proposed Anthropocene term does account for a substantial human footprint on Earth.[2]

2.2.3 The Impacts of the Columbian Exchange (1492–1600) and the Copernican Revolution

Since the sixteenth century the successive scientific revolutions (Renn 2020: 118–142) in astronomy triggered by Copernicus, Kepler and Galileo, and the many new discoveries in the natural sciences by Newton, Bacon, Leibnitz et al. did trigger many turning points in the epochs of Earth's history but they also fundamentally changed the world-view and challenged the orthodoxy of the Catholic Church.

Lewis and Maslin (2016: 9f.) organised their book on the *Human Planet* around four themes relevant for the Anthropocene: "The environmental changes caused by human activity have increased to a point that today human actions constitute a new force of nature." They identified four major transitions related to (a) energy use and its environmental impacts on the Earth System; (b) the gradual start of colonialism since Columbus reached the Americas in 1492, which resulted in a 'Columbian Exchange' of species (animals and plants) and the emergence of a global capitalist economy; (c) the availability and use of coal as a result of mining; and (d) the Great Acceleration after the end of World War II, partly based on the easy access to cheap fossil fuel energy sources. Lewis and Maslin (2016: 13) argued:

> that the start of the Anthropocene should be dated to the Orbis Spike, a trough in carbon dioxide levels associated with the arrival of Europeans in the Americas. Reaching a minimum around 1610, global carbon dioxide levels were depressed below 285 parts per million, largely as a result of sequestration due to forest regrowth in the Americas. This was likely caused by indigenous peoples abandoning farmland following a sharp population decline due to initial contact with European diseases – around 50 million people or 90% of the indigenous population may have succumbed. For Maslin and Lewis, the Orbis Spike represents a *Global Boundary* Stratotype Section and Point [GSSP], a kind of marker used to define the start

[2] "Anthropocene", in: *Wikipedia*; at: https://en.wikipedia.org/wiki/Anthropocene (31 May 2020).

> of a new geological period. They also go on to say that associating the Anthropocene with European arrival in the Americas makes sense given that the continent's colonisation was instrumental in the development of global trade networks and the capitalist economy, which played a significant role in initiating the Industrial Revolution and the Great Acceleration.[3]

In its discussions, the *Anthropocene Working Group* (AWG) later rejected these three turning points in human history – (1) the *Agricultural Revolution*, (2) the *Columbian Exchange*, and (3) the start of the *Industrial Revolution* – as starting points for the Anthropocene that had a significant impact on nature due to their increasing human impact on the natural environment.

The AWG opted for two more recent turning points that were instrumental in triggering more intrusive global environmental and an anthropogenic climate change: (4) the start of the *Nuclear Era or Atomic Age* (since 1945) that caused a fundamental change in military strategy and politics during the Cold War and created for the first time the ability to destroy a significant part of humankind during a major nuclear war; and (5) the *Great Acceleration* (since 1950) that was made possible by the US-planned new world order and its active multilateral global leadership role that has been challenged by its 45th President.

2.2.4 Almogordo: The Nuclear Age with Its Nuclear Fallout as a Major Marker of the Transition from the Holocene to the Anthropocene

The period from 1940 to 1950 was a major turning point in international politics. It has been intensively analysed in international political, military and security history, and also in economic, social and human history. It heralded a fundamentally new international order that was intensively discussed and planned within the US government, most particularly by the US State Department, as well as by the War and Navy Departments since 1939 (Notter 1949). From 1939 until the Japanese attack on Pearl Harbor, the US was officially neutral, but President Franklin Delano Roosevelt was carefully edging towards active US involvement in World War II by supporting the anti-Hitler coalition through Lend-Lease agreements with military armaments and economic support (Brauch 1976, 1977).

After the first nuclear fission by Otto Hahn and Lise Meitner in Berlin (Hahn/Strasssmann 1939, cited in Renn 2020: 231), Albert Einstein and Leo Szilard[4] persuaded President F.D. Roosevelt to launch the top secret "Manhattan District"

---

[3]"The Anthropocene", in: *Wikipedia*; at: https://en.wikipedia.org/wiki/Anthropocene (31 May 2020).

[4]"The Einstein-Szilard Letter – 1939", in: "Atomic Heritage Foundation" website; at: https://www.atomicheritage.org/history/einstein-szilard-letter-1939.

Project (1942–1945),[5] which “produced the atomic bomb” that “created not only a new technology, but also a new kind of socio-economic complex of technological, political, economic, and knowledge structures that seems impossible to abolish” (Renn 2020: 373). For the first time humankind had the military tools to destroy the human species in a major nuclear war. This new ‘super weapon’ was first tested secretly on 16 July 1945 in Almogordo in New Mexico and employed on 6 August 1945 against Hiroshima and on 9 August 2020 against Nagasaki, thus forcing Japan to accept unconditional defeat in World War II.

Between 1940 and 1950 the US role in world politics had fundamentally changed from an inward-centred, largely pacifist, isolationist country during the 1920s and 1930s to a globally orientated political, military, economic and intellectual ‘superpower’ that had gradually adopted an active world leadership role during World War II and was instrumental in crafting, developing and enforcing its new role as the leader of the so-called ‘free world’ and the antipode of the Soviet Union in the post-World War II world. During this decade a fundamental structural change occurred in the determining factors of American foreign, defence, armament, arms control and disarmament policy during World War II and at the beginning of the Cold War (Brauch 1976, 1977: 341).

Inspired by Eisenhower’s (1961) term of a ‘military-industrial complex’ – something he warned against in his farewell speech in 1961 – Brauch called this linkage system of social actors a ‘MIBUC complex’ that combined, in addition to the two classical components of the military (M) and the armaments industry (I), the civilian employees of the War and Navy ministries, which in 1947 became the Department of Defense (DoD), representing the security bureaucracy (B); the military-funded scientific research and development agencies (U); and parts of the US Congress (C) whose Representatives and Senators have benefited from military-funded jobs in the large defence industry (Brauch 1990: 257). Renn (2020: 231) argued more recently that

> The ‘military-industrial-scientific complex’ turned out to be a persistent legacy of World War II and was reinforced, rather than loosened, during the Korean War of the early 1950s, after the Sputnik shock of 1957, and the course of the long Cold War. This particular form of knowledge economy left profound marks on the further development of science and technology, and even the social sciences, which underwent a scientific turn, shaped by the recruitment of humanists and social scientists for military-relevant problems. … The military-industrial-scientific complex constituted a model for the integration of science into large-scale societal structures so that we may speak more generally of ‘socioepistemic complexes’ with a general tendency to persist and enforce the dependence of societies on scientific knowledge. … Public investment and support for research and professional training played an important role in the expansion and further development of this socioepistemic complex.

[5] “Manhattan District History Project: Y The Los Alamos Project, Vol. I: Inception until August 1945”; at: https://digital.library.unt.edu/ark:/67531/metadc841449/; “Manhattan Project”; at: https://en.wikipedia.org/wiki/Manhattan_Project.

However, in 1945 there was no scientific awareness – and consequently no scientific research – on such issues as *global environmental change* (GEC) and anthropogenic *climate change* (CC). Policy-makers were similarly unaware of the 'environment' and 'development' as policy issues to be addressed by the United Nations Charter that was adopted in June 1945 in San Francisco as the framework for the post-World War II global political and security order.

The social construction of reality (Berger/Luckman 1966, 1967), science (Hacking 1999) and scientific knowledge (Mendelsohn 1977) of GEC, CC and the Anthropocene had not yet occurred and could thus have no impact on the framing and development of the Post-World War II order. These ecological problems had not yet been socially constructed and thus had not then existed. Therefore, no ecological reflection in academia on these issues was possible.

Although in 1896 Arrhenius (Rodhe et al. 1997) – and other natural scientists – had already hypothesised a link between the burning of fossil fuels and the concentration of greenhouse gases, especially *carbon dioxide* (CO_2), the scientific debate on climate change did not take off until the 1970s (v. Weizsäcker 1989; Brauch 1996), and the argument of a transition to the Anthropocene Epoch of Earth's history was not suggested until 2000, when Crutzen coined the term. Jan Zalasiewicz,[6] chairman of the AWG, and his colleagues, linked the Anthropocene with the start of the nuclear age when they suggested:

> The entry into the nuclear age (signalled by the detonation of the Trinity A-bomb in New Mexico on July 16, 1945) could be used to mark the beginning of the Anthropocene because of the stratographic evidence from artificial radionuclides but also from the sediment of global industrialisation such as fly ash, concrete and plastic. This was a unique event, comparable to the meteorite impact causing the extinction of the dinosaurs and marking the end of the Mesozoic era (Renn 2020: 360).

Lewis/Maslin (2018: 257–258) supported this argument when they argued:

> The most ubiquitous signature of humans in geological deposits is probably the fallout from nuclear bomb tests. This could provide a very clear 'golden spike' for the Anthropocene, as the fallout was global. The clearest of these signals, and among the best understood scientifically, is the radioactive isotope of carbon, carbon-14, which reaches a maximum just after the 1963 Partial Test Ban Treaty outlawed above-ground weapons testing. The peak, in 1964 in most of the world, has an abundance of radioactive carbon-14 at 190% of 1950 levels. … Measurements of the radioactive isotope plutonium-239 and plutonium-240 also provide a fingerprint of both Cold War nuclear testing from the early 1950s onwards and the Chernobyl accident of 1986. … However, for really long-term preservation, iodine-129, also produced as part of the fallout from nuclear weapons testing, has a half-life of 15.7 million years. This would provide a marker for the duration of the Anthropocene Epoch and beyond, even if the epoch lasted many million years (Waters et al. 2015: 46–57; Lewis/Maslin 2015: 171–180).

[6]Zalasiewicz et. al. (2015); Waters et al. (2018): 379–429.

Thus, in 1945 two turning points – in *geological time*, with the transition from the Holocene to the Anthropocene, and in *political time* of *longue durée* (Braudel 1968, 1969, 1972) – coincided with the start of the new 'glocal era' (Sassen 1991, 2001, 2008) and the end of World War II, which resulted in a fundamental change in the international diplomatic, peace and security order (Brauch 2016). In modern international history, different international orders (Holsti 1991, 2016, 2016a; Osiander 1994; Brauch 2008) may be distinguished:

- the *Hispanic World Order* (1492–1618),
- the order of the *Westphalian State* (1648–1715), based on the peace treaties of Münster and Osnabrück (1648),
- the *Utrecht Settlement* (1715), resulting in an order of Christian princes (1715–1814),
- the *peace settlement of Vienna* (1815) that brought about a European order based on a *Concert of Europe* (1915–1914), with the independence of most Spanish colonies in Latin America and colonial expansion in Africa and Asia.

In the twentieth century three turning points in international order have occurred:

- after World War I (1914–1918) with the *peace treaty of Versailles* (1919);
- after World War II (1939–1945) with the *adoption of the United Nations Charter* (1945); and
- after the fall of the Berlin Wall with the end of the Cold War (1989–1991) that resulted in the reunification of Germany and several expansions of the European Union.

During World War II, President Roosevelt was already discussing the new global order for peace and security with the British Prime Minister, Winston Churchill (1941, 1943), and with Joseph Stalin, the Chairman of the Council of Ministers of the Soviet Union (1945). After meetings in Dumbarton Oaks (1944) and Chapultepec (January 1945) this resulted in the UN Charter (1945), which was weakened from the outset by the veto power Roosevelt granted to Stalin at the Yalta Conference in February 1945.

During the Cold War, the system of the 'collective security' of the UN was paralysed from the outset by the veto power of the five permanent members of its Security Council. Instead, the US's military dominance and security influence was achieved by a series of systems of 'collective self-defence' (Art. 50, UNCh), among them the *North Atlantic Treaty Organization* (NATO). The US's intention of playing the leading role in the post-war global economy resulted in the creation of several international institutions as an outcome of the Bretton Woods Conference (1944): the *World Bank* (WB), the *International Monetary Fund* (IMF) and, in 1948, the *General Agreement on Tariff and Trade* (GATT). This multilateral institutional framework, based on US political, military and economic power, created a stable global order aiming for a free trade policy that fostered a process of economic and financial globalisation. As part of this process, the 'Great Acceleration' has developed since 1950, when, with the acceptance of the security planning paper NSC-68 in June 1950, the

US launched its first global military order, which was instrumental in tripling the US defence expenditure at the outset of the Korean War.

2.2.5 *The Great Acceleration (1950)*

In the USA, after 1945 a new domestic structure for the creation of a national security state was institutionalised via the National Security Act of 1947, which offered a new rationale for US national interests. The US's National Security Council (NSC) justified the Act on the grounds that its purpose was to achieve 'national security' goals. The contest between rival ideologies, systems of rule and visions of world order resulted in a Cold War that combined militarisation in a bipolar world with the establishment of an open door policy based on free trade that superseded the British imperial preference system. The system of Bretton Woods (World Bank, IMF), NATO and the first global security strategy devised by Paul Nitze via NSC-68 in January 1950 was gradually implemented after the start of the Korean War on 25 June 1950. Thus in 1950, NSC-68 became the turning point towards an internationalist, globalist and partly also interventionist foreign, foreign economic (open door) and military (national security) policy (Brauch 1976, 1977). This multiple turning point in US global policy coincided with what was later called the 'Great Acceleration', marking the silent transition in Earth's history from the 'Holocene' to the 'Anthropocene'. Between 1945 and 1950, several major turning points and transitions occurred in the realms of international politics and national policies that fundamentally changed the context for the study of international peace and security in the social sciences.

For the first time in Earth's history, the manifold human economic and military activities have directly interfered with the Earth System, which, according to Crutzen (2000), has triggered a transition in Earth's history from the Holocene to the Anthropocene Epoch. Several natural and social scientists have stated that this silent transition to the Anthropocene was triggered by a 'Great Acceleration' in twelve exponential Earth systems and twelve socio-economic trends that began around 1950 and intensified after the global 'turn' of 1990. Neoliberal globalisation trends have benefited the rapid economic and scientific innovation and expansion of the People's Republic of China since Deng Xiao Ping's economic reforms (1979–1992).

In international relations, with the rapid political and economic decolonisation since the 1950s and during the 1960s, development policy was set up as a new field of global and national policy-making, partly managed by the World Bank, the IMF and the *United Nations Development Programme* (UNDP). In the 1960s and 1970s, national and international environment policies were added, and in the late 1980s and the 1990s several new regimes of "governance without government" (Rosenau/Czempiel 1992) were set up.

Despite these global contextual changes since the 'Great Acceleration' of the 1950s and the 'intensified acceleration' since 1990, only some industrialised countries have implemented the quantitative reductions in GHGs and CO_2 emissions in the atmosphere in accordance with the framework of the Kyoto Protocol (1997–2012).

While the societal impact of the physical consequences of anthropogenic climate change – (a) temperature rise, (b) change in precipitation, (c) sea-level rise, and (d) increase in the number, intensity and cost of natural hazards (storms, floods, forest fires, drought, soil erosion and desertification) – is apparent, the sense of urgency among top policy-makers concerning the need to address them has been inadequate.

A 'window of opportunity' opened with the co-existence of global, regional and national challenges in the realms of peace, international security, and the international political economy. During negotiations, global environmental governance emerged towards the end of the Cold War in autumn 1988. However, momentum stalled somewhat in 1998, when the US Senate refused to ratify the Kyoto Protocol (1997) in case it damaged the US economy.

The conceptual innovation in the discussions on human and environmental security since the 1990s and the policy debate on the possible impacts of climate change on security since 2004 reflected efforts to hightlight the conceptual link between the two policy areas. Unfortunately, most peace researchers, peace research foundations and many peace research institutes on the one hand, and environmental specialists on the other, did not address these contextual factors that directly affected their research and analyses. Instead, disciplinary specialisation has prevailed over multi-, inter-, transdisciplinary and transformative approaches, while broader holistic perspectives which address the links and interactions between different policy fields have declined. During the brief period of conceptual innovation there was hardly any discussion on the need for a 'peace ecology' approach that conceptually and politically links both issue areas.

This institutional framework in the US and the bipartisan foreign policy consensus in the US Congress was the backbone of the US leadership position until 2017, when President Donald Trump started to gradually tear it down and tried to replace the dominant multilateral trade system and liberal global order with his self-centred motto of "Making America great again".

Steffen et al. (2007: 616–618) initially distinguished between two stages of the Anthropocene: (a) the *industrial era* (1800–1945), and (b) the *Great Acceleration* (1946–2015) (documented in Fig. 2.1):

> From 1950 to 2000 the percentage of the world's population living in urban areas grew from 30 to 50% and continues to grow strongly. … The pressure on the global environment from this burgeoning human enterprise is intensifying sharply. Over the past 50 years, humans have changed the world's ecosystems more rapidly and extensively than in any other comparable period in human history … The Earth is in its sixth great extinction event, with rates of species loss growing rapidly for both terrestrial and marine ecosystems … The atmospheric concentrations of several important greenhouse gases have increased substantially, and the Earth is warming rapidly … More nitrogen is now converted from the atmosphere into reactive forms by fertilizer production and fossil fuel combustion than by all of the natural processes in terrestrial ecosystems put together. … The Great Acceleration took place in an intellectual, cultural, political, and legal context in which the growing impacts upon the Earth System counted for very little in the calculations and decisions made in the world's ministries, boardrooms, laboratories … The exponential character of the Great Acceleration is obvious from our quantification of the human imprint on the Earth System, using atmospheric CO_2 concentration as the indicator … Although by the Second World War the CO_2 concentration had clearly risen above the upper limit of the Holocene, its growth rate hit a take-off point

around 1950. Nearly three-quarters of the anthropogenically driven rise in CO_2 concentration has occurred since 1950 (from about 310 to 380 ppm), and about half of the total rise (48 ppm) has occurred in just the last 30 years (Steffen et al. 2007: 616–618).

Steffen et al. (2007: 619–620) identified three philosophical responses – (a) *business as usual*, (b) *mitigation*, and (c) *geoengineering* – any of which "can raise serious ethical questions and intense debate", given "the possibility for unintended and unanticipated side-effects that could have severe consequences. The cure could be worse than the disease." They concluded that "The Great Acceleration is reaching criticality", but pointed to "evidence for radically different directions built around innovative, knowledge-based solutions".

The *International Geosphere-Biosphere Programme* (IGBP)[7] analysed data for 1750–2010 by distinguishing between twelve *socio-economic* (population, real GDP, foreign direct investment, urban population, primary energy use, fertiliser consumption, large dams, water use, paper production, transportation, telecommunications, international tourism) and twelve *Earth systems* (carbon dioxide, nitrous oxide, methane, stratospheric ozone, surface temperature, ocean acidification, marine fish capture, shrimp aquaculture, nitrogen to coastal zone, tropical forest loss, domesticated land and terrestrial biosphere degradation) trends (Figs. 2.1 and 2.2).

According to *Future Earth*, the following observation is apparent: "the synchronous acceleration of trends from the 1950s to the present day … with little sign of abatement".[8] (1 December 2019); Steffen et al. (2004, 2015). The emergence of the Anthropocene reflects the impacts of the US-dominated world order since 1945 of the nuclear era, a world trade system based on 'open doors', cheap fossil energies and the 'American way of life', or a 'Western lifestyle' based on abundance.

Since 1945 humankind has for the first time threatened its own survival through the fallout from atmospheric nuclear tests and a dominant economic system that has resulted in a massive increase in the burning of cheap fossil energy sources (coal, oil and gas). However, there is a major difference between the two existential threats. While the first threat is influenced by rival superpowers or geopolitical camps, the second is the result of our own individual and joint economic behaviour through the individual burning of fossil energy resources.

The *nuclear era* and the *Anthropocene* are two sides of the same coin that gradually emerged between 1940 and 1950. However they are addressed by different disciplines and research programmes: the first has been primarily tackled in the social sciences via security and peace research, while the second has been primarily developed in the natural sciences via environmental research studies focusing on global environmental change. A new peace ecology approach in the social sciences could combine both research programmes by addressing the manifold interlinks.

Carbon dioxide (CO_2) as the key indicator had increased from about 279/280 ppm in 1750 to 310 ppm in 1950, and soared to 354 ppm by July 1990, increasing by

[7]See at: http://www.igbp.net/ (1 December 2019).

[8]See the figures on the twenty-four IGBP categories of the 'Great Acceleration' on the website of Future Earth; at: https://futureearth.org/2015/01/16/the-great-acceleration/ (1 December 2019); Steffen et al. (2004, 2015).

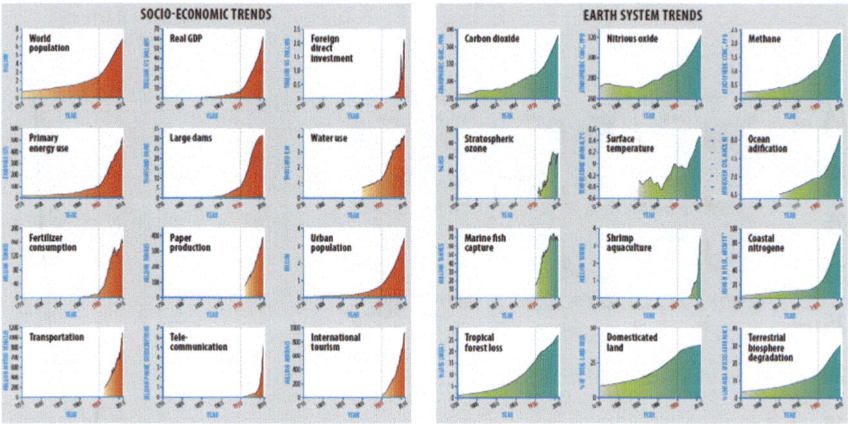

Fig. 2.1 The 'Anthropocene' began around 1950. *Source* Globaia 2011; at: https://globaia.org/great-acceleration; based on Steffens et al. Steffens. (*Source* Globaia, at: https://globaia.org/great-acceleration, based on Steffens et al. 2007, 2015. Design: © Félix Pharand-Deschênes, Globaïa.)

44 ppm in just forty years. Over the next thirty years from July 1990, CO_2 rose to 416 ppm in April 2020 – i.e. by another 62 ppm (Fig. 2.3).

There were also close interactions between the contextual changes in international order with regard to international peace and security and international economic relations. The fundamental structural change in the USA during World War II, its rise to the status of world power, and its acceptance of a world leadership role (Brauch 1977) acted as multiple political triggers that made the launch of the 'Great Acceleration' after 1945 and 1950 possible.

2.2.6 Decision-Making on the Anthropocene Within the Community and Institutions of the Global Organisations of Geologists

To Professional organisations for geologists (the International Union of Geologists, the International Commission on Stratigraphy and the Subcommission on Quaternary Stratigraphy) that are entitled to determine the start and end of epochs of Earth's history set up the *Anthropocene Working Group* (AWG) in 2008. In May 2019 they agreed that the Anthropocene has become a new epoch of Earth's history and replaces the Holocene. On 26 January 2015, twenty-six of the thirty-eight members of the AWG suggested that the first nuclear test (Trinity) on 16 July 1945 was the starting point of the proposed Anthropocene Epoch. In April 2016 the AWG voted to recommend to the International Union of Geological Sciences (IUGS) during its congress in Cape Town in August 2016 "the Anthropocene as the new geological age" and:

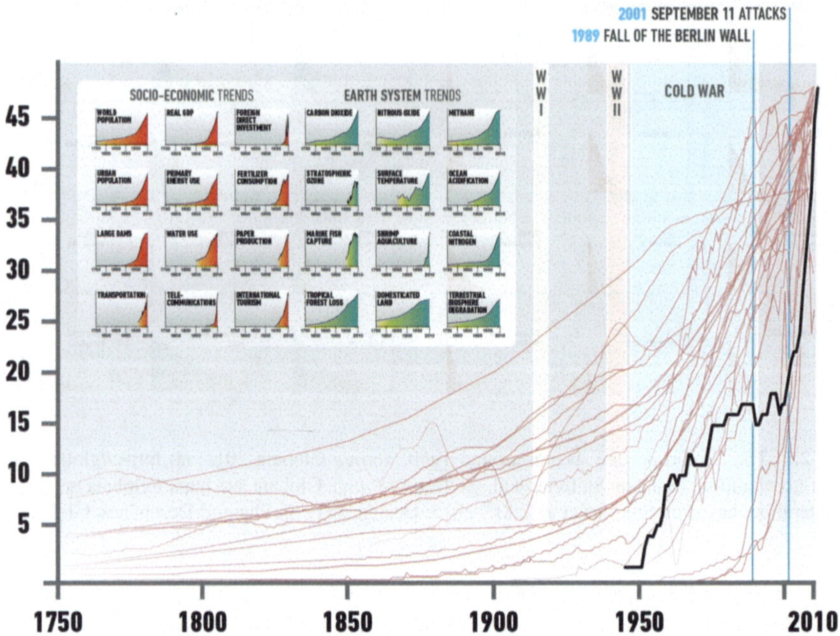

Fig. 2.2 An increasingly walled world and the broader context of the Great Acceleration. *Source Globaia* 2011; at: https://globaia.org/great-acceleration

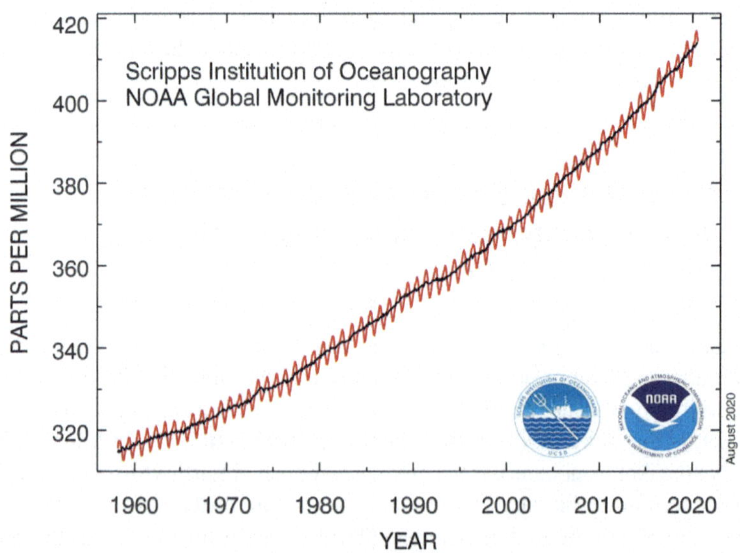

Fig. 2.3 Atmospheric CO_2 at Mauna Loa Observatory (1958–May 2020). *Sources* Dr. Pieter Tans, NOAA/ESRL, at: www.esrl.noaa.gov/gmd/ccgg/trends/ and Dr. Ralph Keeling, Scripps Institution of Oceanography, at: scrippsco2.ucsd.edu/. This figure is in the public domain

proceed towards a formal proposal to define the Anthropocene Epoch in the "Geologic time scale" and presented the recommendation to the International Geological Congress in August 2016. In May 2019, the AWG voted in favour of submitting a formal proposal to the ICS by 2021, locating potential stratigraphic markers to the mid-twentieth century of the Common Era. This time period coincides with the Great Acceleration, a post-WWII time during which socio-economic and Earth system trends started increasing dramatically, and the Atomic Age. … [T]he peak in radionuclides fallout consequential to atomic bomb testing during the 1950s has been more favoured than others, locating a possible beginning of the Anthropocene to the detonation of the first atomic bomb in 1945, or the Partial Nuclear Test Ban Treaty in 1963.[9]

On 21 May 2019, twenty-nine of the thirty-four members of the AWG voted that by 2021 an official proposal on the Anthropocene as a geological epoch with a starting point in the mid-twentieth century should be made to the ICS. The AWG[10] voted on 21 May 2019 that:

- The "Anthropocene [should] be treated as a formal chronostratigraphic unit" …
- The "primary guide for the base of the Anthropocene [should] be one of the stratigraphic signals around the mid-twentieth century of the Common Era" …

In its definition and status of the Anthropocene within the statement cited above, the AWG declares:

- This impact has intensified significantly since the onset of industrialisation, taking us out of the Earth System state typical of the Holocene Epoch that post-dates the last glaciation.
- A proposal to formalise the *Anthropocene* is being developed by the *AWG*. Based on preliminary recommendations made by the AWG in 2016 and a vote of May 2019, this proposal is being developed on the following basis:
  1. It is being considered at the series/epoch level (and so its base/beginning would terminate the Holocene Series/Epoch as well as Meghalayan Stage/Age);
  2. It would be defined by the standard means for a unit of the Geological Time Scale, via a *Global boundary Stratotype Section and Point* (GSSP), colloquially known as a 'golden spike';
  3. Its beginning would be optimally placed in the mid-twentieth century, coinciding with the array of geological proxy signals preserved within recently accumulated strata and resulting from the 'Great Acceleration' of population growth, industrialisation and globalisation;
  4. The sharpest and most globally synchronous of these signals, that may form a primary marker, is made by the artificial radionuclides spread worldwide by the thermonuclear bomb tests from the early 1950s.
- Broadly, to be accepted as a formal geological time term the *Anthropocene* needs to be (a) scientifically justified, i.e. the 'geological signal' currently being produced in strata now forming must be significantly large, clear and distinctive; sufficient evidence has now been gathered to demonstrate this phenomenon (b) useful as a formal term to the scientific community. In terms of (b), the currently informal term *'Anthropocene* has already proven highly useful to the global change and Earth System science research

[9]"The Anthropocene", in: *Wikipedia*; at: https://en.wikipedia.org/wiki/Anthropocene (31 May 2020).

[10]See the statement of the AWG at: http://quaternary.stratigraphy.org/working-groups/anthropocene/ (1 December 2019).

communities and thus will continue to be used. Its value as a formal geological time term to other communities continues to be discussed.

- The *Anthropocene* has emerged as a popular scientific term used by scientists … to *designate* the period of Earth's history during which humans have a decisive influence on the state, dynamics and future of the Earth System. It is widely agreed that the Earth is currently in such a state. The term has also been used in a non-chronostratigraphic context to be an informal term to denote a broader interpretation of anthropogenic impact on the planet that is markedly diachronous, reaching back many millennia. …

According to *Nature*'s news analysis, the AWG:

> is considering ten candidate sites from around the globe, including a cave in northern Italy, corals in the Great Barrier Reef and a lake in China. … They hope to identify a single site to include in their formal proposal. They must also define the type of physical evidence in the sedimentary record that represents the start of the epoch. The group is considering whether to choose the radionuclides that came from atomic-bomb detonations from 1945 until the Limited Nuclear Test Ban Treaty of 1963… The AWG['s] … formal proposal … will be considered by several more groups of the International Commission on Stratigraphy. … Final ratification would come from the executive committee of the International Union of Geological Sciences (Subramanian 2019).

Thus, since 1950 three global trends influenced by political decisions have coincided: (a) the active development of a new global order that was largely framed and established by the USA during and after World War II; (b) the start of the nuclear era with both nuclear weapons and nuclear energy; and (c) the 'Great Acceleration" in environmental history since 1945 (McNeill/Engelke 2014; Steffen et al. 2015) . The consequences of these trends have profoundly affected the global environment (Table 2.1).

In modern human history, major contextual changes in the international order occurred in 1815, 1919, 1945–1950, 1990 and 2020, when the changes in the international political, security, economic and financial order had an increasingly significant impact on the global environmental context through the emissions of GHG and especially CO_2. The French Revolution and the Napoleonic Wars resulted in the Peace of Vienna (1815), heralding a relatively stable international geopolitical security order in Europe in which large wars between the major powers were avoided until 1914. The Peace of Versailles (1919) and the collective security order of the League of Nations were rather fragile from the outset and collapsed within less than two decades. The human impact of both wars on nature and, in particular, the atmosphere was still negligible, amounting to an increase of only 31 ppm in the 200 years between 1750 (279 ppm) and 1950 (310 ppm).

With the increase in the use of fossil fuels, especially coal and gas, as well as electricity for wartime production of military equipment and oil for aeroplanes, military ships, tanks and mobile transportation carriers, the armed forces – especially during major wars and since the Cold War – have increasingly contributed to GHG emissions, particularly CO_2.

The contextual changes that had a major impact on the political and security realm were well observed during World War II, whereas environmental issues (pollution, conservation) remained unnoted until the 1960s. Research on issues of global environmental change (climate change, biodiversity loss, soil erosion, desertification,

Table 2.1 Multiple contextual changes since 1815 to 2019 and the impact of human interference on nature and the atmosphere

Events and turning points in human history (Time)	International political, military and security order		International economic and financial order		Environmental policy and regimes		Environmental impact		Period of earth's history
	International political, military, security, economic and financial context and impacts				Context of environmental *policies* and regimes and their impact on humankind				Contextual changes
	International order	Wars and revolutions	Financial	Trade	Policies	Regime	Key fossil consumers	GHG/CO_2emissions	From the Holocene to the Anthropocene (1950)
French revolution (1789)	Order of Monarchies (Republic)	French Revolution & Napoleonic Wars (1796–1815)			None	None	UK	CO_2emission into the atmosphere (279ppm)	*Holocene* (since the end of the glacial period; about 11,700 years before the present)
Vienna settlement (1815)	System of Negotiation	Colonialism Imperialism	First Economic globalisation (migration, enhanced communications, capital transfers)	Transportation expansion (US open-door, free trade system) in and with China after 1890s	None	None	(Coal mining) UK France Germany USA (oil since 1869) (city and natural gas)		
World war I (1914–18)	Collapse of old international political order	Socialist Revolution in Czarist Russia	Global war economy		None	None	USA, UK, Canada France Germany, Austria	Environmental impact of WWI	

(continued)

Table 2.1 (continued)

Events and turning points in human history (Time)	International political, military and security order		International economic and financial order		Environmental policy and regimes		Environmental impact		Period of earth's history
	International political, military, security, economic and financial context and impacts				Context of environmental *policies* and regimes and their impact on humankind				Contextual changes
	International order	Wars and revolutions	Financial	Trade	Policies	Regime	Key fossil consumers	GHG/CO_2emissions	From the Holocene to the Anthropocene (1950)
Treaty of Versailles (1919)	Fragile new order: League of Nations	Italy: Ethiopia (1935) Japan: China (1931) Germany: Czechoslovakia (1938)	US contributions to reduce debt of defeated countries	Autarky Global Imperial Preference System	None	None	USA Germany Russia Japan UK France		
World War II (1939–45)			Closer economic cooperation between military allies. US economic and financial support for UK and Russia (global war economy).		None	None	USA Japan Germany USSR UK France	Environmental impact of WWII	

(continued)

Table 2.1 (continued)

Events and turning points in human history (Time)	International political, military and security order		International economic and financial order		Environmental policy and regimes		Environmental impact		Period of earth's history
	International political, military, security, economic and financial context and impacts				Context of environmental *policies* and regimes and their impact on humankind				Contextual changes
	International order	Wars and revolutions	Financial	Trade	Policies	Regime	Key fossil consumers	GHG/CO_2emissions	From the Holocene to the Anthropocene (1950)
Post-war (1945–50)	United Nations Charter (collective security)	Expansion of USSR influen-ce and control over Eastern Europe	Bretton Woods World Bank, IMF	Free trade GATT EU Comecon	None	None		Nuclear weapons (testing)	Great Acceleration (1950): start of *Anthropocene* (first anthropogenic intervention on the Earth System)
Cold war (1947–89)	Military alliances (NATO, WP)	Korea (1950-1953) Vietnam War (1961-1976)	Reforms of Deng Xiaoping (1976-1989)	US trade embargos against USSR and China	Pollution ozone layer depletion (Montreal Protocol 1987)	Stockholm international environmental. conference (1972) Introduction of UNEP Brundtland Report (1987)		1950: 310 ppm nuclear weapons testing until 1963 (LTB Treaty (1963) Vietnam War (Agent Orange, 1961–1971)	

(continued)

Table 2.1 (continued)

Events and turning points in human history (Time)	International political, military and security order		International economic and financial order		Environmental policy and regimes		Environmental impact		Period of earth's history
	International political, military, security, economic and financial context and impacts				Context of environmental *policies* and regimes and their impact on humankind				Contextual changes
	International order	Wars and revolutions	Financial	Trade	Policies	Regime	Key fossil consumers	GHG/CO_2emissions	From the Holocene to the Anthropocene (1950)
Post cold war era (1990–2019)	(NATO)			WTO	Global environ-mental and climate change, ozone layer depletion biodiversity	UNFCCC (1992), Kyoto Protocol (1997)		Iraq aggression against Kuwait (1990), oil war	
1990 1991	UN, OSCE, NATO, WP	War of Iraq, against Iraq						1990: 354 ppm	Intensification of the Great Acceleration
Cooperative (1990–1996) G. Bush (1989–1993)	Collective security (UN, OSCE)	Wars in post Yugoslav and Soviet spaces				Rio de Janeiro UNFCC (1992) CBD (1992) UNCCD (1994)			
W.J. Clinton (1993–2001)	Comprehensive Test Ban Treaty					Kyoto Protocol (1997)	China, USA, India		

(continued)

Table 2.1 (continued)

Events and turning points in human history (Time)	International political, military and security order		International economic and financial order		Environmental policy and regimes		Environmental impact		Period of earth's history
	International political, military, security, economic and financial context and impacts				Context of environmental *policies* and regimes and their impact on humankind				Contextual changes
	International order	Wars and revolutions	Financial	Trade	Policies	Regime	Key fossil consumers	GHG/CO_2emissions	From the Holocene to the Anthropocene (1950)
G.W. Bush War on terror (2001–2009)	Unilateralism (NATO)	9/11 (2001) Afghanistan (2001-today) Iraq (2003)							
Obama (2009–2017)	(NATO)	Libya (2011/2013) Syria (2011-today)				Paris climate treaty (2015)			
D. Trump (2017)	Nationalism, Isolationism			End of trade multi-lateralism, sanctions and tariff wars (EU, USA vs. China,			China, USA, EU 28 India, Russia, Japan	May 2020: 416 ppm (exponential increase)	August 2020: 414.38 ppm
Scientific disciplines	Political science: international relations (peace security, international political economy)				Political science and natural sciences (environment policy, ecology, geo-ecology, global environmental challenges, Earth Systems science) studies or analysis				
Objective	Peace Ecology perspective: combining peace and security and environmental (ecological) studies								

Source The Author

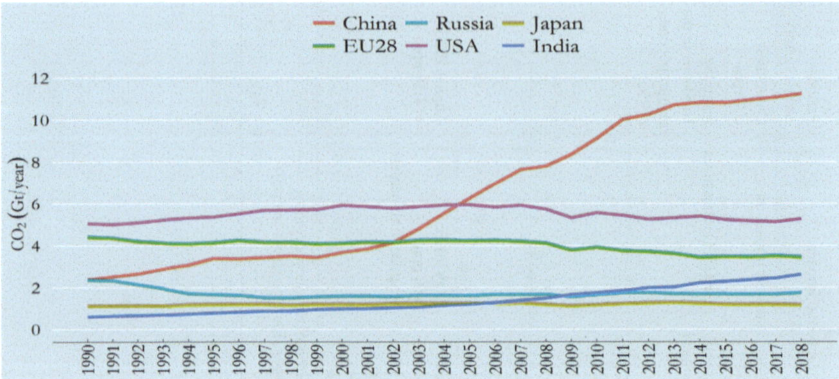

Fig. 2.4 Fossil CO_2 emissions of the six major emitting economies (1990–2018). *Source* Crippa et al. 2019. This figure is in the public domain

water scarcity and pollution) were not politically addressed until the late 1980s, while environmental regimes and governance did not emerge as themes in the social sciences until the 1990s.

While Crutzen initially claimed that the Anthropocene began with the Industrial Revolution in the late eighteenth century, most members of the *Anthropocene Working Group* (AWG) were convinced that the transition from the Holocene to the Anthropocene started with the 'Great Acceleration' around 1950, when numerous nuclear and hydrogen tests took place. These left evidence of their impact in sediments. In the 1950s it would have been difficult to measure GHG emissions in the atmosphere, and even more difficult to identify the impact of the nuclear and hydrogen bomb tests on sediments in the soil and on rocks, but nowadays the effects of the nuclear and hydrogen tests can easily be measured by geologists.

There were two phases of a great acceleration in CO_2: in 1950 and after 1990. China overtook the CO_2 emissions of the EU-28 by 2002 and those of the US in 2004, thereby becoming the major CO_2 emitter. By 2018 Chinese emissions were twice as high as those of the USA. The emissions of India overtook those of Japan in 2007 and those of Russia in 2010 (Fig. 2.4).

2.3 Conceptual Mapping of the Anthropocene Concept

This author (Brauch 2008: 1) argued that the conceptual changes to the 'security' concept since the 1990s occurred due to a threefold contextual change: the end of the Cold War (1989–1991), the globalisation process and global environmental change issues. For the Anthropocene this new concept triggered a search when this new epoch of earth history began (2.2): "The changes in the thinking on security and their embodiment in security concepts are also a semantic reflection of the fundamental

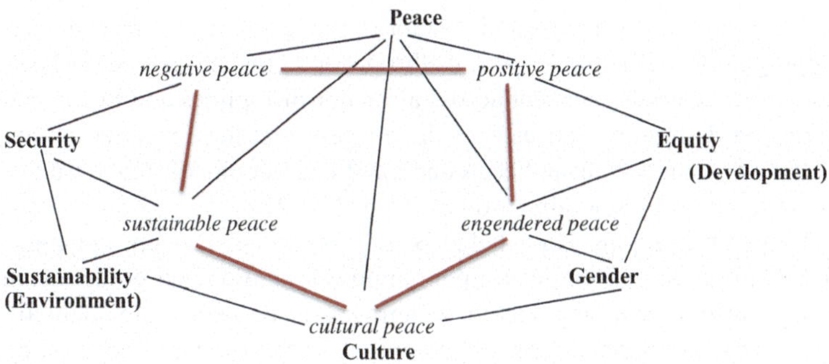

Fig. 2.5 Five Pillars of Peace Ecology and their four linkage concepts of negative, positive, cultural and engendered peace. *Source* Oswald Spring, Brauch and Tidball (2014: 19)

changes as they have been perceived in different parts of the world and conceptually articulated in alternative or new and totally different security concepts" (Brauch 2008: 34).

While an intensive theoretical and methodological discourse on conceptual history (Koselleck 2006; Müller/Schmieder 2016) exists in the literature, the 'conceptual mapping' approach is still underdeveloped. With regard to the security concept, this author (Brauch 2008: 34) used the term to map the different definitions of security in social science disciplines (political science, sociology, economics, law) by different actors in the political realm (governments, parliaments, the public, media, citizens); in society (societal groups); and the business community (banks, industrial and service companies).

> In political science, the security concept has been used in its threefold context of *policy* (field ...), *politics* (process ...), and *polity* (legal norms, laws, and institutions...). The US National Security Act of 1947 and its adjustments have created the legal and institutional framework for the evolution of the 'national security state', sometimes also referred to as a 'military-industrial complex'. This evolution has been encapsulated in the US debate on the concepts of 'national' and, since 2001, also 'homeland' security.

A 'concept map' that presents – according to [11] Wikipedia – "relationships between concepts [as] ... a graphical tool that instructional designers, engineers, technical

[11] Wikipedia "Concept Map"; at: https://en.wikipedia.org/wiki/Concept_map. Wikipedia distinguished between 'topic maps' (developed by Joseph Novak and Alberto J. Cañas, 2008: "The Theory Underlying Concept Maps and How to Construct and Use Them", Technical Report IHMC CmapTools 2006–01, Rev 01–2008, Florida Institute for Human and Machine Cognition; at: https://cmap.ihmc.us/docs/pdf/TheoryUnderlyingConceptMaps.pdf [29 July 2020]), which "make concepts, and propositions composed of concepts, the central elements in the structure of knowledge and construction of meaning" and 'mind maps', "which typically emerge from a single centre". Concept mapping was developed by Joseph D. Novak ... in the 1970s as a means of representing the emerging science knowledge of students. It has subsequently been used as a tool to increase meaningful learning in the sciences and other subjects as well as to represent the expert knowledge of individuals and teams in education, government and business. Concept maps have their origin in the learning movement called constructivism. In particular, constructivists hold that learners actively construct knowledge.

writers, and others use to organise and structure knowledge" differs from the 'conceptual mapping' offered below for the Anthropocene concept as a snapshot of the spread of a new scientific concept across all major disciplines and of the manifold often unrelated definitions and uses of the concept that have resulted within only two decades in about a thousand books and and in December 2020 more than five thousand peer-reviewed scientific articles (Brauch 2021).

The 'Research Mapping Knowledge Base' associated 'concept mapping' with methods and processes of "brainstorming, brainwriting, nominal group techniques, focus groups, affinity mapping, Delphi techniques, facet theory, and qualitative text analysis".[12] Trochim defined 'concept mapping' as "a structured process, focused on a topic or construct of interest, involving input from one or more participants, that produces an interpretable pictorial view (concept map) of their ideas and concepts and how these are interrelated. Concept mapping helps people to think more effectively as a group without losing their individuality. It helps groups to manage the complexity of their ideas without trivialising them or losing detail."

The notion of 'conceptual mapping' used in this chapter significantly differs from both the research tool of a 'concept map' and the research method of 'concept mapping'. Through the proposed 'conceptual mapping technique' this author tries to offer a conceptual tracking and overview of how the Anthropocene concept has been taken up and defined by different academic and scientific disciplines in the natural sciences, the social sciences, the humanities and the arts since the year 2000 (Brauch 2021).

2.3.1 The Anthropocene Concept: Similar Earlier Concepts and Proposed Alternative Concepts

The Anthropocene concept is a combination of *anthropo-* from *anthropos* (ancient Greek: ἄνθρωπος),[13] meaning 'human', and *-cene* from *kainos* (ancient Greek: καινός), meaning 'new' or 'recent'.

2.3.1.1 Earlier Scientific Concepts Addressing Humankind and the Earth System

As early as 1873, the Italian geologist Stoppani (1873) pointed to the increasing power and effect of humanity on the Earth's systems and referred to an *anthropozoic*

[12]See at the 'Research Mapping Knowledge Base': William M.K. Trochim: "Concept Mapping"; at: https:// conjointly.com/kb/concept-mapping/ .

[13]For an oral discussion of the *anthropos* concept in classical Greek, see Frederik Arends (†):"'Anthropos' and 'Politik' as seen by the Greeks", in: "AFES-PRESS Brainstorming in Mosbach: 'Politik' in and for the Anthropocene: Obstacles, Challenges, Opportunities and Tasks for the Social Sciences in the 21st Century, 31 May 2017".

era.[14] Even earlier, in 1864, Marsh (1864, 1965) addressed the growing influence of humankind on nature in his book *Man and Nature*; and in 1926, the Russian geologist, Vernadsky (1929, 1945, 1986, 1998) recognised the increasing power of mankind as part of the biosphere when referring to "the direction in which the processes of evolution must proceed, namely towards increasing consciousness and thought, and forms having greater and greater influence on their surroundings." Later, in 1924, the Jesuit priest Teilhard de Chardin and Le Roy (1927) used the term *noosphere*, the world of thought, for the growing role of humankind's brainpower and technological talents in shaping its own future and environment (Crutzen/Stoermer 2000) through a tenfold increase in human population during the past three centuries and a tenfold increase in urbanisation during the past century, while "30–50% of the land surface has been transformed by human action". Vitousek et al. (1997) stated that "human activity has increased the species extinction rate by thousand to ten thousand fold in the tropical rain forests" (Wilson 1992, 2002). From a biological perspective, Markl (1986, cited by Ehlers 2008) referred to an *Anthropozoikum* due to the dominance of human beings as an essential element of present time. Crutzen and Stoermer (2000) argued in the year 2000 in the IGBP Newsletter:

> Considering these and many other major and still growing impacts of human activities on earth and atmosphere, and at all, including global scales, it seems to us more than appropriate to emphasise the central role of mankind in geology and ecology by proposing to use the term *anthropocene* for the current geological epoch. The impacts of current human activities will continue over long periods. … Without major catastrophes like an enormous volcanic eruption, an unexpected epidemic, a large-scale nuclear war, an asteroid impact, a new ice age, or continued plundering of Earth's resources by partially still primitive technology (the last four dangers can, however, be prevented in a real functioning noosphere) mankind will remain a major geological force for many millennia, maybe millions of years, to come.

Two decades later, the German historian Renn (2020: ix) defined the Anthropocene "as the new geological epoch of humankind, defined by the profound and lasting impact of human activities on the Earth system. The Anthropocene is thus the ultimate context for a history of knowledge and the natural vanishing point for an investigation of cultural evolution from a global perspective." Renn (2020: 5–6) noted that the "the idea of humankind as a planetary force" had already been posited by Comte de Buffon (1778), who stated twelve years prior to the French Revolution that "the entire face of the Earth bears the imprint of human power".

2.3.1.2 Proposed Alternative Concepts to the Anthropocene

Dalby (2016) identified different perspectives for the framing of the Anthropocene: "the good, the bad and the ugly".

> While ecomodernists argue that current circumstances present opportunities and possibilities for a thriving future for humanity, a 'good Anthropocene', critics suggest that the future will

[14]"Anthropocene ", in: *Wikipedia*; at: https://en.wikipedia.org/wiki/Anthropocene (7 June 2020); *Steffen, Grinevald,* Crutzen and, McNeill (2011): 842–67; Stoppani (1873).

> be bad for at least most of humanity as we accelerate the sixth extinction event on the planet. The geopolitics of all this, which may be very ugly in coming decades, requires much further elucidation of the common Anthropocene tropes currently in circulation. … How the Anthropocene is interpreted, and who gets to invoke which framing of the new human age, matters greatly both for the planet and for particular parts of humanity. All of which is now a key theme in the discussions of political ecology that requires careful evaluation of both how geology has recently become so important in global politics, and in discussions of humanity's future, and how scholars from various disciplines might now usefully contribute to the discussion.

The notion of a 'good Anthropocene' has been extensively used by 'cornucopian' (Gledisch 2003) approaches and authors, among them the so-called *ecomodernists* (Breakthrough Institute), who argue that liberal Capitalism can cope with these challenges with technological fixes (geoengineering) and technological innovation without any change in Western lifestyles. The 'bad Anthropocene' is associated with the negative consequences of 'dangerous' or 'catastrophic' climate change (Schellnhuber et al. 2006) and with 'hothouse earth' (Steffen et al. 2018). Dalby also referred to the ugly geopolitical consequences of global environmental change issues on security, conflicts and wars (Scheffran et al. 2012) in his *Anthropocene Geopolitics* (Dalby 2020). For Arias-Maldonado (2019: 137–138), the Anthropocene concept is "in itself as ambitious as it is ambiguous, and the more so in relation to normative concerns". He has proposed "a reframing of the good Anthropocene, so that it is separated from the ecomodernist vision and presented not only as sustainable but as the result of a collective reflection about the way in which socio-natural relations should be organised".

Several alternatives to the geological concept of the Anthropocene have been suggested, primarily in the social sciences, including the *Anthrocene* (Revkin 1992), the *Homogenocene* (Samways 1999), the *Capitalocene*, the *Chthulucene* (Haraway 2016), the *Econocene* (Norgaard 2013), the *Technocene* (Hornborg 2015), the *Misanthropocene* (Patel 2013), the *Manthropocene* (Raworth 2014; di Chiro 2017), the *Necrocene* (McBrien 2016), the *Novacene*, the *Myxocene* (Pauly 2010), the *Plasticene*, the *Plantationocene*, and others (Moore 2016: 6). The geologists Zalasiewicz et al. (2019a: 15) concluded that as a geological unit

> attempts to 'design' a name that might better symbolise its essence … would have little significance. … There is considerable congruence between the meaning of the Anthropocene as originally devised and used in the Earth Systems Science community and the Anthropocene as considered geologically, as a chronostratigraphic unit (Steffen et al. 2016; Zalasiewiecz et al. 2017a). This, together with the way the name has become quickly established in the literature, suggests that the term Anthropocene should be retained with this meaning – with appropriate qualifications as needed when it is necessary to distinguish it from other meanings and interpretations of the word.

Among the concepts that have been suggested as alternatives to the Anthropocene, the *Capitalocene* (Moore 2016; Altvater 2016; Hartley 2016; Parenti 2016) has been discussed most by both Marxists and critical political economists, while some ecologists prefer the *Novacene* (Lovelock/Appleyard 2019), coined by the founder of the *Gaia hypothesis*. In discussing nature, history and the crisis of capitalism, Moore (2016a: 3) noted that "the Anthropocene concept entwines human history and natural history – even if the 'how' and the 'why' remain unclear and hotly debated". He and his collaborators argue "for reconstructions that point to a new way of thinking humanity-in-nature, and nature-in humanity" (Moore 2016a: 5). For Moore the "Capitalocene does not stand for capitalism as an economic and social system". For him this term "signifies capitalism as a way of organising nature – as a multispecies, situated, capitalist world-ecology" (Moore 2016a: 6).

For Dryzek and Pickering (2019: 13–14, 59–60) the *Capitalocene* "ties itself too closely … to [an] aspect of human social organisation". Its proponents "blame certain groups (e.g. capitalists) for global environmental change and its associated injustices". Kelly (2019: 68–69, 74) stated that the adherents of the Capitalocene addressed the "conjoined histories of capitalism and ecological exploitation". Moore and Patel (2018) suggested that the early winners of modern capitalism should repay "a form of ecological reparation for past wrongs." In the view of Hickmann et al. (2019: 4), the proponents of the *Capitalocene* "highlight the causal role of capitalism for irreversible environmental impacts and the emergence of the new epoch". Lewis/Maslin (2016: 444 n.21) agree with Moore

> that the development of the Anthropocene is tied to the modern world and various forms of capitalism. Calling this the Capitalocene is, in our view, not correct because the Anthropocene will last so far into the future – perhaps millions of years – that it may well encompass other future modes of living that are not based on the hallmarks of a capitalist social system.

From a critical Marxist perspective, Angus (2016: 232) stated that "capitalism is a 600-year old social and economic system, while the Anthropocene is a 60-year old Earth System epoch … Insisting on a different word … can only cause confusion, and direct attention away from far more important issues."

The science historian Renn (2020: 382) argued that "the Anthropocene is now a 'Capitalocene', in the sense that the economic, social and political structures of financialised capitalism drive and fundamentally shape current anthropogenic environmental changes, forcing nature into a world external to the logic of capital, where it serves as a provider of resources and a dumping ground for waste and emissions." However, Renn is opposed to relabelling the Anthropocene the Capitalocene because this would tear down "an important bridge between the natural sciences and the social sciences and humanities".

From a legal perspective, Wheeler (2017: 293) prefers 'Anthropocene' to 'Capitalocene', as the latter term "does risk disguising the reality of the Anthropocene. This reality is that the effects of the Anthropocene will be felt by all, albeit unevenly. Thus, '*Anthropos*', emphasising the totality of the crisis, should be the underpinning idea rather than what might be seen as a focus on a limited capital-accruing or benefiting class." Of the many suggestions to replace 'Anthropocene', only 'Capitalocene' has been selectively discussed in the literature, in which a majority argued that the 'Capitalocene' should not be used instead of the 'Anthropocene' that has been increasingly discussed across most scientific disciplines.

In *The Anthropocene as a Geological Time Unit*, Zalasiewicz et al. (2019: 1), the chairman and other active members of the AWG, present evidence for defining the Anthropocene as a geological epoch and "demonstrate the extent to which the term has practical utility in the field of geology, in the field of natural science generally, and to the wider academic community".

Elsewhere, this author (Brauch 2021) published a bibliometric analysis (2000–2020) on the use of the Anthropocene concept in the natural and social sciences, in the humanities and law. Thus, we turn next to 'what' the Anthropocene signified in the social sciences and political realm: a turning point, context and challenge.

2.4 The Anthropocene: A Turning Point, Context and Challenge for Science and Politics

For the purpose of programmatic suggestions for the creation of a 'peace ecology' research approach or research and policy programme in the Anthropocene, a brief discussion of its purpose as a turning point, context and challenge for science and politics will be offered next.

2.4.1 The Anthropocene as a Turning Point *in Earth and Human History*

The Anthropocene was initially proposed by Crutzen (2000) as a turning point in Earth's history that replaced the Holocene Epoch which "began approximately 11,650 cal years before present, after the last glacial period. … The Holocene and the preceding Pleistocene together form the Quaternary period".[15] According to a majority vote and suggestions by the *Anthropocene Working Group* (AWG) of May 2019, the Holocene Epoch ended between 1945 and 1950 (2.2.6) when the Anthropocene Epoch started. It was argued above that this geological turning point in Earth's history coincided with a fundamental transformation in the international political, security and economic order during the emerging Cold War (1945–1950). There was a fundamental change in the global economic order and the lifestyle of many people in the countries of the Global North (industrialised, OECD member states), the joint economic consequences of which can be observed in the proposed 'Great Acceleration'. In their *Environmental History of the Anthropocene since 1945* McNeill and Engelke (2014: 4) offered two reasons for this turning point or transition:

> *First*, that since the mid-twentieth century human action (unintentionally) has become the most important factor governing crucial biogeochemical cycles, to wit, the carbon cycle, the sulphur cycle, and the nitrogen cycle. Those cycles form a large part of what is now called the 'Earth system', a set of interlocking global-scale processes. The *second reason* is that since the mid-twentieth century the human impacts on the Earth and biosphere, measured and judged in several different ways … has escalated. … One cannot say when the Great

[15] Wikipedia: "Holocene"; at: https://en.wikipedia.org/wiki/Holocene (10 June 2020).

Acceleration will end, and one cannot say just how, but it is almost certainly a brief blip in human history, environmental history, and Earth history.

Thus, the years 1945–1950 mark the only joint turning point in Earth and human history when the 'geological time' of Earth history and 'the structural time' (*histoire du longue durée*) of international political, security and economic order have coincided (Brauch 2016). Those who planned the post-World War II international order and drafted the UN Charter (1941–1945) were not aware of the new challenges triggered by the atomic bomb, the processes of decolonisation or the new threats posed by problems of global environmental change, as these were not yet socially constructed and did not exist in the 'world-views' of scientists and policy advisers or the 'mindsets' of policy-makers.

Not surprisingly, neither environmental issues nor ecological perspectives were addressed in most studies by specialists in international relations, security studies, peace research, and development studies. Also not surprisingly, Kalevi J. Holsti's (1992) major book, *Peace and War: Arms Conflicts and International Order 1648–1989*, could not include an environmental dimension in its analysis of the 177 wars in the European and global states system. Holsti showed

how the great peacemaking efforts of 1648, 1713, 1815, 1919, and 1945 involved implicit and explicit theories of international relations. The new orders established in those watershed peace conferences attempted to solve the issues of the past, yet few successfully anticipated those of the future. Indeed, some created the basis of new conflicts.[16]

The same applies to Osiander's (1994) study, *The States System of Europe 1640–1990: Peacemaking and the Conditions of International Stability*. In the dominant world-views of scholars and mindsets of decision-makers that had influenced their idealist, realist or Grotian theories of international relations, environmental issues and problems had not yet existed and thus could not be addressed.

In *The Global Transformation. History, Modernity and the Making of International Relations,* which focuses on the 'long nineteenth century' (1776–1914), Buzan and Lawson (2015) proposed repositioning "the roots of the discipline [International Relations] and establish[ing] a new way of understanding and teaching the relationship between world history and international relations". Buzan and Lawson (2015: 96) noted the start of the processes that have resulted in the 'Great Acceleration' since 1950:

The huge expansion of both the human population and industrial output (and in pollution) that took off during the global transformation makes it the central reference point for the time when human activity began to become a significant factor in climate change. The use of fossil fuels multiplied by a factor of five during the nineteenth century and by a factor of twenty during the twentieth century (Christian 2004: 346–347). Although most people did not know it at the time, the global transformation marks the beginning of contemporary concerns about the environment.

Buzan and Lawson did not refer to the Anthropocene or the Industrial Revolution that was initially suggested by Crutzen as the turning point from the Holocene to the

[16]See the blurb of the book by Holsti (1992).

Anthropocene. Between 1945 and 1950 the following structural changes in the US and in the international order have evolved:

- As the key political planner and architect of post WWII institutions and regimes, by 1950 the US had adopted a bipartisan consensus in the US Congress to maintain the structural elements of a 'national security state' and to assume an active world leadership role in global political, security, economic, ideological and other issues.
- President F.D. Roosevelt took the lead in designing a security order 'with teeth' via the UN Charter and its UN Security Council in order to realise his vision of four 'world policemen'.
- When the Soviet Union blocked and paralysed the new collective security system, the US federal government built up a close network of military alliances as systems of collective self-defence based on Art. 50 of the UNCh via the Rio Treaty of 1947, which created the OAS (1948). This was followed by NATO in 1949; the ANZUS Treaty consisting of the United States, Australia, and New Zealand in 1951; the SEATO Treaty (1954–1977) with the United States, the United Kingdom, France, Australia, New Zealand, Pakistan, the Philippines, Thailand, South Vietnam, and Laos as members; and the CENTO pact (1955–1979) with the United Kingdom, Turkey, Iran, Iraq, and Pakistan as members. After the end of the Cold War (1981–1991) only NATO and the ANZUS Pact survived.
- After 1945 the US ended the Global Preference System of the UK and established its own Global Economic Order based on the so-called Bretton Woods Institutions (the World Bank, the IMF, GATT, WTO etc.). This created a framework for globalisation in which the neo-liberal 'Washington Consensus' has prevailed since the 1980s (Reagan, Thatcher).
- These US-led political, security and economic institutions established a framework in which the processes of a 'Great Acceleration' could evolve and flourish.
- These global institutions have been analysed by many social scientists and IR specialists from a rather narrow institutional perspective that lacked a holistic framework linking results in peace research with security, development and environmental studies.

The period between 1945 and 1950 became instrumental in creating and building up the multilateral international order of the Anthropocene Epoch, whose Western (originally American) way of life with rapidly increasing greenhouse gas emissions benefited from the US-led international order during the Cold War. In 2017 the 45th US President began to undermine this multilateral order and launched his partly isolationist and militarist perspective that denies the existence of anthropogenic climate change and opposes the Paris Climate Change Treaty of 2015.

2.4.2 *The End of World War II, the Anthropocene and the Nuclear Era as a* **Context** *in Earth and Human History*

The Anthropocene coincides with a new threefold political, economic and societal global context:

1. the *end of World War II* on 8 May 1945 in Europe and 2 September 1945 in Asia, and the beginning of the Cold War (1946/1947–1989/1991);
2. the *start of the nuclear age* on 16 July 1945 with the first test of a nuclear weapon in Alamogordo in the desert of New Mexico;
3. the *end of the Holocone Epoch* and the *start of the Anthropocene Epoch* in Earth history.

The first successful secret nuclear weapon test in Alamogordo and the first and so far only use of nuclear weapons in war – against Hiroshima and Nagasaki on, respectively, 6 and 9 August 1945 – for the first time provided political decision-makers with a weapon of mass destruction with which they could severely damage the future of the human species on Earth. However, after the end of World War II, only the beginning of the new global context of international relations was visible to all policy-makers and human beings. As a result of the top secret Manhattan Project (1942–1946) in Los Alamos, the nuclear test in Alamogordo on 16 July 1945 was known only to top scientists, military leaders, and President F.D. Roosevelt and his major aides. The transition from the Holocene to the Anthropocene was not socially constructed until February 2000, when Paul J. Crutzen claimed: "We are now in the Anthropocene". Since the 1970s the causes of *global environmental change* (GEC) and *anthropogenic climate change* have been addressed by scholars in the natural sciences, and this new epoch of Earth and human (political, economic and societal) history has been discussed as a new context in which climate change has been identified as a major threat for the survival of humankind.

The Anthropocene has become the long-term environmental or natural context and the rapidly growing greenhouse gases – especially *carbon dioxide* (CO_2) – have become a determinant for multiple international orders and their much shorter time spans. The start of the nuclear age resulted in critical philosophical reflections on the role of humankind in the atomic era (Anders 1956, 1980, 1981),[17] in which the survival of the human species is threatened by weapons of mass destruction.

[17] See the philosophical discourses in the nuclear age; see at: blog.nuclearphilosophy.org/ (10 June 2020); Anders (1985 [1956], 2002 [1980], 1981).

A new debate on a philosophy for the Anthropocene is emerging (Raffnsøe 2016).[18] The contributors to Polt/Wittrock's (2018) edited volume, *The Task of Philosophy in the Anthropocene*,[19]

> reflect on the responsibilities and possibilities for philosophy today. … [They] explore whether philosophy has meaningful tasks to fulfil in this unparalleled situation. Do philosophers need to reflect on new topics today? Do they need to think in new ways? Do they need new relationships to their own tradition? And are there concrete actions they should take, over and above philosophical reflection? The contributors to this volume thus take on the question of the relevance and responsibility of philosophy, drawing upon diverse legacies, in the current global situation.

However, there is a need to reflect on the Anthropocene as the only joint *turning point* in Earth and human history and the *new context* for the relationship between Earth and humankind. It is also important to reflect on the relationship and mutual influences between human beings and nature in a more holistic way that takes into account the fundamental challenges of the unpredictable 'tipping points' posed by chaotic reactions within the climate system, including the linear physical impacts of climate change, soil erosion, water crises and biodiversity losses and extinctions.

2.4.3 *The Anthropocene as a* **Challenge** *for Present and Future Scientific Research and for Societal, Economic and Political Action*

The growing awareness of anthropogenic climate change requires a deeper understanding of the observable and projected *challenges* which natural variations in climate, anthropogenic climate change and chaotic tipping points may pose for humankind during the Anthropocene, an epoch which is set to stretch far beyond the end of this century. These challenges have been projected using different climate models based on different assumptions about the increases in CO_2 concentration in the atmosphere that were assessed in the first five assessment reports of the IPCCC (1990, 1996, 2001, 2007, 2013).

The physical impacts of anthropogenic climate change have already had multiple societal and economic impacts that may cause societal, economic and political crises, national and regional conflicts, and, in the very worst case, even wars (Scheffran et al. 2012). The unpredictable chaotic interactions within the climate system may bring

---

[18]Raffnsøe (2016) argues that "a critical human turn affecting the human condition is still in the process of arriving in the wake of an initial Copernican Revolution and Kant's ensuing second Copernican Counter-revolution. Within this landscape, issues concerning the human – its finitude, responsiveness, responsibility, maturity, auto-affection and relationship to itself – appear rephrased and re-accentuated as decisive probing questions." His book "explores how the change has ramifications for the kinds of knowledge that can be acquired concerning human beings and for the human sciences as a study of human existential beings in the world.".

[19]See the back cover of Polt/Wittrock (2018).

about dangerous 'tipping points' with unforeseeable and unpredictable geopolitical consequences.

Since the start of the Anthropocene Epoch of earth history (1945–1950) the recognition of the human impact on the environment and especially on global environmental change and climate change has increased and its.

> challenges are real with respect to the sustainability of planet Earth in light of the intensifying human interactions … The 'Anthropocene' is a time when humanity must reconcile its influence as the key driver of climate change and its capacity to develop pathways for climate resiliency and sustainable development (Chin et al. 2017: 1).

A special issue of the Journal *Anthropocene* (2017) includes selective papers that were presented during a symposium of the American Association of Geographers on "Challenges of the 'Anthropocene'". It addressed four themes: (a) the early Anthropocene, (b) the evidence of large-scale human impacts, (c) "couplings and societal responses to human-induced environmental change", and (d) "measuring risk and planning sustainability in an 'Anthropocene' twenty-first century", in which Preiser et al. (2017) outlined:

> four perspectives for framing the problem of the 'Anthropocene' – the eco-modernist perspective, the planetary stewardship paradigm, the pathways to sustainability approach, and the critical post-humanist paradigm – with consequently different paths toward achieving sustainability. The authors suggest that a deeper framing integrating these perspectives is an understanding of the 'Anthropocene as responsibility', which elicits an ethics that comes with being human, required to bring about sustainable futures. Similarly, Berzonsky/Moser (2017) suggest that overcoming the challenges associated with the 'Anthropocene' requires psycho-cultural changes in worldviews, identity, and values (Chin et al. 2017: 2–3).

Problems of global environmental change in the Anthropocene Epoch have already posed multiple challenges for national and international policies, and policy-makers must face and cope with these challenges during the remaining decades of this century. With regard to the political responses to these challenges during the early Anthropocene Epoch, different world-views and mindsets must be distinguished between:

- *Denial* that anthropogenic climate change exists, open opposition to the Paris Agreement, and the so far only withdrawal from it by the Trump Administration on 1 June 2017;
- *A business-as-usual perspective* with modest – and insufficiently implemented – obligations under the *Nationally Determined Contributions* (NDCs) in the Paris Agreement (2015);
- *A cornucopian* approach relying on technological innovation and technological fixes by geoengineering without changing the Western lifestyle, as proposed by eco-modernists;
- *A transformative approach* aiming for a transition towards sustainability in production and consumption. and lifestyle changes that require both structural changes in the economy and society and behavioural changes by a large majority of the population (Brauch et al. 2016).

Problems of global environmental change in the Anthropocene Epoch require a holistic scientific perspective, a multi- and interdisciplinary approach and a transformative design, and thus challenge the tendency of the sciences to favour a high degree of specialisation that makes bridge-building between research programmes difficult and, so far, the exception between peace research and environmental studies during the emerging Anthropocene Epoch.

This requires a rethink about the normative goal of peace and the problems and approaches of peace research (2.5), as well as revised ecology concepts and approaches (2.6). We also need to build on past bridge-building efforts between peace research and ecology (2.7) in order to develop a scientific 'peace ecology' programme which is complemented by a holistic, enlightening and critical scientific programme for the Anthropocene.

2.5 Rethinking Peace and Peace Research in the Anthropocene

Peace has different meanings, but it is globally understood as a state of tranquillity, quietness, and freedom from any disturbance, oppressive thoughts or emotions.[20] Peace also stands for harmony in interpersonal relations based on mutual agreements between the people involved. The term 'peace' (in French *paix*; in Italian *pace*; in Spanish *paz*; and in German *Frieden*) originates from the Latin *pax* and is associated in English with: (1) no war; (2) an agreement that ends a war; (3) a peaceful situation with no unpleasant noise; (4) a feeling of calmness and lack of worry and problems; and (5) a situation in which there is no quarrelling between people who live or work together (Langenscheidt-Longman 1995: 1041). Many dictionaries combine a state of no war with a positive state of harmony. In Russian *mir* refers to both 'peace' and the 'world'. In pre-Hispanic cultures 'peace' implies equilibrium between nature and humans and between gods and humans, as well as between human beings. Peace may also be linked to the oriental concepts of harmony and equilibrium.

2.5.1 The Evolution of the 'Peace Concept' Through History in Different Cultures, Religions, and Scientific Disciplines in the Social Sciences and Humanities

The peace concept has emerged from different roots: in Europe from the Greek *eirene* (*Ειρήνη*), the Roman *pax* and *pax romana,* or peace within the borders (*limes*) of the Roman Empire, and the Christian tradition ('pax Christiana'); in the Middle East from the Hebrew *shalom* and the Arabic 'سلام*salām*'; and in South Asia from

[20]This portion of text heavily relies on Oswald, Brauch and Tidball (2014) and on several texts published in Brauch et al. (2008).

ahimsa (peace with nature). The Jainist *ahimsa*, "no harm", adds the ecological dimension that was missing in the occident. In other cultures, different values, goals and concepts (law, security, justice, harmony with nature) are also associated with 'peace'. The Hindu and Buddhist philosophy of unity and a connection between all living species widened the understanding of non-violence, and for Jains, non-violence is the major religious duty to avoid harmful *karma*. Mahatma Gandhi based his struggle for independence on *ahimsa* (Dadhich 2008), and Martin Luther King promoted the use of non-violence in the campaign against the racial discrimination of Afro-Americans.

In Greek philosophy, Plato argued that war and conflicts should be avoided within the *polis* (city). Aristotle combined peace with politics and emphasised that all political goals may only be realised under conditions of peace, and that war is only acceptable as a means for the defence of the *polis*. During the Roman period, *pax* was closely tied to law and contracts, and, with the emergence of the Roman Empire, the imperial *Pax Romana* relied on contractual subjugation under the emperor in exchange for protection against external intruders.

St Augustine developed a comprehensive Christian concept of peace that distinguished between peace on earth (*pax humana*) and the peace of God *(pax divina*). Thomas Aquinas stressed the close connection of peace with justice (*iustitia*), but also with the love for other human beings (*caritas*). For him, peace is a political good, reflecting goals of the state, and a precondition for a good life. During the fourteenth and fifteenth centuries, several convents called for peace between contentious Christian factions (*pax Christiana*), and the Westphalian Peace of 1648 requested that all parties adhere to the *pax Christiana universalis perpetua* ('universal eternal Christian peace'). After the Peace of Utrecht (1713), Abbé de Saint-Pierre called for a federation of princes to secure *paix perpétuelle* ('perpetual peace') in the tradition of peace proposals from Thomas More's *Utopia* (1516) to William Penn's *Essay towards the Present and Future Peace in Europe* (1693).

Besides the 'peace within the state' that was achieved through its monopoly of the means and use of force, the 'peace between and among states' has become a major concern of modern international law since the seventeenth (de Vitoria, Suárez) and eighteenth centuries (Grotius, Pufendorf). Its authors still considered war between states a legitimate means for the realisation of national interests (*ius ad bellum*), while calling for constraints during war, such as the continuation of diplomacy and the activities of neutral organisations (*ius in bello*). In his treatise advocating *perpetual peace* (1795) Kant went a step further and proposed a ban on war itself. He developed a legal framework for permanent peace based on six preliminary and three definite articles that called for a democratic system of rule, an international organisation (league of nations), and respect for human rights. In the nineteenth and early twentieth centuries many writers glorified war (*bellicists*), while *pacifists* and the peace movement of the late-nineteenth century advocated the condemnation of war. After Mexican independence (1810), Benito Juárez introduced the principle that "respect for the rights of others is peace" in treaties with other countries. In modern theories of hegemonic stability, *Pax Americana* refers to a peace as proposed and implemented by the USA.

After World War I, the Kantian tradition was instrumental in the creation of the League of Nations, while after World War II, Hobbesian lessons were drawn from its collapse. The *United Nations* (UN) gained 'teeth' with the advent of the Security Council, and during the Cold War a bipolar power system based on military alliances and the terror of destructive arms of a *mutual assured destruction* (MAD) prevailed. With the end of the Cold War, war as a social institution returned in the form of resource, ethnic, and religious conflicts, primarily within states but also as pre-emptive wars not legitimised by the UN Security Council and against the expressed preferences of many state members. Since the 1980s there has been an intensive debate in international relations about 'liberal peace' (Paris 2004; Campbell et al. 2011) or 'democractic peace' (Russett 1993; Oren 1995). During the 1990s proposals for a new international order in the Kantian and Grotian traditions were gradually replaced by power-driven concepts of preventive wars and the 'war against terror'.

Peace has been defined as a basic value (Zsifkovits 1973) and as a goal of political action, as a situation of non-war, and as the utopia of a more just world. Schwerdtfeger 2001: 28–29) identified four methods for defining peace: (1) a nominal definition; (2) as a result of a contemplative hermeneutic process; (3) a review of the historic evolution of the concept; (4) a determination reached after analysis of opposite concepts. Galtung (1967, 1968, 1969, 1988, 2013) distinguished between a condition of 'negative peace' (absence of physical or personal violence – or a state of non-war) and 'positive peace' (absence of structural violence, repression and injustice), taking the form of "economic exploitation and/or political repression in intra-country and inter-country class relations". In his mini-theory of peace, Galtung (2007, 2014) argued that "Peace is not a property of one party alone, but a property of the relation between parties". He distinguished between negative (disharmonious), indifferent, and positive (harmonious) relations that often result in *negative peace* (absence of violence, ceasefire, indifferent relations) or *positive peace* (Galtung 1985, 1993: 688–689).

Senghaas (1997, 2013) pointed to five conditions of peace between nations: (1) positive interdependence; (2) symmetry of interdependence; (3) homology; and (4) entropy; that require (5) common softly regulating institutions. In his 'civilisatory hexagon' Senghaas (1994, 1995, 2013) referred to six related aspects: (1) an efficient monopoly over the use of force; (2) effective control by an independent legal system; (3) interdependence of social groups; (4) democratic participation; (5) social justice, and (6) a political culture of constructive and peaceful conflict transformation. Among the many attempts to define peace, no consensus on a generally accepted minimal definition emerged.

According to Czempiel (1986), peace has an institutionalised patterned process of no-war. For Brock (2002: 104f.), peace should be more than the absence of war in the framework of time (eternal peace), space (peace on earth), society (domestic intra-societal peace), procedure (peace as peaceful dispute on peace), and a heuristic dimension to move from the causes of war to the conditions of peace.

Alger (1999: 13–42, 2014, 2014a, b) provided a map of 24 peace tools that can be derived from efforts of peacebuilding during the nineteenth century (2 tools) and the twentieth century (22 tools) which he associated with both the negative (11) and the positive (13) peace concept. Alger (1999: 40–42) filed the 24 peace tools into nine categories:

> (1) words, (2) limited military power, (3) deterrent military power, (4) reducing weapons, (5) alternatives to weapons, (6) protecting rights of individuals and groups, (7) collaboration in solving common economic and social problems, (8) equitable sharing of economic, communications and ecological systems, and (9) involvement of the population at large through peace education and organised participation.

In the UN Charter of 1945, the 'concept of peace' was mentioned among the purposes of the UN in Art. 1(1): "to maintain international peace and security", and "to take effective collective measures for the prevention and the removal of the threats to the peace, and for the suppression of acts of aggression or other breaches of the peace", as well as peaceful conflict settlements. Wolfrum (1994: 50) pointed to both narrow and wide interpretations of peace in the Charter. In Art. 1(2) and 1(3) the UN Charter uses a wider and positive peace concept when it calls for developing "friendly relations among nations" and for achieving "international cooperation in solving international problems of an economic, social, cultural, or humanitarian character." A wider concept of peace was the basis for the "Proclamation of the International Year of Peace" in GA Res. 40/3 of 3 October 1985, which stated that the promotion of international peace and security required continuous and positive action by peoples and states (Wolfrum 1994: 51).

In Chap. 6 on the Pacific Settlement of Disputes, Art. 33 uses a 'negative' concept of peace that is "ensured through prohibitions of intervention and the use of force" (Tomuschat 1994: 508). In Chapter VII of the UN Charter dealing with "Action with Respect to Threats to the Peace, Breaches of the Peace, and Acts of Aggression", in Art. 39, a 'negative' concept of peace prevails, referring to "the absence of the organised use of force between states". In a Security Council meeting on 31 January 1992 the Heads of States and Government "recognised that the absence of war and military conflicts amongst states does not in itself ensure international peace and security" (Frowein 1994: 608). In the framework of Chapter IX on "International Economic and Social Cooperation", Art. 55(3) refers to the "universal respect for, and observance of, human rights and fundamental freedoms". It has been suggested that the Charter should include "the right of self-determination, to peace, development, and to a sound environment" (Partsch 1994: 779) as "human rights of the third generation" (Vasak 1984: 837).

In the UN Charter of June 1945, a narrow or 'negative' concept of peace was at the centre, with a few direct references to 'positive' aspects to be achieved by 'friendly relations among nations' and 'international cooperation'. No reference to 'peace with nature' was included in the Charter, nor were extreme outcomes emerging from global environmental change conceptualised as 'threats to peace'. The 'positive peace' concept refers to peaceful social and cultural beliefs and norms, the presence of economic, social and political justice, and the democratic use of power, including non-violent mechanisms of conflict resolution. 'Sustainable peace', or 'peace with nature', was added later in the debate within the UN, therefore now let us turn first to the ecology concept (see below in Sect. 2.6).

2.5.2 Evolution of Peace Research Since 1919–2020

The international peace activities by the Labour and Women's movements[21] before and after World War I increased awareness of international peace diplomacy and advocated the prevention of the outbreak of violence between states and the avoidance of wars. Several policy advisers at the Versailles Peace Conference (1919) noted the lack of scientific and political knowledge during attempts to create a lasting post-World War I peace order.

In 1919 the two first university Chairs in international relations, focusing to a large extent on issues of international law and organisations from a Wilsonian or *idealist* (ideas-driven) perspective, were established at the Universities of Geneva and Aberystwyth. Despite a few initial scholars and authors (Quincy Wright 1941, L.F. Richardson [Gleditsch 2020], Ludwig Quidde,[22] Emil Julius Gumbel[23] et al.), peace research as a distinct research programme in contrast to *realist* war, security and strategic studies did not emerge until the 1950s. In the US (Kenneth and Elise Boulding, Karl W. Deutsch), the UK (John W. Burton[24]) and Scandinavian countries (Johan Galtung and Gunnar Myrdal) the first peace research institutes, departments and societies emerged in the late 1950s, with Germany, Austria and Switzerland following suit in the early 1970s. In 1964 early pioneers (J.W. Burton, B. Roeling, J. Galtung et al.) set up the *International Peace Research Association* (IPRA).[25]

The institutionalisation of peace research departments and institutes began in 1955 with the *Theodore F. Lentz Institute* in St Louis (Missouri, USA). In 1959 the young Norwegian sociologist Johan Galtung established the *Peace Research Institute Oslo* (PRIO), and in 1967 Gunnar Myrdal was instrumental in setting up the *Stockholm International Peace Research Institute* (SIPRI). In 1969 the Finnish Parliament established the *Peace Research Institute in Tampere* (TAPRI), which in 1994 became part of the University of Tampere (from 2019 Tampere University), and in 2011 TAPRI joined the School of Social Sciences and Humanities and later the Faculty of Social Sciences as an independent research institute.

In Germany, on 1 September 1969 Federal President Gustav Heinemann noted the need for peace research, which resulted in the foundation of the *German Association for Peace and Conflict Research* (DGFK) and the establishment of two peace research institutes: the *Peace Research Institute Frankfurt* (PRIF) in 1970, and the *Institute for Peace Research and Security Policy* (ISFH) in Hamburg in 1971. The German

---

[21] "Peace Movement"; at: https://en.wikipedia.org/wiki/Peace_movement (29 July 2020); "Friedensbewegung"; at: https://de.wikipedia.org/wiki/Friedensbewegung (29 July 2020) .

[22] See Quidde's brief biography at: https://de.wikipedia.org/wiki/Ludwig_Quidde and https://en.wikipedia.org/wiki/Ludwig_Quidde.

[23] See Gumbel (1958) and a brief biography at: https://en.wikipedia.org/wiki/Emil_Julius_Gumbel.

[24] See the brief biography of John W. Burton; at: https://en.wikipedia.org/wiki/John_Burton_(diplomat); Dunn (2004); *see the* Guide to the Papers of John W. Burton; at: https://nla.gov.au/nla.obj-340342029/findingaid#arrangement (6 August 2020).

[25] An extensive body of literature exists on the history of peace research institutions and key pioneers of peace research during and after the Cold War. For an overview, see Brauch (1979); Stephenson (2017).

Association for Peace and Conflict Studies (*Arbeitsgemeinschaft für Friedens- und Konfliktforschung* e.V. – AFK) is a scientific association of peace researchers and institutions from various disciplines, particularly from German-speaking countries.

Founded in 1968, the AFK aims to promote research that contributes to a deeper understanding of the causes of war and peace and thus also provides a basis for political practices orientated towards long-term peace. In spring 2020 the AFK had more than 270 members. The association is represented by its Executive Board. Fourteen German universities offer thirty degree courses on peace research.[26]

In 1970, the private *Berghof Foundation on Conflict Research* was set up by Prof Dr. Georg Zundel as a private limited company with tax-exempt status under German law, and it initially supported critical analyses of the arms race during the Cold War. In 2012 the Berghof Foundation integrated three previously independent institutions: the former Berghof Foundation for Conflict Studies along with its operational arm, Berghof Conflict Research; Berghof Peace Support; and the former Institute for Peace Education in Tübingen.

In October 2000, the *German Foundation for Peace Research* (Deutsche Stiftung für Friedensforschung [DSF]) was established by the *Federal Ministry for Education and Science* (BMBF) with a public contribution of DM 50 million (€25.26 million) that will be increased after 2020.

Later *The Austrian Study Centre for Peace and Conflict Resolution* (ASPR) was founded in 1982 in Schlaining in the Burgenland (Austria). The *European Peace University* (EPU) was established in 1988 in Schlaining, but closed down in 2013 when EPU's accreditation was withdrawn by Austria's accreditation body.

In Switzerland, the Swiss Peace Foundation was founded in 1988 in Bern. In 2020, *Swisspeace* (Goetschel/Hellmüller 2019) is a practice-orientated peace research institute with the mission to contribute to the improvement of conflict prevention and conflict transformation by producing, shaping discourses on international peace policy, developing and applying new peacebuilding tools and methodologies, and supporting and advising other peace actors, as well as providing and facilitating spaces for analysis, discussion, critical reflection and learning. *Swisspeace* is an associated institute of the University of Basel and a member of the Swiss Academy of Humanities and Social Sciences. Its major clients are the Swiss Federal Department of Foreign Affairs, the State Secretariat for Education, Research and Innovation, international organizations, think tanks and NGOs.

In the UK, the Lancaster Peace Research Centre was set up in 1959. It became the *Richardson Institute* in 1969 and moved to Lancaster in 1978, where it has been directed by Michael Nicholson, Paul Smoker, Morris Bradley, Hugh Miall, Feargel Cochrane and Simon Mabon (since 2012). The *Bradford School of Peace Studies* was established in 1973 and has become the Department of Peace Studies and International Development within the School of Social Studies. It collaborates with governments around the world, the United Nations, the Swedish International Development Cooperation Agency, the Africa and the China Development Banks,

[26]See at: https://studieren.de/suche.0.html?sw=frieden&lt=course&rs=tile&term=72791%3Aterm%3Afrieden&sort=best (30 July 2020).

the Aga Khan Foundation, the UK Ministry of Defence, Saferworld, and the Oxford Research Group focusing on peacebuilding, humanitarian assistance and development. In 2020, about thirteen British universities offered degree courses in peace and conflict studies.

In the Netherlands, the peace research institute that was established at the University of Groningen in 1965 by Bert Roeling, the first Secretary General of IPRA, was closed down after the IPRA conference in December 1990. Thirty years later in 2020, Master's degrees were offered at Groningen University in Religion, Conflict and Globalization and International Humanitarian Action; in Conflict, Territories and Identities at Radboud University in Nijmegen; in Conflict Studies and Human Rights at Utrecht University; and in Crisis and Security Management at the University of Leiden.

In North America, in July 2020, the School Directory of Peace Studies & Conflict Resolution Programs compiled by the *Peace and Justice Studies Association* (PJSA) listed 260 college and university programmes in the USA and twenty-nine in Canada in the fields of Peace and Conflict Studies, Conflict Resolution, Social Justice Studies, Conflict Analysis, Conflict Management, Dispute Resolution and a variety of related fields. Of these, ninety-eight programmes offered a Bachelor's and sixty-five a Master's degree, while eleven offered a Ph.D. and four a JD degree.[27]

In 1984 the *US Institute of Peace* (USIP) was established by Congress as a non-partisan and independent institution devoted to the non-violent prevention and mitigation of deadly conflict abroad that provides "research, analysis, and training to individuals in diplomacy, mediation, and other peace-building measures".[28]

In July 1992, when Japan organised the IPRA conference in Kyoto, Japan had a very strong peace research community whose global visibility and academic impact has since gradually declined, but several small institutes or university centres still exist at the universities of Hiroshima, Kyoto and Tokyo. *The Peace Studies Association of Japan* (PSAJ) was established in 1973 to develop peace research in Japan. PSAJ is a member of IPRA, has regularly held an annual convention, and publishes Peace Studies Bulletin. Its membership increased from seventy-two in 1973 to about 880 registered individual members and twelve corporate members.

The *International Christian University Peace Research Institute* (ICUPRI) was founded in Japan in 1991. As a research institute, it has not undertaken any educational activities. The Tokyo University of Foreign Studies offers an interdisciplinary research and study programme on *Peace and Conflict* Studies (PCS), which in 2004 launched the university's first degree programme in English. PCS offers graduate-level research and practical training for students from around the world in response to the need for professionals with expertise in peace and conflict studies, violence, development, and other areas of international affairs. The 'PCS Course to Foster

---

[27] See at: https://www.peacejusticestudies.org/school-directory/ (29 July 2020).

[28] See at: https://www.usip.org/ (29 July 2020); https://en.wikipedia.org/wiki/United_States_Institute_of_Peace (29 July 2020).

Professionals for Post-Conflict Peacebuilding' is an *International Priority Graduate Program* (PGP) with a government scholarship for international students from conflict-affected countries.

The Toda Peace Institute in Japan is an independent, non-partisan institute to advance a more just and peaceful world through policy-orientated peace research and practice informed by evidence-based research, multi-track and multi-disciplinary problem-solving workshops and seminars, and dialogue across ethnic, cultural, religious and political divides. The Toda Peace Institute's current research and practice focus on: (a) Cooperative Security, Arms Control and Disarmament; (b) Social Media and Peacebuilding; (c) Climate Change and Conflict; and (d) Peace and Security in North-East Asia. In September 2018 the Toda Peace Institute and the National Centre for Peace and Conflict Studies (University of Otago, New Zealand) held a workshop on "Climate Change and Conflict in the Pacific: Prevention, Management and the Enhancement of Community Resilience" in Auckland, New Zealand.[29]

The City University of Hiroshima maintains a Graduate School of Peace Studies and in April 1998 it established the Hiroshima Peace Institute as an affiliated research institute. The Institute conducts peace-related education and academic research that aims to contribute to the abolition of nuclear weapons, the establishment of world peace, and the development of local communities. The Graduate School of Peace Studies has a curriculum that fosters analytical academic methodologies in peace studies, international politics, international law and international relations, and wide-ranging skills for analysing real problems. It prepares students for career paths in international relations and human security to propose effective preventive measures aimed at avoiding and solving disputes; trains international officials and international NGO employees who can contribute to the drafting and formulation of international public policy aimed at building and realising peace, and national and regional public servants involved with public policy and internationally related work; and provides journalists and mass-media specialists with the skills to analyse a diversity of issues, including current international disputes and global problems, and the ability to share information with civil society and the international community from a peacebuilding perspective.[30]

In Australia and New Zealand funding for peace research centres has declined and several institutions have disappeared, although courses, clusters and degree programmes are still offered. At the University of Sydney in 1999 a Department for Peace and Conflict Studies was set up to promote non-violent, peaceful ways of resolving conflict that meet human needs and respect human rights from the family to the boardroom, rural villages to cities, and governments to international organisations. Its teaching addresses the causes of conflict and the conditions affecting resolution and transformation. Research projects focus on attaining positive peace in just societies with resources for non-violent responses to conflict. Its staff specialises

[29]On the Toda Peace Institute see at: https://toda.org/.

[30]See at: https://future-students.uq.edu.au/study/programs/master-peace-and-conflict-studies-5644#how-to-apply (30 July 2020).

on peace journalism, transitional justice, genocide studies, human rights, reconciliation, non-violence, the United Nations and the psychology of peace. The University of Queensland in Brisbane offers a Master's degree in peace and conflict studies.[31] The University of Canberra offers a course on Global Peace and Conflict Studies.[32] The University of Melbourne has a cluster on Peace, Development and Justice.[33]

The Institute for Economics and Peace (IEP) is headquartered in Sydney and has offices in five other countries (in New York, Mexico City, Brussels, The Hague and Harare). It aims to bring about a paradigm shift in our thinking about peace by developing global and national indices, calculating the economic cost of violence, analysing country-level risk and fragility, and understanding positive peace. Its work is used extensively by governments, academic institutions, think tanks, non-governmental organisations, and intergovernmental institutions such as the OECD, the Commonwealth Secretariat, the World Bank and the United Nations.[34]

In New Zealand *The National Centre for Peace and Conflict Studies* (NCPACS) was established at the University of Otago in 2009 as New Zealand's first centre to combine global cross-disciplinary expertise on the issues of development, peacebuilding and conflict transformation.[35] At the University of Auckland, a *Conflict, Terrorism and Peace group* (CTAP) works on the causes, dynamics, aftermath and prevention of all types of violent conflict in regions across the globe. It offers a Master of Conflict and Terrorism Studies programme focusing on the causes, dynamics and consequences of political violence, and explores practical and ethical approaches to its prevention and resolution.[36]

In Africa, the *African Peace Research Association* (AFPRA) was founded in 1985. During the IPRA conference in Tampere in 2000 it was renamed the *Africa Peace Research and Education Association* (AFPREA) and registered according to Nigerian law.[37] On 13–15 April 2015 an AFPREA conference occurred in Abuja (Nigeria) in cooperation with IPRA in the facilities of the *Economic Commission of West African States* (ECOWAS).

From 18–20 September 2019 the *African Centre for the Constructive Resolution of Disputes* (ACCORD) in Durban (South Africa) partnered with the Foundation for Sport, Development and Peace to host the second AFPREA Conference. This took place in Cape Town (South Africa). The conference discussed "youth, women,

[31] See at: https://future-students.uq.edu.au/study/programs/master-peace-and-conflict-studies-5644#how-to-apply (30 July 2020).

[32] See at: https://www.canberra.edu.au/coursesandunits/unit?unit_cd=8167 (30 July 2020).

[33] See at: https://arts.unimelb.edu.au/school-of-social-and-political-sciences/research/research-clusters/conflict-development-and-justice (30 July 2020).

[34] See at: https://www.economicsandpeace.org (30 July 2020).

[35] See at: https://www.otago.ac.nz/ncpacs/about/ (30 July 2020).

[36] See at: https://www.arts.auckland.ac.nz/en/about/schools-in-the-faculty-of-arts/school-of-social-sciences/politics-and-international-relations/staff-research/ctap.html (30 July 2020).

[37] For a survey and proposal for "Strengthening Peace Research and Peace Education in African Universities" see: Omeje (2015). For a survey of institutions in the Global North working on peace and development academic programmes with a focus on Africa, see McCandless et al. (2007); at: https://maryking.info/wp-content/PeaceResearchforAfrica.pdf 9 August 2020): 233–249.

climate change, cross-border migration and human trafficking, as well as election justice, peace and development based on case studies from Uganda, South Sudan, Côte d'Ivoire and Nigeria".[38]

Since 1992, ACCORD has been a major non-governmental African peace research organisation that aims

> to influence political developments by bringing conflict resolution, dialogue and institutional development to the forefront as an alternative to armed violence and protracted conflict. ACCORD specialises in conflict management, conflict analysis and conflict prevention. We intervene in conflicts through mediation, negotiation, training, research and conflict analysis. … For the past 28 years, the institution has been building the capacity of Africa's leaders to resolve conflict and to address the underlying political barriers to growth and stability. … ACCORD's current five-year medium-term strategy [1917–2021] seeks to respond to a continental and global context that is characterised by extractive economic systems, socio-economic marginalisation, identity-driven exclusion, social protest, demographic pressures of rapidly growing and more youthful populations, and competition over natural resources. … The medium-term strategy informs ACCORD's intent to use resources and programmes to continue to play a meaningful role in developing mechanisms and capabilities to respond, manage and transform conflicts.[39]

There are many small NGOs and a few university departments in West, Southern and Eastern Africa that address issues of peace, security, development and environment issues in several parts of Africa, but apart from ACCORD there are no permanent funded institutions. Several European peace research institutions have addressed peace issues relating to Africa, primarily through externally funded research projects by PRIO (Oslo),[40] SIPRI (Stockholm)[41] and the Berghof Foundation (Berlin),[42] while the UN University of Peace in Costa Rica (McCandless et al. 2007) has had a limited focus and a few research areas pertaining to Africa.

The *Latin American Council for Peace Research* (CLAIP) was founded in December 1977 in Oaxtepec (Morelos, Mexico) and celebrated its fortieth Anniversary in Mexico City in January 2017 (Oswald Spring et al. 2018). In Latin America there is only one major academic institution focusing on peace: the UN Mandated

[38] See at: https://www.accord.org.za/news/accord-presents-at-afprea-2019-regional-conference-in-cape-town/ (9 August 2020).

[39] See at: https://www.accord.org.za/about/.

[40] See PRIO at: https://www.prio.org/Projects/Project/?x=1407. "DIASPEACE was a large EU-funded cooperation project that studied the transnational political activities of Somali, Ethiopian and Eritrean diaspora organizations based in Europe. The project sought to generate policy-relevant, evidence-based knowledge on how diasporas play into the dynamics of conflict and peace in their countries of origin. It ran from March 2008 to February 2011.

[41] See SIPRI at: https://www.sipri.org/research/conflict-peace-and-security/africa-security-and-governance-project SIPRI conducted an Africa Security and Governance Project (2009–2012) supported by the Open Society Institute (www.soros.org). SIPRI's recent work on Africa spans several regions and cross-cutting themes, including peacekeeping, peacebuilding and military expenditure; at: https://www.sipri.org/research/conflict-peace-and-security/africa (9 August 2020).

[42] See at: https://www.berghof-foundation.org/en/programmes/africa/ (9 August 2020). The Berghof Foundation's "sub-Saharan Africa Unit supports various types of dialogue spanning high-level political dialogue, civil-society forums, and grass-roots dialogue initiatives. The Unit … supports the establishment and strengthening of infrastructures for peace in the East Africa region.".

University of Peace in Costa Rica that was established by the UN General Assembly in December 1980.[43]

A *German-Colombian Peace Institute* (CAPAZ), a Centre of Excellence in Research and Teaching, is financed by the *German Academic Exchange Service* (DAAD), and the German Federal Foreign Office focuses on peacebuilding through the creation of networks between universities, research centres, civil society organisations, and governmental entities that act in the territorial arena. CAPAZ supports research, teaching, and advisory activities that create new approaches to understanding peace and conflict, transmit knowledge to society, and provide responses to the multiple challenges of a society in transition. The network consists of ten Founding Members, ten Associate Members and various Project Partners.[44]

Many peace research centres and university departments – especially in the Global South – are underfunded and lack a solid and lasting economic basis. Since the peace research focus of UNESCO ended, no global survey of peace research centres has been published. Only a few of the existing peace research institutes, centres and departments have systematically addressed peace and ecological issues jointly.

2.5.3 *Rethinking Peace During the Cold War (1945–1990), in the Post-Cold War Era and in the Anthropocene*

When the UN Charter was framed (1943–1945), the scientific and political problems with which the Anthropocene concept has been associated since the year 2000 had not been socially constructed and thus were not perceived to exist. Consequently, policy-makers were not aware of them. During the bipolar Cold War ideological motives and considerations influenced the realist mainstream in the thinking on peace in international relations in the framework of systemic competition. Peace aims were pursued in the West by the 'Free World', while the East tried to instrumentalise the 'peace struggle' to weaken the 'imperialists' militarily.

With the peaceful end of the Cold War (1989–1991), the bipolar political, economic, military and ideological framework for the 'long peace' (Gaddis 1992) did not exist any longer. However, no peace strategy existed to cope with the increasing violence or prevent destructive 'new wars' (Kaldor 1999, 2002; Kaldor/Vashee 1997; Münkler 2002, 2005) in the rapidly disintegrating multi-ethnic states of the Soviet Union and Russia in the Caucasus, or in Yugoslavia's successor states between Serbia and Slovenia and Croatia (1991), within Bosnia-Herzegovina (1995), and between Serbia and Kosovo (1999). While environmental factors did not trigger the outbreak of violence in any of these conflicts, the uranium-enriched ammunition used by the US Army did have environmental consequences. After the aggression of Iraq against Kuwait in summer 1990 and during the Gulf War of 1991 (Westing 2003,

[43] See at: https://www.upeace.org/.

[44] See at: https://www.instituto-capaz.org/en/about-us-2/ and https://www.instituto-capaz.org/en/staff/.

2013, 2013a), burning oilfields and uranium-enriched ammunition (Haavisto 2003)[45] exhibited elements of an 'ecocide'[46] (Broswimmer 2002) that was not triggered by the consequences of global environmental change.

The Anthopocene Epoch (1945/1950-) as proposed by Crutzen (2000) and specified by the *Anthropocene Working Group* (AWG 2019)[47] coincides with the post-World War II era, the atomic age, the Cold War order and the global post-Cold War disorder. The concept of peace and the proclaimed 'peace policy' goals have to be reconsidered, reframed and redefined from a broader peace ecology perspective that analytically brings together the distinct narratives of peace research, security, development and environmental studies. For the past seventy-five years since 1945 the environmental dimension of peace, the environmental causes of conflicts and the environmental consequences of violent domestic conflicts and wars must be added to the "Environmental History of the Anthropocene since 1945" (McNeill/Engelke 2014).

2.5.4 Three International Peace Research Organisations: IPRA, PSS(I), and ISA-PEACE

Since the 1960s three international peace research organisations have evolved in parallel with limited academic interchange: the *International Peace Research Association* (IPRA) since 1964; the *Peace Science Society International* (PSS-I) since 1964; and the *Peace Studies Section* within the *International Studies Association* (ISA-PEACE) since 1959.

IPRA was founded on 1–3 December 1964 in London at a meeting organised by John W. Burton, at which it was decided to form a professional association to increase the quantity of research on world peace and ensure its scientific quality. Its first Executive Committee members included Bert V.A. Röling (The Netherlands) as Secretary General, and John W. Burton (Australia/United Kingdom), Ljubivoje Acimovic (Yugoslavia), Jerzy Sawicki (Poland), and Johan Galtung (Norway).

Since 1964 twenty-seven biennial IPRA General Conferences have been convened. Fourteen General Conferences of IPRA have taken place in Europe – in the Netherlands (1965, 1990), Sweden (1967), Czechoslovakia (1969), Yugoslavia (1971), Finland (1974, 2000), West Germany (1979), Hungary (1983, 2004), UK (1986), Malta (1994), Belgium (2008), and Turkey (2014). So far, five IPRA General Conferences have been held in Asia – in India (1974, 2018), Japan (1992, 2012) and South Korea (2002); four in the Americas– in Mexico (1977), Canada (1981, 2006)

---

[45]On 'depleted uranium' see UNEP's website; at: https://www.unenvironment.org/explore-topics/disasters-conflicts/what-we-do/preparedness-and-response/post-crisis-environmental-0 (29 July 2020).

[46]See 'ecocide law'; at: https://ecocidelaw.com/the-law/what-is-ecocide/ (14 June 2020).

[47]*Anthropocene Working Group* (AWG 2019).

and Brazil (1988); and two each in Australia (1996, 2010) and Africa – in South Africa (1998) and in Sierra Leone (2016).

IPRA has five regional organisations: the *African Peace Research and Education Association* [48](APREA), the *Asia-Pacific Peace Research Association* (APPRA),[49] the *European Peace Research Association* (EUPRA),[50] the North American *Peace and Justice Studies Organisation* (PJSO),[51] and the *Latin American Peace Research Association* (CLAIP).[52] IPRA has represented peace research in the International Social Science Council (ISSC) and, since 2019, the *International Council of Science* (ICS).[53]

IPRA is an underfinanced NGO that does not receive any financial support from any government or international organisation, relies on a small paying membership, and has no permanent secretariat. IPRA publishes a newsletter about its activities and has occasionally published books with selected texts presented at its conferences. The IPRA Foundation has supported IPRA with travel grants for scholars from the Global South.

In 2018 and 2020 IPRA maintained ten research commissions on: (1) Media Conflicts and Journalism; (2) Development, Political Economy and Sustainable Peace; (3) Gender and Peace; (4) Peace Education; (5) Peace Theories and History; (6) Peace and Ecology in the Anthropocene; (7) Human Security and Society; (8) Pathways towards Peace and Justice; (9) Youth, Sports and Peace; and (10) Non-Violence and Peace Movements.

The *Peace Research Society (International)* was established in 1963 in Malmö during a meeting convened by Isard (2000), and its first conference was held in 1964 in Chicago. The participants included Kenneth Boulding, Anatol Rapoport, and many other leading scholars. In 1973, the Society became the *Peace Science Society (International).* PSS (I) "promotes scholarship on the causes of conflict between and within states, conflict resolution, and the promotion of sustainable peace". The PSS (I) is "an association for the advancement of peace and conflict research" that claims to be "an objective, professional, and scholarly organisation without political, ideological, financial, or national bias" and is "committed to promoting work which utilises rigorous scientific methods and analyses on the causes of conflict and the promotion of peace". It "encourages and supports the publication of research, particularly, but not exclusively, quantitative research. PSS(I) does not promote direct political

---

[48]See: *African Peace Researchand EducationAssociation* (APREA); at: https://uia.org/s/or/en/1100011209 (31 July 2020).

[49]See: *Asia-Pacific Peace Research Association* (APPRA); at: https://appra.net/ and https://appra.net/appra-conference-2019/ (31 July 2020).

[50]See: *European Peace Research Association* (EUPRA); at: https://www.euprapeace.org/ (31 July 2020).

[51]See: North American *Peace and Justice Studies Organisation* (PJSO); at: https://www.peacejusticestudies.org/ (31 July 2020).

[52]See: *Latin American Peace ResearchAssociation* (CLAIP); at: https://uia.org/s/or/en/1100057573 (31 July 2020); Oswald Spring and Serrano Oswald (2018).

[53]See: the *International Council of Science* (ICS); at: https://council.science/members/online-directory/ (31 July 2020).

action or polemical discussion." PSS-I (Isard 2000) has been managed by four Executive Secretaries: Walter Isard (1964–1989), Stuart A. Bremer (1989–2002), Glenn Palmer (2002–2019) and Idean Salehyan (2019–). In 2020, PSS-I had thirty institutional members, twenty-nine of which are universities from the US and the remaining one a university from Sweden. PSS(I) publishes the peer-reviewed journal *Conflict Management and Peace Science* (CMPS) in conjunction with SAGE. This contains scientific papers on topics such as international conflict, civil conflict and insurgency, arms races, the effect of international trade on political interactions, foreign policy decision-making, international mediation, and game theoretic approaches to conflict and cooperation. Since 1991 all annual conferences have taken place at US universities. PSS-I offers these awards: (a) the Walter Isard Award for Best Dissertation in Peace Science; (b) the Stuart A. Bremer Award for Best Graduate Student Paper; (c) the Will H. Moore III Prize; and d) The Founders' Medal.[54]

The *Peace Studies Section* within the *International Studies Association* (ISA-PEACE) aims "to seek a better understanding of the causes of war and violence and of the conditions of peace in the international system". The ISA was founded in 1959 and is now "one of the oldest interdisciplinary associations dedicated to understanding international, transnational and global affairs. … It has more than 7000 members comprising academics, practitioners, policy experts, private sector workers and independent researchers, among others". ISA has a permanent secretariat at a US university with a staff led by an Executive Director who "oversees these administrative functions, ensuring that the organization fulfills the mandates of the Governing Council".[55] The Peace Studies Section is the third largest section that "links scholars of various disciplines and methodologies, develops, encourages, and disseminates research, and facilitates research-based teaching in peace and conflict studies".[56] From 2018 to 2019 its membership increased from 791 to 875 members, and in February 2020 it was the third-largest section in ISA. During its last annual convention in 2019 in Toronto (Canada), ISA-Peace managed 102 panels partly co-sponsored by other sections. ISA-Peace offers two annual awards: (i) the *Kenneth Boulding Award* for the best graduate paper presented, and (ii) the *Peace Distinguished Scholar Award* "to honour an individual in the field of peace and conflict studies and, ideally, a record of service to the ISA". Two travel grants for participants from the Global South are sponsored by the Kroc Institute for International Peace Studies and by Lynne Rienner Publishers. The PSS-I is a partner organisation of ISA.

The three international peace research associations have had different goals and functions and a different regional membership and orientation. Two are headquartered and registered in the US: ISA and the PSS-I. PSS-I is primarily an American academic non-political scientific organisation with a preference for quantitative methods. It primarily serves the American academic community with a rather narrow focus on traditional peace research issues that rarely address an environmental dimension. ISA-Peace is the largest rallying point at which scholars primarily from the Global

[54]See at: https://peacescience.unt.edu/ (13 June 2020).

[55]See at: https://www.isanet.org/ISA/About-ISA (13 June 2020).

[56]See at: https://www.isanet.org/ISA/Sections/PEACE/News (13 June 2020).

North and Threshold countries meet annually. IPRA is the smallest and most financially fragile organisation, and to a large extent where the scientific discourse between scholars from the Global South and a few from the Global North takes place.

Since 2014 direct scientific debates between peace researchers from the Global North and the Global South have declined. ISA-Peace has become the largest global meeting point for peace researchers of the Global North whose members can afford to attend ISA meetings. All meetings of PSS-I since 1991 have taken place in the US, and ISA conventions have primarily occurred in the USA, with a few in Canada, one in London (UK) and two in Mexico – one in Mexico City (1983) and the other in Acapulco (1993). Since the end of the Cold War the discourse between peace researchers from the Global North and the Global South has declined, while the internal discussion between scholars from the Global North at ISA-Peace has increased.

This brief global 'snapshot' of peace research institutions and activities in summer 2020 indicates a high degree of professionalisation which follows dominant academic career patterns in OECD countries focusing on highly specialised methodological, theoretical and often quantitative analyses published in specialised journals with a high impact factor and less on scientific monographs addressed to a wider general and policy-orientated audience.

2.5.5 Reconceptualising Peace in the Post-Cold War Era (1990–2020) and the Anthropocene (Since 2000)

In the post-Cold War era the scientific peace concept has gradually expanded beyond Galtung's differentiation of 'negative' and 'positive peace', especially since the release of the Brundtland report (1987), to 'sustainable peace' (OECD 2011). The ecology concept has been taken up – e.g. in economics as 'ecological' economics – and the concept has proliferated as 'deep', 'human', 'social', 'political', 'peace' ecology and 'political geo-ecology.'

Mahatma Gandhi's thinking, inspired by the Hindu and Buddhist traditions and the Jain *ahimsa* concept of non-violence and the environment, has had a significant impact on Arne Naess' environmental philosophy and 'deep ecology', and on Schumacher's (1999) 'small is beautiful' philosophy (Weber 1999: 349–361). However, neither in peace research nor in environmental studies have the manifold links between peace and ecology been systematically explored.

Within peace research only a few scholars have addressed environmental problems and challenges, among them Boulding (1966, 1978) and Elise Boulding (1988, 1989, 1992, 2000). Kenneth Boulding built bridges between both research programmes in *The Economics of the Coming Spaceship Earth* (1966, 1970),[57] and in *Ecodynamics*

[57] See: https://www.jayhanson.us/page160.htm.

(1979)[58]. He combined evolutionary biology, ecology, peace research and Keynesian, socio-economics and environmental economics. Later, Boulding (1983) critically reviewed ecosystems and his concept of the 'empty niche' in biology, societal evolution and artefacts, biological catastrophes, ecological and human interactions, social ecosystems and, finally, evolutionary economics. Elise Boulding, sociologist, feminist and a mother of peace education, linked peace to ecology from practical daily experience, asserting that there is no true peace without ecological links, respect for nature and human ecology (Morrison 2005).

From a peace education perspective, *Hutchinson* (2009) reflected on the links between peace, environmental and futures education, arguing that "the direct and indirect environmental impacts ... of unequal patterns of global consumption and bloated military budgets cannot be adequately understood by staying within the conventional disciplinary, interpretative frame of economics [and that] ... enhanced cross-disciplinary and cross-cultural dialogue are likely to be important". Hutchinson (2009) considers a "cross-cultural and inter-civilisational dialogue crucial, including a greater awareness of Islamic, Confucian, Taoist and Buddhist knowledge traditions on peace and the environment".

Special care must be given to post-conflict peacebuilding processes in places where pollution and resource scarcity may create further conflicts and insecurities. Intra-state and inter-state cooperation is insufficient, and the establishment of early warning systems and good governance are often lacking, which has threatened sustainable peace (Swain/Krampe 2011).

2.6 Evolution and Rethinking Ecology Concepts and Approaches in the Anthropocene

Ecology is based on the two Greek terms *oikos* (οἶκος) – the equivalent of a household, house or family – and *logos* (λόγος), meaning speech, philosophy or science. *Oikos* is the root of both 'economics' and 'ecology'. The *Shorter Oxford English Dictionary* (2002: 789) defines ecology as: "(1) The branch of biology that deals with organisms' relations to one another and to the physical environment in which they live; (the study of) such relations as they pertain to a particular habitat or a particular species; also human ecology; (2) The political movement that seeks to protect the environment, especially from pollution."[59]

The *ecology* concept was coined between 1866 and 1869 by Ernst Haeckel (1834–1919) for the study of living species and their physical and biotic surroundings. The ecology concept was developed by many scientists from different disciplines and world regions, based in part on observations of indigenous cultures in the Americas,

[58]For a summary by Tanya Glaser, see: https://www.colorado.edu/conflict/peace/example/boul7525.htm.

[59]This section is based on several previous texts by the author that have been updated in order to take most recent publications into account: Brauch et al. (2014a: 6–14).

China, India and the Middle East, where knowledge of the use and dangers of plants and animals was crucial for human survival and cultural development. It was deeply influenced by Darwinian theory, which focused on competition as a selective force. During the late nineteenth century the concept was used for the ecology of animals, plants and in hydrobiology, while a modern definition includes (a) the interactions between organisms (individuals, populations, biocoenosis[60]), (b) in their abiotic and biotic environment and (c) the links in the energy, material and information flow.[61]

According to Ellen (1996: 207), the ecology concept "has been centrally concerned with the concept of adaptation and with all properties having a direct and measurable effect on demography, development, behaviour and spatio-temporal position of an organism." *Biological ecology* has been concerned "with population dynamics, energy transfer, systems modelling, nutrient cycles, environmental degradation and conservation; and since the 1970s, especially with the application of neo-Darwinian thinking of socio-ecology".

Human ecology is used in human geography, urban sociology and anthropology. Ellen argued that "the other major impact of ecological concepts in the social sciences has been in the relation of political environmentalism to environment and development. ... Increasing attention is also being paid to the cultural construction of nature, indigenous technological knowledge, the management of collectively owned resources, and environment history" (Ellen 1996: 208).

Vernadsky (1926) defined the biogeochemical cycles as the sum of all ecosystems. Tansley (1935) established an interactive system (biocoenosis) created between the living and their environment (biotope) and ecology, which he transformed into a science of ecosystems that was crucial for the development of ecology as a modern systems science. Odum (1953, 1977, 1993, 1998) defined ecology as the study of the linkages of organisms and groups of organisms with their environment and of their structure and functions (Nentwich et al. 2004: (1) Suess (1875) proposed the term biosphere, including the conditions that promote life on Earth, such as flora, fauna, minerals and different cycles. Theories about the nitrogen cycle, the atmosphere, the hydrosphere and the lithosphere were developed.

The ecology concept was used primarily in biology as 'autoecology' (referring to eco-physiology or biochemical ecology), 'synecology', 'populations (demographic) ecology', 'community ecology', and 'systems ecology'; and in physical geography as 'landscape ecology' (Troll 1968) and 'geoecology' (Huggett 1995). Therefore, historically, the ecology concept was basically related to the biophysical sciences and only after World War II did the concept become more integrated in the social sciences and the humanities.

---

[60]The concept of 'biocoenosis', coined by Karl Möbius in 1877, describes the interacting organisms living together in a habitat (biotope); see at: https://en.wikipedia.org/wiki/Biocoenosis (31 July 2020).

[61]See 'ecology'; at: https://en.wikipedia.org/wiki/Ecology (14 June 2020).

2.6.1 *Widening Ecology*

The ecology concept has been conceptualised by many social scientists as 'deep ecology' (Leopold 1949; Naess 1973, 1989; Pepper 2002), 'human ecology' (Marsh 1864; Young 1974; *Human Ecology*[62]), 'social ecology' (Bookchin 1988, 2005), 'ecofeminism' (d'Eaubonne 1974; Mies 1998; Shiva/Mies 1997), eco-Socialism' (Pepper 2002a), 'political ecology' (Thone 1935) and 'political geoecology' (Brauch et al. 2011).

'Ecological Economics' has evolved since the late 1980s, when the *International Society for Ecological Economics* (ISEE)[63] was established with the statuary goal to advance "our understanding of the relationships among ecological, social, and economic systems and the application of this understanding to the mutual well-being of nature and people, especially of the most vulnerable, including future generations" (Norgaard 2002: 37). ISEE aims to facilitate "understanding between economists and ecologists and the integration of their thinking into a trans-discipline aimed at developing a sustainable world", with a special focus on modelling, equity, indicators, limits to development, trade and development, valuation and policy instruments.

2.6.2 *Human Ecology*

'Human ecology' (Young 1974; Douglas 2002)[64] studies gradually evolved in geography, sociology, psychology, anthropology, zoology, epidemiology, public health, economics, and natural ecology to examine the relationship between humans and their biophysical, social, and built environments. The Geography Department at Lund University defined human ecology as "the study of the interactions between man and nature in different cultures. Human Ecology combines the ideas and methods from several disciplines, including anthropology, sociology, biology, economic history and archaeology", claiming that their "multidisciplinary approach enables us to comprehensively address issues of environmental justice, sustainability and political ecology".[65] The Department of Evolutionary Anthropology in Vienna addresses "various forms of interactions between humans and their environment" and combines

[62]See the websites of the journal *Human Ecology* at: http://www.hunter.cuny.edu/humaneco/ and https://www.springer.com/journal/10745; for a recent selection of texts on human ecology see: "Science Direct", at: https://www.sciencedirect.com/topics/agricultural-and-biological-sciences/human-ecology, and for a review of the history of the concept see at: https://www.encyclopedia.com/philosophy-and-religion/philosophy/philosophy-terms-and-concepts/human-ecology (17 June 2020);

[63]See ISEE's mission on its website at: http://www.isecoeco.org/.

[64]Young (1974); Douglas (2002); see at: https://www.springer.com/journal/10745.

[65]See Lund University, Department of Human Geography, at: https://www.keg.lu.se/en/education/academics/üsubjects/human-ecology (16 June 2020).

"nature and culture, human beings and their surroundings" that are covered "by in-depth views into human environmental history: questions like brain evolution, development of human group sizes or the evolution of speech are within the focus here" and the unique modes of humans' "resource use and resource production" since "the beginning of resource production during the Neolithic transition" up to "the limited resources and sustainable development".[66]

In *Man and Nature* George Perkins Marsh (1864) addressed the ecological relations between humans and their urban environments. In its first edition (1973), the interdisciplinary journal *Human Ecology* referred to "genetic, physiological, and social adaptation to the environment and to environmental change", including "social, cultural, and psychological factors in the maintenance or disruption of ecosystems … interrelations of technological and environmental changes … the relation of food quality and quantity … to demographic change". Today problems facing individuals and how actors deal with their consequences on an individual level are also addressed. Research in the *Human Ecology Review*[67] argues that the discourse has shifted toward applying principles of human ecology.

2.6.3 Political Ecology

The 'political ecology' concept[68] has been used since Thone (1935) coined it and Wolf (1972) revived it in human and development geography, anthropology and ecology to explain "the political dynamics surrounding material and discursive struggles over the environment in the Third World" (Bryant 1998: 80–89). As an interdisciplinary approach political ecology is "constituted by and concerned with, political economy, cultural ecology, social ecology, green socialism, environmental sociology, development ecology, anthropology" etc. (Minch 2011). It studies "the relationships between political, economic and social factors with environmental issues and changes", trying to politicise environmental phenomena. Some people have integrated it "with political economy (Peet/Watts 1996: 6)" on "degradation and marginalisation, environmental conflict, conservation and control, and environmental identities and social movements (Robbins 2004: 14)." Bryant and Bailey (1997) pointed to key assumptions of political ecology, including the unequal distribution of costs and benefits that reinforce existing social and economic inequalities

---

[66] See Vienna University, Department of Evolutionary Anthropology, at: https://www.anthropology.at/research/human-ecology/(16 June 2020).

[67] See the website of the journal *Human Ecology Review*, at: https://www.humanecologyreview.org/.

[68] See 'political ecology', at: https://en.wikipedia.org/wiki/Political_ecology; on 'Political Ecology', see Minch (2011); and see the *Journal of Political Ecology* at: https://journals.uair.arizona.edu/index.php/JPE; for recent critical assessments, see Bridge et al. (2015): 117–126.

that have political implications. It took half a century for humans to understand that global colonisation resulted in a major ecological transformation (human ecology).[69]

When discussing the debate on the linkages between *International Relations Theory and Ecological Thought* since the 1990s, Laferrière and Stoett (1999) distinguished between *utilitarian ecology* (utilitarian anti-environmentalism, conservationism), *authoritarian ecology* (ecofascism, Gaia and misanthropic ecology) and *radical ecology* (deep ecology, social ecology, ecosocialism and ecofeminism), and reviewed the debate in International Relations (IR) on realism, liberalism and critical IR theory and ecology. For Bassett and Peimer (2015):

> Political ecology emerged in the 1970s and early 1980s in the context of political economic critiques of cultural ecology and systems ecology. The critique basically argued that to explain environmental degradation (e.g. soil erosion) one had to situate resource management practices, typically of smallholders in the developing world, within the broader political economy (Watts 1983a; Blaikie 1985). This fusion of political economic and cultural ecological perspectives became known in Anglophone geography as political ecology (Blaikie/Brookfield 1987; Basset 1988). The political ecological critique, initially influenced by Marxist agrarian studies, gave rise to a succession of conceptualisations of nature-society interactions. Paul Robbins identifies three distinct theoretical approaches to nature-society relations in the political ecology literature which he terms the 'destruction', 'production', and 'co-production' of nature (Robbins 2012).

The 'deep ecology' concept was coined by Arne Naess (1973). It considers humans to be part of nature and the earth and critiques the abuse that has prevailed in the Western world-view. Deep ecology calls for the preservation of natural diversity, lifestyles that rely on "simplicity, frugality, self-reliance", and limits on the size of the human population (Pepper 2002: 211).

2.6.4 Social Ecology

Social ecology evolved out of "biological ecology, human ecology, systems theory and ecological psychology" and "studies relationships between people and their environment, often the interdependence of people, collectives and institutions" by taking a "broad, interdisciplinary perspective that gives greater attention to the social, psychological, institutional, and cultural contexts of people-environment relations than did earlier versions of human ecology".[70] The concept has been employed to study a diverse array of social problems and policies within the behavioural and social sciences.

'Social ecology' integrates "the study of human and natural ecosystems" and focuses on the "relationship between culture and nature" (Pepper 2002a). Bookchin

---

[69] For a survey of recent texts on "Political Ecology", see at: https://www.sciencedirect.com/topics/earth-and-planetary-sciences/political-ecology (16 June 2020).

[70] See "Social ecology (academic field)", in: Wikipedia; at: https://en.wikipedia.org/wiki/Social_ecology_(academic_field) (16 June 2020);

(1990), who founded the *Institute of Social Ecology* (ISE)[71] in 1974, conceptualised it "as a critique of current social, political, and antiecological trends, [that] espouses a reconstructive, ecological, communitarian, and ethical approach to society". He argued that "social ecology advocates a reconstructive and transformative outlook on social and environmental issues, and promotes directly democratic, confederal politics. As a body of ideas, social ecology envisions a moral economy that moves beyond scarcity and hierarchy, toward a world that reharmonises human communities with the natural world, while celebrating diversity, creativity and freedom" (Bookchin 2005: 85–87).[72] In Pepper's interpretation (2002a: 484) Bookchin's approach does not prioritise nature over human society and it sees "humankind as evolution's highest expression, e.g. nature's consciousness, so human transformation of nature is natural and desirable".

Research and training, institutes and degree programmes, scientific associations and research programmes on 'social ecology' have been set up in the USA (*Institute on Social Ecology* [ISE][73]) and the UK (Social Ecology London). The School of[74]Social Ecology at the University of California in Irvine (UCI) focuses on:

> human behaviour in a larger social and institutional context, that moves beyond traditional disciplinary boundaries, and that seeks to disseminate knowledge beyond the confines of the university to a broader public. The faculty pursue knowledge production and dissemination in the service of fostering informed social action as they address issues ranging from global poverty to prison overcrowding, from gang violence to healthy child development, from health risks to community empowerment. The School is an internationally recognised pioneer in developing interdisciplinary approaches to social problems that encourage flexibility and independent thinking among faculty and students and nourish collaboration across different fields and with people of many different experiences.[75]

This school was founded in 1970 by Arnold Binder as an interdisciplinary programme that was upgraded to a formal academic school in 1992. In 2020 the school had four departments: (1) Criminology, Law and Society; (2) Environmental Analysis and Design; (3) Psychology and Social Behaviour; and (4) Planning, Policy and Design. In the Conceptual Social Ecology course[76] the UCI School offered: (1) Contemporary definitions; (2) Four assumptions of the social ecology perspective; (3) Core principles of social ecological theory; (4) Development of the ecological paradigm; (5) Historical milestones of development; (6) Application of the social ecology perspective; (7) The potential of transdisciplinary research; (8) The future of interdisciplinarity in the School of Social Ecology.

[71] See at: http://social-ecology.org/wp/ (16 June 2020).

[72] For a discussion of his work see: https://www.socialworkdegreeguide.com/faq/what-is-social-ecology/ (16 June 2020).

[73] See: ISE, Prescott College, NY, USA https://www.social-ecology.org/about/about-the-ise/, which offers an extensive bibliography of Bookchin: http://www.social-ecology.org/learn/published-books/.

[74] See at: http://socialecologylondon.wordpress.com/ and a bibliography at: http://socialecologylondon.wordpress.com/articles-and-reviews/.

[75] For details see: https://socialecology.uci.edu/pages/about-school (16 June 2020).

[76] For details see: https://socialecology.uci.edu/pages/conceptual-social-ecology (16 June 2020).

The concept has also been widely discussed in other countries and languages. In Germany, two major institutes exist: the *Institut für Sozialökologische Forschung* [Institute for Social Ecological Research], Frankfurt (ISOE)[77] has existed since 1989, and the *Institute für Sozialökologie* (ISÖ) in Siegburg was established in 1987 by Michael Opielka, who looked for an alternative to the growth-orientated welfare state model and for answers for eco-social questions.[78]

The *Institut für Sozialökologische Forschung*, Frankfurt (ISOE) has conducted project research and served as a consultant on social ecological themes for government institutions, NGOs and for companies. In 2020, ISOE maintained six research units on Water Resources and Land Use; Water Infrastructure and Risk Analyses; Energy and Climate Protection in Everyday Life; Mobility and Urban Spaces; Biodiversity and People; and Transdisciplinary Methods and Concepts. From 1999–2018, the German Federal Ministry for Education and Research[79] funded a €350 million research programme on social ecological research (Ronzheimer 2019).

2.6.5 Ecofeminism

The term 'ecofeminism' was introduced by Françoise d'Eaubonne in *Le Féminisme ou la Mort* (1974) "to call attention to women's potential to bring about an ecological revolution", and it now refers to "interdisciplinary perspectives on the inextricable inter-connections among human systems of unjustified domination – both of humans and earth others" (Warren 2002:218). Ecofeminism sees "environmentalism, and the relationship between women and the earth, as foundational to its analysis and practice. Ecofeminist thinkers draw on the concept of gender to analyse the relationships between humans and the natural world". Ecofeminism[80] has "varying approaches and analyses, including liberal ecofeminism, spiritual/cultural ecofeminism, and social/socialist ecofeminism (or materialist ecofeminism)." The American ecofeminist Charlene Spretnak (1990) categorised ecofeminist approaches as: (1) through the study of political theory as well as history; (2) through the belief and study of nature-based religions; (3) through environmentalism.

Environmental philosophy with a gender perspective is a novel systemic approach to a complex reality from critical, interdisciplinary and constructivist perspectives.

---

[77]For information about ISOE, Frankfurt, see: http://de.wikipedia.org/wiki/Institut_f%C3%BCr_sozial-%C3%B6kologische_Forschung and: http://www.isoe.de/en/isoe/; for its publications, especially: Becker, Jahn (2006), see at: http://www.isoe.de/en/publications/.

[78]For details see: https://www.isoe.org/institut/profil/ (16 June 2020).

[79]For details see: http://www.sozial-oekologische-forschung.org/ and for a select bibliography of the sponsored research, see: http://www.sozial-oekologische-forschung.org/de/494.php. BMBF (2015).

[80]See "Ecofeminism", in: Wikipedia; at: https://en.wikipedia.org/wiki/Ecofeminism (16 June 2020); Miles, Kathryn: "Ecofeminism", in: *Encylopaedia Britannica*; at: https://www.britannica.com/topic/ecofeminism (16 June 2020).

Alicia Irene Bugallo (1955, 2011, 2015)[81] argues that environmental philosophy and ecophilosophy not only highlight current environmental problems but are associated with values, and cultural and economic patterns and styles, which the more affluent societies or groups should seriously reconsider. Deconstructing the dominant model and reconstructing a comprehensive and sustainable scientific concept from a gender perspective with different theoretical and empirical approaches challenges the positivist androgenic Western liberal vision of the world, such as epistemological feminism, empirical feminism, postmodern feminism and point-of-view feminism. These theoretical approaches refer to empowerment and liberation from the shackles of patriarchy and therefore question the liberal and neoliberal world-view of business-as-usual. Ecofeminists (Mies 1998; Bennhold-Thomsen/Mies 1999; Oswald Spring 1999, 2020) have linked the mechanisms of the oppression of women to the exploitation of nature.

Reardon (1996, 2014) was among the first feminists to analyse the links between patriarchy, a culture of war, authoritarianism and violence. The symbolic distribution space assigned to the predominantly male public and to economic output is the *res publica* and therefore became *homo sapiens*, while women were symbolically located in the private sphere or in the home and considered socially as *homo* (*doña*?) *domesticus* (Lagarde 1992). Reardon analysed these antagonisms, arguing that building security is based on peace education (Reardon/Nordland 1994), exposing the underlying power relations and the exercise of violence. She claimed that the deep roots of often subconscious violence are rooted in gender relations that emerged over thousands of years.

Barad (1999) established a bridge between descriptive and normative epistemology, and between naive realism and social constructivist approaches. She argued that all knowledge has practical, descriptive elements, whereby analytical components explain the phenomena globally. The outcomes of these critical ecophilosophical trends were systematised from a gender perspective in 'deep ecology', ecofeminism, social ecology, and different schools of environmental ethics. All assume some degree of criticism of anthropocentrism, biocentrism between shades, and weak anthropocentrism. Speciesism is questioned (moral discrimination of individuals on the basis of the species to which they belong) by an anthropocentric understanding of nature, cornucopian overexploitation of Mother Earth and humankind.

[81] Alicia Irene Bugallo Finnemann is an environmental philosopher and an anthropologist who specialises in ecological philosophy and environmental ethics in the Department of Practical Philosophy at the University of Business and Social Affairs (Universidad de Ciencias Empresariales y Sociales [UCES]), Buenos Aires; at: http://www.aliciabugallo.com/.

2.6.6 From Landscape Ecology to 'Geoecology'

Influenced by the debate on *landscape ecology*[82] (Troll 1939, 1968), Huggett (1995) introduced 'geoecology' as an interdisciplinary natural science,[83] which he defined as the study of the "structure and function of geoecosystems" at different scales, focusing on the nature and hierarchical structure of geoecological systems, and on ideas about their interdependence and integrity, exploring internal or 'ecological' interactions between geoecosystems and their near-surface environment, and looking at the influence of climate, altitude, topography, insularity, and substrates, including the role of external factors, both geological and cosmic, as agencies disturbing the dynamics of geoecosystems.[84]

For Blumenstein et al. (2000: 9), geoecology "defines structures, functions and modes of action within geosystems or between them and their environment". Geoecology draws on both the spatial sciences (geography, cartography, landscape and regional planning) and the natural sciences, but it does not address the effects of environmental degradation or environmental stress and their outcomes. Blumenstein et al. called for the integration of legal, social science and economic aspects but ignored the political dimension. While Huggett's (2010) definition of geoecology combined geography with ecology, Stüdemann (2008: 9) defined geoecology as a holistic science whose objects are "geoecosystems and the images of all spatial and temporal structures of the geosphere. They are analysed with inventory process and structure analyses regarding systems stability and change and defined for an action maxim". Based on diverse European and North American approaches, Ingegnoli (2002) offered several theories, tested various methods for analysing and diagnosing the ecological state of a landscape, and proposed a new approach to understanding landscape pathology and degradation.

But in this approach the political dimension is limited to environmental planning, reflecting the requirements of sustainable development. It introduces landscape and geoecology into the development of environment policy, law, education and implementation strategies at global level. But like other geoecological approaches, Stüdemann's (2008) lacks a specific political dimension in his horizontal widening of the theoretical scope of geoecology to include sustainability goals.

In 1998 the Latin American Centre of Social Ecology (*Centro Latino Americano de Ecología Social* [CLAES], Gudynas 1999)[85] integrated geoecology with policy, placing great emphasis on economic equality. This Centre strongly supported

[82]See "Landscape ecology", in: Wikipedia; at: https://en.wikipedia.org/wiki/Landscape_ecology (16 June 2020); and the journal "Landscape Ecology", at: https://www.springer.com/journal/10980 (16 June 2020).

[83]See definition of 'geoecology' at: http://www.answers.com/topic/geoecology, based on: *Oxford Dictionary of Geography* (1992, 1997, 2004).

[84]See "Geoökologie ", in: Wikipedia; at: https://de.wikipedia.org/wiki/Geo%C3%B6kologie (16 June 2020). Six German universities in Bayreuth (since 1978), Braunschweig, Potsdam, Tübingen, Freiberg und Karlsruhe offer Bachelor's and Master's degrees in geoecology.

[85]Latin American Centre of Social Ecology (*Centro Latino Americano de Ecología Social* [CLAES], Gudynas 1999.

the World Social Forum in Porto Alegre, and in 2006 the *Centre for Development, Economy, Ecology and Equity* (D3E) in Latin America[86] was set up. Its key themes include geoecology and human ecology. During the Earth Summit in Rio de Janeiro (1992) the precautionary principle was systematically introduced into the geoecological thinking in Latin America (Cafferatta 2004).

2.6.7 Political Geoecology

This author (Brauch 2003, 2017) argued that there is a need for *political geoecology*, a branch of study that combines the divergent approaches of 'geoecology', 'social ecology' and 'human ecology'.[87] The broad canvas of *political geoecology* must combine the complex causes and interactions of key factors of regional environmental change with environmental stress, natural disasters, distress migration, crises, and conflicts. Relying on the results of the natural sciences, *political geoecology* should also use methods of political science and international relations.[88]

Brauch et al. (2011) formulated the concept of 'political geoecology' in response to earlier proposals by Alker and Haas (1993) for a new area of research called *ecopolitics* and by Dalby (2000, 2002, 2002a, 2000b) for *ecological geopolitics.* While Dalby (2000) approached *ecogeopolitics* from critical geopolitics (1998) and challenged the framing of environmental matters in terms of national security, Brauch (2003, 2003a) argued that *political geoecology* needs to combine the regional implications of global change and its potential outcomes – disasters, environmentally-induced migration, crises, and conflicts – and Oswald (2008a, 2008b) stressed the triggering factors and focused on social vulnerability.

Political geoecology suggests a more explicit focus on ecology and also a clear indication that human choices are shaping the world of the future. Recognition of the impact of our actions is emerging in many parts of the world, as the debate on climate change suggests that we and our predecessors have shaped the future condition of the biosphere. This terminology suggests the inapplicability of traditional geopolitical notions of an external environment in discussions about human security in the future. Brauch et al. (2011) have argued that the analysis of the security impacts of global environmental change in the Anthropocene requires the knowledge produced by both the *geoecology* programmes in physical geography, Earth System science and numerous other disciplines, and the *social ecology* and *human ecology* approaches which combine scientific efforts in geography, sociology, psychology and political science.

---

[86] *Centre for Development, Economy, Ecologyand Equity* (D3E).

[87] This section is based on Brauch et al. (2011).

[88] A few scholars in the social sciences have picked up and discussed this concept: Loftus (2018); Burke et al. (2015); Zurita et al. (2017); Kotzé (2018).

Since Haeckel coined the ecology concept in 1866 for biology,[89] the concept has diversified, and since the end of World War II and the start of the Anthropocene it has increasingly been used by the social and political sciences. However, peace and security issues have not been addressed in the framework of human, political, social ecology, ecofeminism, landscape and geoecology. While the concept of 'political geoecology' has been noted by a few geographers, it has not been used to address peace and security issues. Thus, the suggestions for bridge-building between peace and ecology research have all come from the social sciences, especially international relations and peace research.

2.7 Bridge-Building Between Peace Research and Ecology

Between peace and security issues on the one hand and environment and sustainable development problems on the other, various relationships and linkages have been addressed in social science literature since the 1960s, during the emerging Anthropocene Epoch:

- the early conceptual debate on the linkages between peace and the environment (2.7.1);
- the study of the impacts of weapons and wars on the environment (2.7.2);
- the conceptual debate on environmental and ecological security (2.7.3);
- the impact of environmental degradation and stress on environmental conflict (2.7.4);
- environmental peace-making and the role of the environment in post-conflict peacebuilding (2.7.5);
- the emerging discourse and policy debate on climate change and conflicts (2.7.6);
- and the early approaches to peace ecology and their shortcomings (2.7.7).

However, these multiple research projects and studies have not yet resulted in a joint comprehensive peace ecology research programme within the framework of the Anthropocene Epoch in Earth and human history.

2.7.1 From the Bouldings to the Brundtland Report: Addressing Linkages Between Peace, Security and the Environment

The link between peace and environment issues has been addressed by only few social scientists, among them the economist and peace activist, Kenneth Boulding (1966, 1978), and his wife, the sociologist and peace educator Elise Boulding (1988,

[89] See Stauffer (1957), Egerton (2012), and McIntosh (1986).

1989, 1992, 2000). Elise Boulding linked peace to ecology as a result of practical daily experience. She argued that there is no true peace without ecological links, respect for nature and human ecology (Morrison 2005; Boulding 2017, a, b, c).

In *The Economics of the Coming Spaceship Earth* (1966, 1970), Kenneth Boulding[90] used the symbol of cowboys arguing that the world of unlimited resources is coming to an end, and in *Ecodynamics* (1979) he combined evolutionary biology, ecology, peace research and Keynesian, socio- and environmental economics (Khalil 1996). Scott (2015: 100–107) reconstructed the genesis of Boulding's (1966) argument and its impacts on economic thought:

> This paper … marks the beginning of modern ecological economics in that it sees the economy as a subset of the larger ecosystems (or biosphere) and promotes protection of natural capital. … He castigates neoclassical economists for ignoring the environment in their models. … Boulding's article made it clear that the environment is important and that economists can no longer ignore it in their analyses. In the 1970s, due to growing national interest in environmental issues …, mainstream neoclassical economists developed two subfields for studying the environment: (1) environmental economics, and (2) natural resource economics. Today most mainstream economists use the term environmental economics to encompass both subfields. … Boulding's (1966) article was a clarion call for all economists to begin considering the limitations of planet Earth, and to start incorporating the effects of economic decisions on the environment. He argued that Earth has finally reached an exhaustive point where there were no new lands to inhabit.

Later, Boulding (1983) critically reviewed the arguments for and against ecosystems and his concept of the 'empty niche' in biology, societal evolution and artefacts, biological catastrophes, ecological and human interactions, social ecosystems and evolutionary economics. In assessing "Boulding's Place in Economic History", Scott (2015: 185–197) concluded:

> Boulding's entire research output reveals two areas where he was especially prescient and original: First, his work on peace and conflict resolution; and second, his metaphorical Spaceship Earth as an argument for sustainability and controlling rampant consumption and economic growth.

In both research programmes – peace research and ecology – Kenneth Boulding made pioneering and innovative contributions, arguing for the need to overcome traditional narrow disciplinary boundaries, which Wilson (1988) later termed the need for 'consilience'. In Scott's (2015: 185–186) assessment, Boulding's thought-provoking contributions on "peace and Conflict Resolution"

> was unique because it was transdisciplinary – encompassing political science, economics, sociology, mathematics, anthropology, and biology. Boulding was successful at creating a community of scholars who came from a variety of disciplines. They were all studying peace, but approaching it from their own disciplines. …

On the second theme of "Spaceship Earth", Scott (2015: 187) noted that Kenneth Boulding

---

[90] See: http://www.jayhanson.us/page160.htm.

was the first to integrate concepts such as entropy and the second law of thermodynamics into economic thinking. … He argued that we have to change the way we think about economic growth – that our consumer culture leads to resource waste and pollution, quickly leading to untold human suffering. Even today people suffer from anthropocentrism, failure to realise that they (and the economy) are merely a subset of the natural environment (biosphere). … It is imperative that we put intellectual effort into overcoming the destruction we have already caused the planet as well as developing technology and strategies for using more sustainable energy and better use of sustainable resources.

While Boulding's dual intellectual bridge-building from his own discipline, economics, to both peace research and ecology was unique, pioneering and conceptually and methodologically innovative, the Brundtland Report (1987) on *Our Common Future* linked both themes from a policy perspective by discussing the linkage between "Peace, Security, Development, and the Environment" with regard to "environmental stress as a source of conflict", "conflict as a cause of unsustainable development", and "steps towards security and sustainable development". This political agenda-setting addressed the following four linkage problems:

1. Among the dangers facing the environment, the possibility of nuclear war, or military conflict of a lesser scale involving weapons of mass destruction, is undoubtedly the gravest. Certain aspects of the issues of peace and security bear directly upon the concept of sustainable development. …
2. Environmental stress is both a cause and an effect of political tension and military conflict Nations have often fought to assert or resist control over raw materials, energy supplies, land, river basins, sea passages, and other key environmental resources. Such conflicts are likely to increase as these resources become scarcer and competition for them increases.
3. The environmental consequences of armed conflict would be most devastating in the case of thermonuclear war. But there are damaging effects too from conventional, biological, and chemical weapons, as well as from the disruption of economic production and social organisation in the wake of warfare and mass migration of refugees. But even where war is prevented, and where conflict is contained, a state of 'peace' might well entail the diversion into armament production of vast resources that could, at least in part, be used to promote sustainable forms of development.
4. A number of factors affect the connection between environmental stress, poverty, and security, such as inadequate development policies, adverse trends in the international economy, inequities in multi-racial and multi-ethnic societies, and pressures of population growth. These linkages among environment, development, and conflict are complex and, in many cases, poorly understood. But a comprehensive approach to international and national security must transcend the traditional emphasis on military power and armed competition. The real sources of insecurity also encompass unsustainable development, and its effects can become intertwined with traditional forms of conflict in a manner that can extend and deepen the latter (Brundtland Report 1987).

Within the UN system, the *United Nations Environment Programme* (UNEP) focused "the attention of governments on critical environmental problems…, in helping develop many global and regional action plans and strategies …, in contributing to the negotiation and implementation of international conventions …, and in preparing global guidelines and principles for action by governments."

2.7.2 Arthur Westing: A Pioneer in the Study of Environmental Impacts of Conflicts and Wars and UNEP's Case Studies

UNEP funded the research of scientists, such as Arthur H. Westing, and conducted (1995–2008) case studies on the environmental impacts of conflicts and wars in UNEP's Post-Conflict Branch.[91] This later became the *Post-Conflict and Disaster Management Branch* (PCDMB)[92] that operated in twenty-five countries and published eighteen environment assessment reports (1995–2008). In 2008 UNEP worked in Côte d'Ivoire, Afghanistan, Sudan, and Nigeria, conducting environmental assessments, mitigating environmental risk, strengthening institutions for environmental governance, integrating environmental considerations in reconstruction, and strengthening international and regional environmental cooperation.[93]

As a senior researcher at the *Stockholm International Peace Research Institute* (SIPRI) and as project director on the environmental impact of war at the *International Peace Research Institute Oslo* (PRIO (1976–1990), the US forester Arthur H. Westing (1928–2020) published many pioneering books on the environmental impacts of war. These publications were sponsored by the United Nations Environment Programme's Executive Director Mustafa Tolba, who withstood US pressure to stop funding Westing's UNEP work.

With the many books he authored and edited, Westing (2013, 2013a) became the single most important "Pioneer on the environmental impact of war" (Brauch 2013).[94] His pioneering fieldwork in South-East Asia since the late 1960s on the use of defoliants by US and South Vietnamese forces during the Second Indochina War (1961–1975) influenced not only the research and writing of many young scholars in peace research and environmental studies, but also had a significant impact on the internal debate in the USA by increasing public opposition to the Vietnam war (Westing 1976). He addressed both the environmental impact of war and the environmental causes of or impacts on multiple conflicts (Westing 2013, 2013a). He initially framed 'environmental security' as a state-centred concept and gradually moved to what he called "comprehensive human security" (Westing 2013a).

---

[91] See Haavisto (2003); see UNEP's Disasters and Conflicts Programme.

[92] See more on UNEP's PCDMB at: https://postconflict.unep.ch/about.php (17 June 2020).

[93] UNEP's PCDMB documents are at: https://www.preventionweb.net/organizations/3668 (17 June 2020).

[94] See the bibliography on Westing's major writings in Westing (2013: 19–34); Brauer (2009).

Arthur H. Westing was a *conceptual and empirical innovator* during his professional life as a *natural scientist*; a *professor and educator*; an *ecologist and environmentalist*; a *peace researcher*; a *policy consultant*, and a *politically active scientist and citizen* (Brauch 2013).Westing's scientific studies influenced the work of various intergovernmental agencies (UN, UNEP, UNIDIR, UNESCO), and also raised the awareness of major non-governmental organizations (e.g., the *International Committee of the Red Cross* [ICRC], the *International Union for Conservation of Nature* [IUCN], and the *International Campaign to Ban Landmines* [ICBL]). He did not remain silent on the misuse of scientific knowledge in warfare nor did he ignore the longer term effects on the life of present and future generations (Brundtland Report 1987).

2.7.3 The Conceptual Debate on Environmental and Ecological Security: Scientific Discourse and Policy Debate

Since 1989, environmental and security linkages have emerged as a topic of the conceptual and policy debate and for international organisations. Brauch (2003: 92–120) identified four phases of research on environmental security in the social sciences:

- *Phase I*: In the 1970s and 1980s research focused on the environmental impact of wars (Westing 1976, 1980, 1984, 1988), with conceptual contributions by Osborn (1953), Brown (1954), Galtung (1982) and Brock (1991, 1992), and proposals by Ullman (1983), Mathews (1989) and Myers (1989).
- *Phase II*: During the 1990s, two empirical environmental research projects were pursued by the *Toronto Group* (Homer-Dixon 1991, 1994, 1996, 1999, 2000) and the *Bern-Zürich Group* (Bächler 1990, 1995; Bächler/Spillmann 1996a, 1996b; Bächler et al. 1996).
- *Phase III:* Since the mid-1990s there have been comparative studies and conceptual deepening by many research teams, partly relying on modelling and management efforts and focusing on the conflict potential of resource use, state failures, and syndromes.
- A *fourth phase* of environmental security research, suggested by Dalby (2002) and Brauch (2003: 124–134; 2003a: 919–953), combines the structural factors from natural dimensions (climate change, water, soil) and human dimensions (population, urban and rural systems) based on the expertise of both sciences, with outcomes and conflicts (Oswald Spring et al. 2008).

During the first phase "there was a need to redefine security and to include a new range of threats" and "there was an acceptance that the object of security was no longer simply the state, but ranges to levels above and below the level of the state" (Lonergan 2002V: 270–271). During the second phase in the 1990s the research teams led by Homer-Dixon and Bächler and Spillmann focused on the interaction between factors of global change, environmental degradation, scarcity and stress, and how these factors resulted in either environmentally triggered conflicts or environmental cooperation. During the third phase environmental security studies diversified in many directions with little integration (Brauch 2003: 92–120).

Environmental security emerged as a discourse in the US in the waning years of the Cold War. Several authors (Ullman 1983; Mathews 1989; Myers 1989; Prins 1990; Rowlands 1991) proposed including the environment as a US national security issue. Assumptions that environmental disruptions would lead to conflict were common; even more so were fears of wars to control scarce resources, and alarmist accounts of imminent chaos as environmental decay fed numerous social problems. This series of themes later found an especially vivid expression in Kaplan's (1994, 1996, 2000) texts on *The Coming Anarchy*.

In the late 1980s, policy-makers put environment and security concerns on the international agenda. Most prominent were the authors of the *Brundtland Report* (WCED 1987) and Soviet President Gorbachev (1988), who introduced the topic at the United Nations based on previous proposals by think tanks such as the *Worldwatch Institute* (Brown 1977). This scientific discourse and political debate focused on the environment as a threat to national and international security, and hence a matter of high politics which ought to be a priority for states concerned with preventing international conflict. National security was key here; the state was understood as the referent object in need of securing.

A second stage of the discussion emerged in the 1990s as researchers investigated the assumptions of the first phase and tried to operationalise the concepts in field research. Most high profile was a series of case studies conducted by Thomas Homer-Dixon (1991, 1994, 1999; Homer-Dixon/Blitt 1998; Homer-Dixon/Deligiannis 2009) and associates at the University of Toronto. In Switzerland, Günther Bächler (1998). as well as Bächler and Karl Spillmann (1996a, 1996b, 2002). led an interdisciplinary study on environmental conflict that produced complex analyses of environment and development issues. While the Canadian group focused on the linkage between environmental scarcity, stress and conflict, the Swiss group dealt with both environmental scarcity and degradation as causes of environmental conflict, as well as matters of conflict resolution outcomes (Brauch 2005, 2005a).

The inductive case studies of the Toronto and Swiss groups were complemented with deductive approaches. Both focus on the complex interaction between environmental inputs, environmental-societal linkages and extreme outcomes. One difference between the second and third phases was leaving the dependent variable– violent conflict or cooperation – open. During the third phase several research projects were pursued, such as:

- The *Global Environmental Change and Human Security* (GECHS 1999) project, which was launched as a framework for research cooperation and coordination.
- The *ECOMAN*, *ECONILE* and *Environmental Change and Conflict Transformation* projects in Zürich and Bern, which continued the case study approach and

focused on peaceful and cooperative management of renewable resource use in the Horn of Africa, the Nile region and other areas (Arsano 2002).
- The analytical and conflict theoretical syndrome approach of the *Scientific Advisory Council on Global Environment Issues of the German Government*, which focused on the patterned interaction of symptoms of global change with socio-economic processes (WBGU 1996, 1997; Biermann 1999; Biermann et al. 1999) and led to the WBGU-Report: *Security Risk Climate Change* (2007).
- The *Transboundary Freshwater Dispute Database* at Oregon State University and the *Global Assessment of Environment and Security* (GLASS) project at Kassel University.
- Analysis by several research teams (Peluso/Watts 2001) of the causes and intensity of violent conflicts, though only a few studies have focused on environment and conflict linkages.

Other research, for example in geography (Bohle 2002), anthropology (Elwert 1990, 1999, 2003) and hydrology (Biswas 1993, 2004; Bogardi/Castelein 2002) has also produced theoretical and empirical results that are applicable to aspects of the environmental security discussion.

This scientific discourse was gradually picked up by the global peace (Oswald 1990, 1992, 2008a, 2008b; Brock 1991, 1992; Gleditsch 1997, 1998, 2001; Diehl/Gleditsch 2001, 2001a), security (Brown 1977; Renner 1996; Buzan et al. 1998; Wæver 2008, 2008a), environment (Carius/Lietzmann 1999) and development (Bhattacharya/Miller 1999) research communities. A joint NATO/CCMS (Committee on the Challenges of Modern Society) project by the US Department of Defense and the German Ministry of Environment promoted environmental security within NATO countries (Lietzmann/Vest 1999), and since then several NATO-sponsored *Advanced Research Workshops* and *Advanced Study Institutes* have taken place, and environmental security has become a key topic in NATO's Science Programme.[95]

What emerged in this debate, amid numerous methodological and empirical discussions, was the realisastion that environmental change and resource scarcity and degradation were less likely to lead to international war than had been supposed in the first phase. While national security is important, and there are plausible arguments concerning threats of state collapse and internal conflict caused, triggered or intensified at least in part by environmental factors, the focus is more on state capacity and the policy dilemmas of social and environmental change.

The third phase includes analyses of global change linked to larger, more general concerns about human security rather than national security (Brauch 2003, 2003a; Chen et al. 2003, Najam 2003; Dalby 2008). A new approach suggested identifying and mitigating syndromes of global environmental change.[96] Furthermore, several research programmes and projects on global environmental change (IHDP, IGBP, WCRP, Diversitas, GECHS 1999, GWSP) fostered research and scientific networking

[95] See the website of NATO's Science Division, at: https://www.nato.int/science/.

[96] See: WBGU (1996); Petschel-Held et al. (1999); Schellnhuber et al. (1997); Schellnhuber and Wenzel (1998); and for a research review: Brauch (2003, 2003a).

on related issues. Case studies have focused on climate change (Stripple 2002; Page/Redclift 2002; Brauch 2002; Schwartz/Randall 2003; WBGU 2007; Scheffran et al. 2012), soil (Vlek 2005; Brauch et al. 2009), and cooperative resource and conflict management of water (Bogardi/Castelein 2002; Oswald 2005, 2011; Oswald et al. 2009), as well as transboundary water basins[97] and water-related conflicts.[98]

This research has focused on insecurity in many places looking for policy initiatives that can mitigate disruptions caused by environmental change. The community working on hazards and disasters (Wisner/Walker 2005; IFRC 2002, 2005; Cardona 2004, 2009) has identified environmental and social vulnerabilities (Bohle 2002) resulting from natural hazards, storms and droughts. However, so far only a few studies have discussed the linkages between hazards, disasters and conflicts that occur in complex emergencies (Oswald 2008b).

These three phases of environmental security and additional studies on global change, development and conflict suggest a whole series of causes that may be leading to insecurity and violence.

In the fourth phase the focus is on human security rather than environmental factors that may cause concern for national security. The assumption cannot be sustained any longer that environmental change is an independent variable in scientific investigations of security outcomes. Rather the analysis starts with people first and works backward to identify the circumstances, social, cultural and ecological threats, challenges, vulnerabilities and risks that render them insecure and that may contribute to conflicts or cooperation. Focusing on human vulnerabilities and then working backwards to understand the causes of those insecurities is an approach that differs from the earlier phases, although it builds on many of their insights (Oswald et al. 2009).

2.7.4 Impact of Environmental Scarcity, Degradation and Stress on Environmental Conflict: Bächler and Homer-Dixon

The Toronto Group analysed assumed linkages between environmental stress factors and conflicts (Homer-Dixon 1991, 1994). Homer-Dixon (1994: 39) argued: "that environmental scarcity causes violent conflict. This conflict tends to be persistent, diffuse, and sub-national". He assumed that "global warming will probably not have a major effect for several decades, and then mainly by interacting with already existing scarcities" (see figure in Homer-Dixon 1994: 31). Levy (1995: 35–62) commented that it "is of very little importance" that environmental problems constitute security risks for the USA, and argued "that ozone depletion and climate change are the only

[97]See UNESCO (2003); Biswas (1993); Wolf (1995, 1998, 1999, 1999a); Scheumann (2003); Enrique (2004); Wolf et al. (1999).

[98]See Gleick (1993, 1993a, b, 1994, 1998, 2000); Amery/Wolf (2000); Wolf (1995, 1998); Wolf, Hamner (2000); Shiva (2002, 2003); Oswald (2005); Brauch et al. (2009).

significant environmental problems that currently pose a direct physical harm to US interests" (61–62). In reply, Homer-Dixon (1995: 189) argued that climate change "could endanger core American values" and could thus become "direct threats to US security interests", but not in the short term.

A second project analysed the links between environment, population, and security, based on case studies (Homer-Dixon/Blitt 1998) that focus on the social consequences of renewable resource scarcity. They argued that under certain circumstances scarcity of these resources may cause violent conflict. They focused on six types of environmental change: water and land degradation, deforestation, decline in fisheries, and, to a lesser extent, global warming and ozone depletion that can produce scarcities of vital renewable resources. They distinguished between "supply-induced" (environmental change), "demand-induced" (population growth, per-capita consumption), and "structural" scarcity (unequal social distribution of a resource).

They discussed two patterns of interaction among these three types: *resource capture* by powerful groups, and *ecological marginalisation* resulting in a lack of access for the poor segments of society that are often forced to migrate to ecologically fragile and vulnerable regions. Such environmental scarcity is not certain to result in social disruption and violent conflict. But different adaptation failures, such as market failure, social friction, and the lack of capital availability, may produce five types of social effects: "constrained agricultural productivity, constrained economic productivity, migration, social segmentation, and disruption of legitimate institutions". Homer-Dixon/Blitt (1998) considered environmental scarcity to be a cause that interacts with various contextual factors, ranging "from the nature of relations among ethnic groups to the state's degree of autonomy from outside pressure groups".

Homer-Dixon (1999: 5) pursued the linkages between environment, scarcity, and violence, looking to five future types of likely violent conflicts that Third World countries will be less able to prevent: (1) disputes arising from local environmental degradation; (2) ethnic clashes arising from population migration and deepened social cleavages; (3) civil strife (insurgency, banditry, coups d'état); (4) interstate war (on water); and (5) North-South conflicts over global environmental problems (global warming, ozone depletion, biodiversity). He considers the first and last type unlikely, and interstate scarcity wars the least likely. He discussed the scarcity's causal role between: *Environmental Scarcity→Social Effects→Violent Conflict*. While he admits that scarcity, as such, is neither a necessary nor sufficient cause for conflict, he claims "that environmental scarcity is an important cause of these conflicts" (Homer-Dixon 1999: 8) . Among the major determinants of supply-induced environmental scarcity that contribute to the depletion and degradation of renewable resources, Homer-Dixon (1999: 50) referred to (a) *ideational factors* (social relations, preferences and beliefs), (b) *physical availability* of natural resources, and (c) *ecosystem sensitivity* that impact on population and resource consumption (energy global warming→; cropland scarcity→deforestation; fresh water).

With continued population growth, the decrease in quality of renewable resources can either result in (a) *resource capture* (via unequal resource access) or (b) *unequal resource access*. After a decade of research, Homer-Dixon (1999: 177) concluded: "that scarcity of renewable resources … can contribute to civil violence, including insurgencies and ethnic clashes", and he predicted that in the future "such violence will probably increase as scarcities of cropland, fresh water, and forests worsen in many parts of the developing world", where the role of scarcity will be "often obscure and indirect", interacting with political, economic, and many other factors. He further predicted that continued population growth and rising resource demand and persistent inequalities will affect environmentally sensitive regions.

The *Environment and Conflicts Project* (ENCOP), co-directed by Günther Bächler and Kurt R. Spillmann (Bern and Zürich group), started from the premise that environmental transformation does not directly result in conflicts but that it impacts on existing socio-economic conflict potentials that violently escalate. According to its working definition: "environmental conflicts manifest themselves as political, social, economic, ethnic, religious or territorial conflicts over resources or national interests, or any other type of conflict". They are traditional conflicts *induced by environmental degradation*. Environmental conflicts are characterised by the principal importance of degradation in one or more of the following fields: (a) overuse of renewable resources; (b) overstrain of the environment's sink capacity (pollution); (c) improvement of the living-space (Libiszewski 1992: 13; Bächler 1998: 24).

According to ENCOP's analytical framework (Bächler 1993; Libiszewski 1992, 1996: 339–340), the analysis of environmental conflict involved four steps: (1) describing the environmental situation as the backdrop of human activities; (2) deducing the social and economic effects of environmental transformation and degradation; (3) analysing the political implications of these socio-economic effects and conflicts arising from them; and (4) evaluating approaches to peaceful management and resolution on different levels of analysis. ENCOP concluded that besides resource degradation other contextual factors were decisive for conflicts,[99] and "while conflict and environmental change are related in many ways, conflict is more likely to be linked directly to the disruptions of modernity" (Dalby 2002: 97). In a study on Rwanda, Bächler (1999b) clarified the ENCOP model and stressed "that violence was to occur in more remote areas, mountain locations, and grasslands – places where environmental stress coincides with political tensions and unjust access to resources" (Dalby 2002: 97).[100]

Bächler (1998: 24–44) examined "the critical role of transformation regarding causation on environmental conflicts in certain areas of developing countries". To do so, it is necessary "to build on prepared empirical ground to highlight the different patterns of causation, to select types of environmental conflicts in terms of different

[99]Bächler (1998, 1999); Dalby (2002: 97).

[100]The results of the ENCOP project were published in three volumes, the first of which (Bächler et al. 1996) examined the environment as a topic of conflict research, environmental degradation through over-development (*wealth-driven*) and underdevelopment (*poverty-driven*), focusing on actors and the key environmental factors of soil, rivers, and mining. It offered a synthesis of environmental degradation as a cause of war and how environmental conflicts can be solved peacefully. The second volume contains case studies focusing on Bangladesh, Sudan and Darfur, Nigeria, Central Asia, the Jordan Basin, Rwanda and mining in the South Pacific (Bächler/Spillmann 1996a). The third volume includes thirteen country studies by external experts (Bächler/Spillmann 1996b).

pathways leading to violence, as well as to stress the socio-political characteristics of environmental conflicts". Bächler (1998: 24) concluded

> that neither apocalyptic scenarios of environmental catastrophes nor alarmist prognoses of world environmental wars are tenable. Environmentally-caused conflicts escalate across the violence threshold only under certain conditions. Human-induced environmental change can be either a contributing or a necessary factor for both the emergence and/or the intensification of violent conflicts. On one hand, violent conflicts triggered by environmental disruption are due in part to socio-economic and political developments. On the other, social and political maldevelopment, due in part to degradation of natural resources, has become an international peace and security challenge. *Development and security dilemmas* are connected to a syndrome of problems which produces environmental conflicts of varying intensity and nature.

Bächler (1998: 40–44) provided a typology of forty environmental conflicts with different conflict intensity, which he categorised as: (1) *intrastate* (a) ethnopolitical, (b) centre-periphery, and (c) regionalist migration/displacement conflicts; (2) *intrastate* conflicts with a transboundary dimension, caused by a) migration, (b) demographic pressure, and (c) water/river basin conflicts (28–39); and (3) *international* global environmental conflicts.

Both approaches had a significant impact by developing an empirical basis for the primarily policy-orientated discourse that added an environmental dimension to the US national security agenda in the post-Cold War era which succeeded during the Clinton administration but was discontinued during the administration of George W. Bush (Matthew/McDonald 2003, 2009).

Why a critical socio-economic constellation escalated into violent conflict, and when and why such occurrences could be avoided by the bilateral and multilateral co-operation of states, experts, and representatives of civil society could not yet be explained by these studies.[101] After a decade of research a consensus emerged that "environmental stress is rarely considered to be the sole factor in precipitating conflict" either within and between nations. Schwartz (2002: 137) stated:

> In many cases environmental stress is a relatively distant factor, acting in combination with other economic and social factors, such as poverty and weak governments. In other cases, conflict breaks out when rival nations, or rival groups within a nation, battle for diminishing supplies of environmental resources. Although environmental stress is usually only one cause of conflict among many, the evidence suggests that it can play an important role, and that violence may be avoided by addressing environmental problems.

Schwartz considers population growth to be closely linked with environmental stress. Among the wide-ranging environmental factors he includes ozone depletion and global warming, and among the localised ones those environmental factors that affect small areas at different times (desertification, water pollution). He points to "five pathways to indirect internal conflict that involve environmental stress: economic decline, migrations, social fragmentation, erosion of civil society, and curtailment of the state". A major effect of several types of environmental stress is

[101] This approach was criticised by Diehl/Gleditsch (2001); Peluso/Watts (2001); Hartmann (2001: 39–62); Bannon/Collier (2003); see overview in Brauch (2003).

economic decline that will affect the poor more than the rich strata of society and countries. A second pathway from environmental stress to conflict is through migration that may be caused by floods, droughts, locusts, lack of arable land that again is often the result of environmental stress, water and air pollution, or a shortage of fuelwood. The complex interaction of environmental stress and its social, economic, and political ramifications has often resulted in increasing urban violence.

Global and localised environmental stress may in some cases contribute to conflicts. Gleick (1989) noted that global warming could affect fresh water availability and food productivity, and that this would have severe impacts on poorer nations. Some argued that environmental stress coupled with rapid population growth could induce large-scale migration pressures from South to North, and that this may increase tensions between host countries and immigrants. Direct internal conflict has occurred as a result of environmental stress, e.g. in the Sahel (due to drought), where many nomads have clashed with farmers in less affected zones. Schwartz argues that causal linkages between environmental stress and conflict could be shown in individual case studies, but that future research is needed "to estimate the causal effects of environmental stress". Kawashima and Akino (2001), Stripple (2002), and Brauch (2002) have discussed the possible longer-term security implications of climate change.

Hauge and Ellingsen (1998) integrated environmental degradation (soil erosion, deforestation) into a model of civil war. They concluded: "that environmental degradation does stimulate the incidence of conflict, but less so than political, economic, and cultural factors, or previous conflict history". Saltnes (1998) found that "there was indeed a bivariate relationship between the spread of deserts and internal armed conflicts in the period 1980–1990". Gleditsch and Urdal (2002) suggested that resource and environmental aspects of conflict "should be examined within the context of a broader view of armed conflict" with a special focus on politics, economics, cultural factors, and the conflict history.

This first phase of the social science discourse on different approaches to the study of linkages between environmental causes and environmental conflict outcomes based on different case studies did not result in a consensus or a joint 'structured-focused comparison' case study methodology, as has been suggested by Alex George (Caldwell 2019) during the Cold War in the context of the analysis of national security issues. This debate later contributed to the research into the security implications of anthropogenic climate change resulting in environmental crises, conflicts, and, in the very worst case, climate wars.

2.7.5 Environmental Peacemaking and Post-conflict Peacebuilding

Reviewing the emerging global change discourse on linkages between peace and environmental studies, Conca (1994: 20, 2015) suggested an "environmental agenda for peace studies" and a discussion on whether "ecologically desirable futures include concerns for peace and justice", arguing that it is not enough "to place 'sustainable development' and 'ecological security' alongside peace or social justice as 'world-order values'" but that scholars must ask "not only their formal definitions, but also their metaphorical and institutional associations, further the purposes of peace, justice, and community".

Later, Conca (2002: 9) fundamentally challenged a core premise of the debate on environmental and ecological (in)security and conflict by asking "whether environmental cooperation can trigger broader forms of peace" by defining peace "as a continuum ranging from the absence of violent conflict to the inconceivability of violent conflict" by addressing also "problems of structural violence and social inequality" and by "building an imagined security community" based on peaceful conflict resolution. Indirectly inspired by Mitrany's functionalist ideas where environmental collaboration "through cooperation in international environmental agreements" may foster "international peaceful behaviour".

Conca et al. (2005: 150) called for "building peace through environmental cooperation", noting that "most environmental peacemaking initiatives fall into one of the three partially overlapping categories: efforts to prevent conflicts related directly to the environment, attempts to initiate and maintain dialogue between parties in conflict, and initiatives to create a sustainable base for peace". They argued that environmental peacemaking may help "forestall environmentally induced conflict" and "soften group grievances that … are worsened by ecological injustices", which is also identified as 'negative peace', while a second approach "moves beyond conflicts with a specifically environmental component, seeking to build peace through cooperative responses to shared environmental challenges", thus partly aiming for 'positive peace'.

UNEP (2009) provided technical support for the UN Peacebuilding Commission for a study on environmental post-conflict peacebuilding that focused on the role of natural resources and the environment in conflicts, the impact of conflicts on natural resources and the environment, and the role of natural resources and the environment in peacebuilding (Matthew et al. 2009). The report suggested:

> 1) Further develop UN capacities for early warning and early action…; 2) Improve oversight and protection of natural resources during conflicts…; 3) Address natural resources and the environment as part of the peacemaking and peacekeeping process…; 4) Include natural resources and environmental issues in integrated peacebuilding strategies…; 5) Carefully harness natural resources for economic recovery…; [and] 6) Capitalise on the potential for environmental cooperation to contribute to peacebuilding: Every State needs to use and protect vital natural resources such as forests, fertile land, energy and biodiversity. Environmental issues can thus serve as an effective platform or catalyst for enhancing dialogue, building confidence, exploiting shared interests and broadening cooperation between divided groups, as well as states (UNEP 2009: 5).

In 2008, the *Environmental Law Institute* (ELI), the *United Nations Environment Programme* (UNEP), the University of Tokyo, and McGill University launched a five-year global research initiative "to analyse experiences in post-conflict peacebuilding and natural resource management, identify lessons, and raise awareness of those lessons among practitioners, researchers, and decision-makers". Based on this cooperation, between 2011 and 2016 the initiative produced "six edited books with over 150 case studies and other analyses from more than sixty conflict-affected countries and territories, written by 225 researchers, practitioners, and decision-makers from around the world."[102] This series on *Post-Conflict Peace-building and Natural Resource Management* examined the "challenges and opportunities of managing natural resources for post-conflict peace building around the world" that rely:

> upon the combined expertise and field experience of more than seventy researchers, diplomats, military personnel, and practitioners from governmental, intergovernmental, and non-governmental organisations to illustrate the mutually reinforcing relationship between natural resources, good governance, and peace. It examines the theory, practice, and reality of post-conflict governance in fifty conflict-affected countries around the world to explore the opportunity and challenge of effectively and equitably governing the use of natural resources and then converting the subsequent revenues into the jobs, infrastructure, and public service needed to consolidate and sustain peace.[103]

UNEP, ELI and the Universities of California Irvine, Duke, and Columbia have launched a *massive open online course* (MOOC) on Environmental Security and Sustaining Peace which offers an introduction to the roles played by natural resources and the environment in the onset, escalation, resolution of, and recovery from violent conflicts:

> Conflicts over natural resources are among the greatest challenges in twenty-first century geopolitics and present serious threats to human security at both the national and local levels. Natural resources can nonetheless serve as a vehicle for peace if managed in a sustainable and equitable manner. Environmental peacebuilding has emerged as a new frontier in interdisciplinary studies and offers a conceptual and operational framework for researchers and practitioners to understand and maximise the positive peacebuilding potential of natural resources across the conflict lifecycle while mitigating potential risks. … The Steering Committee for the MOOC includes: … Erika Weinthal, Duke University; Richard Matthew, UC-Irvine, Marc Levy, Columbia University; David Jensen, United Nations Environment Programme and Carl Bruch, Environmental Law Institute.[104]

Within the UN system, UNEP has thus performed a pioneering role in supporting research into the environmental impacts of war, environmental peacemaking and post-conflict peacebuilding, in agenda-setting and in awareness-raising that also extended to the policy debate on climate change, security and conflicts (2.7.6). However, it has been up to member states to decide whether to translate these studies and proposals into national strategies, policy projects and financial support for those

---

[102] See at: https://environmentalpeacebuilding.org/about/ (19 June 2020).

[103] See press release at: https://www.unenvironment.org/news-and-stories/press-release/new-study-good-governance-natural-resources-essential-post-conflict (17 June 2020). Six volumes were published: Lujala/Rustad (2011); Jensen/Lonergan (2012); Unruh/Williams (2013); Weinthal et al. (2014); Young/Goldman (2015); Bruch et al. (2016). This six-volume set is available at: https://www.routledge.com/search?kw=Post-Conflict+Peace-building+and+Natural+2499+Resource+Management (17 June 2020).

[104] See at: https://environmentalpeacebuilding.org/education/mooc/ and https://www.epmooc.org/ (19 June 2020).

countries most affected by the negative consequences and opportunities referred to in the scientific case studies. In addition to ideographic case studies, systematic comparable case studies on the challenges posed by the new dangers in the Anthropocene are needed.

The cooperation between the environmental peacebuilding community and UNEP was instrumental in the establishment of the *Environmental Peacebuilding Association*[105] for "fostering knowledge development and exchange, building capacity and awareness, and strengthening relationships between scholars, practitioners, and decision-makers through an integrated series of activities by forming interest groups on water, monitoring & evaluation, law, and disasters, and by organising conferences, webinars, panels, and other events", including an International Conference on Environmental Peacebuilding at the University of California, Irvine, from October 23–25, 2019. Carl Bruch of the *Environment Law Institute* (ELI) in Washington, DC was the founding president and Erika Weinthal and Richard Matthews were elected Vice Presidents of the Environmental Peacebuilding Association. The same group also set up an Environmental Peacebuilding Academy[106] and an Environmental Peacebuilding Working Group.[107]

2.7.6 Scientific Discourse and Policy Debate on Climate Change, Security and Conflicts

Since the global turn with the end of the Cold War in 1989, the most recent policy debate on climate change, security and conflicts (2.7.6.1) has developed on three levels since 2003: (1) as national, regional and global policy debates in the US, the UK, Germany, and later within the *European Union* (EU Council, Commission, Parliament) and in the *United Nations General Assembly* (UN-GA), the *United Nations Security Council* (UNSC) and by the *Secretary General of the United Nations* (UNSG) (2.7.6.2); (2) as UNEP analyses and activities (2.7.6.3); and (3) as an ongoing scientific discourse in the social sciences (2.7.6.4).

2.7.6.1 The Initial Policy-Orientated Scientific Debate (1989–2020)

The policy-orientated scientific debate on anthropogenic climate change and security began in autumn 1989 with a panel at the annual IISS Conference, at which Peter Gleick (1989, 1989a) and Neville Brown (1989, 1991) spoke, a year after Gro Harlan Brundtland (1988) had addressed this 'assumed' causal link at a political conference

---

[105] See at: https://environmentalpeacebuilding.org/about/ (19 June 2020).

[106] See at: https://environmentalpeacebuilding.org/education/academy/ (19 June 2020).

[107] See at: https://www.researchgate.net/project/Environmental-Peacemaking-and-Peacebuilding-Working-group (19 June 2020).

in Washington DC. In November 2002 the German Federal Ministry for the Environment, Nature Conservation and Nuclear Safety (BMU) released a project report on *Climate Change and Conflict* (Brauch 2002), and in early 2003 a leaked contract study for the US DoD (Randal/Schwartz 2002) triggered a policy debate in the UK and later in the US. During the joint German EU and G-8 presidencies in 2007 a book by the German Advisory Council on Climate Change (WBGU 2007), *Climate Change as a Security Risk*, tabled this linkage on the German policy agenda, and also on the EU (2008), the UNGA (2009), and the UNSC (2007, 2011), which triggered a report by the UN Secretary-General on 11 September 2011 (SG 2009).

The year 2007 was a turning point in the process of *securitisation*[108] questions about *global environmental change* (GEC) and *global climate change* (GCC), when several of the highest national policy-makers and high-level fora (UN Security Council) and officials of international organisations addressed global warming (cause) and climate change (effect) as a major objective security danger and subjective security concern that could lead to internal displacements, forced distress migration, crises and conflicts. This emerging scientific discourse, policy debate and political process of the *securitisation* of GEC and GCC focuses on the environmental dimension of security, especially the complex interaction between human beings and humankind as causes, triggers, and victims of the societal consequences of this anthropogenic change. Thus, the *securitisation* of GEC and GCC issues is also closely linked to different referent objects of international, national, and human security.

Along with the *securitisation* of GEC and GCC, the focus of analysis during the past thirty years has been on the process of scientific and political agenda-setting, and the prioritisation of nature-societal issues. While in 1987 both GEC and GCC were still primarily emerging scientific problems for a small group of water, soil, and climate specialists, by 2007, after an intensive process of *politicisation* and *securitisation*, global warming and climate change had moved to the top of the policy agenda as the most urgent security dangers and concerns that require urgent, stringent, and long-lasting policy responses via a fundamental transformation in the global energy system (decarbonisation) and also in human values and consumer patterns.

2.7.6.2 Policy Debates at the National Level, Within the EU and in the United Nations Security Council (2003–2020)

The climate change and security linkage has been addressed by many countries, including the US (2003–2016), the UK, France, Germany, and many small island states and least developed countries in the Global South. It has been taken up and reiterated by the EU since 2008, and discussed by the UN, its Secretary-General, its General Assembly, the Security Council, the Economic and Social Council

[108] The theory of securitisation has been developed by Ole Wæver (1995, 2008, 2008a) and the Copenhagen school: Buzan et al. (1998).

(ECOSOC), the United Nations Environment Programme (UNEP), the United Nations Conference on Disarmament (UNCD), and the Peacebuilding Commission.

Climate change and security were discussed by the UN Security Council in 2007 when the UK put it on its agenda (Brauch 2009). In 2011, a Security Council Presidential Statement tabled by Germany asked the then UN Secretary General, Ban-Ki Moon, to provide contextual information about climate change in his report to the Council. While the US, France and the UK co-operated by keeping climate change on the Council's agenda, Russia and China suggested the topic should be addressed by UNEP and the UN General Assembly (Brauch/Scheffran 2011). Since 2017, the UN Security Council (UNSC) has increasingly addressed natural phenomena, environmental law and climate-related security risks associated with natural disasters (Scott/Ku 2018). In March 2017, UNSC resolution 2349 addressed climate-related risks, including the conflict in the Lake Chad basin; and in July 2018, a debate attended by over seventy member states was held on "understanding and addressing climate-related security risks", with statements by Ministers from Kuwait, Belgium, Indonesia, Germany and Poland.[109] At a UNSC meeting on climate change in January 2019 more than eighty states participated in the debate discussing climate change as a "threat multiplier", and a UNSC Resolution was adopted by fifteen states and organisations, including the EU and the African Union, with the goal of overcoming the dispute between France and the UK on the one side calling for a better use of climate-risk data and more financial support for those most at risk, and China and Russia on the other side. This disagreement avoided the term *threat to peace* for climate change:

> The UNSC … holds more open debates … Second, it allows the UNSC to discuss the issue. Third, while the qualification of a situation as a *threat multiplier* has never been used by the UNSC before … On the one hand, the qualification acknowledges the risks of climate change. Increased migration due to land loss leads, *inter alia*, to a higher food and health insecurity, which overall creates an environment more prone to conflict proliferation. On the other hand, it reflects the controversy over whether or not it is a sufficient basis for the UNSC to act. In this sense, the new qualification *is* a step forward. The UNSC has understood its importance and is carefully assessing ways to open the door to coercive measures. The next step for the UNSC could be to use this terminology as a way to by-pass the need for a Chapter VII resolution to act with regards to climate change. This would pave the way towards a broader understanding of Art. 39 UN Charter by also encompassing *threat multipliers* (Bourghelle 2019).

During its membership in the Security Council 2019/2020, Germany made climate change and security one of its top priorities to raise awareness of this linkage within the UN:

> Climate change is … increasingly becoming a risk multiplier that threatens the stability of countries and societies worldwide. … The German Security Council initiative on climate and security is an overdue security policy addition to climate policy as a whole. The international community needs to take action in places where climate change poses a threat to people's

[109] "Climate change recognized as 'threat multiplier': UN Security Council debates its impact on peace", in: *UN News*, 25 January 2019; at: https://news.un.org/en/story/2019/01/1031322 (18 June 2020).

> livelihoods before conflicts break out or escalate. The Security Council must also be capable of taking action in this regard. The Security Council must therefore be equipped to take concrete action in situations where the impact of climate change exacerbates conflicts. … The Security Council cannot and should not be a substitute for the established instruments of climate policy. … All countries need to step up their efforts in the area of climate protection, also in the interest of global security and stability. As long as the international community's climate goals are not sufficient to limit global warming to an acceptable level, they must address the foreign and security policy impact of climate change. In an effort to advance the discussion on the link between climate and security, Foreign Minister Maas hosted the high-level Berlin Conference on Climate and Security in June 2019. The Call for Action presented at the conference is a road map for concrete, preventive options aimed at tackling security risks caused by climate change at an early stage.[110]

A year later Germany hosted the second *Berlin Conference on Climate and Security*[111] on 23–24 June 2020, with a high-level political segment on 24 June, and from 7 September to 2 October 2020 "on how more comprehensive risk assessments could support forward-looking and preventive foreign and security policy … and offer a space to discuss the role that the inter-national community, and the UN Security Council in particular, can and should take in this respect, including during Germany's membership of the UN Security Council in 2020."[112] The Berlin Climate and Security Conferences in 1919 and 2020 were hosted by the German *Federal Foreign Office* (AA), the *Potsdam Institute for Climate Impact Research* (PIK) and the consulting firm Adelphi, which, with the support of the AA, is conducting "The Climate Diplomacy Initiative". This

> explores the foreign policy dimension of climate change and climate policy, including different regional and thematic approaches to it … [providing] arguments for enhanced international cooperation and diplomatic engagement for climate in multi- and bilateral fora … and includes support for bringing the topic of addressing climate fragility risks into different international fora. Despite the increasing salience of the topic of climate change and security, there is currently no single address to turn to for support and policy advice.[113]

In support of the Berlin Climate and Security Conference 2020, a number of high-level representatives of governments and international organisations have contributed statements.[114] The planned meeting on Climate Change and Security during the German UNSC presidency did not take place as one permanent member of the UNSC blocked it.

In 2003, when the public political debate on the linkage between climate change, security and conflict began in the UK and the US, this claimed relationship also

---

[110] Federal Foreign Office, 2019: "Germany is putting the impact of climate change on the agenda of the UN Security Council" (22 January), at: https://www.auswaertiges-amt.de/en/aussenpolitik/internationale-organisationen/vereintenationen/climate-change-security-council/2179806 (18 June 2020).

[111] See for documentation at: https://berlin-climate-security-conference.de/watch-bcsc-2020 (5 July 2020).

[112] Federal Foreign Office, 2020; at: https://berlin-climate-security-conference.de/ (19 June 2020).

[113] Federal Foreign Office, 2020; at: https://berlin-climate-security-conference.de/co-organisers (19 June 2020).

[114] See at: https://berlin-climate-security-conference.de/high-level-statements (5 July 2020).

had a specific political purpose in the US: to try to sell climate change as an issue of national security and thus overcome the opposition within the Republican Party in the US Congress to joining international agreements and treaties, e.g. the Paris Climate Agreement of 2015. To undercut this strategy of the Democrats, the Republicans in the US Congress have tried to block US partipation in such treaties since 2015, aiming at funding cuts, while the 45th US President stopped the public funding for the Climate Change Center of the US National Intelligence Council of the CIA in 2017 and opposed any discussion of this linkage in the UNSC.[115]

The armed forces in several countries launched policy studies on the linkage between climate change, security and conflict. In *The Securitization of Climate Change: Australian and United States' Military Responses (2003–2013)* Michael Durant Thomas (2017) analysed the debate within the Australian and US military services. Since the end of the Cold War, different environment and security and environmental peace-making scientific discourses and policy debates have evolved without specific contextualisation in the Anthropocene, or a common research programme in peace ecology or on ecological peace policy.

2.7.6.3 From Knowledge to Action: UNEP Analyses and Activities

In 2008 the *United Nations Environment Programme* (UNEP) was asked by Jan Egeland, the UN Special Envoy for Climate Change, to conduct an analysis of climate change and security risks in the Sahel Region. In 2009, UNEP started to work with the *International Organisation on Migration* (IOM), the *United Nations Office for the Coordination of Humanitarian Affairs* (OCHA), the *United Nations University* (UNU), and the *Permanent Interstate Committee for Drought Control in the Sahel* (CILSS) on a report on *Livelihood Security: Climate Change, Migration and Conflict in the Sahel* (2011) that identified nineteen hotspots over the past twenty years. Climate change effects on resource availability have already led to migration and increased competition over scarce resources in some of the hotspots:

> As a follow-up to this initial work, UN Environment was requested in 2009 by the UN Secretary General to provide technical inputs to the drafting of the report to the General Assembly entitled *Climate Change and its Possible Security Implications* (A/64/350). UN Environment's Executive Director was invited to address the Security Council in 2011. This thematic debate resulted in the Security Council Presidential Statement S/PRST/2011/15 on climate change. … The next major international milestone was the report *A New Climate for Peace*, commissioned by G7 foreign ministries, which was launched in New York in June 2015. Based in part on substantive contributions by UNEP, the report identifies seven key compound climate and fragility risks that should form the basis for united action. These include local resource competition, livelihood insecurity and migration, volatile food prices

[115] In the US, on 25 September 2009, the Central Intelligence Agency opened a Center on Climate Change and National Security (see: https://www.cia.gov/news-information/press-releases-statements/center-on-climate-change-and-national-security.html), but in 2015 the Republicans requested severe funding cuts on climate change both for the CIA and the US Department of Defense; see: Foran, Clare, 2015: "Republicans Push Climate Change Cuts at CIA, Defense Department", in: *Atlantic* (17 March).

> and provision, transboundary water management, and unintended effects of climate change policies. As a direct follow-up to the G7 report, UN Environment has established a partnership with the EU to address the security implications of climate change in two pilot countries. … A combination of different approaches to build resilience will be tested ranging from ecosystem restoration and improved resource management, to the development of social capital and early warning mechanisms, to training, monitoring and local institution building. UN Environment will either provide additional funding to help existing climate change adaptation projects understand and address security risks, or help scale-up existing good practices.[116]

In a Briefing Paper Achim Steiner (2009), as UNEP's Executive Secretary General, summarised UNEP's analyses on the climate change and security linkage and listed among the proposed response strategies: (1) ecosystem-based adaptation; (2) green economy, (3) strengthening national and regional governance; (4) strengthening global governance, offering *the Environmental Security Initiative* (ENVSEC) as an example.[117]

2.7.6.4 Scientific Discourses on Climate Change and Security

Since 2007, a scientific discourse has gradually emerged in the social sciences on climate change and security issues (Cabot 2016; Harrington/Shearing 2017) with a focus on:

1. *conceptual aspects* framing the assumed linkage as issues of climate change and state-centred security (planetary, global, international, regional, national) or *economic* and *societal* or *human* security[118];
2. *policy-orientated* problems that are addressed by government institutions (foreign ministries, defence ministries), private policy research think tanks and policy consultants[119];

---

[116]UNEP: "Climate change and security risks"; at: https://www.unenvironment.org/explore-topics/disasters-conflicts/what-we-do/risk-reduction/climate-change-and-security-risks; *Guidance Note and Tools to Address Climate-Fragility Risks* (UNEP, EU, Adelphi); *Livelihood Security: Climate Change, Migration* and Conflict in the Sahel (UNEP, IOM, OCHA, UNU, CILSS, 2011); UN-EU Guidance Note: *Managing and Preventing Conflicts over Renewable Resources* (UNEP with EU-UN Partnership on Land, Natural Resources and Conflict, 2012); DFID (2014): *Topic Guide: Conflict, Climate and Environment;* "Security Council, in Statement, Says 'Contextual Information' on Possible Security Implications of Climate Change Important When Climate Impacts Drive Conflict", 20 July 2011.

[117]See Achim Steiner (2009): https://www.nato.int/strategic-concept/pdf/UNEP.pdf (18 June 2020).

[118]See: "Climate Security", at: https://en.wikipedia.org/wiki/Climate_security (2 August 2020); Adger et al. (2014): 755–791; Detraz/Betsill (2009); Brauch (2002, 2006); Brauch/Scheffran (2010); McDonald (2013); Taylor (2013); Brauch (2014): Boas/Rothe (2016); Selby (2014); Trombetta (2014); Baldwin et al. (2014); Floyd (2008); Brzoska (2016); von Lucke (2014); Warner, Boas (2017); Risbey (2008).

[119]See: EU position paper of March 2008; at: https://www.consilium.europa.eu/media/30862/en_clim_change_low.pdf (2 August 2020); Ministry of Defence (2010a); US Department of Defense (2015); UN Security Council (2007); Kurtz (2012).

3. *qualitative* national security or socio-political *case studies*[120]; and
4. *quantitative trend analyses.*[121]

This review, by analysing the linkages between climate change, human security, and violent conflict, aims to contribute new knowledge based on a plurality of scientific disciplines, approaches, theories, and methods in order to develop Beck's 'principle of *precaution through prevention*'; contribute to the discussion on geopolitical aspects of the required transformation to 'manage' conflicts; 'prevent' them escalating into violence and wars; and contribute to a discussion on a scientific and policy agenda for a 'sustainability transition' (Grin et al. 2010) by addressing a policy priority to 'avoid' future climate-induced violent conflicts and move towards sustainable peace.

Climate change may also result in an increase in hydro-meteorological hazards. This aspect has barely been addressed in the initial environmental security research (de Wilde 2009; Dalby et al. 2009) but it is key to the debate on the *securitisation* of climate change. Environmental stress may increase the impact of hazards (for the highly socially vulnerable) and cause or contribute (with natural hazards and conflicts) to internal displacement, urbanisation, and transboundary forced migration.

Whether these factors result in domestic crises, disasters, or violent conflicts, or can be avoided, depends on many specific factors and activities resulting from the interaction between the three actors representing the state, society, and the business community. It also depends on the use of both traditional and modern technical and organisational knowledge and knowledge-based response strategies by governments and international organisations and transnational societal and economic organisations.

According to Scheffran et al. (2016), there are various possible pathways between climate change and conflict that are influenced by a number of contextual conditions, intermediate variables, and intervening responses.[122] The societal and political levels are not only affected *by* climate change; they also affect it themselves, indicating that circular feedbacks are relevant. Within this context, the analysis is embedded into an integrated assessment framework of climate-society interaction that represents the causal links between climate change, natural resources and environmental stress, human values and needs, and the societal consequences and instabilities:

1. Changes in the climate system, such as increases in greenhouse gas concentration, temperature, and precipitation, affect environmental systems and natural resources (e.g. soil, water, ecosystems, forests, biodiversity) through a sequence of complex interactions.
2. Changes in natural resources can have adverse impacts on human values and capabilities, which may provoke human responses that can affect social systems.

---

[120] Allan (2017); Ide (2017); Gilmore et al. (2018).

[121] Buhaug (2015); Buhaug et al. (2014); Buhaug/Rudolfsen (2015); Burke et al. (2015); Gleditsch (2012); Hsiang et al. (2013); Ide/Scheffran (2014); Detges (2014, 2016, 2017, 2018).

[122] See for more details and publications by the CLISEC team at: https://www.clisec.uni-hamburg.de/staff/current-members-of-clisec/scheffran.html and https://www.clisec.uni-hamburg.de/publications.html (5 September 2020).

3. Depending on the degree of vulnerability, socio-economic stress increases as a result of water and food insecurity, health problems, migration, economic degradation, the weakening of institutions, diminishing economic growth, and eroding societies.
4. Interdependencies between these factors may lead to societal instability that can manifest itself in violent forms such as riots, insurgencies, urban violence, and armed conflict.
5. A feedback loop allows human beings and societies to adapt to the changing situation and mitigate climate stress through strategies, institutions, and governance mechanisms that may apply technology or human and social capital to adjust the economy and the energy system to the altered environmental conditions.

The significance of the impacts of climate change on society and security can be deduced from the links between the variables and how events spread along the causal chain or cascade, which is a function of the sensitivities between variables.

2.7.7 Research on Sustainability Transition and Sustainable Peace in the Anthropocene

From an interdisciplinary perspective, William C. Clark of the Kennedy School of Government, Harvard University, Nobel laureate Paul C. Crutzen of the Max Planck Institute for Chemistry in Mainz, and Hans Jochen Schellnhuber, former director of the Potsdam Institute of Climate Impact Research, have called for a "second Copernican revolution" in science that will result in "a new paradigm for sustainability" (Clark et al. 2004) to meet a 50% reduction in global GHG or an 80% reduction for the G8 countries within only three decades (2020–2050). They argue that the guidance system for a sustainability transition requires "systems for adaptive management and social learning" whose elements require "appropriate information, incentives, and institutions". Such reform efforts should mobilise "the right knowledge", "integrate knowledge", balance "flexibility and stability", and address the required "infrastructure and capacity". They concluded by calling for a new 'social contract' between society and the scientific community, an idea that was developed further in a WBGU (2011) report.

The debate on 'sustainability transition' emerged first in the US in the 1970s and was taken up in a report by the US Academy of Science (NRC 1999; Raskin et al. 2002; Johnson 2016). It has looked forward to the processes of a long-term system transformation necessary to contain and reduce the effects of the dominant *business-as-usual* paradigm and to reduce GHG emissions through both *multilateral* quantitative emission reduction obligations and *unilateral* transformations. From 2005, a specific 'sustainability transition' research paradigm emerged from the *Dutch Knowledge Network on Systems Innovation and Transition* (KSI)[123] and the Amsterdam Conference in 2009, where the *Sustainability Transition Research Network* (STRN),[124] focusing on long-term transformations in culture, values, behaviour and lifestyle, was founded.

From the mid-1970s and the late 1990s there have been several initiatives aiming at 'sustainability transition'. In 1976 the *Tellus Institute* became an early pioneer of sustainability transition research. Kates (2001: 15325) noted "an effort to re-engage the scientific community around the requirements for a sustainability transition" in the US during the 1990s. This sought "a transition towards a state of sustainable development", which "was studied as a series of interlinked transitions, as a process of adaptive management and social learning, and as a set of indicators and future scenarios". This effort was influenced by "the widespread perception that the continuation of current trends in both environment and development would not provide for the desired state of sustainable development. To the extent that transitions represent breaks in such trends, there was growing interest in identifying the needed or desired transitions in the relationship between society and the environment."

In a first major conceptualisation of 'sustainability transition', Speth (1992) specified five "inter-linked transitions (demographic, technological, economic, social and institutional) as collective requirements for a sustainability transition". He proposed "a transition towards technologies that were environmentally benign and reduced sharply the consumption of natural resources, and the generation of waste and pollutants. For the economy, the desired transition was towards one in which prices reflected their full environmental costs. A social transition would move towards a fairer sharing of economic and environmental benefits both within and between countries." As a fifth process, he proposed "a transition in the institutional arrangements between governments, businesses and people that would be less regulation-driven and more incentive-led, less confrontational and more collaborative."

In order to study these necessary transitions researchers jointly organised "a major, albeit uncompleted, research effort known as the '2050' project (World Resources Institute, Brookings Institutions, Santa Fe Institute) that sought to track these transitions and understand their interactions". Kates (2001) noted two different approaches by the US National Academy of Sciences in its report *Our Common Journey: A Transition toward Sustainability* (NRC 1999), and a process of social learning (Social Learning Group 2001). The NRC's research agenda was "based on both current scientific understanding and the requirements of an emerging sustainability science".

---

[123] For details of KSI see: http://www.ksinetwork.nl/home; on its participants see: http://www.ksinetwork.nl/what-is-ksi/participants and for its publications see: https://www.ksinetwork.nl/output/publications. See also Routledge Studies in Sustainability Transition: http://www.routledge.com/books/series/RSST/

[124] For STRN, see: http://www.transitionsnetwork.org/; and for a reference list of ST literature see: http://www.transitions-network.org/files/Reference%20list%20to%20transition%20publications.pdf. STRN has held conferences in Amsterdam (2009), Lund (2011), Copenhagen (2012) and Zürich (2013); selected results are published in: *Journal on Environmental Innovation and Sustainability Transition*; at: http://www.journals.elsevier.com/environmental-innovation-and-societal-transitions/.

The report "identified concrete goals and appropriate next steps to accelerate major transitions underway or needed in each sector" and tried "to integrate global and local perspectives to shape a 'place-based' understanding of the interactions between environment and society; and initiate focused research programs on a small set of understudied questions that are central to sustainability transition". A fourth group focused on long-term scenarios of alternative pathways to transition, and assessed "how the global system might unfold". These scenarios reflected "current trends and reform proposals, various forms of social and environmental breakdown, and more fundamental transitions or transformations (Raspin et al. 1998, Hammond 1998)". The NRC (1999) report concluded that

> although the future is unknowable a successful transition toward sustainability is possible over the next two generations. This transition could be achieved without miraculous technologies or drastic transformations of human societies. What will be required, however, are significant advances in basic knowledge, in the social capacity and technological capabilities to utilise it, and in the political will to turn this knowledge and know-how into action (Kates 2001: 15328).

The Tellus Institute focuses on "global scenarios, planetary phase of civilization, sustainable development, climate change, eco-efficiency, globalization, information technology", and since 2005 it has addressed

> a *Great Transition* to a sustainable, just, and liveable global civilization. To attain this vision, the world must navigate toward ways of producing, consuming, and living that balance the rights of people today, future generations, and the wider community of life. The prospects for such a transition rest with the ascendance of new values, a planetary consciousness, and a sense of global citizenship. These aims will lie at the heart of the Institute's program of research, education, and network-building in the coming years.[125]

In 2014, Paul Raskin, the Tellus President, identified three global models: *Conventional Worlds* (business-as-usual), a model which assumes structural continuity of present trends and actors; *Barbarization* (worst case), which assumes "a deluge of instability swamps society's adaptive capacity, leading to a general global crisis and the erosion of civilized norms"; and *Great Transitions*, which imagines "how the imperatives and opportunities of the Planetary Phase might advance more enlightened aspirations", envisioning the new values of "human solidarity, quality-of-life, and an ecological sensibility" instead of "individualism, consumerism, and domination of nature", and aiming for "institutions that support democratic global governance, well-being for all, and environmental sustainability".

A totally different approach to sustainability transition emerged from a large research project by the *Dutch Knowledge Network on Systems Innovation and Transition* (KSI) in the Netherlands, in which eighty-five researchers participated (2005–2010). Grin et al. (2010) combined "three perspectives on transitions to a sustainable society: complexity theory, innovation theory, and governance theory". The authors

> seek to understand transitions dynamics, and how and to what extent they may be influenced. ... They do so from the conviction that only through drastic system innovations and transitions

[125]For Tellus, see at: http://www.tellus.org/about/ (11 February 2015).

> does it become possible to bring about a turn to a sustainable society to satisfy their own needs, as inevitable for solving a number of structural problems on our planet, such as the environment, the climate, the food supply, and the social and economic crisis. Among other things this implies that our world has to overcome the undesirable side-effects of the ongoing 'modernization transition', which began around 1750. However, the transition to sustainability has to compete with other developments, and it is uncertain which development will gain the upper hand. In *Transitions to Sustainable Development* the authors ... closely address the need for transitions, as well as their dynamics and design (Grin et al. 2010: xvii–xix).

A new discourse on 'sustainability transition' has evolved since 2009 within the *Sustainability Transition Research Network* (STRN).[126] It focuses on sustainability problems in the energy, transport, water and food sectors from different scientific perspectives on the ways "in which society could combine economic and social development with the reduction of its pressure on the environment. ... There is a need for transformative change at the systems level, including major changes in production, consumption that were conceptualised as 'sustainability transitions'. The STRN defined transitions research as a "new approach to sustainable development" that

> is also developing its own core set of questions and theories. Major research efforts ... have advanced knowledge of transitions to sustainability, particularly in the field of a broad understanding of how major, radical transformations unfold and what drives them. ... Technical changes need to be seen in their institutional and social context, generating the notion of 'socio-technical (s-t) systems', which are often stable and path dependent, and therefore difficult to change. Under certain conditions and over time, the relationships within s-t systems can become reconfigured and replaced in a process that may be called a system innovation or a transition.

The STRN defined its mission as coordinating its scientific capacity "towards the production of foresight reports on strategic sustainability policy questions, ... to support the development of a sustainability transitions research community internationally, and provide an independent, authoritative and credible source of analysis and insight into the dynamics and governance of sustainability transitions".[127] Their multilevel perspective on transitions was influenced by Geels's (2002: 1263) model that distinguished between technological niches, the sociotechnological regime and landscape developments that start with ... interrelated and mutually reinforcing technological innovations in 'niches'. Such multiple events initially face opposition in the respective sociotechnical regime, and once they are overcome they often result in a structural change by exploiting windows of opportunity, and lead to changes in the landscape.[128]

[126] See at: http://www.transitionsnetwork.org/ (22 January 2015).

[127] See the opening page of the STRN, at: http://www.transitionsnetwork.org/ (22 January 2015).

[128] The *Environmental Innovation and Sustainability Transitions* (EIST) Journal "offers a platform for reporting studies of innovations and socio-economic transitions to enhance an environmentally sustainable economy and thus solve structural resource scarcity and environmental problems, notably related to fossil energy use and climate change." See the *Environmental Innovation and Sustainability Transitions* (EIST) Journal at: http://www.journals.elsevier.com/environmental-innovation-and-societal-transitions/.

The research focus of KSI and the STRN influenced a policy report by the *German Advisory Council on Global Change* (WBGU 2011) on a 'Social Contract for Sustainability' (2011). The WBGU (2011: 93) report adapted Geels's model and added several megatrends in both the Earth System (climate, biodiversity, land degradation, water, raw materials) and the human system (development, democratisation, energy, urbanisation, food). Innovative changes in the regime may directly affect the megatrends. The report argued that the realisation of a low-carbon economy and society is overcoming the many barriers and exploiting the favourable factors (WBGU 2011: 6). The WBGU report stated that "… a low-carbon transformation can only be successful if it is a common goal, pursued simultaneously in many of the world's regions" (WBGU 2011). It discussed (WBGU 2011: 5) the global "remodelling of economy and society towards sustainability as a 'Great Transformation'. Production, consumption patterns and lifestyles in all of the three key transformation fields must be changed in such a way that global greenhouse gas emissions are reduced to an absolute minimum over the coming decades, and low carbon societies can develop."

The transformation towards a climate-friendly society requires many existing change agents to lead to a "concurrence of multiple change" (Osterhammel 2009), which "can trigger historic waves and comprehensive transformations". This transformation aims to produce a low-carbon society that starts with the decarbonisation of energy systems. For this to be achieved, "the proactive state and the change agents" will need to be "the key players". Within the next decade, they should jointly initiate a transformation towards a sustainable energy path. This transformation should permit a major decarbonisation of the economy by 2050, avoiding both a rebound effect and a climate crisis.

To achieve a major transformation towards a low-carbon economy and society, the WBGU has proposed specific measures for the energy sector, land-use changes and global urbanisation that could accelerate and extend the transition to sustainability. The WBGU Report suggested that "research and education are tasked with developing sustainable visions, in co-operation with policy-makers and citizens; identifying suitable development pathways, and realizing low-carbon and sustainable innovations". The WBGU argued "that transformation costs can be lowered significantly if joint decarbonisation strategies are implemented in Europe. The transformation also represents a great chance for Europe to make innovation-driven contributions to a globalisation process that has a viable future" (WBGU 2011).

A report by the *International Social Science Council* (ISSC 2012: 16) on *Transformative Cornerstones of Social Science Research for Global Change* defined transformation as "a process of altering the fundamental attributes of a system, including in this case structures and institutions, infrastructures, regulatory systems, financial regimes, as well as attitudes and practices, lifestyles, policies and power relations". It argued that the necessary additional social science research will contribute to change by calling "for an additional response to global change and to climate change; additional to and building on the enduring focus in this field on adaptation and mitigation; … a critical questioning of the systems and paradigms that have created climate change and on which climate change rests".

So far none of the studies on sustainability transition in North America (by the Tellus Institute, Speth, Kates and the NRC), Europe (by the KSI, STRN and WBGU) or by the ISSC have addressed the possible linkage of the suggested transitions to questions of international and national peace and security. This leads to the second key goal of a Handbook (Brauch/Oswald Spring/Grin/Scheffran 2016): 'peace with nature' or 'sustainable peace'.

UNEP (2011, 2014), OECD,[129] the EU and several governments have promoted an alternative vision and outlined alternative policies for a global green deal, green growth, and a decoupling of economic growth from energy consumption. These policy proposals were partly taken up by the European Commission and the European Council in its longer-term goals and policy papers on climate change, energy (EU 2011), resource (Happaerts 2016; EU 2011c, 2011d) and transport policies[130] (EU 2011a), and in its "Roadmap for moving to a competitive low carbon economy in 2050" (EU 2011b),[131] in which the European Commission addressed the goal "of reducing greenhouse gas emissions by 80–95% by 2050 compared to 1990". The Commission's Roadmap study for moving towards a low-carbon economy indicated "that a cost effective and gradual transition would require a 40% domestic reduction in greenhouse gas emissions compared to 1990 as a milestone for 2030, and an 80% reduction for 2050" (EU 2011b: 14).

By the end of 2015, at COP 21 in Paris, a majority of state parties agreed on legally binding voluntary commitments defined by the member countries themselves. They undertook to reduce their GHG emissions in the decades to come. Moreover, they would fully implement these commitments by gradually moving towards a low-carbon economy, by achieving a sustainability transition in the energy and production sectors, and adopting national policies that supported sustainable consumption by their citizens. However, no global targets and or sanctions were approved. While the EU, UN, UNEP and OCED have suggested a transition towards a green economy, so far only a few countries have initiated detailed policy programmes for a sustainable energy transition.[132]

On 19 December 2019, the new President of the European Commission, Ursula von der Leyen "presented The European Green Deal – a roadmap for making the EU's economy sustainable by turning climate and environmental challenges into opportunities across all policy areas and making the transition just and inclusive

[129] See publications on: "Green growth and sustainable development".

[130] See "White paper 2011: Roadmap to a Single European Transport Area: Towards a Competitive and Resource Efficient Transport System", at: http://ec.europa.eu/transport/themes/strategies/2011_white_paper_en.htm.

[131] European Commission, Climate Action, "Roadmap for moving to a Low-Carbon Economy in 2050"; This report is not online any more but instead is accessible in 2020: "2050 long-term strategy"; at: http://ec.europa.eu/clima/policies/roadmap/index_en.htm.

[132] See: "The Energy Transition in Europe: Initial Lessons from Germany, the UK and France" see also the "Berlin Energy Transition Dialogue" https://www.dena.de/veranstaltungen/berlin-energy-transition-dialogue-2016.html (24 February 2016).

for all".[133] As part of its long-term climate strategy the new EU Commission aims to be carbon–neutral by 2050 and it requested the Council "to take forward the work on the European Green Deal", which is to provide a roadmap with multiple actions, including the efficient use of resources in a clean, circular economy, restoring biodiversity and cutting pollution with the goal to be "climate neutral in 2050". The EU Commission President "proposed a *European Climate Law* turning the political commitment into a legal obligation and a trigger for investment" with the goal of:

- investing in environmentally-friendly technologies
- supporting industry to innovate
- rolling out cleaner, cheaper and healthier forms of private and public transport
- decarbonising the energy sector
- ensuring buildings are more energy efficient
- working with international partners to improve global environmental standards.[134]

In 2020, a few weeks prior to the COVID-19 pandemic, the EU Commission launched its long-term strategy and repeatedly updated its proposed long-term budget (2021–2027) for implementing a sustainability transition in the framework of a European Green Deal:

> The European Commission adopted its first proposal for the EU's long-term budget, the 2021–2027 multiannual financial framework (MFF) package, on 2 May 2018. To respond to the economic and social fallout of the COVID-19 pandemic, the Commission proposed a *revamped long-term EU budget* on 27 May 2020. The proposal includes an emergency recovery instrument, Next Generation EU, to help repair the immediate damage brought by the coronavirus pandemic and kick-start the recovery. *The Commission's proposal* is a seven-year EU budget of *€1,850 billion*; a revised long-term EU budget of *€1,100 billion* for 2021–2027; a temporary reinforcement of *€750 billion* for the Next Generation EU.[135]

As a result of the controversy within the European Council on the EU long-term budget (2021–2027) and especially on the temporary reinforcement of *€750 billion* for the Next Generation EU, the compromise achieved by the twenty-seven EU members and the European Commission at a special summit in Brussels on 17–21 July 2020 on a budget of *€1,824.3 billion* also implied cuts in the budget for climate action and for The European Green Deal.[136] The final conclusions of the joint meeting of the European Council with the European Commission noted:

> A21. Climate action will be mainstreamed in policies and programmes financed under the MFF [*Multiannual Financial Framework*] and NGEU [*Next Generation EU*]. An overall

[133] European Commission, 19 December 2019; at: https://ec.europa.eu/commission/presscorner/detail/e%20n/ip_19_6691.

[134] Europena Green Deal; at: https://ec.europa.eu/info/strategy/priorities-2019-2024/european-green-deal_en and https://www.consilium.europa.eu/en/policies/green-deal/ (12 December 2020).

[135] European Council, "Long-term EU budget 2021–2027"; at: https://www.consilium.europa.eu/en/policies/the-eu-budget/long-term-eu-budget-2021-2027/ (12 December 2020).

[136] EUCO 10/20 EN, European Council, Brussels, 21 July 2020 (OR. en), EUCO 10/20CO EUR 8 CONCL 4; see "2020 climate & energy package"; at: https://ec.europa.eu/clima/policies/strategies/2020_en; see: "2030 climate & energy framework"; at: https://ec.europa.eu/clima/policies/strategies/2030_en (12 December 2020).

climate target of 30% will apply to the total amount of expenditure from the MFF and NGEU and be reflected in appropriate targets in sectoral legislation. They shall comply with the objective of EU climate neutrality by 2050 and contribute to achieving the Union's new 2030 climate targets, which will be updated by the end of the year. As a general principle, all EU expenditure should be consistent with Paris Agreement objectives.[137]

18. Reflecting the importance of tackling climate change in line with the Union's commitments to implement the Paris Agreement and the United Nations Sustainable Development Goals, programmes and instruments should contribute to mainstream climate actions and to the achievement of an overall target of at least 30% of the total amount of Union budget and NGEU expenditures supporting climate objectives. EU expenditure should be consistent with Paris Agreement objectives and the 'do no harm' principle of the European Green Deal. An effective methodology for monitoring climate-spending and its performance, including reporting and relevant measures in case of insufficient progress, should ensure that the next MFF as a whole contributes to the implementation of the Paris Agreement. The Commission shall report annually on climate expenditure. In order to address the social and economic consequences of the objective of reaching climate neutrality by 2050 and the Union's new 2030 climate target, a Just Transition Mechanism, including a Just Transition Fund, will be created.

84. Funding in this Heading {on *Natural Resources and Environment*} focuses on delivering added value through a modernised, sustainable agricultural, maritime and fisheries policy as well as by advancing climate action and promoting environmental and biodiversity protection. The mainstreaming of climate across the budget and enhanced integration of environmental objectives gives this Heading a key role in reaching the ambitious target of at least 30% of EU expenditure contributing to climate objectives.

The final EU long-term budget (2021–2027) will be a result of a tough bargaining between the European Parliament, the European Commission and of the twenty-seven EU members represented in the European Council. Most likely the projected budget expenditures for COVID-19-related government assistance in coping with the economic impact of the pandemic may result in significant budget cuts for measures to cope with climate change impacts and support policies for a sustainability transition.[138]

However, the weak performance and implementation of quantitative GHG emissions reductions by most countries will most severely affect the highly socially and environmentally vulnerable developing and least developed countries. These have already seen the highest number of deaths and people affected. The economic losses from climate-induced hazards were highest in developed countries because of insurance. It is projected that many countries with a continuing high level of population growth and a high level of people below the poverty line will also have a low

---

[137] See at: https://www.consilium.europa.eu/media/45109/210720-euco-final-conclusions-en.pdf (1 August 2020); see: "European Council conclusions, 10-11 December 2020. On 10 and 11 December 2020, the European Council adopted conclusions on the MFF and Next Generation EU, COVID-19, climate change, security and external relations", at: https://www.consilium.europa.eu/en/press/press-releases/2020/12/11/european-council-conclusions-10-11-december-2020/; see the text; at: https://www.consilium.europa.eu/media/47296/1011-12-20-euco-conclusions-en.pdf (12 December 2020).

[138] See for details on the final version of the "Long-term EU budget 2021–2027: A budget powering the recovery from COVID-19"; at: https://www.consilium.europa.eu/en/policies/the-eu-budget/long-term-eu-budget-2021-2027/ (9 December 2020)

level of resilience and limited capabilities for adaptation and mitigation during the twenty-first century. This again may trigger public protest, unrest, crises and violent conflicts.

The above review focused on seven problem areas that were put on the national and international political agendas and became objects of the policy debate and international diplomatic activities. They have increasingly been addressed by specialists of international relations from the perspective of peace, security, development and environmental studies. To what extent do the proposals for a peace ecology approach reflect this reviewed research and the political debates and activities on the linkages between the environment and peace and security?

2.7.8 Early Approaches to Peace Ecology and Their Shortcomings

A 'peace ecology' concept was proposed by Kyrou (2007) as an "integrative, multi-contextual, and case sensitive approach in identifying resources for conflict and violence transformation" with the goal "to include issues of conflict analysis and peacebuilding" in environmental studies. The concept emerged from the debate on 'environmental security' and 'environmental peacemaking'. A shortcoming of this approach is "the lack of a common world-view and the absence of a shared philosophical space in relating ecology with peace". Some researchers have applied 'environmental peacemaking' to 'peace parks' (Ali et al. 2005; Ali 2008, 2009), eco-museums (Davis 1999; Vietnam 2006), ecovillages (Bang 2005), peace museums (Yamane 2007) and cultural heritage sites as either a peacebuilding or an environmental problem-solving tool.

Kyrou argues that peace ecology "applies a world-view approach to environmental peacemaking" whereby both the people and the environment benefit from overcoming cultural violence against the environment resulting from "monolithic, mechanistic and unsustainable methods of resource extraction", a business-as-usual approach or ecological violence that refers "to the direct injury to the environment through pollution, degradation, overexploitation …, especially in cases of severe or irreversible damage". In his view, "peace ecology values the preservation and harmonious interaction of societies with the nature of peace; at the same time, it values a society striving to maintain positive peace as an ecological asset". Peace ecology, he claims, links the value of biodiversity with that of cultural diversity and aims to protect the environment and to maintain the peace far into the future. Other elements of his peace ecology approach are bioregionalism, and the 'do-no-harm' principle that aims at the "preservation of positive peace in society while maintaining ecological integrity". He claims that "peace ecology places the environmental peace-making activities within the context of bioregions and examines their impact on various forms of violence".

Kyrou (2007a) also introduced a *sustainable peace assessment method* (SPAM) for evaluating "the effectiveness of environmental peace-making projects" in terms of (a) the project's "potential for reducing or removing violence", (b) "the qualitative and quantitative transformations that it may generate for each form of violence", and (c) "the degree to which the project may contribute to or reinforce a balanced and sustainable coexistence and interdependence between people and nature". His SPAM includes five tools: (i) conflict assessment maps/guides, (ii) conflict-specific investigative tools, (iii) a SPAM archive, (iv) engagement with local communities, and (v) a violence assessment matrix.[139]

From a peace activist perspective, Amster (2015) developed the concept further, but neither Kyrou nor Amster contextualised peace ecology in the framework of the Anthropocene, the new era of Earth and human history that coincided with the nuclear era, the end of World War II and the 'Great Acceleration'.

So far this 'peace ecology' approach has not been taken up and discussed by schools of peace and environment studies for the issue areas reviewed above. In the view of this author, peace ecology should be reconceptualised and this approach must be developed further by linking it to the 'political geoecology' approach (Brauch et al. 2011).

2.8 Peace Ecology: A Holistic, Enlightening and Critical Scientific Programme for the Anthropocene

Since the 1990s, environmental security research has shifted from environmental scarcity, degradation and conflict to the dangers posed by global environmental and climate change (Sygna et al. 2013). With the direct impacts of humans upon ecosystems in the Anthropocene (Crutzen 2002, 2011) and with the progressive securitisation of issues of GEC since 2003, these anthropogenic changes are increasingly threatening human lives and livelihoods. Worldwide, the destruction of key ecosystem services, the pollution of air, water and soil, land use change and extreme events are creating new 'anthropogenic challenges' for humankind, although these do not pose a threat of violent conflict or war.

Conca (1994: 20) suggested an "environmental agenda for peace studies" and a discussion on whether "ecologically desirable futures include concerns for peace and justice". He argued that it is not enough "to place 'sustainable development' and 'ecological security' alongside peace or social justice as 'world-order values'" but that scholars must ask whether "not only their formal definitions, but also their metaphorical and institutional associations, further the purposes of peace, justice, and community". Conca (2002: 9) fundamentally challenged a core premise of the debate on environmental (in)security and conflict by asking "whether environmental cooperation can trigger broader forms of peace defined as a continuum ranging from

139 Since giving presentations at ISA conferences in 2006 and 2007, Christos Kyrou (2007, 2007a, b) has not developed his papers on peace ecology any further.

the absence of violent conflict to the inconceivability of violent conflict" by also addressing "problems of structural violence and social inequality" and by "building an imagined security community" based on peaceful conflict resolution.

Partly building on Conca's work, Christos N. Kyrou (2007) introduced the concept of 'peace ecology'[140] in environmental studies. Peace ecology calls philosophically for "peace with nature" (Meyer-Abich 1979, 2003), both a domestic and an international task where human behaviour has to be brought back into line with the wholeness of nature, where increasing hazards and disasters are an expression of the disharmony and absence of humankind's peace with nature. How humans respond to these new threats to their survival depends not only on the world-view of scientists but also on the mindset of the elites. Business-as-usual perspectives prevail when the political, economic and military elites are unwilling or unable to act to address the root causes of GEC and climate change in the Anthropocene.

Peace ecology in the Anthropocene was conceptualised by Oswald Spring et al. (2014: 18–19) within the framework of six conceptual pillars of peace, security, equity, sustainability, culture and gender, in which *negative peace* (non-war) is defined by the linkages between peace and security, while for the relationship between peace and equity the concept of *positive peace* is defined by peace with social justice and global equity; for interactions between peace, gender and environment the concept of *cultural peace* was proposed, and for the relations between peace, equity and gender the concept of an *engendered peace* was suggested (Oswald Spring 2020).

These five pillars of peace ecology point to different conceptual features of peace. The classic relationship between 'international peace and security' in the UN Charter refers only to a narrow *negative peace* without war and violent conflict. It aims for the prevention, containment and resolution of conflicts and violence and the absence of 'direct violence' in wars and repression. To achieve peace with equity or *positive peace* requires the absence of 'structural violence'. This is achieved by overcoming social inequality, discrimination, marginalisation and poverty where there is no access to adequate food, water, health or educational opportunities.

Sustainable peace involves the manifold links between peace, security and the environment, where humankind and the environment as the two interdependent parts of global Earth face the consequences of destruction, extraction and pollution (Oswald Spring 2008). The concept of sustainable peace also includes the processes of recovering from environmental destruction, reducing the human footprint in ecosystems through less carbon-intensive processes, and in the long term possibly carbon-free and dematerialised production, so that future generations still have the opportunity to decide on their own resource and development strategies.

Policies aiming for 'sustainability transition' (2.7.7) are thus part of a positive strategy that addresses possible new causes of instability, crises, conflicts and, in the worst case, even war. These causes may be either the scarcity of fossil energy sources or the possible security consequences of anthropogenic global environmental

[140] This section partly relies on a previous co-authored chapter (Oswald Spring, Brauch et al. 2014), with the permission of the co-authors.

and climate change, either of which may be triggered by linear trends as well as by chaotic tipping points (Lenton et al. 2008).

The relationship between peace, the environment and gender may result in *cultural peace* that facilitates the creation of peace in the minds and actions of humankind. It socialises people so that religious and social discrimination can be overcome by establishing human rights granted equally to all people. This enables them to develop the ability to negotiate solutions to present and future conflicts peacefully and to share political, economic, social and cultural powers. The rights also respect different ecosystems by taking into account their vulnerability to human actions.

Finally, the links between peace, equity and gender introduce an *engendered peace*, creating relationships between men and women, children, young people and elderly people based on equity. This builds up the capacity for negotiation, exchange, sharing, bargaining and developing tools, with the most vulnerable empowered through development processes, non-violent conflict management, and disaster risk reduction (Oswald Spring 2016). For Randall Amster (2015: 8–9):

> peace ecology is more than merely a conceptual synthesis of peace and ecology. In essence, it contemplates the ways in which the same environmental processes that often drive conflict – e.g. resource depletion, anthropogenic climate change, food and water shortages – can also become opportunities for peaceful engagement. … When confronted with a crisis … recent research shows that people are actually more likely to cooperate than they are to compete, and that instances of violence in such situations are the exception rather than the norm (e.g. Solnit 2009). … People around the world who strive to manage scarce essential resources (as with sensitive watersheds) often find that their mutual reliance on the resource transcends even profound cultural and political differences – and in some cases even warring parties have found ways to work together positively on such issues. Informed by schools of thought including *social ecology* … and *deep ecology*, peace ecology is concerned equally with the human–human and human-environmental interfaces as they impact the search for peace at all levels (Amster 2015: 8–9).

Amster (2015: 143ff.) reviewed different concepts of and approaches to environmental cooperation, including *environmental dispute resolution* (Caplan 2010), *environmental peacemaking* (Conca/Dabelko 2002; Ali 2007, 2009) and *environmental peacebuilding* (Ecopeace 2016).[141] However, he did not frame his argument in the context of the Anthropocene, nor did he discuss issues of sustainability transition or the long-term global transition and transformation of national and global economies that need to be achieved through strategies aiming for gradual decarbonisation and dematerialisation of the economy (Brauch et al. 2016).

[141] See "EcoPeace Middle East"; at: https://en.wikipedia.org/wiki/EcoPeace_Middle_East (6 August 2020); see direct link, at; https://ecopeaceme.org/.

2.8.1 Environmental Security and Peace Studies in the Anthropocene: Fragmentation of Scientific and Political Knowledge

This author suggests a wider 'peace ecology' concept to develop a research framework and programme that can conceptually and methodologically comprise the many research projects that have emerged since the end of the Cold War, primarily in the social sciences (Table 2.2). Since 1989, the political issues of global environmental change were addressed in several global environmental regimes that were set up after the first Rio Earth summit in 1991 through the adoption of several international conventions (UNFCCC, CBD in 1992 and UNCCD in 1994).

It is appropriate that the temporal context for this 'peace ecology' research programme is the Anthropocene (Crutzen/Stoermer 2000; Crutzen 2002), since this epoch of Earth and human history began when a turning point in the geological time of Earth history coincided with the long-term structural turning points in human history after the end of World War II, with the emerging nuclear era and the order of the Cold War (1947–1989) and post-Cold War disorder (1990-present).

The environmental dimensions of this change could not be socially constructed until the late 1980s, when the exponential increase in greenhouse gases, particularly CO_2, since the Great Acceleration in the 1950s made the issues relating to global environmental change and climate change all too apparent. Whereas the CO_2 concentration in the atmosphere increased from 279 ppm in 1750 to 310 ppm in 1950 – i.e. by 31 ppm in 200 years – between 1950 and 1990 it leapt up by 44 ppm in just 40 years (from 310 to 354 ppm), and in the 30 years between 1990 and May 2020 it increased by a further 62 ppm (from 354 to 416 ppm – 137 ppm higher than the level in 1750).

Thus, with the exponential increase of *greenhouse gases* (GHG) in the atmosphere, scientific knowledge of the physical impacts of global climate change has increased since the 1980s, initially in the natural sciences. Several new research and training programmes and research institutes have been set up in the newly emerging areas of *Earth Systems Sciences* (ESS) and *Earth Systems Analyses* (ESA), and new degree courses in geoecology are being offered in the sphere of physical geography. After the build-up of new scientific capabilities, the scientific knowledge of GEC has expanded rapidly. Its peer-reviewed research output was assessed by the *Intergovernmental Panel on Climate Change* in its first five assessment reports (IPCC 1990, 1996, 2001, 2007, 2013, 2014) (process of *scientisation*). Among policy-makers in governments and parliaments political awareness of anthropogenic climate change emerged in 1988, and after only four years resulted in the first two framework agreements on climate change and biological diversity at the Rio Earth Summit (process of *politisation*). In 1988 analysis of the security implications of climate change gradually started. In spring 2003 the national public debate and studies in think tanks in several OECD countries (UK, US, Germany) took off (process of *securitisation*), and after 2007 related research in the social sciences also began.

Table 2.2 Different research approaches on linkages between the environment, security and peace, and five pillars of an extended peace ecology approach. *Source* The author

Research Programmes	Science Disciplines	Authors Organisations	Earth Systems Science (ESS)	Earth Systems Analysis (ESA)	Geoecology	Political Geoecology	Anthro-pocene studies	
Earth systems science (ESS)	Natural sciences (NS)	Lenton et al. (2008)	x					
Earth systems analysis (ESA)		Schellnhuber/Wenzel (1998) Schellnhuber (1999)		x				
Geoecology		Hugget (1995, 1999)			x			
Political Geoecology	Natural & social sciences	Brauch (2003)				x		
Anthropocene studies		Crutzen (2002)					x	
Research projects of a peace ecology programme			Five pillars of an extended peace concept					
			negative	cultural	sustainable	engendered	positive	

(continued)

Table 2.2 (continued)

Research Programmes	Science Disciplines	Authors / Organisations	Earth Systems Science (ESS)	Earth Systems Analysis (ESA)	Geoecology	Political Geoecology	Anthro-pocene studies	
			Peace & security	Sustainable & engendered peace	Peace & environment	Peace & gender	Peace & justice	
Peace Ecology	Social sciences (SS)	Kyrou (2007) Amster (2015)			x			P E A C E E C O L O G Y // An thropocene
1. Environmental impacts of war	NS & SS (political science)	Westing (1976) UNEP	x		x			
2. Environmental security (ENVSEC)	SS, NATO German BMU	Multiple (1989) NATO, UNDP, UNEP,	x		x			
3. Environmental degradation & conflict	SS (PS, IR)	Bächler (1990s)	x		x			
4. Environmental stress & conflict/war	SS (PS, IR)	Homer Dixon (1990s)	x		x			
5. Environmental dispute resolution	SS (PS, IR)	Caplan (2010)		x	x	x	x	
6. Environmental peace-making	SS (PS, IR)	Conca (1994) UNEP	x		x			
7. Environmental post-conflict peacebuilding	SS (PS, IR)	Bruch UNEP			x		x	
8. Climate change & security & conflict	SS (PS, IR)	Multiple, (1989) EU, UNSG UNSC, UNGA	x		x			
9. Sustainability transition & transformation	SS & NS	Grin/Rotmans/Schot, 2010 WBGU, EU			x			
10. Global Green Deal	SS & NS, Engineering	EU, UNEP UNDP	x		x			

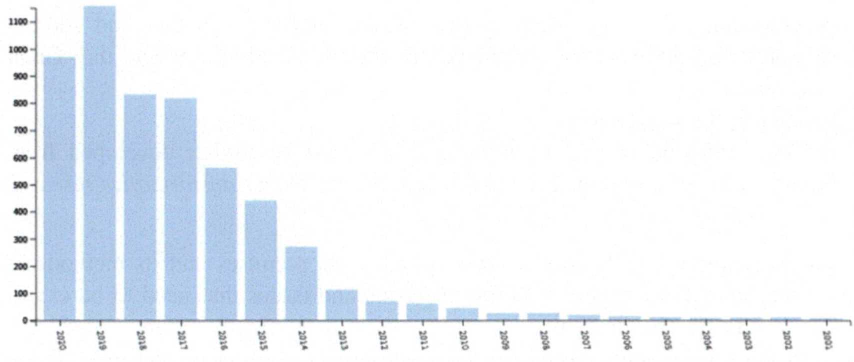

Fig. 2.6 'Anthropocene' in publication titles by year (2000–2020). *Source:* Web of Science, Core Collection (9 December 2020)

The number of peer-reviewed publications on the Anthropocene has slowly risen since 2010. On 9 December 2020, *Amazon.com* counted over 1,000 results on the Anthropocene, the *World Catalogue* based on title data listed 14,284 publications, the *Web of Science* for all databases offered 6,941 results and the *Web of Science Core Collection* had 5,342 records included (Fig. 2.6) and *Scopus* listed 5,220 document results (Brauch 2021).

Nevertheless, in their studies on 'peace ecology' both Kyrou (2007) and Amster (2015) avoided any reference to the Anthropocene, as did most studies in the framework of the ten research themes on linkages between environmental issues and problems of security or peace.

2.8.2 *Peace Ecology in the Anthropocene: The Need for Holistic Perspectives and Transformative Approaches*

Neither peace ecology nor political geoecology have so far been widely recognised research programmes in either the social or the natural sciences. In *Anthropocene Geopolitics* Dalby (2020: 169–187, 184) briefly discussed the previously proposed interdisciplinary research programme of political geoecology (Brauch et al. 2011), which brings natural science perspectives into primarily social science-orientated research.

In the perspective of this author, a peace ecology approach should:

- overcome the overspecialisation in both the social and natural sciences;
- offer *holistic* approaches that address interlinkages between themes addressed by peace research and environmental studies;

- be contextualised in the Anthropocene Epoch of Earth's history and address scientific and political linkages since the end of World War II and the 'Great Acceleration':
- be interdisciplinary; and
- *methodologically*, in the Anthropocene it should be further developed from interdisciplinary and transdisciplinary approaches into a *transformative research programme.*

Each discipline has its specific epistemology, its premises and its methods of generating new knowledge.[142] As the problems and issues that need to be examined scientifically become more complex, multidisciplinarity offers a first step towards analysing complex problems from different disciplinary perspectives. These multidisciplinary studies rely on the methodologies of their respective disciplines.

The Swiss scholar Jean Piaget,[143] who worked simultaneously in different disciplines, e.g. developmental psychology, cognitive theory and genetic epistemology, pioneered a new transdisciplinary scientific approach, which resulted in educational reforms. Piaget promoted communication between different disciplines, and in the 1960s he proposed using the term 'interdisciplinary'. He applied interdisciplinarity to pedagogic units or modules in order to integrate knowledge from different disciplines. This interdisciplinary approach was taken up by new approaches and fields, such as bioengineering and brain sciences.

Given the complexity of the Anthropocene, global environmental change and resource scarcity, several research centres and think tanks have proposed transdisciplinarity as a new scientific approach to overcome the narrow disciplinary boundaries of specialised subfields and epistemic schools of knowledge creation. Darbellay (2015: 165) has argued that

> the increasing interest in ITD [inter- and transdisciplinarity] reflects a growing awareness of the multidimensional complexity of research contexts and objects and, hence also, of societal issues that require the greater synergization of institutionalised disciplinary skills. Today, the implementation of this type of approach responds to a need within the community of researchers, who must provide answers on a daily basis to theoretical and practical questions that are highly complex and not reducible to a single disciplinary point of view.

For Hadorn et al. (2008) and Jaeger and Scheringer (1998), transdisciplinarity refers to "the cause of the present problems and their future development (*system*

[142] This section relies on Oswald Spring et al. (2016: 43).

[143] Piaget/Inhelder (1969) argued that meta-thinking and behaviour are not simply acquired from the outer world but emerge from within, in relation to the outside. Piaget's constructivism is embedded in social, economic, environmental, political, gender and cultural structures. According to Piaget's epistemology, the development of a child's knowledge comes about through understanding the chronological sequences and structural interrelations between each of these processes, their advances and sometimes also their regressions. He used empirical data to show how a simple structure becomes more complex, then global, and how this generalisable structure may facilitate a sustainable transition process. From this epistemology comes the idea of complex structures, processes, and social interactions in which a sustainable-engendered peace may help overcome greed, social differentiation, discrimination, and violence.

knowledge)"; to the "values and norms … [to] be used to form goals of the problem-solving process (*target knowledge*)"; and to "how a problematic situation can be transformed and improved (*transformation knowledge*)". They argue that "transdisciplinarity requires adequate [ways of] addressing … the complexity of problems and the diversity of perceptions of them, that abstract and case-specific knowledge are linked, and that practices promote the common good" (Hadorn et al. 2008; Pohl/Hadorn 2007; Jaeger/Scheringer 1998; Mittelstrass 2003; Brand et al. 2004). While "*multidisciplinarity* draws on knowledge from different disciplines but stays within their boundaries",[144] a definition of transdisciplinary and interdisciplinary research[145] states:

> ***Transdisciplinary Research*** is defined as research efforts conducted by investigators from different disciplines working jointly to create new conceptual, theoretical, methodological, and translational innovations that integrate and move beyond discipline-specific approaches to address a common problem. ***Interdisciplinary Research*** is any study or group of studies undertaken by scholars from two or more distinct scientific disciplines. The research is based upon a conceptual model that links or integrates theoretical frameworks from those disciplines, uses study design and methodology that is not limited to any one field, and requires the use of perspectives and skills of the involved disciplines throughout multiple phases of the research process.

In short, *transdisciplinarity* refers to a research strategy that establishes a common research objective that crosses *disciplinary* boundaries. The goal is to create a holistic approach by addressing complex problems that require close cooperation between several disciplines, such as brain or cancer research or issues of global environmental change, in which medical, behavioural, environmental, economic and political sciences work together. Funtowicz and Ravetz (1993) argued that "transdisciplinarity can help determine the most relevant problems and research questions involved" (Funtowicz/Ravetz 1993, 2003).

Holistic system analysis (Güvenen 2015) also contributed to *transdisciplinary* research, which includes all possible aspects and focuses on the interaction between different elements. *Transdisciplinarity* takes a structural approach (Nicolescu w/d) and distinguishes between different levels of analysis. The surrounding conditions facilitate dynamic adjustment of undesirable disturbers. Of particular interest is a systemic dissipative and self-regulating approach, based originally on Ilya Prigogine and the thermodynamic understanding of processes (Prigogine/Stengers 1977: 184) and Haken's (1983) *Synergetics*.

These 'dissipative and open systems' operate outside and mostly far from a thermodynamic equilibrium in a setting where energy and matter are exchanged. Prigogine characterised 'dissipative structures' as having the spontaneous appearance of anisotropy or symmetry-embracing processes in which complex and often chaotic structures interact and create unpredictable new system formations. These dissipative systems are part of a permanent dynamic process which creates a new

---

[144] Choi/Pak 2006: "Multidisciplinarity, Interdisciplinarity and Transdisciplinarity in Health Research, Services, Education and Policy: 1: Definitions, Objectives, and Evidence of Effectiveness", in: *Clinical and Investigative Medicine*, 29,6 (December): 351–364.

[145] "Definitions"; at: https://www.hsph.harvard.edu/trec/about-us/definitions/ (4 February 2016).

equilibrium between the existing structures and substructures, and also between the flows at different levels. The outcomes are permanently changing processes and new structures, which are far from equilibrium but able to maintain some dynamic functionality within the global system.

Luhmann (1991) applied dynamic system analysis to sociology and used the term 'autopoiesis', which originally described and explained the nature of a living system (Maturana 1996), e.g. a biological cell consisting of various biochemical components such as nucleic acids and proteins which are organised and bound in a cell nucleus. Luhmann's term 'autopoiesis' refers to the complexity of dynamic systems which interact with the complexity of the environment. Luhmann insisted on the radical nature of the concept and assessed five key characteristics: autonomy, emergency, operative closure, self-structuration and autopoietic reproduction. These elements are essential for the analysis of new risks and uncertainties caused by changes in the environment and social behaviour in the Anthropocene.

Schneidewind et al. (2016) proposed moving from a 'transdisciplinary' approach to a 'transformative science', while Swilling (2016) suggested an 'anticipatory science'. The concept of 'transformative research' or 'science' has been used since the 2000s for a new approach that cuts across the dominant scientific paradigms. The US National Science Board (2007) adopted the following working definition of 'transformative research':

> [it] involves ideas, discoveries, or tools that radically change our understanding of an important existing scientific or engineering concept or educational practice or lead to the creation of a new paradigm or field of science, engineering, or education. Such research challenges current understanding or provides pathways to new frontiers.

Building on this approach, in *World in Transition – A Social Contract for Sustainability*, the WBGU (2011: 21–23, 321–356) referred to "four transformative pillars of the knowledge society": 'transformation research' and 'transformation education', as well as 'transformative research' and 'transformative education'. It asserted (WBGÜ 2011: 21) that transformation research "specifically addresses the future challenge of transformation realisation" by exploring "transitory processes in order to come to conclusions on the factors and causal relations of transformation processes" and should "draw conclusions for the transformation to sustainability based on an understanding of the decisive dynamics of such processes, their conditions and interdependencies. … Transformative research supports transformation processes with specific innovations in the relevant sectors and it should encompass, for example, "new business models such as the shared use of resource-intensive infrastructures, and research for technological innovations like efficiency technologies" by aiming at a "wider transformative impact". Schneidewind/Mandy (2013) and Göpel (2017) have developed this transformative approach further for climate policy and research on sustainability transition.

In its report on the *Transformative Cornerstones of Social Science Research for Global Change*, the *International Social Science Council* (ISSC 2012: 21–22) identified six cornerstones: (1) historical and contextual complexities; (2) consequences; (3) conditions and visions for change; (4) interpretation and subjective sense-making;

(5) responsibilities; and (6) governance and decision-making. The report concluded that

> the transformative cornerstones framework speaks to the full spectrum of social science disciplines, interests and approaches – theoretical and empirical, basic and applied, quantitative and qualitative. By not fashioning a global change research agenda around a substantive focus on concrete topics – water, food, energy, migration, development, and the like – the cornerstones are not only inclusive of many social science voices but, perhaps most importantly, show that climate change and broader processes of global environmental change are organic to the social sciences, integral to social science preoccupations, domains par excellence of social science disciplines. … The transformative cornerstones of social science function not only as a framework for understanding what the social sciences can and must contribute to global change research. They function as a charter for the social sciences, a common understanding of what it is that the social sciences can and must do to take the lead in developing a new integrated, transformative science of global change.

Various initiatives by the US *National Science Board* (NSB 2007), the ISCC (2012), and the STRN (2016) have called for a new scientific paradigm for research into both global environmental change and sustainability transitions. The policy dimension should be included in the research design by moving from knowledge creation to action, policy initiatives, development and implementation.

These efforts are still highly dependent on the top-down efforts of governments and multinational enterprises. In the Global South, meanwhile, many excluded stakeholders have for decades put into practice transformative education, e.g. by implementing the pedagogy of liberation inspired by Freire (1970, 1975). These excluded social groups promote transformative processes from their daily situation of marginalisation, violence and exclusion, and promote sustainable livelihoods not for elites, but for wider social groups (Benavenides 2015; Silvia/Tavares 2015).

2.8.3 From Knowledge to Action: Addressing the Challenges for Peace and Sustainable Development in the Anthropocene

A transformative research programme implies moving "from knowledge to action" by addressing the challenges for peace and sustainable development arising from the impacts of global environmental and climate change in the Anthropocene as a result of anthropogenic interventions in multiple Earth Systems processes. Peace ecology – as introduced above– is action research, in which the transformative action is already reflected in the research design.

2.9 Towards an Ecological Peace Policy in the Anthropocene

'Peace ecology' and 'ecological peace policy' are two sides of the same coin. The tasks of an ecological peace policy are to project the probable ecological consequences of the physical effects of global environmental change; to launch early preventive actions to avoid environmental crises, conflicts or wars; and to solve these environmental consequences peacefully without any outbreaks of violence.

This book's publication date – 2020 – marks roughly the halfway point between two phases of human and Earth history which offer current researchers a reasonably long-term perspective of the Anthropocene:

1. the past eighty years of world history since the start of World War II in 1939 and the Japanese attack on Pearl Harbor (1941) up to the year 2020; and
2. the next eighty years of world history from the present to the end of the twenty-first century (2020–2100).

In 2020, midway between these two phases, we can analyse and interpret the first eight decades of the Anthropocene and project trends for the next eight decades – i.e. until 2100, which is the horizon of most climate models that were assessed by the IPCC. The outcome for the planet – and humankind – depends on the political strategies and programmes that are launched now or later or not at all.

It is impossible to be sure how countries which have previously failed to implement their obligations under climate change agreements will behave in the future. However, deductions can be made about societal and political outcomes by using climate models which correlate projections of future population levels with the production and consumption levels associated with different lifestyles.

An ecological peace policy should aim to provide a preventive strategy so that (1) the projected physical effects of anthropogenic climate change can be contained, resolved and structurally prevented through multilateral diplomacy and cooperation, and (2) domestic and international violent crises, conflicts and wars can be avoided. Jointly and proactively addressing resource scarcity, resource pollution and stress (of water, soil, air) will make it possible to enhance the health and security of the affected people.

2.9.1 The Anthropocene: Its Genesis (1940–2020) and Policies of Dealing with Projected Strategic Alternatives (2020–2100)

The German aggression against Poland on 1 September 1939 triggered a series of structural changes (Brauch 1976):

1. The US totally changed its global policy and its perception and definition of its role from an isolationist and pacifist country to a globalist nation with the strongest military and arms industry and the most powerful economy in the world.
2. The successful nuclear fission by Otto Hahn, Lise Meitner, Fritz Strassmann and Otto Robert Frisch in Berlin in 1938 persuaded Albert Einstein to send a letter to President Roosevelt warning of a possible German nuclear weapon. Roosevelt launched the top-secret Manhattan Project in Los Alamos in New Mexico to develop and build the first three nuclear weapons. This triggered a nuclear arms race once the Soviet Union and several other countries had caught up. The radiation from the subsequent nuclear testing in the atmosphere and, since 1963, underground created the 'golden spike' which led to the *Anthropocene Working Group* (AWG) proposing the year 1945 as the turning point in Earth's history from the Holocene to the Anthropocene.
3. The new international political, economic and societal order that emerged after 1945 and the 'Great Acceleration' that could be observed since 1950 in both environmental and socio-economic trends were proposed as the starting point of the Anthropocene.
4. The politics of the cheap oil and natural gas protected by US military power made the Western way of life, with its profligate attititude towards resources, possible.
5. The result was a rapid increase in the concentration of six greenhouse gases, particularly *carbon dioxide* (CO_2), in the atmosphere (global warming) due to the exponential increase in the consumption of coal, oil and gas and the burning of these fossil fuels.

The next eighty years are far more difficult to foresee and impossible to predict, especially whether chaotic changes in the climate system will occur as "tipping points" (Lenton et al. 2006) that may have catastrophic geopolitical impacts, e.g. if the Gulf Stream would turn in the mid North Atlantic, if the Amazon basin would dry out or the Indian monsoon would fundamentally change challenging the food security for the survival of several billion people in South, Southeast and East Asia.

2.9.2 Addressing the Challenges Posed in the Anthropocene

This author argued above that the proposed new Anthropocene Epoch in Earth's history since 1945/1950 coincided with severe and long-term changes in political and in human history. As there was no confirmed knowledge among natural and social scientists on anthropogenic environmental and climate change until the 1980s and 1990s, policy-makers were both unaware of and totally ignored environmental factors in their political planning, strategies and policies.

Thus, political scientists, international relation specialists, strategic analysts and peace researchers ignored the environmental dimension of policy-making almost completely until environmental historians addressed environmental impacts and ecological concerns after the end of World War II (McNeill 1992, 2000, 2010; McNeill/Unger 2010; McNeill/Engelke 2014; Radkau 2009). Only nuclear weapons experts, arms control specialists and a few peace scholars who worked on the emissions of nuclear testing took their impact primarily on human health into account, but they ignored the research on global environmental and climate change as a challenge for international peace and security.[146]

In 1989 a debate emerged, initially in strategic studies, on the possible security impacts of climate change on global, international and national security (Brown 1989; Gleick 1989). The topic was gradually also addressed by peace researchers (Brauch 2002; Scheffran et al. 2012; Brauch et al. 2017, 2018), the EU (2008, 2008a), the UNGA (2009), the UNSG (2009) and the UNSC (2007, 2020).

2.9.3 Strategies for Coping with the Anthropocene Challenges

Problems of climate change have been addressed as issues of domestic and foreign policies, particularly national, regional and international environmental and economic policies in the framework of the international climate change regime within the boundaries of the UNFCC, the Kyoto Protocol (2007) and the Paris Agreement (2015). Despite these multiple diplomatic efforts since 1992 to slow down the growth of GHG, the greenhouse gas and especially the CO_2 concentration increased by 62 ppm between 1990 and May 2020 from 354 to 416 ppm, a substantial rise compared with the 44 ppm increase during the previous four decades of the Cold War (1950–1990). In most countries there has been a huge gap between declaratory commitments and the political will to implement these commitments.

[146]"Nuclear Testing 1945 – today"; at: https://www.ctbto.org/nuclear-testing/history-of-nuclear-testing/nuclear-testing-1945-today/ (12 December 2012).

2.9.4 *Outlook for a Peace Ecology Research Programme and an Ecological Peace Policy in the Anthropocene*

While several environmental linkages between peace and security issues have increasingly been addressed in the social sciences – in political science, international relations, strategic studies and peace research – a peace ecology approach and an integrated comprehensive peace ecology programme have only been slowly emerging since 2007 (Kyrou 2007; Oswald Spring et al. 2014; Amster 2014). Le Billion/Duffy (2019) pointed to these impediments to building bridges between political ecology and peace and conflict studies:

> Conflict is at the core of many political ecology studies. Yet there has been limited engagement between political ecology and the field of peace and conflict studies. This lack of connection reflects in part the broader disciplinary context of these two fields. Whereas political ecology research mostly comes from disciplines that eschewed environmental determinism, such as human geography, much of peace and conflict studies is associated with political science using positivist approaches to determine the causal effects of environmental factors on conflicts. Yet greater connections are possible, notably in light of political ecology's renewed engagement with 'materialism', and peace and conflict studies' increasingly nuanced mixed-methods research on environment-related conflicts. Furthermore, political ecology's emphasis on uneven power relations and pursuit of environmental justice resonates with the structural violence approaches and social justice agenda of peace and conflict studies (Le Billion/Duffy 2019).

However, both authors have remained within the mainstream debates on human geography and political science and international relations, and have not so far addressed the proposed concepts of 'political geoecology' and 'peace ecology', nor have they contextualised their analysis in the Anthropocene.

Although between 2000, when Crutzen coined the term 'Anthropocene', and 2020 nearly a thousand books and several thousand scholarly articles were published, a debate on linkage concepts between political science, international relations and peace research and environmental studies has remained underdeveloped within the social sciences, and the much-needed integration of knowledge derived from climate and Earth Systems science and analysis in the natural sciences with analysis in the social sciences has remained rare despite the many pleas for interdisciplinary research.

The American biologist Wilson (1988) noted a growing *consilience* (interlocking of causal explanations across disciplines), in which the "interfaces between disciplines become as important as the disciplines themselves", that would "touch the borders of the social sciences and humanities." The key problem that will be addressed in a peace ecology approach is the complex linkage between anthropogenically-induced changes in the climate system and their societal outcome as multiple forms of conflicts that sometimes lead to violence or societal instability, and how the latter can be managed, prevented, or avoided. To this end, reactive or proactive political strategies, policies, and measures can address the cause by reducing *greenhouse gas* (GHG) emissions, and can address the impacts by political adaptation and mitigation measures to avoid escalation into violent conflicts.

In the Anthropocene there is a complex dual causal relationship between human beings, the Earth System and the social system. During the Anthropocene, humankind has for the first time directly intervened in the Earth System through the burning of fossil fuels, resulting in an anthropogenic increase in greenhouse gases, most particularly carbon dioxide (CO_2), in the atmosphere. This global warming has triggered four major physical effects: (a) an increase in global average temperature, (b) variations in precipitation, (c) low-onset sea-level rise, and (d) an increase in the probability and intensity of extreme weather events that have been assessed in so far five assessment reports (IPCC 1990, 1996, 2001, 2007, 2013/2014) and several special reports (IPCC 1997, 2011, 2012). The physical effects of anthropogenic climate change have had multiple societal impacts that may cause severe domestic or international crises, conflicts, and, in the very worst case, even violent wars (Brauch 2009).

A peace ecology approach or programme in the Anthropocene must combine and integrate the results of peace research in the social sciences with the research in Earth Systems science and analyses which has primarily been conducted from the perspective of natural sciences. A peace ecology approach must thus cross the narrow disciplinary and research programme boundaries and must move from multidisciplinary perspectives to interdisciplinary assessments. A peace ecology research programme that wants to contribute to an ecological peace policy should be transformative by including the transformation of the status quo into the research design.

References

Adger, W.N.; Pulhin, J.M.; Barnett, J.; Dabelko, G.D.; Hovelsrud, G.K.; Levy, M.; Oswald Spring, Ú.; Vogel, C.H., 2014: "Human Security", in: Field, C.B.; Barros, V.R.; Dokken, D.J.; Mach, K.J.; Mastrandrea, M.D.; Bilir, T.E.; Chatterjee, M.; Ebi, K.L.; Estrada, Y.O.; Genova, R.C.; Girma, B.; Kissel, E.S.; Levy, A.N.; MacCracken, S.; Mastrandrea, P.R.; White, L.L. (Eds.), 2014: *Climate Change 2014: Impacts, Adaptation, and Vulnerability. Part A: Global and Sectoral Aspects. Contribution of Working Group II to the Fifth Assessment Report of the Intergovernmental Panel on Climate Change* (Cambridge – New York: Cambridge University Press): 755–791.

Alger, Chadwick F., 1999: "The Expanding Tool Chest for Peacebuilders", in: Jeong, Ho-Won (Ed.): *The New Agenda for Peace Research* (Aldershot – Brookfield – Singapore – Sydney: Ashgate): 13–42.

Alger, Chadwick F., 2014: *Chadwick Alger: Pioneer in the Study of the Political Process and on NGO Participation in the United Nations* (Cham: Springer International Publishing).

Alger, Chadwick F., 2014a: *Chadwick Alger: The UN System and Cities in Global Governance* (Cham: Springer International Publishing).

Alger, Chadwick F., 2014b: *Chadwick Alger: Peace Research and Peacebuilding* (Cham: Springer International Publishing).

Ali, Saleem H., 2008: "The Siachen Peace Park Proposal: Moving from Concept to Reality", in: *Environment*, 50,3 (May): 43.

Ali, Saleem H., 2009: "Extractive Industries and the Environmental Aspects of International Security", in: Brauch, Hans Günter et al. (Eds.), 2009: *Facing Global Environmental Change: Environmental, Human, Energy, Food, Health and Water Security Concepts* (Berlin – Heidelberg – New York: Springer-Verlag): 335–342.

Ali, Saleem; Besançon, Charles; Kim, Ke Chung; Ponce, Carlos F.; Duffy, Rosaleen; Swatuk, Larry; Dabelko, Geoffrey; Williams, Alison, 2005: "Parks for Peace or Peace for Parks? Issues in Practice and Policy"; in: *Environmental Change and Security Program of the Woodrow Wilson International Program for Scholars 2005.*

Alker, Hayward R.; Haas, Peter M., 1993: "The Rise of Global Ecopolitics", in: Chucri, Nazli (Ed.): *Global Accord: International Challenges and International Responses* (Cambridge, MA: MIT Press): 133–171.

Allan, Bentley B., 2017: "Producing the Climate: States, Scientists, and the Constitution of Global Governance Objects", in: *International Organizations,* 71,1 (Winter): 131–162.

Altvater, Elmar, 2016: "The Capitalocene, or Geoengineering against Capitalism's Planetary Boundaries", in: Moore, Jason M. (Ed.): *Anthropocene or Capitalocene? Nature, History, and the Crisis of Capitalism* (Oakland, CA: PM Press): 138–153.

Amery, Hussein, A.; Wolf Aaron T., 2000: *Water in the Middle East: A Geography of Peace* (Austin, Texas: University of Texas Press)

Amster, Randall, 2014: *Peace Ecology* (Boulder, CO: Paradigm).

Anders, Günther, 1985 [1956]: *Die Antiquiertheit des Menschen,* vol. I (Munich: C.H. Beck).

Anders, Günther, 2002 [1980]: *Die Antiquiertheit des Menschen,* vol. II (Munich: C.H. Beck).

Anders, Günther, 1981: *Die Atomare Drohung: Radikale Überlegungen* (Munich: C.H. Beck).

Angus, Ian, 2016: *Facing the Anthropocene: Fossil Capitalism and the Crisis of the Earth System* (New York: Monthly Review Press).

Arends, Frederik, 2017: "'Anthropos' and 'Politik' as seen by the Greeks", in: *AFES-PRESS Brainstorming in Mosbach: 'Politik' in and for the Anthropocene: Obstacles, Challenges, Opportunities and Tasks for the Social Sciences in the 21st Century,* 31 May 2017; at: http://www.afes-press-books.de/html/PDFs/2017/From%20the%20Brainstorming%20to%20the%20Book%20Schedule_22-6-2017.pdf (9 June 2020).

Arias-Maldonado, Manuel; Trachtenberg, Zev (Eds.), 2019: *Rethinking the Environment for the Anthropocene: Political Theory and Socio-Natural Relations in the New Geological Epoch* (London – New York: Routledge).

Arsano, Y., 2002: *Conflict Management over Water Rights in Ethiopia: The Case of the Woiyto Valley in Southern Ethiopia* (Bern: Swisspeace).

Bächler, Günther, 1998: "Why Environmental Transformation Causes Violence: A Synthesis", in: *Environmental Change and Security Project Report 4* (Washington, DC: Woodrow Wilson International Center for Scholars): 24–44.

Bächler, Günther, 1999: "Environmental Degradation and Violent Conflict: Hypotheses, Research Agendas and Theory-Building", in: Suliman, Mohamed (Ed.): *Ecology, Politics and Violent Conflict* (London: Zed): 76–112.

Bächler, Günther; Spillmann, Kurt R.; Suliman, Mohamed (Eds.), 2002: *Transformation of Resource Conflicts: Approach and Instruments* (Bern: Peter Lang).

Bächler, Günther; Böge, Volker; Klötzli, Stefan; Libiszewski, Stephan; Spillmann, Kurt R. (Eds.), 1996: *Kriegsursache Umweltzerstörung: Ökologische Konflikte in der Dritten Welt und Wege ihrer friedlichen Bearbeitung,* ENCOP Final Report, vol. 1 (Chur – Zürich: Rüegger).

Bächler, Günther, 1990: *Ökologische Sicherheit und Konflikt.* Arbeitspapiere der Schweizerischen Friedensstiftung, No. 5 (Bern: Schweizerische Friedensstiftung).

Bächler, Günther, 1994: *Umweltflüchtlinge: Das Konfliktpotential von Morgen* (Münster: agenda).

Bächler, Günther, 1995: "Desertification and Conflict: The Marginalization of Poverty and of Environmental Conflicts", in: Puigdefábregas, Juan; Mendizabal, Teresa (Eds.): *Desertification and Migrations – Desertificacion y Migraciones: International Symposium on Desertification and Migrations* (Madrid: Ministerio de Asuntos Exteriores – Logroño: Geoforma Ediciones): 185–224.

Bächler, Günther, 1998: "Why Environmental Transformation Causes Violence", in: *Environmental Change and Security Report,* No. 4: 24–44.

Bächler, Günther, 1999a: "Environmental Degradation in the South as a Cause of Armed Conflict", in: Carius, Alexander; Lietzmann, Kurt (Eds.): *Environmental Change and Security: A European Perspective* (Berlin: Springer): 107–130.

Bächler, Günther, 1999b: *Violence Through Environmental Discrimination – Causes: Rwanda Arena, and Conflict Model* (Dordrecht – Boston – London: Kluwer Academic Publishers).

Bächler, Günther; Böge, Volker; Klötzli, Stefan; Libiszewski, Stephan; Spillmann, Kurt R. (Eds.), 1996: *Kriegsursache Umweltzerstörung. Ökologische Konflikte in der Dritten Welt und Wege ihrer friedlichen Bearbeitung*, vol. I (Chur-Zürich: Rüegger).

Bächler, Günther; Spillmann, Kurt R. (Ed.), 1996a: *Kriegsursache Umweltzerstörung: Länderstudien von externen Experten/Environmental Degradation as a Cause of War: Country Studies of External Experts*, vol. III (Chur: Rüegger).

Bächler, Günther; Spillmann, Kurt R. (Ed.), 1996b: *Kriegsursache Umweltzerstörung. Regional- und Länderstudien von Projektmitarbeitern/Environmental Degradation as a Cause of War: Regional and Country Studies of Research Fellows*, vol. II (Chur: Rüegger).

Bächler, Günther; Spillmann, Kurt R.; Suliman, Mohamed (Eds.), 2002: *Transformation of Resource Conflicts: Approach and Instruments* (Bern – Berlin: Peter Lang).

Baldwin, Andrew; Methmann, Chris; Rothe, Delf, 2014: "Securitizing 'Climate Refugees': The Futurology of Climate-Induced Migration", in: *Critical Studies on Security*, 2,2: 121–130.

Bang, Jan Martin, 2005: *Ecovillages: A Practical Guide to Sustainable Communities* (London: New Society).

Bannon Ian; Collier, Paul, 2003: *Natural Resources and Violent Conflict. Options and Actions* (Washington, DC: World Bank).

Bassett, Thomas J.; Peimer, Alex W., 2015: "Political Ecological Perspectives on Socio-Ecological Relations", in: *Dans Natures Sciences Sociétés*, 23,3: 157–165.

Barad, Karen, 1999: "Agential Realism: Feminist Interventions in Understanding Scientific Practices", in: Biagioli, Mario (Ed.): *The Science Studies Reader* (New York: Routledge): 1–11.

Becker, Egon; Jahn, Thomas (Eds.), 2006: *Soziale Ökologie: Grundzüge einer Wissenschaft von den gesellschaftlichen Naturverhältnissen* (Frankfurt a.M., New York: Campus 2006).

Benavenides de Pérez, Raquel Amada, 2015: "Estrategia para el Desarrollo Integral Adolescente y Juvenil", in Serrano, S. Eréndira; Oswald Spring, Úrsula; de la Rúa, Diana (Eds.): *América Latina en el Camino hacia una Paz Sustentable: Herramientas y Aportes* (Guatemala: ARP, FLACSO, CLAIP): 142–162.

Bennholdt-Thomsen, Veronika; Mies, Maria, 1999: *The Subsistence Perspective: Beyond the Globalized Economy* (London: Zed).

Berger, Peter L.; Luckmann, Thomas, 1966, 1967: *The Social Construction of Reality: A Treatise in the Sociology of Knowledge* (Garden City, NY: Anchor).

Berzonsky, Carol L.; Moser, Susanne C., [2017]: "Becoming Homo Sapiens: Mapping the Psycho-Cultural Transformation in the Anthropocene", in: *Anthropocene*, 20: 15–23.

Bhattacharya, A.; Miller, M., 1999: "Coping with Crisis: Is there a 'Silver Bullet?'", in: Agenor, R.; Miller, M.; Vines, D.; Weber, A. (Eds.): *The Asian Financial Crisis: Causes, Contagion and Consequences* (Cambridge, UK: Cambridge University Press): 357–385.

Biermann, Frank, 1999: "Syndromes of Global Change: A Taxonomy for Peace and Conflict Research", in: Carius, Alexander; Lietzmann, Kurt M. (Eds.), 1999: *Environmental Change and Security. A European Perspective* (Berlin – Heidelberg: Springer): 131–146.

Biermann, Frank, 1998: *Weltumweltpolitik: Zwischen Nord und Süd* (Baden-Baden: Nomos).

Biermann, Frank; Petschel-Held, Gerhard; Rohloff, Christoph, 1999: "Environmental Degradation as a Cause of Conflict? Theoretical Conceptualisation and Empirical Analysis of the Relationship of 'Environment' and 'Security'", in: *Law and State*, 59/60: 158–195.

Biser, E., II, 1972: "Friede", in: Ritter, Joachim (Ed.): *Historisches Wörterbuch der Philosophie* (Basel – Darmstadt: Wissenschaftliche Buchgesellschaft): 1,113–1,115.

Biswas, Asit K., 1993: "Management of International Waters: Problems and Perspectives", in: *International Journal of Water Resources Developments*, 9,2: 167–188.

Biswas, Asit K., 2004: "Dams: Cornucopia or Disaster?", in: *Water Resources Development*, 20,1 (March): 3–14.

Blumenstein, Oswald; Schachtzabel, Hartmut; Barsch, Heiner; Bork, Hans-Rusolf; Küppers, Udo, 2000: *Grundlagen der Geoökologie: Erscheinungen und Prozesse in unserer Umwelt* (Heidelberg-Berlin: Springer).

BMBF, 2015: *Sozialökologische Forschung Förderkonzept für eine gesellschaftsbezogene Nachhaltigkeitsforschung 2015/2020* (Berlin: BMBF)

Boas, Ingrid; Rothe, Delf, 2016: "From Conflict to Resilience? Explaining Recent Changes in Climate Security Discourse and Practice", in: *Environmental Politics*, 25,4: 613–632.

Bogardi, Janos; Castelein, Saskia (Eds.), 2002: *Selected Papers of the International Conference: From Conflcit to Co-operation in International Water Resources Management: Challenges and Opportunities. UNESCO-IHE Delft, The Netherlands, 20–22 November 2002* (Paris: UNESCO).

Bohle, Hans-Georg, 2001: "Vulnerability and Criticality: Perspectives from Social Geography", in: *IHDP Update*, 2/01: 3–5.

Bohle, Hans-Georg, 2002: "Land Degradation and Human Security", in: Plate, Erich (Ed.): *Environment and Human Security: Contributions to a Workshop in Bonn* (Bonn: mimeographed).

Bohle, Hans-Georg; O'Brien, Karen, 2007: "The Discourse on Human Security: Implications and Relevance for Climate Change Research: A Review Article", in: Climate Change and Human Security. Special issue of: *Die Erde*, 137,3: 155–163.

Bookchin, Murray, 1988: "Social Ecology versus Deep Ecology", in: *Socialist Review*, 18,3: 9–29.

Bookchin, Murray, 1990: *Remaking Society: Pathways to a Green Future* (Boston: South End Press).

Bookchin, Murray, 2005: *The Ecology of Freedom: The Emergence and Dissolution of Hierarchy* (Oakland: AK Press).

Boulding, Elise. 1988: *Building a Global Civic Culture: Education for an Interdependent World* (New York, NY: Teachers College Press).

Boulding. Elise. 1989: *One Small Plot of Heaven: Reflections on Family Life by a Quaker Sociologist* (Philadelphia, PA: Pendle Hill Press, 1989).

Boulding, Elise (Ed.), 1992: *New Agendas for Peace Research: Conflict and Security Re-examined* (Boulder, Co: Lynne Rienner).

Boulding, Elise, 2000: *Cultures of Peace: The Hidden Side of History* (New York: Syracuse University Press).

Boulding, Kenneth E. 1966: "The Economics of the Coming Spaceship Earth", in: Jarrett, H. (Ed.): *Environmental Quality in a Growing Economy* (Baltimore: Johns Hopkins Press).

Boulding, Kenneth E., 1970: "The Economics of the Coming Spaceship Earth", in: Boulding, Kenneth E., (Ed.): *Beyond Economics: Essays on Society, Religion, and Ethics* (Ann Arbor, University of Michigan Press): 275–287.

Boulding, Kenneth E., 1978: *Ecodynamics* (Beverly Hills, CA: Sage).

Boulding, Kenneth E., 1983: "Ecodynamcis", in: *Interdisciplinary Science Reviews*, 8,2 (June): 108–113.

Boulding, J. Russell (Ed.), 2017: *Elise Boulding: A Pioneer in Peace Research, Peacemaking, Feminism and the Family: From a Quaker Perspective*, with a Foreword by Betty Reardon and a Preface by Úrsula Oswald Spring (Cham – Heidelberg – New York – Dordrecht – London: Springer International Publishing).

Boulding, J. Russell (Ed.), 2017a: *Elise Boulding: Writings on Peace Research, Peacemaking and the Future* (Cham – Heidelberg – New York – Dordrecht – London: Springer International Publishing).

Boulding, J. Russell (Ed.), 2017b: *Elise Boulding: Writings on Feminism, the Family and Quakerism* (Cham – Heidelberg – New York – Dordrecht – London: Springer International Publishing).

Boulding, J. Russell (Ed.), 2017c: *Elise Boulding: Autobiographical Writings and Selections from Unpublished Journals and Letters* (Cham – Heidelberg – New York – Dordrecht – London: Springer International Publishing).

Bourghelle, Valentine, 2019: "Climate Change in the Security Council: On the Road to Qualifying Climate Change as 'Threat Multiplier'", in: *Völkerrechtsblog* (9 December); at: https://voelkerrechtsblog.org/climate-change-in-the-security-council/ (18 June 2020).

Brand, Frank; Schaller, Franz; Völker, Harald (Eds.), 2004: *Transdisziplinarität. Bestandsaufnahme und Perspektiven: Beiträge zur THESIS-Arbeitstagung im Oktober 2003 in Göttingen* (Göttingen: Universitätsverlag).

Brauch, Hans Günter, 1976: *Struktureller Wandel and Rüstungspolitik der USA (1940–1950): Zur Weltführungsrolle and ihren innenpolitischen Bedingungen* (Inaugural doctoral dissertation, Ruprecht-Karl University of Heidelberg, Faculty of Philosophy and History, 30 June).

Brauch, Hans Günter, 1977: *Struktureller Wandel and Rüstungspolitik der USA (1940–1950): Zur Weltführungsrolle and ihren innenpolitischen Bedingungen* (Ann Arbor – London: University Microfilms).

Brauch, Hans Günter, 1979: *Entwicklungen and Ergebnisse der Friedensforschung (1969-1978): Eine Zwischenbilanz and konkrete Vorschläge für das zweite Jahrzehnt* (Frankfurt: Haag + Herchen).

Brauch, Hans Günter, 1990: "Theorien zur Rüstungsdynamik im Kontext des Ost-West-Konflikts", in: Rittberger, Volker (Ed.): *Theorien der Internationalen Beziehungen – Bestandsaufnahme und Forschungsperspektiven*, in: *PVS Sonderheft*, 21,31 (Opladen: Westdeutscher Verlag): 240–262.

Brauch, Hans Günter (Ed.), 1996: *Klimapolitik: Naturwissenschaftliche Grundlagen, internationale Regimebildung and Konflikte, ökonomische Analysen sowie nationale Problemerkennung and Politikimplementation* (Berlin – Heidelberg: Springer-Verlag).

Brauch, Hans Günter, 2002: "Climate Change, Environmental Stress and Conflict: AFES-PRESS Report for the Federal Ministry for the Environment, Nature Conservation and Nuclear Safety", in: Federal Ministry for the Environment, Nature Conservation and Nuclear Safety (Ed.): *Climate Change and Conflict: Can Climate Change Impacts Increase Conflict Potentials? What is the Relevance of this Issue for the International Process on Climate Change?* (Berlin: Federal Ministry for the Environment, Nature Conservation and Nuclear Safety, 2002): 9–112; at: http://www.afes-press.de/pdf/Brauch_ClimateChange_BMU.pdf.

Brauch, Hans Günter, 2003: "Security and Environment Linkages on the Mediterranean Space: Three Phases of Research on Human and Environmental Security and Peace", in: Brauch, Hans Günter; Liotta, P.H.; Marquina, Antonio; Rogers, Paul; Selim, Mohammed El-Sayed (Eds.): *Security and Environment in the Mediterranean: Conceptualising Security and Environmental Conflicts* (Berlin – Heidelberg: Springer): 35–144.

Brauch, Hans Günter, 2003a: "Towards a Fourth Phase of Research on Human and Environmental Security and Peace. Conceptual Conclusions", in: Brauch, Hans Günter; Liotta, P.H.; Marquina, Antonio; Rogers, Paul; Selim, Mohammed El-Sayed (Eds.): *Security and Environment in the Mediterranean. Conceptualising Security and Environmental Conflicts* (Berlin – Heidelberg: Springer): 919–954.

Brauch, Hans Günter, 2003b: "Worldviews and Mindsets: American vs. European Perspectives on Mediterranean (Environmental) Security Policy", in: Brauch, Hans Günter; Liotta, P.H.; Marquina, Antonio; Rogers, Paul; Selim, Mohammed El-Sayed (Eds.): *Security and Environment in the Mediterranean: Conceptualising Security and Environmental Conflicts* (Berlin – Heidelberg: Springer): 237–266.

Brauch, Hans Günter, 2005: *Environment and Human Security: Freedom from Hazard Impact*, InterSecTions, 2/2005 (Bonn: UNU-EHS).

Brauch, Hans Günter, 2005a: *Threats, Challenges, Vulnerabilities and Risks in Environmental Human Security*. Source, 1/2005 (Bonn: UNU-EHS); at: https://www.ehs.unu.edu/index.php?module=overview&cat=17&menu=36.

Brauch, Hans Günter, 2008: "Introduction: Globalization and Environmental Challenges: Reconceptualizing Security in the 21st Century", in: Brauch, Hans Günter; Oswald Spring, Úrsula; Mesjasz, Czeslaw; Grin, John; Dunay, Pal; Behera, Navnita Chadha; Chourou, Béchir; Kameri-Mbote, Patricia; Liotta, P.H. (Eds.): *Globalization and Environmental Challenges: Reconceptualizing Security in the 21st Century* (Berlin – Heidelberg: Springer-Verlag): 27–43.

Brauch, Hans Günter, 2009: "Securitizing Global Environmental Change", in: Brauch, Hans Günter; Oswald Spring, rsula; Grin, John; Mesjasz, Czeslaw; Kameri-Mbote, Patricia; Behera, Navnita Chadha; Chourou, Béchir; Krummenacher, Heinz (Eds.), 2009: *Facing Global Environmental*

Change: Environmental, Human, Energy, Food, Health and Water Security Concepts (Berlin – Heidelberg – New York: Springer-Verlag): 65–102.

Brauch, Hans Günter, 2013: "Preface", in: Westing, Arthur H.: *Arthur H. Westing: Pioneer on the Environmental Impact of War*. SpringerBriefs on Pioneers in Science and Practice No. 1 (Heidelberg – Dordrecht – London – New York: Springer-Verlag): vii–x.

Brauch, Hans Günter, 2014: "From Climate Change and Security Impacts to Sustainability Transition: Two Policy Debates and Scientific Discourses", in: Oswald Spring, Úrsula; Brauch, Hans Günter; Tidball, Keith G. (Eds): *Expanding Peace Ecology: Peace, Security, Sustainability, Equity and Gender* (Cham: Springer International Publishing); at: https://doi.org/10.1007/978-3-319-00729-8_2.

Brauch, Hans Günter, 2016: "Historical Times and Turning Points in a Turbulent Century: 1914, 1945, 1989 and 2014?", in: Brauch, Hans Günter; Oswald Spring, Úrsula; Bennett, Juliet; Serrano Oswald, Serena Eréndira (Eds.): *Addressing Global Environmental Challenges from a Peace Ecology Perspective* (Cham – Heidelberg – New York – Dordrecht – London: Springer International Publishers): 11–54.

Brauch, Hans Günter, 2017: "Conceptualizing Sustainable Peace in the Anthropocene: A Challenge and Task for an Emerging Political Geoecology and Peace Ecology", in: Brauch, Hans Günter; Oswald Spring, Úrsula; Grin, John; Scheffran; Jürgen (Eds.): *Handbook on Sustainability Transition and Sustainable Peace* (Cham – Heidelberg – New York – Dordrecht – London: Springer): 187–236.

Brauch, Hans Günter, 2021: "Review of the Use of the Anthropocene Concept in the Natural and Social Sciences and in the Humanities – A Conceptual Mapping (2000–2020)", in: Benner, Susanne; Lax, Gregor; Lelieveld, Jos; Pöschl, Ulrich (Eds.): *The Anthropocene – A New Phase of Earth History* (Cham: Springer International Publishing).

Brauch, Hans Günter; Dalby, Simon; Oswald Spring, Úrsula, 2011: "Political Geoecology for the Anthropocene", in: Brauch, Hans Günter; Oswald Spring, Úrsula; Mesjasz, Czeslaw; Grin, John; Kameri-Mbote, Patricia; Chourou, Béchir; Dunay, Pal; Birkmann, Jörn (Eds.): *Coping with Global Environmental Change, Disasters and Security – Threats, Challenges, Vulnerabilities and Risks* (Berlin – Heidelberg – New York: Springer-Verlag, 2011): 1,453–1,486.

Brauch, Hans Günter; Oswald Spring, Úrsula; Mesjasz, Czeslaw; Grin, John; Dunay, Pal; Behera, Navnita Chadha; Chourou, Béchir; **Kameri-Mbote, Patricia;** Liotta, P.H. (Eds.), 2009: *Globalization and Environmental Challenges: Reconceptualizing Security in the 21st Century* (Berlin – Heidelberg – New York: Springer-Verlag).

Brauch, Hans Günter; Oswald Spring, Úrsula, 2011: "Introduction: Coping with Global Environmental Change in the Anthropocene", in: Brauch, Hans Günter; Oswald Spring, Úrsula; Mesjasz, Czeslaw; Grin, John; Kameri-Mbote, Patricia; Chourou, Béchir; Dunay, Pal; Birkmann, Jörn (Eds.): *Coping with Global Environmental Change, Disasters and Security – Threats, Challenges, Vulnerabilities and Risks* (Berlin – Heidelberg – New York: Springer-Verlag, 2011): 31–60.

Brauch, Hans Günter; Dalby, Simon; Oswald Spring, Úrsula, 2015: "Political Geoecology for the Anthropocene" [in Chinese], in: Brauch, Hans Günter; Oswald Spring, Úrsula; Mesjasz, Czeslaw; Grin, John; Cheng, Liu (Eds.): *Coping with Global Environmental Change, Disasters and Security – Threats, Challenges, Vulnerabilities and Risks* [in Chinese] (Nanjing: PR China: Nanjing Press Company).

Brauch, Hans Günter; Oswald Spring, Úrsula; Mesjasz, Czeslaw; Grin, John; Dunay, Pal; Behera, Navnita Chadha; Chourou, Béchir; **Kameri-Mbote, Patricia;** Liotta, P.H. (Eds.): *Globalization and Environmental Challenges: Reconceptualizing Security in the 21st Century* (Berlin et al.: Springer-Verlag).

Brauch, Hans Günter; Oswald Spring, Úrsula, 2009: *Securitizing the Ground: Grounding Security* (Bonn: UNCCD).

Brauch, Hans Günter; Oswald Spring, Úrsula; Bennett, Juliet; Serrano Oswald, Serena Eréndira (Eds.), 2017: *Addressing Global Environmental Challenges from a Peace Ecology Perspective* (Cham: Springer International Publishers).

Brauch, Hans Günter; Oswald Spring, Úrsula; Collins, Andrew E.; Serrano Oswald, Serena Eréndira (Eds.), 2019: *Climate Change, Disasters, Sustainability Transition and Peace in the Anthropocene* (Cham: Springer International Publishers).

Brauch, Hans Günter; Oswald Spring, Ursula, 2017: "Sustainability Transition and Sustainable Peace: Scientific and Policy Context, Scientific Concepts and Dimensions", in: Brauch, Hans Günter; Oswald Spring, Úrsula; Grin, John; Scheffran, Jürgen (Eds.): *Handbook of Sustainability Transition and Sustainable Peace* (Heidelberg – New York – Dordrecht – London: Springer-Verlag): 1–66.

Brauch, Hans Günter; Oswald Spring, Úrsula; Grin, John; Scheffran, Jürgen, (Eds.), 2017: *Handbook of Sustainability Transition and Sustainable Peace* (Heidelberg – New York – Dordrecht – London: Springer-Verlag).

Brauch, Hans Günter; Scheffran, Jürgen, 2012: "Introduction", in: Scheffran, Jürgen; Brzoska, Michael; Brauch, Hans Günter; Link, Peter Michael; Schilling, Janpeter (Eds.): *Climate Change, Human Security and Violent Conflict: Challenges for Societal Stability* (Berlin – Heidelberg – New York: Springer-Verlag).

Braudel, Fernand, 1966: *La Méditerranée et le Monde Méditerranéen à l'époque de Philippe II* (Paris: Armand Colin).

Braudel, Fernand, 1969: "Histoire et Science Sociales: La Longue Durée", in: *Écrits Sur l'Histoire* (Paris: Flammarion): 41–84.

Braudel, Fernand, 1972: *The Mediterranean and the Mediterranean World in the Age of Philip II,* 2 volumes (New York: Harper & Row).

Brauer, Juergen, 2009: *War and Nature: The Environmental Consequences of War in a Globalized World* (Lanham, MD: AltaMira Press).

Bridge, Gavin; McCarthy, James; Perreault, Tom (Eds.), 2015: "Editor's Introduction", in: Perreault, Tom; Bridge, Gavin; McCarthy, James (Eds: *The Routledge Handbook of Political Ecology* (London: Routledge): 3–18.

Brock, Lothar, 1991: "Peace Through Parks: The Environment on the Peace Research Agenda", in: *Journal of Peace Research*, 28,4: 407–423.

Brock, Lothar, 1992: "Security through Defending the Environment: An Illusion?", in: Boulding, Elise (Ed.): *New Agendas for Peace Research: Conflict and Security Re-examined* (Boulder, Co: Lynne Rienner).

Brock, Lothar, 1997: "The Environment and Security: Conceptual and Theoretical Issues", in: Gleditsch, Nils-Petter (Ed.), 1997: *Conflict and the Environment* (Dordrecht – Boston – London: Kluwer): 17–34.

Brock, Lothar, 2002: "Was ist das 'mehr' in der Rede: Friede sei mehr als die Abwesenheit von Krieg?", in: Sahm, Astrid; Sapper, Manfred; Weichsel, Volker (Eds.): *Die Zukunft des Friedens. Band 1: Eine Bilanz der Friedens- und Konfliktforschung* (Opladen: Westdeutscher Verlag, 2002): 95–114.

Broswimmer, Franz J., 2002: *Ecocide: A Short History of the Mass Extinction of Species* (London: Pluto Press).

Brown, H., 1954: *The Challenge of Man's Future* (New York: Viking).

Brown, Lester R., 1977: *Redefining National Security*. Worldwatch Paper No. 14 (Washington, DC: Worldwatch Institute).

Brown, Lester R., 2004: "Troubling New Flows of Environmental Refugees", in: Earth Policy Institute (Ed.): *Update 33* (28 January): 2.

Brown, Lester R., 2006: *Plan B 2.0: Rescuing a Planet under Stress and a Civilization in Trouble* (London – New York: Norton).

Brown, Neville, 1989: "Climate, Ecology and International Security", in: *Survival*, 31,6: 519–532.

Brown, Neville, 2001: *History and Climate Change. A Eurocentric Perspective* (London – New York: Routledge).

Bruch, Carl; Muffett, Carroll; Nichols, Sandra S. (Eds.), 2016: *Governance, Natural Resources and Post-Conflict Peacebuilding* (London: Routledge).

Brundtland Commission (World Commission on Environment and Development), 1987, 1988: *Our Common Future. The World Commission on Environment and Development* (Oxford – New York: Oxford University Press).

Bryant, Raymond L., 1998: "Power, Knowledge and Political Ecology in the Third World: A Review", in: *Progress in Physical Geography*, 22,1: 79–94.

Bryant, Raymond L.; Bailey, Sinéad, 1997: *Third World Political Ecology* (London: Routledge).

Brzoska, Michael; Fröhlich, Christiane J., 2016: "Climate Change, Migration and Violent Conflict: Vulnerabilities, Pathways and Adaptation Strategies", in: *Migration and Development*, 5,2: 190–210.

Buffon, Georges-Louis Leclerc de, 1778: *Histoire Naturelle, Générale et Particulière,* supplement 5: *Des Époques de la Nature* (Paris: Imprimerie Royale).

Bugallo Alicia Irene, 1995: *De Dioses, Pensadores y Ecologistas* (Buenos Aires: Grupo Editor Latinoamericano).

Bugallo, Alicia Irene, 2011: *La Filosofía Ambiental en Arne Naess: Influencias de Spinoza y James* (Río Cuarto: Ediciones del ICALA).

Bugallo, Alicia Irene, 2015: *Filosofía Ambiental y Ecosofías: Arne Naess, Spinoza y James* (Buenos Aires: Prometeo).

Buhaug, Halvard, 2015: "Climate-Conflict Research: Some Reflections on the Way Forward", in: *Wires Climate Change 6*: 269–275.

Buhaug, Halvard; Nordkvelle, Jonas; Bernauer, Thomas; Böhmelt, Tobias; Brzoska, Michael; Busby, Joshua W.; Ciccone, Antonio; Fjelde, Hanne; Gartzke, Erik; Gleditsch, Nils P.; Goldstone, Jack A.; Hegre, Håvard; Holtermann, Helge; Koubi, Vally; Link, Jasmin S. A.; Link, Michael P.; Lujala, Päivi; O'Loughlin, John; Raleigh, Clionadh; Scheffran, Jürgen; Schilling, Janpeter; Smith, Todd G.; Theisen, Ole M.; Tol, Richard S.J.; Urdal, Henrik; von Uexkul, Nina, 2014: "One Effect to Rule Them All? A Comment on Climate and Conflict", in: *Climatic Change*, 127,3: 391–397.

Buhaug, Halvard; Rudolfsen, Ida, 2015: *A Climate of Conflicts?* (Oslo: Peace Research Institute Oslo [PRIO])

Burke, Anthony; Dalby, Simon; Fishel, Stefanie; Levine, Daniel; Mitchell, Audra, 2015: "Planet Politics – A Manifesto for the End of IR", Presented at the Millennium Conference: Failure and Denial in World Politics, London, October 2015.

Burke, Marshall; Hsiang, Solomon M.; Miguel, Edward, 2015: "Climate and Conflict", in: *Annual Review of Economics*, 7,1: 577–617.

Buzan, Barry; Lawson, George, 2015: *The Global Transformation: History, Modernity and the Making of International Relations* (Cambridge: Cambridge University Press).

Buzan, Barry; Wæver, Ole; Wilde, Jaap de, 1997, 1998, 2004: *Security. A New Framework for Analysis* (Boulder – London: Lynne Rienner).

Cabot, Charlène, 2016: *Climate Change, Security Risks, and Conflict Reduction in Africa: A Case Study of Farmer-Herder Conflicts over Natural Resources in Côte d'Ivoire, Ghana, and Burkina Faso* (Berlin – Heidelberg – New York: Springer).

Cafferatta, Néstor, 2004: "El Principio Precautorio", in: *Gaceta Ecológica*, 73 (October-December); at: https://www.isla.org.mx/documentos/444[1].pdf.

Caldwell, Dan (Ed.), 2019: *Alexander L. George: A Pioneer in Political and Social Sciences* (Cham: Springer International Publishing).

Caldwell, Dan; McKeown, Timothy J. (Eds.), 1993, 2020: *Diplomacy, Force, and Leadership Essays in Honor of Alexander L. George* (London: Routledge).

Campbell, Susanna; Chandler, David; Sabaratnam, Meera, 2011: *A Liberal Peace? The Problems and Practices of Peacebuilding* (London: Zed).

Caplan, James A., 2010: *The Theory and Principles of Environmental Dispute Resolution* (San Bernardino, CA: edrusa.com).

Cardona, Omar D., 2004: "The Need for Rethinking the Concepts of Vulnerability and Risk from a Holistic Perspective: A Necessary Review and Criticism for Effective Risk Management", in: Bankoff, Greg; Frerks, Georg; Hilhorst, Dorothea (Eds.): *Mapping Vulnerability, Disasters, Development & People* (London – Sterling, Va.: Earthscan): 37–51.

Cardona, Omar D., 2009: "Disaster Risk and Vulnerability: Concepts and Measurement of Human and Environmental Insecurity", in: Brauch, Hans Günter; Oswald Spring, Úrsula; Mesjasz, Czeslaw; Grin, John; Kameri-Mbote, Patricia; Chourou, Béchir; Dunay, Pal; Birkmann, Jörn (Eds.): *Coping with Global Environmental Change, Disasters and Security – Threats, Challenges, Vulnerabilities and Risks* (Berlin – Heidelberg – New York: Springer-Verlag, 2011): 107–122.

Carius, Alexander; Lietzmann, Kurt M. (Eds.), 1999: *Environmental Change and Security: A European Perspective* (Berlin – Heidelberg: Springer).

Chakrabarty, Dipesh, 2018: "Anthropocene Time – The Seventh *History and Theory Lecture*", in: *History and Theory*, 57,1 (March): 5–32.

Chen, Lincoln; Fukuda-Parr, Sakiko; Seidensticker, Ellen (Eds.), 2003: *Human Insecurity in a Global World* (Cambridge, MA: Harvard University Press).

Chin, Anne; Beach, Timothy; Luzzadder-Beach, Sheryl; Solecki, William D., 2017: "Editorial: Challenges of the 'Anthropocene'", in: *Anthopocene*, 20 (December): 1–3.

Choi, B.C.; Pak, A.W., 2006: "Multidisciplinarity, Interdisciplinarity and Transdisciplinarity in Health Research, Services, Education and Policy, 1: Definitions, Objectives, and Evidence of Effectiveness", in: *Clinical and Investigative Medicine* 29,6 (December): 351–364; at: https://www.ncbi.nlm.nih.gov/pubmed/17330451.

Christian, David, 2004: *Maps of Time* (Berkeley: University of California Press).

Clark, William C.; Crutzen, Paul J.; Schellnhuber, Hans Joachim, 2004: "Science and Global Sustainability: Toward a New Paradigm", in: Schellnhuber, Hans Joachim; Crutzen, Paul J.; Clark, William C.; Claussen, Martin; Held, Hermann (Eds.): *Earth System Analysis for Sustainability* (Cambridge, MA; London: MIT Press): 1–28.

Conca, Ken, 1994: "In the Name of Sustainability: Peace Studies and Environmental Discourse", in: Käkonen, Jyrki (Ed.): *Green Security or Militarized Environment* (Dartmouth: Aldershot): 7–24.

Conca, Ken, 2002: "The Case for Environmental Peacemaking", in: Conca, Ken; Dabelko, Geoffrey (Eds.): *Environmental Peacemaking* (Baltimore: Johns Hopkins University Press): 1–22.

Conca, Ken, 2015: *An Unfinished Foundation: The United Nations and Global Environmental Governance* (Oxford: Oxford University Press).

Conca, Ken; Carius, Alexander; Dabelko, Geoffrey D., 2005: "Building Peace through Environmental Cooperation", in: The Worldwatch Institute (Ed.): *State of the World 2005: Redefining Global Security* (New York – London: W.W. Norton): 144–157.

Crippa, M.; Oreggioni, G.; Guizzardi, D.; Muntean, M.; Schaaf, E.; Lo Vullo, E.; Solazzo, E.; Monforti-Ferrario, F.; Olivier, J.G.J.; Vignati, E., 2019: *Fossil CO_2 and GHG Emissions of All World Countries – 2019 Report*, EUR 29849 EN (Luxembourg: Publications Office of the European Union).

Crutzen, Paul J., 2002: "Geology of Mankind", in: *Nature*, 415,3 (January): 23.

Crutzen, Paul J., 2011: "The Anthropocene: A Geology of Mankind", in: Brauch, Hans Günter; Oswald Spring, Úrsula; Mesjasz, Czeslaw; Grin, John; Kameri-Mbote, Patricia; Chourou, Béchir; Dunay, Pal; Birkmann, Jörn (Eds.), 2011: *Coping with Global Environmental Change, Disasters and Security – Threats, Challenges, Vulnerabilities and Risks* (Berlin– Heidelberg– New York: Springer-Verlag): 3–4.

Crutzen, Paul J.; Stoermer, Eugene F., 2000: "The Anthropocene", in: *IGBP Newsletter*, 41: 17–18.

Czempiel, Ernst-Otto, 1986: *Friedensstrategien* (Paderborn – München – Wien – Zürich: Schöningh).

Dadhich, Naresh, 2008: "Thinking on Security in Hinduism: Contemporary Political Philosophy and Ethics in India", in: Brauch, Hans Günter; Oswald Spring, Úrsula; Mesjasz, Czeslaw; Grin, John; Dunay, Pal; Behera, Navnita Chadha; Chourou, Béchir; Kameri-Mbote, Patricia; Liotta, P.H. (Eds.): *Globalization and Environmental Challenges: Reconceptualizing Security in the 21st Century* (Berlin – Heidelberg: Springer-Verlag): 243–252.

Dalby, Simon, 1998: "Ecological Metaphors of Security: World Politics in the Biosphere", in: *Alternatives*, 23,3: 291–319.

Dalby, Simon, 2000: "Geopolitics and Ecology: Rethinking the Contexts of Environmental Security", in: Lowi, Miriam R.; Shaw, Brian R. (Eds.): *Environment and Security. Discourses and Practices* (Basingstoke – London: Macmillan; New York: St. Martin's Press): 84–100.

Dalby, Simon, 2002: "Environmental Change and Human Security", in: *Canadian Journal of Policy Research*, 3,2: 71–79.

Dalby, Simon, 2002a: "Security and Ecology in the Age of Globalization", in: Woodrow Wilson International Center for Scholars (Ed.): *Environmental Change & Security Project Report*, Issue No. 8 (Summer): 95–108.

Dalby, Simon, 2002b: *Environmental Security* (Minneapolis: University of Minneapolis Press).

Dalby, Simon, 2008: "Security and Environment Linkages Revisited", in: Brauch, Hans Günter; Oswald Spring, Úrsula; Mesjasz, Czeslaw; Grin, John; Dunay, Pal; Behera, Navnita Chadha; Chourou, Béchir; Kameri-Mbote, Patricia; Liotta, P.H. (Eds.): *Globalization and Environmental Challenges: Reconceptualizing Security in the 21st Century* (Berlin – Heidelberg: Springer-Verlag): 165–172.

Dalby, Simon, 2009: *Security and Environmental Change* (Cambridge: Polity).

Dalby, Simon. 2013: "Peace in the Anthropocene" in: *Peace Review*, 25,4 (October):

Dalby, Simon, 2016: "Framing the Anthropocene: The Good, the Bad and the Ugly", in: *Anthropocene Review*, 3,1: 33–51.

Dalby, Simon, 2020: *Anthropocene Geopolitics*: *Globalization, Security, Sustainability Politics and Public Policy* (Ottawa: University of Ottawa Press)

Dalby, Simon; Brauch, Hans Günter; Oswald Spring, Úrsula, 2009: "Environmental Security Concepts Revisited During the First Three Phases (1983–2006)", in: Brauch, Hans Günter; Oswald Spring, Úrsula; Grin, John; Mesjasz, Czeslaw; Kameri-Mbote, Patricia; Behera, Navnita Chadha; Chourou, Béchir; Krummenacher, Heinz (Eds.), 2009: *Facing Global Environmental Change: Environmental, Human, Energy, Food, Health and Water Security Concepts* (Berlin – Heidelberg – New York: Springer-Verlag): 781–790.

Darbellay, Frédéric, 2015: "Rethinking Inter- and Transdisciplinarity: Undisciplined Knowledge and the Emergence of a New Thought Style", in: *Futures*, 65: 163–174.

Davis, Peter, 1999: "Ecomuseums: A Sense of Place", in: Pearce, Susan M. (Ed.): *Leicester Museum Studies* (New York: Leicester University Press).

d'Eaubonne. Françoise, 1974: *Le Féminisme ou la Mort* (Paris: Pierre Horay).

Desfor, Gene; Keil, Roger, 2004: *Nature and the City: Making Environmental Policy in Toronto and Los Angeles* (Tucson, AZ: The University of Arizona Press): 231.

Detges, Adrien, 2014: "Close-up on Renewable Resources and Armed Conflict: The Spatial Logic of Pastoralist Violence in Northern Kenya", in: *Political Geography*, 42: 57–65; at: https://doi.org/10.1016/j.polgeo.2014.06.003.

Detges, Adrien, 2016: "Local Conditions of Drought-Related Violence in Sub-Saharan Africa: The Role of Road and Water Infrastructures", in: *Journal of Peace Research*, 53,5: 696–710; at: https://doi.org/10.1177/0022343316651922.

Detges, Adrien, 2017: "Droughts, State-Citizen Relations and Support for Political Violence in Sub-Saharan Africa: A Micro-Level Analysis", in: *Political Geography*, 61: 88–98; at: https://doi.org/https://doi.org/10.1016/j.polgeo.2017.07.005.

Detges, Adrien, 2017a: *Climate and Conflict: Reviewing the Statistical Evidence: A Summary for Policy-Makers* (Berlin: Adelphi).

Detges, Adrien, 2018: *Drought, Infrastructure and Conflict Risk in Sub-Saharan Africa* (PhD dissertation, Free University of Berlin).

Detraz, Nicole; Betsill, Michele M., 2009: "Climate Change and Environmental Security: For Whom the Discourse Shifts", in: *International Studies Perspectives,* 10,3 (August): 303–320; at: https://www.jstor.org/stable/44218603 (2 August 2020).

de Wilde, Jaap H., 2008: "Environmental Security Deconstructed", in: Brauch, Hans Günter et al. (Eds.), 2008: *Globalization and Environmental Challenges: Reconceptualizing Security in the 21st Century* (Berlin et al.: Springer-Verlag): 595–602.
DFID, 2014: Topic Guide: Conflict, Climate and Environment; Security Council, in Statement, Says 'Contextual Information' on Possible Security Implications of Climate Change Important When Climate Impacts Drive Conflict, 20 July 2011.
di Chiro, Giovanna, 2017: "Welcome to the White (M)Anthropocene? A Feminist-Environmental Critique", in: MacGregor, Sherilyn (Ed.): *Routledge Handbook of Gender and Environment* (London: Routledge): 487–505.
Diehl, Paul F.; Gleditsch, Nils Petter (Eds.), 2001: *Environmental Conflict* (Boulder – Oxford: Westview).
Diehl, Paul F.; Gleditsch, Nils Petter, 2001a: "Controversies and Questions", in: Diehl, Paul F.; Gleditsch, Nils Petter (Eds.): *Environmental Conflict* (Boulder – Oxford: Westview): 1–9.
Dombrowsy, Ines, 2003: "Water Accords in the Middle East Peace Process: Moving towards Cooperation?", in: Brauch, Hans Günter; Liotta, P.H.; Marquina, Antonio; Rogers, Paul; Selim, Mohammed El-Sayed (Eds.): *Security and Environment in the Mediterranean. Conceptualising Security and Environmental Conflicts.* Hexagon Series on Human and Environmental Security and Peace, vol. 1 (Berlin – Heidelberg – New York: Springer): 729–744.
Douglas, Ian, 2002: "Human Disturbance of the Earth System: Dynamics and Complexities", in: Douglas, Ian (Ed.): *Encyclopedia of Global Environmental Change*, vol. 3: *Causes and Consequences of Global Environmental Change* (Chichester: John Wiley): 1–11.
Dryzek, John S.; Pickering, Jonathan, 2019: *The Politics of the Anthropocene* (Oxford: Oxford University Press).
Dunn, David J., 2004: *From Power Politics to Conflict Resolution: The Work of John W Burton*, (London: Palgrave Macmillan).
Dupont, Claire, 2019: "The EU's Collective Securitisation of Climate Change", in: *West European Politics*, 42,2.
Ecopeace [Kool, Jeroen], 2016: *Sustainable Development in the Jordan River Valley: Final Report of the Regional NGO Master Plan.* Hexagon Series on Human and Environmental Security and Peace, vol. 13 (Berlin – Heidelberg – New York: Springer).
Egerton, Frank N., 2012: *Roots of Ecology: Antiquity to Haeckel* (Berkeley: University of California Press).
Ehlers, Eckart, 2008: *Das Anthropozän: Die Erde im Zeitalter des Menschen* (Darmstadt: Wissenschaftliche Buchgemeinschaft).
Elwert, Georg, 1990: "Markets of Violence. The Violence Trade and the Goal Rationality of Violence", in: *Law and State*, 59/60: 40–60.
Elwert, Georg, 1999: "Markets of Violence", in: Elwert, Georg; Feuchtwang, Stephan; Neubert, Dieter (Eds.): *Dynamics of Violence: Processes of Escalation and De-Escalation in Violent Group Conflicts: Sociologus Supplement 1* (Berlin: Duncker & Humblot): 85–102.
Elwert, Georg, 2003: "The Socio-Anthropological Interpretation of Violence", in: Heitmeyer, Wilhelm; Hagan, John (Eds.): *International Handbook on Violence Research* (Boston –London – Dordrecht: Kluwer Academic Publishers).
Ellen, R.F., 1996: "Ecology", in: Kuper, Adam; Kuper; Jessica (Eds.): *The Social Science Encyclopedia* (London – New York: Routledge): 207–209.
Enrique Mazlum, Davis, 2004: "Hidrodiplomacia y la Funci6n de la COCEF y BDAN en el Desarrollo Sustentable en la Zona Fronteriza de México y Estados Unidos" (Master's thesis, Monterrey Institute of Technology and Higher Education (ITESM), Mexico).
EU [European Commission; Council], 2008: *Climate Change and International Security*. Doc 7249/08 (Brussels: European Commission, 14 March).
EU, 2008a: *Report on the Implementation of the European Security Strategy – Providing Security in a Changing World*, S407/08 (Brussels: European Council, 11 December).
European Commission, 2010: *Roadmap to a Resource Efficient Europe (Impact Assessment Roadmap)* (Brussels: European Commission).

European Commission, 2011a: *Roadmap to a Resource Efficient Europe* (Brussels: European Commission).
European Commission, 2011b: *A Roadmap for Moving to a Competitive Low Carbon Economy in 2050* (Brussels: European Commission).
European Commission, 2011c: *Analysis associated with the Roadmap to a Resource Efficient Europe: Part I* (Brussels: European Commission).
European Commission, 2011d: *Energy Roadmap 2050* (Brussels: European Commission).
Falcon-Lang, H.: "Anthropocene: How Humans Created a New Geological Age"; at: *BBC website*, 11 May.
Floyd, Rita, 2008: "The Environmental Security Debate and its Significance for Climate Change", in: *The International Spectator*, 43,3.
Freire, Paulo, 1975: *Conscientization* (Geneva: World Council of Churches).
Frowein, Jochen Abr., 1994: "Chapter VII: Actions with Respect to Threats to the Peace, Breaches to the Peace, and Acts of Aggression, Art. 39", in: Simma, Bruno (Ed.): *The Charter of the United Nations: A Commentary* (Oxford: Oxford University Press): 605–616.
Funtowicz, S.; Ravetz, J.R., 1993: "Science for the Post-Normal Age", in: *Future,* 25: 735–755.
Funtowicz, S.; Ravetz, J.R., 2003: "Post-Normal Science", in: International Society for Ecological Economics, *Internet Encyclopaedia of Ecological Economics*; at: http://isecoeco.org/pdf/pstnormsc.pdf (6 February 2016).
Gaddis, John Lewis, 1986: "The Long Peace: Elements of Stability in the Postwar International System", in: *International Security*, 10,4 (Spring): 99–142.
Gaddis, John Lewis, 1987, 1989: *The Long Peace: Inquiries into the History of the Cold War* (New York – Oxford: Oxford University Press).
Gaddis, John Lewis, 1992: *The United States and the End of the Cold War* (New York – Oxford: Oxford University Press).
Galtung, Johan, 1967: "Peace Research: Science, or Politics in Disguise?", in: *International Spectator,* 21,19: 1573–1603.
Galtung, Johan, 1968: "Peace", in: *International Encyclopedia of the Social Sciences* (London – New York: Macmillan): 487–496.
Galtung, Johan, 1969: "Violence, Peace and Peace Research", in: *Journal of Peace Research*, 3: 167–191.
Galtung, Johan, 1982: *Environment, Development and Military Activity: Towards Alternative Security Doctrines* (Oslo: Norwegian University Press).
Galtung, Johan, 1985: "Twenty-Five Years of Peace Research: Ten Challenges and Some Responses", in: *Journal of Peace Research*, 22,2: 141–158.
Galtung, Johan, 1988 [1987]: "What is Meant by Peace and Security? Some Options for the 1990s", in: Galtung, Johan (Ed.): *Transarmament and the Cold War: Peace Research and the Peace Movement*, Essays in Peace Research, vol. 6 (Copenhagen: Christian Ejlers): 61–71.
Galtung, Johan, 1993: "Peace", in: Krieger, Joel (Ed.): *The Oxford Companion to Politics of the World* (Oxford: Oxford University Press): 688–689.
Galtung, Johan, 2007: "A Mini Theory of Peace"; at: http://www.oldsite.transnational.org/Resources_Treasures/2007/Galtung_MiniTheory.html.
Galtung, Johan, 2012: "Positive and Negative Peace", in: Webel, Charles P.; Johansen, Jørgen (Eds.): *Peace and Conflict Studies: A Reader* (New York: Routledge): 75–78.
Galtung, Johan; Fischer, Dietrich, 2013: *Pioneer of Peace Research* (Heidelberg – New York – Dordrecht – London: Springer).
Galtung, Johan, 2014: "A Mini Theory of Peace"; at: https://www.galtung-institut.de/wp-content/uploads/2014/11/Mini-Theory-of-Peace.pdf.
GECHS [Global Environmental Change and Human Security], 1999: *GECHS Science Plan.* IHDP Report No. 11 (Bonn: IHDP).
GECHS, 2005: *Annual Report 2004/2005* (Oslo: GECHS).
Geels, Frank W., 2002: "Technological Transitions as Evolutionary Reconfiguration Processes: A Multi-Level Perspective and a Case Study", in: *Research Policy*, 31, 8–9: 1257–1274.

George, Alexander L., 1979: "The Causal Nexus between Cognitive Beliefs and Decision Making Behavior: The 'Operational Code'", in: Falkowski, L. (Ed.): *Psychological Models in International Politics* (Boulder: Westview): 95–124.

George, Alexander L., 1988: "Case Studies and Theory Development: the Method of Structured, Focused Comparison", in: Lauren, P.G. (Ed.): History: *New Approaches* (New York: Free Press).

German Advisory Council on Global Change (WBGU), 2011: *Social Contract for Sustainability* (Berlin: German Advisory Council on Global Change, July 2011).

Gilmore, Elisabeth A.; Herzer Risi, Lauren; Tennant, Elizabeth; Buhaug, Halvard, 2018: "Bridging Research and Policy on Climate Change and Conflict", in: *Current Climate Change Reports* 4,3: 313–319.

Gleditsch, Nils Peter (Ed.), 1997: *Conflicts and the Environment* (Dordrecht – Boston – London: Kluwer Academic Publisher).

Gleditsch, Nils Petter, 1998: "Armed Conflict and the Environment: A Critique of the Literature" in: *Journal of Peace Research*, 35,3 (May): 381–400.

Gleditsch, Nils Petter, 2001: "Armed Conflict and the Environment: A Critique of the Literature", in: Diehl, Paul F.; Gleditsch, Nils Petter (Eds.): *Environmental Conflict* (Boulder, CO: Westview): 251–272.

Gleditsch, Nils Petter, 2003: "Environmental Conflict: Neomalthusians vs. Cornucopians", in: Brauch, Hans Günter; Liotta, P.H.; Marquina, Antonio; Rogers, Paul; Selim, Mohammed El-Sayed (Eds.): *Security and Environment in the Mediterranean: Conceptualising Security and Environmental Conflicts* (Berlin – Heidelberg: Springer 2003): 477–485.

Gleditsch, Nils Petter, 2012: "Whither the Weather? Climate Change and Conflict", in: *Journal of Peace Research*, 49,1: 3–9.

Gleditsch, Nils Petter (Ed.), 2020: *Lewis Fry Richardson: His Intellectual Legacy and Influence in the Social Sciences* (Cham: Springer International Publishing).

Gleditsch, Nils Petter; Urdal, Henrik, 2002: "Ecoviolence? Links between Population Growth, Environmental Scarcity and Violent Conflict in Thomas Homer-Dixon Work", in: *Journal of International Affairs*, 56,1: 283–302.

Gleick, Peter H., 1989: "Climate Change and International Politics: Problems Facing Developing Countries", in: *Ambio*, 18,6: 333–339.

Gleick, Peter H., 1989a: "The Implications of Global Climate Changes for International Security", in: *Climate Change*, 15: 303–325.

Gleick, Peter H., 1993a: "Water and Conflict: Fresh Water Resources and International Security", in: *International Security*, 18,1: 79–112.

Gleick, Peter H., 1993b: *Water in Crisis: A Guide to the World's Fresh Water Resources* (New York: Oxford University Press).

Gleick, Peter H., 1994: "Water, War and Peace in the Middle East", in: *Environment*, 36,3: 6–15.

Gleick, Peter H., 1998: *The World's Water: The Biannual Report on Fresh Water Resources* (Washington, DC – Covelo, CA: Island Press).

Gleick, Peter H., 2000: *The World's Water 2000–2001: The Biennial Report on Freshwater Resources* (Washington, DC: Island Press).

Göpel, Maja, 2017: *The Great Mindshift: Why We Need a New Economic Paradigm for Sustainability Transformations* (Cham – New York – Heidelberg – Dordrecht – London: Springer).

Goetschel, Laurent; Hellmüller, Sara, 2019: "Friedens- und Konfliktforschung in der Schweiz: ein facettenreiches Patchwork", in: *Zeitschrift für Friedens- und Konfliktforschung*, 8: 79–98.

Gorbachev, Mikhail, 1988: "The Problem of Mankind's Survival", speech before the United Nations General Assembly (New York: UN, 8 December).

Grin, John; Rotmans, Jan; Schot, Johan, 2010: *Transitions to Sustainable Development: New Directions in the Study of Long Term Transformative Change* (New York – London: Routledge).

Gudynas, E., 1999: *La Reforma del Estado para la Sustentabilidad del Desarrollo* (Montevideo, Uruguay: Centro Latino Americano de Ecología Social [CLAES]).

Gumbel, Emil Julius, 1958: *Statistics of Extremes* (New York: Columbia University Press).

Güvenen, Orhan, 2016: "Transdisciplinary Science Methodology as a Necessary Condition in Research and Education", in: *Transdisciplinary Journal of Engineering & Science*, 7: 69–78.

Haavisto, Pekka, 2003: "Environmental Post-Conflict Assessments: A New UN Tool Developed by UNEP", in: Brauch, Hans Günter; Liotta, P.H.; Marquina, Antonio; Rogers, Paul F.; Selim, Mohammad El-Sayed (Eds.): *Security and Environment in the Mediterranean: Conceptualising Security and Environmental Conflicts* (Berlin – Heidelberg – New York – Hong Kong – London – Milan – Paris – Tokyo: Springer-Verlag): 535–562.

Hacking, Ian, 1999: *The Social Construction of What?* (Cambridge, MA: Harvard University Press).

Hahn, Otto; Strassmann, Fritz, 1939: "Über den Nachweis und das Verhalten der bei der Bestrahlung des Urans mittels Neutronen entstehenden Erdalkalimetalle", in: *Naturwissenschaften*, 27,1: 11–15

Haken, Hermann, ^{3}1983: *Synergetics: An Introduction: Nonequilibrium Phase Transition and Self-Organization in Physics, Chemistry, and Biology* (Berlin – Heidelberg: Springer-Verlag).

Hammond, A., 1998: *Which World? Scenarios for the 21st Century* (Washington, DC: Island Press).

Hamner, J.; Wolf, Aaron T., 1998: "Patterns in International Water Resource Treaties: The Transboundary Freshwater Dispute Database", in: *Colorado Journal of International Environmental Law and Policy 1997 Yearbook.*

Happaerts, Sander, 2017: "Discourse and Practice of Transitions in International Policy-Making on Resource Efficiency in the EU": in: Brauch, Hans Günter: Oswald Spring, Úrsula; Grin, John; Scheffran, Jürgen (Eds.): *Handbook on Sustainability Transition and Sustainable Peace* (Cham – Heidelberg – New York – Dordrecht – London: Springer): 869–884.

Haraway, Donna J., 2016: "Staying with Trouble: Anthropocene, Capitalocene, Chthulucene", in: Moore, Jason M. (Ed.): *Anthropocene or Capitalocene? Nature, History, and the Crisis of Capitalism* (Oakland, CA: PM Press).

Harrington, Cameron; Shearing, Clifford, 2017: *Security in the Anthropocene – Reflections on Safety and Care* (Bielefeld: Transcript).

Hartley, Daniel, 2016: "Anthropocene, Capitalocene, and the Problem of Culture", in: Moore, Jason M. (Ed.): *Anthropocene or Capitalocene? Nature, History, and the Crisis of Capitalism* (Oakland, CA: PM Press): 154–165;

Hartmann, Betsy, 2001: "Will the Circle be Unbroken? A Critique of the Project on Environment, Population and Security", in: Peluso, Nancy Lee.; Watts Michael (Eds.): *Violent Environments* (Ithaca – London: Cornell University Press): 39–62.

Hauge, Wenche; Ellingsen, Tanja, 1998: "Beyond Environmental Scarcity: Causal Pathways to Conflict", in: *Journal of Peace Research*, 35,3 (May): 299–317.

Hickmann, Thomas; Partsch, Lena; Pattberg, Philipp; Weiland, Sabine (Eds.), 2019: *The Anthropocene Debate and Political Science* (Abingdon – New York: Routledge).

Hirsch Hadorn, Gertrude; Hoffmann-Riem, Holger; Biber-Klemm, Susette; Grossenbacher-Mansuy, Walter; Joye, Dominique; Pohl, Christian; Wiesmann, Urs; Zemp, Elisabeth (Eds.), 2008: *Handbook of Transdisciplinary Research* (Berlin – Heidelberg: Springer).

Holsti, Kalevi J., 1991, 1992: *Peace and War: Arms Conflicts and International Order 1648-1989* (Cambridge: Cambridge University Press).

Holsti, Kalevi, 2016: *Kalevi Holsti: A Pioneer in International Relations Theory, Foreign Policy Analysis, History of International Order, and Security Studies* (Cham – Heidelberg – New York – Dordrecht – London: Springer, 2015).

Holsti, Kalevi J., 2016a: *Kalevi Holsti: Major Texts on War, Peace, and International Order* (Cham – Heidelberg – New York – Dordrecht – London: Springer, 2015).

Homer-Dixon, Thomas F., 1991: "On the Threshold: Environmental Changes as Causes of Acute Conflict", in: *International Security*, 16,2 (Fall): 76–116.

Homer-Dixon, Thomas F., 1994: "Environmental Scarcities and Violent Conflict: Evidence From Cases," in: *International Security*, 19,1 (Summer): 5–40.

Homer-Dixon, Thomas F., 1995: "Correspondence: Environment and Security", in: *International Security*, 20,3: 189–194.

Homer-Dixon, Thomas F., 1996: "Strategies for Studying Causation in Complex Ecological-Political Systems", in: *The Journal of Environment and Development,* 5,2 (June): 132–48.
Homer-Dixon, Thomas F., 1999: *Environment, Scarcity, and Violence* (Princeton, NJ: Princeton University Press).
Homer-Dixon, Thomas F., 2000, 2001: *The Ingenuity Gap: How Can We Solve the Problems of the Future?* (New York – Toronto: Alfred A. Knopf).
Homer-Dixon, Thomas F.; Blitt, Jessica (Eds.), 1998: *Ecoviolence: Links among Environment, Population, and Security* (Lanham, Md.: Rowman and Littlefield).
Homer-Dixon, Thomas F.; Deligiannis, Tom, 2009: "Environmental Scarcities and Civil Violence", in: Brauch, Hans Günter; Oswald Spring, Úrsula; Grin, John; Mesjasz, Czeslaw; Kameri-Mbote, Patricia; Behera, Navnita Chadha; Chourou, Béchir; Krummenacher, Heinz (Eds.): *Facing Global Environmental Change: Environmental, Human, Energy, Food, Health and Water Security Concepts* (Berlin – Heidelberg – New York: Springer-Verlag): 309–324.
Hornborg, Alf, 2015: "The Political Ecology of the Technocene", in: Hamilton, Clive; Gemenne, Francois (Eds.): *The Anthropocene and Global Environmental Crisis* (New York: Routledge).
Hsiang, Solomon M.; Burke, Marshall; Miguel, Edward, 2013: "Quantifying the Influence of Climate on Human Conflict", in: *Science*, 341,6151: 1–17,
Huber, Wolfgang, II, 1972: "Friedensforschung", in: Ritter, Joachim (Ed.): *Historisches Wörterbuch der Philosophie* (Basel – Darmstadt: Wissenschaftliche Buchgesellschaft): 1,119–1,122.
Huggett, Richard John, 1995: *Geoecology. An Evolutionary Approach* (London – New York: Routledge).
Huggett, Richard, 2000: "Geoecology", in: Hancock, Paul; Skinner, Brian J. (Eds.): *The Oxford Companion to the Earth* (Oxford: Oxford University Press).
Huggett, Richard John, 2010: *Physical Geography: The Key Concepts* (London: Routledge).
Hutchinson, Francis P., 2009: "Reflections on the Connections between Peace Education, Environmental Education, Futures Education" (March).
Ide, Tobias, 2017: "Research Methods for Exploring the Links between Climate Change and Conflict", in: *Wiley Interdisciplinary Reviews: Climate Change*, 8,3.
Ide, Tobias; Link, Peter Michael; Scheffran, Jürgen; Schilling, Janpeter, 2016: "The Climate-Conflict Nexus: Pathways, Regional Links, and Case Studies", in: Brauch, Hans Günter; Oswald Spring, Úrsula; Grin, John; Scheffran, Jürgen (Eds.): *Handbook on Sustainability Transition and Sustainable Peace* (Cham – Heidelberg – New York – Dordrecht – London: Springer): 285–304.
Ide, Tobias; Scheffran, Jürgen, 2014: "On Climate, Conflict and Cumulation: Suggestions for Integrative Cumulation of Knowledge in the Research on Climate Change and Violent Conflict", in: *Global Change, Peace & Security*, 26,3: 263–279.
IFRC, 2002: "Assessing Vulnerabilities and Capacities – During Peace and War", in: IFRC (Ed.): *World Disasters Report 2002: Focus on Reducing Risk* (Geneva: IFRC): 128–147.
IFRC, 2005: *World Disasters Report 2005: Focus on Information in Disasters* (Bloomfield, CT: Kumarian – London: Eurospan).
Ingegnoli, Vittorio (Ed.), 2002: *Landscape Ecology: A Widening Foundation* (Berlin – Heidelberg: Springer-Verlag).
International Social Science Council (ISSC), 2012: *Transformative Cornerstones of Social Science Research for Global Change* (Paris: ISSC).
IPCC, 1990: *Climate Change: The IPCC Impacts Assessment* (Geneva: WMO; UNEP; IPCC).
IPCC, 1996: *Climate Change 1995: The Science of Climate Change: Contributions of Working Group I to the Second Assessment Report of the Intergovernmental Panel on Climate Change* (Cambridge: Cambridge University Press).
IPCC, 1997: *Special Report: The Regional Impacts of Climate Change: An Assessment of Vulnerability* (Cambridge: Cambridge University Press, March).
IPCC, 2001: *Climate Change 2001: The Scientific Basis* (Cambridge – New York: Cambridge University Press).

IPCC, 2007: *Climate Change 2007: The Physical Science Basis: Working Group I Contribution to the Fourth Assessment Report of the IPCC* (Cambridge – New York: Cambridge University Press).

IPCC, 2011: *Special Report: Renewable Energy Sources and Climate Change Mitigation* (Cambridge: Cambridge University Press, April).

IPCC, 2012: *Managing the Risks of Extreme Events and Disasters to Advance Climate Change Adaptation* (Cambridge: Cambridge University Press, April).

IPCC, 2013: *Climate Change 2013: The Physical Science Basis: Working Group I Contribution to the Fifth Assessment Report of the Intergovernmental Panel on Climate Change* (Cambridge – New York: Cambridge University Press); at: https://www.ipcc.ch/report/ar5/wg1/.

IPCC, 2014: *Climate Change 2014: Synthesis Report: Summary for Policymakers* (Geneva: IPCC).

IPCC, 2018: *Global Warming of 1.5°C* (Cambridge: Cambridge University Press, October).

IPCC, 2019: *The Ocean and Cryosphere in a Changing Climate* (Cambridge: Cambridge University Press, September).

IPCC, 2019: *Climate Change and Land* (Cambridge: Cambridge University Press, August).

Isard, Walter, 2000: "Formative and Early Years of the Peace Science Society (International)", in: *Conflict Management and Peace Science,* 18,1: 1–48.

Jaeger, Jochen; Scheringer, Martin, 1998: "Transdisziplinarität: Problemorientierung ohne Methodenzwang", in: *GAIA*, 7,1: 10–25.

Janssen, Wilhelm, 1975: "Friede", in: Brunner, Otto; Conze, Werner; Koselleck, Reinhart (Eds.): *Geschichtliche Grundbegriffe*, vol. 2 (Stuttgart: Klett-Cotta): 543–591.

Jensen, David; Lonergan, Stephen, 2012: *Assessing and Restoring Natural Resources in Post-Conflict Peacebuilding* (London: Routledge).

Johnson, Twig, 2016: "Policy, Politics and the Impact of Transition Studies", in: Brauch, Hans Günter; Oswald Spring, Úrsula; Grin, John; Scheffran, Jürgen (Eds.): *Handbook on Sustainability Transition and Sustainable Peace* (Cham – Heidelberg – New York – Dordrecht – London: Springer): 481–492.

Kaldor, Mary, 1999: *New and Old Wars: Organized Violence in a Global Era* (Cambridge: Polity; Stanford: Stanford University Press).

Kaldor, Mary, 2002: *New and Old Wars: Organized Violence in a Global Era* (Cambridge: Polity).

Kaldor, Mary; Vashee, Basker (Eds.), 1997: *New Wars* (London: Pinter).

Kaplan, Robert, 1994: "The Coming Anarchy", in: *The Atlantic Monthly*, 273,2: 44–76.

Kaplan, Robert D., 1996: *The Ends of the Earth: A Journey at the Dawn of the 21st Century* (New York: Random House).

Kaplan, Robert D., 2000: *The Coming Anarchy: Shattering the Dreams of the Post Cold War* (New York: Random House).

Kaplan, Robert, 2009: "The Revenge of Geography", in: *Foreign Policy*, 172 (May/June): 96–105.

Kates, R.W., 2001: "Queries on the Human Use of the Earth", in: *Annual Review of Energy and the Environment*, 26: 1–26.

Kawashima, Yasuko; Akino, Sayaka, 2001: "Climate Change and Security: Regional Conflict as a New Dimension of Impact of the Climate Change", in: *Global Environmental Research*, 5,1: 33–43.

Kelly, Duncan, 2019: *Politics and the Anthropocene* (Cambridge: Polity Press).

Khalil, Elias L., 1996: "Kenneth Boulding: Ecodynamicist or Evolutionary Economist?", in: *Journal of Post Keynesian Economics,* 19,1 (Autumn): 83–100.

Kimmenich, Otto, 1972, II: "Friede", in: Ritter, Joachim (Ed.): *Historisches Wörterbuch der Philosophie* (Basel – Darmstadt: Wissenschaftliche Buchgesellschaft): 1,115–1,119.

Koselleck, Reinhart, 2006: *Begriffsgeschichten: Studien zur Semantik und Pragmatik der politischen und sozialen Sprache* (Frankfurt am Main: Suhrkamp).

Kotzé, Louis, 2018: "Reflections on the Future of Environmental Law Scholarship and Methodology in the Anthropocene", in: Pedersen, O. (Ed.): *Perspectives on Environmental Law Scholarship: Essays on Purpose, Shape and Direction* (Cambridge: Cambridge University Press): 140–161.

Kurtz, Gerritt, 2012: "Securitization of Climate Change in the United Nations 2007–2010", in: Scheffran, Jürgen; Brozska, Michael; Brauch, Hans Günter; Link, Peter Michael; Schilling, Janpeter (Eds.): *Climate Change, Human Security and Violent Conflict* (Berlin – Heidelberg: Springer-Verlag): 669–684.

Kyrou, Christos, 2007: "Peace Ecology: An Emerging Paradigm in Peace Studies", in: *The International Journal of Peace Studies*, 12,2.

Kyrou, Christos, 2007a: "Peace Ecology: Theoretical and Methodological Reflections", Paper presented at conference on Cutting Edge Theories and Recent Developments in Conflict Resolution, Syracuse, NY, 27–28 September.

Kyrou, Christos, 2007b: "Peace Ecology: Evaluative and Methodological Reflections", Paper presented at the annual meeting of the ISA 48th Annual Convention, Chicago, 28 February 2007.

Laferrière, Eric; Stoett, Peter J., 1999: *International Relations Theory and Ecological Thought: Towards a Synthesis* (London: Routledge).

Lagarde, Marcela, 1992: *Los Cautiverios de las Mujeres: Madresposas, Monjas, Putas, Presas y Locas,* Coordinación General de Estudios de Posgrado (Mexico City: UNAM).

Langenscheidt-Longman, 1995: *Longman Dictionary of Contemporary English* (Munich: Langenscheidt-Longman).

Le Billon, Philippe; Duffy, Rosaleen, 2018, 2019: "Conflict Ecologies: Connecting Political Ecology and Peace and Conflict Studies", in: *Journal of Political Ecology*, 24: 239–260.

Le Roy E., 1927: L'Exigence Idealiste et le Fait de l'Evolution [Idealistic Exigency and the Feat of Evolution] *(Paris: Boivin).*

Lenton, Timothy; Held, Hermann; Kriegler, Elmar; Hall, Jim W.; Lucht, Wolfgang; Ramstorf, Stefan; Schellnhuber, Hans Joachim, 2008: "Tipping Elements in the Earth's Climate System", in: *Proceedings of the National Academy of Science* (PNAS), 105,6 (12 February): 1,786–1,793.

Lenton, Timothy M.; Rockström, Johan; Gaffney, Owen; Rahmstorf, Stefan; Richardson, Katherine; Steffen, Will; Schellnhuber, Hans Joachim, 2019: "Climate Tipping Points – Too Risky to Bet Against: The Growing Threat of Abrupt and Irreversible Climate Changes Must Compel Political and Economic Action on Emissions", in: *Nature*, 575 (27 November): 592–595.

Leopold, Aldo, 1949: *A Sand County Almanac* (Oxford: Oxford University Press).

Levy, Marc A., 1995: "Is the Environment a National Security Issue?", in: *International Security,* 20,2: 35–62.

Lewis, Simon L.; Maslin, Mark A., 2015: "Defining the Anthropocene", in: *Nature*, 519, 7542: 171–180.

Lewis, Simon L.; Maslin, Mark A., 2018: *The Human Planet: How We Created the Anthropocene* (New Haven: Yale University Press).

Libiszewski, Stephan, 1992: *What is an Environmental Conflict? Environment and Conflicts Project:* ENCOP Occasional Paper No. 1, 2, 6 (Zürich: Swiss Federal Institute of Technology [ETH], Center for Security Studies and Conflict Research; Bern: Swiss Peace Foundation).

Libiszewski, Stephan, 1995: *Water Disputes in the Jordan Basin Region and their Role in the Resolution of the Arab-Israeli Conflict:* ENCOP Occasional Paper No. 13 (Zürich: ETH Zürich; Bern: Swiss Peace Foundation).

Libiszewski, Stephan, 1996: "Water Disputes in the Jordan Basin and their Role in the Resolution of the Arab-Israeli Conflict", in: Bächler, Günther; Spillmann, Kurt R. (Eds.): *Kriegsursache Umweltzerstörung: Environmental Degradation as a Cause of War, vol. 2: Regional and Country Studies of Research Fellows* (Chur – Zürich: Rüegger): 337–460

Libiszewski, Stephan, 1996a: "Internationale Konflikte um die Nutzung fliessender Gewässer", in: Baechler, Günther; Böge, Volker; Klötzli, Stefan; Libiszewski, Stephan; Spillmann, Kurt R. (Eds.): *Kriegsursache Umweltzerstörung – Ökologische Konflikte in der Dritten Welt und Wege ihrer friedlichen Bearbeitung:* ENCOP Final Report, vol. 1 (Chur – Zürich: Rüegger): 117–166.

Libiszewski, Stephan, 1997: "Integrating Political and Technical Approaches: Lessons from the Israeli-Jordanian Water Negotiations", in: Gleditsch, Nils-Petter (Ed.): *Conflict and the Environment* (Dordrecht – Boston – London): 385–402.

Lietzmann, Kurt M.; Vest, Gary D., 1999: *Environment & Security in an International Context* (Brussels: NATO, CCMS, March).
Loftus, Alex, 2018: "Political Ecology II: Whither the State?", in: *Progress in Human Geography*.
Lonergan, Steve, 2002: "Environmental Security", in: Timmerman, Peter (Ed.): *Encyclopedia of Global Environmental Change*, vol. 5: *Social and Economic Dimensions of Global Environmental Change* (Chichester: John Wiley): 269–278.
Lovelock, James E.; Appleyard, Bryan, 2019: *Novacene. The Coming Age of Hyperintelligence* (London: Penguin, Allen Lane).
Luhmann, Niklas, 1991: *Soziale Systeme* (Frankfurt am Main: Suhrkamp).
Lujala, Päivi; Rustad, Siri Aas, 2011: High-Value Natural Resources and Post-Conflict Peace-building (London: Routledge),
Markl, Hubert, 1986: *Natur als Kulturaufgabe: Über die Beziehung des Menschen zur lebendigen Natur* (Stuttgart: Deutsche Verlagsanstalt).
Marsh, George Perkins, 1864: *Man and Nature* [reprinted in 1965 as]: *The Earth as Modified by Human Action (*Cambridge, Massachusetts: Belknap Press).
Matthew, Richard A.; Brown, Oli; Jensen, David, 2009: *From Conflict to Peacebuilding: The Role of Natural Resources and the Environment* (Nairobi: UNEP).
Matthew, Richard A.; Gaulin, T., McDonald, B., 2003: "The Elusive Quest: Linking Environmental Change and Conflict", in: *Canadian Journal of Political Science,* 36,4: 857–878.
Matthew, Richard A.; McDonald, Bryan, 2009: "Environmental Security: Academic and Policy Debates in North America", in: Brauch, Hans Günter; Oswald Spring, Úrsula; Grin, John; Mesjasz, Czeslaw; Kameri-Mbote, Patricia; Behera, Navnita Chadha; Chourou, Béchir; Krummenacher, Heinz (Eds.), 2009: *Facing Global Environmental Change: Environmental, Human, Energy, Food, Health and Water Security Concepts* (Berlin – Heidelberg – New York: Springer-Verlag): 791–802.
Mathews, Jessica Tuchman, 1989: "Redefining Security", in: *Foreign Affairs,* 68,2 (Spring): 162–177.
McBrien, Justine, 2016: "Accumulating Extinction: Planetary Catastrophism in the Necrocene", in: Moore, Jason M. (Ed.): *Anthropocene or Capitalocene? Nature, History, and the Crisis of Capitalism* (Oakland, CA: PM Press): 78–115.
McCandless, Erin; Bangura, Abdul Karim [King, Mary E.; Sall, Ebrima (Eds.)], 2007: *Peace Research for Africa: Critical Essays on Methodology* (Addis Ababa: University for Peace, Africa Programme).
McDonald, Matt, 2013: "Discourses of Climate Security", in: *Political Geography*, 33 (March): 42–51.
McIntosh, Robert P., 1986: *The Background of Ecology: Concept and Theory* (Cambridge: Cambridge University Press).
McNeill, John R. 1992. *The Mountains of the Mediterranean World: An Environmental History* (New York: Cambridge University Press).
McNeill, John R., 2000: *Something New Under the Sun: An Environmental History of the 20th-Century World* (New York: Norton).
McNeill, John R., 2009: "The International System, Great Powers, and Environmental Change since 1900", in: Brauch, Hans Günter; Oswald Spring, Úrsula; Grin, John; Mesjasz, Czeslaw; Kameri-Mbote, Patricia; Behera, Navnita Chadha; Chourou, Béchir; Krummenacher, Heinz (Eds.), 2009: *Facing Global Environmental Change: Environmental, Human, Energy, Food, Health and Water Security Concepts* (Berlin – Heidelberg – New York: Springer-Verlag): 43–52.
McNeill, John R., Engelke, Peter, 2014: *The Great Acceleration; An Environmental History of the Anthropocene since 1945* (Cambridge, MA – London: The Belknap Press).
McNeill. J.R.; Unger, Corinna R., 2010: *Environmental Histories of the Cold War* (New York: Cambridge University Press).
MEA, 2005: *Ecosystems and Human Well-being: Synthesis Report* (Washington, DC: Island Press).
MEA, 2005a: *Ecosystems and Human Well-being: Desertification Synthesis* (Washington, DC: World Resources Institute).

Mendelsohn, Everett, 1977: "The Social Construction of Scientific Knowledge", in: Mendelsohn, Everett; Weingart, Peter; Whitley, Richard (Eds.): *The Social Production of Scientific Knowledge: Sociology of the Sciences: A Yearbook*, vol. 1 (Dordrecht: Springer).
Meyer-Abich, Klaus M. (Ed.), 1979: *Frieden mit der Natur* (Freiburg – Basel – Wien: Herder).
Meyer-Abich, Klaus M., 2003: "Frieden mit der Natur", in: Simonis, Udo E. (Ed.): *Ökolexikon* (Munich: Beck): 84.
Mies, Maria, 1998: *Patriarchy and Accumulation on a World Scale* (Melbourne: Zed).
Miles, Kathryn: "Ecofeminism", in: *Encylopaedia Britannica.*
Minch, Michael, 2011: "Political Ecology", in: Chatterjee D.K. (Eds.): *Encyclopedia of Global Justice* (Dordrecht: Springer).
Ministry of Defence (MOD), 2010a: *MOD Climate Change Strategy 2010* (London: MOD).
Ministry of Defence (MOD), 2010b: *MOD Sustainability and Environmental Appraisal Tool Handbook: Section 7: Climate Impacts Risk Assessment Method* (London: MOD).
Mittelstrass, Jürgen, 2003: *Transdisziplinarität: Wissenschaftliche Zukunft und institutionelle Wirklichkeit* (Konstanz: Universitätsverlag Konstanz).
Möllers, Nina; Schwägerl, Christian; Trischler, Helmuth (Eds.), 2014: *Welcome to the Anthropocene: The Earth in Our Hands* (Munich: Deutsches Museum).
Moore, Jason M. (Ed.), 2016: *Anthropocene or Capitalocene? Nature, History, and the Crisis of Capitalism* (Oakland, CA: PM Press).
Moore, Jason W., 2016a: "Introduction: Anthropocene or Capitalocene? Nature, History, and the Crisis of Capitalism", in: Moore, Jason M. (Ed.): *Anthropocene or Capitalocene? Nature, History, and the Crisis of Capitalism* (Oakland, CA: PM Press): 1–13.
Moore, Jason M.; Patel, Raj, 2018: *A History of the World in Seven Cheap Things: A Guide to Capitalism, Nature, and the Future of the Planet* (Oakland: University of California Press; London: Verso).
Morrison, Mary Lee, 2005: *Elise Boulding: A Life in the Cause of Peace* (Jefferson, NC: McFarland).
Müller, Ernst; Schmieder, Falko, 2016: *Begriffsgeschichte und historische Semantik: Ein kritisches Kompendium* (Frankfurt am Main: Suhrkamp).
Münkler, Herfried, 2002: *Über den Krieg: Stationen der Kriegsgeschichte im Spiegel ihrer theoretischen Reflexion* (Weilerswist: Velbrück Wissenschaft).
Münkler, Herfried, 2005: *The New Wars* (Cambridge: Polity).
Myers, Norman, 1989: "Environment and Security", in: *Foreign Policy*, 74 (Spring): 23–41.
Myers, Norman, 1989a: "Environmental Security: The Case of South Asia", in: *International Environmental Affairs*, 1,2: 138–154.
Naess, Arne, 1973: "The Shallow and the Deep, Long-Range Ecology Movement. A Summary", in: *Inquiry*, 16: 95–100.
Naess, Arne, 1989: *Ecology, Community and Lifestyle: Outline of an Ecosophy* (Cambridge: Cambridge University Press).
Najam, Adil, 2003: "Environment and Security: Exploring the Links", in: Najam, Adil (Ed.): *Environment, Development and Human Security: Perspectives from South Asia* (Lanham – New York – Oxford: University Press of America): 1–24.
Nentwich, Wolfgang; Bacher, Sven; Beierkuhnlein, Carl; Brandl, Roland; Grabherr, Georg, 2004: *Ökologie* (Heidelberg: Spektrum der Wissenschaft).
Norgaard, Richard B., 2002: "Ecological Economics", in: Timmerman, Peter (Ed.): *Encyclopedia of Global Environmental Change,* vol. 5: *Social and Economic Dimensions of Global Environmental Change* (Chichester: John Wiley): 37–48.
Norgaard, Richard B., 2013: "The Econocene and the Delta", in: *San Francisco Estuary and Watershed Science,* 11,3: 1–5.
Notter, Harley, 1949: *Postwar Foreign Policy Preparation, 1939–1945* (Washington, DC: US Government Printing Office).
Novak, Joseph; Cañas, Alberto J., 2008: "The Theory Underlying Concept Maps and How to Construct and Use Them", Technical Report IHMC CmapTools 2006–01, Rev 01–2008, Florida

Institute for Human and Machine Cognition; at: https://cmap.ihmc.us/docs/pdf/TheoryUnderlyingConceptMaps.pdf (29 July 2020),
NRC (US National Research Council), Policy Division, Board on Sustainable Development, 1999: *Our Common Journey: A Transition toward Sustainability* (Washington, DC: National Academy Press); at: https://www.nap.edu/openbook.php?record_id=9690& page=1.
NSB [US National Science Board], 2007: *Enhancing Support for Transformative Research at the National Science Foundation* (Washington, DC: NSF).
Odum, Eugene P., 1953, 1971: *Fundamentals of Ecology* (Philadelphia: W.B. Saunders).
Odum, Eugene P., 1975: *Ecology, the Link between the Natural and the Social Sciences* (New York: Holt, Rinehart and Winston).
Odum, Eugene P., 1977: "The Emergence of Ecology as a New Integrative Discipline", in: *Science*, 195: 1289–1293.
Odum, Eugene P., 1993: "Biosphere 2: A New Kind of Science", in: *Science*, 260,5110 (14 May 1993): 878–879.
Odum, Eugene P., 1998: *Ecological Vignettes: Ecological Approaches to Dealing with Human Predicaments* (Amsterdam: Harwood).
OECD, 2011: From Power Struggles to Sustainable Peace: Understanding Political Settlements (Paris: OECD); at: https://www.keepeek.com/Digital-Asset-Management/oecd/development/from-power-struggles-to-sustainable-peace_9789264116498-en.
Omeje, Kenneth, 2015: "Strengthening Peace Research and Peace Education in African Universities", in: *African Sociological Review*, 19,1: 19–36.
Oren, Ido, 1995: "The Subjectivity of the 'Democratic' Peace: Changing US Perceptions of Imperial Germany", in: *International Security*, 20,2: 147–184.
Osborn, Fairfield, 1953: *Our Plundered Planet* (Boston: Little Brown).
Osiander, Andreas, 1994: *The States System of Europe 1640-1990: Peacemaking and the Conditions of International Stability* (Oxford: Clarendon Press).
Osterhammel, Jürgen, 2009: *Die Verwandlung der Welt: Eine Geschichte des 19. Jahrhunderts* (München: C. H. Beck).
Oswald Spring, Úrsula (1990): "Pobreza Perversa y Estrategias de Supervivencia", *Revista Pueblo, Democracia Social y Política*, 152 (January): 28–29.
Oswald Spring Úrsula, 1999: *Fuenteovejuna o Caos Ecológico* [Fuenteovejuna or Ecological Chaos] (Cuernavaca: CRIM/UNAM).
Oswald Spring, Úrsula, 2005: *El Valor del Agua: Una Visión Socioeconómica de un Conflicto Ambiental* [The Value of Water: A Socio-Economic View of an Environmental Conflict] (Mexico City: Coltlax, CONACYT, Gov. of Tlaxcala, SEFOA).
Oswald Spring, Úrsula, 2008: "Sustainable Development", in: de Rivera, Joe (Ed.): *Culture of Peace* (New York: Springer-Verlag): 211–227.
Oswald Spring, Úrsula, 2008a: "Peace and Environment: Towards a Sustainable Peace as Seen from the South", in: Brauch, Hans Günter; Oswald Spring, Úrsula; Mesjasz, Czeslaw; Grin, John; Dunay, Pal; Behera, Navnita Chadha; Chourou, Béchir; Kameri-Mbote, Patricia; Liotta, P.H. (Eds.): *Globalization and Environmental Challenges: Reconceptualizing Security in the 21st Century* (Berlin – Heidelberg – New York: Springer-Verlag): 113–126.
Oswald Spring, Úrsula, 2008b: *Gender and Disasters: Human, Gender and Environmental Security: A HUGE Challenge.* Source 8/2008 (Bonn: UNU-EHS).
Oswald Spring, Úrsula, 2016: "Sustainability Transition in a Vulnerable River Basin in Mexico", in: Brauch, Hans Günter; Oswald Spring, Úrsula; Grin, John; Scheffran, Jürgen (Eds.): *Handbook on Sustainability Transition and Sustainable Peace* (Cham – Heidelberg – New York – Dordrecht – London: Springer): 675–704.
Oswald Spring, Úrsula 2020: *Earth at Risk in the 21st Century: Rethinking Peace, Environment, Gender, and Human, Water, Health, Food, Energy Security, and Migration* (Cham: Springer International Publishing).
Oswald Spring, Úrsula; Brauch, Hans Günter, 2009: "Securitizing Water", in: Brauch, Hans Günter; Oswald Spring, Úrsula; Grin, John; Mesjasz, Czeslaw; Kameri-Mbote, Patricia; Behera, Navnita

Chadha; Chourou, Béchir; Krummenacher, Heinz (Eds.): *Facing Global Environmental Change: Environmental, Human, Energy, Food, Health and Water Security Concepts*. Hexagon Series on Human and Environmental Security and Peace, vol. 4 (Berlin – Heidelberg – New York: Springer-Verlag): 175–202.

Oswald Spring, Úrsula; Brauch, Hans Günter; Tidball, Keith G. (Eds.), 2014: *Expanding Peace Ecology: Security, Sustainability, Equity and Peace: Perspectives of IPRA's Ecology and Peace Commission* (Heidelberg – Dordrecht – London – New York: Springer-Verlag).

Oswald Spring, Úrsula; Brauch, Hans Günter; Dalby, Simon, 2009: "Linking the Anthropocene, HUGE and HESP in a Fourth Phase of Human, Environmental and Gender Security and Peace Research", in: Brauch, Hans Günter; Oswald Spring, Úrsula; Grin, John; Mesjasz, Czeslaw; Kameri-Mbote, Patricia; Behera, Navnita Chadha; Chourou, Béchir; Krummenacher, Heinz (Eds.): *Facing Global Environmental Change: Environmental, Human, Energy, Food, Health and Water Security Concepts* (Berlin – Heidelberg – New York: Springer-Verlag): 1277–1294.

Oswald Spring, Úrsula; Brauch, Hans Günter; Tidball, Keith G., 2014a: "Expanding Peace Ecology: Peace, Security, Sustainability, Equity and Gender", in: Oswald Spring, Úrsula; Brauch, Hans Günter; Tidball, Keith G. (Eds.): *Expanding Peace Ecology: Peace, Security, Sustainability, Equity and Gender: Perspectives of IPRA's Ecology and Peace Commission* (Heidelberg – Dordrecht – London – New York: Springer-Verlag, 2014): 1–30.

Oswald Spring, Úrsula; Serrano Oswald, Serena Eréndira, 2018: *Risks, Violence, Security and Peace in Latin America: 40 Years of the Latin American Council for Peace Research* (CLAIP) (Cham: Springer International Publishing).

Page, Edward A.; Redclift, Michael (Eds.), 2002: *Human Security and the Environment: International Comparisons* (Cheltenham: Edward Elgar).

Parenti, Christian, 2016: "Environment-Making in the Capitalocene: Political Ecology of the State", in: Moore, Jason M. (Ed.): *Anthropocene or Capitalocene? Nature, History, and the Crisis of Capitalism* (Oakland, CA: PM Press): 166–184.

Paris, Roland, 2004: *At War's End – Building Peace after Civil Conflict* (Cambridge: Cambridge University Press).

Partsch, Karl-Josef, 1994: "Art. 55 (c)", in: Simma, Bruno (Ed.): *The Charter of the United Nations. A Commentary* (Oxford: Oxford University Press): 776–793.

Patel, Raj, 2013: "Misanthropocene?", in: *Earth Island Journal*, 28,1.

Pattberg, Philipp; Zelli, Fariborz (Eds.), 2016: *Environmental Politics and Governance in the Anthropocene – Institutions and Legitimacy in a Complex World* (London – New York: Routledge).

Pauly, D., 2010: *5 Easy Pieces: The Impact of Fisheries on Marine Systems* (Washington, DC: Island Press).

Peace Trees Vietnam, 2006: "Former Battleground Dedicated as Peace Trees Friendship Village.

Peck, Connie, 1998: *Sustainable Peace: The Role of the UN and Regional Organizations in Preventing Conflicts* (Lanham – Boulder – New York: Carnegie Commission on Preventing Deadly Conflict).

Peet, R.; Watts, M.J., (Eds.), 1996: *Liberation Ecologies: Environment, Development, Social Movements* (London – New York: Routledge).

Peichert, Henrike, 2003: "The Nile Basin Initiative: A Catalyst for Cooperation", in: Brauch, Hans Günter; Liotta, P.H; Marquina, Antonio; Rogers, Paul; Selim, Mohammed El-Sayed (Eds.): *Security and Environment in the Mediterranean: Conceptualising Security and Environmental Conflicts* (Berlin – Heidelberg: Springer): 761–774.

Peluso, Nancy Lee; Watts, Michael (Eds.): 2001: *Violent Environments* (Ithaca – London: Cornell University Press).

Pepper, David, 2002: "Deep Ecology", in: Timmerman, Peter (Ed.): *Encyclopedia of Global Environmental Change*, vol. 5: *Social and Economic Dimensions of Global Environmental Change* (Chichester: John Wiley): 211.

Pepper, David, 2002a: "Social Ecology", in: Timmerman, Peter (Ed.): *Encyclopedia of Global Environmental Change*, vol. 5: *Social and Economic Dimensions of Global Environmental Change* (Chichester: John Wiley): 484.

Petschel-Held, Gerhard; Lüdeke, Matthias K.B.; Reusswig, Fritz, 1999: "Actors, Structures and Environments: A Comparative and Transdisciplinary View on Regional Case Studies of Global Environmental Change", in: Lohnert, B.; Geist, H. (Eds.): *Coping with Changing Environments* (London: Ashgate): 255–291.

Piaget, Jean; Inhelder, Barbara, 1969: *The Psychology of the Child* (New York: Basic Books).

Polanyi, Karl, 1944: *The Great Transformation: The Political and Economic Origins of Our Time* (Boston: Beacon Press).

Polt, Richard; Wittrock, Jon, 2018: *The Task of Philosophy in the Anthropocene: Axial Echoes in Global Space* (London: Rowman & Littlefield).

Preiser, Rika; Pereira, Laura M.; Biggs, Reinette (Oonise), 2017:, in: *Anthropocene*, 20: 83–87; at: https://www.sciencedirect.com/science/article/pii/S2213305417300474.

Prigogine, Ilya; Stengers, Isabelle, 1984: *Order out of Chaos: Man's New Dialogue with Nature* (London: Flamingo).

Prigogine, Ilya; Stengers, Isabelle, 1997: *The End of Certainty* (Glencoe: The Free Press).

Prins, Gwyn, 1990: "Politics and the Environment", in: *International Affairs*, 66: 711–730.

Radkau, Joachim, 2002: *Natur und Macht: eine Weltgeschichte der Umwelt* (Munich: C.H. Beck).

Radkau, Joachim, 2009: *Nature and Power: A Global History of the Environment* (Washington, DC: German Historical Institute).

Radkau, Joachim, 2011: *Die Ära der Ökologie: Eine Weltgeschichte* (Munich: C.H. Beck).

Raffnsøe, Sverre, 2016: *Philosophy of the Anthropocene: The Human Turn* (London: Palgrave).

Raskin, P.; Banuri, T.; Gallopin, G.; Gutman, P.; Hammond, A.; Kates, A.; Swart, R., 2002: *Great Transition: The Promise and the Lure of the Times Ahead: A Report of the Global Scenario Group: Pole Start Series Report 10* (Stockholm: Stockholm Environment Institute).

Raskin, P.; Gallopin, G.; Gutman, P.; Hammond, A.; Swart, R., 1998: *Bending the Curve: Toward Global Sustainability: A Report of the Global Scenario Group* (Boston: Stockholm Environment Institute).

Raworth, Kate, 2014: "Must the Anthropocene be a Manthropocene?", in: *The Guardian* (20 October).

Reardon, Betty, 1996: *Sexism and the War System* (New York: Syracuse University Press).

Reardon, Betty, 2014: *Betty Reardon, A Pioneer in Peace Education, on Sexism and War* (Heidelberg – New York – Dordrecht – London: Springer-Verlag).

Reardon, Betty; Nordland, Eva, 1994: *Learning Peace: The Promise of Ecological and Cooperative Education* (Albany, NY: State University of New York Press).

Renn, Jürgen, 2020: *The Evolution of Knowledge: Rethinking Science for the Anthropocene* (Princeton – Oxford: Princeton University Press).

Renner, Michael, 1996: *Fighting for Survival: Environmental Decline, Social Conflict, and the New Age of Insecurity* (New York: Norton).

Revkin, A.C., 1992: *Global Warming: Understanding the Forecast* (New York: Abbeville Press).

Risbey, James S., 2008: "The New Climate Discourse: Alarmist or Alarming?", in: *Global Environmental Change*, 18: 26–37.

Robbins Paul, 2004: *Political Ecology: A Critical Introduction. Blackwell Critical Introductions to Geography* (London: Blackwell).

Robbins, Paul, 2nd edn., 2012: *Political Ecology: A Critical Introduction* (London: Blackwell).

Rodhe, Henning; Charlson, Robert; Crawford, Elisabeth, 1997: "Svante Arrhenius and the Greenhouse Effect", in: *Ambio*, 26,1 (February): 2–5.

Ronzheimer, Manfred, 2019: "Wissenschaft und Nachhaltigkeit: Sozial-Ökologie auf Erfolgskurs", in: *TAZ* (23 December).

Rosenau, James N.; Czempiel, Ernst-Otto, 1992: *Governance Without Government: Order and Change in World Politics* (Cambridge: Cambridge University Press).

Rowlands, Ian H., 1991: "Ozone Layer Depletion and Global Warming: New Sources for Environmental Disputes", in: *Peace and Change*, 16,3 (July): 260–284.
Ruddiman, William F., 2003: "The Anthropogenic Greenhouse Era Began Thousands of Years Ago", in: *Climatic Change*, 61,3: 261–293.
Ruddiman, William F., 2005: *Plows, Plagues, and Petroleum: How Humans Took Control of Climate* (Princeton, NJ: Princeton University Press).
Ruddiman, William F., 2016: "Late Holocene Climate: Natural and Anthropogenic", in: *Review of Geophysics*, 54: 93–118.
Russett, Bruce, 1993: *Grasping the Democratic Peace: Principles for a Post-Cold War World* (Princeton, NJ: Princeton University Press).
Saltnes, Turi, 1998: *Desertification and Conflict: A Quantitative Analysis of the Relationship between the Environment and Civil War* [in Norwegian] (Cand. Polit. thesis, Norwegian University of Science and Technology, Trondheim, Norway).
Samways, M., 1999: "Translocating Fauna to Foreign Lands: Here Comes the Homogenocene", in: *Journal of Insect Conservation*, 3: 65–66.
Sassen, Saskia, 1991, 2001: The *Global City* (Princeton: Princeton University Press).
Sassen, Saskia, 2008: "The Global City", in: Nugent, David; Vincent, Joan (Eds.): *A Companion to the Anthropology of Politics* (Hoboken, NJ: Wiley-Blackwell, March): 168–178; at: DOI: https://doi.org/10.1002/9780470693681.ch11.
Scott, Robert, 2015: *Kenneth Boulding* (Basingstoke – New York: Palgrave Macmillan).
Scott, Shirley V.; Ku, Charlotte, 2018: *Climate Change and the UN Security Council* (Cheltenham: E. Elgar).
Scheffran, Jürgen, 1999: "Environmental Conflicts and Sustainable Development: A Conflict Model and its Application in Climate and Energy Policy", in: Carius, Alexander; Lietzmann, Kurt M. (Eds.): *Environmental Change and Security: A European Perspective* (Berlin – Heidelberg: Springer): 95–218.
Scheffran, Jürgen; Brzoska, Michael; Brauch, Hans Günter; Link, Peter Michael; Schilling, Janpeter (Eds.), 2012: *Climate Change, Human Security and Violent Conflict: Challenges for Societal Stability* (Berlin – Heidelberg – New York: Springer-Verlag)
Schellnhuber, Hans Joachim, 1999: "'Earth System' Analysis and the Second Copernican Revolution", in: *Nature*: 402: C19-C23 (2 December); doi: https://doi.org/10.1038/35011515.
Schellnhuber, Hans-Joachim; Block, Arthur; Cassel-Gintz, Martin; Kropp, Jürgen; Lammel, Gerhard; Lass, Wiebke; Lienenkamp, Roger; Loose, Carsten; Lüdeke, Matthias K.B.; Moldenhauer, Oliver; Petschel-Held, Gerhard; Plöchl, Matthias; Reusswig, Fritz, 1997: "Syndromes of Global Change", in: *Gaia*, 6,1: 19–34.
Schellnhuber, Hans Joachim; Cramer, Wolfgang; Nakícénovic, Nebojsa; Wigley, Tom; Yohe, Gary (Eds.): *Avoiding Dangerous Climate Change* (Cambridge: Cambridge University Press).
Schellnhuber, Hans-Joachim; Crutzen, Paul; Clark, William C.; Hunt, Julian, 2005: "Earth System Analysis for Sustainability", in: *Environment*, 47,8 (October): 11–25.
Schellnhuber, Hans-Joachim; Wenzel, Volker (Eds.), 1998: *Earth System Analysis: Integrating Science for Sustainability* (Berlin – Heidelberg: Springer).
Scheumann, Waltina, 2003 "The Euphrates Issue in Turkish-Syrian Relations", in: Brauch, Hans Günter; Liotta, P.H.; Marquina, Antonio; Rogers, Paul; Selim, Mohammed El-Sayed (Eds.): *Security and Environment in the Mediterranean: Conceptualising Security and Environmental Conflicts* (Berlin – Heidelberg: Springer): 745–760.
Schneidewind, Uwe; Singer-Brodowski, Mandy, 2013, 2014 (revised edn.): *Transformative Wissenschaft. Klimawandel im deutschen Wissenschafts- und Hochschulsystem* (Marburg: Metropolis).
Schneidewind, Uwe; Singer-Brodowski, Mandy; Augenstein, Karoline, 2016: "Transformative Science for Sustainability Transitions", in: Brauch, Hans Günter; Oswald Spring, Úrsula; Grin, John; Scheffran, Jürgen (Eds.): *Handbook of Sustainability Transition and Sustainable Peace* (Heidelberg – New York – Dordrecht – London: Springer-Verlag): 123–136.

Schumacher, E.F., 1999: *Small Is Beautiful: A Study of Economics As If People Mattered* (Vancouver, BC: Hartley & Marks Publishers.

Schwägerl, Christian, 2015: "A Concept with a Past", in: Möllers, Nina; Schwägerl, Christian; Trischler, Helmuth (Eds.), 2014: *Welcome to the Anthropocene: The Earth in Our Hands* (Munich: Deutsches Museum): 128–129.

Schwartz, Daniel M., 2002: "The Environment and Violent Conflict", in: Timmerman, Peter (Ed.): *Encyclopedia of Global Environmental Change*, vol. 5: *Social and Economic Dimensions of Global Environmental Change* (Chichester: John Wiley): 137–149.

Schwartz, Peter; Randall, Doug, 2003, 2004: *An Abrupt Climate Change Scenario and Its Implications for United States National Security*, Contract study for the US Department of Defense (Washington, DC: DoD, Net Assessment; Emeryville, CA: Global Business Network).

Schwerdtfeger, Johannes, 2001: *Begriffsbildung und Theoriestatus in der Friedensforschung* (Opladen: Leske + Budrich).

Selby, Jan, et al., 2014: "Rethinking Climate Change, Conflict and Security. Introduction", in: *Geopolitics*, 19,4.

Senghaas, Dieter, 1994: "Frieden als Zivilisationsprojekt", in: Senghaas, Dieter: *Wohin driftet die Welt?* (Frankfurt/M.: Suhrkamp).

Senghaas, Dieter, 1995: "Frieden als Zivilisationsprojekt", in: Senghaas, Dieter (Ed.): *Den Frieden Denken* (Frankfurt/M.: Suhrkamp).

Senghaas, Dieter, 1997: "Frieden – ein mehrfaches Komplexprogramm", in: Senghaas, Dieter (Ed.): *Den Frieden Machen* (Frankfurt/M.: Suhrkamp): 560–575.

Senghaas, Dieter, 2013: *Dieter Senghaas: Pioneer of Peace and Development Research* (Heidelberg – New York – Dordrecht – London: Springer-Verlag, 2013).

Shiva, Vandana, 2002: *Water Wars: Privatization, Pollution, and Profit* (New Delhi: India Research Press).

Shiva; Vandana, 2003: *Water Wars* (Melbourne: Zed).

Shiva, Vandana; Mies, Maria, 1997: *Ecofeminism* (Melbourne: Zed).

Silva Diniz, Barbara, Tavares Beleza, Flávia, 2015: "La Mediación Social en la Escuela: Espacio de Construcción de Paz", in: Serrano Oswald, S. Eréndira; Oswald Spring, Úrsula; de la Rúa, Diana (Eds.): *América Latina en el Camino hacia una Paz Sustentable: Herramientas y Aportes* (Guatemala: ARP, FLACSO, CLAIP): 181–200.

Speth, James Gustavo, 1992: "The Transition to a Sustainable Society", in: *Proceedings of the National Academy of Sciences of the United States of America*, 89: 870–872.

Spretnak, Charlene, 1990: "Ecofeminism: Our Roots and Flowering", in: Diamond, Irene; Ornstein, Gloria (Eds.): *Reweaving the World: The Emergence of Feminism* (San Francisco, CA: Sierra Club Books): 3–14.

Stauffer, Robert C., 1957: "Haeckel, Darwin, and Ecology", in: *The Quarterly Review of Biology*, 32,2 (June): 138–144.

Steffen, Will, 2013: "Commentary: Paul J. Crutzen and Eugene F. Stoermer", in: Robbin, Libby; Sörlin, Sverker; Warde, Paul (Eds.): *The Future of Nature. Documents of Global Change* (New Haven: Yale University Press): 486–490.

Steffen, Will; Andreae, Meinrat O.; Bolin, Bert; Cox, Peter M.; Crutzen, Paul, J.; Cubasch, U.; Held, Herman; Nakicenovic, Nebojsa; Talaue-McManus, Liana; Turner II, Billie L., 2004: "Group Report: Earth Systems Dynamics in the Anthropocene", in: Schellnhuber, Hans Joachim; Crutzen, Paul J.; Clark, William C.; Claussen, Martin; Held, Hermann (Eds.): *Earth System Analysis for Sustainability* (Cambridge, MA; London: MIT Press): 313–340.

Steffen, Will; Broadgate, Wendy; Deutsch, Lisa; Gaffney, Owen; Ludwig, Cornelia, 2015: "The Trajectory of the Anthropocene: The Great Acceleration", in: *Anthropocene Review*, 2,1: 81–98.

Steffen, Will; Crutzen, Paul J.; McNeill, John R., 2007: "The Anthropocene: Are Humans Now Overwhelming the Great Forces of Nature", in: *Ambio: A Journal of the Human Environment*, 36,8 (December): 614–621.

Steffen, Will; Grinevald, Jacques; Crutzen, Paul; McNeill, John, 2011: *"The Anthropocene: Conceptual and Historical Perspectives"*, in: *Philosophical Transactions of the Royal Society A*, 369,1938: 842–867.

Steffen, Will; Leinfelder, Reinhold; Zalasiewicz, Jan; et al., 2016: "Stratigraphic and Earth Systems Approaches in Defining the Anthropocene", in: *Earth's Future*, 8, 324–345.

Steffen, Will; Rockström, Johan; Richardson, Katherine; Lenton, Timothy M.; Folke, Carl; Liverman, Diana; Summerhayes, Colin P.; Barnosky, Anthony D.; Cornell, Sarah E.; Crucifix, Michel; Donges, Jonathan F.; Fetzer, Ingo; Lade, Steven J.; Scheffer, Marten; Winkelmann, Ricarda; Schellnhuber, Hans Joachim, 2018: "Trajectories of the Earth System in the Anthropocene", in: *Proceedings of the National Academy of Sciences PNAS*. DOI: https://doi.org/10.1073/pnas.1810141115.

Steffen, Will; Crutzen, Paul. J.; McNeill, John R., 2007: "The Anthropocene: Are Humans Now Overwhelming the Great Forces of Nature?"; in: *Ambio*, 36: 616–618.

Steffens, Will; Broadgate, Wendy; Deutsch, Lisa; Gaffney, Owen; Ludwig, Cornelia, 2015: "The Trajectory of the Anthropocene: The Great Acceleration", in: *Anthropocene Review*, 2,1 (April): 81–98.

Steffen, Will; Sanderson, R. Angelina; Tyson, Peter D.; Jäger, Jill; Matson, Pamela A.; Moore III, Berrien; Oldfield, Frank; Richardson, Katherine; Schellnhuber, Hans Joachim; Turner, B.L.; Wasson, Robert J., 2004: *Global Change and the Earth System: A planet under pressure* (Berlin – Heidelberg: Springer-Verlag).

Stephenson, Carolyn M., 2016: "Paradigm and Praxis Shifts: Transitions to Sustainable Environmental and Sustainable Peace Praxis", in: Brauch, Hans Günter; Oswald Spring, Úrsula; Grin, John; Scheffran; Jürgen (Eds.), 2016: *Handbook on Sustainability Transition and Sustainable Peace* (Cham – Heidelberg – New York – Dordrecht – London: Springer): 89–104.

Stoppani, Antonio., 1873: *Corsa di Geologia (*Milan: G. Bernadom; E.G. Brigola).

Stripple, Johannes, 2002: "Climate Change as a Security Issue", in: Page, Edward A.; Redclift, Michael (Eds.), 2002: *Human Security and the Environment: International Comparisons* (Cheltenham: Edward Elgar).

Stüdemann, Otto, 2008: *Aspekte der Geoökologie* (Berlin: Weißensee Verlag).

Subramanian, Meera, 2019: "Anthropocene Now: Influential Panel Votes to Recognize Earth's New Epoch", in: *Nature* (19 May); at: doi: https://doi.org/10.1038/d41586-019-01641-5 (31 May 2019)

Suess, Eduard, 1875: *Die Entstehung Der Alpen* [*The Origin of the Alps*] (Vienna: W. Braunmüller).

Sundberg, Juanita, 2015: "Ethics, Entanglement and Political Ecology", in: Perreault, Tom; Bridge, Gavin; McCarthy, James (Eds.): *The Routledge Handbook of Political Ecology* (London: Routledge): 117–126.

Swain, Ashok; Krampe, Florian, 2011: "Stability and Sustainability in Peace Building: Priority Area for Warfare Ecology", in: Machlis, G.E.; Hanson, T.; Špirić, Z.; McKendry, J.E. (Eds.): *Warfare Ecology* (Dordrecht: Springer): 199–210.

Swilling, Mark, 2016: "Preparing for Global Transition: Implications of the Work of the International Resource Panel", in: Brauch, Hans Günter; Oswald Spring, Úrsula; Grin, John; Scheffran, Jürgen (Eds.): *Handbook of Sustainability Transition and Sustainable Peace* (Heidelberg – New York – Dordrecht – London: Springer-Verlag): 391–418.

Sygna, Linda; O'Brien, Karen; Wolf, Johanna (Eds.), 2013: *A Changing Environment for Human Security. Transformative Approaches to Research, Policy and Action* (London – New York: Routledge).

Tansley, Arthur G., 1935: "The Use and Abuse of Vegetational Terms and Concepts", in: *Ecology*, 16,3: 284–307.

Taylor, Charles Lewis, **Russett**, Bruce M. (Eds.), 2020: *Karl W. Deutsch: Pioneer in the Theory of International Relations* (Cham: Springer International Publishing).

Taylor, Chris, 2013: "The Discourses of Climate Change", in: Cadman, T. (Eds.): *Climate Change and Global Policy Regimes: International Political Economy* (London: Palgrave Macmillan); at: https://doi.org/https://doi.org/10.1057/9781137006127_2.

Thomas, Michael Durant, 2017: *The Securitization of Climate Change: Australian and United States' Military Responses (2003-2013)* (Cham: Springer International Publishing).
Thone, Frank, 1935: "Nature Rambling: We Fight for Grass", in: *The Science Newsletter*, 27,717 (5 January): 14.
Tomuschat, Christian, 1994: "Chapter VI: Pacific Settlement of Disputes, Art. 33", in: Simma, Bruno (Ed.): *The Charter of the United Nations. A Commentary* (Oxford: Oxford University Press): 505–514.
Trischler, Helmuth (Eds.): *Welcome to the Anthropocene: The Earth in Our Hand*s (Munich: Deutsches Museum): 128–129.
Tritschler, Helmuth, 2016: "The Anthropocene: A Challenge for the History of Science, Technology, and the Environment", in: *NTM Zeitschrift für Geschichte der Wissenschaften, Technik und Medizin*, 24: 309–335; at: https://doi.org/10.1007/s00048-016-0146-3.
Troll, Carl, 1939: "Luftbildplan und ökologische Bodenforschung" [Aerial Photography and Ecological Studies of the Earth], in: *Zeitschrift der Gesellschaft für Erdkunde (Berlin): 241–298.*
Troll, Carl, 1968: *Geo-Ecology of the Mountainous Regions of the Tropical Americas: Proceedings of the UNESCO Mexico Symposium, 1–3 August 1966* (Bonn: Dümmler).
Trombetta, Maria Julia, 2014: "Linking Climate-Induced Migration and Security within the EU: Insights from the Securitization Debate", in: *Critical Studies on Security*, 2,2.
Ullman, Richard, 1983: "Redefining Security", in: *International Security* 8,1 (Summer): 129–153.
UNEP, International Resource Panel, 2011: *Decoupling Natural Resource Use and Environmental Impacts from Economic Growth* (Paris: UNEP).
UNEP, International Resource Panel, 2014: *Decoupling 2: Technologies, Opportunities and Policy Options* (Paris: UNEP).
UNEP: "Climate Change and Security Risks"; at: https://www.unenvironment.org/explore-topics/disasters-conflicts/what-we-do/risk-reduction/climate-change-and-security-risks.
UNESCO, 2003: *Combating Desertification: Rehabilitation of Degraded Drylands and Biosphere Reserves*. UNESCO-MAB Dryland Series No. 2 (Paris: UNESCO).
UNESCO, 2006: *UN World Water Development Report II: Water a Shared Responsibility* (Paris: UNESCO 2006).
UNGA (General Assembly), 2009: "Climate Change and its Possible Security Implications". Resolution adopted by the General Assembly, A/RES/63/281 (New York: United Nations General Assembly, 11 June).
UN-EU Guidance note: Managing and Preventing Conflicts over Renewable Resources (UNEP with EU-UN Partnership on Land, Natural Resources and Conflict, 2012).
Unruh, Jon; Williams, Rhodri, 2013: *Land and Post-Conflict Peacebuilding* (London: Routledge);
UNSC (Security Council), 2007: Letter dated 5 April 2007 from the permanent representative of the United Kingdom of Great Britain and Northern Ireland to the United Nations addressed to the President of the Security Council: *Meeting Records*, S/PV. 5663, 5,663rd meeting (Part I) (New York: UN, 17 April).
UNSC (Security Council), 2011: Statement by the President of the Security Council on "Maintenance of Peace and Security: Impact of Climate Change", S/PRST/1011/15, 20 July 2011.
UNSG (Secretary General), 2009: *Climate Change and its possible security implications, Report by the Secretary General*, UN Doc. A/64/350 (New York: UN, 11 September).
US Department of Defence (DOD), 2015: *National Security Implications of Climate-Related Risks and a Changing Climate* (Washington, DC: US DOD).
US National Academy of Sciences, 1999: *Our Common Journey: A Transition toward Sustainability* (Washington, DC: NRC 1999),
US National Science Board, 2007: *Enhancing Support of Transformative Research at the National Science Foundation* (Washington, DC: National Science Foundation, 7 May).
von Weizsäcker, Ernst Ulrich, 1989: *Erdpolitik: Ökologische Realpolitik an der Schwelle zum Jahrhundert der Umwelt* (Darmstadt: Wissenschaftliche Buchgesellschaft).
Vasak, K., 1984: "Pour une Troisième Génération des Droits de l'Homme", in: Swinarski, Christophe (Ed.): *Studies in Honour of Jean Pictet* (The Hague: Martinus Nijhoff Publishers).

Vernadsky, Vladimir I., 1929 [1st edn. 1926]: *La Biosphere* (Paris: Librairie Felix Alcan, Nouvelle Collection Scientifique).
Vernadsky, Vladimir. I., 1945: "The Biosphere and the Noosphere", in: *American Scientist,* 33,1: 1–12
Vernadsky, Vladimir, 1986: *The Biosphere* [reprinted] (Oracle AZ: Synergetic Press).
Vernadsky, Vladimir I., 1998: *The Biosphere* (New York: Springer Science+Business Media).
Vitousek, P.M.; Mooney, H.A.; Lubchenco, J.; Melillo, J.M., **1997: "**Human Domination of Earth's Ecosystems", in: *Science*, 277: 494–499.
Vlek, Paul, 2005: *Nothing Begets Nothing: InterSecTions*, 1/2005 (Bonn: UNU-EHS).
von Lucke, Franziskus; et al., 2014: "What's at Stake in Securitising Climate Change? Towards a Differentiated Approach", in: *Geopolitics*, 19,4;
Wæver, Ole, 1995: "Securitization and Desecuritization", in: Lipschutz, Ronnie D. (Ed.): *On Security* (New York: Columbia University Press): 46–86.
Waever, Ole, 2008: "Peace and Security: Two Evolving Concepts and their Changing Relationship", in: Brauch, Hans Günter; Oswald Spring, Úrsula; Mesjasz, Czeslaw; Grin, John; Dunay, Pal; Behera, Navnita Chadha; Chourou, Béchir; Kameri-Mbote, Patricia; Liotta, P.H. (Eds.): *Globalization and Environmental Challenges: Reconceptualizing Security in the 21st Century* (Berlin – Heidelberg – New York: Springer-Verlag): 99–111.
Wæver, Ole, 2008a: "The Changing Agenda of Societal Security ", in: Brauch, Hans Günter; Oswald Spring, Úrsula; Mesjasz, Czeslaw; Grin, John; Dunay, Pal; Behera, Navnita Chadha; Chourou, Béchir; Kameri-Mbote, Patricia; Liotta, P.H. (Eds.): *Globalization and Environmental Challenges: Reconceptualizing Security in the 21st Century* (Berlin – Heidelberg – New York: Springer-Verlag): 581–593.
Wæver, Ole; Buzan, Barry; de Wilde, Jaap, 2008, 2009: *Politics of Security: A Comprehensive Framework of Analysis* (Boulder, CO: Lynne Rienner).
Warner, Jeroen; Boas, Ingrid, 2017: "Securitisation of Climate Change: The Risk of Exaggeration", in: *Ambiente & Sociedade*, 20,3 (July/Sept.).
Warren, Karen, 2002: "Karen Warren's Ecofeminism", in: *Ethics & the Environment,* 7,2 (September): 12–26; at: doi:https://doi.org/10.2979/ETE.2002.7.2.12.
Waters, Colin N.; Syvitski, James P. M.; Gałuszka, Agnieszka; Hancock, Gary J.; Zalasiewicz, Jan; Cearreta, Alejandro; Grinevald, Jacques; Jeandel, Catherine; McNeill, John R.; Summerhayes, Colin; Barnosky, Anthony, 2015: "Can Nuclear Weapons Fallout Mark the Beginning of the Anthropocene Epoch", in: *Bulletin of the Atomic Scientists*, 71 (1 May): 46–57.
Waters, Colin N.; Zalasiewicz, Jan, 2008: "Our Geological Footprint", in: Wilkinson, C. (Ed.): *The Observer Book of the Earth* (London: Observer Books).
WBGU [German Advisory Council on Global Change], 1996: *Welt im Wandel: Herausforderungen für die deutsche Wissenschaft: Jahresgutachten 1996* (Berlin: Springer).
WBGU, 1997: *World in Transition: The Research Challenge* (Berlin: Springer).
WBGU, 1998: *World in Transition: Ways Towards Sustainable Management of Freshwater Resources* (Berlin: Springer-Verlag).
WBGU, 2007: *Welt im Wandel: Sicherheitsrisiko Klimawandel* (Berlin – Heidelberg: Springer-Verlag).
WBGU, 2008: *World in Transition: Climate Change as a Security Risk* (London: Earthscan); at: https://www.wbgu.de/en/publications/publication/welt-im-wandel-sicherheitsrisiko-kli mawandel.
WBGU, 2011: *World in Transition: A Social Contract for Sustainability* (Berlin: German Advisory Council on Global Change, July 2011).

WCED (World Commission on Environment and Development or: Brundtland Commission), 1987, 1988: *Our Common Future. The World Commission on Environment and Development* (Oxford – New York: Oxford University Press).

Weber, Thomas, 1999: "Gandhi, Deep Ecology, Peace Research and Buddhist Economics", in: *Journal of Peace Research*, 36,3: 349–361.

Weinthal, Erika; Troell, Jessica J.; Nakayama, Mikiyasu, 2014: *Water and Post-Conflict Peacebuilding* (London: Routledge).

Weizsäcker, Ernst Ulrich von, 1989: *Erdpolitik: Ökologische Realpolitik an der Schwelle zum Jahrhundert der Umwelt* (Darmstadt: Wissenschaftliche Buchgesellschaft).

Westing, Arthur H., 1976: *Ecological Consequences of the Second Indochina War* (Stockholm: Almqvist & Wiksell).

Westing, Arthur H., 2003: "Environmental Dimension of the Gulf War", in: Brauch, Hans Günter; Liotta, P.H.; Marquina, Antonio; Rogers, Paul F.; Selim, Mohammad El-Sayed (Eds.): *Security and Environment in the Mediterranean: Conceptualising Security and Environmental Conflicts* (Berlin – Heidelberg – New York – Hong Kong – London – Milan – Paris – Tokyo: Springer-Verlag): 523–534.

Westing, Arthur H., 2013: *Arthur H. Westing: Pioneer on the Environmental Impact of War*. (Heidelberg – Dordrecht – London – New York: Springer-Verlag).

Westing, Arthur H., 2013a: *Texts on Environmental and Comprehensive Security* (Heidelberg – New York – Dordrecht – London: Springer-Verlag).

Wheeler, Sally, 2017: "The Corporation and the Anthropocene", in: Kotzé, Louis J. (Ed.): *Environmental Law and Governance for the Anthropocene* (Oxford – London – New York – New Delhi – Sydney: Hart): 289–308.

Wilson, Edward O., 1998: *Consilience* (New York: Knopf).

Wilson, Edward O., 1992: *The Diversity of Life (London:* Allen Lane; Harmondsworth: The Penguin Press).

Wilson, E.O., 2002: *The Future of Life* (New York: Alfred A. Knopf).

Wisner, Ben; Walker, Peter, 2005: "Beyond Kobe: A Proactive at the World Conference on Disaster Reduction".

Wolf, Aaron T., 1995: *Hydropolitics along the Jordan River: Scarce Water and its Impact on the Arab-Israeli Conflict* (Tokyo – New York – Paris: United Nations University Press).

Wolf, Aaron T., 1998: "Conflict and Cooperation along International Waterways", in: *Water Policy*, 1,2: 251–265.

Wolf, Aaron T., 1999: "'Water Wars' and Water Reality: Conflict and Cooperation along International Waterways", in: Lonergan, Steve (Ed.): *Environmental Change, Adaption, and Security* (Dordrecht: Kluwer): 251–265.

Wolf, Aaron T., 1999a: "International Basins of the world", in: *International Journal of Water Resources Development*, 15,4 (December).

Wolf, Aaron T., 1999b: "The Transboundary Freshwater Dispute Database", in: *Water International*, 24,2 (Spring): 29–60.

Wolf, Aaron T., 1999c: "Criteria for Equitable Allocations: The Heart of International Water Conflict", in: *Natural Resources Forum*, 23,1 (February): 3–30.

Wolf, Aaron T., 2001: "Transboundary Waters: Sharing Benefits, Lessons Learned", Thematic Background Paper to the International Conference on Freshwater "Water – A Key to Sustainable Development", Bonn, FRG, 3–7 December.

Wolf, Aaron T., 2001a: "Water and Human Security", in: Bennett, Lynne Lewis (Ed.): *Complexities with Transboundary Water Resource Management: Progress and Stumbling Blocks*, Universities Council on Water Resources, Issue No. 118, January.

Wolf, Aaron T. (Ed.), 2002: *Conflict Prevention and Resolution in Water Systems* (Cheltenham – Northampton, MA: Edward Elgar).
Wolf, Aaron T., 2002a: "Conflict Prevention and Resolution in Water Systems", in: Howe, Charles W. (Ed.): *The Management of Water Resources Series* (Cheltenham: Elgar).
Wolf, Aaron T.; Hamner, J., 2000: "Trends in Transboundary Water Disputes and Dispute Resolution", in: Lowi, Miriam R.; Shaw, Brian R. (Eds.): *Environment and Security: Discourses and Practices* (London: Macmillan – New York: St. Martin's Press): 123–148.
Wolf, Aaron T.; Natharius, J.; Danielson, J.; Ward, B.; Pender, B., 1999: "International River Basins of the World", in: *International Journal of Water Resources Development*, 15,4 (December): 387–427.
Wolf, Aaron T.; Stahl, Kerstin; Macomber, Marcia F., 2003: "Conflict and Cooperation within International River Basins: The Importance of Institutional Capacity, in: *Water Resources Update*, 125 (Universities Council on Water Resources).
Wolf, Aaron T.; Yoffe, Shira B.; Giordano, Mark, 2003: "International Waters: Identifying Basins at Risk", in: *Water Policy*, 5,1 (February): 29–60.
Wolf, Eric R., 1972: "Ownership and Political Ecology", in: *Anthropological Quaterly*, 35. 201–205.
Wolfrum, Rüdiger, 1994: "Preamble and Art: 1", in: Simma, Bruno (Ed.): *The Charter of the United Nations: A Commentary* (Oxford: Oxford University Press): 45–56.
Wright, Quincy, 1942, 1965: *A Study of War* (Chicago – London: University of Chicago Press).
Yamane, Kazuyo, 2007: "Peace Education Through Peace Museums", *International Security, Peace, Development and Environment, EOLSS,* vol. 2 (Cambridge: EOLSS Press).
Young, Gerald L. 1974: "Human Ecology as an Interdisciplinary Concept: A Critical Inquiry", in: Macfadyen, Amyan (Ed.): *Advances in Ecological Research*, vol. 8 (New York: Academic Press): 1–105.
Young, Helen; Goldman, Lisa, 2015: *Livelihoods, Natural Resources, and Post-Conflict Peacebuilding* (London: Routledge).
Zalasiewicz, Jan, 2008: "Our Geological Footprint", in: Wilkinson, C. (Ed.): *The Observer Book of the Earth* (London: Observer Books).
Zalasiewicz, Jan, 2008a: "Are We Now Living in the Anthropocene?", in: *Geological Society of America Today*, 18: 4–8.
Zalasiewicz, Jan; Steffen, Will; Leinfelder, Reinhold, et al. 2017a: "Petrifying Earth Process: The Stratigraphic Imprint of Key Earth System Parameters in the Anthropocene", in: Clark, N.; Yusuff, K. (Eds.): *Theory Culture and Society, Special Issue: Geosocial Formations and the Anthropocene, Theory, Culture and Society,* 34: 83–104.
Zalasiewicz, Jan; Summerhayes, Colin; et al., 2018: "Global Boundary Stratotype Section and Point (GSSP) for the Anthropocene Series: Where and How to Look for Potential Candidates", in: *Earth Science Reviews*, 178: 379–429.
Zalasiewicz, Jan; Waters, Colin N.; Williams, Mark; Summerhayes, Colin P.; Head, Martin J.; Leinfelder, Reinhold, 2019a: "A General Introduction to the Anthropocene", in: Zalasiewicz, Jan; Waters, Colin N.; Williams, Mark; Summerhayes, Colin P. (Eds.), 2019: *The Anthropocene as a Geological Time Unit: A Guide to the Scientific Evidence and Current Debate* (Cambridge: Cambridge University Press).
Zalasiewicz, Jan; Waters, Colin N.; Williams, Mark; Summerhayes, Colin P. (Eds.), 2019: *The Anthropocene as a Geological Time Unit: A Guide to the Scientific Evidence and Current Debate* (Cambridge: Cambridge University Press).
Zalasiewicz, Jan; Waters, Colin N.; Head, Martin J.; Poirier, Clément; Summerhayes, Colin P.; Leinfelder, Reinhold; Grinevald, Jacques; Steffen, Will; Syvitski, Jaia; Haff, Peter; McNeill, John R.; Wagreich, Michael; Fairchild, Ian J.; Richter, Daniel D.; Vidas, Davor; Williams, Mark; Barnosky, Anthony D.; Cearreta, Alejandro, 2019: "A Formal Anthropocene is Compatible With but Distinct From its Diachronous Anthropogenic Counterparts: A Response to W.F. Ruddiman's 'Three Flaws in Defining a Formal Anthropocene'", in: *Progress in Physical Geography: Earth and Environment*; at: https://doi.org/10.1177/0309133319832607.

Zalasiewicz, Jan; Waters, Colin N.; Williams, Mark; et al., 2015: "When Did the Anthropocene Begin? A Mid-Twentieth-Century Boundary Level is Stratographically Optimal", in: *Quaternary International*, 183: 196–203;

Zsifkovits, Valentin, 1973: *Der Friede als Wert: Zur Wertproblematik der Friedensforschung* (Munich – Wien: Olzog).

Zurita, Melo; de Lourdes, Maria; Munro, Paul George; Houston, Donna, 2017: "Unearthing the Subterranean Anthropocene". in: *Area*, 50,3 (September): 298–305.

Chapter 3
Transformative and Participative Peace: A Theoretical and Methodological Proposal of Epistemology for Peace and Conflict Studies

Esteban A. Ramos Muslera

Abstract Among the most recent theoretical and methodological proposals in peace studies that have transcended the foundational paradigm of a science initially disconnected from social participation, the proposal of 'Transformative Peace' epistemologically justifies the need to integrate the population in the processes of research, education and action for peacebuilding and conflict transformation as subjects – and not objects – of study and/or beneficiaries of the actions designed by others. This chapter describes the conceptual evolution that has enabled the development of an epistemological perspective in peace and conflict studies that is concerned with and occupied by participatory action-reflection from a socio-praxical perspective.

Keywords Peace · Peacebuilding · Participation · Social research · Peace and conflict studies · Peace theory

3.1 From Negative Peace to Transformative and Participatory Peace

The conceptualisation of peace as an antithesis of war and synonym of order, control and tranquillity within 'ourselves' that gave rise to the Studies of Peace and Conflict in the 1940s restricted the field of peace by limiting the role of the human being in research and peace-building processes. Negative Peace (Galtung 1985) – based on the conception of Greek *eirene* and Roman *pax* – contributed to the crystallisation of an idea according to which it was the states and their military forces who should act as the protagonists of peace – and conflict – and not, in this way, citizenship. Consequently,

Esteban A. Ramos Muslera, Peace Studies Programme Coordinator, *Instituto Universitario en Democracia Paz y Seguridad* (IUDPAS), *Universidad Nacional Autónoma de Honduras* (UNAH); Co-Secretary General *Consejo Latinoamericano de Investigación para la Paz* (CLAIP). *Email*: esteban.ramos@gmail.com, esteban.ramos@unah.edu.hn. *Telephone*: +504 97862636. This chapter summarises the work developed by the author (Ramos 2013, 2015, 2019) on the epistemological foundation of the 'transformative peace' approach and its participatory method for peacebuilding and conflict transformation.

Ú. Oswald Spring and H. G. Brauch (eds.), *Decolonising Conflicts, Security, Peace, Gender, Environment and Development in the Anthropocene*, The Anthropocene: Politik—Economics—Society—Science 30,
https://doi.org/10.1007/978-3-030-62316-6_3

the first studies of the discipline focused on "discovering the causes of war to avoid its outbreak and maintain the Status Quo, but without really considering the need to change the structures of the international system" (del Arenal 1987: 560).

From the 1960s onwards, the concern for the study of war between states shifted to the study of conflicts (Frankel 1973). The new theories of conflict and, above all, the conceptualisation of positive peace proposed by Galtung (1985) led to the development of a second phase in the history of the discipline that would end up shaping the Studies of Peace and Conflict (del Arenal 1987) as a multidisciplinary social science, explicitly orientated by values, that would combine the use of quantitative and qualitative methodologies, and that would focus on the analysis of conflicts with the aim of propitiating their transcendence for the construction of peace (Braillard 1977; del Arenal 1987; Jiménez 2009).

The conception of positive peace managed to overcome the dominant paradigm to date by understanding peace as a social order linked to two necessary and complementary conditions: on the one hand, the absence of the three typologies of violence distinguished by Galtung himself (1985) – direct violence; structural violence; and cultural violence – and, on the other, the effective presence of their opposites: direct peace (absence of violent confrontations between human beings and presence of conflict regulation mechanisms); structural peace (absence of inequality and material presence of social, economic and political justice); and cultural peace (absence of cultural traits that legitimise the use and reproduction of violence, and presence of cultural traits and universal values that impel direct peace and structural peace), which would together form the three dimensions of positive peace.

Such an approach would allow the discipline of Peace and Conflict Studies to enter previously ignored fields of knowledge, such as development, social movements, poverty, inequality, and human rights, among others (Jiménez 2009), increasing, likewise, the interest in social participation (Lederach 1997).

The absoluteness of positive peace, which conditioned the existence of peace to the conquest of a goal – "there cannot be Positive Peace if there are relations characterized by domination, inequality and non-reciprocity, even if there is no open conflict" (Lederach 2000: 35) – led some authors to consider that positive peace constitutes an unattainable utopia rather than a palpable reality dependent on human action (Muñoz 2004). This approach allowed Francisco Muñoz (2001) to develop a new epistemological perspective of peace, known as 'imperfect peace', under which peace would be understood as an unfinished social reality not defined by the absence of any violence. According to this conception, violence, conflicts and peace are independent realities that coexist: the presence of violence and conflicts does not imply the non-existence of peace, nor does it preclude peaceful action, nor the presence of peace actors, nor the existence of spaces or moments stimulated by peace. From the imperfect peace perspective, it is understood that human beings, who are subject to a culture of belonging acquired during socialisation processes, opt for one behaviour or another, contributing to peace, violence or conflict.

The epistemological shift proposed from positive peace to the imperfect peace perspective contributed decisively to the development of the constructivist approach to the discipline. This was evident in the fact that peace was conceptualised as a palpable reality, entirely dependent on human action, and not on the conquest of a predetermined result by the presence or absence of direct, structural and cultural violence. Conceiving peace as a process linked to human action justified seeing peace not as the goal to be achieved in an eternal obstacle course, but as the means and the end at the same time: a sort of 'dynamic balance' in permanent advance and retreat produced by the interactions of spaces, actors, practices, feelings and values of peace present in every culture (Muñoz 2004).

The imperfect peace approach deepened the existing relationship between peace, conflict and coexistence. Understanding that, as human beings, we opt for the realities of peace, conflict and violence in our coexistence, and sustaining that such realities are external to human beings but relative – since they penetrate with greater or lesser intensity, in one way or another, in the being of each one of us – situated the discipline in the antechamber of the constructivist socio-praxic perspective (Montañés 2012; Ramos 2015: 48–61). This complex constructivist-materialist perspective differs from the abstract-constructivist approach insofar as it does not consider the constructed reality to be a simple capricious representation produced by the particular inventiveness of each one – nor a relative reality that penetrates from 'real reality' into the being of each one – but a praxical internal construction of the subject and not dissociable from him or her (Varela 2002), that is socially compatibilised at the same time that it is constructed[1] (being, therefore, dependent on the social practices and the given material conditions at the same time that it transforms the subject, the social practices and the given material conditions). Reality is a social construction that emerges within the framework of the compatibilisation process of senses inferred to the received stimuli derived from the sensorimotor links that, among human beings, are produced in the networks in which we participate to address our needs.

> It is the material conditions that force us to construct one or another reality. That is why this constructivism is called materialistic. The constructed reality does not respond (…) to a historicist determinism in which the person is programmed to serve a certain purpose. We are not in an existential emptiness (…), we are beings submitted to a sociohistorical context, which also inhabits us, being the singularity of each subject the entity where the objective reality is produced (Montañés 2012: 453).

Thus, the socio-praxical perspective allows us to conceive peace, conflict and violence as praxical social realities that emerge from the process of compatibilising realities in each space of human interaction to which we give life in a transductive way (Villasante 2006), constructing and transforming, uninterruptedly, both the social realities we build and all those responsible for their construction. This socio-praxical approach, applied to peace and conflict studies by Manuel Montañés and Esteban A. Ramos (2012, 2019; Ramos 2013a, 2015), crystallised into the theoretical and methodological proposal of transformative (and participatory) peace.

---

[1] Reality as a permanent inference of sense made on the ongoing flow of experiences (Varela 2002). "We usually believe, naively, that there is a world independent of us. Due to a lack of attention training, we do not realize the co-dependence and dynamics of the inside and the outside, in an original and indissoluble unity in permanent flow" (Montero 2009: 151).

3.1.1 *Conceiving Peace from the Socio-Praxic Perspective: Transformative Peace*

The emergence of the imperfect peace approach, as well as other approaches that criticise the dominant paradigm of 'Liberal Peace',[2] have emphasised the importance of social participation in research and action for peace. The concept of transformative science (Schneidewind et al. 2016a, 2016b) agglutinates multiple disciplines and epistemological approaches that debate and propose the involvement of stakeholders in research processes with the clear intention of contributing to the transformation of problems through transdisciplinary collaboration and through the integration of the knowledge of researchers and non-scientists. However – given that the relevance of the aforementioned trend is growing but not yet hegemonic – there are still very few approaches that go beyond considering participation as merely optional or only convenient,[3] depending on the interest of the researcher or the specific characteristics of the peace-building process.

The 'transformative peace' proposal is based on the fact that promoting integral social participation in peace research and peacebuilding processes is an intrinsic need for the discipline. This approach addresses the relationship between peace, coexistence, human needs and social participation, and argues that the practice of limiting participation in research and peacebuilding processes has a flawed epistemological basis: the consideration of human beings as mere subjects who are subject to the influences of external forces that determine their existence.

According to the transformative (and participatory) peace approach, peace, conflict or violence are not understood as substantive external realities (absolute or relative), but as internal realities praxically constructed and socially crystallised when human beings "act to satisfy some needs or others, in one way or another, with some people or another, in one space or another, build internally, project, make compatible and crystallise peaceful, conflictive and violent social realities, conditioned by networks, cultures, needs and social realities – which are also praxically constructed" (Ramos 2015: 257). Thanks to the reflective nature of human beings, which allows us to see others seeing the reality that we see in the process of making realities compatible (Montañés 2006), we are capable of crystallising some or other coexistence models according to the means we use and the way in which we address our needs within the framework of the relationships we establish; as previously discussed, each human being constructs his or her own non-transferable realities and makes them compatible with other human beings. As Glaserfeld (1994: 138) suggests: "I have not built this table, but I adapt to the table by not crossing it". The practice

---

[2]The book *Pax Crítica: Theoretical Contributions to the Post-Liberal Peace Perspectives*, coordinated by K. Pérez de Armiño and Zirion (2019), presents several critical theoretical and methodological approaches to the 'Liberal Peace' perspective.

[3]The study of action research as a tool for peacebuilding made by Geoff Harris and Sylvia Kaye in their book *Building Peace Via Action Research* (2017) is a good example of this. The book analyses several case studies of participatory peacebuilding processes made over brief periods of time, involving small populations or restricted geographical areas.

of fitting realities allows us to act socially in order to meet our needs, transforming the environment in which we interact, our own constructed realities, ourselves and the social realities that we crystallise, in the process. The open (Bertalanffy 1978), praxic and autopoietic (Maturana/Varela 1990), reflective (Montañés 2006) and not trivial (Foerster 1991) human nature enables socially crystallised realities to change by the articulation of the newly sensed inferences that we permanently apply to the received stimuli over the previously constructed and compatible realities in order to address our needs. In this way, we build (and transform) models of coexistence: by being in a permanent relationship with other human (and non-human) systems and finding ourselves obliged to address our own needs through the implementation of actions, we impact on other systems, cause the emergence of different coexistence models in permanent transformation, and depend on the practices we implement to meet our needs and the assessments we make of them. In order to study human needs, Max-Neef et al. (1986) took into account the actions that are carried out to address human needs – means – and the way in which these are met – modes – conceiving five types of satisfiers.

From the sociopraxic perspective it is understood that, according to the satisfiers that we human beings build, apply, reproduce or impose in the cohabitation spaces, the construction and crystallisation of five coexistence models for attending to needs is propitiated.[4] The synergic coexistence model is considered to be the pacific one: when human beings put into practice actions that allow them to meet their own needs while promoting responsiveness to the needs of others, producing responses valued as non-violators, inhibitors or pseudo-satisfiers of the needs of the whole population, they propitiate the construction of peaceful coexistence models. That is, they build 'transformative peace'. When, on the contrary, we act to meet our own needs by putting into practice actions viewed as harmful to the needs of others, we propitiate the crystallisation of violating or inhibiting coexistence models, we generate violence and, potentially, conflicts.

[4]Coexistence models for attending to needs (Ramos 2016: 519):

1. *Violator coexistence model*: articulated from the production, reproduction and imposition of satisfiers that not only annihilate the possibility of addressing the need they seek to meet, but also make it impossible to address other needs (one's own and those of others);
2. *Inhibitor coexistence model*: articulated from satisfiers through which coverage is given to a specific need, making it difficult to meet other needs (one's own and those of others);
3. *Pseudo-satisfactory coexistence model*: articulated from the construction, reproduction and/or imposition of pseudo-satisfiers that feed the false sensation of addressing a certain need;
4. *Singular coexistence model*: articulated from attention to a single need without taking into account the rest of one's own needs or those of others, although not inhibiting or violating possible attention to those other needs.
5. *Synergic coexistence model*: articulated from the construction and reproduction of satisfiers that propitiate attention to a need, being able, in turn, to contribute to attention to other needs and the needs of others.

The multiplicity over time of action-responses and assessments on the part of the different human systems, as well as the involvement of different actors and sociocultural networks[5] in the process, is a constituent part of peace. A peace that, as has been suggested, is formed according to the following cycle (Montañés/Ramos 2012, 2015):

1. As human beings, we interact in spaces of cohabitation or interrelation.
2. We receive stimuli through our sensory system, we value them, and we infer particular meaning from them, constructing internal realities in accordance with the principle of reflectivity: to see the other seeing what one sees.
3. We proceed to the organisational closure in search of the internal balance and we emit responses in accordance with the realities constructed to meet our own needs.
4. We make the realities constructed between human systems compatible by fitting them together, and we crystallise objectified social realities.
5. We encourage the concatenation of action-responses and assessments that promote the linkage of new actors and networks to the process, generating:

   5.1 Concerns, resistances and a multiplicity of action-responses that perpetuate or deepen discomfort in human beings and sociocultural networks, causing social conflict to emerge and encouraging the rise of pseudo-satisfactor, inhibitor or violator coexistence models of the attention of needs; and/or,
   5.2 Synergic attention to the needs of human beings and sociocultural networks that promote the emergence of synergic coexistence models.

Both conflictive and peaceful realities are processes in permanent transformation due to the redefinition, reorganisation and/or re-prioritisation of particular needs, as well as variations in the action-responses (and in the assessments of the stimuli received from them) that human beings and sociocultural networks put into practice.

Transformative peace is therefore conceived as a process of a praxical nature in whose development three dimensions intervene: the contextual dimension, which refers to the grammatical, psychological, situational and socio-economic-historical-cultural context that human beings construct in the coexistence-interrelation framework; the logical-complex-strategic dimension, which is related to the ability, skill and way in which human beings reason and act socially; and the evaluative-creative dimension, which has to do with the human capacity to value and construct diverse realities in the process of compatibilisation. Each of these dimensions is associated with a series of elements that make up a complex structure, and that function as potential dynamos that amplify, buffer and/or diffuse peace (and conflict), taking into account the following principles (Ramos 2015: 117–120):

---

[5] Social networks are configured from the links (or nodes) existing between a set of elements – persons, actors (Garrido 2001). There are different types of networks or social groups (family networks, production networks, networks that are articulated in leisure time or in the neighbourhood area, personal networks, telematics networks, etc.) and all of them influence the behaviour of their elements, as they are constituted by as well as constituent of each element.

Table 3.1 Elements of peace and conflict from the socio-praxic perspective. *Source*: Prepared by the author based on Ramos (2015: 117–118)

Dimensions	**Constituent elements, potentially amplifiers, dampers or diffusers**
Contextual dimension	Actors – networks (parts, groups or sociocultural networks).
	Relationships.
	Well-being, joy or annoyances, worries, dissatisfactions.
	Nuclear issue(s) and polarising positions.
	Practices: action-responses.
	Needs.
	Positive aspects or socio-political-economic problems.
	Power.
	Historical experiences and origin.
	Situation-related scope or scale.
	Formal and informal rules and regulations.
	Cultures or uses and customs.
	Wishes, feelings, passions.
Logical-complex-strategic dimension	Methods (action logics, strategies and tactics).
	Objectives or goals.
	Interests and expectations.
	Roles and positions.
	Lifestyles or regulating styles.
Valorative-creative dimension	Assessments or sense inferences.
	Redefinition, re-prioritisation of needs and reorganisation of the system: transformations in elements of the contextual dimension and the logic-complex-strategic dimension.

1. The elements that potentially shape peace (and conflict) have a more or less significant influence on its dynamics. They are interdependent variables, some of which acquire greater relevance on some occasions while others become more relevant on other occasions.
2. The elements that potentially shape peace (and conflict) are endowed with precise content in each specific process, and depend on the actors and socio-cultural networks that foster their development in each cohabitation or inter-relational space (Table 3.1).

3.1.2 Methodological Implications of Transformative Peace in the Discipline of Peace and Conflict Studies

The recognition of peace as a socio-praxic process allows us to lay the theoretical and methodological foundations of an epistemology of peace conceived from a complex paradigm: the conception of transformative peace, considering the human being as the practice builder of social reality – and neither as a mere object nor as a subject whose behaviour is fully or partially determined by one or other pre-existing external realities – makes it possible to configure an epistemological framework in the discipline of peace and conflict studies that requires the full involvement of the protagonists of peace, violence and conflict, i.e. with human beings as subjects in the processes of research and action for peace and conflict transformation.

Indeed, if the production and reproduction of more synergic and recursive coexistence models of peace – or more violent and conflictive ones – depend on the realities, action-responses and assessments that as human beings we are capable of constructing and putting into practice, collectively, to address our own needs and those of others, it is possible to conclude that engaging in peace and conflict as subjects of study and action is not a discretionary matter. To conceive human beings as subjects, and not as objects of study or beneficiaries of the action designed by others, is an epistemological need. How else could we discover the problems and conflicts that crystallise socially, ascertain how they arise, define what actions to implement in order to address them, and build proposals that generate synergic coexistence models?

Proceeding – as is often done – in accordance with the classic paradigms of research and social intervention based on the configuration of a diagnosis that tells only the story of the technician – and does not include the secondary reflections that emerge in any participatory process (Montañés 2006) – makes it difficult to guarantee that the proposals put forward are effective when it comes to meeting the needs of the population (or transforming them) in a synergic way. That is the main disadvantage faced by those peace programmes designed from large offices or 'elite hotels' (top-down) (Lederach 1997: 84) , or those that seek to impose the use of prefabricated peace formulas.[6] If citizens are not included as a subject in the analysis and configuration of a self-diagnosis that makes it possible to understand what the needs, problems or conflicts are and how they evolved, it is unlikely that prefabricated proposals will be effective when it comes to addressing their needs, problems or conflicts.

Authors who subscribe to the conceptualisation of negative peace by denying the praxic nature of the human being are unaware of the relevance of social participation; they understand that subjects are mere automated objects constrained by the mandate of external realities that determine our being. This is why they proceed in the classic

[6]This is, indeed, the fundamental problem that has manifested itself recurrently in the so-called 'peacebuilding operations' promoted by the United Nations and conducted in accordance with the amalgam of currents of 'Liberal Peace' that ends up assimilating peace with governance (peace-as-governance) (Mateos 2011).

non-participatory way: why should the people be involved if peace is signed by states? The authors who transcend this premise, but still understand human beings as a subject exposed to the influence of external forces – absolute or relative – that to a greater or lesser extent permeate the essence of being and thus determine its behaviour, prioritise methods for research and/or action for peace that restrict participation and obstruct the co-construction of peaceful coexistence models. If human beings are (totally or partially) determined by external forces, it would be pointless to waste resources and efforts in promoting the integral participation of all citizens.

Those who conceive human beings as subjects more (or less) determined by the degree of penetration of one or another culture into their being, for example Muñoz (2004) from the perspective of imperfect peace, opt for methods that favour the participation of the representatives of the cultural identity postulates, but not of the population as a whole. The same happens when, as Zapata (2009: 26) – following Lederach (1997) – proposes from a positive peace approach, human beings are considered more (or less) determined by their basic needs: "thinking about peace building activities is a luxury that [people from the social base] cannot assume" as they have to focus their attention, according to the author, on security, housing, health or food needs. It is clear, therefore, that it would make no sense to design peacebuilding actions with human beings who are unable to carry them out, to the point that it could even be considered reckless to propose strategies aimed, as Zapata (2009: 26) continues, at something more than "impacting on high levels of leadership […] in eventual negotiation processes".

As opposed to these conceptions, the socio-praxic perspective argues that the only way to build compatible realities that allow synergic attention to needs is to promote a dialogical strategy that fosters a *sentipensante*[7] (Fals Borda/Brandao 1987; Moraes/Torre 2002) dialogue between human beings as well as the collective construction of knowledge and proposals from the particular logics and resources of each one. Therefore, from the perspective of transformative peace, a participatory method of research and action for peace and conflict transformation is proposed – one that is capable of promoting the integral participation of the whole population: the *participatory construction of peaceful coexistence method*. This method, in which human systems are involved as subjects of study and action, is conceived by Ramos (2015: 211–212) as:

- A participatory method of action research designed to promote the transformation of coexistence models that inhibit and violate attention to needs into synergic and recursive coexistence models for attending to these needs.
- A participatory method of action research that is flexible, complex and dialogical and in which all of us, collectively, have the possibility of building, deconstructing

[7]Neologism coined by Fals Borda; translated as sensing-thinking by Laura Rendón in her book *Sentipensante (Sensing/thinking) pedagogy: educating for wholeness, social justice and liberation* (2009) and conceived by Eduardo Galeano, a delicious writer, as "the language that speaks the truth, the one able to think by feeling and feel by thinking".

Table 3.2 Distinctive features of theoretical approaches to peace. *Source*: Prepared by the author based on Ramos (2015)

	Negative Peace	Positive Peace	Imperfect Peace	Transformative Peace
Principle defining peace	The law, order, control and absence of war.	Absence of violence (direct, structural and cultural), and presence of its opposites: direct peace, structural peace and cultural peace. Attention to basic needs. Social justice.	Interaction of pacifist practices, feelings, ideas and expressions, generating 'dynamic balances'.	Participatory praxis of coexistence models of synergic attention to needs.
Nature of Peace	Result: end of confrontation/ establishing a socio-political order not necessarily legitimate.	Social order determined by the absence of violence and the presence of social justice.	Option/ negotiation process in which the presence of violence is minimised, and social justice is maximised.	Socio-praxic process of construction and implementation of synergic coexistence models for attending to needs: social construct in permanent transformation that acquires particular characteristics in each case.
Style of action for peace and conflict transformation	Force: imposition, top-down.	Collaborative: top-down negotiator or bottom-up.	Collaborative: negotiator.	Participatory: action-reflection-action at all levels and scales.
Key actors in conflict transformation and peacebuilding	Military forces, state, inter-national organisations.	Organised civil society, NGOs, state, international organisations, military forces.	Social movements, organised civil society, NGOs, state, international organisations, military forces.	Everyone: grass-roots citizenship, social movements, organised civil society, NGOs, state, international organisations, military forces.

and reconstructing realities aimed at promoting the articulation of peaceful coexistence models; of asking, reflecting and making proposals that make it possible to extend the spaces of peaceful coexistence; of limiting the presence and reproduction of violence; of promoting the emergence of networks, powers and cultures valued as contributors to peace; and of transforming our realities and ourselves in the process.

- A participatory method of action research in which all of us, especially the ones made most vulnerable, have the possibility of participating integrally, and of making their voices, thoughts, needs, demands, logics, stories and visions count.
- A method of participatory action research that adapts tools, techniques and even classical methods to a participatory way of proceeding.

In short, a participatory method of research and action that generates transformative peace, which makes peace and conflict studies the discipline of collective action-reflection-action for the construction of coexistence models of synergic response to needs.

The following Table 3.2 shows some of the distinctive features of theoretical approaches to peace in relation to research and action for peace, and conflict transformation.

3.2 A Participatory Peacebuilding Method

As argued by Bourdieu et al. (1989: 27), the division between scientific-theoretical knowledge and opinion-driven practical knowledge is not an irrefutable universal truth, but rather a complementary duality. To address peacebuilding and conflict-transformation processes, given that theoretical and practical knowledge are of equal value, it is necessary to resort to methods of research and social transformation in which all human beings are considered subjects and not mere objects to be studied or simply consulted. As Villasante et al. (2003) state, this entails the need for deeper development of a critical-transformative social research approach; theoretical and methodological contributions to participatory action-research made by Latin American authors since the 1970s has greatly assisted the development of this line of thought initially developed by Lewin (1946): Fals Borda (1993), Vio Grossi (1981), Molano (1978), Oquist (1978), Rojas (1977), Briones (1975), or Sanguineti Vargas (1981). These proposals were characterised by the active commitment in defence of those excluded by hegemonic powers and were inspired by the emancipation pedagogy of Paulo Freire (1971) and the liberation philosophy of Enrique Dussel (1977), and are today being enriched by contributions made by Zemelman (2005), Castro-Gómez/Grosfoguel (2007), and Quijano (1992), among others, within the framework of the decolonial epistemological perspective, or from a socio-praxic approach that states that the production of scientifically valid knowledge, whether it relies on the observation of facts and the construction of postulates and theories, or whether it is based on deductive reasoning, ends up being dependent on the theoretical framework of the research subject: first, because every theory contains premises that, being true, are non-demonstrable (Gödel 1962); and, second, because it is the theory itself that constructs that which is observed according to a purpose in relation to the individual's conception of the world (Montañés 2006). Thus both the rationalist-deductive method and the empiricist-inductive method are conditioned by a certain theoretical framework: the first one directly, because it develops from the general to the particular; and, the second one indirectly, because it is impossible to observe the facts with which one elaborates hypotheses, verifies them and constructs theories if one does not start from some observation criterion.

This compels us to rethink the role of the object and subject of study in social research, leading to substantial methodological consequences. Among these is the fracture of the research subject-researched object binomial in favour of a research-action paradigm that stimulates the collective construction of knowledge among subjects (the researchers and the researched) endowed with reflective capacity. The importance of this fact is not minor for any of the social sciences, especially a discipline in which the 'for what' and the 'for whom' of knowledge production are intimately related to the construction of peaceful coexistence among, by and for human beings – in other words, among, by and for subjects responsible for peacebuilding and conflict transformation. In this sense, the following question should be answered: Is it convenient – or even possible – to build peace and transform conflicts without the active participation of the protagonists of peacebuilding and conflict in the definition of the 'for what' and the 'for whom'?

The knowledge produced in classical research processes is based and conditioned by a predetermined 'what for' and 'for whom', so it is highly probable that, as mentioned above, the results will not address the needs of everyone synergistically. In this sense, the 'what for' of the research and action processes for transformative peace should not be established as a pre-existing goal to be achieved, defined by the research subject according to a certain theoretical frame of reference – nor according to the opinions of instances outside the research-action process itself. If this were the case, a classic methodological paradigm would be reproduced. The 'what for' of the *participatory peacebuilding method* should state the purpose of changing a given situation by establishing a guiding starting point for the action-research process for peace; this starting point should be redefined in a participatory manner throughout the process.

The same goes for the 'for whom'. In an action-research process for transformative peace, the 'for whom' must be made explicit, defined and redefined in a participatory manner, in such a way that all the subjects involved have the opportunity to guide the production of knowledge: having the occasion to define what, when, how, why, for what, with whom and for whom at the time of building knowledge and models for addressing needs synergistically.

Since there is no external instance to which to turn in order to discern whether the knowledge produced is true or not, the legitimacy of the scientific method is not found in the theoretical framework on which it is based – deductive method – nor in the empirical facts – inductive method – but in the process constructed in a participatory manner proposed by the abductive research method. This method promotes the participation of the research subject and of the population under study in a process of social planning that fosters action-reflection-collective action in and between networks, permanently, as a mechanism capable of rescuing "the buried knowledge of erudition and those disqualified by the hierarchy of knowledge and science" (Foucault 1991: 130). Therefore, it is considered necessary to opt for a participatory abductive method and for a *sentipensante* and dialogical strategy for the construction of transformative peace.

3.2.1 An Abductive Method and a Dialogical Strategy

Accepting the logics that structure the thoughts, reflections and positions of subjects, and considering them equally valid and equally enriching realities for the production of knowledge and the construction of peaceful coexistence, requires opting for a method that overcomes the classic Platonic separation between scientific knowledge (theoretical) and popular knowledge (practical). The abductive method is appropriate, as it allows the aggregation of discourses and reflections of the research subjects to produce new scenarios inserted transductively through the implementation of a conversational process (Montañés/Martín 2017).

The 'abductive' method of the *participatory construction of peaceful coexistence* is developed according to a dialogical strategy that allows the formalisation of a representation of the realities constructed by the subjects and their networks through sociological-discursive analysis. Based on what has been lived and experienced in everyday life, this proposal promotes a reflective process with which to transform social conflicts and build peaceful coexistence, involving all social actors in the territory.

This key to this dialogical strategy is to proceed by analysing the discourses and the inferred senses, bearing in mind that it is not possible to separate the product from those who produce the product. The suggested strategy uses the discourses of human beings participating in the process of action-research for peace as a means and unit of analysis, achieving the knowledge of socially crystallised realities to submit them to collective debate and to their later transformation or impulse.

This is feasible because with words we not only say – semantic dimension – but also do things – pragmatic dimension. And if what has been said so much designates things – referential component – as it allows us to relate some words with others – structural component – the study of the doing of what has been said allows us to know (or more accurately, allows us to infer) an image of who says what he says; given that when saying we are said in what has been said – sociolinguistic commitment (Montañés 2009: 51–74).

In fact, by expressing ourselves with some words and not others, we reflect ourselves, at the same time as we reflect our networks and our cultures. For this reason, the method initially considers the collection and sociological-discursive analysis of what people say to interpret the existing relationship between inferred senses and compatibilised signifiers in the social networks. Subsequently, the method proposes the collective debate in and between the networks of the relation established between what has been said and the inferred senses, according to an abductive logic. In this way, discursive interpretation becomes useful for the construction of peaceful coexistence: discursive interpretation facilitates the transformation of coexistence models that violate the attention of needs into coexistence models of synergic attention of the needs of the whole population.

To do so, it is necessary to foster debates among the producers of sayings about what was said (why and for what purpose it was said, and why and for what purpose what was not said was not said) and to configure a self-diagnosis on which the problems, conflicts or confrontations that occur in the territory are considered; which elements are involved in the crystallisation of which conflicts, and the role they play; how they are produced and which actors reproduce peaceful coexistence spaces, considering how networks and social groups, powers and cultures are articulated, as well as the set of potentially intervening elements in the cycle of conflict and peace in its three dimensions.

The proposed participative modus operandi suggests the development of conversational processes of and between the different group realities, in such a way that it is possible to configure a discursive *trencadís*[8] (Ramos 2013) from which to design, implement and follow up on proposals for consensual action built from the

[8]This term is taken from the ornamental technique created by Antonio Gaudí, consisting of the configuration of mosaics from fragments of ceramic tiles. In their likeness, the discursive *trencadís* is configured from the discursive fragments and the debate on the observed inferred senses that were collected for the configuration of the self-diagnosis.

creative feedback of self-diagnosis and the consequent compatibility of inter-reticular realities.[9]

The step taken from a self-diagnosis to a discursive *trencadís* is made possible through the promotion of a second-order reflective process regarding the sayings of the networks (Villasante 1987) among the networks. To do this, it is necessary to inquire about the results, generating questions that elicit new responses that facilitate the development of new paths to follow, in an endless process of answers-questions that articulate collective deliberation and the transformation of the initial positions of the networks.

The process of *participatory construction of peaceful coexistence* assumes the challenge of generating personal transformations aimed at strengthening new internal balances (personal dimension), equitable relations in and between networks (relational dimension), horizontal, equal and participatory power structures (structural dimension), and life formulas that are respectful to nature (environmental dimension[10]) by means of the empowerment of the different actors and social networks, the empowerment of new structures of social participation, the strengthening of management capabilities (co-responsibility) and action for peace (beyond the strategies based on negotiation) in and between the networks. In this sense, the proposed method constitutes a model of social dynamisation designed to facilitate processes of reflection, self-training, programming and collective action similar to the one that, according to Galtung (2008: 36–37), Gandhi raised in relation to *satyagraha* when he conceived it as an instrument of power that "can be managed by all, and particularly by those of the bottom assuming that a sufficient number has embarked on it (…). People trained in *satyagraha* are also able to fight against the abuse of power from the top."

[9]Note that if we only proceeded by means of the configuration of a classic technical diagnosis – based, for example, on surveys – without taking into account the reflections of human beings that give sense to the lived reality, there would be no guarantee that the suggested action proposals would attend to the needs of the population. This is essentially because the configuration of a technical diagnosis based on the distributive method "makes grouped decisions taken by the ones taken by those who have the power to group, be considered as group decisions" (Montañés 2006: 262). In a survey, the answers are implicit in the questions and issues addressed by the questionnaire, preventing even the intermediate positioning between alternatives (Ortí 1986).

[10]Linkage with Sustainable Peace proposal is evident, as it refers to the "processes of recovering from environ-mental destruction, reducing the human footprint in nature through less carbon-intensive – and in the long term possibly carbon-free – and increasingly dematerialised production processes that future generations may still be able to decide on their own resources and development strategies" (Brauch/Oswald Spring 2016: 8).

3.2.2 Purpose, Scope and Methodological Structure of the Peaceful Coexistence Participatory Construction Process

The purpose of the *participatory construction of peaceful coexistence method* is to build co-existence models of synergic and recursive attention to needs; the scope of such models is specified in the process itself and related to the participation and involvement of the different actors. In this sense, and in accordance with what Montañés (2009: 86) acknowledges for participatory research, a process of *participatory construction of peaceful coexistence* could consider the following possible scopes that would condition the magnitude of the process as a whole:

1. A process that is orientated towards recognising and taking action on the socio-cultural problematic in general by addressing the conflicts and the aspects that promote the peaceful coexistence of a specific population segment.
2. A process that is orientated towards recognising and taking action on a specific problem, conflict and/or aspect which affects, worries or benefits a specific population segment in order to promote peaceful coexistence.
3. A process that is orientated towards recognising and taking action on some problem, conflict and/or aspect that promotes the peaceful coexistence of the population in general.
4. A process that is orientated towards recognising and acting on each and every sociocultural problem, conflict and aspect that affects, worries or benefits the population as a whole in order to promote peaceful coexistence.

Depending on the initial scope, the methodological design of the process acquires its particular characteristics, although the approach to what is considered to be the object of analysis must always be carried out as a "total social phenomenon" (Mauss 1979: 260); that is, in such a way as to facilitate action-reflection-action in and between the networks with respect to the potentially intervening elements in the three dimensions of peace and conflict.

Any process of *participatory construction of peaceful coexistence* should favour the permanent creation of new scenarios, new realities, new questions and new answers in an open process,[11] with a design in permanent reconstruction, adapted so that the participating population influences the analysed topics and the objectives, reflection on the problems, and the definition of socio-communitarian action for the development of coexistence models of synergic attention to needs. In this sense, the coordinated and simultaneous development of at least the following three strategic lines[12] is considered appropriate (Ramos 2016: 525):

[11]However, this does not mean that the process is at the expense of improvisation. The way of proceeding must be organised according to the principles of "complex, dialogic, hologrammatic and recursive logic" (Morin 1994: 106).

[12]Depending on the specific context in which the process is carried out, these strategic lines acquire some or other specific contents, and it is advisable to incorporate others if considered pertinent.

- Construction and impulse of actions and strategies that contribute to the synergic attention of needs.
- Regulation and constructive transformation of crises, controversies, problems and conflicts.
- "Education for Transformative Peace" (Montañés/Ramos 2019: 467).

This methodological process considers the proposals from participatory action-research made by Orlando Fals Borda and Carlos Brandaö (1987), and of socio-praxis made by Manuel Montañés (2006, 2009) and Tomás R. Villasante (2006); and adapts to a participatory approach many techniques and methods conceived from the paradigms of positive peace and 'imperfect peace' – the Transcend method (Galtung 2006), the integrated conceptual framework (Lederach 1997), mediations for peace (Muñoz 2004) – from the Gandhian *satyagraha* (Sharp/Finkelstein 1973) and from the liberating praxis of Freire's awareness processes (Freire 1971). These contributions have been integrated into a five-phase process in which dialogue and participatory action-reflection-action are promoted at various levels and in various spheres. For this purpose, a series of techniques, dynamics and tools for research and education are used to facilitate the development of the above-mentioned strategic lines. These phases are conceived as non-sealed departments, interlinked with each other, and in permanent collective reconstruction.

The proposed methodological process develops like an expansive spiral, abductively enabling knowledge to be built, deconstructed and transformed through successive moments of information-gathering and collective construction of knowledge in progressively concurrent spaces, in and between networks. The interaction with the first socio-cultural actors and networks of the territory is considered the starting point. In this initial phase, the configuration of a pre-diagnosis is proposed, as well as the preliminary definition of the components that will be developed later. During the second phase, the process requires the identification and involvement of new networks, the adjustment – or redefinition – of the assessment carried out in the first phase, the collection of primary information, and the elaboration of a network self-diagnosis. In the third phase, the self-diagnosis information is communicated among the networks in order to obtain feedback. The results of these second collective reflections allow the construction of the discursive *trencadís* on which, in the fourth phase, the planning tasks are based; and in the fifth phase, the execution, follow-up and evaluation of the actions that contribute conflict transformation and peacebuilding in the defined scopes and scales are carried out. Once these five phases have been completed, the methodological process will resume its development from a second starting point with a new scope and particular characteristics. In such a manner, the suggested methodological course promotes the participation of and among the networks sequentially, establishing points of closure or conclusion that function as anchors between the phases of the process. In essence, these points that close and open the participation and reflection within and between groups aim to promote qualitative leaps between the phases, on the one hand allowing the process to be re-opened in a new moment, and on the other serving as an antidote to the potentially eternal abstraction or undefinition of the debates.

3.3 Conclusions

- The conceptualisation of peace from a socio-praxical perspective and its associated peacebuilding proposal (the method for the *participatory construction of peaceful coexistence*) is offered as a contribution to the theoretical and methodological corpus of the discipline of peace and conflict studies.
- The development of this proposal, which is the outcome of a dialogue initiated in 2010 by a group of researchers committed to participatory methodologies and enriched by the organisational processes articulated in the *Minga* of social and community resistance of the south-west node of the People's Congress in Colombia, has stimulated participatory action-reflection-action for peace and conflict transformation in both the academic and the popular action spheres: The approach of transformative peace has been the foundation for some research studies, publications and educational processes promoted in Latin American academic spaces. At the same time, it has underlined the need for integral and binding participation in many popular organisational processes.
- Transformative peace goes beyond the consideration of social participation as merely optional by demonstrating that it is necessary and pertinent when promoting peace-building processes, and by understanding that peace cannot be ordered or decreed by fitting prefabricated or imported formulas or by imposing agreements signed in negotiations carried out between actors who arrogate to themselves the representativeness of the people's feelings and thoughts.
- The broad and decisive involvement of the entire population is considered key to peacebuilding processes. In this sense, it is argued that the method for the Participatory Construction of Peaceful Coexistence favours peacebuilding and conflict transformation, especially at local and community level. The process developed in *Las Palmeras, Córdoba*, Spain (Ramos, 2013) is a good example of this: the inhabitants of this neighbourhood, historically excluded and marginalised, were able to promote collective transformations based on their needs and the means and modes of addressing them, on the networks of relationships, on uses and customs, and on the powers and conflicts crystallised in the socio-cultural environment. Similarly, multiple organisational processes in Colombia, articulated around the Peoples' Congress, constitute a paradigmatic example that shows that the participatory construction of Plans for a Dignified Life (*Planes de Vida Digna*) and the promotion of participatory decision-making mechanisms contribute to the expansion of models of synergistic attention to needs (Ramos 2013b, 2016) and should be taken into account for the design of effective peacebuilding processes.
- The further development of this approach by research centres and peace activists would contribute to the disruption of the equivalence that, from the hegemonic approach of liberal peace, is made between peace and governance. It would also promote the emergence of new participatory processes of peacebuilding, more viable and with a greater impact. In this sense, the promotion of research and

action for peace efforts inspired by the Sustainable Transition paradigm is considered appropriate, especially when taking into account that the global and long-term transformation of the systems of production, consumption and governance proposed by this paradigm requires "a change in culture (thinking on the human-nature interface), world views (thinking on systems of rule, e.g. democracy vs autocracy, on domestic priorities and policies, and on inter-state relations in the world), mindsets (the strategic perspectives of policymakers), and new forms of national and global sustainable governance" (Brauch/Oswald Spring 2016: 4), and that, for this purpose, the active action-reflection of the population is essential.

References

Arenal del, Celestino, 1987: "La Investigación sobre la Paz: Pasado, Presente y Futuro", in: *Congreso Internacional sobre la Paz*, Tomo II (Mexico City: Universidad Nacional Autónoma de México): 549–586.

Bertalanffy, Ludwig, 1969: *General System Theory: Foundations, Development, Applications* (New York: G. Braziller) [1978: *Tendencias en la Teoría General de Sistemas* (Madrid: Alianza)].

Bourdieu, Pierre; Chamboredon, Jean-Claude; Passeron, Jean-Claude, 1968: *Le métier de sociologue* (Mouton/EHESS) [1989: *El Oficio del Sociólogo* (Madrid: Siglo XXI)].

Braillard, Philippe, 1977: *Théories des relations internationals* (París: Presses universitaires de France).

Brauch, Hans Günter; Oswald Spring, Úrsula, 2016: "Sustainability Transition and Sustainable Peace: Scientific and Policy Context, Scientific Concepts and Dimensions", in: Brauch, Hans Günter; Oswald Spring, Úrsula; Grin, John; Scheffran, Jürgen (Eds.): *Handbook on Sustainability Transition and Sustainable Peace* (Cham: Springer International Publishing): 3–66.

Briones, Guillermo, 1975: *Metodología de la Investigación Evaluativa* (Bogotá: Centro de Investigaciones de la Universidad Pedagógica Nacional).

Castro-Gómez, Santiago; Grosfoguel, Ramón, 2007: *El Giro Decolonial: Reflexiones para una Diversidad Epistémica Más Allá del Capitalismo Global* (Bogotá: Siglo del Hombre Editores).

Dussel, Enrique, 1977: *Filosofía de la Liberación* (México: Edicol).

Fals Borda, Orlando; Brandaö, Carlos, 1987: *La Investigación Participativa* (Montevideo: Instituto del Hombre).

Foucault, Michael, 1991: *Saber y verdad* (Madrid: La Piqueta).

Fals Borda, Orlando, 1993: "La Investigación Participativa y la Intervención Social", in: *Documentación Social*, 92, 9–21.

Foerster, Heinz von, 2003: *Understanding Understanding: Essays on Cybernetics and Cognition* (New York: Springer). [Spanish edition, 1991: *Las Semillas de la Cibernética* (Barcelona: Gedisa)].

Freire, Paolo, 1971: *Pedagogía del Oprimido* (Buenos Aires: Siglo XXI).

Galtung, Johan, 1985: *Sobre la Paz* (Barcelona: Fontamara).

Galtung, Johan, 2006: *La Transformazione dei Confliti con Mezzi Pacifici (il Método Trascend)* (Turín: United Nations Disaster Management Training Programme).

Galtung, Johan, 1987: *Der Weg ist das Ziel. Gandhi und die Alternativbewegung.* Hammer [2008: *La Meta Es el Camino: Gandhi Hoy* (México: Trascend Peace University)].

Garrido, Francisco, 2001: "El Análisis de Redes en el Desarrollo Local", in: Montañés, Manuel; Villasante, Tomás R.; Martín, Pedro (Eds.): *Prácticas Locales de Creatividad Social* (Madrid: El Viejo Topo): 67–89.

Geoffrey, Harris; Sylvia, Kaye, 2017: *Building Peace via Action Research: African Case Studies* (Addis Ababa: University for Peace Africa Programme).

Glasersfeld, Ernst von, 1994: "La Construcción del Conocimiento", in: Schnitman, Dora (Ed.): *Nuevos Paradigmas, Cultura y Subjetividad* (Buenos Aires: Paidós): 115–141.

Gödel, Kurt, 1962: *On formally undecidable propositions of «Principia Mathematica» and related systems* (Valencia: Universidad de Valencia).

Ibáñez, Jesús, 1986: *Más Allá de la Sociología. El Grupo de Discusión* (Madrid: Siglo XXI).

Jiménez, Francisco, 2009: "Hacia un Paradigma Pacífico: la Paz Neutra", in: *Convergencia Revista de Ciencias Sociales,* vol. 16 (Universidad Autónoma del Estado de México): 141–189.

Lederach, John Paul, 1997: *Building Peace: Sustainable Reconciliation In Divided Societies* (Washington DC: USIP Press) [1998: *Construyendo la Paz, Reconciliación Sostenible en Sociedades Divididas* (Gernika: Catarata-Bakeaz)].

Lederach, John Paul, 2000: *El Abecé de la Paz y los Conflictos: Educación para la Paz* (Madrid: Catarata).

Lewin, Kurt, 1946: "Action Research and Minority Problems", in: *Journal of Social Issues*, 2,4: 34–46.

Martín, Pedro, 2010: "Elaboración de la Muestra Reticular en un Proceso de Investigación-Planificación Participativa", Paper for the 10th Spanish Congress of Sociology – Spanish Federation of Sociology, Pamplona, 1–3 July.

Mateos, Óscar, 2011: *La Construcción de Paz Posbélica: Análisis de los Debates Críticos a través del Caso de Sierra Leona,* doctoral thesis, Universitat Autónoma de Barcelona.

Maturana, Humberto; Varela, Francisco, 1990: *El Árbol del Conocimiento* (Madrid: Debate).

Mauss, Marcel, 1979: *Sociología y Antropología* (Madrid: Tecnos).

Max-Neef, Manfred; Elizalde, Antonio; Hopenhayn, Martín, 1986: *Desarrollo a Escala Humana: una Opción para el Futuro* (Barcelona: Icaria).

Molano, Alfredo, 1978: *Crítica y política en ciencias sociales* (Bogotá: Punta de Lanza).

Montañés, Manuel, 2006: *Práxis Participativa Conversacional de la Producción de Conocimiento Sociocultural,* doctoral thesis, Universidad Complutense de Madrid.

Montañés, Manuel, 2009: *Metodología y Técnica Participativa. Teoría y Práctica de una Estrategia de Investigación Participativa* (Barcelona: Universitat Oberta de Catalunya).

Montañés, Manuel 2012: "Verdad e Incertidumbre: Una Mirada desde la Perspectiva Sociopráxica", *Prisma Social Revista de Ciencias Sociales*, 8, 440–469.

Montañés, Manuel; Ramos, Esteban A., 2012: "La Paz Transformadora: Una Propuesta para la Construcción Participada de Paz y la Gestión de Conflictos desde la Perspectiva Sociopráxica", in: *OBETS. Revista de Ciencias Sociales* (Universidad de Alicante), 7,2: 241–269.

Montañés, Manuel; Martín, Pedro, 2017: "De la IAP a las Metodologías Sociopráxicas", in: *Hábitat y Sociedad*, 10, 35–52; at: http://dx.doi.org/https://doi.org/10.12795/HabitatySociedad.2017.i10.03

Montañés, Manuel; Ramos, Esteban A., 2019: "La Participación del Estudiantado en el Diseño Curricular de la Educación para la Paz Transformadora", in: *OBETS. Revista de Ciencias Sociales,* 14,2: 437–470; doi: https://doi.org/10.14198/OBETS2019.14.2.06.

Montero, Jaime 2009: "Enacción, Filosofía de la Mente y el Yo", in: *Franciscanum: Revista de las Ciencias del Espíritu*, LI, 152: 133–179.

Moraes, Maria; Torre, Saturnino, 2002: "Sentipensar bajo la Mirada Autopoiética o cómo Reencantar Creativamente la Educación", in: *Revista Creatividad y Sociedad*, 2: 41–56; at: https://bit.ly/2JZGu3w (9 October 2018).

Morin, Edgar, 1994: *Introducción al Pensamiento Complejo* (Barcelona: Gedisa).

Muñoz, Francisco, 2001: "La Paz Imperfecta ante un Universo en Conflicto", in: Muñoz, Francisco (Ed.): *La Paz Imperfecta* (Granada: Instituto de la Paz y los Conflictos Universidad de Granada): 21–66.

Muñoz, Francisco, 2004: *Manual de Paz y Conflictos* (Granada: Universidad de Granada).

Ortí, Alfonso, 1986: "La Apertura y el Enfoque Cualitativo o Estructural: la Entrevista Abierta y la Discusión de Grupo", in: García, Manuel; Ibáñez, Jesús; Alvira, Francisco (Eds.): *El Análisis de la Realidad Social: Métodos y Técnicas de Investigación* (Madrid: Alianza): 153–186.

Oquist, Paul, 1978: Epistemología de la Investigación-Acción (Bogotá: Punta de Lanza).

Pérez de Armiño, Karlos.; Zirión, Iker 2019: *Pax Crítica. Aportes Teóricos a las Perspectivas de Paz Posliberal* (Madrid: Tecnos)

Quijano, Aníbal, 1992: Colonialidad y modernidad/racionalidad, in: *Perú Indígena*, 13,29: 11–20.

Ramos, Esteban A., 2012: "Metodologías Participativas desde la Perspectiva Sociopráxica", in: *Revista Encrucijadas* (Universidad de Salamanca USAL), 4: 95–114.

Ramos, Esteban A., 2013a: *Teoría y Práctica del Conflicto Social y la Paz desde la Perspectiva Sociopráxica: La Construcción Participada de Convivencias Pacíficas en Entornos Socialmente Vulnerables*, doctoral thesis, Universidad de Valladolid.

Ramos, Esteban A., 2013b: *El Conflicto Sociopolítico Colombiano y la Construcción de Paz Transformadora y Participativa: Una Mirada desde el Movimiento Social* (Cali: PcD-Unverteilen).

Ramos, Esteban A., 2015: *Paz Transformadora (y Participativa): Teoría y Método de la Paz y el Conflicto desde la Perspectiva Sociopráxica* (Tegucigalpa: IUDPAS-Universidad Nacional Autónoma de Honduras).

Ramos, Esteban A., 2016: "El Proceso de Construcción de Paz Colombiano Más Allá de la Negociación: una Propuesta desde la Paz Transformadora y Participativa", in: *Revista Ágora*, 16,2 (Universidad San Buenaventura): 513–532.

Ramos, Esteban A., 2019: "Práxis Participativa de la Paz Transformadora: Teoría y Método", in: Pérez de Armiño, K.; Zirión, Iker (Eds.): *Pax Crítica. Aportes Teóricos a las Perspectivas de Paz Posliberal* (Madrid: Tecnos): 425–471.

Rendón, Laura, 2009: *Sentipensante (Sensing/thinking) Pedagogy: Educating for Wholeness, Social Justice and Liberation* (Sterling: Stylus).

Rojas, Fernando, 1977: "El Apego de las Ciencias Sociales a lo Real: Crítica y Política en Ciencias Sociales: El Debate sobre Teoría y Práctica", paper for the Simposio Mundial sobre Investigación Activa y Análisis Científico, Cartagena, 18–24 April 1977.

Sanguineti Vargas, Yolanda, 1981: "La Investigación Participativa en los Procesos de Desarrollo de América Latina", in: *Revista de la Asociación Latinoamericana de Psicología Social,* 1,1: 221–238.

Sharp, Gene; Finkelstein, Marina, 1973: *The politics of nonviolent action*, vol. 3 (Boston: P. Sargent Publisher).

Schneidewind, Uwe; Singer-Brodowski, Mandy; Augenstein, Karoline; Stelzer, Franziska, 2016a: "Pledge for Transformative Science", in: *Wuppertal Papers*, 191: 4–19.

Schneidewind, Uwe; Singer-Brodowski, Mandy; Augenstein, Karoline 2016b: "Transformative Science for Sustainability Transitions", in: Brauch, Hans Günter; Oswald Spring, Úrsula; Grin, John; Scheffran, Jürgen, (Eds.): *Handbook on Sustainability Transition and Sustainable Peace* (Cham: Springer International Publishing): 123–136.

Sharp, Gene, 1973: *The Politics of Nonviolent Action, Part 3: The Dynamics of Nonviolent Action* (Boston: Porter Sargent Publisher).

Varela, Fernando, 2002: *El Fenómeno de la Vida* (Santiago de Chile: Dolmen).

Vío Grossi, Francisco, 1981: "Investigación Participativa: Precisiones de Ayacucho", in: Vio Grossi, Francisco; Gianotten, Vera; Wit, Ton de (Eds.): *Investigación Participativa y Praxis Rural: Nuevos Conceptos en Educación y Desarrollo Comunal* (Lima: CIPCA): 69–80.

Villasante, Tomás R., 1987: "La Ciudad Más que Dual: Pobrezas y Alteraciones", in: *Documentación Social: Revista de Estudios Sociales y Sociología Aplicada* (Barcelona: Icaria), 67: 83–104.

Villasante, Tomás R., 2006: *Desbordes Creativos: Estilos y Estrategias para la Transformación Social* (Madrid: La Catarata).

Villasante Tomás, R.; Montañés, Manuel; Martín, Pedro, 2001: *Prácticas Locales de Creatividad Social. Construyendo ciudadanía,* 2 (Barcelona: El Viejo Topo).

Villasante Tomás, R.; Martín, Pedro; Hernández, Dolores, 2003: "Estilos y coherencias en las metodologías creativas", in: Villasante Tomás, R.; Garrido, Francisco (Eds.): *Metodologías y presupuestos participativos* (Madrid: ICARIA-CIMAS): 17–42.

Zapata, María, 2009: *Acción sin Daño y Reflexiones sobre Prácticas de Paz. Una Aproximación desde la Experiencia Colombiana. Construcción de paz y transformación de conflictos* (Bogotá: Universidad Nacional de Colombia).

Zemelman, Hugo, 2005: *Voluntad de Conocer: el Sujeto y su Pensamiento en el Paradigma Crítico* (México: Centro de Investigaciones Humanísticas Universidad Autónoma de Chiapas).

Chapter 4
Peaceful Societies Through Social Cohesion? The Power of Paradigms for Normative and Interdisciplinary Research

Katarina Marej

Abstract Agenda 2030 sets the goal of building peaceful, just societies – but how do we get there? More and more sociopolitical questions are being addressed to science, and peace research should get involved in order to create an impact. While positive peace is rarely on the political agenda, social cohesion is an issue of growing interest because of the need to find new ways of living together in increasingly diverse societies. However, current policies tend to be exclusionary, due to their underlying paradigms. Yet there is no elaborated paradigm of positive peace that could be contrasted with this. Therefore, existing paradigmatic positions are examined. Because they mostly deal with ontological and axiological aspects, suggestions for suitable supplements concerning methodology and epistemology are made. This will contribute to the development of an integrative paradigm of positive peace which facilitates interdisciplinary communication and enables suggestions to be made for shaping social cohesion in a more peaceful way. Furthermore, indications are given on what to consider as a researcher in interdisciplinary and normative research settings in order to meet social responsibility.

Keywords Agenda 2030 · Social cohesion · Positive peace · Research paradigms · Transformation · Narratives · Interdisciplinary communication · Socially responsible research

4.1 Social Cohesion as a Synonym for Positive Peace?

The UN Agenda 2030 calls upon all states and people to engage in the question of a desirable future society. Peace plays a central role in this process, because the goal to be striven for is "peaceful, just and inclusive societies" (UN 2015). In the Sustainable Development Goals (SDGs) which form part of the Agenda, No. 16 is to "promote peaceful and inclusive societies", and one objective of education should

Ms. Katarina Marej, M.A., Ph.D. candidate; cultural anthropologist, sociologist, historian; currently working on issues of citizenship education in diverse societies; University of Münster, Institute of Political Science, Germany. *Email*: katarina.marej@uni-muenster.de.

Ú. Oswald Spring and H. G. Brauch (eds.), *Decolonising Conflicts, Security, Peace, Gender, Environment and Development in the Anthropocene*,
The Anthropocene: Politik—Economics—Society—Science 30,
https://doi.org/10.1007/978-3-030-62316-6_4

be the "promotion of a culture of peace and non-violence" (UN 2015, SDG 16; SDG 4.7). This task also calls upon the Western and Northern states and stakeholders and civil society to further develop and improve: "All countries and all stakeholders, acting in collaborative partnership, will implement this plan" (UN 2015, preamble).

This idea of a global transformation illustrates the realisation that the modernist idea that a 'catch-up development' along the lines of Western democracies would represent a global patent remedy has become unconvincing. Not only has the notion of 'catch-up development' been criticised politically and theoretically for a long time, but empirical evidence within these states that would justify a model function has also been crumbling for a number of years. Western states are constantly confronted with crisis diagnoses, from financial ones to multicultural fragmentation; the very stability of democracy is being called into question. In order to address this variety of problems, the concept of intra-societal cohesion has come into focus. Since the 1990s attempts have been made to promote cohesion through political measures and to 'activate' the population to achieve this goal. This poses political and academic questions regarding the basis of existing and future policies aimed at the attainment of intra-societal cohesion.

The discourses of social cohesion and positive peace have so far run parallel to and independent of each other. An exchange would be useful and constructive, since both pose the question of a desirable society – although both have conceptual weaknesses, as will be discussed below. While social cohesion was primarily politically initiated, positive peace is rooted in academic tradition. Yet the scope and disciplinary orientation of peace research is not undisputed (see recently e.g. Bernshausen/Bonacker 2015; Wintersteiner/Graf 2015). In academic reality, the dominant focus is on political sciences and international relations (Brunner 2015: 212), and therein on conflict rather than peace research. Here I follow the view and concept of Karlheinz Koppe, who describes peace research as an 'interface research' [*Schnittstellenforschung*]. According to this, everything that "endangers external and/or internal peace" such as "economic, ecological or social phenomena and processes" is relevant to peace research (Koppe 1990: 109; see also Eberwein/Reichel 1976: 60; Heitkämper 1989: 130f.). Therefore, it is indispensable to engage in inter-, possibly transdisciplinary research and to deal with thematically related discourses and "bring together hitherto unconnected theoretical strands" (Graf/Wintersteiner 2016: 79).

This will be pursued here with regard to social cohesion and positive peace. While 'peace' is frequently used at the UN and global level, the term is unusual at the level of nation states (i.e. the future "peaceful and inclusive societies"), at least in Western countries. Instead, 'social cohesion' is a popular term there. Since the UN goals can only be implemented through states, the question arises as to whether these national policies orientated towards social cohesion are simply using a different terminological label and already represent developments towards the goals of Agenda 2030 – or, if not, what the reasons for this are and what should be changed.

Both the requirements of Agenda 2030 and the demand for social cohesion at national level are highly topical and charged. The trend in politics towards using academia to solve sociopolitical challenges or legitimise measures leads to a new level of responsibility for researchers. Fundamental to the academic treatment of normative questions is the need for methodological transparency and effective interdisciplinary communication.

The following chapter tries to constructively address the above-mentioned challenges. *First*, I analyse the theoretical concept of social cohesion and then relate it to strategies and outcomes of practical policies. Then, I try to explain this knowledge and policy production by their underlying paradigms. In the *second part* I explore positive peace as a paradigmatic alternative. As the discussions in this field mostly concentrate on ontological and axiological dimensions, I additionally discuss the transformative and the narrative paradigm in order to add methodological and epistemological aspects which may complete an integrative peace paradigm. *Finally*, I suggest alternatives for cohesion research. The overall aims are to broaden and combine the discussions, to emphasise the importance of transparency in paradigmatic positioning, and to offer orientation by proposing paradigmatic systematisations that provide value-driven objectives.

4.2 Social Cohesion: Origins, Intentions, and Critique

There are two main perspectives on social cohesion in the academic discussion. These correspond to the perceived challenge to be mastered:

1. post-conflict countries which are on the way to reconciliation
2. (Western) democracies which are attested to be in crisis.

Within the first functional frame, social cohesion is seen to help reconciliation processes by creating stronger bonds within and across different groups, and by fostering greater trust in the institutions of government. It aims at decreasing hostility or mistrust between different identity groups or the state and its citizens; the role of change and education/learning is emphasised (e.g. Sayed/Novelli 2016; Fonseka/Glaister 2017; Rubagiza et al. 2016; Ministry of Education Sri Lanka 2008).

In this chapter, I will concentrate on the second strand of social cohesion: approaches in and for Western democracies, especially in Europe. Unfortunately, this strand does not make recourse to approaches from the first one, especially when it comes to the dimension of change and learning, a deficit which will be discussed later.

4.2.1 Origins and Intentions

The dominant narrative in Western literature is a perceived deterioration of social cohesion (e.g. Council of Europe 2005; Jenson 1998; Schmeets/te Riele 2014), caused by:

- globalisation, economic changes;

- global migration, ethno-cultural diversity;
- new information and communication technologies.

These tendencies are said to be challenging national identities – at least in the 'traditional' way they were conceptualised (more or less homogeneous entities, referring to the same history or values etc.). There is concern that without corrective measures, further development will lead to disintegration, asymmetry, competition, and ultimately to the disappearance of society and the dissolution of the social (e.g. Heitmeyer 1997).

As Fischer (2006: 25) succinctly puts it, the problem behind those "crises of cohesion" is political ambiguity. On the one hand, there is a growing ambition to fend off any risk of illiberal ideologies gaining acceptance; on the other hand, there is the open question of 'ties that bind' alongside entrenched notions about the political community. It is not surprising that these tensions are coming to the surface at a time when globalisation and migration challenge to proof the proclaimed plurality, tolerance, and democracy.

The discourse on social cohesion thus arose in concrete sociopolitical situations and with a focus on solutions and applications. At the same time, such discourse provides an example of the way that research is increasingly being used as an advisory or legitimising instance for political measures. The resulting dilemmas and quality requirements for scholars are discussed in the second part of this chapter, but approaches to the concept of social cohesion will be elaborated upon first.

The term social cohesion is characterised by a vagueness or multiplicity that arises from the diverging intentions of the various research and policy concepts related to it. Having unconnected roots in the 1980s, the current discourse gained popularity in the 1990s. The collapse of the Soviet Union led to the re-invention of nation states in the East and the disappearance of an enemy in the West. In addition, there were questions of globalisation, migration and economic adjustment. Linked to this are the questions: What holds us together? Who are we and where do we want to go? The aim was to find ways to deal with heterogeneous populations – in other words, ways to govern diversity.

In Canada, a Social Cohesion Network (SCN) was founded in 1997 (Bittle 2001) with the political participation of the Department of Justice Canada and Canadian Heritage. One year later, this procedure of creating a research task force was adopted by the Council of Europe and led to social cohesion becoming a strategic goal of the Council of Europe (CoE 2004, 2010). As Schiefer/van der Noll (2017: 581) point out, these approaches are mainly aimed at preventing or reducing economic instabilities. The close connection to the concept of social capital becomes clear – social cohesion is to be achieved primarily through integration into the labour market. The same can be said for OECD conceptions (2011). Generally, the concerns of agents from particular fields lead to different consequences. For example, the World Bank focuses on economic development and poverty reduction, the UK on increasing cultural diversity, and in any democracy different political parties pursue their respective goals (Schiefer/van der Noll 2017: 581).

The research on social cohesion is manifold and accesses or connects with various research strands: "the relation of social cohesion to other concepts, integration, harmony, social order or stability, is not clearly established" (Dobbernack 2010: 5; cf. Mekoa/Busari 2018). In addition to those already mentioned, there are further research directions: inclusion, belonging, social identity, coherence, citizenship, welfare, governance, solidarity, social responsibility, security, labour flexibility, migration, multiculturalism, well-being, health – this list is not exhaustive (cf. e.g. CoE: *Trends of Cohesion*). Within these various conceptions, social cohesion reveals a special feature: it can be seen either as an outcome or as a prerequisite for other outcomes or even both at the same time.

Last but not least, a desired outcome is to increase the stability of the political system. Social cohesion can generally function as the cause of and cure for all kinds of conflicts: "significant social problems were re-described as resulting from a lack of social cohesion" (Dobbernack 2014: 9). Due to the scope and orientation of this chapter, it is not possible to discuss the various approaches here, but it is particularly worth mentioning the article by Schiefer/van der Noll (2017), which makes helpful attempts at sorting, and Oxoby (2009), who describes the close connection of social cohesion to labour market integration and 'social capital' from an economic point of view. The following sections are intended to help to clarify and comprehend the different approaches from a paradigmatic angle.

However, this vagueness of the term seems to be one of its advantages, especially in political discourses, because its semantic vacuum can be filled with very different associations and goals. It serves as an 'empty' or 'floating signifier', it "represent[s] an indeterminate value of signification, in itself devoid of meaning and thus susceptible of receiving any meaning at all" (Lévi-Strauss 1950/1987: 55). This means that the term does not refer to a clearly identifiable 'outside', but also addresses unconscious ideas and the affective levels on which people communicate. Thereby, it becomes a representative of the unconscious ideals that enter the various concepts which assign particular semantic values to the term. Therefore, the minimum conceptual consensus is that social cohesion is a desirable feature of a social entity and that it may promote several positively connotated attitudes or practices.

4.2.2 Critique: Exclusive Directions and Outcomes of Social Cohesion Policies

So far, even if the concept itself is pretty open, the objectives associated with social cohesion appear to be all positive. But what are the political realities? Though more research is necessary to be able to provide well-founded diagnoses, the following examples of France, the UK, and Germany already raise concern. Nevertheless, these countries are particularly suitable examples insofar as they are perceived as established democracies and are often regarded as role models in politics and research. Therefore, I analyse their policy approaches to social cohesion in the following subsections. The presentation is based on three identified critical aspects: homogenising and harmonising, re-mystifying, and neoliberal tendencies. These are particularly relevant as they touch the core dimensions of the policy initiatives and they are characterised by inner ambiguity – while actually aiming at cohesion, they produce cleavages themselves. Only a few examples can be given here to illustrate the criticism. For a comprehensive analysis see e.g. Dobbernack (2010, 2014) or Osler (2011), and for an analytical comparison of cohesion-orientated education policies see Marej (in preparation).

4.2.2.1 Community and the Problem of Homogenising and Harmonising Tendencies

The idea that a group of people on a certain territory should form a unit was promoted particularly through the formation process of European states. This claimed commonality could be based, for example, on ethnic or religious grounds, and the most successful way to formalise it is the concept of a 'nation' (Anderson 1983). How a nation can be conceptualised in a democratic and diverse society without reference to logics of exclusion is something most approaches to social cohesion struggle with. The goal is to create a "social imaginary" (Taylor 2004) that connects the individuals with each other and at the same time with the state. This image should be shared by all inhabitants or at least citizens, regardless of their further group affiliations and identity aspects. The aimed imaginary therefore refers to 'community' rather than to 'society' (see Tönnies 1887 for theories about the different organisational structures and ranges).

A widespread approach is that of a "community of values", populated by "good citizens, law-abiding and hardworking members of stable and respectable families" (Anderson 2013: 3). 'Values' and 'morals' are being loaded – the desired political community is not a bureaucratic or formalistic one, but (also) an emotional one. A lack of cohesion can then be seen as a 'moral crisis', 'a mental state'. In his candidacy for presidency Jacques Chirac resorted to the widespread narrative that the "technocratic orientation had made French society cold and inhumane". The community of values should be led by the President as a "moral centre" with the task of "providing a sense of psychological security" (Dobbernack 2014: 172). In contrast, in the UK, the concept of a community of values is based primarily on individuals. They shall build "vibrant communities", "united by a vision, a sense of civic pride and diversity knowledge" (Dobbernack 2014: 172). The communities' futures lie "in their capacity for self-regeneration", their parts shall be "restored and linked up through the tonic of community cohesion, local trust and active citizenship" (Amin 2005: 614).

Local communities and neighbourhoods are often seen as a basic dimension for the functioning of democracy as well as social cohesion (e.g. Putnam's "civic communities" 1993 or LGA 2004). This perspective focuses on co-operation in everyday life, the pragmatic negotiation of difference and community, the solving of real-life problems, and increased security. For Lockwood (1999), the local community is the actual entity which is capable of social cohesion in contrast to (national) "civic integration".

Though both levels are connected, they follow different logics and dissolution factors (corruption in the case of civic integration, deviant behaviour in the case of social cohesion).

Many cohesion policies mix these levels, ignoring the fact that the commonality of a shared life world provides different prerequisites to a unit based primarily on imagination (Anderson 1983). They aim to influence behaviour in order to guarantee civic integration. In the UK, the focus is on labour-market-related codes of conduct: "to shape up and become active, self-reliant and responsible" (Dobbernack 2014: 160). In Germany, notions of community are sometimes expressed in terms of attitudes: "education and performance are values in themselves", "we are part of the West". Other notions relate to interpersonal interaction and behaviour in public: "…we say our names. As a welcome gesture, we shake hands. […] We show our faces. We are not burka" (Minister of the Interior, de Maizière 2017). Social cohesion would work "if only the correct set of attitudes would be adopted" (Dobbernack 2014: 172) and the addressed minority groups would assume new identities (Modood/Dobbernack 2013).

In addition to the one-sided harmonistic construction of community, the problem here is the transfer of principles of community to those of society. The approach of conceiving cohesion not vertically but horizontally means that "greater emphasis is placed on the dimension of inter-personal unity" (Dobbernack 2010: 5). This promotes the emphasis on cultural differences instead of social (socio-economic) similarities or challenges.

The problem of this narrative is that it works with homogenised protagonists and that it is orientated towards the past. The good citizens and their community face two enemies: the 'migrant' or the non-citizen, and the 'failed citizen'. If read benignly, it tries to create an offer for a common cultural orientation. However, the unequal access possibilities criticised by Merton (1968) are not taken into account, nor is the emergence of these 'common' values questioned.

Accordingly, plurality (which is actually constitutive for democracy) and diverging interests are faded out of this image. Instead, there is a "re-description of society as an essentially harmonious and non-conflictual endeavour" (Dobbernack 2010: 3). This notion tends to limit the right to dissent: protest and deviation are at risk of being criminalised. A prominent example is the discussion about limiting academic freedom and freedom of speech in the name of security (De Wijze 2018; Gayle 2019). The state, on the other hand, is conceived as a 'finished' place, and its role in creating vulnerability as well as structural problems in general is denied.

This is critical from both a democratic and a peace perspective, because conflictual deliberation is disavowed as the basis for social decisions. Furthermore, historical developments are negated as well as the hegemonic structures in which this search for harmonious unity is negotiated. At the same time, such a stance promotes authoritarian understandings of democracy, because it defines, "normalizes and pathologizes (i.e., it defines what constitutes a 'social problem' that is worthy of governmental intervention), and it predisposes government as regards the type of political interventions that appear necessary to rectify social problems thus defined" (Dobbernack 2010: 3).

This approach supports those who see social cohesion as a way to prevent radicalisation. In Germany, this argument is mainly brought up to avert the path to the radical Right by portraying a moderate ('normal') national image. In France and UK, due to the experiences of terror attacks and riots, there is a particular fear of Islamistic radicalisation. In both countries, for example, schools were required to have a cohesive and preventive impact. In France, the school subject *Éducation morale et civique* (Moral and Civic Education) was introduced in 2015, and in June 2018 a handbook for teachers was provided to ensure laicity in schools. It contains information on how long a skirt and how wide a hairband is allowed to be without being considered a religious sign. The focus is on repression, not on education and emancipation, and the idea of laicism is abused for securing hegemonic positions and anti-Muslim arrangements (cf. Lindner 2017; Almeida 2017). In England, the school subject 'Citizenship Education' was introduced in 2002 and since 2014 'Fundamental British Values' are to be taught. While actually comprising the values of democracy, rule of law, individual liberty and mutual respect for religious beliefs, they have to be understood in the context they arose. The term was developed as part of the 'Prevent Strategy', which in turn is part of the counter-terror 'Contest Strategy'. This includes further-reaching monitoring mechanisms, and schools are inspected to evaluate whether they meet the requirements. At the National Union of Teachers Conference, this approach was overwhelmingly rejected, among other things because it "disproportionately targeted Muslim students" and the "training given to many teachers was crude and often involves loads of stereotypes" (Adams 2016). In the name of social cohesion, people in schools and other public places are urged to monitor each other; the focus is on suspicion. Both educational policy approaches work with images of an enemy and demarcate them from 'the right and normal'.

Social cohesion in this sense is anti-pluralistic and mostly addressed towards newcomers and established minorities, and expresses itself in schools as "disciplinary forms of citizenship training" (Osler 2011: 201). This also entails the danger of making authoritarian interpretations of democracy socially acceptable, corresponding to the global trend (cf. Foa/Mounk 2017). In addition to the issue of the new authoritarianisms, questions arise about the quality and the social and cultural design of democracy under the conditions of globalisation – topics of current peace research desiderata (Segert 2016).

Following this logic of social cohesion, the aim is not to pursue further processes of democratisation or structural improvements, but to make people agree to the existing systems and to be grateful for being allowed to live within them. This applies in particular to the treatment of migrants, but also to established minorities, because these are the ones who are perceived as disturbing "the assumed calmness of the hosting society" (Bekerman 2019: 3).

Therefore, the aim is the alignment, the assimilation of the deviants to the preferred idea of the good citizen. This consists of "good character" requirements and specifies which kind of family relations, work, skills, and identity are desirable, and which are not. For this purpose, in many countries in the 2000s not only binding requirements for language skills, but also citizenship courses and naturalisation tests were introduced. Here, citizenship is presented as "a prized possession that is to be earned and can be lost if not properly cultivated" and the state is "sacralized by an emphasis on the national community" – two trends which can be grasped as "neoliberal communitarianism" (van Houdt et al. 2011).

Yet, as mentioned above, 'the others' are not only non-citizens but also (this has been dialectically reinforced by immigration regulations) 'failed citizens'. This classification is based on a "hierarchy of character types", targeting especially "disadvantaged social groups", namely "welfare recipients, ethnic minority populations, inhabitants of sink estates, unemployed" (Dobbernack 2014: 177). Thus, certain 'disturbers' of social cohesion are identified. This image, however, is merely the liberal self-narration of the dominant White elites, who confirm that they are "rational, self-owning and independent" (Anderson 2013: 3), and consequently the "privileged groups face no call to evince any particular cohesiveness" (Dobbernack 2014: 177). Social cohesion in this understanding does not promote deliberation but the assimilation of 'the others' to the ideas of the elite:

> Our approach is based on the belief that people must take responsibility for their own choices but that government has a responsibility to help people make the right choices. (SJPG 2007)

4.2.2.2 The Nation and the Problem of Re-mystification

Closely connected with the trend described above is the observed need for re-mystification. In some adaptations, especially when it leaves the economic sphere and enters social, cultural or political space, the cold rationality and fragmentation of the world is lamented and the longing for a warm, emotional, holistic belonging is being expressed. Frequently, the solution is to be found within the concepts of 'nation' and 'patriotism'.

This is one of the fundamental issues in the German debate. The widespread approach of constitutional patriotism is devalued as cold, artificial, rationalistic; it would be a "frail [dünnblütige] professorial fiction" (Schwarz 1990: 3). Instead, an emotional connection should be established, the central feeling in the debate being 'pride'. In addition, there is a "grown culture", which, including its unconscious elements, should serve as a *Leitkultur* ['leading culture' – i.e. the orientation point for those wishing to belong German society] (first in this line of argument at the national level: Merz 2000). Culture and identity are thereby mixed and heightened to form a collective body: "Beyond language, constitution and respect for fundamental rights, there is something that holds us together in our innermost being, something that makes us special and distinguishes us from others" (de Maizière 2017). Generally said, those determinations of the German culture often serve to distinguish it from other (especially Muslim) cultures, whose "pre-modern and archaic values" (AfD 2018) are judged as contrary to Germany's own professed culture of freedom and enlightenment.

While in Germany the argument is 'culture', in France and UK the debate revolves more explicitly around the nation. The nation is not only present on the level of banal nationalism (Billig 1995), but nowadays through "blatant nationalism", the overt "(re-)production of national identity" (Benwell 2014). In France, it was decided in February 2019 that the EU flag, the national flag and the national anthem must be hung in every classroom (Johnstone/Fischer 2019).[1] The 'Fundamental British Values' are depoliticised and conveyed as a "touristic vision", "including red phone boxes, fish and chips, teapots and the Queen" – "there is confusion between values such as democracy and symbols that have become associated with an essentialized and nostalgic view of Britain" (Starkey 2018: 156). Teaching Britishness is glorified as a national initiation process, "as a means to end young people's political disenfranchisement" (Habib 2018: 3). Colonialist discourses are thereby reproduced (cf. Winter 2018), and the nation's culture and history are beyond all criticism: "Lessons should celebrate rather than denigrate Britain's role through the ages, including the Empire. Guilt about Britain's past is misplaced" (Gove 2010, two months before becoming Minister of Education).

What 'nation, as well as 'Leitkultur', promises is an "intuition of fullness that cannot be granted by the reality of the present" (Laclau 1990: 63; 2005: 71, 116); it "alludes to apocalyptic scenarios of social disintegration that are tied up to utopian ideas of a wholesome, peaceful and harmonious society" (Dobbernack 2010: 3). This utopia may be appealing, but the way of trying to achieve it (the basic misconception is the idea that it existed in the past, so it should be possible to 'restore' it) is problematic precisely because it seeks in the past and in old concepts instead of daring to develop something new. For this, deliberative processes would be necessary, even changes for the social majority.

As mentioned above, the most common narrative concerning social cohesion in Western countries is about threat and decay. Interestingly, this is accepted without hesitation by many people. The question, however, is whether this diagnosis goes beyond 'in former times everything used to be better' and actually refers to real times and states of greater social cohesion. In the light of real socio-historical developments, this can be strongly doubted (especially from the perspective of non-members of the hegemonic groups). What has certainly become more difficult is the acceptance of hierarchies due to more information and the increasing diversification and democratisation of societies. Furthermore, there is rarely any discussion about what exactly this sense of cohesion is that is to be awakened on the individual side; what is it good for, and what is the normative impetus?

It seems plausible that this is ultimately a search for orientation in times of rapid changes. What is criticised would then be the lack of stability provided by clear rules and norms and functional assignments within society. Durkheim (1897) argued that this 'anomie' disturbs and hinders social order and integration. As Merton elaborated further, the basis is shared culturally accepted goals, but the possibility of achieving them depends on social conditions which are not accessible to all (Merton 1968: 216). The question arises as to what these unifying culturally accepted goals are and who actually determines them. Another question is why future goals should be based on concepts from the past when their disadvantages are widely known. It appears to be

[1] Jean-Michel Blanquer, France's Education Minister, told *Le Parisien* that this was because: "Civic spirit must be strengthened. Everywhere in the world, knowing the symbols of your country is the most natural thing in the world." See at: https://www.bbc.com/news/world-europe-49553523.

an attempt to conceal the fact that the future is uncertain and that far-reaching changes will come anyway, at least due to globalisation, digitalisation, and climate change. This kind of 'anti-enlightening identitarian security policy' prevents a constructive approach to present and future challenges.

4.2.2.3 The Active Citizen and the Exclusivity of Neoliberal Values

In addition to these above all identity-related aspects, there is another dimension that runs through the entire discourse: the adaptation of people and society to the needs of the economy. Economic growth is frequently seen as a prerequisite, sometimes as a motor and most often as the desired outcome of successful social cohesion policies. The economic system itself and its growth ideology are not questioned; on the contrary, they are invoked as a means to solve all other problems. Closely connected to this emphasis of economic integration is the concept of the active citizen. This assumes that the processes of integration require only individual responsibility and an active commitment to the common good:

> Social cohesion can't be guaranteed by the top-down action of the state or by appeal to tradition. We have to make our lives in a more active way than was true of previous generations, and we need more actively to accept responsibilities for the consequences of what we do and the lifestyle habits we adopt. (Giddens 1998: 37)

The approaches described as 'New Governance' may be seen as a willingness to share the power to define and shape public debates. Social cohesion could provide a framework for "new modes of socio-economic governance" (CoE 2004: 5), "a new relationship between governed and governing" (Blunkett 2002). The connection of social cohesion to concepts of citizenship thus becomes clear, because questions about the distribution of responsibilities are aimed at structures and responsibilities (Jeannotte et al. 2002: 12ff.). This might suggest an emancipatory, future-orientated, creative social project – but, as shown above, in existing policies activity is unilaterally conceptualised as a shift in responsibility, the scope and direction of activity is limited, and citizens are conditioned by 'nudging'. To the extent that this "libertarian paternalism" (Thaler/Sunstein 2009) is even considered a legitimate steering procedure by the state, the question arises as to what constitutes right choices. What is the normative and ethical evaluative framework? Here it quickly becomes obvious that (neo-)liberal logics dominate.

Social cohesion approaches point towards "inactivity, immobility and inflexibility as pertinent causes of social problems", so "regulatory regimes" are sometimes built in response. These "activation policies" were initially aimed at labour markets and welfare regimes, but then developed into blueprints for a generally desirable "realm of behaviour and conduct" (Dobbernack 2014: 174f.). Those programmes did "not ha[ve] the effect of stabilizing society" but "they facilitated the adoption of new responsibilities and requirements", especially for those "social groups that are less able to comply" (Dobbernack 2014: 2).

The current developments illustrate the ambiguity of liberal conceptions. Joppke distinguishes between: liberalism according to Rawls, "with its emphasis on equality, individual rights and neutrality" and promotion of emancipation efforts; and Foucault's liberalism focusing on "power and disciplining", which is about the maximum use of resources and shifting responsibility:

> the contemporary state, hollowed out by economic globalization, is coercing individuals, as well as the communities of which they are a part, into releasing their self-producing and self-regulating capacities as an alternative to the redistribution and public welfare that fiscally diminished states can no longer deliver. (Joppke 2007: 23)

Here it is the principles of liberalism itself that lead to discrimination and support illiberalism, because "contemporary civic integration or workfare policies [...] involve putting illiberal means to the service of liberal goals" (Joppke 2007: 24). The notion of non-optimal resource use leads to group-specific policies, which are not intended to promote their self-determination or emancipation but their assimilation to the needs of economy, as its growth is the ultimate normative direction.

4.2.3 The Demise of Social Cohesion?

The discussion so far has shown that:

- social cohesion is an empty signifier, which can be used to address various issues;
- policy approaches are aligned with current needs and interests;
- a top-down procedure tends towards lack of transparency and uncritical self-affirmation of the social majority;
- the search for concepts of cohesion in the past of a particular society tends to lead to excluding approaches;
- neo-liberal attributions of responsibility are unilaterally addressed to specific individuals and groups;
- consequently, the existing social cohesion policies do not meet the requirements of Agenda 2030.

Therefore, the question arises as to whether social cohesion inherently contains exclusive moments or whether it could be conducive to the goals of global sustainable development. Are the exclusionary policies based on the idea of social cohesion itself or is it due to (respective and general) national adaptations? If one does not regard the overcoming of nation-states as a necessary prerequisite for sustainable peaceful developments, but accepts their relevance out of conviction or pragmatic consideration, it is important to work within the logic of states. This raises the fundamental question about which aspects are to be considered in researching and politically promoting cohesion in order to avoid falling into the trap of backward-looking, exclusive policies – in other words, how can cohesion be promoted without referring to an 'enemy', a question which is also raised in the discourse of *Culture of Peace* (cf. Wintersteiner 2006: 99). How can the goals of Agenda 2030 be adapted to specific contexts? Can the concept of social cohesion be used to promote inclusion, sustainability and maybe even peace, all the more since its core dimensions are:

- social relations,
- a sense of belonging,
- orientation towards the common good (Schiefer/van der Noll 2017).

To conceptualise social cohesion differently, it is necessary to take a step back and delve a level deeper. What strikes one's eye is that the dimension of peace, sustainability and inclusion is not yet deeply reflected within the discussion. Since the discourse is already pretty much accepted and even tries to combine economic and 'somehow social' aspects, this is a great opportunity for peace research to enrich and redirect the discussion, especially if understood as an interdisciplinary endeavour. While the social cohesion discourse is mainly led by economists and political scientists who tend to be focused on mechanistic, cybernetic understandings, peace research could provide deeper analyses and strengthen humanitarian and social perspectives. It is interesting to note that the advocates of cohesion-orientated measures are to be found mostly in the majority society. This concerns both society and politics as well as academia with its disciplinary traditions, in which political sciences and economics, for example, tend to take a majority social view, while social, cultural and cognitive sciences more often explore minority perspectives and point to the dangers of oppression, overwhelming, and exclusion. If initiatives for social cohesion are aligned with peace, the necessary perspective is not to fight against or ignore each other's position, but to achieve constructive understanding.

Changing the perspective in that way might lead to the result that social cohesion is not a mere 'catchword' (Chan et al. 2006: 277) for all types of social challenges that society faces but a visionary goal for social development. Besides the tasks of developing and spreading future-orientated visions, an important desideratum of research would be how to facilitate the acceptance of change. This does not just concern the suppressed or underprivileged who gain power and become emancipated. Rather, it is the majority which (also) has to change but will probably resist sharing power and resources etc. if there are no incentives. Yet, before considering how to achieve change, fundamental considerations about clarifying goals and appropriate research are necessary.

4.3 Paradigmatic Foundations and the Role of Research(ers)

As mentioned above, social cohesion as well as peace involve researchers in normative and social issues. Therefore, research methodology and transparency become crucial factors in the evaluation of scholarly statements. A central factor in normative research is that of research perspective and attitude. On the one hand, this is influenced by structural and institutional aspects. For instance, most studies on social cohesion have been commissioned by politics and the question arises as to whether this serves scholarly informed politics or politically informed research. In order to understand research not as a critical or evaluative instance of politics but as a co-operation partner, one has to bear in mind that academia has other justification requirements than politics. Normative research needs to deal with underlying paradigms in order to counteract the danger of fulfilling expectations or reproducing unconscious subjective assumptions. The following sections may help to identify critical aspects.

Until now the focus of research has been mainly on two questions:

(1) How/with which operationalisations can the degree of social cohesion be measured?
(2) What can be done to increase social cohesion?

Unfortunately, this means that the question 'what is social cohesion?' is quickly relegated to the background in favour of quantitative concepts, and that academic activities are prematurely orientated towards (post-)positivist approaches. Despite the obvious normative orientation, framing, and objective, normative foundations and positioning are rarely clarified or reflected, but rather adapted to practical requirements: "Its definition and operationalisation are inseparable from the context of its invocation and not least from the crises it is seen to respond to" (Dobbernack 2010: 6). In such instrumental approaches, quantitative studies quickly present graphs and diagrams that are intended to enable comparability between different countries or longitudinal developments. The approach itself and the selection of indicators are presented as logical or obvious; the usefulness for the political discourse is in the foreground: "As a quasi-concept, social cohesion is judged not only by its analytical rigour but also by its utility" (Beauvais/Jenson 2002: 30). Of course, important data is collected in this work and correlations are established, tested and checked. However, a key question remains: from which perspective and on which basis are these concepts considered 'useful'?

In order to explain the knowledge production of social cohesion and compare it with positive peace approaches, I proceed in three steps. *First*, I will examine paradigms that may underlie social cohesion, *then* I will analyse paradigms of peace, and *finally* I will complement them with methodological and epistemological approaches which will contribute to the improved substantiality and professionalisation of peace research.

4.3.1 Paradigms of Social Cohesion

Academic research is organised by paradigms. By consensus on a paradigm, disciplines emerge which focus on a content-related field and contain agreements on methodological procedures: "Paradigms [provide] the scientists not only with a map [...], but also with some essential guidelines for the production of the map" (Kuhn

1962: 116; 122). Paradigms thus create an overarching order and direction, differentiate between relevant and irrelevant, between acceptable and unacceptable, and offer further selection criteria and directions of interpretation within this order.

Kuhn's considerations referred to natural sciences, but were transferred to other research areas (Hoyningen-Huene 1989). Lincoln and Guba (2003) have developed a method of systematising paradigms, according to which a paradigm makes ontological, axiological, methodological and epistemological statements. Based on this approach, the following explanations and considerations are an attempt to organise the resultant knowledge.

When analysing and systematising research on social cohesion, it is striking that the underlying paradigm is often multidimensional (cf. Mekoa/Busari 2018). The multidimensional approach to exclusion is

> naïvely heuristic and tautological in that it identifies social problems and then labels them as aspects of exclusion. It is not guided by any particular social science paradigm or theorisation of what either exclusion or inclusion is. (O'Reilly, 2005: 81)

The advantage is high openness or connectivity. The disadvantage is a certain arbitrariness or unquestioned positioning. It is assumed that social cohesion is a desirable condition and a diagnosis is made that it is 'threatened'. It is then empirically measured, but rarely critically analysed or constructively addressed. The analytical approaches are based on moral premises; epistemological and ontological levels are blended without theoretical reflection or empirical verification. This may lead to the above-mentioned exclusive cohesive practices. Therefore, further systematisation of possible paradigmatic approaches will be attempted here.

O'Reilly (2005: 81f.) builds on the works of Silver (1994) and Levitas (1998) and identifies five further possible paradigmatic orientations in his systematics, which I apply to the foundation of social cohesion:

(1) *Organic*: Social cohesion "based on a naturalistic conception of the organic integration". This is structured on supposedly natural categories and is founded on an 'identity' (also understood as natural), to which moral 'correctness' is subordinated and which distinguishes itself from identities conceived as different.

(2) *Specialisation*: Social cohesion based on individual freedom and rational choices; this follows (neo-)liberal intentions. Individual freedom may be limited by discrimination which causes exclusion (mostly related exclusively to the labour market). Yet, the moral discourse around this paradigm deals mainly with the 'character' of the excluded themselves (which is identified as the reason for exclusion) and is not necessarily logically but symbiotically connected (Levitas 1998). While the paradigm addresses (only) individuals for the production of cohesion, the moral discourse works with collective character attributions. Structural aspects are completely left out of the equation.

(3) *Monopoly*: Social cohesion based on the prevention of monopolies. "Groups that are able to assert a monopoly on the means of production receive 'rents' that perpetuate their unequal economic position". This refers both to the vertical level

of the hierarchical structure of production and to the horizontal level on which centres and corresponding peripheries are formed. This approach is initially related to economic, structural and material aspects in general, but with regard to the question of social cohesion it also applies to questions of discourse power and other instances of social recognition.

(4) *Republican*: Social cohesion based on the fundamental equality of citizens. Ruptures of this social bond are a failure of the republican state. The problem that arises here is the necessary equality – is it (only) related to legal and civic issues, or does it involve, in a broader sense, the requirement of cultural equality? The above-mentioned examples of existing cohesion policies point to the latter, the direction being assimilatory rather than dialectical.

(5) *Postmodernis*t: Social cohesion goes beyond inclusion into the labour market, which is the main focus for paradigms (2), (3) and (4) (Yépez del Castillo 1994). Access to not only production, but also consumption must be guaranteed, since the former connection between the two dimensions has now been eliminated (Geddes/Benington 2002). In relation to social cohesion, this paradigm can be extended beyond the material side and include questions of discourse power and political and cultural representation with regard to their degree of inclusion as well as to agency in production and consumption (who produces, who consumes?).

As this brief summary is intended to show, all paradigms have their justification as well as their specific strengths and weaknesses but usually focus on only one aspect without relating to each other constructively. Existing policies then multidimensionally select and combine single elements, echoing political intentions. In the following sections, I present and discuss alternative paradigms that may be helpful for anti-exclusive cohesion research and policies. I start with positive peace, which may be a conceptual equivalent though generated by academia.

4.3.2 Peace Paradigm(s)

The main idea of positive peace is that it is 'more' than the absence of war/direct violence and also addresses structural and cultural violence. According to Galtung (2012: 52), basic features are: co-operation, equality, equity, culture of peace and dialogue. However, these principles have also been translated and supplemented into various models without it always being transparent how these were achieved (e.g. IEP, latest report 2019). Here, too, the impression of an 'empty signifier' is evoked, which is narrowed and shortened, especially in the course of quantitative measurability. It is therefore necessary to clarify what peace actually means and which paradigms are used in this field. What makes peace research become peace research?

There is a large amount of literature covering various perspectives which cannot be exhaustively presented here. Peace research works either on a disciplinary basis and then adopts the paradigms of the respective discipline or on an interdisciplinary and multiparadigmatic basis; an integrative approach is lacking (Graf/Wintersteiner 2016: 74ff.). Instead of an excessive literature analysis, I will therefore attempt to provide an ordering overview, including further theoretical and methodological implications, which will serve as a basis for discussion. The consequences regarding social cohesion are presented in the conclusion.

Richmond (2005; 2008) has identified four metatheories or paradigmatic generations of peace research for the Anglo-Saxon world in research and politics:

(1) *Realistic paradigm*: aims at negative peace; negative image of humans (violence as part of nature); state as central actor; conflict management through intervention.
(2) *Structuralist paradigm*: aims at elimination of structural violence and at positive peace (understood as the fulfilment of human needs); active role of civil society; possible positive function of liberal states; based on 'universal' norms, such as democracy, pluralism, human rights, justice.
(3) *Liberal paradigm*: aims at ending open violence, structural reforms, democratisation, free market economy; civil society subordinated to the ideals of the liberal state; elite-driven.
(4) *Post-structural, emancipatory paradigm*: aims at emancipation and true self-determination (life chances, feelings, culture, learning); criticism of existing approaches, especially the liberal approach as a hegemonic power project; questions implicit norms (security, development) and basic values (universality, rationality, sovereignty); emphasises the situation of individuals.

When it was noted above that social cohesion is conceptualised multidimensionally, this complementary scheme reveals a focus on liberal approaches, which places a strong focus on power. Though this is also one of the strands of peace research, the other approaches show that aspects such as capitalism's potential for violence and the interveners' own interests are ignored. Addressing deeper structural causes of conflict is limited by the liberal agenda (Graf/Wintersteiner 2016: 71). Richmond's system originates from the field of International Relations and offers a valuable overview at a descriptive level. However, it suggests an either/or mind-set, as if it were necessary to decide on one of the approaches.

Funk (2002) goes beyond the Anglo-Saxon world and offers a different sorting approach. Conceptually, it is more analytically orientated and better suited to intra-societal peace. It distinguishes five paradigms:

(1) *Power politics*: Peace through coercion.
Starting from a negative image, human nature is characterised by egoism, greed, competition and violence. The "ever-present threat of violence" and "chronic insecurity" necessarily lead to "acquisition of material power and military capability to compel and deter others". Since there are no universally shared values, the world is a "self-help system" and policies "serve the private good of their immediate 'national interests'". Peace is understood here as "temporary suspension of hostilities secured by military power".

This paradigm resembles the 'realistic' paradigm of Richmond and refers to 'negative peace' (i.e. the absence of war or direct violence). The following four paradigms refer to positive notions of peace.

(2) *World order*: Peace through value-maximizing institutions.
Rejecting coercive power as "a form of disorder", this paradigm highlights the necessity and possibility of co-operation and collaboration. Humans are capable of both selfishness and altruism. Hence 'world order' "emphasizes human choice and intentionality" and "the roles of concerned citizens and ethical values in politics". Global, national and local "loyalties, values and interests" should be mitigated in "value-maximizing institutions to move beyond fearful and reactive behaviour". Collaboration includes not only states, but also NGOs and social movements. Necessary "values, norms, and programmes" are: non-violent conflict resolution, human dignity, development, ecological balance, political participation, consensus, justice-orientation, opportunity-creation, multilateral frameworks. Peace is here the "presence of certain value conditions that are required for human flourishing and for long-term survival within a global context".

(3) *Conflict resolution*: Peace through communication.
This paradigm is more pragmatic and aimed at developing skills to deal with conflicts in a constructive, proactive and coordinated way. It emphasises that "conflict is natural at all levels of human interaction and organisation", but that "conflict not inevitably leads to violence and is often necessary for change". Instead, the "empathy, creativity, and 'shared positive power'" should be used and "the underlying interests and human needs" analysed in order to enable "processes of dialogue and engagement" (also non-formal). Peace is created here "through direct interaction with 'the other'" and "skills for communication and coexistence" need to be developed.

(4) *Non-violence*: Peace through will and decision.
Non-violence does not stand for passivity, but "action animated by principle and informed [...] that means and ends are inseparable". Power arises through the "consent of the people", who define their "own behaviour as moral agents irrespective of external norms and pressures". The "value of human life" serves as the benchmark to avoid or reduce "insult to human needs (and [...] to the balance of nature)". This needs willpower and solidarity which should be (further) developed for "a revolution of the human spirit" and "a shift in human consciousness". Peace here is closely linked to freedom and justice, as "justice entails an absence of oppression". The task is "overcoming their own fear and anger" in order to "work for justice – justly".

(5) Transformation: Peace through love.
Focusing on cultivating the self, "the transformation of the individual becomes a metaphor for and instrument of broader changes". "Education, cultural change, and spirituality" are central. Building on the notion that "sacred ideals are personalized for application", "consciousness and character" are the prerequisites for "'internal disarmament' and personal integration". A "universal loyalty" inspires "actions born of loving commitment to the wholeness and

integrity of creation". Uniting "doing with being", the mantra is: "If you want peace, be peace. Be an instrument of peace."

These paradigms for positive peace are built on the notion that power is not (only) a category of coercion, but is supplemented by other dimensions: law and institutions, communication, will, love. The last three dimensions, in particular, have so far been discussed rarely or only very superficially in social cohesion debates (e.g. as 'love of the country', which is usually limited to national loyalty or gratitude; or as 'social justice', which is reduced to welfare).

Other factors also play a role. Lutz (2004) characterises peace research as "normative, interdisciplinary, practice-oriented" and points out that peace must not be thought of as a status quo but as a continuous process. One special feature, he says, is that it is not only about restraining violence, but about overcoming it. Therefore, non-personal, structural elements have to be taken into account and 'positive peace' should be filled with meaning. Peace research often emphasises the human being as a cultural being. This focus is not reduced to the control of instincts; it deals with the shaping of being, with cultivation, changeability and the ability to learn. "People are capable of peace" and "peace is feasible" are consequently central guiding principles (Lutz 2004: 26).

Therefore, learning specific skills is central as well as the 'unlearning' or reflection of unconsciously acquired dispositions, and the creative search for ways to satisfy the various human needs. Chappell (2018) lists the following needs: purpose and meaning, belonging, self-worth, explanations as well as nurturing relationships, expression, inspiration, challenge, and transcendence. He underlines that their non-fulfilment leads to trauma. The need for reflection and learning refers to both the individual and the collective levels: "Peace […] must be an activity of culture" and therefore "we must locate the cultural sources of conflict as a foundation for transformation" (Knight 1988: 17, 26).

It is important to understand culture as a universal anthropological disposition and not as a terminological substitute for 'race' or 'ethnicity' (see Hall 1989; Balibar/Wallerstein 1990). In view of the "renaissance of culturalised difference" (Brunner 2010: 86), the anthropological principle reminds us that not only 'the others' have 'culture' (possibly while 'we' are 'civilised'), but that we are all cultural beings.

It is crucial not to simplify problems and conflicts, but, on the contrary, to acknowledge and deal with them in their complexity. Graf et al. (2010) therefore propose "complexity thinking" as a meta-framework. This is based on the "Dynamic Systems Theory" (Coleman 2006) and "generalised complexity" (Morin 2006), which is characterised by dialogical, recursive, and holographic understandings.

This meta-framework and the above-mentioned paradigms relate primarily to ontological and axiological levels, i.e. the view of the world and the values that are central to the objectives and the conduct of research, but without really referring to each other. Bringing these (and other) aspects together for a paradigm for positive peace is an exciting task for future peace research. The following considerations may serve as preparation for this. According to the systematics of Lincoln/Guba (2003), additional statements on methodology and epistemology have to be made in order to produce a fully-fledged paradigm. For this purpose, I would like to introduce the transformative and the narrative paradigm into the discussion.

4.3.3 The Transformative Paradigm

For legitimising and organising normative research, Mertens developed the transformative paradigm as a "framework for addressing inequality and injustice in society using culturally competent, mixed methods strategies" (Mertens 2007: 212). Following critical theorists and evolving ideas of Critical Systems Thinking (CST) (cf. Romm 2015), the aim is to open up hegemonic discourses and bring marginalised voices into the discussion in order to get below the surface of the epidemic and address its root causes in an action-orientated way.

Besides its explicitly normative positioning, this has a strong effect on the self-image of the researcher as the one who not only 'recognises' but also "strives to challenge the status quo" (Mertens 2007: 212); the relation between theory and practice is seen as one of mutual influence. Like the approaches of participatory research, inquiry should not be done *about* people or groups, but *with* them. However, while the aim of participatory research is to enable disadvantaged groups to participate in society and politics, the transformative approach can (in a broader reading) address even stronger the structures that prevent participation. Thus, the aim of research would not only be to support the struggle for participation in a competitive system, but also to establish equal access as a basic value within the system.

The expression *transformative paradigm* serves as an umbrella term for emancipatory, participatory, pragmatic and inclusive perspectives which "link the results of social inquiry to action, and […] to wider questions of social inequality and social justice" (Mertens 1999: 4). While including much from critical, constructivist and participatory approaches, its extraordinary quality is its ability to bring together different specialised perspectives to strengthen the focus on social justice, including the dimensions of culture, power, and privilege.

Mertens specifies the transformative approach on the basis of the categorisation of Lincoln and Guba (2003: 265) (Fig. 4.1).

As a consequence, normative research has to be participatory; the researcher needs to establish an interactive link with community members. This means researchers must proactively and constructively address the understanding of "differing versions of reality and how they are synergistically related to power issues" (Mertens 2012: 6). It is essential to acknowledge the differences in power and to work towards a "formation of trusting relationships" (Mertens 2012: 6) in a respectful way, appreciating the expert knowledge of the participants.

If the question of social cohesion goes beyond the ambitions of economic logics in terms of measurability, it cannot be predetermined top-down, but only co-constructed

Ontological	Emphasises the existence of multiple versions of reality while rejecting cultural relativism. Explains how certain perspectives on reality become privileged over others and how researchers can serve to undercut undue privileging views. Focuses on making versions of reality visible that have the potential to further human rights and social inclusion.
Axiological	Stresses the need to respect cultural diversity and to develop cultural competency and culturally sensitive interventions. Concerning the underlying values of research, explicitly acknowledges the "[connection] between the process and the outcomes of research and furtherance of a social justice agenda" (Mertens 2007: 216).
Epistemological	Problematises the nature of knowledge related to power and privilege issues. Distinguishes between postpositivist approaches (observing from a distant and dispassionate standpoint in order to get 'objective' results) and constructivist approaches (knowledge only created through interaction); emphasises the attitude and manner in which researchers relate to participants; aims at an interactive link between researcher and participants and at a fair understanding of key viewpoints, and relates them to theories.
Methodological	Open to qualitative, quantitative, or mixed methods. It is crucial to involve 'affected' people in the research process, and to be culturally informed and responsive, and dialogue-orientated. Collecting data is not enough; the results must be processed into action-orientated outcomes.

Fig. 4.1 The transformative paradigm. *Source* Compilation of Mertens (1999, 2007, 2012) and adapted by the author

participatively. Not all participative research is transformative; an important characteristic is a procedure that goes beyond practical considerations, is culturally competent and socially sensitive, and considers all elements of the transformative paradigm (Mertens 2007: 220; Whitmore 1998, see also Arnstein 1969).

Furthermore, it is necessary for researchers to engage in personal self-reflection and be willing to learn. The co-operative basic attitude alone, as well as the awareness of ontological polyphony, makes questioning oneself as a researcher unavoidable. This applies not only to one's own role and 'practical questions of power' in the research process, but also to wider socio-political affiliations, imprints and preferences. Located in a very discourse-powerful, possibly hegemonic group (such as academics), it is particularly crucial that scholars critically address possibilities of pretentiously or unconsciously fading out divergent realities (thematically focused approaches in, for example, Tißberger 2017; Rüsen 2007; practice-orientated insights in, for example, Malthouse et al. 2015).

The need for reflection includes not only (possibly unconscious) individual-biographical, but also collective-cultural imprints. Interesting here are Galtung's reflections on deep culture, i.e. collectively subconscious practices, norms, discourses, and cosmologies (see Graf 2009: 47). Galtung (2010) identifies two pathologies, one cognitive and one emotional. The *cognitive pathology* (DMA syndrome) is characterised by thinking in terms of either/or (Dualism), good vs. evil (Manichaeism), and embracing the idea of a final decisive battle (Armageddon). The *emotional pathology* (CGT syndrome) is fed by the idea of being Chosen, to carry Glory from the past and to strive for glory in the future, but also to have experienced injuries and defeats, which have an enduring effect (Trauma). Trauma as violence from the past requires reconciliation and dialogue in order to be resolved.

This awareness of biographical and cultural influences and the search for and design of alternatives can be subsumed under the understanding of 'civilisation' as a central category of peace research (Vogt 1996: 113). Yet, the very term 'civilisation' has a colonialist bias and may lead to it being understood as an elite-driven project. However, since no one is immune to subjectivity and limited knowledge, but at the same time expert knowledge and specialisations need to be acknowledged, permanent self-reflection is required, especially on normative questions (cf. Jaberg 2011: 61). If "learning peace refers to societal learning processes" (Wintersteiner/Gruber 2014: 9), one of the tasks is to understand academia and politics as parts of society as well, and thus as addressees of the call for learning. The connection to the axiological claim for non-violence – the alliance of means and outcomes – is here spelled out in epistemological and methodological terms.

Consequently, there is particular sensitivity to the data-related empirical orientation: (post)positivist approaches use "primarily quantitative methods that are interventionist and decontextualized"; constructivist approaches use "qualitative methods in a hermeneutical and dialectical manner" (Mertens 1999). Hence, there is a strong case for using mixed methods, whereby not only a mix of qualitative and quantitative methods can be employed, but also, for example, different qualitative methods can be combined. The decisive factor is to capture the contextual complexity.

The approach could be criticised with the argument that research should strive for 'neutrality' and 'objectivity', which would be undermined by the normative objective. However, following the aforementioned arguments of subjectivity always playing a role, and the at least unconscious articulation of normative statements, one can conclude that it is better to make them transparent and to direct them towards social justice and peace.

4.3.4 The Narrative Paradigm

The narrative paradigm offers further implications for methodology and epistemology. First, it states that human consciousness is not (only) rational and follows arguments, but that it uses stories to understand and create meaning (Fisher 1984). The very process of linking elements, of trying to bring order and meaning to the overwhelming 'outside', is structured narratively. This is true on individual as well as collective levels (Müller-Funk 2007).

Despite its enormous significance for various disciplines, the term 'narrative' is used very inconsistently. In order to be able to work productively with it, interdisciplinary understanding and interest are necessary. So far, according to my research, there is no comprehensive overview that takes up different disciplinary approaches and traditions, especially since the term is also used differently in different countries and languages. For the following draft of a systematisation, 'narrative' was

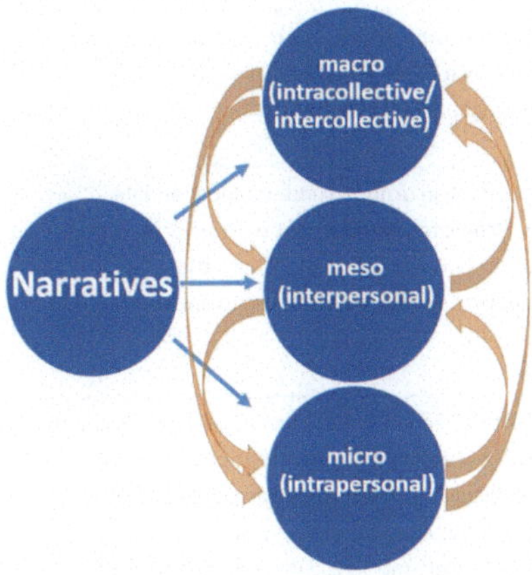

- Grand narratives (e.g. theories, ideologies, religions, history) (H. White 1973; Lyotard 1979; Müller-Funk 2007; Assmann 1992)
- Collective identities ("We" and "the others") (Barthes 1957; Anderson 1983; Hammack 2008; Mana et al. 2012)

- Disseminations and connectivity of narrative truths (Kristeva 1972; Green/Brock 2000; Bekerman/Zembylas 2012)
- Mediation (Cobb 1993; Salomon 2004)
- Change/development/solutions (M. White/Epston 1989; Grossmann 2003; de Shazer/Berg 2003)
- Empowerment (Stark 1992; Holland/Kilpatrick 1993; Rappaport 1995)

- Organising experiences, creating sense and engagements (Fisher 1984; Bruner 1990; Adams 1996; Hammack/Pilecki 2012)
- Identities, biographies (Sarbin 1986; Somers 1994; Polkinghorne 1996)

Fig. 4.2 Systematics of narrative approaches at different levels with a tiny selection of literature. *Source*: The author

distinguished from 'narration' (story). Narration means that (at least) two events or situations are connected by an explaining connection. Narratives are a sub-category of narrations and create legitimations for realities and claim to be "factual" (Genette 1992)[2] (Fig. 4.2).

With regard to the peace paradigms presented above, the narrative paradigm is particularly applicable to post-structuralist approaches (Richmond) and the focus on communication, will, and love (Funk). Translating it to the four levels of Lincoln and Guba results in these thoughts (Fig. 4.3).

The interesting feature of narratives as an analytical category, e.g. in delineation to discourses, is the emphasis on structure and context, while equally stressing the question of content, power, and hegemony. Narratives assign agency to actors, and the sequential order of events is relevant for the meaning. It is analytically interesting that 'the narrator' emphasises 'the unusual and the exceptional', which enables an understanding of the underlying 'ordinary and right' (Patterson/Monroe 1998: 316). This brings the unsaid or presupposed into focus and leads particularly to cognitively and culturally relevant aspects. Consequently, authorship and context of genesis and circulation are pointed out.

Hayden White (1986: 102) asserted that even academic writing is "linguistic fiction, whose content is as created as it is found". This serves as a reminder that insights transferred through language (i.e. narratives) are always limited. As the term 'narrative' thus suggests that truth depends on perspective, sources, and expressions, it is a suitable means to bring together different views, judgements, and methods. Narratives can be analysed linguistically, psychologically, sociologically, and politically, which helps us to understand their logics and developments. This is important

[2]To illustrate this by examples: "The king died" is a fact, not a narration. "The king died and then the queen died of grief" is a narration (Forster's minimalist narration). "Because all monarchs were dead, we developed democracy" would be a narrative if presented as factual.

Ontological	'Truth' and 'reality' are always narrated and captured by words, and are therefore location-bound and limited. Narratives play a special role as social, political, or academic grand narratives as well as individual variations of truth.
Axiological	It is necessary to really engage with the story in order to understand it and its fundamental elements and structures. Promotes engagement with several narrations; self-limitation to one predictably narrows insight. Implications for processes of change and learning; relativises holistic claims to truth and thus facilitates dialogue.
Epistemological	Cognitively, we only know 1) what we have been told and 2) what we have assumed or invented by making connections and thereby creating narratives by ourselves. Narratives connect the outside and the inside. New narratives are judged by their inner coherence, integrability to other narrations (including the narrative of rationality), and ties to experienced reality and emotions. Offers indicators for differentiating between fictitious and factual narratives.
Methodological	A wealth of systematic research approaches from linguistics to psychology, history, cultural anthropology and more. Focuses on actors, events and structures as well as authorship and the production context. Provides reminders about the influence of researchers' narrative dispositions in gaining insights.

Fig. 4.3 The narrative paradigm. *Source*: The author

for conveying and handling political issues as well as for dialogues between academic disciplines. Narrative work therefore combines analytical potential and integrating abilities.

Furthermore, the concept of narratives has the advantage of enabling analytical links between society and the individual. They allow the relationship between master narratives and individual narratives to be reconstructed. Narratives provide access to the current structure of identity (personal or collective), value a developmental perspective (Hammack 2008) and may turn "societal abstractions" into concrete situations (cf. Müller-Funk 2008). In contrast to other analytical categories, narratives include an element of time and space, and as meaning is always historically situated, this is an important aspect for the legitimisation of "social imaginaries" (Taylor 2004).

This, of course, also entails risks, one being the irrelevance of 'reality' or 'truth' – if the world is understood as purely narrative, it appears to be a toy of cognitive preferences. For this reason, especially with regard to the truth claim of academic research, there are many studies on the question of the delineation of fictional and non-fictional/factual texts (e.g. Friedländer 1992; Cohn 1999; Kiesow/Simon 2000). However, at least the insight that it is impossible for everything to be captured and depicted is ontologically convincing: "To believe that there is actually a story 'out there' in the world waiting to be discovered – as something that exists independently of the process of narrative construction of meaning – is what we call an ontological fallacy" (Brockmeier/Harré 2005: 42).

Looking at this point from the perspective of mobilisation, e.g. in the context of the fake news debate, it does indeed appear that the distinction between fictional and non-fictional texts does not necessarily determine action. If 'true' arguments were effective, there would be no unreasonable actions. More important is the inner coherence of the narrative and its ability to link to existing texts (Kristeva 1972). Reason and rationality are bound to underlying narratives. This is not a plea for fictionalisation, but for taking the relevance of unreasonable or untrue narratives action seriously and for responding to the logic they contain. A further danger is the legitimisation of cultural relativism – if all narratives are equally valid, none is 'better' than the other. In the context of the argumentation of this chapter, however, this is resolved by the fact that 'peace' is given as a basic axiological orientation (though needs to be specified for concrete endeavours).

An advantageous feature of the narrative paradigm is the emphasis on dynamics and development while at the same time stressing complexity. It makes it easier to understand the social construction of language, politics, and thought, the need for collective solidarity through shared meaning, and the need for coherence through narrative identity. It captures the human being in his/her need for culture and meaning. Due to its mediational capabilities, narrative approaches are especially suitable for sociopolitical activity and practice (Hammack/Pilecki 2012). These are properties which also make narratives a promising tool for learning contexts as they can be used to increase empathy for 'the others' as well as to work with one's 'own' narrative in order to empower or integrate change (Bekerman/Zembylas 2012; Bar-On 2006; Gergen/Gergen 2006; Pennebaker/Seagall 1999; Holland/Kilpatrick 1993; Salomon 2004).

4.4 Summary and Conclusions

The article has shown that social cohesion is an empty signifier which can be connected to a variety of concepts and which shares some dimensions of positive peace. However, in empirical reality, it is connected to exclusionary policies, due, among other things, to the underlying multidimensional paradigm. Furthermore, I have shown that positive peace is also not entirely defined and that the different paradigmatic orientations are mainly on ontological and axiological levels. In order to develop a 'fully-fledged' paradigm, I propose the transformative and narrative paradigms as complements, especially for the methodological and epistemological dimensions.

What does this mean for the research and design of social cohesion with respect to peaceful, just communities? First of all, that a development in the direction of Agenda 2030 is a question of will and consciousness. The fact that social cohesion policies have hitherto contained arbitrary, divisive, majority-based, exclusive, and unjust elements is not necessarily due to malicious intent, but rather to the unconscious reproduction of the familiar. To counteract this, participation, professionalisation and, of course, peace are necessary.

Participation facilitates democratic negotiations that are orientated towards the community as a whole, future-orientated, and creative entity. As the considerations of the transformative paradigm indicate, it is necessary to involve not just selected stakeholders but a broad spectrum of 'affected people' in the research process if its impact is to be long-lasting and fair. It is also necessary to design the research process as peaceful/non-violent. Quantitative surveys seem to be based primarily on a 'narrow choice approach', in which variables are derived within exclusive contexts and then measured using a small number of indicators whose origins cannot be traced or are justified by their prior usage. With such an approach, academia engages in socially relevant epistemic violence (see e.g. Brunner 2016). This is based, among other things, on simplified thinking, which "has become the barbarism of science. This is the specific barbarism of our civilization" (Morin 1977/2010: 448). Even though this may not be overcome completely, a transparent and participatory approach at least contributes to its reduction.

Professionalisation primarily entails transparency with regard to normative goals, perspectives, methodology, and structural conditions (Marej 2020). Precisely because socio-political issues are increasingly being addressed by scholars, peace research needs to participate in these discourses, also in order to "convincingly prove its legitimacy towards practical politics" (Lutz 2004: 131). Since such research projects are increasingly interdisciplinary and the number of publications is exploding, communication between disciplines must be facilitated. Peace research needs a paradigmatic foundation that is not limited to political sciences and International Relations. It is therefore necessary to develop an approach that integrates the presented paradigmatic perspectives and extends them by epistemological and methodological dimensions. The transformative and narrative paradigms offer starting points for this discussion. These, in turn, have an impact on the self-image of the researcher. Structured self-reflection can serve as a moment of deceleration, which on the one hand helps to clarifying the researcher's own position, and on the other hand helps researchers to recognise their own presumptions and limitations, leading, if appropriate, to self-correction and learning.

Peace research emphasises that it is generally necessary to acknowledge the sociopolitical responsibility of academic activity and to reflect on and question the effects of its work. Ontology plays a role not only as a background, but also as a product: which world views should be created, which narrative linkages and solutions offered? Approaches without socio-critical potential that undermine the subject and do not encourage self-determined action appear to be historically outdated (cf. Laucken 1995: 298). Peace research serves initially as a 'watchdog' for this – "in this respect it initially has a 'courier function': it brings 'bad news' quickly and in concise form, without always being able to offer solutions at the same time" (Lutz 2004: 30). However, since the central categories of social cohesion are connectedness, social relations, and orientation towards the common good (Schiefer/van der Noll 2017), it shares with positive peace the search for a desirable societal vision. There are numerous points of intersection that can only be briefly touched on here. Fundamentally, "de-securitizing the social imaginary" (Dobbernack 2014: 182) is necessary in order to avoid "new, 'catastrophist' politics of societal management forged out of deep insecurity about a future perceived as calamitous and out of control" (Amin 2013: 5). In short, it is about peace logic instead of security logic (Birckenbach 2012).

For this purpose, it is helpful to think beyond the hitherto dominant liberal paradigm and the dichotomy between liberal and non-liberal (cf. Kahl 2019) and to include other dimensions. Institutions can be conceived not only for control and punishment (or nudging), but also as places of dialogue and negotiation. Their role becomes clear, for example, in the civilising hexagon (Senghaas 1997). Conflicts can be perceived not only as disturbing, but also as a stimulus for necessary improvement. They should be dealt with in a dialogical and non-violent way. The precondition is the will to do so. Instead of wanting to assimilate 'the other', direct exchange should be fostered, including the willingness to learn something new and to correct structural and cultural injustices. This willingness is closely linked to the dominant underlying emotion. In the social cohesion discourse 'pride' is addressed and considered desirable; in the peace discourse it is 'love'. While pride is connoted with ego, victory, and demarcation, love overcomes these. A society that collectively shares the motto 'if you want peace, be peace' will take a different stance from one that propagates pride.

Of course, the individual is then also required to reflect on him/herself and his or her actions and, if necessary, to change them – which is more difficult than a group membership that allows actions to be ascribed to others without personal effort. In general, a change of perspective away from conservation to creation seems necessary. This also concerns the self-image of the West as peaceful. This is only true within a narrow ('negative') concept of peace. If cultural and structural factors such as "the production and sale of military weapons, the consumption of energy and the ecological footprint, or the economic profit of crude exploitation" (Brunner 2010: 94) are taken into account, it becomes clear that Western states are still far from the ideal of peaceful, just societies. In order to promote developments in this direction, the individual as an 'active citizen' is also required here, albeit with a different orientation. It is not activity as a value itself and not activity limited to the labour market but societal activity towards a (not yet specified) utopia or vision.

At the same time, citizens should be respected and recognised in their diversity rather than being homogenised. Attempts at homogenisation do not work anyway if the narrative of the macro level is not mirrored in the reality of life; they may even be counter-productive to substantial social cohesion. If the aim is to create an inclusive collective identity at macro level, the narratives of all, or at least as many social groups as possible, should be narratively connectable and also enable resonance with individual life worlds. This may require the renegotiation or modification of narratives, symbols, anniversaries, etc. at macro level – which is nothing new (Hobsbawm/Ranger 1983), but should be initiated consciously and democratically.

When aiming to create a common narrative for social cohesion, a critical analysis is necessary to prevent imperial aspirations or side-effects, because "[t]he power to narrate, or to block other narratives from forming and emerging, is very important to culture and imperialism, and constitutes one of the main connections between them" (Said 1993: xiii). Instead, if working creatively and deliberatively with them, thinking in narratives allows access to different realities and solutions. At the same time, it promotes empathy, self-reflection, and the acceptance of plurality in a more playful and relaxed way than the attempt to 'convince' through 'reason'. Working with narratives in this context is not "'story-telling' as a form of governmental intervention [...] working upon and shaping the ideational circumstances around given policy issues" (Sørensen 2006: 109f.) – but a societal learning process and means of communicative exchange. Social cohesion could then serve as a reference point towards which different narratives are directed. The term could be emptied and refilled or, if too spoiled, replaced by another (e.g. peaceful, just society). Due to the practices of earlier hegemonic articulations, signifiers must compete with each other to represent the whole (cf. Laclau 1996). The semantic power of terms and concepts must not be underestimated.

Another suggestion is to move the perspective away from the state and its self-preservation towards the people and their needs. Instead of wanting to regulate them out of fear and 'force' identification, the question from the perspective of peace logic is the other way around: what leads to alienation and how can this be reduced (Graf 2009: 42)? Stability and cohesion are then not created by perseverance, but by correction and improvement. The connection between anomie and social integration can be considered from different directions: does integration not work because of a lack of cultural agreements or are those agreements missing because of pathological forms of integration?

Diagnoses could be made more systematically and causes of crises could be perceived less intuitively. An analytic approach is more useful than binary or dichotomous ethnic and/or moral conceptions of 'we and the others' when addressing the social fault lines of structural and cultural violence which can be found at micro, meso, and macro level, on the actor and on the structural side. On this basis (cf. Galtung 1998: 115ff., 174; Graf 2009: 46) I suggest a more complex model for the analysis of social cohesion (Fig. 4.4).

By applying these categories to a matrix of cohesion, more complex (and more realistic) pictures emerge of the points where tensions and divisions exist, where cohesion can be developed and where it is already relatively established. Instead of one-sided explanations and accusations, causes and interdependencies can be recognised and addressed. Identifying violent relationships is the first step towards reducing them. This approach is, of course, much more complicated than simple binary ascriptions, but if really peaceful, just societies are the goal, these aspects have to be dealt with.

In the end, it is all about humanity moving together in a globalised world. While Huntington (1996), for example, wants to counter this with struggle and separation, Galtung, for instance, stands for civilising self-reflection and intercultural dialogue. Attitude and will decide which path is preferred, but at least the options should be made transparent and it is one of the tasks of peace research to broaden the discussion. The whole potential of social cohesion only becomes clear when understood as a project, a social task, not as a product of conservation. The goal of stabilising democratic systems is shared by a peace perspective, but this does not need to lead to 'freezing' (Suvarierol 2012) existence at a certain point and the current (or somewhat

Fig. 4.4 Model for more complex diagnoses of social crises and social cohesion. *Source* The author

premature) state being declared a 'Golden Age'. It is important to recognise that sometimes transformations (at various levels) are necessary to stabilise the system and to pursue democratisation further and more consistently. Instead of declaring social decline and promoting national ideologies, it seems more promising to design social cohesion as a mutual task, a shared search, with invitations to engage while emphasising the common as well as the individual good – which is what also Agenda 2030 aims to achieve.

References

Almeida, Dimitri, 2017: *Laizität im Konflikt. Religion und Politik in Frankreich* (Wiesbaden: Springer).

Amin, Ash, 2005: "Local Community on Trial", in: *Economy and Society* 34,4: 612–633.

Amin, Ash, 2013: "Land of strangers", in: *Identities – Global Studies in Culture and Power*, 20,1: 1–8.

Anderson, Benedict, 1983/2005: *Die Erfindung der Nation: Zur Karriere eines folgenreichen Konzepts* (Frankfurt a.M.: Campus).

Anderson, Bridget, 2013: *Us & Them? The Dangerous Politics of Immigration Control* (Oxford: Oxford University Press).

Arnstein, Sherry, 1969: "A ladder of citizen participation", in: *Journal of the American Institute of Planers*, 4: 216–224.

Assmann, Jan, 1992: *Das kulturelle Gedächtnis. Schrift, Erinnerung und politische Identität in frühen Hochkulturen* (München: Beck).

Balibar, Étienne; Wallerstein, Immanuel, 1990: *Rasse, Klasse, Nation* (Hamburg: Argument).

Barthes, Roland, [1957] 2010: *Mythen des Alltags* (Berlin: Suhrkamp).

Baur, Nina; Knoblauch, Hubert; Akremi, Leila; Traue, Boris, 2018: "Qualitativ – quantitativ – interpretativ: Zum Verhältnis methodologischer Paradigmen in der empirischen Sozialforschung", in: Akremi, Leila et al. (Eds.): *Handbuch Interpretativ forschen* (Weinheim: Beltz Juventa): 246–284.

Bekerman, Zvi; Zembylas, Michalinos, 2012: *Teaching Contested Narratives. Identity, Memory and Reconciliation in Peace Education and Beyond* (Cambridge: Cambridge University Press).

Bekerman, Zvi, 2019: "Epistemological Difficulties in the Development of Civic Identities in Western Education", in: de Toro, Alfonso; Tauchnitz, Juliane (Eds.): *The World in Movement: Performative Identities and Diasporas* (Leiden: Brill): 88–105.

Benwell, Matthew C., 2014: "From the Banal to the Blatant: Expressions of Nationalism in Secondary Schools in Argentina and the Falkland Islands", *Geoforum*, 52: 51–60.

Bernshausen, Sirin; Bonacker, Thorsten, 2015: "Ist die Friedens- und Konfliktforschung auf dem Weg zur Disziplin? Ein wissenschaftssoziologischer Blick auf die Bedeutung von Studiengängen für die Ausdifferenzierung von Disziplinen", in: Bös, Matthias et al. (Eds.): *Konflikte vermitteln? Friedens- und Konfliktforschung* (Wiesbaden: Springer): 247–268.

Billig, Michael, 1995: *Banal Nationalism* (London et al.: Routledge).

Birckenbach, Hanne-Margret, 2012: "Friedenslogik statt Sicherheitslogik. Gegenentwürfe aus der Zivilgesellschaft", in: *Wissenschaft & Frieden*, 2: 42–47.

Bittle, Steven, 2001: *Inclusion for All: A Canadian Roadmap to Social Cohesion: Insights from Structured Conversations* (Department of Justice Canada, Cultural Heritage Canada); at: https://www.justice.gc.ca/eng/rp-pr/csj-sjc/jsp-sjp/tr01-rt01/tr01.pdf.

Brockmeier, Jens; Harré, Rom, 2005: "Die narrative Wende: Reichweite und Grenzen eines alternativen Paradigmas", in: *Psychologie und Gesellschaftskritik*, 29,3–4: 31–57.

Bruner, Jerome, 1990: *Acts of Meaning* (Cambridge: Harvard University Press).

Brunner, Claudia, 2010: "Knowing Culture, Knowing Peace? Epistemological and/as Political Aspects of the 'Culture of Peace'-Initiative, Concept and Programme", in: Ratković, Viktorija; Wintersteiner, Werner (Eds.): *Yearbook Peace Culture Bd. 5: Culture of Peace. A Concept and a Campaign Revisited* (Klagenfurt: Drava), 82–103.

Brunner, Claudia, 2015: "Frieden lernen, Frieden lehren: Zehn Jahre transdisziplinäre Friedensstudien an der Alpen-Adria-Universität Klagenfurt", in: Wintersteiner, Werner; Wolf, Lisa (Eds.): *Jahrbuch Friedenskultur Bd. 10: Friedensforschung in Österreich. Bilanz und Perspektiven* (Klagenfurt: Drava): 211–226.

Brunner, Claudia, 2016: "Gewalt weiter denken in der Kolonialität des Wissens", in: Ziai, Aram (Ed.): *Postkoloniale Politikwissenschaft. Theoretische und empirische Zugänge* (Bielefeld: transcript): 91–108.

Chan, Joseph; To, Ho-Pong; Chan, Elaine, 2006: "Reconsidering Social Cohesion: Developing a definition and analytical framework for empirical research", in: *Social Indicators Research*, 75,2: 273–302.

Chappell, Paul, 2018: *A New Peace Paradigm. Our Human Needs and the Tangles of Trauma*; at: https://www.Peaceliteracy.org/.

Cobb, Sara, 1993: "Empowerment and Mediation: A Narrative Perspective", in: *Negotiation Journal*, 9,3: 245–259.

CoE (Council of Europe) (Ed.), 2001–2012: *Trends in Social Cohesion*, 1–25; at: https://www.coe.int/t/dg3/socialpolicies/socialcohesiondev/trends_en.asp.

CoE (Council of Europe) (Ed.), 2004: *A New Strategy for Social Cohesion: Revised Strategy for Social Cohesion*, approved by the Committee of Ministers of the Council of Europe on 31 March 2004; at: https://www.coe.int/t/dg3/socialpolicies/socialcohesiondev/source/ Revised-Strategy_en.pdf.

CoE (Council of Europe) (Ed.), 2010: *New Strategy and Council of Europe Action Plan for Social Cohesion*, approved by the Committee of Ministers of the Council of Europe on 7 July 2010; at: https://www.coe.int/t/dg3/socialpolicies/socialcohesiondev/source/2010Strategy_ActionPlan_SocialCohesion.pdf.

Cohn, Dorrit, 1999: *The Distinction of Fiction* (Baltimore, London: John Hopkins University Press).

Coleman, Peter, 2006: "Conflict, Complexity, and Change: A Meta-Framework for Addressing Protracted, Intractable Conflicts – III", in: *Peace and Conflict: Journal of Peace Psychology*, 12,4: 325–348.

de Shazer, Steve; Berg, Insoo Kim, 2003: *Lösungen (er-)finden: Das Werkstattbuch der lösungsorientierten Kurztherapie* (Dortmund: Modernes Lernen).

Dobbernack, Jan, 2010: "'Things Fall Apart': Social Imaginaries and the Politics of Cohesion", in: *Critical Policy Studies*, 4,2: 146–163.

Dobbernack, Jan, 2014: *The Politics of Social Cohesion in Germany, France and the United Kingdom* (Basingstoke: Palgrave Macmillan).

Eberwein, Wolf-Dieter; Reichel, Peter, 1976: *Friedens- und Konfliktforschung. Eine Einführung* (München: Piper).

Fischer, Karsten, 2006: *Moralkommunikation der Macht: Politische Konstruktion sozialer Kohäsion im Wohlfahrtsstaat* (Wiesbaden: VS Verlag).

Fisher, Walter, 1984: "Narration as a Human Communication Paradigm: The Case of Public Moral Argument", in: *Communication Monographs*, 51: 1–22.

Foa, Roberto Stefan; Mounk, Yascha, 2017: "The Signs of Deconsolidation", in: *Journal of Democracy*, 28,1: 5–15

Fonseka, Dilrukshi; Glaister, Isla, 2017: *Social Cohesion Framework: Social Cohesion for Stronger Communities: Knowledge, Skills, Understanding*; at: https://www.sfcg.org/wp-content/uploads/2017/02/SC2_Framework-copy.pdf.

Friedländer, Saul (Ed.), 1992: *Probing the Limits of Representation: Nazism and the "Final Solution"* (Cambridge/Mass.: Harvard University Press).

Funk, Nathan C., 2002: "Peace Paradigms: Five Approaches to Peace", *Gandhi Marg*, 24,3: 298–306.

Galtung, Johan, 1998: *Frieden mit friedlichen Mitteln. Friede und Konflikt, Entwicklung und Kultur* (Münster: Agenda).

Galtung, Johan, 2010: *A Theory of Conflict – Overcoming Direct Violence* (Oslo: Transcend University Press).

Galtung, Johan, 2012: *A Theory of Peace – Building Direct-Structural-Cultural Peace* (Oslo: TRANSCEND University Press).

Genette, Gérard, 1992: *Fiktion und Diktion* (München: Wilhelm Fink).

Giddens, Anthony, 1998: *Der dritte Weg: die Erneuerung der sozialen Demokratie* (Frankfurt a.M.: Suhrkamp).

Graf, Wilfried, 2009: "Kultur, Struktur und das soziale Unbewusste: Plädoyer für eine komplexe, zivilisationstheoretische Friedensforschung: Johan Galtungs Gewalt- und Friedenstheorie kritisch-konstruktiv weiterdenken", in: Isop, Utta et al. (Eds.): *Spielregeln der Gewalt. Kulturwissenschaftliche Beiträge zur Friedens- und Geschlechterforschung* (Bielefeld: transcript): 27–66.

Graf, Wilfried; Kramer, Gudrun; Nicolescou, Augustin, 2010: "Complexity Thinking as a Meta-Framework for Conflict Transformation. In Search of a Paradigm and a Methodology for a Transformative Culture of Peace", in: Ratković, Viktorija; Wintersteiner, Werner (Eds.): *Yearbook Peace Culture Bd. 5: Culture of Peace. A Concept and a Campaign Revisited* (Klagenfurt: Drava), 58–81.

Graf, Wilfried; Wintersteiner, Werner, 2016: "Friedensforschung. Grundlagen und Perspektiven", in: Diendorfer, Gertraud et al. (Eds.): *Friedensforschung, Konfliktforschung, Demokratieforschung. Ein Handbuch* (Köln et al.: Böhlau): 35–86.

Green, Melanie; Brock, Timothy, 2000: "The Role of Transportation in the Persuasiveness of Public Narratives", in: *Journal of Personality and Social Psychology*, 79,5: 701–721.

Grossmann, Konrad P., 2003: *Der Fluss des Erzählens* (Heidelberg: Carl-Auer).

Habib, Sadia, 2018: *Learning and Teaching British Values: Policies and Perspectives on British Identities* (Cham: Palgrave Macmillan).

Hall, Stuart, 1989: "Rassismus als ideologischer Diskurs", in: *Das Argument*, 178: 913–921.

Hammack, Phillip L., 2008: "Narrative and the Cultural: Psychology of Identity", in: *Personality and Social Psychology Review*, 12,3: 222–247.

Hammack, Phillip L.; Pilecki, Andrew, 2012: "Narrative as a Root Metaphor for Political Psychology", in: *Political Psychology*, 33,1: 75–103.

Heitmeyer, Wilhelm, 1997: "Auf dem Weg in eine desintegrierte Gesellschaft", in: Heitmeyer, Wilhelm (Ed.): *Was treibt die Gesellschaft auseinander? Bundesrepublik Deutschland: Auf dem Weg von der Konsens- zur Konfliktgesellschaft* Bd. 1 (Frankfurt a.M.: Suhrkamp): 9–26.

Hobsbawm, Eric; Ranger, Terrence (Eds.), 1983: *The Invention of Tradition* (Cambridge: Cambridge University Press).

Holland, Thomas; Kilpatrick, Allie, 1993: "Using Narrative Techniques to Enhance Multicultural Practice", in: *Journal of Social Work Education* 29,3: 302–308.

Hoyningen-Huene, Paul, 1989: *Die Wissenschaftsphilosophie Thomas S. Kuhns* (Wiesbaden: Springer).

IEP (Institute for Economics & Peace), 2019: *Positive Peace Report 2019: Analyzing the Factors that Sustain Peace*; at: https://visionofhumanity.org/app/uploads/2019/10/PPR-2019-web.pdf.

Jaberg, Sabine, 2011: "Friedensforschung", in: Gießmann, Hans J.; Rinke, Bernhard (Eds.): *Handbuch Frieden* (Wiesbaden: VS-Verlag): 53–69.

Jeannotte, M. Sharon et al., 2002: *Buying In or Dropping Out: The Public Policy Implications of Social Cohesion Research* (Ottawa: Strategic Research and Analysis Directorate, Department of Canadian Heritage, Strategic Planning and Policy Coordination).

Joppke, Christian, 2007: "Immigrants and Civic Integration in Western Europe", in: Banting, K. et al. (Eds.): *The Art of the State III – Belonging? Diversity, Recognition, Citizenship in Canada* (Montreal: Institute for Research on Public Policy): 321–350.

Kiesow, Rainer; Simon, Dieter (Eds.), 2000: *Auf der Suche nach der verlorenen Wahrheit: zum Grundlagenstreit in der Geschichtswissenschaft* (Frankfurt a.M.: campus).

Knight, Kathy, 1988: "Cultural Roots of a Peace Paradigm", in: *The Personalist Forum*, 4,1: 13–26.

Koppe, Karlheinz, 1990: "Exkurs zum Friedensbegriff in der Friedenswissenschaft", in: Koppe, Karlheinz; Senghaas, Dieter (Eds.): *Friedensforschung in Deutschland. Lagebeurteilung und Perspektiven für die neunziger Jahre* (Bonn: Arbeitsstelle Friedensforschung Bonn): 106–110.

Kristeva, Julia, 1972: "Probleme der Textstrukturation", in: Blumensath, Heinz (Ed.): *Strukturalismus in der Literaturwissenschaft* (Köln: Kiepenheuer & Witsch): 243–262.

Kuhn, Thomas, 1962: *The Structure of Scientific Revolutions* (Chicago, London: University of Chicago Press)

Laclau, Ernesto, 1990: *New Reflections on the Revolution of our Time* (London: Verso).

Laclau, Ernesto, 2005: *On Populist Reason* (London: Verso).

Laucken, Uwe, 1995: *Sozialpsychologie: Geschichte, Hauptströmungen, Tendenzen* (Oldenburg: BIS).

Lévi-Strauss, Claude, 1950/1987: *Introduction to the Work of Marcel Mauss* (London: Routledge & Kegan Paul).

Levitas, Ruth, 1998: *The Inclusive Society? Social Exclusion and New Labour* (Basingstoke: Macmillan).

LGA (Local Government Association) (Ed.), 2004: *Community Cohesion – an Action Guide: Guidance for Local Authorities* (London: LGA Publications); at: https://www.london.gov.uk/sites/default/files/communitycohesionactionguide.pdf.

Lincoln, Yvonna; Guba, Egon, 2003: "Paradigmatic Controversies, Contradictions, and Emerging Confluences", in: Denzin, Norman; Lincoln, Yvonna (Eds.): *The Landscape of Qualitative Research: Theories and Issues* (London: Sage): 253–291.

Lindner, Kolja, 2017: *Die Hegemoniekämpfe in Frankreich. Laizismus, politische Repräsen-tation und Sarkozysmus* (Hamburg: Argument).

Lockwood, David, 1999: "Civic Integration and Social Cohesion", in: Gough, Ian; Olofsson, Gunnar (Eds.): *Capitalism and Social Cohesion. Essays on Exclusion and Integration* (London: Palgrave Macmillan): 63–84.

Lutz, Dieter, 2004: "Friedensforschung – normativ, interdisziplinär, praxisorientiert", in: Eckern, Ulrich et al. (Eds.): *Friedens- und Konfliktforschung in Deutschland. Eine Bestandsaufnahme* (Wiesbaden: VS-Verlag), 23–32.

Mana, Adi; Sagy, Shifra; Srour, Anan; Mjally-Knani, Serene, 2012: "Perceptions of Collective Narratives and Identity Strategies: The Case of Palestinian Muslims and Christians in Israel", in: *Mind & Society*, 11,2: 165–182.

Malthouse, Richard; Watts, Mike; Roffey-Barentsen, Jodi, 2015: "Reflective Questions, Self-Questioning and Managing Professionally Situated Practice", in: *Research in Education*, 94: 71–87.

Marej, Katarina, 2020: "Creating Impact in Civic Education by Transformative Research: Indications for Professionalisation", in *Journal of Social Science Education* 19,2.

Mekoa, Itumeleng; Busari, Dauda, 2018: "Social Cohesion: Its Meaning and Complexities", in: *Journal of Social Sciences*, 14,1: 107–115.

Mertens, Donna, 1999: "Inclusive Evaluation: Implications of Transformative Theory for Evaluation", in: *American Journal of Evaluation*, 20,1: 1–14.

Mertens, Donna, 2007: "Transformative Paradigm: Mixed Methods and Social Justice", in: *Journal of Mixed Methods Research*, 1,3: 212–225.

Mertens, Donna, 2012: "Transformative Mixed Methods: Addressing Inequities", in: *American Behavioral Scientist*, 56,6: 802–813.

Merton, Robert K., 1968: *Social Theory and Social Structure* (New York: Free Press).

Ministry of Education Sri Lanka (Ed.), 2008: *National Policy and a Comprehensive Framework of Actions on Education for Social Cohesion and Peace (ESCP)* (Battaramulla: Ministry of Education, Social Cohesion and Peace Education Unit).

Modood, Tariq; Dobbernack, Jan, 2013: "Accepting Multiple Differences: The Challenge of Double Accommodation", in: Dobbernack, Jan; Modood, Tariq (Eds.): *Tolerance, Intolerance and Respect: Hard to Accept?* (Basingstoke: Palgrave Macmillan): 186–207.

Morin, Edgar, 1977/2010: *Die Methode: Die Natur der Natur* (Wien, Berlin: Turin Kant).

Morin, Edgar, 2006: "Restricted Complexity, General Complexity", paper presented at the Colloquium on "Intelligence de la complexité:´epistémologie et pragmatique", Cerisy-La-Salle, France, 26 June 2005.

Müller-Funk, Wolfgang, 2007: *Die Kultur und ihre Narrative. Eine Einführung* (Wien, New York: Springer).

Newman, Ines; Ratcliffe, Peter, 2011: "Conclusion: Towards a Theory of Change for Social Cohesion", in: Radcliffe, Peter; Newman, Ines (Eds.): *Promoting Social Cohesion: Implications for Policy and Evaluation* (Bristol: Policy Press): 261–276.

O'Reilly, Dermot, 2005: "Social Inclusion: A Philosophical Anthropology", in: *Politics* 25,22: 80–88.

OECD (Organisation for Economic Co-operation and Development) Development Centre, 2011: *Perspectives on Global Development* (Paris: ebrary).

OECD (Organisation for Economic Co-operation and Development) Development Centre, 2012: *Social Cohesion in a Shifting World* (Paris: ebrary).

Osler, Audrey, 2011: "Education Policy, Social Cohesion and Citizenship", in: Radcliffe, Peter; Newman, Ines (Eds.): *Promoting Social Cohesion. Implications for Policy and Evaluation* (Bristol: Policy Press): 185–206.

Oxoby, Robert, 2009: "Understanding Social Inclusion, Social Cohesion, and Social Capital", in: *International Journal of Social Economics*, 36,12: 1,133–1,152.

Polkinghorne, Donald E., 1996: "Explorations of Narrative Identity", in: *Psychological Inquiry*, 7,4: 363–367.

Putnam, Robert O., 1993: *Making Democracy Work: Civic Traditions in Modern Italy* (Princeton: Princeton University Press).

Rappaport, Julian, 1995: "Empowerment Meets Narrative: Listening to Stories and Creating Settings", in: *American Journal of Community Psychology*, 23,5: 795–807.

Richmond, Oliver, 2008: *Peace in International Relations* (London: Routledge).

Romm, Norma, 2015: "Review", in: *Systemic Practice and Action Research*, 28: 411–427.

Rubagiza, Jolly; Umutoni, Jane; Kaleeba, Ali, 2016: "Teachers as Agents of Change: Promoting Peacebuilding and Social Cohesion in Schools in Rwanda", in: *Education as Change*, 20,3: 202–224.

Rüsen, Jörn, 2007: "How to Overcome Ethnocentrism? Approaches to a Culture of Recognition by History in the 21st Century", in: Lange, Dirk; Himmelmann, Gerhard (Eds.): *Demokratiebewusstsein: Interdisziplinäre Annäherungen an ein zentrales Thema der Politischen Bildung* (Wiesbaden: VS Verlag): 229–242.

Said, Edward, 1993: *Culture and Imperialism* (London: Chatto & Windus).

Salomon, Gavriel, 2004: "A Narrative-Based View of Coexistence Education", in: *Journal of Social Issues* 60,2: 273–287.

Sarbin, Theodore, 1986: *Narrative Psychology: The Storied Nature of Human Existence* (New York: Praeger).

Sayed, Yusuf; Novelli, Mario, 2016: *The Role of Teachers in Peacebuilding and Social Cohesion. Synthesis Report on Findings from Myanmar, Pakistan, South Africa and Uganda* (Sussex: Research Consortium on Education and Peacebuilding); at: https://s3.amazonaws.com/inee-assets/resources/15-Role-of-teachers-in-Peacebuilding-synthesis-report-final.pdf.

Schiefer, David; van der Noll, Jolanda, 2017: "The Essentials of Social Cohesion: A Literature Review", in: *Social Indicators Research*, 132,2: 579–603.

Segert, Dieter, 2016: "Demokratieforschung", in: Diendorfer, Gertraud et al. (Eds.): *Friedensforschung, Konfliktforschung, Demokratieforschung. Ein Handbuch* (Köln et al.: Böhlau): 97–123.

Senghaas, Dieter, 1997: "Frieden – Ein mehrfaches Komplexprogramm", in: Senghaas, Dieter (Ed.): *Frieden Machen* (Frankfurt a.M.: Suhrkamp): 560–575.

Silver, Hillary, 1994: "Social Exclusion and Social Solidarity: Three Paradigms", in: *International Labour Review*, 133,5–6: 531–578.

SJPG (Social Justice Policy Group), 2007: *Breakthrough Britain: Ending the Costs of Social Breakdown: Overview: Policy Recommendations to the Conservative Party*; at: https://www.centreforsocialjustice.org.uk/core/wp-content/uploads/2016/08/BBChairmansOverview.pdf.

Somers, Margaret R., 1994: "The Narrative Constitution of Identity: A Relational and Network Approach", in: *Theory and Society*, 23,5: 605–649.

Sørensen, Eva, 2006: "Metagovernance: The Changing Role of Politicians in Processes of Democratic Governance", in: *The American Review of Public Administration*, 36,1: 98–114.

Stark, Wolfgang, 1992: "Gemeindepsychologische Geschichte(n): Zur Bedeutung von Geschichten für eine gemeindepsychologische Perspektive. Fünf Annäherungen", in: Böhm, Ingrid et al. (Eds.): *Gemeindepsychologisches Handeln: Ein Werkstattbuch* (Freiburg: Lambertus): 28–44.

Starkey, Hugh, 2018: "Fundamental British Values and Citizenship Education: Tensions between National and Global Perspectives", in: *Geografiska Annaler: Series B, Human Geography*, 100,2: 149–162.

Suvarierol, Semin, 2012: "Nation-Freezing: Images of the Nation and the Migrant in Citizenship Packages", in: *Nations and Nationalism*, 18,2: 210–229.

Taylor, Charles, 2004: *Modern Social Imaginaries* (Durham: Duke University).

Thaler, Richard; Sunstein, Cass, 2009: *Nudge. Improving Decisions about Health, Wealth and Happiness* (London: Penguin).

Tißberger, Martina, 2017: *Critical Whiteness: Zur Psychologie hegemonialer Selbstreflexion an der Intersektion von Rassismus und Gender* (Wiesbaden: Springer).

UN (United Nations), 2015: *Transforming our world: The 2030 Agenda for Sustainable Development* (A/RES/70/1); at: https://sustainabledevelopment.un.org/content/documents/21252030%20Agenda%20for%20Sustainable%20Development%20web.pdf.

van Houdt, Friso; Suvarierol, Semin; Schinkel, Willem, 2011: "Neoliberal Communitarian Citizenship: Current Trends towards 'Earned Citizenship' in the United Kingdom, France and the Netherlands", in: *International Sociology* 26,3: 408–432.

Vogt, Wolfgang, 1996: *Frieden durch Zivilisierung? Probleme, Ansätze, Perspektiven* (Münster: agenda).

White, Hayden, 1986: *Auch Klio dichtet oder die Fiktion des Faktischen: Studien zur Tropologie des historischen Diskurses* (Stuttgart: Klett-Cotta).

White, Hayden, 1980: "The Value of Narrativity in the Representation of Reality", in: *Critical Inquiry*, 7,1: 5–27.

White, Michael; Epston, David, 1989: *Literate Means to Therapeutic Ends* (Adelaide: Dulwich Centre Publications).

Whitmore, Elizabeth (Ed.), 1998: *Understanding and Practicing Participatory Evaluation* (San Francisco: Jossey-Bass).

Winter, Christine, 2018: "Disrupting Colonial Discourses in the Geography Curriculum during the Introduction of British Values in Schools", in: *Journal of Curriculum Studies*, 50,4: 456–475.

Wintersteiner, Werner, 2006: "Kultur des Friedens – ein neuer Leitbegriff der Friedensforschung?", in: Zentrum für Friedensforschung und Friedenspädagogik (Ed.): *Jahrbuch Friedenskultur 1* (Klagenfurt: Drava): 94–108.

Wintersteiner, Werner; Graf, Wilfried, 2015: "Friedensforschung in Zeiten des Umbruchs. Plädoyer für (selbst-)kritische Neuorientierung und transdisziplinäre Zusammenarbeit", in: Wintersteiner, Werner; Wolf, Lisa (Eds.): *Jahrbuch Friedenskultur 10: Friedensforschung in Österreich. Bilanz und Perspektiven* (Klagenfurt: Drava): 16–37.

Wintersteiner, Werner; Gruber, Bettina, 2014: "Building Peace: Social Transformation as a Collective Learning Process", in: Gruber, Bettina; Wintersteiner, Werner (Eds.): *Yearbook Peace Culture 9: Learning Peace: An Integrative Part of Peace Building: Experiences from the Alps-Adriatic Region* (Klagenfurt: Drava), 7–20.

Other Literature

Adams, Richard, 2016: "Teachers back motion calling for Prevent strategy to be scrapped", in: *The Guardian* (28 March); at: https://www.theguardian.com/politics/2016/mar/28/teachers-nut-back-motion-calling-prevent-strategy-radicalisation-scrapped.

AFD (Alternative für Deutschland, political party) 2018: "Leitkultur, Identität, Patriotismus. Ein Positionspapier der AfD-Fraktion im Thüringer Landtag als Beitrag zur Debatte um die deutsche Leitkultur (Erfurt)".

Blunkett, David, 2002: *Speech on Social Capital.*

de Maizière, Thomas, 2017: "Leitkultur für Deutschland – Was ist das eigentlich?", in: *Bild am Sonntag*, 1 May, Bundesministerium des Innern, für Bau und Heimat; at: https://www.bmi.bund.de/SharedDocs/interviews/DE/2017/05/namensartikel-bild.html.

De Wijze, Stephen, 2018: "Why Norman Geras's essay 'Our Morals' should be essential reading for politics students – not a subversive threat", in: *The Conversation* (16 November); at: https://theconversation.com/why-norman-gerass-essay-our-morals-should-be-essential-reading-for-politics-students-not-a-subversive-threat-106871.

Johnstone; Fischer 2019: "Französische Abgeordnete stimmen für obligatorische EU-Flaggen in Klassenzimmern", in: *Euronews* (12 February); at: https://de.euronews.com/2019/02/12/abgeordnete-obligatorische-eu-flaggen-klassenzimmern-europa-frankreich.

Gayle, Damien, 2019: "UK's Prevent guidance to universities unlawful, court rules", by Gayle, in: *The Guardian* (8 March), at: https://www.theguardian.com/uk-news/2019/mar/08/uks-prevent-guidance-to-universities-unlawful-court-rules.

Gove, Michael, 2010: "Gove unveils Tory plan for return to 'traditional' school lessons", in: *The Times* (6 March); at: https://www.thetimes.co.uk/article/gove-unveils-tory-plan-for-return-to-traditional-school-lessons-wnkhqw0k3mz.

Merz, Friedrich, 2000: "CDU/CSU-Fraktionschef Friedrich Merz fordert: Zuwanderer sollen sich deutscher Kultur anpassen", in: *rp online* (18 October); at: https://rp-online.de/politik/zuwanderer-sollen-sich-deutscher-kultur-anpassen_aid-8244647.

Schwarz, Hans-Peter, 1990: "Das Ende der Identitätsneurose", in: *Rheinischer Merkur/Christ und Welt*, 30 (7 September): 3f.

Chapter 5
The National and Universal Importance of the Non-violent Policy of Mohandas K. Gandhi

Egbert Jahn

Abstract It is a matter of dispute as to whether the attainment of national independence is primarily the result of non-violent actions by the Indians, or of the moderate British colonial policies. Numerous domestic political factors in India and Britain, and above all the weakening of the British Empire as a result of the World Wars, paved the way for the success of the largely non-violent independence movement. Its most important political leader was Mohandas Karamchand Gandhi. He only rarely took up official positions in the Indian National Congress. Nevertheless he enjoyed authority over the masses. The core element of Gandhi's way of life is an awareness of one's responsibility not only for one's actions, but also of one's failure to put up resistance against injustice in one's own environment. Non-cooperation is the primary legal means of non-violent action. Civil disobedience is a stage of escalation of non-violent policies that requires careful preparation and consideration of the risks to the common good that it entails. The basic principles of non-violent social behaviour and Gandhi's policies have universal significance, particularly in liberal democracies and in dictatorships that are losing their legitimacy in the eyes of a population suffering from injustice.

Keywords Civil rights · Gandhi M. K. · Global peace movement · India · Indian national movement · King M. L. · Non-violence · South Africa

5.1 The International Mystification of Mohandas K. Gandhi the Politician

Even before Mohandas K. Gandhi began to play a role in the Indian national independence movement, he had become highly famous between 1893 and 1915 as a leading figure in the resistance movement among the Indian minority in British South Africa against the beginnings of the Apartheid system. On his return to India in January

Prof. Dr. Egbert Jahn (Em.), University of Mannheim and Senior Prof., Goethe University of the Third Age, Frankfurt on Main, Germany; *Email*: e.jahn@em.uni-frankfurt.de. The text was translated from German into English by Anna Güttel-Bellert.

Ú. Oswald Spring and H. G. Brauch (eds.), *Decolonising Conflicts, Security, Peace, Gender, Environment and Development in the Anthropocene*,
The Anthropocene: Politik—Economics—Society—Science 30,
https://doi.org/10.1007/978-3-030-62316-6_5

1915, he was welcomed as a Mahatma (or 'great soul' in Sanskrit) by the philosopher and winner of the Nobel Prize for literature, Rabindranath Tagore (1861–1941). Not long afterwards, his first name was replaced by 'Mahatma'. Most non-Indians are unaware of the fact that several other people have also been given the title 'Mahatma', which is approximately the same as the Christian term 'saint'. The extent to which Gandhi was also honoured outside India was so great that in a book publication of his essays in the *Young India* newspaper, in 1924, John Haynes Holmes (USA) equated him with Christ returned to earth.[1] The declaration of Gandhi as an ahistorical saintly figure[2] obscures his real-life political and societal role in achieving the independence of India and Pakistan in August 1947 after decades of non-violent protest, thereby also contributing to the shattering of the British Empire. According to Gandhi's German biographer, Dietmar Rothermund: "Unfortunately, he tends to be remembered as a saint who became unwittingly involved in politics, rather than as a politician whose ideas were and still are of relevance to the future" (Rothermund 1997: 15).

When analysed through an objective, socio-historical lens, Mohandas K. Gandhi should be regarded as an individual who, like no other in global history, steered the political behaviour of countless millions of people during his lifetime without having access to the force and power of a major state or a party, and solely through his charisma, without recourse to violence. No Lenin or Stalin, Mao or Hitler had such personal power during the twentieth century. Only in rare cases was Gandhi's charisma reinforced by the official authority of the President (1924) or by another leading position in the *Indian National Congress party* (INC). In 1934 he left the INC, although as the party's 'advisor' he probably had even greater influence over it than before. He played a key role in the conversion of the institutional structures of the INC from an elite organisation into a mass organisation of farmers. In independent India, he did not aspire to hold an official office, as a result of which he was never confronted with the problematic issue of having to exert state force.

Even before the partition of British India into the independent states of India and Pakistan, murders between Hindus, Sikhs and Muslims were committed on an extensive scale in the years prior to 1948. Up to one million of people lost their lives as a result; 20 million people were resettled from India and Pakistan, or were driven out or fled from their homes. Before the Partition, Gandhi succeeded in ending the acts of violence in several provinces through his personal presence and through hunger strikes. The last British Viceroy, Lord Louis Mountbatten, wrote: "In the Punjab, we have 55,000 soldiers and large-scale rioting on our hands. In Bengal,

---

[1]"If I were to believe in reincarnation, then – with all respect – I would see Christ in Mahatma Gandhi, who had returned to Earth. If I were to believe in the principle of the second coming, I would claim that this event had already occurred in India. ... He, too, moves us most deeply with his writings, and like Jesus in the gospels, he lifts us to the greatest heights," in the introduction to: Gandhi (1924, p. XII, XVI).

[2]Gandhi frequently complained about being honoured as a Mahatma, or saint. "I think that word 'saint' should be ruled out of present life. It is too sacred a word to be lightly applied to anybody, much less to one like myself, who claims only to be a humble searcher after Truth..." See Gandhi (1924: 71, 78).

our forces consist of one man, and there is no rioting. As a serving officer, as well as an administrator, may I be allowed to pay my tribute to the One-Man Boundary Force…" (Rothermund 1997: 472).

In the history and society of India, violence plays no less a role than in other countries (Singh 2017). The independent Republic of India waged several wars after Gandhi, and has become an atomic power. It can hardly be claimed that Indian society has become less violent than societies in other countries since Gandhi's historical impact, although smaller groups of Gandhiists who live according to his teachings continue to exist in India. The question therefore arises as to whether this was simply a one-off historical episode in which – under extremely favourable social and international constellations – a mainly non-violent national independence movement was able to succeed in a decades-long struggle against a major global power at the time. And this was achieved with around 8000 political casualties[3] among a population of far more than 300 million. How much bloodier the battles for independence were among other, far smaller groups of people! How was Gandhi able, therefore, to rise to his outstanding political role? What were the methods of combat that were in many respects historically new? What was the basis for his authority and ability to assert his demands within Indian society? Why did the British then ultimately allow their empire to be destroyed by a 'semi-naked fakir', as the colonial politician Winston Churchill disparagingly described him in the 1930s?

First I will focus exclusively on the role played by Gandhi in the Indian national independence movement. Then I will discuss his universal, global historical importance.

5.2 The Theory of the Singularity of the Success of Non-violent Policy in the Indian National Movement

While admirers of Gandhi and the proponents of the principle of non-violent politics[4] throughout the world assume that the Indian resistance movement against British colonial rule is of universal significance and that it acts as a model, in current political thinking the dominant attitude is that the success of the non-violent politics of the Indian national movement was a unique historical event, and that it could not, and cannot, be replicated in other political situations.

Several reasons have been cited for this view. The most fundamental argument is that of Max Weber (1864–1920), that politics is, in essence, the leadership or influence of the leadership of a state (in general terms: a political association), whereby the state is a "human community that (successfully) claims the monopoly of the legitimate

[3]I quote these numbers from memory from one of my earlier lectures, although I am currently unable to find written evidence for them.

[4]In the German version of this text, Ebert (1969: 34) proposed a differentiation between non-violence as the avoidance of violent behaviour on principle and non-violence as the avoidance of violence based on circumstances.

use of physical force within a given territory" (Weber 1988: 506). The ethics of non-violence, of "do not resist evil with force", is exclusively the way of the saint (Weber 1988: 550). However, politics does not just consist of leading and influencing a state – in other words, of exerting governmental and state power – but is also the process of attaining the power to rule, which in parliamentary democracies usually means using non-violent methods. The process of obtaining the power to rule one's own state can also be non-violent or even violence-free when achieved through a national independence movement, which Max Weber had not considered as a possibility when writing in 1919.

Most authors explain the success of non-violent politics in India in terms of a specific historical constellation of colonial rule and a national independence movement. Here, some emphasise that in pre-industrial Indian society, and above all in the Hindu religion, there was a particular inclination among large swathes of the Indian population to be humble followers of a charismatic, non-violent leader.

Others see the main reason for the development of the leadership role of Gandhi, the non-violent politician, in the Indian national movement as being a specifically liberal British colonial policy, which became democratically legitimised and controlled over the course of time, and which had become more humane, which began as a movement to reform the Empire, before radicalising to become an independence movement. One proponent of this view was Karl Jaspers, who initially assumed that "in its essence, politics is a way of dealing with violence",[5] while claiming that Gandhi, quite unlike the pacifists, had successfully pursued 'supra-political' religious non-violence: "...he did not detach himself from the world in solitude, like a saint indifferent to the world. He also did not act alone, but together with the mass of people who believed in him." According to Jaspers, this success was only possible because Britain was not willing to pursue reckless violence: "Only under England and only under this uniquely liberal rule in the history of empires was Gandhi able to be successful. Such non-violent politics would never have led to such a result in former times, and in the future, it would only be able to be successful under conditions similar to those under England with regard to liberalism, openness and legality. The liberation of India through Gandhi's politics of non-violence is to a far greater extent the consequence of a struggle in England with itself than of an Indian act." And Jaspers concludes that "In a battle with the totalitarian regimes, Gandhi's approach would no longer have been a political path, but a sure route to destruction."[6]

[5]These and the following quotes can be found in the section "The Notion of Non-Violent Politics" in Jaspers (1960: 63, 65, 67, 68). The translator could not find published translations of the cited phrases. Therefore, the translations are those of the translator of this text.

[6]Note by the translator: This text was taken from the English original.

5.3 Discussion Surrounding the Combat Methods of the Indian Movement to Reform British Colonial Rule in South Africa

British colonial rule of the Indian subcontinent began with the establishment of the trading centres of the British East India Company on the coasts of India in the early seventeenth century. Later, the trading company, which had been invested with political and military rights by the King of England, took control of the interior of the subcontinent and finally subjugated around half it, along with two-thirds of the population, to direct rule, and 562 princely states to supreme British colonial authority. Nominally, India was still ruled by the Muslim Grand Moguls, who in 1526 had conquered the sultanate of Delhi (in existence since 1206), which was also Muslim, but had lost its power over the course of time to regional rulers, many of them Hindus, and then to the British. In 1857 there was an uprising against British rule among Indian soldiers in the British forces in north-eastern and central India, which was joined by civilians, mostly from poorer castes, and a small number of princes as well as the Grand Mogul. Shortly before he was deposed in 1858, the Grand Mogul assumed the title of Emperor. The British only succeeded in crushing the revolt after two years, with the help of allied Indians, and after many atrocities on both sides that led to the deaths of over 10,000 people. The uprising led the United Kingdom to dissolve the trading company and to assume direct rule over the subcontinent. From that time onwards, the Governor General bore the title of Viceroy. He was assigned command over the executive and legislative council, the members of which were initially only British, although later they were joined by Indians elected from the upper castes. The territory of British India covered not only today's Republic of India, Pakistan and Bangladesh, with a total area of 4.2 million square kilometres (by comparison, the EU currently covers an area of 4.4 million square kilometres), but from 1886–1937 also Burma, known today as Myanmar.

From 1856–1947, India as a colony of the Crown was the 'jewel' of the British Empire, which in 1922 covered a quarter of the land surface of the earth, and included a quarter of the world's population. In 1875, 301 million people lived in British India, with just 33 million living in the United Kingdom.[7] The Kingdom succeeded in exerting its rule over the entire subcontinent with just 150,000 or so British officials and military personnel,[8] cleverly incorporating numerous princes and the upper-caste Indians into the colonial administrative system. In 1876, Queen Victoria was

[7]Historical population.

[8]I quote these numbers from memory from one of my earlier lectures, although I am currently unable to find written evidence to support them. In 1921, Gandhi (1924: 227) spoke of 300 million Indians, including 70 million Muslims, and 100,000 British. In the final decades of the nineteenth century, 140,000 Indians and 70,000 British served in the British-Indian Army, with the British alone occupying the higher officers' positions, according to Kulke and Rothermund (2006: 325).

crowned Empress of the Indian Empire by the British Parliament, in response to the establishment of the German Empire.[9]

The *Indian National Congress* (INC), an organised national movement, was not created until 1885. For a long time, only Western-educated elites, Hindus and Muslims alike, who wanted to participate in the administration of the colony were actively involved. They attempted to assert liberal citizens' rights and social reforms. After the Hindus had gained the upper hand in the INC, a separate All-India Muslim League was formed in 1906, to which Muhammad Ali Jinnah (1876–1948) was elected President in 1916.

Mohandas Karamchand Gandhi was born on 2 October 1869, far away from the areas of influence of the national movement, in Porbandar, a small coastal town in what is now the federal state of Gujarat in the westernmost region of the Republic of India. At that time, the town was the centre of one of the princely states of British India, most of which were very small. The Gandhi family belonged to the merchant class (Vaishya), which held a great deal of influence in society; the name 'Gandhi' means 'grocer' in English. At the age of eight, Gandhi was engaged to Kasturbai Makthaji,[10] who was the same age, and they married when they were thirteen. The couple had four children.

His strictly religious Hindu mother (who belonged to the group of Hindus who worship the god Vishnu) exerted an important influence over him. In her religion, fasting and the making of vows played a significant role. She was also influenced by Jainism. This religion, which emerged at about the same time as Buddhism, and which with its strong ascetic beliefs has only about four million followers, espouses, among other things, the radical command of *ahimsa*, the non-killing and non-injuring of even the smallest life forms; accordingly, the Jains are vegetarians. Another tenet of Jainism is *satya*, or truth. "The search for truth was at the centre of Gandhi's life, but this was not the truth that can be founded on methods of critical scholarship. For him, the truth had to be tested in well thought-through action, supported by a vow" (Rothermund 2011: 7–8). As a result, Gandhi was able to understand religiosity not merely as contemplation, but also as something that had to prove its worth through practical deeds, on a personal level as well as for the general good, in other words, in politics. In his politics, Gandhi never referred to religious commands from holy scriptures, but only regarded those writings as being a religious command that appeared to be of sensible use to him. In his public prayer meetings, he recited prayers from all the major religions. Therefore, he had a clearly rational understanding of religion and politics based on his understanding of the truth, which whenever possible excluded violent physical and mental behaviour.

Mohandas K. Gandhi's father and grandfather were prime ministers in the small princely state of Porpandar. Mohandas, the youngest son, was expected to follow

[9] Austria had already been an empire since 1804 (in response to the proclamation of Napoleon as Emperor of France and in anticipation of the dissolution of the Holy Roman Empire of the German Nation), while Russia had been an empire since 1721 (as a means of increasing its status over the Holy Roman Empire).

[10] Gandhi Arun (1981). Kasturbai died before Gandhi, in 1944.

in their path and for this reason was sent to London to study (1888–1891). There, he received important inspiration for his future life, such as from the non-violent suffragist movement, constitutional thinking, conflict mediation in labour disputes, theosophy and the Sermon on the Mount in the New Testament. Following his return from London, he worked as a lawyer in Mumbai (Bombay) with moderate success. In 1893, he took up an offer to represent the legal interests of a Muslim trader from Gujarat in Durban, in Natal/South Africa.[11] After experiencing the racial discrimination of Indians in person during a train journey from Durban to Pretoria/Transvaal – he was ejected from the train when he refused to leave his reserved seat in the first-class carriage following a complaint made to the guard by a white passenger – he was inspired not only to protest against his personal discrimination, but also to mobilise the Indians living in Pretoria to demonstrate against the racial discrimination against them in general.

The British had brought many Indians to their sugar plantations as contract workers. Many of them also remained in South Africa as free workers following their five-year tenure to the individual plantation owners.[12] Their increasing number led to fears of foreign infiltration among the Boers and British, which engendered the beginnings of the legally underpinned policy of Apartheid. When Gandhi learned, just before he was due to return to India, that in Natal a law was being drafted that would have deprived all Indians of their voting rights, he organised protest rallies and became Secretary General of a newly founded Natal Indian Congress.

Strongly influenced by the book *Unto this Last* (1860), by the Calvinist art historian, painter and social reformer John Ruskin (1819–1900),[13] who in his critique of the principles of national economy sharply criticised not only capitalism and industrialisation, but also Marxism, espousing instead a simple life, manual work and an equal wage for all, in 1906 Gandhi converted to a life of frugality, abstinence and harsh discipline. He was also motivated by his experience of aiding the British by serving in the Natal Indian Ambulance Corps during the crushing of the Zulu uprising. On a farm outside Durban he founded a community with the staff of his newspaper, *Indian Opinion*, after deciding to leave his law practice and dedicate himself entirely to political and social work.

Gandhi had no interest in socio-economic and political theories such as capitalism or socialism, parliamentary democracy, realism or liberal institutionalism, and is also difficult to classify politically. Gandhi wrote no systematic political theory books, and very few more extensive works.[14] The name 'Gandhi' therefore hardly appears at all in political and sociological works, and only occasionally in passing in

[11] The four British Crown colonies of Natal, Transvaal, the Orange Free State and the Cape Colony were not combined to form the South African Union until 1910.

[12] Similar working conditions still exist today in Qatar and other Gulf states.

[13] See Ruskin (2011). The title refers to the New Testament parable in which the day labourer most recently employed in a vineyard, and who has therefore worked there for the shortest time, is paid the same wage as those who were employed there earlier (Matthew 20). Gandhi translated Ruskin's book into Gujarat, calling it *Sarvodaya* ('the ascent of man').

[14] In addition to the autobiography, these include *Satyagraha in South Africa* (Gandhi 1924, 2017), and *Hind Swaraj or Indian Home Rule* (Gandhi 1909, 2010).

political-philosophical writings.[15] Even so, his political attitudes and behaviours can be described as being extremely rationally, systematically thought-through concepts within the framework of a theory of non-violent politics.

In 1907, Gandhi read The short essay, "Civil Disobedience",[16] by Henry David Thoreau (1817–1862), in which civil disobedience was propagated, for example through refusal to pay taxes or against unjust laws that justify slavery. According to Thoreau, punishment by imprisonment for violating the law should be accepted. Gandhi first called for the law to be broken when he organised a boycott of a new law in Transvaal that required Indians to register themselves. The purpose of this law was to prevent unwanted immigration by more Indians. Due to his refusal to register, he was sentenced to two months in prison. During the course of his life, Gandhi spent over six years[17] in total in prison.

Gandhi called his new combat strategy by a new name, *satyagraha*, i.e. adherence to the truth, whereby the truth must be proven through one's own non-violent behaviour. Truth therefore emerges through interaction, and is not something that is discovered in itself.[18] In order to consolidate the policy of non-violent resistance, those taking non-violent resistance action, the *satyagrahis*, were required to take an oath. Many of them and their families lived with Gandhi on the Tolstoy Farm (a precursor of the later ashrams in India), where they led a simple, hard-working communal life, eating only vegetarian food. When a court declared that only marriages conducted in a Christian church and registered at a registry office were valid, it was the Indian women above all who were outraged, along with their children, who were now declared illegitimate. They mobilised the Indian mineworkers to take part in a long-term strike. When they were subsequently sentenced to forced labour in their mines, solidarity strikes were called among the plantation workers. When the government of South Africa also came under pressure from a railway strike by the whites, it finally relented and passed a law that recognised Indian marriages, abolished the poll tax on Indians and permitted the immigration of qualified Indians. Gandhi then felt that his work in South Africa was complete, and left the country for good at the end of 1914. The leader of a minority movement would then become the leading personality of a majority movement.

In 1896, when he travelled to India to bring his family to South Africa, Gandhi had already become well known for his report about the situation of the Indians in South Africa, and had met the leading members of the INC. The national reformer and professor, Gopal Krishna Gokhale (1866–1915), supported union between Hindus and Muslims, education for ordinary people and public promotion of health, while being against the caste system and discrimination against the 'Untouchables'. The reformers also wanted to implement constitutional reforms in the Crown colony,

---

[15]For example, Karl Jaspers (1960) occasionally examined Gandhi's political thinking.

[16]Thoreau (2014).

[17]In total, 2089 days in Indian prisons and 249 days in South African prisons, according to Fischer (1962: 147).

[18]Probably the best book about Gandhi is the psychoanalytical study by Erikson (1993). For the religious historical background, see Mühlmann (1950).

and to abolish child marriages and the immolation of widows. Gokhale, who was just three years older than Gandhi, became his most important adviser. Following Gandhi's return to India at the start of 1915, Gokhale urged him to first spend a year travelling through India in order to get to know the country that was the size of an entire continent, and while doing so, to refrain from giving political speeches or issuing political statements.

As well as the national reformers, there was a strong national-revolutionary group, whose key political representative was Bal Gangadhar Tilak (1856–1920). It propagated achieving independence through violence as a prerequisite to social reforms, and encouraged the uprising against British rule. At times, it formed its own organised wing in the INC. Violence and war certainly played an important role in Indian history and in the history of the national movement. The national-revolutionary group gained the upper hand in the INC during the First World War, and Tilak was elected its president in 1918. However, after his death, it lost a considerable amount of influence when Gandhi rose to become the de facto leader of the INC, which was increasingly moving away from being a broad movement organisation and towards becoming a political party, alongside which other parties were also emerging, albeit with a lesser degree of influence.

5.4 Gandhi's Role in the Indian National Independence Movement

At the start of the 1920s, after the death of Gokhale and Tilak, Gandhi quickly became the main leader of the Indian national independence movement, which was at the same time a social reform movement. Gandhi had already written down his ideas for the social and political development of India on a journey from London to South Africa in 1909, in the form of a Socractic dialogue with a violent extremist. He called this work *Hind Swaraj* (Gandhi 2010), or 'Indian home-rule'. This home-rule, he claimed, could only be achieved through the self-discipline of each individual. Simply replacing British rule by force with an Indian equivalent would not bring India freedom. The document contained radical criticism of Western civilisation, and in so doing, went far beyond the ideas, in terms of its propagation of comprehensive social reform, of the moderate Indian liberals and the national revolutionaries, who simply attacked British foreign rule.

Apart from Ruskin's book, Gandhi obtained important ideas for his new sociopolitical thinking from Socrates' *Apology* (Fuhrmann 1989) and Leo Tolstoy's (2017) ideas of how to organise a Christian society. It was more through his successes in South Africa than his writings that Gandhi quickly became an influential figure in India. However, later, his autobiography was received with a great deal of interest, both internationally and in India. In typical fashion, he called it *The Story of my Experiments with Truth*, which he understood to be a way of coming closer to the truth of a non-violent individual and societal life explored through his own activity.

The autobiography appeared as a sequel to journal articles written from 1925–1929, when Gandhi was already aged 46–49 (Gandhi 1983). In this book, he discloses his highly personal, even intimate problems and weaknesses, and gives his views on all possible subjects, such as health, children's upbringing, gender relations, and also socio-economic and political events and fundamental questions with which he was engaged. Social classes, nations, humanity and other collective bodies had no meaning for Gandhi. For him, they were essentially associations of self-responsible individuals in communities and social contexts. He had become a journalist at an early age and had published his own newspapers. And he was a prolific letter-writer. His collected works comprised 100 volumes[19] and run to 45,000 pages – far more than the works of Marx and Engels, Lenin, or other political thinkers.

Gandhi was not a socialist, and did not reject the concept of private property of the means of production in principle. Instead, he regarded owners of capital as trustees with an obligation to pay a fair wage. He largely rejected industrialisation and advocated a concept of village farmer communities. He became famous when he turned against socially unjust and ethnically discriminatory laws that made life very difficult for farmers, while at the same time also opposing the suppression of workers' strikes. He always stood up for those in a weaker social position, although at the same time he was willing to reach a compromise with social and political rulers. The most accurate way to describe Gandhi, with certain constraints, is as an anarchist, who wanted to achieve non-hierarchical social self-organisation, preferably in village communities, although he by no means wanted to do away with the State, the judiciary, the police or the military overall.

Communist contemporaries found it difficult to evaluate Gandhi: "It is, as it were, as though Gandhi embodies two different people, two class truths", even when they also adjudge him harshly: "This same Gandhi is a smooth political entrepreneur, one of the most crafty, slippery, cunning professional politicians, a master of the smooth compromise, the great master of lies and deception.... The class countenance of Gandhi is the countenance of the Indian bourgeoisie, the countenance of the civic national movement in India" (Reissner 1930: 1130). "The Communist Party and the working class are themselves leading the movement to a higher level, by forcing back the miserable ideology of Gandhiism, exposing Gandhi himself and his counter-revolutionary ideas, clearly revealing the irreconcilable dichotomy of interests of the revolutionary proletariat and the farmers and the interests of the national-reformist block." "Gandhi hoped to provide distraction from the fundamental issues of the agricultural revolution through a pompous rally against the salt monopoly of the government… The final exposure of the betrayal of the National Congress and its leaders, who flee to the prisons before this exposure, is one of the most important tasks of the Communist Party of India" (Reissner 1930: 1138).

Gandhi was by no means a pacifist, and himself volunteered to participate in wars three times, albeit in the unarmed medical service: in the Boer War, during the suppression of the Zulu uprising, and at the start of the First World War. During the

[19] *Collected Works of Mahatma Gandhi* (CWMG), 100 (Delhi: Government of India, Publications Division, 1958–1994).

First World War, in which 1.3 million Indian soldiers fought on the side of the British, Gandhi campaigned for the recruitment of volunteers, since he expected the British to grant India self-government as a mark of gratitude for Indian participation in the war. However, while India became a founding member of the League of Nations, it remained under rigid British colonial rule.

As was the case in South Africa, it was both laws perceived to be discriminatory that were imposed by the British Crown and the social harassment of farmers and workers by property owners and companies, as well as the British authorities, that motivated Gandhi to conduct several spectacular, non-violent individual and mass actions of non-cooperation and civil disobedience. Here, he often used limited or even unlimited fasting as a means of urging both the authorities and his supporters to accept his demands or to negotiate compromises. Gandhi's demonstrative failure to comply with an existing law or police order often led to his imprisonment. In many cases, his internment led to massive, sometimes also violent, mass protests, as a result of which Gandhi was repeatedly prematurely released from prison, or his imprisonment was avoided by a re-interpretation of the law.

A turning point in the history of the Indian national movement and Gandhi's attitude towards the British Empire came when the Rowlatt Act was passed in March 1919, which aimed to uphold, in an amended form, the emergency war legislation that had been imposed during the First World War. The law made it possible to imprison individuals suspected of being involved in terrorist activity without a prior court trial. Gandhi organised a *hartal*, a traditional religious, one-day general strike in opposition to this law, with fasting and demonstrations, in several parts of India. In Delhi and elsewhere, the government ordered that shots should be fired at the demonstrators, leading to violent conflicts. It was only after the massacre ordered by the British general Reginald Dyer at a large gathering of non-violent demonstrators in Amritsar in the Punjab region, which led to the deaths of around 400[20] people and which went unpunished, that Gandhi became convinced that India must gain independence from British rule. However, after the campaign of civil disobedience degenerated into a series of violent disputes, Gandhi brought it to a halt and admitted committing a 'Himalayan blunder', since, unlike previously in South Africa, he had not trained a sufficient number of *satyagrahis* to take non-violent action in a controlled manner. Later, he placed great value on training disciplined cadres for civil disobedience. He tended to call on the mass population to commit acts of legal non-cooperation, which could take on huge dimensions. This began with the return of medals and awards and the refusal to participate in official events, but could also involve boycotting elections and goods or relinquishing official state posts. Occasionally, people were called on to interrupt their studies in colonial administration schools and universities.

In parallel with the campaign against the Rowlatt Act, Gandhi also organised a campaign of solidarity with the Ottoman caliph (caliphate campaign), from whom the British had withdrawn the power of rule over the holy Muslim sites after the war. This was met with outrage among numerous Indian Muslims, which Gandhi

[20]Rothermund (2002: 74). According to other sources, the number of casualties was 600 (Mann 2005: 87).

now wanted to exploit for his policies opposing British rule. However, the campaign was brought to a halt by the Turkish Grand National Assembly under Atatürk, who disbanded the caliphate in March 1924.

One of the most famous campaigns was the boycott of English textile goods. In order to avoid exerting excessive pressure on the import traders, Gandhi called on people to publicly burn clothing personally owned by them that had been produced by the British. At the same time, he summoned the population to spin thread for new clothing themselves on simple spinning wheels, and to weave new clothes. From that time on, the membership fee for the INC would have to be paid in the form of self-spun, later also purchased thread. This campaign had a dual impact. On the one hand, it made a severe dent in the British textile industry, leading to the loss of tens of thousands of jobs. On a visit to the centres of the British textile industry, Gandhi was, however, able to gain understanding for his campaign among the textile workers. On the other hand, the campaign mobilised many millions of simple Indians, who through their own legal action not only made a contribution to the weakening of British rule, but at the same time also strengthened their personal and political self-assurance. It was for this reason that the Indian National Congress incorporated a spinning wheel into its flag in 1931. When India gained independence, the spinning wheel was replaced by the old Indian 'wheel of the law'. However, even today, the flag must still be made of hand-woven thread.

In 1929, the INC passed a resolution demanding Indian independence. Gandhi initiated a campaign of civil disobedience in support of it, which formed the basis for his global fame. In the spring of 1930, he organised a salt march to the west coast of India with a small number of followers (*satyagrahis*) trained in civil disobedience. He notified the Viceroy in advance of this action. The march took several weeks, and when he arrived at the coast, he and his supporters, in front of the global press, collected several grains of salt from the sea. This was a punishable act, since the British colonial authorities had a monopoly on salt and drew a considerable portion of their funds from the salt tax. Thousands of Indians followed Gandhi's example in different parts of India. During one march to a salt depot, countless *satyagrahis* were bludgeoned with bamboo rods filled with lead, leading to several fatalities. Tens of thousands were imprisoned. Finally, Gandhi, too, was interned for nine months. The campaign against the salt tax went hand in hand with a boycott, mainly by women, of shops selling alcohol, in order to achieve a prohibition on alcohol and thus also the loss of the tax on alcohol, which was a considerable source of revenue. When the global economic crisis led to a drastic drop in prices for wheat, millet and rice, a campaign was launched among the farmers of refusal to pay the land tax that placed a severe strain on their resources. When the British Labour government finally declared that it was willing to enter into round-table negotiations with the Indian parties, and in particular with the INC and Gandhi, concerning reform of the constitution and the economic and other problems in India, Gandhi brought the sweeping campaign of civil disobedience to an end. However, the negotiations in London did not produce Gandhi's desired result.

Gandhi regarded his inability to prevent India being split apart as one of his greatest failures. Muhammad Ali Jinnah (1876–1948), who was three years older than Gandhi, had become his political opponent in the national movement at an early stage. He left the INC in 1920 after being actively involved in the party for two decades. However, many Muslims remained as members of the INC. In 1916, the INC and the Muslim League demanded in a joint declaration, the Lucknow Pact, that India should attain the status of a dominion. In response, the British government agreed to a gradual transition to self-government. In 1909, in order to weaken the influence of the INC, the British colonial authorities had strengthened the political clout of the Muslims by introducing separate constituencies for Hindus and Muslims for the Viceroy's legislative council. Without intending so, this paved the way for the partition of India in 1947.

5.5 Turning Points at Which the Non-violent Movement for an Independent India Failed

Looking back, the outstanding role played by Mohandas K. Gandhi and the non-violent strategy in the Indian national independence movement appears to be undisputed. However, there were repeated phases in the history of the Indian national movement after the violent uprising of 1857–59 in which politicians wanted to apply a strategy implemented in many other countries, including India, involving the use of force. Additionally, during the course of the following century, there were repeated cases of unrest, with many casualties. In 1907, the INC split into two branches, one of which was led by Bal Gangadhar Tilak, who propagated the national revolutionary uprising. After the death of Tilak, the movement's importance declined and it reunited with the moderate reformist branch. Under Gandhi's influence, this branch became politically radicalised, although it adopted the approach of non-violent mass mobilisation after 1920. Many leading members of the INC, such as Jawaharlal Nehru (1889–1964), who was President of the INC for a time, and who would later become the first prime minister of independent India from 1947–1964, followed Gandhi less from fundamental conviction than because his non-violent individual and mass actions were successful (Rothermund 2010).

Several times, there was violent unrest when Gandhi was interned or threatened. This explains why the British avoided arresting him whenever possible, or even released him from prison prematurely, in order to prevent outbreaks of violence. Gandhi was also always willing to enter into dialogue, and was also prepared to compromise on many issues. To this extent, the threat and option of violent protest and resistance was a key factor in the success of non-violent protest among a large minority in Indian society. When Gandhi made compromises, he was frequently deceived by his partners in South Africa and India alike, who broke the agreements they made with him. This repeatedly led to a considerable loss of face for Gandhi and gave impetus to politicians within and beyond the INC to take measures that were just as brutal as those employed by the colonial government. Thus, the terrorist

Bhagat Singh, who threw a bomb into the parliament in New Delhi, and who was executed in March 1931, was just as popular as Gandhi (Rothermund 1997: 281).

During the Second World War, around two million Indians fought on the British side. However, some of them regarded Germany and Japan as their natural allies in the fight against British colonial rule. Gandhi and the INC were not willing to enter into an alliance with the two dictators, however. Subhas Chandra Bose (1897–1945), who was a long-term supporter of Gandhi, and who was president of the INC in 1937 and 1939, initially founded an Indian Legion in Germany, consisting of 3500 prisoners of war. However, they were hardly ever used, due to Hitler's racist prejudice. In 1943, Bose was taken to Japan by submarine, where he founded an Indian National Army with 87,000 members, which fought side by side with the Japanese against the British in Burma, and which advanced towards India after the British had conquered Singapore and Malaysia. Bose also formed a government in exile, Free India (Azad Hind).

In Bengal, a major famine broke out as a result of the war. Under these circumstances, Gandhi pressed for full independence for India, claiming that Japan was unlikely to be interested in conquering India, but was only waging war against the British Empire. For this reason, he caused the INC to pass a Quit India resolution in August 1942, whereupon he was sent to prison for two years by the British, together with the entire leadership of the INC. They remained incarcerated until the end of the war.

Soon after the Second World War, the British came to the conclusion that they could no longer maintain their colonial regime in India without risking a costly war against the insurgent Indians. The strong political influence of the Muslim League under Muhammad Ali Jinnah, which since 1940 had taken up the proposal, with its Pakistan Resolution, of partitioning India into two nations, and which now demanded the establishment of a separate Muslim nation state, moved the British to agree to the founding of two independent states: Pakistan and India. For a long time, Gandhi attempted to prevent this from happening; however, ultimately, he was forced to bend to willingness of the INC to see the country divided. In the tense, as yet unresolved political situation of August 1946, the prime minister of Bengal organised a massacre of 5000 Hindus by Muslims in Calcutta, which was then followed by mass murders of Muslims perpetrated by Hindus in those areas in which Muslims were in the minority. While Gandhi travelled to Bengal, where he was able to contain the acts of violence through his speeches and public prayer meetings (which led Viceroy Mountbatten to write the letter quoted at the beginning of this chapter), in other regions new violent conflicts broke out. These caused the British to withdraw from India and led to the partition of the country after attempts at federalisation failed. When Gandhi voiced his support for a fair separation of the British-Indian state finances between Pakistan and India, he was shot by a Hindu extremist who, like many others, regarded Gandhi as being a traitor to the Hindu cause, since Pakistan was able to use the money it received to wage its war against India in Kashmir.

The border between Pakistan and India was not negotiated between the Muslim and Hindu contingents, but was determined by the British. The border was drawn on the basis of a religious state map, creating areas with a Muslim and a Hindu majority, without taking the Sikhs into account. The task was assigned by the British to Cyril Radcliffe, who had never been in India and who also had no personal relationship with the country, and who was therefore considered to be non-partisan.[21] The precise course of the border was not announced until the date the two states became independent, a move that contributed significantly to the extreme violence that followed.

5.6 The Favourable Historical Constellation for Gandhi's Partial Political Successes

Shortly before he was assassinated, Gandhi recognised that he had failed to achieve a fundamental reform of Indian society in a shared state in a spirit of non-violence and religious plurality and tolerance. He was successful in overcoming British colonial rule, albeit only through the establishment of an Islamic Republic of Pakistan and of a secular, multi-religious Indian state, in which 40 million Muslims remained. There is no doubt that Gandhi also gave important, sustained impetus to the social recognition and equality of both the 'Untouchables' (those who belonged to no caste, and who were assigned the lowest tasks) and to women, although they were still a long way from full emancipation. Finally, his historic example still has an impact today on non-violent and peaceful movements throughout the world.

Gandhi's successes and the peaceful national movement must be seen in the light of a historical constellation that was favourable in many respects. First, this was a movement supported by a huge national majority against colonial rule that was upheld only by a relatively small colonial bureaucracy and military. Second, there were only a small number of weapons circulating in Indian society. Third, despite repeated infringements of the law and a brutal use of force, the British rulers abided by fundamental legal norms, and opened up significant educational opportunities and access to the judicial system to an Indian elite. Over time, they created regional and central parliamentary institutions through several constitutional reforms (1909, 1919, 1935). They endeavoured to avoid violent uprisings wherever possible, and were therefore willing to compromise with the INC and Gandhi. Under a totalitarian regime, Gandhi would never have had the opportunity to develop his strategy as he had in South Africa or India, and would very likely have been killed at an early stage in one way or another. Fourth, Britain as a global power had been decisively weakened by two world wars. Fifth, the upcoming global power, the USA, which under President Roosevelt had sent its own troops to India to defend the British Empire against the Japanese, who were advancing into the subcontinent from Burma, exerted pressure on the British to end colonial rule. Finally, in July 1945, the coalition government led by the Conservatives under Winston Churchill, who had not been willing to give up

[21] On the treatment of the princely states and above all Kashmir, see Jahn (2015a).

British colonial rule, was replaced by a Labour government under Clement Attlee,[22] who finally agreed to the termination of British colonial rule in India.

5.7 The Global Inspiration for Non-violent Political Movements Arising from the Intellectual Stimuli and Successes of Mohandas K. Gandhi

During the 1920s, his ideas and, above all, his political successes inspired civil rights and peace movements throughout the world, including in Germany (Jahn 1993). In particular, the model of Gandhi's non-violent movement was a key source of inspiration for the Afro-American civil rights movement under Martin Luther King (1929–1968). As a student, King was fascinated by a lecture or sermon about India and Gandhi, and began to study his activities intensively (King 1968: 74; Oates 1984: 50; Lewis 1970: 34). In 1959, he visited the places where Gandhi campaigned in India. King himself said: "It was in this Gandhian emphasis on love and nonviolence that I discovered the method for social reform that I had been seeking" (Scott King 1979: 54; Carson 2000: 24). Structurally, the US civil rights movement, as the movement of an ethnic-racial minority, had more in common with the Indian movement in South Africa against the beginnings of the Boer-English policy of Apartheid from 1894–1914 than with the independence movement in India.

The example of the Indian minority movement and the *Indian National Congress* (INC) was also a major influence on the *African National Congress* (ANC) and the South African-Indian Congress and its defiance campaign against unjust laws under the leadership of Nelson Mandela (1918–2013) in 1952 and 1953. It ended with the arrest of 8500 activists and trials for high treason against Mandela and others. Mandela was not a supporter of non-violent policies in general, but merely accepted it as a tactic, while at times approving of violent unrest: "I do not consider non-violence according to the Gandhian model to be an inviolable principle, but a tactic to be employed depending on the situation. The principle was not so important that one should pursue the strategy even if it was self-destructive, as Gandhi believed."[23]

Lesser known campaigns were inspired by Gandhi, such as that of Danilo Dolci (1924–1997), who initiated a non-violent reformist movement in Sicily (Dolci 1969) and was popularly known as the 'Gandhi of Sicily', in the way that the press clearly enjoys recreating new, regional 'Gandhis' every so often. Ibrahim Rugova (1944–2006), the chairman of the Democratic League of Kosovo and then the first president of independent Kosovo from 2002–2006, was lauded as the 'Gandhi of the Balkans' (Birukoff 2006), but did not refer to Gandhi himself in his policies of non-violence

[22]During their brief period of rule in 1924 and from 1929–1931 the British Social Democrats were already inclined to agree to the Indian demands.

[23]Mandela (2013: 147, 182). The political context in which Mandela's views changed is discussed by Limb (2008: 33–62, in comparison with Gandhi: 50). As an example of the uncritical glorification of Mandela, see also Sharma (2014: 25, 138–142).

and the institutionalisation of a shadow state with numerous functioning underground administration, education and health facilities during the 1990s.[24] The list of leaders of regional and national non-violent civil rights, reformist and independence movements throughout the world who were inspired by Gandhi could be extended even further. The aim here is not to provide a detailed history of the impact of Gandhi's ideas and actions on the policies of non-violence in a very large number of countries. The purpose here is rather to discuss the key reasons for non-violent policies as presented by Gandhi in many scattered texts, most of them newspaper articles and letters, as well as the fundamental objections to these policies. Finally, an assessment will be made of the possible future prospects for success throughout the world of both general non-violent policies in line with Gandhi's thinking and also non-violent, pragmatic-tactical policies that are willing to shift towards limited violence in specific situations.

5.8 Gandhi's Concept of Religion and Politics and Fundamental Objections to It

From Gandhi's statements scattered over the many newspaper articles and letters, and in the few longer works, a systematically reflected, consistent concept of religion and policy, in other words, one with a theoretical foundation, can be discerned, which is based on the ethical norms explicitly referred to by him. Gandhi described himself as a Hindu[25] who follows the particular orientation of the faith towards the god Vishnu. However, he does not differentiate between various gods, but regards the ideas of god in all religions as being an approach to comprehending the godly, which he also describes as truth. To this extent, it became a matter of course for him to use ritual prayers from all the major religions in his public prayer meetings. Despite his roots in the Hindu tradition, Gandhi generally had a pan-religious, polyreligious outlook, which attempted to learn from all the holy scripts. In his view, there was no holy script that was an exclusive epiphany of god; rather, such writings were an attempt by humankind to understand the godly. Therefore, Gandhi clearly did not perceive god to be a male entity, but as a normative principle of truth, love and respect for all life, not just human life. In this sense, for Gandhi, atheists could also be religious.

[24]Clearly, the pragmatic consideration that the Albanians were in the minority compared to the Serbs, and that above all, they had no weapons, was a decisive factor in Rugova's policy, as emphasised by Prorok (2004: 93–96). See also Ahmeti (2017).

[25]"It is not the Hindu religion, which I certainly prize above all other religions, but the religion which transcends Hinduism, which changes one's very nature, which binds one indissolubly to the truth within and which ever purifies. It is the permanent element in human nature which counts no cost too great in order to find full expression and which leaves the soul utterly restless until it has found itself, known its Maker and appreciated the true correspondence between the Maker and itself." CWMG (1999), vol. 20: 304.

For Gandhi, truth is not something that one can possess, but that one can only strive to achieve without ever succeeding. The goal of this striving is non-violence[26] as "a perfect state. It is a goal towards which all mankind moves naturally though unconsciously" (CWMG 1999, vol. 26: 292). In Gandhi's view, a man would not become divine but perhaps simply a true man, whereas in the present state he is only partly man and partly beast because:

> we deliver blow for blow and develop the measure of anger required for the purpose. We pretend to believe that retaliation is the law of our being, whereas in every scripture we find that retaliation is nowhere obligatory but only permissible...Restraint is the law of our being. For, highest perfection is unattainable without highest restraint. Suffering is thus the badge of the human tribe. The goal ever recedes from us. The greater the progress, the greater the recognition of our unworthiness. Satisfaction lies in the effort, not in the attainment. Full effort is full victory. (CWMG 1999, vol. 26: 262–263)

For Gandhi, religion is not just a matter of belief, realisation and thought, but above all of behaviour – in other words, action and omission of action with regard to other people and living things. As a result, for Gandhi, religion is naturally and of necessity social and also political. "Those who say that religion has nothing to do with politics do not know what religion means" (Gandhi 1983: 453). It cannot be purely contemplative, observing the world; it is expressed through activity in the world, through a life-protecting mode of living that entails loving one's neighbour; it strives to achieve justice, not just for oneself, but also for one's own environment and potential field of impact. To this extent, individuals are not only responsible for their actions, but also for their omissions to act in order to help people in their environment who are being treated unjustly. That is why, for Gandhi, religion is always political. For him, the yardstick by which he judges his actions is his "inner voice", as he calls it; in other words, his conscience.[27] This is not present in the form of religious imperatives or prohibitions that he has taken from one holy scripture or another, and to which he refers as the highest authority, for example in relation to secular legislators. He only recognises those religious imperatives that make sense to him, and for which he can provide sensible reasons.[28] There is therefore no fundamental difference between Gandhi's socio-political ideas and any secular, rational, humanistic thought. It should be analysed according to the same measures and adjudged on the basis of its practicability in society and in politics.

---

[26] Among the supporters of non-violence, there is general agreement that the negative term is not satisfactory, and can also hardly be replaced by the Sanskrit neologism *satyagraha*. Martin Arnold (2011: 89–94, 36–39) attempts to replace it with the word *Gütekraft* and to prove that it is appropriate across different cultures and world-views, in different religious and atheistic concepts alike.

[27] "The only tyrant I accept in this world is the 'still small voice' within me." CWMG (1999), vol. 26: 260.

[28] "For I do not believe in the exclusive divinity of the Vedas. I believe the Bible, the Koran, and the Zend Avesta to be as much divinely inspired as the Vedas. My belief in the Hindu scriptures does not require me to accept every word and every verse as divinely inspired. Nor do I claim to have any first-hand knowledge of these wonderful books. But I do claim to know and feel the truths of the essential teaching of the scriptures. I decline to be bound by any interpretation, however learned it may be, if it is repugnant to reason or moral sense" CWMG (1999), vol. 24: 371.

Naturally, Gandhi knows that not everyone's conscience tells them the same thing that his does, although he often also deals with people who he assumes are doing something that they themselves regard as being unjust. It is in this latter group of people that he wishes to mobilise the conscience through his non-violent resistance and his suffering, in order to motivate and urge them to change their behaviour. At the same time, for him, suffering to which one consciously submits oneself also means self-purification and examining one's own conscience. Certainly, he acknowledges the decision made by those whose conscience motivates them to behave in a different way from him. And as a politically minded lawyer, he also acknowledges the necessity of restricting the anarchy of individual consciences through an overall, binding law and statutes. As a consequence, he generally recognises the existing statutory law and the legal order (he usually only indirectly considers the hierarchy of constitutional and simple laws) and the state institutions of the legislator, the government and administration and jurisprudence, and in an apparently schizophrenic way proclaims and demands fundamental respect for the law on the one hand, while at the same time, in an act of civil disobedience which he has publicly announced, targetedly and in a controlled manner breaks individual laws that he regards as unjust. The reconciliation of the contradiction between respecting the law and breaking the law emerges from the fact that Gandhi not only accepts the punishments meted out by the court, but even explicitly demands them.[29] This requires consciously provocative, active suffering that has been thoroughly mentally prepared for, which differs fundamentally from the suffering passively endured by most people. This philosophy of suffering fundamentally differentiates Gandhi's thinking from all common political ideology. It clearly alienates most political theorists from Gandhi's ideas, and may perhaps also require political thinking as a precondition.

Gandhi knew as a matter of course that there are different notions of what is just and unjust. For Gandhi, conscience was not a fixed entity, but the intellectual conclusion drawn from a situation analysed by him, in which he balanced the interests of all those involved against each other, following a communicative process with these people, and above all also with those who opposed his ideas. The inner voice of his conscience was, therefore, not a dogmatic, a priori doctrine that was set in stone, that he had concocted himself or drawn from holy scriptures, but the result of a social process of reflection. Accordingly, he could regard his inner voice as being a voice from the people when he experienced support for his activities from a section of the population that was underpinned by active participation, such as for the Salt March of 1930 or the spinning and weaving of self-made clothing. However, he was repeatedly

[29] When Gandhi was sentenced to six years' imprisonment in 1922 due to his call in three newspaper articles for people to join a campaign of non-cooperation and civil disobedience against the existing governmental system, during which terrible acts of violence were committed by Indians against policemen and other representatives of the authorities, he declared that: "The only course open to you, the Judge, is either to resign your post and thus dissociate yourself from evil, if you feel that the law you are called upon to administer is an evil and that in reality I am innocent; or to inflict on me the severest penalty if you believe that the system and the law you are assisting to administer are good for the people of this country and that my activity is, therefore, injurious to the public weal." CWMG (1999), vol. 26: 385.

forced to recognise that he had entirely misjudged the actual response of parts of the population to his campaigns, as a result of which he abandoned them, and did not prolong them with an overblown adherence to what he felt was right and wrong. This communicative understanding of an acknowledgement of the existing legal order and breaking the law repeatedly also enabled him to reach compromises with opponents whom he regarded as being guided by genuine conviction. This also enabled him to recognise his own errors. Gandhi was only dogmatic when it came to his own non-violent behaviour. He was thoroughly aware that most of his supporters only avoided violence for opportunistic reasons, and could also accept that others used violence when they felt that only violent resistance against injustice was possible.

In his view, it was always necessary to differentiate between violence used for a just cause and violence used to implement injustice. For him, violence was not always the same, and neither was war.[30] He declared unequivocally that resistance with violence against an injustice was preferable to tolerance of the injustice out of cowardice.[31] So in his view, the armed resistance put up by a large number of Poles against the German expansionist policy was 'almost non-violent'.[32]

The main argument against a policy of non-violence on principle as opposed to unjust or even illegal, unconstitutional, state or private-societal violence is the pragmatic argument that it is usually unsuccessful. A second frequent argument states that non-violent behaviour towards people who use violence requires an 'inhumane' willingness to suffer and, in extreme cases, even willingness to die, which most people do not have and which should not be demanded of them in the first place. It is natural, so the argument goes, that people should use force to defend themselves against unjust private-societal and also state violence, when other peaceful, legal means have been exhausted.

---

[30]"My non-violence does recognize different species of violence-defensive and offensive. It is true that in the long run the difference is obliterated, but the initial merit persists. A non-violent person is bound, when the occasion arises, to say which side is just. Thus I wished success to the Abyssinians, the Spaniards, the Czechs, the Chinese and the Poles, though in each case I wished that they could have offered non-violent resistance." *Harijan* (9 December 1939); at: https://www.mkgandhi.org/momgandhi/chap29.htm). Quote found in Kraus (1957: 266).

[31]"I do believe that where there is only a choice between cowardice and violence I would advise violence." For this reason, Gandhi advised the following: "I advocate training in arms for those who believe in the method of violence. I would rather have India resort to arms in order to defend her honour than that she should in a cowardly manner become or remain a helpless witness to her own dishonour. But I believe that non-violence is infinitely superior to violence, forgiveness is more manly than punishment. CWMG (1999), vol. 21: 133.

[32]"If a man fights with his sword single-handed against a horde of dacoits armed to the teeth, I should say he is fighting almost non-violently. Haven't I said to our women that, if in defence of their honour they used their nails and teeth and even a dagger, I should regard their conduct non-violent? She does not know the distinction between Himsa and Ahimsa. She acts spontaneously. Supposing a mouse in fighting a cat tried to resist the cat with his sharp beak, would you call that mouse violent? In the same way, for the Poles to stand valiantly against the German hordes vastly superior in numbers, military equipment and strength, was almost non-violence. … You must give its full value to the word 'almost'." *Harijan* (25 August 1940); CWMG (1999), vol. 79: 121–122.

The neutral, social and historical question relates to the conditions under which politically relevant groups among the population have found the strength to offer non-violent resistance against holders and users of violence, and in some cases have even been successful in their efforts, and under which conditions non-violent resistance collapses and perhaps then transforms into violent resistance, or perhaps does not come into being in the first place. Also, are only religious people, and Hindus in particular, capable of non-violent resistance with a high level of willingness to suffer and die, since they believe in a life after death, or even in reincarnation in another form? Another question relates to the possible indications that non-violent resistance against unjust conditions could play an increasingly important role in global society in the future.

5.9 Gandhi's Strong Impact on Civil Rights and Freedom Movements and His Low Degree of Influence on International Politics

Above I named several leaders of civil rights movements, mostly from Western democracies, who were influenced by Gandhi's historical example, his writings and his actions. However, many activists in these movements and in the international peace movement felt inspired by Gandhi even though most of them only knew his name at best, and referred to other sources of inspiration for their work, such as directly to holy scriptures and in particular to the Sermon on the Mount in the New Testament,[33] or to Henry David Thoreau (1817–1862; Thoreau 2014).

Gandhi also exerted an indirect, strong influence, first over publications covering non-violent policies, and second over the generally marginal branch of peace research that aims to develop theoretical and conceptional conclusions from Gandhi's thoughts and actions, as well as from other historical examples of non-violent policies. In Germany, these authors include, in particular, Theodor Ebert[34] (born 1937) and his former colleagues, such as Gernot Jochheim (1984, 1986; born 1942), who worked for the journal *Gewaltfreie Aktion* ("Non-Violent Action") from 1969–2010. The most influential person on the international political stage was and remains Gene Sharp (1928–2018) from the US, who developed guidelines for non-violent policies based not only on the non-violent movement in India but also on many other historical events.[35] There is evidence that Sharp influenced the civil rights movements in Serbia, Georgia, Ukraine, Kyrgyzstan, Myanmar and Egypt.[36]

[33] See, for example, Alt (1983). More than twenty editions had been published by 2000.

[34] Ebert (1969, 1984). See also Steinweg and Laubenthal (2011).

[35] Sharp (1973). Brief instructions for action without any reference to Gandhi are provided by Sharp (2010). This booklet has been translated into thirty-one languages.

[36] On these movements, see also the lectures "Der zweite Demokratisierungsversuch, in Serbien, Georgien und der Ukraine" (Jahn 2008: 149–165) and "Democratisation or the Restoration of Dictatorship as the Outcome of the Arab Rebellion" (Jahn 2015b: 171–186).

Gandhi's influence on non-Indian national independence movements appears to have been fairly weak, although evidence can be found here and there of his impact, for example in Ghana (on the young Kwame Nkrumah 1909–1972; see Hountondji 1983: 135, 146), other countries in Africa,[37] and Georgia (at times on Zviad Gamsakhurdia, 1939–1993; see Gamsakhurdia 1995: 148). This influence was and remains far greater on civil rights movements that demand individual and social freedoms, human rights, the rule of law and liberal democracy. Even under National Socialist rule in Germany, non-violent resistance was successful in some cases, such as against the deportation of the Jewish partners of non-Jews from Rosenstrasse in Berlin (Jochheim 2002; Gruner 2002, 2004). Attempts were also made to assert non-violent resistance against the conquest and occupation policies of other states, such as Norway (1940–1945, see Mez 1976) and Czechoslovakia in 1968 (Horský 1975). Gandhi himself considered the issue of possible non-violent resistance of Jews against the violent National Socialist rule (Rothermund 1997: 395–397), as well as a non-violent defence of India against Japan and of Britain and Poland against the German Reich, which could not constitute a defence of the border and which required toleration of the military occupation. At the end of his life, he was also concerned with the question of how non-violent resistance could be asserted against the use of atomic weapons.[38]

5.10 Contentious Basic Assumptions with Regard to Gandhi's Political Understanding

Gandhi never claimed that violent resistance against unjust, subjugating violence is illegitimate on principle. He only said that he himself did not wish to use violence, and that he also wished that his fellow political activists and other people in general would waive violent measures in favour of non-violent ones. What was his justification for this stance? And how can sociologists and political scientists explain that non-violent resistance is successful in some cases and not in others, either because it transmogrifies into violent resistance, or because it collapses entirely and a violent regime succeeds in suppressing any form of resistance over a longer period of time?

Gandhi never criticised or combated violence or the use of physical and mental violence per se. His motivation for taking social and political action was always the experience of social, and later also political, injustice, which led him to protest and resist, either alone or with other people, in practice nearly always with Indians. He never became involved in resistance campaigns or solidarity campaigns for

[37] See the examples of Julius Nyerere in Tanzania, Kenneth Kaunda in Zambia, Albert Luthuli in South Africa and Sam Nujoma in Namibia. It is noticeable that all these cases occurred in Anglo-Africa and not in Franco- or Ibero-Africa. Kaunda once said that if he were to be forced to select a colonial power, he would choose the British, "because I would be in a position to go to their country and lead a campaign against their own government" (Sutherland/Meyer 2000: 110).

[38] See also the Gandhi quote in Erikson (1978: 515) and Rothermund (1997: 503–504).

suppressed people in other countries,[39] but, when asked by foreigners about possible resistance in other countries, stated on principle that a universal human approach to the issue of resistance against unjust living conditions was possible.

Non-violent resistance can take two basic forms: a legal one that is in compliance with the law, and an illegal, unlawful one, each of which requires different disadvantages to be taken into account.[40] Non-cooperation, such as resigning from one's post in the state administration or in a company that offers miserable working conditions or produces unacceptable goods, may entail a drastic reduction in one's own standard of living, while boycotting certain goods, on the other hand, may sometimes only mean doing without certain conveniences. Generally, boycotting certain goods does not have any direct impact on individuals, aside from parliamentary candidates seeking election. However, non-cooperation can lead to a violent response among those who had counted on the individual's continued co-operation. In such instances, non-cooperation initially entails suffering from acts of violence without responding with violence oneself, even if legal and political means should be used against the perpetrators of the violence.

The most contentious form of non-violent resistance is civil disobedience (Braune 2017), in other words, consciously and openly breaking the law and openly announcing this fact to one's opponent. Either a law is broken that is regarded as being unjust, or another law that represents it is broken in its place. For example, the Rowlatt Act of 1919 in India was an emergency law which could not be disobeyed as long as it was not applied, and which incidentally never was applied in practice. However, Gandhi mobilised his followers to civil disobedience in defiance of it. The systematically organised infringement of the prohibition on privately taking salt from the sea during the famous Salt March in 1930 was not motivated by the fact that this law was perceived as being unjust, but rather, it was chosen by Gandhi as a symbol of foreign rule by the British when he began a campaign for independence, for which he presented eleven specific demands, including amendments to or the abolition of certain laws (Rothermund 1997: 248–272).

The state can respond in various different ways to a breach of the law. It can ignore it and attempt to undermine the intention of the perpetrators to lead the state to impose repressive measures. It can also impose financial penalties (which only rarely occurred in India, since many of those who broke the law were simply unable to pay them; this approach tended to be applied in Europe), or it can attempt to initiate court proceedings that lead to imprisonment, and finally, it can even use direct, physical force that either causes injury or is fatal. The acceptance of the financial penalty or imprisonment and the tolerance of the use of repressive private or state force without responding in kind with force, but merely with legal and political means, is more deeply embedded in the fundamental acceptance, mentioned above, of the state legal order and the right of the state organs to penalise breaches of the law.

---

[39]The purpose of the Khilafat, i.e. the caliphate movement, supported by Gandhi from 1919–1924, was not primarily to support the Ottoman caliph and his Turkish supporters, but above all to offer solidarity with Indian Muslims. See in detail Rothermund (1997: 133–140, 182–183).

[40]Sharp (1973, vol. 2) compiled a detailed list of 198 forms or methods of non-violent action.

To this extent, the refusal on principle to pay taxes (as opposed to rejecting a particular type of tax) can be only the most extreme means of non-violent policy, which should generally be avoided, since when applied systematically and en masse, it would endanger the existence of the state and thus also the entire legal order. For this reason, unlike Thoreau (2014), Gandhi never propagated this instrument of resistance. A publicly announced, non-violent breach of the law was intended to force the state legislative bodies to abolish or amend the law that was regarded as being unjust. The voluntary willingness to suffer is aimed at making it clear to the holders of state power that the demands for just laws are serious, after petitions and protest rallies (in autocratic systems) or regular, institutional procedures such as claims made before the constitutional court, elections, involvement in the formation of policies by political parties, and demonstrations (in liberal, democratic constitutional states) have all had no impact. Non-violent resistance and a limited willingness to suffer, which is also required in liberal democracies, are appropriate reactions to legislative policy conflicts that go beyond the calmer forms of parliamentary conflict and court proceedings.

In times of violent conflict and war, far more suffering is usually involved than when non-violent campaigns are brutally suppressed. However, most people, probably 95% or more, tend to be willing to participate, either actively or with moral support, in civil and state wars when they regard their vital interests as having been violated, or to subjugate themselves to those in power, than to offer non-violent resistance in such cases. There is little to suggest that these attitudes will change within the space of just a few decades, if at all. How else is the greater willingness to suffer in wars than through non-violent resistance to be explained?

On the one hand, there are different reasons for a willingness to suffer, and to suffer in specific ways. While in many wars a hero's death and the willingness to die in this way are glorified (*dulce et decorum est, pro patria mori* or to die for another just cause), this is not the case when it comes to the non-violent approach. However, the suffering of people who have been severely injured or raped, or who are languishing in prisons and concentration camps, is usually not mentioned, although it is accepted. While it may be regarded as being "sweet and honourable" to die for a just cause, the same does not apply to being beaten and tortured, and lying in a military hospital with severe wounds. By contrast, with the non-violent approach, physical and mental suffering, which more rarely does not last for life, is addressed in far greater depth and is acknowledged as the price for resistance against injustice. For Gandhi it is also even occasionally regarded as a means of self-purification.

A second difference appears to lie in the fact that one's own suffering and, if necessary, one's own death appear more tolerable when one has inflicted greater suffering, if possible, on one's enemy. Violent heroism in war includes the heroic act of boldly injuring and killing enemies, not just one's own brave death. The exertion of violent power appears to make it easier to countenance one's own powerlessness. The *satyagrahi*, the non-violent resister, must do without such direct compensation for their own powerlessness. This requires considerable mental preparation, training and self-discipline. The non-violent resister must, in the most extreme case, be more courageous than a warrior or soldier. For the non-violent resister, the justification for one's own suffering lies in the fact that their resistance will in the longer or shorter term contribute to a change of attitude among those in power to rectify an injustice. In democracies, this means the holders of legislative power, the courts or the government and administration.

Thus, in the history of the labour movement, hundreds of thousands of workers were willing to accept millions of accumulated years in prison, death, severe injuries and humiliation by participating in non-violent strikes in order to achieve key improvements in their working and living conditions and ultimately also the legal right to strike.

A third difference, as well as the advantages and disadvantages of non-violent resistance mentioned above, is that consistent non-violent resistance, which can lead to civil disobedience, in other words, a violation of the law and the acceptance of the resulting penalties or also the illegal use of force by those in power, is far less frequently implemented than violent resistance and military service. A soldier goes to war and into battle with an awareness of the risk of dying or of returning to peaceful life injured, although one of these scenarios is more or less likely. In most wars, the large majority of soldiers survive. In some battle situations, the chances of survival for individual soldiers is close to zero, however. The justification for one's own death, for the individual and for the fighting community, is that it possibly contributes to a better life for one's surviving relatives and the political community (of a country, a people, a social class, religious community or party, etc.). Violent battle and war only make sense in terms of their unpredictable character as long as victory or defeat, survival or death are uncertain. If defeat is certain, capitulation in an individual battle or in the entire war and the conclusion of a peace accord with the most tolerable conditions for the vanquished side are advisable. The willingness of soldiers to fight in a war in which their death is certain (kamikaze attack) is an extremely rare exception in all countries and among all peoples, and requires a vast amount of indoctrination, which must usually be supported by alcohol and drugs. In most battle situations, the individual soldier has a chance of survival through their own skill and luck.

By contrast, the *satyagrahi* consciously resists the legal or illegal use of force by their political opponent, be it the state or another societal organisation or group that exerts force, which they do not wish to avoid as a matter of principle, and which they do not wish to prevent through their own use of force or through flight. To this extent, they surrender entirely to the good or bad will of their opponent, thereby displaying elements of doctrinaire self-sacrifice such as are not usually experienced by a soldier. This individual mental factor, which makes one's own potential but uncertain suffering appear more tolerable when suffering is also caused to one's opponent, could explain why most people tend to be more willing to accept an actual or alleged defensive war with a very large number of casualties and injuries than non-violent resistance with far fewer deaths and injuries in society as a whole, which makes individual suffering inevitable and which denies one the satisfaction of having caused suffering to one's opponent.

However, the non-violent resister attempts to use all possible measures, such as the early announcement of resistance campaigns, negotiations with the opponent, influencing public opinion through the media, and finally also through their public suffering, in order to mitigate the violence inflicted by the opponent or even to motivate or force them to change their behaviour. Ideally, the opponent will realise that they are in the wrong and will change their behaviour or modify the law. In reality, non-violent acts of resistance are also successful when they are widely approved of by society and when the holders of power recognise that, for them, violent assertion of their interests and their interpretation of the law are costlier than simply giving way on the matter. The certainty that massive suppression of non-violent resistance may under certain circumstances lead to violent unrest and uprising can also contribute to the success of non-violent resistance.

5.11 The Limited but not Fully Exploited Scope for Action of Non-violent Policies, Now and in the Near Future

Experience since the Second World War and the death of Gandhi has shown that non-violent movements are playing an increasingly important role, both in toppling autocratic regimes and in creating change in democratic societies. Thus, for example, autocratic regimes were overcome by mainly non-violent mass movements in Iran in 1979, in the Philippines in 1986, on repeated occasions in Latin America, in communist Europe from 1989–1993 and in several post-communist neo-autocracies in the decades that followed, and also to a certain extent in the Arab uprisings after December 2010.[41] There are two reasons for this. On the one hand, some autocratic regimes may be able to gain broad approval within society under certain conditions, but sooner or later lose this approval due to their arbitrary actions, which even breach the laws they themselves have decreed, and also due to economic ineffectiveness, corruption and extreme social injustice. Autocratic regimes are clearly less able to adjust to changing societal attitudes and socio-economic challenges than most democracies. Secondly, however, the potential for use of force by the police and military, which has grown enormously over the years, acts as a deterrent against violent resistance in many cases.[42] However, massive and determined suppression, particularly of ethno-national groups and their non-violent protests and demonstrations,

[41] A comprehensive analysis of the success and failure of 323 non-violent and violent resistance movements from 1900 to 2006 is provided by Chenoweth/Stephan (2013: 6–7). According to this analysis, during this period, the success rate of non-violent movements has increased, while that of violent ones has declined.

[42] Some social-geographical factors, such as sparse settlement in mountains and jungles, are doubtless more suited to guerrilla warfare than densely populated locations in urban regions. However, the long civil wars in Beirut, Aleppo and Mosul demonstrate that, even in major cities, war can be conducted over a longer period of time.

can still repeatedly lead to a shift from non-violent movements towards movements that are willing to countenance civil war, as was the case, for example, in Kosovo and Syria in recent decades.

In some non-violent movements, non-violent activists have played an important pioneering role, even though most of their members – as was the case during the Indian independence movement – are not opponents of all application of force on principle. Generally, they support armed defence of the country against foreign aggression and accordingly are in favour of the military preparation that appears to be required for the purpose, in other words, the maintenance of military forces. And all people accept the need in principle for a police force that uses violent measures against criminals within the scope of what is legally permitted.[43]

Therefore, following the spectacular one-week non-violent resistance in Czechoslovakia against the Warsaw Pact intervention troops in 1968, all attempts by a small number of peace researchers during the 1970s to use the combat methods used by non-violent resistance for systematic, state-organised preparation for non-violent defence of the country against foreign aggressors, without a military, remained entirely unsuccessful (Roberts 1969; Ebert 1981; Jochheim 1988), and will probably remain so in the future. At the same time, it is possible that in the future, non-violent resistance will be offered when a country is occupied by an aggressor who is far superior militarily. While this may have almost no chance of direct success, in the long term, the resistive spirit will be encouraged and when changes occur to the international constellation, success can be achieved, as was the case with the inner erosion of the Soviet occupation morale in Eastern Europe at the end of the 1980s.

During the development of the liberal democracies, non-violent resistance also took on an important role in reforming the law, which by itself is not sufficiently observed by institutionalised policy formation in the parliaments and democratic parties. Non-violent resistance with narrowly limited violation of the law can, in liberal democracies, express the seriousness of the will to change the law articulated initially by societal minorities. This resistance does not have to invoke extra-statutory religious or other ethical moral requirements, but can often refer to central constitutional basic rights. Legalisation of civil disobedience cannot by definition be provided for in a legal system. However, a moralistic de-dramatisation of the contradictions between the ruling and oppositional, sometimes innovative, sometimes confusing legal interpretations is certainly possible in the political culture of liberal democratic constitutional states.[44] Thus, the fact cannot be overlooked that civil disobedience

[43] Gandhi certainly did have ideas about a police force that acted non-violently on principle, which countered violent crime without applying violence itself. However, these ideas were never put into practice and will probably have no chance of success in the future either. However, it is possible that the methods used by police for non-lethal suppression (water cannon, police truncheons instead of guns) will be further developed – for example the use of stun guns when dealing with dangerous wild animals.

[44] Thus, Jürgen Habermas, for example, said: "Every constitutional democracy that is secure in its existence considers civil disobedience as being a normalised, since necessary, part of its political culture." He also wrote that the constitutional state "must protect and keep alive mistrust against injustice that occurs in legal forms, even though it cannot adopt an institutionally secured form"

plays a very important role in achieving key civil rights among the Afro-American population, as it has done in the past in the attainment of suffrage among workers and women and, more recently, in the fight for equal rights for homosexuals and the right to exist for animals and plants threatened with extinction. In many countries, the abolition of the death penalty was also preceded by a non-violent movement. Overall, a historical trend towards the civilisation of conflict behaviour can be observed, at least in the liberal democracies. In recent decades, the arming of the police with guns has been increasingly supplemented with truncheons and water cannon, so that suppressive state force has been able to replace older, lethal, police and military force when dealing with non-violent, or at least low-violence, oppositional movements.

What conclusions can be drawn from these considerations for future global-human peace policy? A one-dimensional peace policy, which focuses only on the dimension of action – such as non-violent action, diplomatic negotiation and military deterrence – does not promise success. Most people will continue to support the readiness for military defence of their country, in other words, a certain degree of military deterrence. This only has a chance of succeeding when it is incorporated into a functioning system of collective security of the United Nations, when an attack against an individual country is likely to lead to the threat of military countermeasures by the UN Security Council and possibly by all other countries. At the same time, this includes the will, particularly in a major power and in a country that is a permanent member of the UN Security Council, to refrain from attempting to create a military change to the territorial and political status quo, and also to prevent such attempts from being made in other countries through non-military means.

Actions by governments that take responsibility for maintaining peace remain the cornerstone of all global peace policies and cannot be replaced by the oppositional activities of peace movements. Non-violent movements therefore have the task of motivating the majority of society to only bring parties and governments to power that wish to maintain or achieve peace, and to initially combat those parties and governments that do not promote peace with the existing institutional means of attempting policy change available in the respective political system, and if necessary also non-cooperation and civil disobedience. Non-violent resistance forces are therefore no alternative to armies and police units, as was claimed for a long time by many of their activists and proponents. Instead, they are an important, partially complementary, partially antagonistic, supplement to the traditional instruments and methods of domestic and foreign policy.[45] Accordingly, there is much to suggest that

(Habermas 1985: 81, 87). The contradictions between individual legal norms and in the legal interpretations in the hierarchy of the courts and in public opinion are, according to Dworkin (1984), signs of a very different way of handling breaches of the law in cases of civil disobedience than for criminals and those who violate civil and human rights. Cf. also Dreier (1983) and Rawls (1999: 421): "A general willingness to commit acts of civil disobedience brings stability to a well organised or almost just society".

[45] The difficulty of also practising non-violent policies as democratically responsible statesmen following the attainment of national independence has been very clearly demonstrated by Kenneth Kaunda and Julius Nyerere (Sutherland/Meyer 2000: 95–113, 69–89).

non-violent resistance offers an alternative to civil war, although hardly ever to a willingness to conduct a state defensive war.

5.12 Unavoidable Setbacks of Non-violent Policies

We can be certain that numerous non-violent movements will arise in the future, in which activists who are non-violent on principle play an important, albeit rarely decisive, role. Such movements will in particular play an increasingly important role in toppling autocratic regimes, with the goal of establishing the rule of law, personal freedoms, liberal and social citizens' and human rights, democratic institutions and procedures, and the dismantling of corrupt structures. For several years, the tendencies towards global liberalisation and democratisation have declined, however. It remains to be seen as to what extent non-violent movements contribute to a reverse in developments, which mainly depends on a loss of legitimacy among the autocratic regimes in socio-economic crises. According to the New York-based Freedom House, in 2016, 39% of people out of the total global population (7.4 billion) lived in free states, 25% lived in partially free states, and 36% lived in non-free states. In 2017, of the 195 countries, 87 (45%) were categorised as free states, 59 (30%) as partially free states and 49 (25%) as non-free states.[46] It can be assumed that only a small number of regimes that assert illegitimate, autocratic or imperialist rule will succeed in becoming stabilised over a long period of time. Additionally, the unavoidable increase in problems that can only be regulated internationally, such as climate change, environmental destruction, migration, and terrorism, will in the future lead to the creation of global-humane movements that will exert pressure on the states to find a way to solve these problems.

Since the process of forming new nation states has by no means been completed, even though most larger nations have in the interim achieved their own statehood within multinational federal states or in the form of independent nation states,[47] we can assume that, in the future, several non-violent national autonomy and independence movements will be formed against imperial foreign rule, even though – as past examples have shown – they will remain exposed to the risk of adopting violent strategies if they are severely repressed. Only a small number of nations, as was the case with the Norwegian, Slovak and Montenegrin states, will be in a position in the future to establish their own state without bloodshed, thanks to favourable international constellations.

[46] Freedom House 2017: *Freedom of the World 2017*; at: https://freedomhouse.org/report/freedom-world/freedom-world-2017. As well as the 193 UN member states, Kosovo and the Republic of China (Taiwan) have been included in the list of countries by Freedom House.

[47] For more detail, see the three lectures on the relationship between the state and the nation, and on nationalism (Jahn 2015b: 13–68).

We can expect that liberal-democratic, anti-autocratic and national, anti-imperial movements will shift towards a position of supporting civil war in the short or long term, following an initial non-violent strategy, if they are opposed with ruthless violence by those in power. In Kosovo, this transition took several years following the dissolution of Yugoslavia in 1991, while in Syria it took just a few months after the first non-violent protest rallies were held in March 2011.

Even in liberal-democratic states that are not satisfied with the historically achieved degree of liberalisation and democratisation in their societies, and that do not wish to simply accept the existing institutional procedures regarding policy and legislative change, it is likely that an increasing number of oppositional movements will emerge. In a secularised world, they will probably invoke religious norms less and less, and will increasingly base their actions on fundamental norms enshrined in the constitution and in international law which are not met or are even contradicted by many decisions made by governments, major commercial companies and existing laws. As well as the constitutional forms of protest, legal and political influence and demonstrative non-cooperation, it is likely that, to an increasing extent, acts of civil disobedience – in other words, politically moderate and controlled infringements of the rules and of the law – will play a role in communicating to the public, and to the parliaments and governments, the urgency of certain political and legislative changes. Even if civil disobedience by definition cannot be legally institutionalised, during the course of further liberalisation and democratisation in many countries, it could become regarded in many countries as part of a largely tolerated constitutional reality during the course of further liberalisation and democratisation, and be recognised as an important tool of continuous, dynamic self-correction in liberal-democratic systems when faced with new sociopolitical problems. Such progress in social policy and constitutional law will inevitably have to overcome severe setbacks in which attempts are made to regulate the respective status quo or even to restore one that has already been historically overcome.

References

Ahmeti, Zef, 2017: *Ibrahim Rugova: Ein Leben für Frieden und Freiheit im Kosovo* (Hamburg: Tradition).

Alt, Franz, 1983: *Frieden ist möglich: Die Politik der Bergpredigt* (Munich: Piper).

Arnold, Martin, 2011: *Gütekraft: Ein Wirkungsmodell aktiver Gewaltfreiheit nach Hildegard Goss-Mayr, Mohandas K. Gandhi und Bart de Ligt* (Baden-Baden: Nomos).

Birukoff, André, 2006: "Kosovo-Präsident Rugova. Der ‚Gandhi des Balkans' ist tot", in: *Spiegel-Online*; at: https://www.spiegel.de/politik/ausland/kosovo-praesident-rugova-der-gandhi-des-balkans-ist-tot-a-396663.html.

Blume, Michael, 1987: *Satyagraha: Wahrheit und Gewaltfreiheit: Yoga und Widerstand bei Gandhi* (Gladenbach: Hinder & Deelmann).

Braune, Andreas (ed.), 2017: *Ziviler Ungehorsam: Texte von Thoreau bis Occupy* (Stuttgart: Reclam).

Carson, Clayborne (ed.), 2000: *The Autobiography of Martin Luther King, Jr.* (London: Abacus).

Chenoweth, Erica.; Stephan, Maria J., 2013: *Why Civil Resistance Works: The Strategic Logic of Non-Violent Conflict* (New York: Columbia Universtiy Press).
CWMG, 1999: *The Collected Works of Mahatma Gandhi* (Electronic Book), 98 volumes, (New Delhi: Government of India, Publications Division),
Dolci, Danilo, 1969: *Die Zukunft gewinnen: Gewaltlosigkeit und Entwicklungsplanung* (Bellnhausen: Hinder & Deelmann).
Dreier, R., 1983: "Widerstandsrecht und ziviler Ungehorsam im Rechtsstaat", in: Glotz, Peter (Ed.): *Ziviler Ungehorsam im Rechtsstaat* (Frankfurt am Main: Suhrkamp).
Dworkin, Ronald, 1984: *Bürgerrechte ernstgenommen* (Frankfurt am Main: Suhrkamp).
Dworkin, Ronald, 1977: *Taking Rights Seriously* (Cambridge, MA: Harvard University Press).
Ebert, Theodor, 1969: *Gewaltfreier Aufstand: Alternative zum Bürgerkrieg*, 2nd edn. (Freiburg im Breisgau: Rombach).
Ebert, Theodor, 1981: *Soziale Verteidigung*, 2 vols. (Waldkirch: Waldkircher Verlag).
Ebert, Theodor, 1984: *Ziviler Ungehorsam: Von der APO zur Friedensbewegung.* (Waldkirch: Waldkircher Verlag).
Erikson, Erik H., 1993: *Gandhi's Truth: On the Origin of Militant Nonviolence* (New York/London: W.W. Norton).
Fischer, Louis, 1962: *Mahatma Gandhi: Sein Leben und seine Botschaft an die Welt* (Berlin: Ullstein).
Fischer, Louis, 1951, 2015: *The Life of Mahatma Gandhi* (New York: Vintage).
Freedom House, 2017: *Freedom of the World 2017*; at: https://freedomhouse.org/report/freedom-world/freedom-world-2017.
Freier 1930: "Der revolutionäre Aufschwung in Indien", in: *Die Kommunistische Internationale*, XI,17: 951–955.
Fuhrmann, Manfred (Ed.), 1989: *Platon: Apologie des Sokrates* (Stuttgart: Reclam).
Gamsachurdia, K., 1995: *Swiad Gamsachurdia. Dissident – Präsident – Märtyrer* (Basel: Perseus).
Gandhi, Arun, 1981: *Kasturbai und Mahatma Gandhi* (Gladenbach: Hinder und Deelmann).
Gandhi, M, 1924: *Jung Indien: Aufsätze aus den Jahren 1919 bis 1922* (Erlenbach-Zürich: Rotapfel).
Gandhi, Mahatma K., 1983: *An Autobiography or The Story of My Experiments with Truth* [1927/29] (Penguin, Harmondsworth).
Gandhi, Mahatma K., 2010: *Hind Swaraj* (Cambridge: Cambridge U.P.).
Gandhi, Mahatma K., 2017: *Satyagraha in South Africa* [1928] (Ahmedabad: Indian ebooks).
Gruner, Wolf, 2002: "Die Fabrik-Aktion und die Ereignisse in der Berliner Rosenstraße: Fakten und Fiktionen um den 27. Januar 1943", in: *Jahrbuch für Antisemitismusforschung*, 11: 137–177.
Gruner, Wolf, 2004: "Ein Historikerstreit? Die Internierung der Juden aus Mischehen in der Rosenstraße 1943. Das Ereignis, seine Diskussion und seine Geschichte", in: *Zeitschrift für Geschichtswissenschaft*, 52: 5–22.
Gunturu, Vanamali, 1999: *Mahatma Gandhi: Leben und Werk* (Munich: Diederichs).
Habermas, Jürgen, 1985: "Ziviler Ungehorsam – Testfall für den demokratischen Rechtsstaat: Wider den autoritären Legalismus in der Bundesrepublik", in: *Die Neue Unübersichtlichkeit* (Frankfurt am Main: Suhrkamp).
Historical Population of United Kingdom, *43 AD to Present*, at: https://chartsbin.com/view/28k.
Horský, Vladimír, 1975: *Prag 1968: Systemveränderung und Systemverteidigung* (Stuttgart/Munich: Klett/Kösel).
Hountondji, Paulin J., 1983: "The end of 'Nhrumaism' and the (re-)birth of Nkrumah", in: *African Philosophy: Myth and Reality* (London: Hutchinson University Library for Africa).
Jahn, Beate, 1993: *Politik und Moral: Gandhis Herausforderung für die Weimarer Republik* (Kassel: Weber, Zucht & Co.).
Jahn, Egbert, 2008: *Politische Streitfragen* (Wiesbaden: Verlag für Sozialwissenschaften).
Jahn, Egbert, 2015a: "Kashmir: Flashpoint for a Nuclear War or Even a Third World War", in: *World Political Challenges: Political Issues Under Debate*, vol. 3 (Heidelberg: Springer): 205–220.
Jahn, Egbert, 2015b: *World Political Challenges: Political Issues Under Debate*, vol. 3., (Heidelberg: Springer).

Jaspers, Karl, 1960: *Die Atombombe und die Zukunft des Menschen*, 2nd edn. (Munich: Piper).
Jochheim, Gernot, 1984: *Die gewaltfreie Aktion: Idee und Methoden, Vorbilder und Wirkungen* (Hamburg/Zürich: Rasch und Röhring).
Jochheim, Gernot, 1986: *Länger Leben als die Gewalt: Der Zivilismus als Idee und Aktion* (Stuttgart: Weitbrecht).
Jochheim Gernot (Ed.), 1988: *Soziale Verteidigung – Verteidigung mit einem menschlichen Gesicht* (Düsseldorf: Patmos).
Jochheim, Gernot, 2002: *Frauenprotest in der Rosenstraße Berlin 1943* (Berlin: Hentrich & Hentrich).
King, Martin Luther, 1968: *Freiheit! Der Aufbruch der Neger Nordamerikas* (Munich: Heyne).
King, Martin Luther, 1958: *Stride Toward Freedom* (New York: Harper).
Kraus, Fritz (Ed.), 1957: *Vom Geist des Mahatma: Ein Gandhi-Brevier* (Zürich: Schweizer Druck- und Verlagshaus).
Kulke, Hermann; Rothermund, Dietmar, 2006: *Geschichte Indiens: Von der Induskultur bis heute* (Munich: Beck).
Lewis, David Levering, 1970: *King: A Critical Biography* (Baltimore: Penguin).
Limb, P., 2008: *Nelson Mandela: A Biography* (London/Westport, CT.: Greenwood).
Mandela, Nelson, 2013: *Long Walk to Freedom* (London: Abacus).
Mann, Michael, 2005: *Geschichte Indiens: Vom 18. bis zum 21. Jahrhundert* (Paderborn: Schöningh).
Mez, Lutz, 1976: *Ziviler Widerstand in Norwegen* (Frankfurt am Main: Haag und Herchen).
Mühlmann, Wilhelm E., 1950: *Mahatma Gandhi: Der Mann, sein Werk und seine Wirkung. Eine Untersuchung zur Religionssoziologie und politischen Ethik* (Tübingen: J.C.B. Mohr).
Oates, Stephen B., 1984: *Martin Luther King: Kämpfer für Gewaltlosigkeit: Biographie* (Hamburg: Ernst Kabel).
Prorok Christiane, 2004: *Ibrahim Rugovas Leadership. Eine Analyse der Politik des kosovarischen Präsidenten* (Frankfurt am Main: Lang).
Rawls John, 1999: *A Theory of Justice* (Cambridge: Harvard U.P.).
Reissner 1930: "Gandhi, der Prophet der indischen Bourgeoisie", in: *Die Kommunistische Internationale*, XI,20: 1,130–1,138.
Roberts, Adam (Ed.), 1969: *Civilian Resistance as a National Defence: Non-Violent Action Against Aggression* (Harmondsworth: Penguin).
Rothermund, Dietmar, 1997: *Mahatma Gandhi: Eine politische Biographie*, 2nd edn. (Munich: Beck).
Rothermund, Dietmar, 2010: *Gandhi und Nehru: Zwei Gesichter Indiens* (Stuttgart: Kohlhammer).
Rothermund, Dietmar, 2011: *Gandhi: Der gewaltlose Revolutionär*, 2nd edn. (Munich: Beck).
Ruskin, John, 2011: *Unto this Last* [1860] (Lamar: Dixie Publishing).
Scott King, Coretta, 1979: *Mein Leben mit Martin Luther King* (Gütersloh: Mohn).
Sharma, Gopal, 2014: *Nelson Mandela: The African Gandhi* (New Delhi: Ruby Press).
Sharp, Gene, 1973: *The Politics of Nonviolent Action*, 3 vols (Boston: Porter Sargent).
Sharp, Gene, 2010: *From Dictatorship to Democracy: A Conceptual Framework for Liberation*, 4th edn. (Boston: Porter Sargent).
Singh, Upinder, 2017: *Political Violence in Ancient India* (Cambridge, MA: Harvard University Press).
Steinweg, Reiner; Laubenthal, Ulrike (Ed.), 2011: *Gewaltfreie Aktion: Erfahrungen und Analysen* (Frankfurt am Main: Apsel & Brandes).
Sutherland, Bill; Meyer, Matt, 2000: *Guns and Gandhi in Africa: Pan African Insights on Nonviolence, Armed Struggle and Liberation in Africa* (Trenton, N.J.: Africa World Press).
Thoreau, Henry David, 2014: *Civil Disobedience* [1849] (CreateSpace Independent Publishing Platform).

Tolstoy, Leo, 2017: *The Kingdom of God is Within You* [1894] (CreateSpace Independent Publishing Platform)

Weber, Max, 1988: "Politik als Beruf" [1919], in: *Gesammelte Politische Schriften*, 5th edn., (Tübingen: J. C. B. Mohr): 505–560

Chapter 6
Disobedient Peace: Non-cooperation with Inhuman Orders

Pietro Ameglio Patella

To Lito Marín, an outstanding epistemologist and a great human being

Abstract This chapter presents a practical theory of formal and informal education, and of direct action, based on three decades of experience with non-violent civil resistance. It represents an attempt to conceptualise and systematise the reflection and action processes which have enabled us – and many others – to build forms, on different scales, of 'due disobedience to inhuman orders'. We consider this to be a key concept and practice for building something we call 'disobedient peace', within the broader objective of contributing a new conceptualisation to enrich the already ample panoply of similar input to be found in studies on peace. The complexity involved in 'disobedience' in the social order has enabled us to develop a comparable epistemology.

Keywords Disobedient peace · Inhuman orders · Non-co-operation · Non-violent civil resistance · Peace epistemology

6.1 Introduction

The act of living in Latin America, particularly in Mexico, in the midst of so much social injustice and violence, has led us to constantly reflect on what 'building peace' really means. This entails approaching the issue from individual as well as collective and mass viewpoints, respecting different identities and cultural histories, and assessing how to do it and on which scale this is possible in the short and medium perspectives in terms of active non-violence. We begin these reflections by examining the pertinence of using the word 'peace' as a concept which embraces that which society, in its vast majorities, aspires to as the ideal form of social, human, political

Pietro Ameglio Patella, tenured professor, chair of "Culture of Peace and Non-Violence", National Autonomous University of Mexico (UNAM), Faculty of Philosophy and Letters, Email: serpajc@lanet.apc.org.

Ú. Oswald Spring and H. G. Brauch (eds.), *Decolonising Conflicts, Security, Peace, Gender, Environment and Development in the Anthropocene*, The Anthropocene: Politik—Economics—Society—Science 30,
https://doi.org/10.1007/978-3-030-62316-6_6

and productive relationships, as governments have abused and manipulated this idea continually, depriving it of content or associating it with militarised and repressive systems.

The Mexican government of President Andrés Manuel López Obrador asserts that its primary aim is to realise a "Mexico in Peace", which was stated to be something which already exists. However, every year – at least in the past decade – "acts of war" (Centro DDHH Fray Bartolomé 2015; Libera México 2013; Kloster 2017; Olivares 2019) have intensified, with all their attendant related human costs. At present, this country is riddled with this type of action under many different forms and characteristics, the common denominator being the high incidence of deaths, forced disappearances, displaced persons, kidnappings, extortions, and many other diverse crimes. All social classes are affected.

Official policy promotes a model of "armed peace",[1] which is a big international business deal that has prevailed historically in our species, and which implies permanent forms of militarised public security as well as the militarisation of local, state and federal security organisations. This entails violent tactics and strategies, and ignores human rights in combating organised crime. To achieve this, the authorities have *installed insecurity and social terror* on a permanent basis, with the pretext of a 'war against drugs', which overrides the means as long as it achieves the grand social end of this type of peace: 'security'. They have managed to subordinate – subsume might be a better word – the idea of 'peace' to the notion of 'security', which is a category and conceptualisation laced with military overtones.

In real terms, it is a transnational intercapitalist war over the monopoly of an illegal commodity, and many other crimes. In each opposing band we observe similar social identities in different bodies: authorities of all levels, business persons, legal and illegal armed forces, organised crime, and a sector of civilian society employed by these criminal forces (Bourbaki 2011).

It is, therefore, an excellent business, based on the exploitation of the bodies and natural resources in the whole national territory. A disorganised population, surrounded, confined, terrorised (which is not the same as frightened[2]), tamely surrender their bodies and reflections to an authority – even at the price of waiving rights that have cost decades to obtain – in "anticipated obedience to exert a punishment when authority demands it", where the so-called punishment really involves a confrontation made to appear as an "act of justice" (Marín 2009; Ameglio 2002: 129). All this is exacted under the pretext of guaranteeing minimal conditions for human survival, posing the dramatic paradox that 'peace with security' is being provided precisely by the same band which disseminates and reproduces 'war with insecurity'.

---

[1] This categorisation is linked to the idea of 'negative peace', in the sense that it "ascribes a dynamic and important role to war" (Lederach 1986: 21) and is the one which the Mexican government has favoured, putting all its emphasis and power behind the national construction of a militarised peace imposed by the armed forces and the police, including legal recognition to do so, a unique clause of the recently enacted Internal Security Law.

[2] "While fear is the form in which subjectivity organizes its defenses for self- preservation, terror knows what it fears but has no way of protecting itself from what it fears" (Bleichmar 1995).

There is a certain similarity between the idea of non-violence and the notion of peace in terms of the paradox we have described, but it is harder to approach. The term 'non-violence' – which encompasses a style of life, philosophy, struggle and spirituality – is not easily understood in Mexico, as we have experienced for years, and is prone to false controversies, debates and prejudices. Something similar has also happened in other countries, where social movements have culturally defined their form of non-violent struggle under different names: "Strength of Truth" (the Gandhian quest for the independence of India); 'force of love' (the struggle for human and political rights by the Afro-American population of the US); 'people power' (resistance in the Philippines against the dictatorship of Ferdinand Marcos); 'power of the powerless' (struggle in Czechoslovakia against USSR rule); 'permanent firmness' (social actions in Brazil). In Mexico, we use concepts such as 'civil resistance' and 'civil and peaceful' struggle.

In the hope of transmuting them into knowledge, this chapter attempts to share and *think aloud* on some considerations and experiences culled from the last three decades of individual and collective work in the fields of culture, education and the construction of peace. These three areas are intimately linked, although different – according to the priority of each type of action undertaken – in terms of their work towards non-violent civil resistance.

I will also share part of the theoretical base on which I have built these experiences, and will add new theoretical conceptualisations which I have developed. This last point is important, as the authors I draw from did not always conceive the greater part of their theories in terms of peace and non-violent civil resistance. The *adjustments* are my own, and I hope they do not affect the essence of those basic theoretical inquiries.

What we wish to stress is that the following assertions are born from practical experience –many times repeated in concrete actions – not from speculations on an abstract theory. I believe that this confers a certain value to this exercise of sharing and enables me to provide copious examples. This chapter thus comprises a theoretical reflection based on long practice and 'experimenting'.[3] It is drawn from Mexican actions and experiences in the fields of the social struggle for human rights, against war, for justice and in defence of the environment, as well as on the issue of autonomous, popular and formal education. In most of these actions I have been an apprentice of the people, their social movements and social activists.

Considering reflection to be the main weapon – alongside the body – in the struggle for social justice and peace, I have reached the conclusion, from the viewpoint of the epistemology of peace and non-violence, that the key strategy in the construction of or education for peace based on non-violence is the construction of a corpus of knowledge with ethical action plans based on "Due Disobedience to Inhuman

[3]The title of Gandhi's autobiography – a fundamental book to read, along with Mandela's autobiography – has been a lifelong guide for understanding the construction of a social identity: *My Experiences with Truth* (some colleagues have translated experiences as "experiments"). I believe that approaching actions of peace and non-violence from the logic of an experiment is very helpful in building peace without fear of being mistaken.

Orders".[4] Initially, and as a proposal for collective discussion designed to enrich in some way the already fertile conceptualisations of the subject, I will call this 'disobedient peace'. In other words, "To disobey an inhuman order, an order that produces damage to *the other* and to oneself, is a moral weapon and the major challenge of a non-violent action" (Fracchia 2014: 11). The next question is: *How can we build this humanising knowledge of disobedient and non-violent peace, and from what logical mindset and mode of action can we operate it?*

6.2 Epistemology of Disobedient Peace

I feel it is important to state that when I talk about peace, I do so initially from a perspective of what Braudel called "history of long duration", in the sense of a history which builds on social processes of a "humanisation of the species". Konrad Lorenz was right in pointing out that the present human species is still the "missing link" (Lorenz 1994).

In biological terms, we could define ourselves as a relatively rounded and evolved species during the last hundreds of thousands of years. But from a cultural point of view, we are more like a *hope* than a reality: two out of five individuals do not know if they will eat tomorrow; there are genocides – the most inhuman act of our species – televised worldwide. In view of this, being 'fully human' seems to be more like an end point than the starting point we would like to believe.

> For human beings as a whole, it is not yet clear that they constitute a species; that's why there are still massive killings, and an organisational social relationship for the procurement of food which is hugely unfair, which means that we have not yet achieved the phase in which any human being is a human being for any other human being (Marín 2014: 45–46).

Thus, peace is still a distant reality for a major part of our species, and it is in this struggle of long historical duration that all those who strive for a fairer and more human world are now involved. This perspective enables us to achieve an awareness of our work in a broader sense, because it compels us from the beginning to define clearly and objectively the starting points and the end points in our own processes and in the social, cultural and educational endeavours in which we are

---

[4]This idea was coined initially by Dr Juan Carlos Marín, a noted Argentinian epistemologist and social activist, (Marin 1995: 25–26) and taken up in the Final Declaration of the XXII Congress of the Latin American Association of Sociology (ALAS in Spanish), carried out in Concepcion, Chile, in October 1999:

> We unanimously express that, in the ethical practice of our profession, social scientists cannot restrict themselves to diagnosing their societies, without knowing and facing the multiple dimensions in which the legal monopoly of violence is inhumanly and arbitrarily practiced in our Continent. We therefore propose urgent collaboration in the construction of a moral judgment capable of enabling the rupture of all forms of uncritical obedience to authority, making observable and promoting lifelong a due disobedience to any inhuman order.

involved. Superimposing, inverting or confusing these points is a permanent cause of frustration and disappointment, or erroneous assessments of actions undertaken, as we lose sight of processes and scales.

On the other hand, as Gandhi said: "We can't wait thirty years for the adversary to change, so we must resort to non-violent direct action".

Fortunately, from the second half of the twentieth century, "peace studies" have developed in a big way, linked to those on violence, as these realities are indivisible, systemic and intertwined in a complex way, so it is imperative to be familiar with both. Multiple approaches to peace have been explored, from the viewpoints of many cultures and experiences: peace as 'positive' and 'negative' (J. Galtung and J. P. Lederach) ; as 'imperfect' (F. Muñoz); as 'neutral' (F. Jimenez); as 'intercultural' (R. Pannikar); as "*lekil kuxlejal* [good living]" (Tseltal Maya peoples from Chiapas, Mexico); as 'transpersonal' (Dietrich); as 'spiritual' (Thich Nhat Hanh); as 'holistic' (C. Martinez); as 'engendered' (Ú. Oswald); as 'structural' (Economic Council for Latin America) or 'sustainable'...

Associating the idea of peace with that of non-violence,[5] I have been able to observe in recent years – with the purpose of developing reflections and actions which are really operable in individual and collective situations involving social change – some principles (to be built from certain epistemic axes, that I will share as I go along) that I believe synthesise the essence of this culture and practice:

1. We need to build an "original way of thinking" (Fromm 2005: 231–245) – this is one of the most difficult tasks – embracing personal and collective autonomy, enabling us to "disobey any inhuman order".
2. It is necessary to attempt to "humanize the other and the adversary". A starting point would be to become familiar, with the greatest possible depth and complexity, with the constituent processes of his/her identity and interests. Thus, it is possible to 'disarm' his/her reality principle we consider dehumanising, by means of reflection, not prejudice, stigma or actions that seek to destroy him/her. Comprehension of history is central to this process.
3. We must prioritise the "search for truth", with a lower case 't', as Gandhi used to say, but with clear parameters concerning "absence of injustice", rather than "absence of conflict", procuring the benefits and reparation for victims, regardless of differences.
4. Avoid exacerbating the "spiral of violence" with our reflections, words or actions; seek, at all times, to de-process it and make it regress as much as possible to levels of dialogue and negotiation.
5. Maintain coherence in the proportion between means and ends: social order is constructed – in almost all its aspects – according to the principle so well expressed by Machiavelli: "the end justifies the means". Gandhi used to say: "The means are like the seed, and the end is like the tree... There is an unbreakable

[5]Clearly, non-violence is not the only way to build peace; I respect and have supported many others, but I believe that this culture is a millenarian tradition which has contributed to the long process of the humanisation of our species. However, in order to avoid prejudice, stigma or 'anticipated defeats', it is crucial to distinguish between the end points and the starting points.

relationship (between them). A good tree will never grow from a rotten seed" (Gandhi 1985: 114).

6.2.1 *Start from a 'Reality Principle'*

The epistemic axis which links these five central principles in any work based on disobedient peace governed by non-violence is the construction of a 'reality principle'[6] which is as objective and empirical as possible. In order to handle a social event or actor appropriately according to the form of obedience or disobedience, it is helpful to be familiar with it in all its complexity; therefore, the initial phase of the process is the integration of appropriate knowledge. If this initial approach to the observable factors of the social event which affects us is not grounded in some form of empirical 'recording' which reflects hard facts, the entire process of actions and reflections that follows will probably be rooted in some kind of 'logical empiricism' in which reality is described more in terms of the 'relationships between discourses on that reality'[7] than on the recording of real facts. This causes the perception of reality to be grounded more in fantasy or wishful thinking than in its true material base.

I have observed this repeatedly in social struggles and in the construction of individual and collective peace, and this results in actions that do not respond directly to the targeted conflict, and thus do not transform it. In some cases, they can even exacerbate it, leading to further violence.

Social order imperceptibly sets the epistemic trap that its reality principle is the only one which is valid and feasible, because it is not impossibly utopian, and with this device it sets in place a dominant hegemonic culture. This culture is frequently transmitted by a static vision of history that feeds on a vast variety of social stigma and prejudices – which mask multiple forms of social confrontation and discipline – disseminated and multiplied by the media, laws and regulations, pulpits and many sorts of orders and instructions in institutions, schools and families. There cannot be actions for peace without clashing, to some degree, with the prevailing social order if its reality principle presents an invidious anomalous situation as normal. For this reason, it is vital to understand the mechanisms involved in the construction and reproduction of social order from a dual perspective: history and the recording of present events.

---

[6]This concept was adapted to sociology by Juan Carlos Marín, taking it from Freud and others; it has subsequently been applied to and enriched with new experiences and conceptualizations. In this case, for example, it is being applied to the culture of peace, education and civil resistance.

[7]The important part here is the logic within and without the discourse, not the empirical-experimental correlation according to some type of objectivised register of the material reality observed. Power constructs 'discourses on reality' permanently, and 'operates' them on bodies with enormous strength, as if they were 'truthful' or 'real', because whoever accepts them does not observe reality, but limits his/her attention to the entity which pronounces the discourse, in obedience with an ingrained attitude of deference previously constructed by the 'authority principle'.

In this sense, it is necessary to distinguish between 'information' and 'knowledge': the construction of the latter leads us to reflection, while information on its own can never achieve this; information must be transformed into knowledge – not a straightforward or easy process – registering it as some form of fact, a fundamental epistemic unit in the field of thought[8] which, when analysed – measured and compared – becomes some type of knowledge that can be 'installed'[9] in the identity of the person who builds or receives it. Furthermore, it is important to develop the awareness that it is possible to acquire a certain degree of knowledge concerning a social event, but that this knowledge will not necessarily lead to adequate reflection.

Finally, it is also possible to ponder in an acute and complex way, but still not necessarily perform an action in accordance with the moral identity which is being transformed. In other words, the path from information to knowledge, from knowledge to reflection, and from reflection to action is not a series of mechanical operations but an 'epistemic art', which grows from the principle that, from greater knowledge, reflection and action a greater moral, individual and social identity develops. Thus, the construction of 'moral judgment', according to Piaget and Marín, is an epistemic process rather than an intellectual one, although there is plenty of feedback between the two. It follows that, to influence social order and individual identities, the principal educational task is to build knowledge which is as close to reality as possible, and from there create new conceptualisations which enable us to reinforce reflection and action.

6.2.2 'Becoming Aware' of Concepts and Processes

The epistemology of disobedient peace shared here, centred on the construction of autonomous identities, is based – apart from a reality principle – on 'becoming aware': "…an action pattern transforms the former into a concept, and this becoming aware essentially comprises a conceptualization" (Piaget 1976: 254). In this sense, it is essential that the conceptualisation should be complex and transdisciplinary, because it must enable us to pass from less to more knowledge concerning those social events which, as elements of a social process, simultaneously and in parallel intersect several aspects of each human identity at the same time that they reproduce social order.

The 'becoming aware' – centred on its 'relation with' and not on some generic abstraction – will move forward in each social identity, on the basis of new conceptualisations that enable us to discover new social 'observables', 'unobserved' and

[8]"…discussion and reflection, that is cooperation in the field of thought, increasingly override unsupported statements and intellectual egocentrism" (Piaget 1985: 37). 'Proof' and 'data' are what really enable co-operation, or really 'operate' with others in the intellectual and epistemic spheres.

[9]The 'installation' of knowledge is one of the most complex, unobservable and accepted as normal arts in epistemic work, and it is also a core task in any process of constructing peace and autonomy.

'unobservables',[10] which will help us to learn more, and more deeply, about the social event in question and its indispensable contextualisation in the social order under study. Hence, to know, in a complex manner, means to approach a social event or identity with a successively broader range of concepts and 'epistemic perceptions'.

I consider it is essential for the epistemic construction of peace and non-violent action –either from a strategic or tactical point of view – to always bear in mind this aspect concerning the social observables, unobserved and unobservables in their different dimensions, which we could associate respectively with the intellectual, epistemic and moral areas of identity, in that ascending order in terms of depth.

Resorting to a more graphic form of explanation, we could liken this epistemic process to a flower: the 'observables' would be the flower we see above the ground; the 'unobserved' would be the roots; and the 'unobservables' would be the soil below the roots. The flower cannot be explained in itself; neither is it sufficient to add the roots, as it would not grow without the soil that harbours them. This is somewhat similar to an epistemology – of complex and autonomous thought – which seeks to go beyond the flower – the tip of the iceberg – as Galtung would say in his definition of 'direct violence', to explore what allows the emergence of this dynamic, thus to build reflections and actions which would enable us to increase our influence on real changes aimed at processes of peace, non-violence, autonomy and humanisation. At bottom, it is also a more realistic way of working towards the construction of 'hope' instead of misguidedly focusing on an 'illusion' of reality, which would be extremely damaging to cultural and educational processes and social struggle.

This process of achieving awareness fosters the construction of personal 'rupture principles'[11] at the intellectual, epistemic and moral levels, which could – although not automatically – lead to new actions, both individually and socially. These ruptures, caused by the increasing awareness of something in particular, such as the 'rearrangement' of some previous knowledge (either erroneous or prejudiced)

---

[10]The 'social observables', according to J.C. Marín, are qualities and traits of social events and actors which can be observed at first glance; the 'unobserved' can be discerned with the help of some adequate theory or conceptualisation; on the other hand, the 'unobservables' cannot be observed directly and require the application of indirect observation instruments to bring them to the surface. An example of this could be 'fear', which, to be observed, requires us to ask the individual about certain behaviours and practices.

[11]'Rupture' is a fundamental epistemic and moral category which must be built permanently in us and in others, both individually and collectively. It is indispensable for casting doubt on previous knowledge with the aim of creating new knowledge concerning a determined social event, based on an increase in awareness. This text is based, to a great extent, on how we have been able to *advance* along this epistemic axis, from theory – acquired and built upon – to experiences-experiments in which we ourselves participated. Returning to Fracchia (2018), we can see how, for Bachelard, scientific rupture is that which "contradicts common experience", and in which, as Bourdieu points out, "familiarity with the social universe is the primary epistemological obstacle". In other words, for a rupture to exist, it is necessary to cast doubt upon accepting as normal much knowledge drawn from "common experience" which, as we will see, is really the installation of part of the dominant heteronomy in social order, based on some "authority principle". Thus, wherever knowledge is catechistic, voluntarist and dependent on absolute values, there cannot be "fissures" or autonomous ruptures. A rupture is, therefore, a prerequisite for the construction of acts of disobedience, because it is an "opening towards what is new".

towards a new conceptualisation – on the strength of empirical evidence – which allows us to correct that 'peripheral factor' that generates ignorance, are for Piaget: "maladaptation factors which…enable the achievement of awareness, which would be useless if individual functioning…were normally adapted" (Piaget 1976: 255).

The epistemology of disobedient peace, then, faces the constituent processes of actions – or beliefs – from the point of view of 'how' rather than 'why', bearing in mind that the historical experience of knowledge in our species indicates that the most adequate way to stop suffering something, or to exacerbate it, is not by means of voluntarism or catechistic-dogmatic-absolute transference of information, but rather by means of greater knowledge and reflection.

Finally, a central tenet in every order of our lives is that the achievement of awareness travels from the periphery to the centre, where the initial two peripheral observables are "the awareness of the objective to be accomplished" and "the awareness of its resolution either as success or failure" (Piaget 1976: 256). This is how the culture in which we live operates most of the time, evaluating and judging people and social events principally on the merits of their most external peripheral – the final result – without giving much consideration to the constituent processes of the event involved.

Consequently, a complex approach to the areas of peace mentioned above points towards the construction of a greater conceptualisation of the processes involved in social change, and not simply to changing the final actions, even when these might be positive but occur as a consequence of greater observance of discipline or advanced obedience to authority. We seek, therefore, the construction of autonomous and reflective subjects and forms of disobedience – not any form of disobedience for its own sake – and not 'disobedient obedient subjects', even when the peripheral factor may seem just and legitimate. This is one reason why we must build knowledge and actions aimed not at the periphery of the problem, but at deciphering its constituent process. This is why the achievement of awareness "…should be directed towards the central regions of the action when it proposes to reach its internal workings: recognising the means employed, the reasons for choosing them, or their modification during the exercise, etc." (Piaget 1976: 256).

6.3 From 'Social Infantilism' to 'Co-operation'

Another central point in this epistemology of disobedient peace is linked to factors which need to be deciphered in the actors and actions of violence and non-violence: their logic and construction of thought, because these are the building blocks of moral identity and consequent actions. Unjust or inhuman behaviour chiefly stems from ignorance[12] rather than 'inherent evil'; therefore deciphering thought processes

[12]Ignorance – which is among the principal causes of violence – is associated with lack of knowledge concerning a fact or a trait of some social identity, and not intelligence or intellectual identity. It is counteracted by building and installing greater knowledge.

makes it easier to change inhumane attitudes. For this reason, it is advisable to take the epistemic approach, and not simply focus on increasing information and compliance with certain catechistic-voluntarist-absolute values, or *good* leaders. In this sense, based on studies on 'moral criteria' in children – in terms of practice and awareness of rules in games – but referring also to adults,[13] Piaget himself proposes that the superior moral phase is 'co-operation' (Piaget 1985: 36–90), rooted in two fundamental traits that should be encouraged: the "equalisation principle" and "mutual respect", both emerging from consensus among equals, and not from asymmetry of power (seized, for instance, by an "older adult" who assumes that age and experience automatically endow him or her with greater authority). Reflection and co-operation are thus the first steps on the path to achieving any substantial change in the culture of peace that could empower people to autonomously disobey inhuman orders.

It is necessary to start from a basic achievement of awareness: social order is based on the massive construction of an 'egocentric stadium': hierarchical power and legality, sacralised and born outside historic time and with no human origin, unidirectional, with unilateral respect and asymmetry of power. Any attempt to equalise power or establish mutual respect will be inhibited, punished or annihilated in whatever field it appears. These efforts – incommensurable and perennial in terms of their logical, epistemic and moral construction – launched from all the fields and forms of social order (hierarchic authoritarian institutions of all types, media in blind obedience to power, systematic spreading of fear and terror, schooling without education, corporal imposition of discipline rather than real discipline, authoritarianism[14] without authority...), comprise what I propose calling 'social infantilism or infantilisation', without any negative connotations of that which is infantile; on the contrary, I am trying to point out that, under the form of 'the logic of thought', adults suffer great regression compared with the evolutionary process of the child.[15] This is not 'natural'. It is, on the contrary, a highly sophisticated construction of the whole apparatus regulating social order. It is the effort of the mechanisms and power of social order to obstruct, deliberately or not, any attempt – in whatever degree – by social or individual bodies to evolve from the 'egocentric' stadium to that of 'co-operation', the only way in which it is possible to autonomously disobey inhuman orders. Social order tries to impose the 'authority principle', built upon submission, mystic sacrality and timelessness, to the detriment of the 'consensus among equals' principle.

---

[13]"The whole adult can be found in the child, and all the child survives in the adult" (Piaget 1985: 70).

[14]Foucault adroitly describes the effects of corporal 'discipline': "...the more obedient he is, all the more useful... Thus, discipline makes submissive and well trained bodies, 'docile' bodies. Discipline increases the strength of the body (in terms of economic usefulness) and diminishes that same strength (in terms of political obedience)" (Foucault1976: 160). Antón and Damiano round off the idea: "Thus it seems to be the 'target' of power: to correct bodies to obtain more docile and useful individuals, incapable of reflecting on their own actions" (Antón/Damiano 2010: 24).

[15]"How is it possible that the practice of democracy is so advanced in a game of marbles between boys from 11 to 13 years old, and is so unfamiliar to adults in many fields?" (Piaget 1985: 62).

For autonomy to succeed heteronomy and for laws to lose their sacred nature as the results of consensus processes, it is important to remember that "co-operation between equals… will make the mystic nature of authority disappear" (Piaget 1985: 51–53). Therefore, the construction of disobedience in the face of inhuman orders should generate processes of social equalisation and autonomy, so that individuals or social movements might feel empowered to challenge and face authority.

The other process which must be built, with the objective of breaking the asymmetry of power and unilateral respect, is that of becoming aware of one's own power and its proportion with that of the adversary. Gandhi and the theory of civil resistance see co-operation from the point of view of the relationship between each individual and power, based on the given that social reproduction of injustice is enabled by the co-operation, voluntary or not, with subjects or actions that build it or benefit from it.

This form of non-violent struggle and construction of peace is based on the principle that all of us have some type of power which we can exert against unjust or inhuman situations, reflected upon by our autonomous thought that determines our moral identity, and if we do not exert this power, inhumanity will reproduce, but if we disobey, then we humanise ourselves – as well as many more who suffer the consequences of that order – and the social order in its long historical process.

Gandhi correctly pointed out in his *Constructive Programme for India* that:

> We have become accustomed, over a long time, to think that power should come only from legislative assemblies. I have come to consider that this belief is a serious mistake, caused by inertia or by a sort of hypnosis. A superficial study of history, of English history, makes us think that all power reaches the people through parliaments. The truth is that power rests in the people, and is entrusted provisionally to those whom the people choose as their representatives. Parliaments have no power, not even existence, independently from the people. Civilian disobedience is the repository of power (Ameglio 2002: 305–306).

In other words, the logic of thought linked to co-operation is an action that confers power on those who apply it, because it triggers the degree of power an individual actually possesses, but which social order renders invisible, and expropriates it in the name of the social value involved in obeying authority.

6.4 Due Disobedience to Inhuman Orders

6.4.1 Indignation in the Face of What is Inhuman

Taking into account the axes stated in relation to social order and its compulsory reality principle, authority and its sacralised hierarchisation, power and its exercise of unilateral respect, the achievement of awareness and its emphasis on peripheral results, I believe that the disobedient peace that we need to build – not only in response to our war in Mexico, which makes it indispensable – and that many groups and individuals, nationally and internationally, have termed "with justice and dignity", contains in its basic epistemology, as stated at the beginning, the imperative to build individually and collectively a 'due disobedience to any inhuman order'.

What do we disobey? The inhumanity of social order, in terms of actions, reflections, values, laws, regulations and orders.

Adding depth to this definition of an epistemology of disobedient peace and non-violence, there are various elements and different operators which should be conceptualised: inhuman-order-duty-disobedience.

The social order which pervades us is a machine for creating the orders and processes of inhumanity, injustice or social violence – as well as those of humanisation – regulated according to set phrases, such as: 'it's for your own good', 'it's always been that way', 'everybody does it'... According to this, violence, abuse of power, depredation, negation of the identity or resources of enormous sectors of humanity, and excessive punishment without any proportion to the offence, are the operators that expropriate the humanity of others, in the name of the dominant heteronomy of 'order' and, many times, of peace too.

Harold Laski points out that: "Civilization consists, above anything else, of a lack of disposition to inflict unnecessary suffering... It is our duty, if we do not want to live a life totally bereft of sense and meaning, to not accept anything in contradiction with our basic experience for the simple reason that it represents tradition, convention or authority" (2011: 1–10). To face this inhumanity, or to de-process the inhumanity within us, demands, first, getting to know it and making it observable; this is why the degree and the actions of inhumanity we observe and face depend on the awareness we have of the social order.

There is, likewise, a link with the installation of that which is inhuman and terrifying: "The first principle of humanisation (is)... to begin to de-process that mechanism of terror on which the authority principle has been grounded, not because all authority terrorizes us, but because the mechanism of the given order (inhuman) is based on that force" (Marin 2014: 42). We must be familiar with this mechanism (authority principle-punishment-terror), and how it is set up, to be able to begin to disobey. However, it is possible to be aware of certain inhuman orders without necessarily being able to face them. These are two different actions and epistemologies.

Simultaneously, to be able to advance towards the disobedience of inhuman orders, it is fundamental to build, not only knowledge of the inhuman content in social order, but also the capacity for 'indignation' (Hessel 2010), 'rage', 'anger', and 'rebelliousness' towards this social fact. Hannah Arendt said correctly that: "...the clearest sign of dehumanization is not rage or violence, but the notable absence of both. Rage is not in the slightest a reaction against poverty and suffering as such; nobody reacts with rage in the face of an incurable illness, an earthquake or, in what concerns us, social conditions that seem to be unchangeable. Rage emerges where there are reasons to suspect that those conditions could change, but are not changing" (Arendt 2005: 85). Indignation or rage, then, are a complex construction stemming from experience transformed into knowledge, achievement of awareness and ruptures; they are not some mechanical result of knowledge; they need a particular type of reflection for peace to become disobedient.

6.4.2 The Order and the Duty to Obey It

'Order(s)' and 'punishment(s)'[16] based on threat are the great operators of social order for the enforcement of obedience to authority. Therefore, we must make the set of orders we have accepted as normal and obeyed as observable as possible, and decide which of them are dehumanising us and dehumanising others around us, with or without them being aware, and then disobey them. Many of these orders are accepted due to ignorance, others to affection or emotions, or to fear of punishment, according to their infinite social forms.

It is of central importance, then, to grasp this 'epistemic unit' which operates so strongly within us. Order is associated with heteronomy, with that which is imposed from outside, and consequently "…in each order with which we comply, we are renewing an ancient victory… The power of whoever is in charge grows incessantly. … Power issues orders as a cloud of magic arrows" (Canetti 1983: 301). Canetti enriches the description of the mechanism for installing 'egocentrism' or 'social infantilism', which operates within us by means of a permanent stream of orders that reinforce sacralised authority, unilateral respect, and the necessary asymmetry of power.

Orders are also linked to the dehumanisation process: "Any order comprises an *impulse* and a *barb*… But the barb sinks deep into the individual who has complied with an order, and there it stays, unalterably… Any individual, with the passage of time, finally accumulates a huge amount of barbs… It can happen that somebody is so filled with barbs that he is no longer sensitive to anything else: beyond that, he feels nothing" (Canetti 1983: 302, 318).[17] That sort of regulation dehumanises us, and this is precisely why we must practise disobedience to certain orders as a sort of *daily workout* which humanises us.

Orders, with the passage of time, have become more remote from their biological origins, associated with flight, and have been used in multiple social relationships, pervading a mode of 'domestication' rooted in certain forms of 'bribe', 'submission', 'reward' and 'voluntary captivity' (Canetti 1983: 303–34), by he who holds the greater power and gives something to the person who complies with the order. This shows that, at least in part, there is a relationship of convenience, reciprocal interests, and voluntary captivity in exchange for obedience.

The most complete and transparent model of obedience to authority and its orders is the soldier: "A soldier on duty only acts according to orders… A good soldier is

---

[16]The whole social order is built, in Juan Carlos Marín's words, on "anticipated obedience to exert a punishment when an authority demands it, where the punishment, in reality, masks a confrontation and is presented as an act of justice" (Ameglio 2002: 129). More in Piaget (1985: 167–272), Martínez (2012: 109–118) and Glover (2013: 448–561).

[17]In his research, Milgram claims that, for a soldier in Vietnam, "It is enough hard work to get through the present day alive; there's no time to think about moral problems" (1980: 169). The aim is to transform individuals into "a pure state of agency [in which] moral judgments are to a great extent suspended" (1980: 146). For different reasons, this construction of the "soldier" and "beings in agency state" is in some ways similar to the "Muslim" described by Giorgio Agamben in *Homo Saccer*.

always in a state of conscious *wait for orders*" (Canetti 1983: 308). Also, at another level, but following a similar logic, he is also the model for the 'good citizen',[18] the 'good student', or even the 'good son/daughter'.

For Canetti, orders "are the singular, most dangerous element in the society of humans. We must have the courage to oppose them" (1983: 329), because, being heteronomous, they remove responsibility from the person who executes them and transfer it to an authority which is generally diffuse,[19] and for this very reason activate another great operator of inhumanity: 'duty', bringing into play "something much more dangerous, the human being's capacity to set aside his humanity, or even worse, the inevitability of behaving in this way when he subsumes his unique personality in broader institutional structures". The individual is co-opted in his autonomy "without the hindrances of individual morality, free from any human inhibition, only paying attention to the penalties imposed by authority" (Milgram 1980: 174).

In her study on the *Banality of Evil*, Arendt states that Eichmann: "was performing his duty: he not only followed *orders*, but he also complied with the *law*… a strange notion very prevalent in Germany, in the sense that complying with the law was not limited simply to obeying it, but it was required that one should act as if one were the author of the law being obeyed" (2003: 83–84). She demonstrated, by means of a very detailed study of the behaviour of this high-ranking Nazi official, charged with operating a considerable part of the Jewish genocide, that he was not a sadist or a madman, nor was he an "inhuman exception of our species", but rather a simple bureaucrat who strictly and proudly did his duty by following orders from a higher sacred authority, regardless of the consequences of his acts. He simply separated his actions from their effects.

According to Milgram (1980: 19, 175), "…Hannah Arendt's conception of the *banality of evil* is much closer to the truth than anybody would dare to imagine… The most ordinary people… can become agents of a terribly destructive process. A very large percentage of people do what they are told, regardless of the contents of their actions, and without obstacles provided by their conscience, as long as they perceive that the order originated in some legitimate authority."[20]

So obedience is the great 'epistemic-moral obstacle' to building disobedient peace. To become aware of this component of inhumanity which pervades part of our identity and body, in which we have probably been built by the regulation and violence of the social order, implemented by 'orders', 'duties' and 'a priori obedience to authority', is therefore a central prerequisite for developing disobedient peace.

---

[18]"…the historic construction of what is called a citizen, and if we wanted greater precision we would call it a soldier-citizen. Because we must not forget that before being a citizen, one must be a soldier" (Marín 2009: 68).

[19]Among the causes of obedience, Milgram (1980: 20) pinpoints a fundamental one: "…to be able to consider oneself as not responsible for one's actions. The individual is free from any responsibility when he ascribes the initiative to the experimenter" who gave the order.

[20]Moore (1989: 97) analyses this issue when he compares the experiments of Milgram and Asch.

6.4.3 The Difficult Mission of Disobeying

Finally, how do we achieve the most difficult part: the action of 'disobeying' those orders which dehumanise us and others, that is, how do we face authority?

This is something that nobody – that is, nobody who is close to authority, the imposition of discipline, or the reproduction of social order – will want to make observable or share with us, because it would place them and us in a sort of co-operation towards the 'equalisation of power', 'mutual respect', and 'peace with justice and dignity' for all. Social order is not built so that somebody – or some groups or movements – can say: 'No!' or 'Enough!' to one or many of authority's inhuman orders, and carry this 'conscientious objection' to the extreme of disobeying the order in question.

Experience coincides with what Marín posits about authority: "One of the hardest things for a person is to defy authority, because from the very moment he/she is born, it begins to be implanted, it becomes part of his/her corporal anatomy. It suffices that somebody imbued with some sort of authority should look at us, for that mandate of authority to be exerted on our bodies" (Marín 2014: 38). And Milgram adds:

Disobedience is the ultimate means with which to end a tension. It is an act which is anything but easy… A totally unknown perspective of the relationship follows after the rupture. For many individuals there is a certain apprehension concerning what is going to happen as a result of the act of disobedience… The price of disobedience is a sentiment that gnaws at us from within, a notion that we have not been faithful. Even when one has chosen the morally correct act, the individual remains dazed by the upheaval in the social order he/she has caused… It is he/she, not the obedient subject, who experiences the impact of his/her action (Milgram 1980: 152–153).

It is also true that in many inhuman historic situations "…moral voices were raised against concrete actions, but the usual reaction of the man in the street was to submit to orders" (Milgram 1980: 167).

Disobedience is, therefore, one the most difficult human acts to carry out, because it pits us against the social order, against our own identity, past and present, and against a sacralised authority. It is, therefore, an end point and not a starting point – as in all social actions – and must be viewed as a construction process. Its 'peripheral' (final action) should not be considered the principal thing to observe, because this inevitably leads to frustration and discouragement. I have noted this many times.

In intellectual, epistemic and moral terms, the first element in this construction process – in each one of us and collectively – is a 'tension' (some order which is not completely assimilated to one's own epistemological and moral identity), a 'rupture', a 'discomfort' or a 'maladaptation'.[21] In the first part of this chapter I have attempted to reflect, from an epistemic point of view, which I consider fundamental,

[21] For Piaget (1976: 255) "…the factors of maladaptation would be the triggers of the achievement of awareness… what triggers the achievement of awareness … is the fact that automatic regulations… are no longer enough… and it becomes important to seek new means for a more active adjustment". For Marín (2014), 'discomfort' is something from previous knowledge that does not agree with the new action we are being asked to perform.

on the construction of this tension or rupture, which, based on concrete experiences, I consider to be indispensable for carrying out actions under the form of autonomous disobedience, and I stated the following theoretical observables: the objectivatable empirical reality principle; the achievement of awareness that observes processes and new conceptualisations; the social infantilisation as a consequence of the logic of egocentrism, instead of co-operation among equals; the construction of what constitutes inhumanity; moral indignation; a priori obedience to authority and to any form of punishment it may decree.

Let us now move forward to the step from tension to disobedience, which is the final phase of disobedient peace. To start with, we must take into account Milgram's assertion that: "Tension leads, if it is strong enough, to disobedience. But initially it causes lack of agreement.... Many dissenting individuals, capable of expressing their disagreement with authority, continue recognizing the right of authority to override the opinion they have expressed" (Milgram 1980: 151). As stated earlier, I can be aware of an inhuman order, but it does not necessarily follow that the reflection or reality principle that follows will lead me to oppose it with action. This is very important. A tension or a rupture does not lead mechanically to the disobedience of that order. For this to happen another turn of the screw is necessary: the achievement of moral and epistemic awareness, personally or collectively.

Using the data culled during his experiment, Milgram has attempted to describe and analyse the passage from tension to active disobedience. Starting with the construction of this tension, a process is triggered whereby the sequence of disobedience initiated by 'internal doubt' (as a result of the tension) begins as something private but rapidly becomes something 'external' when it is socialised or shared with another, to whom the individual confides their doubt concerning the legitimacy of the order issued. If the other does not agree or correct, the relationship becomes one of dissent, resulting in "a gradual transition towards rupture". If there is no agreement, then dissent becomes a 'threat' that the individual will refuse to comply with the order. "Finally, the individual, having tried all other measures, comes to the conclusion that he must go to the very root of the relationship... (and) disobeys" (Milgram 1980: 152–153).

"Internal doubt (as a result of the tension or rupture), externalization of doubt, dissent, threat, disobedience: this is a hard road to take, which only a minority is capable of following it to its conclusion. ... Swim deliberately against the current. ... The psychological cost of this road is considerable" (Milgram 1980: 153). So the epistemic process of building a disobedient peace and empowering individuals to put it into practice is comprised of the construction of tensions and ruptures, which evolve from individual to social phases until they reach disobedience. This sequence must be achieved by increasing knowledge of the inhumane social order, and the awareness of greater autonomy, and not by issuing other 'orders', even if their content is more humanising and just, since they would still be heteronomous and 'infantilising'.

It follows that disobeying is intimately linked with 'resistance', active or passive, individual or social. It is a resistance against inhuman orders, against the heteronomy that seeks to annul our autonomy. For Canetti (1983: 304, 302), "The 'free' man is only he who has learned to dodge orders, and not he who only rids himself of them after (passing the arrow/barb to somebody else)". Disobedience is linked, then, to liberty, autonomy and humanisation.

How Can We 'dodge' Orders and Free Ourselves?

In the shooter of the arrow/order, "...There is always something like the perception of a counterattack; what has been done has also become imprinted on oneself, not only on the victim. Many of these counterattacks are accumulated and they generate fear, *fear of ruling,* a conception of the risk (for the shooter) if the many victims were to rise up against him" (Canetti 1983: 305).

One of the key elements is the mass which unites to face and resist authority, as will be seen in the coming paragraphs. An order to the mass is lost in abstract generality, "...because there is a liberation from all the barbs, even the most monstrous ones, but this liberation is in the mass. ... The turnaround mass is made up of many individuals who unite and turn against another group, in which they see the cause of all the orders they have had to endure for such a long time..." (Canetti 1983: 325). Concerning these mass processes of disobedience and resistance, I feel that Canetti and Arendt seem to be rather more optimistic than reality has shown, at least in the short term, when they say: "The general process of a liberation from barbs, once it is initiated, continues unstoppably" (Canetti 1983: 326), and "Where orders are no longer obeyed, the means of violence are rendered useless" (Arendt 2005: 67).

6.5 Disobedient Peace in Action

Considering the social perspective, we still must ask ourselves: how do we build social struggle actions for this 'disobedient peace'?

As stated earlier when considering the empirical reality principle, experience has shown that, within strategic reflection on a social struggle, it is essential to develop some type of measurement and categorisation not only of one's own forms of action (tactics), but also, with equal precision, those of the adversary(ies), who also struggles with all its (their) resources. This is of critical importance because, in different forms and fields, a certain proportionality in the scale of the actions must be planned[22] between both dynamics in conflict, considering of course one's own forces and resources. If not, there will be no real possibility of advancing one's own claims, with the attendant risk of sliding into some type of simulation or self-delusion.

At this point it is important not only to concentrate our attention on the peripheral of the action, its end point, but also on the different variables that comprise it – time, space, actors, context – which could contribute greater or lesser intensity, beyond the action itself. For example, certain actions carried out by bodies with a certain 'social

[22]"*Civil resistance* is a method of collective political struggle based on the primary idea that governments depend ultimately on the collaboration, or at least the obedience, of the majority of the population" (Randle 1998: 25). Some profiles of civil resistance actions can be found in Randle (1998:25–32), Sharp (1986) and Ameglio (2002: 117–18).

power' or 'moral strength' can also contribute their indignation and determination to the cause, although if they are not actions that go 'beyond the law and co-operation', they can face power with considerable force, as they can 'unmask the truth publicly', as Gandhi would say, and this creates a public moral fracture in the legitimacy of power. Another central element in this construct of actions towards disobedient peace is the site of the action, which radicalises its effects when it pinpoints exactly where power resides: a demo-march to a central square is not the same as a demo-march to a military barracks, although both are demos and may comprise the same number of people.

In this sense, and bearing in mind the proportional relationship between actions, in the present Mexican situation riddled with warlike social events, I cannot abstain from encouraging actions involving non-cooperation and civil disobedience. The former is within the bounds of law, but I am seeking to withdraw any form of co-operation, be it economic, political or social, in which people or resources might be used for the reproduction of this injustice against individuals, activities, institutions, regimes, or countries with which we are in conflict.

One of my starting points is that the adversary remains in power due to some sort of consent of our own, explicit or not. Non-cooperation keeps within the limits of the law, but the usual forms of co-operation are withdrawn. In other words, from the point of view of the epistemology of non-violent civil resistance, we no longer defer to the sources which reproduce or provide legitimacy to injustice. We start from the principle and achievement of awareness of how our own bodies, material resources and social identities are the basis of 'consent' (explicit, silent or not) by means of which the adversary maintains him/herself in power (Sharp 1986: 72, 101–102, 162).

Gandhi continuously encouraged this form of life and struggle; his best known action is the boycott against English clothing (to the extent of burning it in great bonfires in public squares), simultaneously with the promotion of homespun textiles (*khadi*) as a symbol of the Hindu struggle for autonomy and self-government (*swaraj*). But he also asked Hindus to refuse to work for the British government, return their titles and honours of any type, and refuse to serve in the police or the army. His "equation for active resistance or *satyagraha*" was: "How is it possible that 100,000 English can control 300 million Hindus? They didn't take India from us, we gave it to them. Their control depends on Hindu collaboration, not on British control." And he added: "If all of us were to withdraw our co-operation, the English would have nothing left to do in India." In consequence, his strategy for increasing confrontation was: "We must stop doing what the English want us to do" (Ackerman/Duvall 2000: 61–102; York 2000).

At the same time, civil disobedience is one of the major instruments for the advancement of the human species. Without it we would still be stuck in a moral Stone Age. I propose openly and publicly violating a law that is considered illegitimate and immoral. Thus, civilian disobedience is not a destructive act resulting in social chaos but, on the contrary, is profoundly creative and ordered in terms of humanity.

For Gandhi, who was an important systematiser and contemporary actor in the conceptualisation and practice of civilian disobedience – which could be individual or collective, offensive or defensive – this "…is a stimulus for those who struggle and a challenge for the adversary. It should be clear that civilian disobedience in terms of Independence, without the co-operation of the masses by means of constructive efforts, is purely and simply bravado and worse than useless" (Ameglio 2002: 328).

In terms of content, Gandhi asserted that "the civilian violation of oppressive moral laws … the disobedience of these laws as a matter of conscience, not from fear of punishment… is an unalienable right of every citizen. To renounce this right means we stop being human" (Ameglio 2002: 155).

Both types of action – non-cooperation and civil disobedience – which are central to the construction of disobedient peace imply a much deeper comprehension of the relationship between legality and legitimacy or, in Gandhi's words: "What comes first? The law or conscience?"[23] To face this epistemic and moral dilemma, it is fundamental to know the historical processes which led to the enactment of the legal instrument, and how the dominant sectors have always striven to socially reinforce the "sacred nature" of the law (Ameglio 2018), masking the processes involved in its construction which, in many cases, were the result of a "victory of some over others", and even more so in its application which frequently represents, directly or indirectly, an "armed device" of those who exercise power.

The relationship between people and the law/regulations/institutions is one of the most complex issues we face in the construction of due disobedience to inhuman orders, because it represents a great epistemic obstacle which later becomes reflected in moral identity. Social order, as stated while describing social infantilism, builds the sacred character and the timelessness of the law, and whatever may emerge from it, as pillars to uphold obedience to authority. This is why, when we become aware of this, face it and challenge it, it is perceived as a very strong rupture, whether individually or collectively. So the construction of civil disobedience acts an end point and never a starting point, because it is necessary to have previously built all these epistemic reflections on rupture with the egocentric stadium in order to move on to the stage of co-operation, in which the law and its application are the fruit of consensus among equals. On many historic occasions, civil disobedience and non-cooperation – the principal radical actions in the non-violent civil resistance struggle – have lost effectiveness and have fallen into public disrepute as simple acts of violent provocation. This has come about because the actors and public opinion have not been adequately prepared by means of reflection on co-operation designed to undermine the sacralised character of the law. Omitting consideration of this massive epistemic obstacle has been the central reason for not being able to advance in the magnitude of necessary actions to build a disobedient peace capable of curbing at least part of the war in Mexico.

I also believe that the so-called 'disobedient' or 'globaliphobic' movements, initiated at least partially with the Zapatista uprising on 1 January 1994, have been important in terms of the search for updates on the meaning and practice of civil disobedience against neoliberalism in this century and within a non-Gandhian culture.

[23] Thoreau is very clear on this point: "The only obligation I have the right to shoulder is to do at all times what I consider proper", and he adds: "Thus, in the name of order and civilian government, we are made to honor our own baseness, and even to uphold it" (1997: 25, 31).

Zapatismo has undertaken to build a human experience on a social scale which strives to rule itself according to non-capitalist principles. This is what the Zapatista call 'autonomy', in which all 'good government' decisions are made by consensus and community co-operation, according to the principle of 'ruling while obeying'.[24]

The movements born in Seattle to oppose the WTO (November 29 – December 3, 1999) advocate active disobedience, bringing together direct action and civil disobedience to counteract dominant economic powers. By means of these acts, they proposed to undertake "public challenges conceived to show growing social support for our rebellion and the justice of our cause… Break the halo of naturalness that envelops the existing order, its institutions and results"[25] (Iglesias 2011: 16, 247).

6.6 Non-cooperation and Civil Disobedience in Mexico Today

To avoid simply falling into the same logical empiricism I have criticised, it is important to conclude this work with a couple of powerful examples of social struggles and disobedient peace which I have observed in present-day Mexico. These exemplify an updated version of the reality principle. The experiences of three decades form the total context of these reflections, which are both theoretical and practical. Two examples of non-cooperation in the sense of not obeying inhuman orders could, I believe, show the way to stop in part this war that pervades Mexico, and build a disobedient peace.

The first case concerns 3360 teachers belonging to the *National Coordination of Education Workers* (CNTE in Spanish); they are tenured teachers and live and work in twenty-eight states in Mexico. In 2015 they refused to sit for the Evaluation Exam and the Reposition Test as a clear conscientious objection to the inappropriately named 'Educational Reform' violently forced by the government, and for this reason they were fired on 1 March 2016. This repressive governmental act was a totally disproportionate punishment and without any direct link with the transgression committed, to which the dissident teachers responded by convening a 'national strike' from 15 May to 19 September 2016. Conscientious objections, boycotts and strikes are actions historically representative of non-cooperation. 'Conscientious objection', particularly, is a type of action originated in reflection concerning an injustice and subsequent public construction of a 'moral frontier' which is not susceptible to be *moved or crossed*. It is one of the central virtues of the education for and the epistemology of peace: to say 'Enough!' and to be capable of building an action which makes this challenge openly observable.

[24] See: https://es.wikisource.org/wiki/Discurso-del-subcomandante-Marcos-%22Mandar-obedeciendo%22.

[25] More in Negri (2006: 40–82) and Rabinovich/Magrini (2011).

A second action of exemplary non-cooperation in Mexico has been that of the National Brigades for the Search of Disappeared Persons[26] in clandestine graves since 2016, organised by their families in the whole north of the country, and in the states of Morelos, Veracruz, Jalisco, Sinaloa, Coahuila, Tamaulipas. After years of formal, legal efforts, the family organisations and their allies decided to go their own way, without waiting for the authorities – in many cases with a certain degree of involvement in the inhuman event under investigation – to provide the action and support that never seemed to arrive. They decided to exercise *their power*, as Gandhi would say, and in the words of Zapatista commander David, in Oventic, during the creation of the *Juntas* for 'Good Government' "without asking for permission" (August 2003). This is a very important form of 'autonomous non-cooperation'.

Concerning actions of civil disobedience in Mexico, I feel that the highway blockades carried out by the dissident teachers between May and September 2016 represented an indispensable form of struggle, given the level of official violence against them.[27] In particular, they enabled civilian society and the government to measure the level of support their struggle garnered in different communities – especially rural communities in the South East – based on the solidary support of students and their parents. This moral force placed its bodies on the highways, thus expressing their radicalisation, indignation and moral determination.

Another form of civil disobedience in Mexico is represented by what is known as 'duality of power', or 'parallel power' (Sharp 1986: 281–290), exercised openly in the Zapatista communities since 2003, in the fields of 'good government', health, education and production. It is one of the most radical forms. This mode of action has been followed in some other communities, especially in indigenous and rural populations in the South, with particular emphasis on issues related to public security and community guards or police forces which have been created (Horta/Aburto 2016; Fernandez 2014).

The challenge to non-violent civil resistance in the construction of a real disobedient peace involves working out: *How to activate the spiral of non-violent civil resistance without increasing the spiral of violence.*

One important non-violent weapon is the "moral reserve" (Ameglio 2011; Ameglio/Ramírez 2016) that resides in a given society, and gives this society the strength to draw a "moral frontier".[28] This social moral reserve can comprise very

[26] These were reinforced in 12–22 May 2017 by an International Caravan for the Search of Disappeared Persons. More at: https://aristeguinoticias.com/0709/mexico/suman-1307-las-fosas-clandestinas-halladas-de-2007-a-la-actualidad-cndh/.

[27] The Nochixtlán massacre, which involved the murder of teachers during demos, the jailing of leaders, and mass firings.

[28] The construction of epistemic and moral 'frontiers' in a dynamic, collective or individual way, and in permanent evolution according to new knowledge and reflection acquired, is fundamental for defining one's own identity and the degrees of 'disobedience' or 'obedience' one is prepared to face. In preparation for peace (which includes education, culture and construction) this is one of the most complex tasks to install, in oneself and in others. The difficulty is directly related with the degree of knowledge available concerning inhumanity –establishment of violence as a normal state of affairs—contained in the social order. See more on this in Ameglio (2002: 197–210).

diverse social identities – including some which are at odds with each other – capable of mobilising themselves in specific historic situations in which they consider that, in some sense, a certain moral line of the given social order has been crossed. In many societies this shifting of the "moral frontier" in the face of situations of gross inhumanity which people are not willing to accept as normal has historically caused great masses of people to take to the streets, or individuals whose bodies and identities concentrate a considerable 'social power' (academic and religious leaders, peasant, indigenous or labour leaders, intellectuals, artists) also to emerge to pressure authority, or the rest of the population to express their massive cry of moral and material indignation: 'Enough!' (1 January 1994), "We are fed up!" (13 April 2011), "I am #132" (11 May 2012), "It was the State!" (September 2014), and "Stop gasoline hikes!" (January 2017). These five have been cries of moral and material indignation by Mexican society in recent years, in the face of highly inhuman situations, such as the condition of the indigenous peoples of Chiapas, the victims of the war of extermination, the violations of democracy, the forty-three students who disappeared during a genocidal action in Iguala (State of Guerrero) on the night between 26 and 27 September 2014, and the serial increases in the price of gasoline. These events triggered an important and radical set of reactions conceived from non-violence, and created the conditions for the construction of a disobedient peace.

It is also true that another factor which partly explains how the spiral of war and social violence in Mexico has grown during the past decade has been that – in the face of other social events revealing brutal inhumanity which, in my view, pushed back the moral frontier towards a greater acceptance of barbarianism and impunity – a considerable part of the national moral reserve did not take to the streets with their bodies in non-violent actions (non-cooperation and civil disobedience) proportional to the level of violence. These actions should exhibit an important trait of disobedient peace and non-violence which could be termed "permanent firmness" (Barbé 1977): "We are not going to move until there is truth, justice and reparation". In other words, "put our bodies on the street for an undetermined time", which would transform these social subjects into a powerful "moral weapon".[29]

I will mention only the events I consider of exceptional inhumanity and which should not be allowed to happen without radical consequences (truth, justice, reparation, resignations, jail…) for power and its representatives: the massacre of forty-nine infants at the ABC Daycare Center in Hermosillo (State of Sonora) on 5 June 2010); the massacre of sixteen young people at a party in Villas de Salvárcar, Ciudad Juarez (State of Chihuahua) on 31 January 2010; the massacre of seventy-two migrants in San Fernando (State of Tamaulipas) on 22 August 2010; the massacre of fifty-two people at the Casino Royale in Monterrey (State of Nuevo Leon) on 26 August 2011; the execution of twenty-two people in Tlatlaya (State of Mexico) on 30 June 2014 by government forces; the disappearance of the forty-three students from the *Ayotzinapa*

[29]"The way to fully interpose a body is to make it a *thinking* body, so that the strength of this body is multiplied, because each part of it is going to behave according to what the environment, as defined by reflection, requires. It will be hugely consistent: the moment of reflection and action coincide. … Everything works. This is a *moral weapon*" (Marín 1995: 26).

Rural School in Iguala (State of Guerrero) on 26–27 September 2014. Of course, the list of inhumanities in Mexico in the past decade is unfortunately much longer, but I wished to highlight the previous examples as significant cases of levels and quantities that a society cannot accept as normal in their impunity without consequent changes in the power structure, without truth, justice and reparation.

Thus, apart from the epistemology of disobedience outlined at the beginning of this article, the individual and collective conceptions and construction of force, reserve, frontier, rupture and moral weapons are also central factors for building, reflecting and taking action for a disobedient peace based on the principles of non-violent civil resistance. It is a task for millennia, as is the improvement of the human species, but there are also multiple events in the short and medium term which provide opportunities to start building it. This is the quest of many more than we think, and it is my hope that we can increasingly *think aloud,* and *think together*. This is my hope, which stems from a reality principle, humble but genuine.

References

Ackerman, Peter; Duvall, Jack, 2000: *A Force more Powerful: A Century of Nonviolent Conflict* (New York: St. Martin's Press).

Ameglio, Pietro, 2002: *Gandhi y la desobediencia civil: México hoy* [*Gandhi and Civil Disobedience: Mexico Today*] (Mexico City: Plaza & Valdés).

Ameglio, Pietro, 2011: "La reserva moral mexicana sale a la calle" ["Mexican Moral Reserve Takes to the Streets"]; at: https://elperiodicodemexico.com/nota.php?id=472627 18 April 2011.

Ameglio, Pietro, 2016: "¿Cómo construir la paz y reflexionar sobre ella en medio de la guerra en México?" [How to Build Peace and Reflect upon it in the Midst of War in Mexico?], in: Sicilia, J.; Vázquez, E. (Eds.): *El movimiento por la paz con justicia y dignidad* [*The Movement for Peace with Justice and Dignity*] (México: Edics. Era).

Ameglio, Pietro, 2018: "México: la Ley como arma del poder y la Guerra" [Mexico: The Law as Weapon of Power and War]; at: https://desinformemonos.org/mexico-la-ley-arma-del-poder-la-guerra/ (9 February 2018).

Ameglio, Pietro; Ramírez, Tania (Eds.), 2016: *¿Cómo construir la paz en el México actual? Textos, autores y preguntas sobre construcción, educación y cultura para la paz* [*How to Build Peace in Mexico Today? Texts, Authors and Questions on Construction, Education and Culture for Peace*] (Mexico City: Plaza & Valdés).

Antón, Gustavo; Damiano, Franco, 2010: *El cuerpo, territorio del poder* [*The Body as a Territory of Power*] (Buenos Aires: P.I. Ca. So).

Arendt, Hanna, 2005: *Sobre la violencia* [On Violence] (Madrid: Alianza Edit.).

Arendt, Hanna, 2005: *Eichmann en Jerusalén: un estudio acerca de la banalidad del mal* [*Eichmann in Jerusalem: A Study on the Banality of Evil*] (Barcelona: Ed. Lumen).

Bourbaki Team, 2011: "El costo humano de la guerra por la construcción del monopolio del narcotráfico en Mexico (2008-2009)" ["The Human Cost of the War for the Construction of the Monopoly of Drug Trafficking in Mexico (2008-2009)"] (Cuernavaca: SERPAJ)

Barbé, Domingo, 1977: *A Firmeza Permanente. A Forca da Nao-Violencia* [*Permanent Firmness: The Strength of Non-Violence*] (Sao Paulo: Edit. Loyola-Vega).

Bleichmar, Silvia, 1995: *Pensar en voz alta* [Thinking Aloud] (Buenos Aires: P.I.Ca.So).

Canetti, Elias, 1983: *Masa y poder* [The Masses and Power] (Madrid: Alianza Editores).

Centro de Derechos Humanos Fray Bartolomé de las Casas, 2015: *Informe: La Realidad, contexto de Guerra* [*Inform: Reality: The Context of War*] (San Cristóbal, México).

Fernández, Paulina, 2014: *Justicia Autónoma Zapatista: Zona Selva Tzeltal* [Autonomous Zapatista Justice: Tzeltal Jungle Zone] (Mexico City: Edics. Autónomas).
Foucault, Michel, 1976: *Vigilar y Castigar: Nacimiento de la Prisión* [*Guard and Punish: The Birth of Prisons*] (México: Siglo XXI Edits.)
Fracchia, Myriam, 2014: "Juan Carlos 'Lito' Marín y la desobediencia debida a toda orden inhumana" [*Juan Carlos 'Lito' Marín and Due Disobedience to any Inhuman Order*], in: Marin, J.C., 2014: *Conocimiento y desobediencia a toda orden inhumana* [*Knowledge and Disobedience of Any Inhuman Order*] (Cuernavaca: Morelos State Autonomous University).
Fracchia, Myriam, 2018: "Propuesta de un enfoque basado en aportes socio-psicogenéticos para la formación en investigación-intervención sobre la relación entre violencia escolar y violencia social" ["Proposal for an Approach based on Socio-Psychogenetic Contributions on the Relationship between School Violence and Social Violence"], in: Velázquez, Guadalupe (Ed.): *El paradigma de la complejidad y la investigación de los problemas sociales, educativos y culturales del siglo XXI* [*The Complexity Paradigm and Research on Social, Educational and Cultural Problems in the Twenty-First Century*] (Mexico City: Universidad Pedagógica Nacional): 139–152.
Fromm, Erich, 2006: *El miedo a la libertad* [Fear of Liberty] (Buenos Aires: Ed. Paidós).
Gandhi, Mohandas, 1985: *En lo que yo creo* [*What I Believe In*] (México: Dante).
Horta, Juan; Aburto, Sabas, 2016: *CRAC-PC: El origen de la Policía Comunitaria: Montaña y Costa Chica de Guerrero* [*CRAC-PC: The Origin of Community Police*] (Mexico City: The authors)
Iglesias, Pablo, 2011: *Desobedientes: De Chiapas a Madrid* [*Disobedient: From Chiapas to Madrid*] (Madrid: Editorial Popular).
Kloster, Karina; Fracchia, Myriam, 2017: "El costo humano en la construcción de los dominios territoriales" ["The Human Cost in the Construction of Territorial Domains"], in: Kloster, Karina (Ed.): *Conflictividad y violencias en América Latina* [*Conflict and Violence in Latin America*] (Mexico City: UACM).
Laski, Harold, 2011: *Los peligros de la obediencia* [*The Dangers of Obedience*] (Madrid: Edics. Sequitur).
Lederach, John Paul, 1986: *Educar para la paz* [*Education for Peace*] (Barcelona: Fontamara).
Libera México, 2013: *México, la guerra invisible: Historias, cifras y negocios de los cárteles criminales y la impunidad de las mafias mexicanas* [*Mexico, the Invisible War: Stories, Figures and Businesses of Criminal Cartels and the Impunity of Mexican Mafias*] (Mexico City: Libera).
Marín, Juan Carlos, 1995: *Conversaciones sobre el poder: una experiencia colectiva* [*Conversations on Power: A Collective Experience*] (Buenos Aires: Buenos Aires University).
Marín, Juan Carlos, 2009: *Cuaderno 8: Leyendo a Clausewitz* [*Copybook 8: Reading Clausewitz*] (Buenos Aires: Buenos Aires University).
Marín, Juan Carlos, 2014: *Conocimiento y desobediencia a toda orden inhumana* [*Know-ledge and Disobedience to any Inhuman Order*] (Cuernavaca: Universidad Autónoma del Estado de Morelos).
Martínez, Carlos, 2012: *De nuevo la Vida: El poder de la Noviolencia y las transformaciones culturales* [*Life Again: The Power of Non-Violence and Cultural Transformations*] (Bogotá: Uniminuto).
Milgram, Stanley, 1980: *Obediencia a la autoridad: un punto de vista experimental* [*Obedience to Authority: An Experimental Point of View*] (Bilbao: Desclée de Brouwer).
Moore, Barrington, 1989: *La injusticia: bases sociales de la obediencia y la rebelión* [*Injustice: Social Bases of Obedience and Rebellion*] (Mexico City: UNAM).
Negri, Antonio, 2006: *Goodbye Mr. Socialism* (Milán: Feltrinelli).
Olivares, Emir, 2019: "Michelle Bachelet: La cifra de muertes violentas en México, de un país en guerra" [*Michelle Bachelet: The Number of Violent Murders in Mexico, a Country at War*]; at: https://www.jornada.com.mx/2019/04/10/politica/003n1pol (10 April 2019).
Piaget, Jean, 1976: *La toma de conciencia* [*Becoming Aware*] (Madrid: Edics. Morata).
Piaget, Jean, 1985: *El criterio moral del niño* [*Moral Criteria in Children*] (Mexico City: Rocas Edics.).

Rabinovich, Eleonora; Magrini, Ana Lucía (Eds.), 2011: *"Vamos a portarnos mal": Protesta social y libertad de expresión en America Latina ["Let's Misbehave": Social Protest and Freedom of Speech in Latin America]* (Bogotá: Centro de Competencia en Comunicación para América Latina).

Randle, Michael, 1998: *Resistencia civil* [*Civil Resistance*] (Barcelona: Paidós).

Sharp, Geene, 1986: *Politica dell'Azione Nonviolenta 2: Le techniche* [*Policy of Non-Violent Action 2: Techniques*] (Turin: Edizione Gruppo Abele).

Thoreau, Henry, 1997: *Del deber de la desobediencia civil* [*On the Duty of Civil Disobedience*] (Buenos Aires: Dissur Edics.).

York, Steve, 2000: "A Force more Powerful: Marcha de la Sal"; online video at: file:///C:/Users/W10x64/Desktop/Gandhi%20y%la%20marcha%20de%la%20sal%20-%20youTube.html.

Part II
Conflicts, Peace, Gender, Families and Vulnerable People

Chapter 7
Risks, Mitigation and Adaptation to Urban Climate Change Impacts in the Global South from a Gender Perspective

Úrsula Oswald Spring

Abstract Confronted with the uncertainty of climate change impacts and the complexity of urban development in the Global South, megacities and urban agglomeration in coastal areas are at high risk of losing lives, wealth and infrastructure. The present chapter explores the Talanoa Dialogue to reduce impacts and increase peacebuilding in regions where climate change effects will increase disasters in urban areas and often oblige people to leave high-risk areas and settle elsewhere. It further analyses how governments and city-dwellers could promote multi-level governance actions to reduce conflicts and promote disaster risks reduction, mitigation and adaptation among extremely exposed social groups, especially women, children and the elderly. Talanoa Dialogue also promotes the interchange of experiences among cities in the Global South and proposes multisectorial climate governance for achieving successful political changes with participative governance that are able to improve resilience. Finally, the chapter examines how cities could promote clean energy, sustainable transport and a radical carbon sink of greenhouse gases to prevent conflicts and disasters by involving all stakeholders within a frame of cultural and socio-environmental diversity.

Keywords Talanoa dialogue · Climate change · Disasters · Gender · Global south · Liveable megacities · Socio-environmental vulnerability · Greenhouse gas emissions · Resilience · Mitigation · Adaptation · Multi-level participative governance · Local climate governance management

7.1 Introduction and Research Questions

Urbanisation has become a major development of the twentieth and twenty-first centuries. In 2019, 55% of people lived in urban settlements and this number is predicted to increase to 68% by 2050. UN DESA (2018) notes that the future increase

---

Prof. Dr Úrsula Oswald Spring, research professor, CRIM-UNAM, Cuernavaca, Morelos, Mexico. Email: uoswald@gmail.com.

Ú. Oswald Spring and H. G. Brauch (eds.), *Decolonising Conflicts, Security, Peace, Gender, Environment and Development in the Anthropocene*,
The Anthropocene: Politik—Economics—Society—Science 30,
https://doi.org/10.1007/978-3-030-62316-6_7

is expected to be highly concentrated in India, China and Nigeria, which will account for 35% of the estimated urban growth. Globally, the urban population has grown rapidly since 1950 from 751 million to 4.2 billion in 2018. This concentration of people also has serious implications for the risks from climate change and for adaptation. The IPCC Cities and Climate Change Science Conference in Edmonton, Canada in 2018 estimated that cities emit 70%yof *greenhouse gases* (GHG) related to food, electricity, clothing, electronics, local and air travel, construction materials, and other factors.

The UN Climate Action Summit 2019 estimated further that cities consume more than 78% of energy, and noted that climate change has greater impacts in urban areas due to population density. However, energy, fuel, transport, food and waste are managed differently by men and women, and vulnerable women and children, especially new immigrants, are highly exposed to climate risks (UNRISD 2019). Further, 90% of the global urban areas are situated on coastlines and are at risk from short-term (hurricanes, storms, floods, bushfires) and long-term impacts of climate change (sea level rise, coastal erosion, intrusion of salt water into coastal aquifers, etc.). On the other hand,

> Wealthy 'consumer cities', such as London, Paris, New York, Toronto and Sydney, that no longer have large industrial sectors have significantly reduced their local emissions. However, when the emissions associated with their consumption of goods and services are included, these cities' emissions have grown substantially and are among the highest in the world on a per person basis… Meanwhile, 'producer' cities in India, Pakistan, or Bangladesh generate lots of industrial pollution and carbon emissions in the manufacture of products that will be sold and consumed in Europe and North America (Leahy 2018:2).

In most Southern countries prospective studies on climate change, if they exist, are generated in a fragmented way, with little synergy between resource availability, actors, potential threats, resource conflicts and governance. These inconsistencies in the collection of knowledge, combined with insufficient incorporation of GHG monitoring and inadequate direct measurement of the current state of risks, increase the threats in marginal areas. Prospective analyses of complex interactions require serious strengthening in terms of the production, accessibility and quality of primary data sources and information, both nationally and internationally. The creation of robust local scenarios in terms of both potential implications (IPCC 2013) and efficient solutions is further limited because research is still insufficient, especially with regard to interdisciplinary efforts around the interface of, on the one hand, the biophysical, and on the other, the socio-economic, political, gender and cultural aspects (IPCC 2014). Research into all aspects is crucial for the advancement of transversal and transectoral public policies that are capable of reducing vulnerability and increasing the resilience of everybody, including the vulnerable. Similarly, we need more data and discussions about what could happen prospectively, and what the implications would be for megacities and countries with a temperature rise scenario of 1.5, 2 °C or more.

In the Global South, informal urban settlements are often not integrated or considered in public policies of mitigation and adaptation to climate change (Delgado et al. 2018). The dominant climate change agenda in the Global North tends to focus more on smart and exposed cities (Leal-Arcas 2018) by exploring expensive technological solutions – drones, sensors, big data and other cutting-edge technologies – but does not reflect the complex nexus between governance, *disaster risk reduction* (DRR) and the reality of the informal cities that are also growing daily in poor neighbourhoods in developed countries (Sander/Amling 2019). In the Global South, with the exception of Singapore and Hong Kong and some better-off neighbourhoods, the reality is not a 'smart city', but an 'informal city' with a lack of or deficient public services, situated in high-risk zones in steep mountains (e.g. La Paz), on river beds (e.g. Mexico City) or in low-lying coastal areas (e.g. Mumbai; Rahman/Hickey 2019). These adverse environmental conditions are aggravated by socio-environmental vulnerability (Oswald Spring 2016), which limits possible routes of resilience-building and DRR capacity-development for these marginal people (IPCC 2012). Therefore, there is a need for new approaches that are capable of reducing the existing socio-environmental risks by promoting adaptation capacities and resilience to the existing and upcoming climate change threats (O'Brian et al. 2008) while taking into account cultural diversity across the globe (Arizpe 2019). Thus, these adaptation processes need to be tailored to suit different regions, particularly in developing countries further limited by scarce public funds and often insufficiently prepared public authorities (Victoria State Government 2018).

Confronted with the uncertainty of climate change impacts and the complexity of global urban development, during the **COP23 of UNFCCC**, the Presidencies of COP22 and COP23 presented the outcome of their consultations on the dialogue developed and made it available to all Parties. The COP welcomed the design of the facilitative dialogue known as the 'Talanoa Dialogue', and the interchange between its participants was launched globally, starting in January 2018. For COP25 in 2020 in Glasgow, the dialogue asserted that "multilateralism and co-operation will enable us to address problems together, find solutions, and build consensus for the common good. Only a global coalition of actors—including Parties, national and sub-national governments, private sector companies, the investment community, civil society and all non-Party stakeholders—can take us there."[1]

7.1.1 Research Questions

Confronted with these complex processes of urbanisation in the Global South within a scenario of a hotter climate and unpredictable impacts, the present chapter explores the following research questions:

1. How can Fiji's *Talanoa Dialogue* (TD) reduce climate change impacts and increase peacebuilding by dialogue, when climate change effects will increase disasters in urban areas and oblige people to leave high-risk areas and settle down elsewhere?

[1] See at: https://unfccc.int/sites/default/files/resource/Talanoa%20Call%20for%20Action.pdf.

2. How could governments and city-dwellers promote multi-level actions to reduce the impacts of climate change and promote DRR and adaptation among highly exposed women and children?
3. How could cities finance the mitigation costs of clean energy, transport and a radical carbon sink to reduce the impacts of climate change, and how could involving all stakeholders, especially women, in participative governance avoid conflicts when seeking to achieve these goals within a diverse cultural and socio-environmental setting?

7.1.2 Organisation of the Chapter

After this brief introduction, the chapter introduces the Talanoa approach (Sect. 7.2) and analyses the most recent data on *global greenhouse gas* (GHG) emissions (Sect. 7.3). Section 7.4 links the evolution of (mega-)cities with population growth, informal settlements, gender and youth relations and the rise in GHG emissions. Without any doubt, cities and especially megacities produce most of the GHG emissions and have the potential to mitigate them substantially, but cities require local emission policies and financial support to implement these strategies, especially in the least developed countries (Sect. 7.5). The Talanoa approach explores the dialogue between cities in the same country, neighbour states and other regions (Sect. 7.6) and proposes multi-sectorial and multi-level climate governance (Sect. 7.7). Finally, some conclusions integrate the research questions and propose some concrete local and regional practices to achieve efficient DRR and a successful climate change policy with participative governance to reduce the risks of climate change impacts (Sect. 7.8).

7.2 Introduction to the Talanoa Approach

Urban planning is diverse, with visions of experts and levels of government that repeatedly enter into tension or conflict (Delgado et al. 2018). As a result of the clash of interests, cities in the Global South have been poorly planned, but are mostly exposed to anarchic, dispersed and disconnected growth, which increases the costs of public services (Strohbach et al. 2019). In most developing cities, basic public services lack quality, are intermittent or do not exist. They are often destroyed during an extreme event, thus increasing the risks for the residents. Poor dwellers, who live in high-risk settlements, have little resilience to the growing impacts of climate change (Brewera/Granta 2015). Thus, the Talanoa approach promotes collaboration between cities in the Global South to exchange and learn from positive experiences.

Talanoa is a term used in Fiji and across the Pacific to promote inclusive, participatory and transparent dialogues in which ideas, skills and experiences are shared. This process should increase empathy, understanding and collaboration when different observations or actions are discussed with mutual respect. Potential actions are based on consensual decision-making processes that support everybody, and they may be able to increase the resilience of the poorest people in the most adverse socio-environmental conditions. The Talanoa approach was launched under the Presidency of Fiji to help poor countries enhance their Nationally Determined Contributions (NDC) by 2020 and to achieve the goal to limit the rise in average global temperature to well under 2 °C above pre-industrial levels, thus to pursue efforts to limit the increase to 1.5 °C. The Talanoa Dialogue was created as a political momentum for COP25 in 2020, at which political leaders can agree on what to do and how to achieve the global fight against the increase of GHG and the impacts of climate change. The Talanoa Dialogue:

> is a mandated process requested by Parties to take stock of collective efforts to reduce emissions and build greater resilience, in line with the long-term goals of the Paris Agreement and to prepare updated or new NDCs by 2020. Ultimately, the goal is to help Parties achieve maximum ambition in implementing and improving their NDCs…The central goal of the Paris Agreement is to achieve net zero emissions as soon as possible and in the second half of the century, and to contain the rise in average global temperatures to well below 2 degrees Celsius, while pursuing efforts to achieve a rise no greater than 1.5 degrees Celsius. The Fijian Presidency is calling for the formation of a Grand Coalition to build momentum to achieve the more ambitious target of 1.5, and to reduce net carbon emissions to zero as soon as possible."[2]

7.3 Global GHG Emissions (GT CO_{2e}/year)

Cities are the greatest users of energy and produce the most GHG. The promotion of actions may reduce the impacts of climate change and support DRR and multi-level adaptation among highly exposed social groups, especially women and children. To achieve these goals, urban authorities and citizens need to streamline their use of energy and embrace an efficient transport system that is able to take into account the different travel requirements of men and women.

The IEA (2019) estimates that global GHG emissions in 2018 rose 1.7% to a historic 33.1 $GtCO_{2e}$. Emission from all fossil fuels increased and the power sector accounted for almost two-thirds of the emission growth. Especially serious is the use of coal (10 $GtCO_{2e}$), mostly emitted in China and India. With regard to keeping the increase in GHG emissions below 2 °C by 2030, there is a gap of 17 $GtCO_{2e}$ and regarding the *Nationally Determined Commitment* (NDC) made by countries, there still exists a gap of 14 $GtCO_{2e}$ (Fig. 7.1). Both gaps indicate further that the agreement of less than 2 °C will not be achieved, if drastic actions are not taken now. With an average increase in temperature of 2 °C, coral reef will be lost across the globe, and urban coastal cities and most of the *small islands developing states* (SIDS) will lose a significant part of their territories due to sea level rise and seawater intrusion into their aquifers. Several SIDS will be covered by salt water and people will be obliged to emigrate, losing their country and their nationality.

[2]See at: https://cop23.com.fj/talanoa-dialogue/.

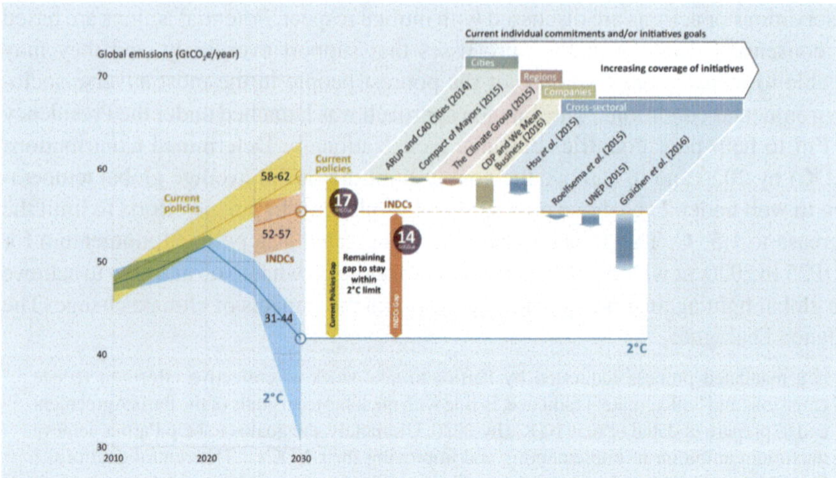

Fig. 7.1 Estimated global GHG emissions (GT CO_{2e}/year). *Source* Höhne et al. (2016: 29)

Thus, there is an ethical problem resulting from the dissonance between the greatest emitters (industrialised countries) and the most affected least developed countries in the Global South. Therefore, the present situation of gaps in the Paris agreements requires a greater understanding of climate pathways and impacts in different timescales and cities in order to curb the emissions of GHG. Including all urban stakeholders, a climate policy could enable Parties in the COP25 to establish long-term climate strategies to maintain the temperature rise to 1.5 °C. This information could also support Parties to develop NDC or adjust medium-term strategies to align their commitments with the long-term strategies, in order to avoid climate catastrophes and maintain the temperature far under 2 °C.

In order to reach the low concentration stabilisation level of 450 ppm CO_{2e} agreed in Paris in 2015, regional reduction targets differ significantly for each effort-sharing approach. Based on the emissions in 1990, the Organisation for Economic Co-operation and Development (OECD 2011) proposed a new regional approach and emphasised the equity principles of responsibility, capability and need. Equal cumulative per capita emissions carbon budgets may lead to significant emissions reduction targets for 2030. These "would be approximately half of the emissions of 2010 with a large range, roughly two-thirds in the *Economies in Transition* (EIT), roughly at the 2010 emissions level or slightly below in Asia, slightly above the 2010 level in the Middle East and Africa and well below the 2010 level in Latin America" (Höhne et al. 2014: 122). This equity approach includes fairness, for which many different effort-sharing approaches have been proposed to support the economically weaker developing countries (Fig. 7.1).

Van den Berg et al. (2019) suggested equal cumulative per capita emissions, contraction and convergence, grandfathering, greenhouse development rights and ability to pay. These authors observed that when we:

(i) calculate required reduction targets in carbon budgets (relative to baseline budgets) and/or (ii) take into account historical emissions when determining carbon budgets this can lead to (large) negative remaining carbon budgets for developed countries. This is the case for the equal cumulative per capita approach and especially the greenhouse development rights approach. Furthermore, for developed countries, all effort-sharing approaches except grandfathering lead to more stringent budgets than cost-optimal budgets, indicating that cost-optimal approaches do not lead to outcomes that can be regarded as fair according to most effort-sharing approaches" (Van den Berg et al. 2019: 1).

With effort-sharing the difference between carbon budget and emission pathways could be avoided. Calculations differ mostly as a function of the different effort-sharing approaches. Van den Berg et al. (2019) also proposed an allocation of negative emissions allowances or budgets, which may stimulate emitters to reduce their GHG emissions further. These global emission balances will help to make the global measurement of GHG for each industrial, public and social sector more transparent.

7.4 Cities, Population, Women, Youth and GHG Emissions

Between now and 2050, 90% of urban growth will happen in developing countries, predominantly in Africa and South Asia. The World Bank (2018) noted that cities generate 80% of global wealth, but consume more than two-thirds of the world's energy. Rapid growth in the Global South necessitates affordable housing, public transportation systems and basic services, including the creation of new jobs for the younger generations. The yearly investment projected to build climate-smart and resilient cities is about 6 billion USD/year.

Women produce very different profiles of GHG emissions, due to their multiple task activities as workers, mothers, carers and providers of food, fuel and water. They have genuine need of transport, water, fuel for cooking and sewerage. However, in most poor countries, public transport is underdeveloped and mostly orientated to male workers, who travel to their daily work. This transport system rarely takes into account the multiple tasks a women with children and extended family has to solve daily, and public transport in the Global South is often chaotic and insecure for women and children (UNRISD 2019). Women face further structural barriers, entrenched in laws that keep them out of jobs and prevent them from owning a business (IBRD/WB 2019). The World Bank's *Women, Business and the Law Report* (2019) found that 104 countries prevent women from working in certain jobs, 58 have no prevention against sexual harassment and in 18 economies husbands can prevent women working. The regions with higher discrimination against women are East Asia and the Pacific, the Middle East and North Africa, South Asia and Sub-Saharan Africa. The second highly vulnerable groups are young female and males in all countries, but especially in the Global South. They not only have the highest rate of under- and unemployment, but young men are also exposed to criminal violence and young women and girls to femicide, rape and trafficking (UNICEF 2018). Diwakar et al. (2019) also asserted that any poverty alleviation programme in Kenya and

India should tackle the structural factors of chronic poverty, avoid trained youths falling back into poverty traps due to lack of employment, and ensure that adaptive capacities include resilience-building against environmental shocks and stresses.

7.4.1 Population and Youth Will Be Urban

By 2050, two-thirds of the planet's population will live in urban centres. Nearly 90% of the 2.5 billion new urban dwellers will live in Africa and Asia. The world's urban areas are responsible for more than 70% of GHG emissions. A crucial theme is youth migration to cities. People in their teens and early twenties prefer to live in cities because they are stimulating, full of energy and offer greater prospects for jobs and income (UNICEF 2018). Cities provide youth opportunities for personal development and a shift towards modernity, but urban areas are also sites of a catastrophic form of youth marginalisation, violence and exclusion. Africa is the fastest changing region towards urbanisation and is predicted to remain rural only until 2030 (UN DESA 2018). Sub-Saharan Africa currently remains one of the least urbanised regions in the world. Caraël and Glynn (2008: 124) point out that "urban populations of Sub-Saharan Africa have increased by 600% in the last 35 years: a growth rate which has no precedent in human history." As Lubuva (2004: 4), a Tanzanian government official, asserted: "urban local authorities have very little revenue of their own, far less than what they would require to keep pace with the rate of urbanization."

On the other hand, in developing countries the private economy is mostly tiny, racked by corruption and nepotism, and does not provide economic opportunities for everybody, but only for a few people from the upper class and governmental families. This is not just true in Africa, but in most other cities of the Global South, where, as a result, informality "has become a vital facet of African urban life in the sense that it is predominantly driven by informal practices in such areas as work, housing, land use, transportation and a variety of social services" (Konings et al. 2006: 3).

Most megacities and medium-sized urban settlements are under-served in the Global South and confront fiercely competitive economic environments further impacted by the neoliberal development policies imposed on developing countries by the IMF due to their high indebtedness, often as a result of corruption and capital flight to tax havens. The pressures and dangers that face young men and women can be extreme, and Latin America is the most violent region without war, due to the organised crime and high level of femicides. To reduce this violence, the needs of the marginalised youth majorities in the urban context should be addressed more effectively. Two in three urban residents obtain their livelihoods from the informal economic sector, which is thought to be growing at an annual rate of 7%. Sommers (2010) estimates that in the near future more than 90% of jobs will be provided through informal economies in Africa, and elsewhere there exist similar conditions for the marginal youth.

Manhood is difficult in towns in the aftermath of economic crises. From 1980 onwards, "many young men lost even the possibility to establish themselves as adults, by building a house or getting married—though they continued to become fathers of children for whom they could not provide" (Utas 2005: 150). The African conditions are similar in Latin America and South Asia, where young people try to survive in precarious life conditions in urban marginal settlements, often highly exposed to climate disasters, especially landslides, floods and related diseases. Further, USAID (2005: 7) asserted that "urbanisation concentrates precisely that demographic group most inclined to violence: unattached young males who have left their families behind and have come to the city seeking economic opportunities."

To reduce this urban violence and offer both male and female youngsters the opportunity to thrive in an urban context, development strategies in the Global South should focus first on those who are most in need. Generally, they represent the majorities in the Global South. Ensuring better education and special programmes that give these youths the potential to develop their own capacities may help to overcome the existing violence by gangs or organised crime, and also reduce intrafamiliar violence through training courses and peacebuilding training.

7.4.2 Women in Marginal Urban Areas in Developing Countries

Women tend to have significantly fewer economic options than their male youth counterparts, and often the sole solution for survival is prostitution. They "…become pregnant by their boyfriends in the hope that this would cement the bonds between them. They saw pregnancy as one of the only ways to escape the stigma that marked their lives… many men reported they were more likely to marry a woman after she had a child by him and proved that she was fertile…." (Moyer 2006: 186). Early pregnancy among girls is also related to their lack of social role inside the urban marginal family. Once she has a child, generally the family accepts the girl as mother and integrates her into the daily activities of caring, fetching water and fuel, cooking and looking after the elderly, thus she receives some protection from the extended family in urban contexts (Menkes Bancet/Sosa-Sánchez 2018).

Gender violence has the highest rate in Latin America and is extremely high in Brazil and Mexico. During 2019, 2833 women were victims of femicide in Mexico (at least eight women a day), but only 25.6% were investigated as femicide and most of the perpetrators (more than 80%) are former or present intimate partners of the victims (Sistema Nacional de Seguridad Pública 2019). The Pan-American Health Organization (Bott et al. 2012) asserted that gender violence has negative consequences for women because of physical injury, unwanted pregnancy, abortion, sexually transmitted infections (including HIV/AIDS), maternal mortality, post-traumatic stress disorder, depression and suicide. When the assassinated women is a mother, children are often left without proper care, and sometimes grandmothers take on the role of raising these traumatised children.

However, gender violence is common all over the world and affects 35% of women globally, with 38% of murders of women committed by an intimate partner. It is the result of lack of legal reinforcement against intrafamiliar violence, tribunals without gender sensibility, and cultures where a violent manhood is promoted among young boys. This violence "is not only devastating for survivors of violence and their families, but also entails significant social and economic costs. In some countries, violence against women is estimated to cost countries up to 3.7% of their GDP—more than double what most governments spend on education."[3] These scarce public resources used in the health system could be invested in prevention, climate mitigation and adaptation.

7.5 Cities Concentrate Most of the GHG Emissions

If current trends continue, in 2050 the GHG emissions in urban areas could grow by 50%. Thus, while urban areas not only pollute seriously, they also offer the potential to substantially reduce GHG emissions through collaboration with all stakeholders: citizen groups, governments, enterprises, scientists and technological improvements. Therefore, cities are key actors for reducing GHG emissions. Cities, and especially megacities in developing countries, are also highly exposed to climate change impacts, but, at the same time, they are the first-responders in a disaster and they are also the first to experience trends of growing risks.

The fifty largest cities in the world combined rank third in both population and GHG emissions, and second in GDP when compared with the largest and wealthiest countries. However, in per capita emissions large cities are quite efficient. It is further estimated that 89% of the increase in CO_{2e} from energy use will be produced in developing countries (IEA 2014). In synthesis, richer cities, extended cities with low density, and cities that produce energy predominantly with coal all emit more greenhouse gases than smart-green cities.

A crucial polluter in cities and megacities is the transportation system (Delgado 2012). Citizens' bicycle ownership and a supportive culture for cycling are essential for the success of cycling across different city administrations. To be successful, projects of this importance should be included in and supported by broader urban development and mobility strategies, such as landscape planning, sustainable management of waste and sewage, and a substantial reduction in public violence in the streets. The support by high-level local government officials, particularly mayors of committed cities, is key to the success of ambitious infrastructure projects, such as the Quinto Centenario Project in Colombia. Bogotá is consolidating its role as a leader on cycling and sustainable mobility in Latin America and a model for the world.

[3]See at: https://www.worldbank.org/en/topic/socialdevelopment/brief/violence-against-women-and-girls.

The creation and updating of an overarching document that outlines the project's progress is a crucial project management tool. Cities interested in replicating projects ought to carefully consider the objectives of any new institutional framework before implementing their own efforts. Without any doubt, successful climate change mitigation and adaptation in cities requires not only the involvement of all stakeholders, but also the full support of the local authorities and a long-term planning process that is independent on the political agenda of changing authorities. Independent academic evaluation of the advance of the projects is crucial to achieve sustained success and the possibility of adapting errors and modifying inconsistencies.

7.5.1 Cities Require Local Climate Change Management

The German government proposed to its municipalities a local climate change management approach embedded into a national climate protection initiative through a process of collective analysis and planning by all stakeholders and local and federal investment in the implementation of the concrete actions. The federal government gives further advice and support for specific actions, and funding is offered after the application of concrete local projects. The availability of enough funds is crucial for the success of climate strategies at local level. In developing countries, the obstacles are related to weak links between federal, state and local governments and lack of funds for promoting at local level new models of climate change mitigation and adaptation that involve all stakeholders.

The World Bank (2019a) stated that the fifty largest cities (based on population alone) have combined economies second only to the US and larger than the entire GDP of China or Japan. These fifty cities with 500 million people generate about 2.6 billion tons of CO_{2e} annually – more than all countries except the US and China. The 10 top GHG emitters have about the same emissions as Japan. New York has the highest global total of GHG emissions, but per capita its GHG emissions are 40% lower than Houston's. Well planned and managed cities, such as Hong Kong, Paris, São Paulo, Tokyo, Dhaka and London, are able to reduce their GHG emissions per person. It is crucial for every city to use its energy more efficiently. Transport, economic growth and urbanisation do not automatically signify greater GHG emissions, although rich cities generally produce more GHG emissions (Fig. 7.2).

Urban density and the spatial organisation of a city decisively influence GHG emissions. Private cars, lack of efficient public transport and no alternatives for bicycles and other non-fossil transport mean a substantial increase in GHG emissions. Prosperity in urban areas was one factor of urban sprawl that has increased the demand for land and the increase in prices. However, if urban growth continues, population density will increase and this will reduce some of the GHG emissions in suburbs. There are also efforts to bring people back into the so-called "dead" centres of cities to reduce the level of commuting from suburban areas. Thus urban planning is crucial to bring down GHG emissions globally.

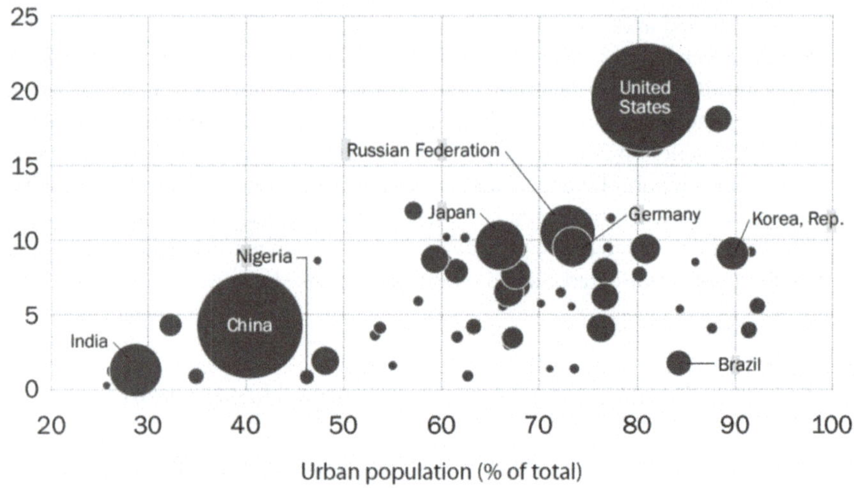

Fig. 7.2 Development and CO_2 emissions. *Source* World Bank (2019a: 19)

7.5.2 *How to Finance Liveable Cities*

The financing of climate mitigation and adaptation projects at local level depends on the decentralisation of each municipality (Fig. 7.3). While in Switzerland local finances are very stable thanks to local taxes and service incomes, in developing countries the federal government collects most of the local income from each area

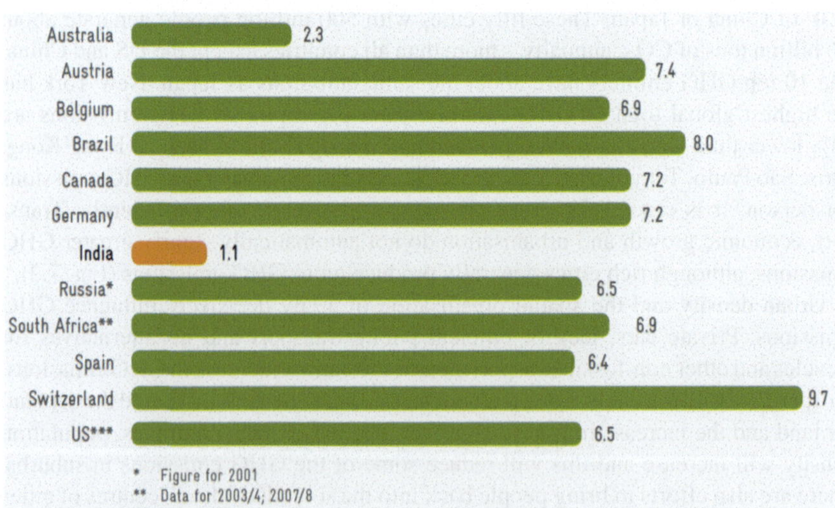

Fig. 7.3 Municipal expenditures in different cities. *Source* UN Habitat (2016: 11)

and redistributes it poorly to local authorities through the Ministry of Finance. A crucial exception is Brazil, with a high level of local financial autonomy, while countries such as Australia have a weak local financial system. Extremely weak is the local availability of funds for native development in India, a country that is highly exposed to climate change impacts. The recent Niño Year in 2019 and a cyclone in the Arabian Sea weakened the trade winds of the monsoon, and multiple cities were running out of drinking water. Chennai, with 4.6 million inhabitants, was in a critical situation, due to the former overexploitation of the aquifer (Cassella 2019). The 820,000 slum dwellers, who could not afford private water tankers, received just 30–40 L of water a day, about thirty times less than an average American family.

In Cape Town, South Africa, similar meteorological conditions have obliged the local authorities to regulate the drinking water availability and force people to use water sparingly (Joubert/Ziervogel 2019). The predictions of less precipitation in drylands, more hurricanes and changes in the trade winds will increase the lack of water, and cities and especially megacities must take these predictions seriously and promote rainwater capture and infiltration into the aquifer during the rainy season (IPCC 2014).

UN Habitat (2016) analysed the climate change investments in different regions, distinguishing between mitigation, adaptation and cross-cutting projects, but omitting the gender perspective. Mitigation is still dominant in the decision-making process; however co-benefits and cross-cutting projects additionally include adaptation to new situations of greater climate change impacts. Africa and Asia–Pacific are leaders in mitigation investment, and Africa and the *Least Developed Countries* (LDC) are also spending funds on cross-cutting activities. About 47% of the investment comes from grants to compensate for the higher impact on the Global South, 42% from loans, 9% from equity and 2% from guarantees. Again, the countries with the lowest emission level of GHG get into debt in order to compensate for the high emission rate of GHG in the Global North. Without any doubt investments are necessary to protect population and infrastructure; however, a gender approach could improve the outcome of the investment and reduce the costs, due to women being directly linked to the well-being of the family, food provision, fuel and safe water.

In Asia, nine years ago an *Asian Cities Climate Change Resilience Network* (ACCCRN) began operating in ten cities in India, Indonesia, Thailand and Vietnam. It has implemented effective processes to reduce urban climate vulnerabilities. Sharma et al. (2013) worked out the contextual factors in each city. These included existing governance structures, industrial make-up, population and demographic conditions, as well as the implementing partners' prior experience and level of comfort with quantitative and qualitative assessments. The authors asserted that any urban resilience activity in cities requires support from the state government and budgetary allocations from the national government. Their research revealed that local governments did not perceive climate change threats as a local problem, but as an additional burden.

> The city respondents agreed to the feasibility of introducing toolkits/capacity building programmes to build the capacity of the cities to take up such an exercise. The partners had also emphasised the importance of toolkits for guiding cities on planning for resilience, engaging various stakeholders, risk assessments and vulnerability assessments. In order to

> offer a plethora of options to cities, the development of as many tools as possible was recommended. For this purpose, the customisation of tools for cities and their various sectors is required (Sharma et al. 2013: 46).

Confronted with these complex and multiple problems of mitigation, adaptation, cross-cutting activities and resilience-building during the COP23, the Fiji government, along with other stakeholders, proposed the Talanoa Dialogue.

7.6 The Talanoa Dialogue

Launched on 9 February 2018 at the ninth World Urban Forum, the Cities and Regions Talanoa Dialogues initiative was facilitated by ICLEI (Local Governments for Sustainability) with the Global Covenant of Mayors for Climate and Energy (GCoM 2018) and UN-Habitat as special partners. More than fifty sessions of Cities and Regions Talanoa Dialogues were scheduled for 2018 in thirty-eight countries, mostly in the Global South, and especially in regions highly affected by the impacts of climate change. More than half of these Dialogues took place in August 2018. Their outcomes demonstrate that the active engagement of ministries of urbanisation, housing, public works, or their equivalent as appropriate, was facilitated by the Dialogues. In particular, cities which had not been adequately involved in national and global climate efforts used the Talanoa approach to acquire toolkits and learn from the experiences of similar cities. If the Dialogues were to include a gender perspective, their impact could increase without additional investment while simultaneously strengthening the resilience of the most vulnerable populations.

7.6.1 Why Talanoa?

Local governments, the private sector, and civil society lack expertise and resources to take effective climate action, but are directly affected by extreme events. Local governments, the private sector and civil society also lack the authority to take effective climate action due to the federal government generally being in charge of climate change laws, strategies and investments. Further, local governments, especially smaller municipalities, lack the knowledge to implement an effective early warning programme, safe preventive evacuation and other actions to protect the lives of the people, their belongings and the public infrastructure. Further, gaps exist between transnational climate action networks, multilateral climate organisations and domestic climate actions. These have limited the learning process even though there is clear recognition of the existing and upcoming threats related to climate change impacts. However, the local approach involving all stakeholders, including scholars, women and teachers, could improve the outcomes of mitigation and adaptation at local level.

The Talanoa effort is also a new and different instrument for multi-level climate governance that should support the realisation of local and regional climate mitigation and adaptation potentials. Through the dialogue it stimulates intermunicipal and regional collaboration, the integration of municipalities belonging to a megacity, support during and after an extreme event, DRR, and local collaboration over climate change mitigation and adaptation. The key proposal is to create local, regional and global networks, city-twinning, and internal and external partnerships which stimulate and help to strengthen possible climate change actions, success stories and financial support.

The weak horizontal and vertical collaboration from global level to people is articulated at local level, with people communicating through modern networks with other persons in similar conditions living in the same place or in similar conditions. Interchanges of experiences also place greater pressure on the local, regional and national government interact with similar partners in other parts of the country or the world. Talanoa further represents a competence framework for allocating more climate-related responsibility to lower levels of government with direct links to threatened people, especially the most vulnerable women, children and elderly. Without any doubt, the first ten minutes during a disaster are crucial for saving lives, and these activities can only be done by locally well-trained people. If, during a disaster, between 65 to 79% of dead people are women, there is a local necessity to empower these exposed social groups and train them in resilience-building.[4] Thus, preparation and adaptation to more serious climate events are crucial to reduce human losses, empower vulnerable women and avoid the destruction of infrastructure and personal belongings. DRR, disaster-risk management, prevention and resilience-building are now intimately related and may motivate people to get together locally and promote climate change adaptation at local level (UNISDR 2005).

Policy activities that are orientated towards introducing policy reforms in priority sectors to enable climate-actions are related to land use planning, transportation systems, building codes, waste management, energy reduction, industries and food culture. Public management with local laws for green buildings, renewable energy, water management and green purchasing of food are factors that may reduce at individual level the emissions of GHG, waste and ecological footprint. Most of these activities are in the hands of women. Finally, financial instruments are necessary to mainstream climate actions into the local financial system, such as carbon pricing or accounting for climate-risk in asset valuation through insurance policies. Confronted with a low fulfilment of NDC in the Global North, China and India, and a dramatic increase of GHG at global level, the Talanoa Dialogue is also reinforcing the achievement of the Paris Agreement. By explaining at local level the dialogues on climate change to people who suffer from its impacts, cities suddenly get involved and people find ways to collaborate and to strengthen their own resilience, often without additional investments.

[4]In 1991, Cyclone Gorky hit Bangladesh and caused around 140,000 deaths; 72% were women – i.e. 14 times more women than men died. During the Indian Ocean tsunami in 2004, more than 70% of the 250,000 fatalities were women.

New opportunities for fulfilling the national commitment during the Paris Agreement and a better understanding of the existing urban mitigation challenges mean that partner cities can learn lessons from other cities and understand the complexity and uncertainty of climate change impacts in the future. Better local mitigation management may also release additional funds for promoting specific activities, which are also preconditions for getting international, national and private credits to ensure efficient climate change adaptation, capable of reducing the risks in urban areas.

An alternative to the Talanoa approach is the 'carbonn Climate Registry', a global platform for cities, towns and regions to create transparent and credible accountability for GHG emissions, with an equity principle included. The participants represent 9% of the global population and, between them, will reduce 5.5 $GtCO_{2e}$ by the end of 2020 and 26.8 $GtCO_{2e}$ in 2050. This platform is open to newcomers, and together these committed cities, regions and towns have developed or are in the process of promoting 7,083 actions of mitigation and adaptation with 1,909 climate targets in eighty-six countries. These types of bottom-up or medium-size agreements could be more effective at reducing GHG than global agreements, as sharing experience and interacting with other cities and social groups may enable the goal to maintain the temperature far beyond 2 °C to be achieved.

The Talanoa Dialogue also addresses concrete problems. Local governments often lack skilled personnel to combat climate change threats efficiently and install renewable energy and energy-saving tools. Links with local and regional universities, technological institutions and training programmes for women and youths are gradually overcoming the lack of trained personal at local level. To improve institutional capacities, governmental personal also require training, often linked to the potential for a career in the public service in which job and social security are guaranteed, independently of the political party in power. Motivating public functionaries could promote and develop locally appropriate adaptation and mitigation strategies, which may be consolidated by effective participation with local citizens, especially young students from technological centres.

Megacities also need collaboration between different levels of government and special administrative and financial instruments, as well as consultation and collaboration between all the governmental structures involved. Greater climate change impacts may also oblige federal, state and municipal authorities to collaborate more closely with private stakeholders to reduce concrete threats and create resilience among the vulnerable. For instance, the Environmental Commission of the Megalopolis of Mexico City (CAMe in Spanish) is represented by the Ministry of Environment and Natural Resources, the government of Mexico City and the state governments of Hidalgo, Mexico State, Morelos, Puebla, Tlaxcala and recently also Querétaro. This integrated approach facilitates the coordination of activities for 45 million inhabitants in the central part of Mexico, creating plans for emergency management (flash floods, earthquakes, hurricanes, volcanic eruptions, bushfires, air pollution) and collectively training local authorities in the professional management of DRR and NDC. In financial terms, the CAMe manages a collective budget. Its reforestation activities may benefit states located up and downstream, and especially improve the air quality and the infiltration of rain into the aquifers in the highlands and the megacity.

7.7 Climate Governance and Conflict Resolution

Uncertainty over climate change impacts obliges authorities to prevent potential conflictive outcomes related to forced migration due to loss of shelter, food, water and livelihood. Multi-level climate change governance may be able to reduce local tensions and maintain political stability (IPCC 2014).

7.7.1 Multi-level Climate Change Governance

Without doubt, there is a strong interconnection between climate instruments and governance capacities. First, it is necessary to provide reliable information about GHG emissions. The data must be obtained via a transparent system of monitoring and reporting in accordance with the agreed methodology of UNFCCC. A certification and award system may stimulate the reduction of GHG emissions by businesses, households and government, especially when supported by a national system of laws and norms. Talanoa specifically promotes regional and local involvement in the NDC. Intersectorial collaboration between those involved in urban planning, the transportation system, natural reserves and productive urban areas is crucial (Fig. 7.4). An efficient weather forecast at regional and local level may improve early warning protocols and reduce loss of life and damage to the infrastructure.

Even the best instrument of measurement does not enable the existing human, institutional and financial resources to be optimised if a stable institutional capacity is missing (Bader/Bleischwitz 2009). Governance capacity occurs at horizontal and vertical level when key actors are identified and opportunities promoted to increase the reduction of GHG emissions. City networks, city twinning and partnerships simultaneously promote local activities and raise the motivation of citizens to participate. Training human resources and retaining qualified people in local government are crucial for responding effectively to the unknown complexities of climate change impacts and potential negative socio-political outcomes.

Small and marginal municipalities often lack the necessary funds to implement the agreed plans. Internationally, funding for adaptation and mitigation activities is limited, but locally there are multiple opportunities to increase the budget through taxes and fees. Carbon taxes are a globally promoted system to reduce the use of fossil energy, and temporary subsidies may promote renewable energies at household level. Without doubt, federal and state governments should help local administrations to consolidate their institutional capacities and foster intermunicipal and regional collaboration.

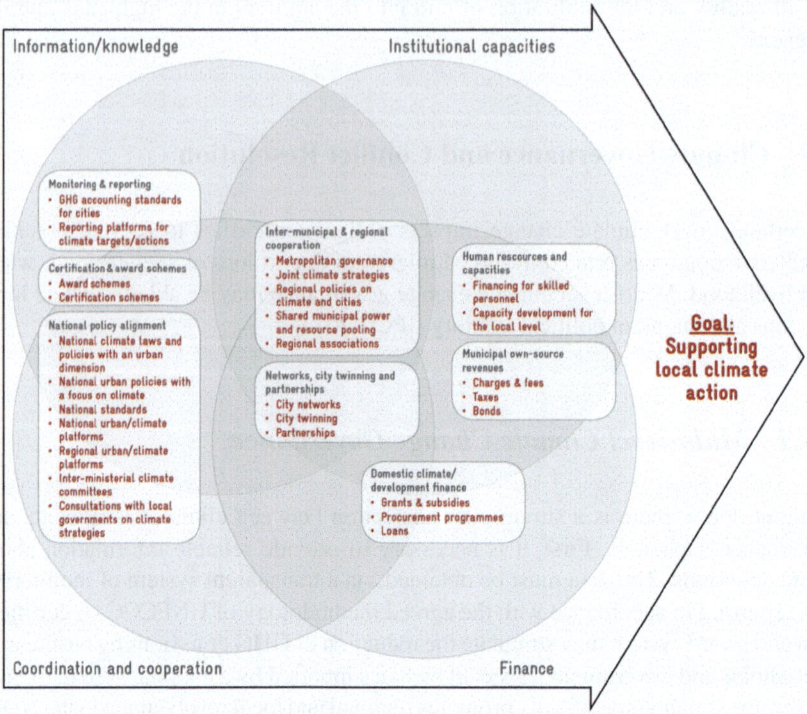

Fig. 7.4 Instruments and governance capacities. *Source* GIZ (2018: 27)

7.7.2 Multi-level Climate Change Actions Involved in Talanoa

The Talanoa Dialogue promotes multi-level climate change actions. These are crucial urban areas to diminish GHG emission. Most of the NDC include urban commitments (Fig. 7.5). There is a correlation between GHG emissions and health impacts. Clean energy in every household substantially reduces respiratory diseases and at the same time mitigates GHG emissions. The WHO (2018) reported that 93% of the world's children breathe toxic air every day, which puts their health and development at serious risk. The WHO estimates that in 2016, 600,000 children died from acute lower respiratory infections caused by polluted air. The report states that: "More than 40% of the world's population – which includes 1 billion children under 15 – is exposed to high levels of household air pollution mainly from cooking with polluting technologies and fuels." In poor countries, women frequently rely on coal

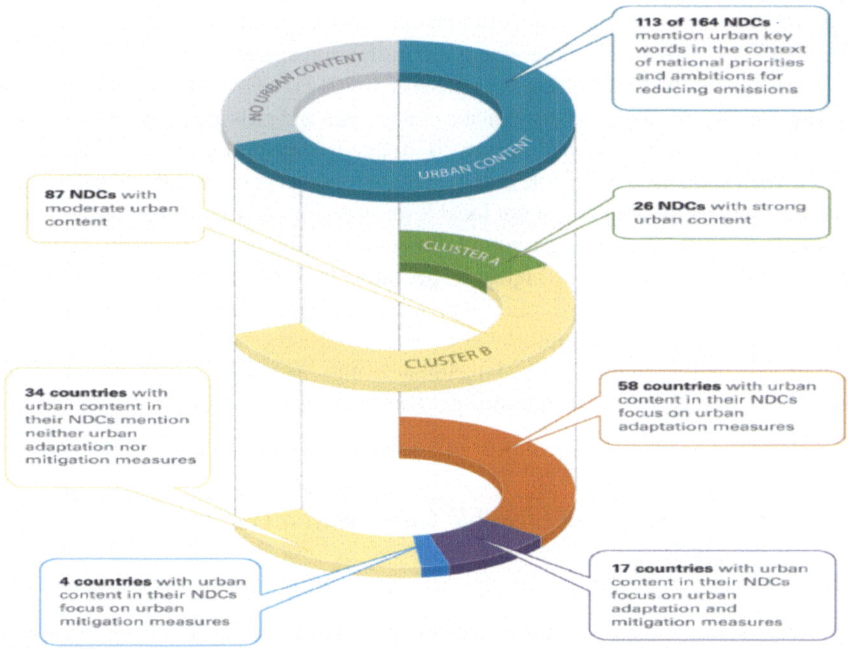

Fig. 7.5 Multi-level climate change actions with Talanoa. *Source* Bainimarama (2018: 1)

and biomass fuels for cooking and heating, which puts them and their children at higher risk of the effects of inside pollutants[5] and of premature death.

Because of population density, disasters strike harder in urban conglomerations (Hoornweg et al. 2011). Megacities are too big for preventive evacuation, thus the authorities need to develop plans to augment the adaptation and resilience capacities of highly exposed social groups (Oswald Spring 2018), provide special training, and

[5]The WHO 2018 Report synthesised the following key findings: "Air pollution affects neurodevelopment, leading to lower cognitive test outcomes, negatively affecting mental and motor development. Air pollution is damaging to children's lung function, even at lower levels of exposures. Globally, 93% of the world's children under 15 years of age are exposed to ambient fine particulate matter (PM2.5) levels above WHO air quality guidelines, which include 630 million children under 5 years of age, and 1.8 billion children under 15 years. In low and middle-income countries around the world, 98% of all children under 5 are exposed to PM2.5 levels above WHO air quality guidelines. In comparison, in high-income countries, 52% of children under 5 are exposed to levels above WHO air quality guidelines. More than 40% of the world's population – which includes 1 billion children under 15 – is exposed to high levels of household air pollution from mainly cooking with polluting technologies and fuels. About 600,000 deaths in children under 15 years of age were attributed to the joint effects of ambient and household air pollution in 2016. Together, household air pollution from cooking and ambient (outside) air pollution cause more than 50% of acute lower respiratory infections in children under 5 years of age in low- and middle-income countries. Air pollution is one of the leading threats to child health, accounting for almost 1 in 10 deaths in children under five years of age" https://www.who.int/news-room/detail/29-10-2018-more-than-90-of-the-world%E2%80%99s-children-breathe-toxic-air-every-day.

undertake preventive mitigation work to reduce the potential risks of extreme events (Marván/López 2018). The combination of mitigation and adaptation also increases the resilience of the most affected people (IPCC 2014).

The statistics of past disasters indicate that women and girls are especially exposed during extreme events (Oswald Spring 2016) due to lack of training in DRR (IPCC 2012). Their socialised social role to provide care for others is an additional risk factor, therefore special training courses at local level in regions of frequent extreme events may reduce deaths and the destruction of personal belongings and public services. At the same time, women are highly co-operative before, during and after extreme events, which often get transformed into a disaster due to lack of prevention, early warning and DRR. Thus the multi-level, intersectorial and gender aspect may increase resilience in regions highly exposed to climate events, and might simultaneously also reduce GHG emissions (Oswald Spring 2018).

7.7.3 How to Promote Liveable Cities with a Gender Perspective

Over the last year Talanoa identified the key issues for cities, especially in rethinking the mobility paradigm so that it is efficient, with low emissions, accessible, affordable and safe for men, women and children. This approach needs to break the existing monopolies of the car industry and the paradigm of private automobiles for all. Megacities are currently overloaded with private cars, often occupied by only one person. To overcome this, Talanoa instead recommends sustainable transport systems featuring motorised public vehicles with low emissions (de Alba 2017). In multiple cities, there are now electric buses (e.g. in China) and electric trams. For personal transport, cities may promote non-motorised mobility. Urban planning does not just cover transport, but should ideally lead to improvements in air quality, natural areas, parks and intermodality. Using city centres as business places and suburbs as sleeping residences increases GHG emission, due to long-distance travel and daily commuting. New paradigms for urban management are emerging, with novel proposals that facilitate the redesign of the urban landscape. At the same time, there are individual efforts to promote multigenerational housing to avoid isolated retirement houses and depression among the elderly.

At local level, the adjustment of street designs, favouring mixed land use, brings together the points of origin and destination as part of an effort to move towards polycentric cities. The transport needs of men and women are very different due to their traditional daily tasks. Therefore, it is crucial to create safer cities for women, especially in public transport at rush hour. Robust criminal laws and policing reforms, and ending impunity for sexual violence inside the household and outside in the jobs and streets are crucial ways to improve the personal security of women, girls, disabled people and minorities (Lecompte/Bocarejo 2017). All these efforts increase conflict resolution and promote peace-building at household and local level up to the national and international scale.

In megacities of the Global South there is significant migration from rural areas to urban settlements, due to better job, health, education and leisure opportunities (Oswald Spring et al. 2014). It is vital for governments to promote collaborative actions to transform informal settlements into dignified colonies and safe neighbourhoods. The involvement of local authorities at grass-roots level may help settlers find solutions to complex problems, improve service facilities and provide previously marginalised people with better lives and livelihood conditions (de Alba 2017).

A key mitigation process is strengthening energy efficiency and transitioning towards sustainable energy as a starting point for achieving regional and national energy security (Winzer 2011). This policy implies a commitment to:

1. the diversification of sources – with the emphasis on those with the least socio-environmental impacts;
2. adjust the budget allocated to research and development in hydrocarbons and transfer these resources to renewable energies;
3. improve the viability of decentralised generation of energy at household level and enterprises with mart grids and energy back-ups;
4. advance in energy efficiency;
5. sustainable development at city level, including planning for energy transition and self-sufficiency in energy, water and sewage (for example, the elaboration of energy profiles and the metabolic profiles of cities to reduce waste and sewage and to produce some basic food items locally simultaneously reduces GHG);
6. reduce the present wasteful consumption patterns, promote local food culture and recycle and compost organic waste (Delgado 2012);
7. manage water and recycle grey water and sewage in gardens, WCs, parks and agriculture;
8. promote urban transport orientated towards low carbon mobility;
9. grant informal settlements descentralised systems of electricity generation from renewable energies (REN21 2019)

7.8 Conclusions

7.8.1 The Talanoa Approach is a Multisectorial Approach Without a Gender Perspective

Beyond doubt and with widespread support from international organisations, a crucial issue for efficient local government is a clear diagnosis that identifies what has to be done. Participative governance and intergovernmental collaboration facilitate the establishment of achievable goals that are also related to the SDG 2030. It is crucial to agree where we want to be, what the priorities are, and by when these targets will be achieved. A further question is how to implement the decided plan and understand what is lacking: skills, finances, technologies and capacities to achieve the agreed goals?

It is also crucial to mobilise all stakeholders. How can the public, the private and the societal sectors be encouraged to get involved, and which concrete actions inspire universal support for better climate governance? Finally, it is crucial to have a route map that indicates which political, technical, financial and participative actions are required to increase the political willingness for collaboration. Within this process, it is necessary to know how local governments, the private sector, women and civil society can work together to reduce GHG emissions domestically and deliver the NDC. Warmer temperatures in oceans and on land create fiercer extreme events which are often transformed into disasters. Thus at local level it is crucial to define the requisite mitigation activities that will lead to a drastic reduction in GHGs and also improve the adaptation skills of the most vulnerable people, orientating actions towards co-benefits that increase resilience-building at household level.

There are crucial intersections between actions, goals, planning and management, whereby meshed networks may promote or limit the success of a sustainable mitigation, adaptation and resilience strategy. Experiences in Asia and Europe have shown that using cycle ways as a primary transport network plays a crucial role in improving the air quality in cities and the health of the people. In traffic management, electronic signals, pavement indicators and physical or painted land which divides cars from pedestrians and cyclists increases the safety of non-motorists. Rental and bicycle sharing and sufficient, safe, affordable car parks outside cities are crucial to promote a different culture of healthy, non-fossil transport. Additionally, an integrated public transport system with bicycles improves the air quality in megacities, limits traffic accidents and substantially reduces the use of private cars and thus the emissions of GHG and urban traffic congestion. Singapore has instrumented very heavy costs for parking in the city centre, enough parking lots in the suburbs for cars to be left safely, and an efficient public transport system which allows everybody to travel in comfort to their work or leisure place.

7.8.2 Granting Gender Equality and Life Quality for Children and Everybody

As mentioned before, men and women have different transport needs, and safety for women and their mobility is crucial. Men and women use public and private transport differently, and engendered-sustainable policies, whereby women can move publicly with their children, may improve livelihoods and the success rate of schemes to reduce poverty and GHG emissions (UNRISD 2019).

City administrations can efficiently support the NDC and help to achieve the targets by reducing GHGs through specific actions and new laws for buildings, urban planning and waste management. Urban gardens on roofs and plants on balconies or walls increase the capture of CO_2. Parks can be reforested with fruit trees for human consumption, which may reduce under- or malnourishment. Multiple Asian countries organise public exercise sessions in their parks. These help to reduce obesity, and when combined with safer cycling networks, air quality and health also improve.

Innovative planning which focuses on the high-emitting urban sector is required to reduce GHG emissions. Crucial issues are reliable data collection and access to safe, cheap and improved technologies for energy efficiency, renewable energy, recycling organic waste, integral management of sewage, bioclimatic houses and urban agriculture. All these efforts, especially when women get actively involved, also help to develop DRR infrastructure.

Without doubt, mass production of sustainable technology reduces the costs of installation and maintenance. Women and young people could actively participate in learning the processes for improving energy efficiency and installing renewable energy systems, which would increase income and jobs for under- and unemployed people. Recycling is a cultural phenomenon, and traditionally older people have been in the habit of reusing, reducing and recycling their waste, especially organic waste. Composting and using the resultant substance in parks and gardens improves soil quality, sequesters more GHG emissions, reduces the evaporation of the water in the soil and can aid the production of organic food in gardens or on balconies.

Finally, it is crucial that all the coordinated urban efforts are systematically monitored. A transparent financial and monitoring system by internet increases confidence in local authorities, reduces the costs of GHG sequestration and produces multi-level governance, better health, improved living standards and a green city, where, additionally, an economy of solidarity (Collin 2014) or circular economy may simultaneously improve the income of self-employed people. New service processes also create extra job opportunities, making it possible for even the most vulnerable to earn a dignified living, which increases social cohesion. As Pickett and Wilkinson (2018) maintain, urban inequality 'is a key catalyst for nonconformity and violence. Greater equity, sustainable management of land and waste, and integrated DRR may improve living conditions and well-being everywhere, including the megacities of the poor Global South.

References

Anderson, Sarah E.; Bart, Ryan R.; Kennedy, Maureen C.; MacDonald, Andrew J.; Moritz, Max A., 2018: "The dangers of disaster-driven responses to climate change. Low-probability, high-consequence climate change events are likely to trigger management responses that are based on the demand for immediate action from those affected. However, these responses may be inefficient and even maladaptive in the long term", in: *Nature Climate Change,* 8: 651–653; at: https://www.nature.com/articles/s41558-018-0208-8.

Bader, Nikolas; Bleischwitz, Raimund, 2009: "Measuring Urban Greenhouse Gas Emissions: The Challenge of Comparability", in: *S.A.P.I.EN.S*; at: https://journals.openedition.org/sapiens/854.

Bainimarama, H.E. Frank (President of Fiji), 2018: "Talanoa and Beyond: Raising Ambition with Cities and Regions" (Joint Policy Briefing Note by GIZ, ICLEI and UN-Habitat), UNFCCC, COP23.

Bott, Sarah; Guedes, Alessandra; Goodwin, Mary; Adams Mendoza, Jennifer, 2012: *Violencia contra las mujeres en América Latina y el Caribe* (Washington: PHO [Pan-American Health Organization]).
Brewera, Kirk; Granta, Jill L., 2015: "Seeking Density and Mix in the Suburbs: Challenges for Mid-Sized Cities", in: *Planning Theory & Practice,* 16,2: 151–168.
Caraël, Michel; Glynn, Judith R., 2008: "HIV Infection in Young Adults in Africa: Context, Risks and Opportunities for Prevention", in: Caraël Michel; Glynn, Judith R., (Eds.): *HIV, Resurgent Infections and Population Change in Africa*, International Studies in Population vol. 6 (New York: Springer).
Cassella, Carly, 2019: "A Major Indian City Has Nearly Run Out of Water, And No One Is Talking About It", in: *ScienceAlert,* 21 June; at: https://www.sciencealert.com/this-indian-city-has-all-but-run-out-of-water-and-no-one-is-talking-about-it.
Chan, Sander; Amling, Wanja, 2019: *Does Orchestration in the Global Climate Action Agenda Effectively Prioritize and Mobilize Transnational Climate Adaptation Action?* (Cham: Springer International).
Collin Harguindeguy, Laura, 2014: *Economía Solidaria: Local y Diversa* (Tlaxcala: Coltlax).
De Alba, Felipe (Ed.), 2017: *Un Debate Actual a Distintas Voces* (Mexico City: Cámara de Diputados/LXIII Legislatura, Centro de Estudios Sociales y de Opinión Pública).
Delgado Ramos, Gian Carlo (Ed.), 2012: *Transporte, Ciudad y Cambio Climático* (Mexico City: CEIICH-UNAM).
Delgado Ramos, Gian Carlo; van Staden, Maryke; Villaseñor Franco, Edgar (Eds.), 2018: *Diálogo Mexicano de Talanoa* (Mexico City: CEIICH-UNAM).
Diwakar, Vidya; Lovell, Emma; Opitz-Stapleton, Sarah; Shepherd, Andrew; Twigg, John, 2019: *Child Poverty, Disasters and Climate Change: Investigating Relationships and Implications over the Life Course of Children* (London: Overseas Development Institute [ODI]).
GIZ, 2018: *Multi-Level Climate Governance Supporting Local Action* (Bonn: GIZ).
Höhne, Niklas; den Elzen, Michel; Escalante, Donovan, 2014: "Regional GHG Reduction Targets Based on Effort Sharing: A Comparison of Studies", in: *Climate Policy*, 14,1: 122–147; DOI: https://doi.org/10.1080/14693062.2014.849452.
Höhne, Niklas; Drost, P.; Bakhtiar, F.; Chan, S.; Gardiner, A.; Hale, T.; Hsu, A.; Kuramoch, T.; Puig, D.; Roelfsema, M.; Sterl, S., 2016: "Bridging the Gap – The Role of Non-State Action", in: *The Emission Gap Report 2016: A UNEP Synthesis Report* (Nairobi: UNEP).
Hoornweg, Daniel; Sugar, Lorraine; Trejos Gómez, Claudia Lorena, 2011: "Cities and Greenhouse Gas Emissions: Moving Forward", in: *Environment & Urbanization*, 23,1: 207–227; DOI: https://doi.org/10.1177/0956247810392270.
IBRD [International Bank for Reconstruction and Development/The World Bank], 2019: *Women, Business and the Law 2019* (Washington: IBRD/WB).
IEA [International Energy Agency], 2014: *CO_2Emissions from Fuel Combustion Highlights* (Paris: OECD/IEA).
IEA, 2019: *Global Energy & CO_2Status Report* (Paris: IEA).
IPCC [Intergovernmental Panel on Climate Change], 2012: *Managing the Risks of Extreme Events and Disasters to Advance Climate Change Adaptation* (Cambridge: Cambridge University Press).
IPCC [Intergovernmental Panel on Climate Change], 2013: *Climate Change 2013: The Physical Science Basis. Working Group I Contribution to the IPCC Fifth Assessment Report* (Cambridge: Cambridge University Press).
IPCC [Intergovernmental Panel on Climate Change], 2014: *Climate Change 2014: Impacts, Adaptation, and Vulnerability: Working Group II Contribution to the IPCC Fifth Assessment Report* (Cambridge: Cambridge University Press).
Joubert, Leonie; Ziervogel, Gina, 2019: "One Day. City's Response to a Record-Breaking Drought, Cape Town"; at: https://www.africancentreforcities.net/wp-content/uploads/2019/07/Day_Zero_Joubert_Ziervogel_2019.pdf.

Konings, Piet; van Dijk, Rijk; Foeken, Dick, 2006: "The African Neighbourhood: An Introduction", in: Konings, Piet; Foeken, Dick (Eds.): *Crisis and Creativity: Exploring the Wealth of the African Neighbourhood* (Leiden & Boston: Brill).

Leahy, Stephen, 2018: "Cities Emit 60% More Carbon than Thought: A new analysis finds that city planners have been undercounting greenhouse gas emissions from a key contributor", in: *National Geographic*, https://news.nationalgeographic.com/2018/03/city-consumption-greenhouse-gases-carbon-c40-spd/.

Leal-Arcas, Rafael, 2018: "Re-Thinking Global Climate Change: A Local, Bottom-Up Perspective", in: *Seton Hall Journal of Diplomacy and International Relations, 20,1 (Autumn/Winter): 4–20 (London: Queen Mary University).*

Lecompte, María Carolina; Bocarejo S., Juan Pablo, 2017: **"Transport Systems and Their Impact on Gender Equity", in: *Transportation Research Procedia*,** 25: 4,245–4,257.

Lubuva, John M., 2004: "Community Approach to Security, Social Inclusion and Development in Tanzania", Paper presented at the Dialogue on Promoting Coexistence and Security in the Information Society, Barcelona, 9–11 September 2004 (unpublished paper).

Marván, Ma. Luisa; López-Vázquez, Esperanza (Eds.), 2018: *Preventing Health and Environmental Risks in Latin America* (Cham: Springer International Publishing).

Menkes Bancet, Catherine; Sosa-Sánchez, Itzel A., 2018: *Desigualdades Sociales, Género y Embarazo Adolescente en México y en el Estado de Morelos* (Cuernavaca: CRIM-UNAM).

Moyer, Eileen, 2006: "Not Quite Comforts of Home: Searching for Locality among Street Youth in Dar es Salaam", in: Konings, Piet; Foeken, Dick (Eds.): *Crisis and Creativity: Exploring the Wealth of the African Neighbourhood* (Leiden & Boston: Brill).

O'Brien, Karen; Sygna, Linda; Leichenko, Robin; et al., 2008: *Disaster Risk Reduction, Climate Change Adaptation and Human Security* (Oslo: GECHS, University of Oslo).

Oswald Spring, Úrsula; Serrano, S. Eréndira; Estrada Álvarez, Adriana; et al., 2014: *Vulnerabilidad Social y Género entre Migrantes Ambientales* (Cuernavaca: CRIM-DGAPA-UNAM).

Oswald Spring, Úrsula, 2016: "Perspectiva de Género ante el Cambio Climático y la Doble Vulnerabilidad, in: *Revista Internacional de Ciencias Sociales y Humanidades, SOCIOTAM,* 26,2: 137–162.

Oswald Spring, Úrsula, 2018: "Vulnerabilidad Social Frente a Eventos Hidro-Meteorológicos Extremos y Políticas de Prevención y Mitigación", in: Calva, J.L. (Ed.): *Derechos Social y Bienestar Social* (Mexico City: Consejo Nacional de Universitarios, JP Ed.): 397–419.

Pickett, Kate; Wilkinson, Richard, 2018: "The Enemy Between Us: How Inequality Erodes our Mental Health", in: *Transformation,* at: https://www.opendemocracy.net/en/transformation/enemy-between-us-how-inequality-erodes-our-mental-heal/.

Rahman H.M. Tuihedur; Hickey, Gordon M., 2019: "What Does Autonomous Adaptation to Climate Change Have to Teach Public Policy and Planning About Avoiding the Risks of Maladaptation in Bangladesh?", in: *Policy and Practice Reviews* (22 January 2019); doi: https://doi.org/10.3389/fenvs.2019.00002.

REN21, 2019: *Global Status Report: Renewable Energy in 2019* (Paris: REN21).

Sharma, Divya; Singh, Raina; Singh, Rozita, 2013: *Urban Climate Resilience: A Review of the Methodologies adopted under the ACCCRN Initiative in Indian Cities* (London: IIED).

Sommers, Marc, 2010: "Urban Youth in Africa", in: *Environment & Urbanization*, 22,2: 317–332.

Strohbach, Michael W.; Döring, Anneke O.; Möck, Malte; Sedrez, Maycon; Mumm, Olaf; Schneider, Anne-Kathrin; Weber, Stephan; Schröder, Boris, 2019: "The 'Hidden Urbanization': Trends of Impervious Surface in Low-Density Housing Developments and Resulting Impacts on the Water Balance", in: *Frontiers in Environmental Science*, 7, Article 29.

The Global Covenant of Mayors for Climate and Energy (GCoM), 2018: "Global Covenant of Mayors for Climate and Energy – April submission to the Talanoa process"; at: https://www.globalcovenantofmayors.org/wp-content/uploads/2018/04/Talanoa-April-Submission-GCoM_.pdf.

UN Climate Action Summit, 2019; at: https://www.un.org/en/climatechange/.

UN DESA, 2018: *World Urbanization Prospects* (New York, UN DESA).

UN Habitat, 2016*: Finance for City Leaders Handbook* (Nairobi: UN Habitat).

USAID, 2005: *Youth and Conflict: A Toolkit for Intervention, Office of Conflict Management and Mitigation* (Washington: United States Agency for International Development [USAID]).
UNICEF, 2018: *Annual Report 2018: For Every Child, Every Right* (New York: UNICEF).
UNRISD, 2019: *Climate Justice from Below: Local Struggles for Just Transition(s)* (Geneva: UNRISD).
Utas, Mats, 2005: "Building a Future? The Reintegration and Remarginalization of Youth in Liberia", in: Richards, Paul (Ed.): *No Peace No War: An Anthropology of Contemporary Armed Conflicts* (Athens: Ohio University Press; Oxford: James Currey): 150.
Van den Berg, Nicole J.; van Soest, Heleen L.; Hof, Andries F.; et al., 2019: "Implications of Various Effort-Sharing Approaches for National Carbon Budgets and Emission Pathways", in: *Climatic Change*; at: https://doi.org/10.1007/s10584-019-02368-y.
Victoria State Government, 2018: *Regional Adaptation Snapshot*.
World Bank, 2010: *Cities and Climate Change: An Urgent Agenda* (Washington: International Bank for Reconstruction and Development/The World Bank).
World Bank, 2018: "Year in Review: 2018 in 14 Charts"; at: https://www.worldbank.org/en/news/feature/2018/12/21/year-in-review-2018-in-14-charts.
World Bank, 2019: *World Bank's Women, Business and the Law Report 2019* (Washington: International Bank for Reconstruction and Development/The World Bank).
World Bank, 2019a: *A Decade of Reform* (Washington: International Bank for Reconstruction and Development/The World Bank).
WHO [World Health Organization], 2016: "An Estimated 12.6 Million Deaths Each Year are Attributable to Unhealthy Environments"; at: https://www.who.int/mediacentre/news/releases/2016/deaths-attributable-to-unhealthy-environments/en/.
WHO [World Health Organization], 2018: *Air Pollution and Child Health: Prescribing Clean Air* (Geneva: WHO).
Winzer, Christian, 2011: *Conceptualizing Energy Security*, EPRG Working Paper 1123, Cambridge Working Paper in Economics 1151 (Cambridge & New York: Cambridge University Press).

Chapter 8
Conflicts in Kenya: Drivers of Conflicts and Assessing Mitigation Measures

Charles Ndalu Wasike

Abstract Modern threats to security differ in both form and dynamics and therefore it is important to appreciate their origins, trends and evolution within the context of the political history of the country in order to deal with the underlying issues surrounding them. Against this backdrop, this chapter delves into the nature and drivers of insecurity in Kenya with the aim of enhancing understanding of the context within which realistic, sustainable policy options can be sought. Kenya is rife with violence, as exemplified by its high levels of sexual and gender-based violence, intercommunal violence, cycles of election-related violence, and increasing numbers of terrorist attacks. The incidence, gravity and intensity of the phenomena have been increasing for years. The conflicts the country experiences are multiple and overlapping, with the Rift Valley, Nairobi, the peripheral pastoralist drylands, and the coastal region being among the areas most affected. The high levels of violence are the result of a range of factors, including: ethnic intolerance, border conflicts, political party zoning, competition over land and other resources, the proliferation of small arms, weak security and poverty, underdevelopment, and marginalisation. Intercommunal violence risks being increased by competition over the fruits of devolution and the elite manipulation of local communities. Violent Islamist activities have tended to be clustered in the north-eastern region of the country (which borders Somalia and Ethiopia), the coastal region in the South East, and Nairobi. Terrorist attacks have pervasively increased since the Kenyan government sent its defence forces to Somalia for the military operation dubbed 'Operation Linda Nchi' in 2011. The response of the government has been characterised mainly by the politicisation of the counterterrorism response and the scapegoating of certain ethnic and religious groups that have also played into Al-Shabaab's hands. The proliferation of small arms in the African Great Lakes Region and Kenya provides more opportunities for violence and wider insecurity. There have been attempts to avert conflicts in Kenya, but failure to address various problems embedded in Kenya's social structures makes the nurturing of peaceful coexistence a mirage. In order to mitigate the foregoing, there is an urgent need to deal with: economic inequalities, provision of basic essentials, culture of violence

Charles Ndalu Wasike (Kenya), Ph.D., head, *African Peace Research and Education Association* (AFPREA), lecturer, Moi University, Kenya; *Email*: ndaluwasike2006@yahoo.com.

Ú. Oswald Spring and H. G. Brauch (eds.), *Decolonising Conflicts, Security, Peace, Gender, Environment and Development in the Anthropocene*,
The Anthropocene: Politik—Economics—Society—Science 30,
https://doi.org/10.1007/978-3-030-62316-6_8

as manifested in the formation of gangs used by politicians to settle political scores, gender inequalities and unequal power relationships between women and men, and the creation of an effective and accessible justice system that has made women and girls in the *internally displaced person* (IDP) camp susceptible to multiple levels and various forms of violence.

Keywords Conflicts · Drivers · Marginalisation · Mitigation · Peacebuilding · Proliferations · Mitigation · Violence

8.1 Introduction

The *Kenya National Bureau of Statistics* (KNBS) 2019 shows that Kenya has a population of 47.6 million people, of which 23.34 million are male, while 24.014 million are female. Further, there are 12 million households with a population density of 82 per square kilometre. Kenya has a youthful rural population. In this case, 35.7 million Kenyans (75.1%) are below 35 years, while 32.73 million (68.9%) live in rural areas. KNBS (2019) data shows that the population grew by 26% from 37.7 million in 2009 to 47.6 million in 2019, translating to an additional 9.9 million people.

Kenya is largely a multi-ethnic country comprised of 43 ethnic groups. According to the 2019 Population and Housing Census, the most populous ethnic groups are the Kikuyus (8,148,668) followed by the Luhya (6,823,842), Kalenjin (6,358,113), Luo (5,066,966), Kamba (4,663,910), Kenyan Somali (2,703,235), Mijikenda (2,488,691), Meru (1,975,869) and Maasai (1,189,522). Christianity is the major religion. The religious population distribution is as follows: Protestants: 15,777,473; Catholics: 9,726,169; Evangelical Churches: 9,648,690; Muslims: 5,152,194; African Churches: 3,292,573; Traditionalists: 318,727; Atheists: 755,750; Hindus: 60,287; Orthodox: 201,263; and other Christian faiths: 1,732,911 (Fig. 8.1).

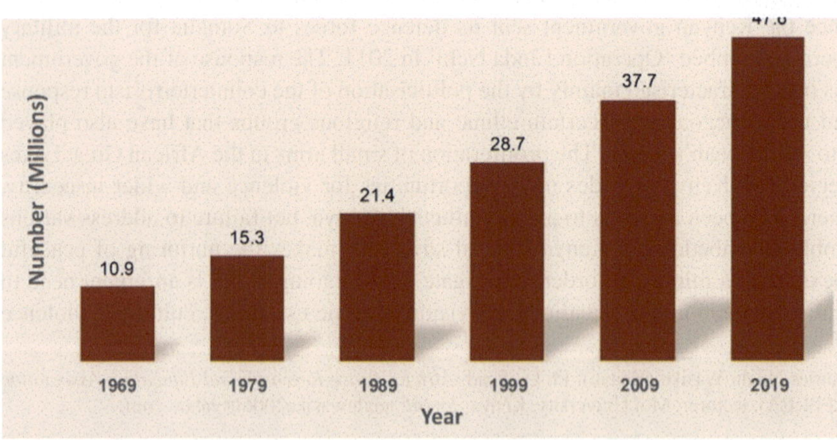

Fig. 8.1 Kenya's population growth rate. *Source* KNBS (2019)

Kenya's experience since independence paints a picture of a country at war with itself, going by the types of intra-state conflicts. Historically, the Somali-speaking peoples of the Horn of Africa have always regarded themselves as one people and have, therefore, never excused the colonial powers for the Balkanisation of the Somali Nation into separate entities. This was a source of tension between the new Republic of Somalia, created in 1960 after the amalgamation of the former British Somaliland and Italian Somaliland and their neighbours, Kenya and Ethiopia, which had large populations of Somali in the *Northern Frontier District* (NFD) and the Ogaden regions respectively.

Kenya is rife with violence which manifests itself in various forms and different dynamics. The violence manifests in the high levels of sexual and gender-based violence, intercommunal violence, cycles of election-related violence, and increasing numbers of terrorist attacks. The cyclical incidences, gravity and intensity of the violence have been on an upward trend for years. Another significant feature of the conflicts is that they are normally multiple and overlapping. For instance, the Rift Valley, Nairobi, the peripheral pastoralist drylands, and the coastal region are among the areas most affected. The high levels of violence have been attributed to: ethnic intolerance, border conflicts, political party zoning, competition over land and other resources, the proliferation of small arms, weak security structures and systems, poverty, underdevelopment and marginalisation. Inter communal violence is exacerbated by increased competition over resources and the political manipulation of local communities.

Terrorism has tended to be clustered in the north-eastern parts of the country which border Somalia, the coastal province in the South East, and Nairobi. The increased incidences of terrorism have been attributed to Kenya's ongoing military involvement in Somalia. However, there was a relative time of peace during the 2013 elections attributable largely to United Nations involvement through the International Criminal Court. The main players who are actively involved in various ways in Kenya's conflicts include: (i) the national government; (ii) county governments; (iii) the security forces; (iv) politicians and elites; (v) political militias; (vi) communal militias; (vii) pastoralist communities; (viii) Kenyan Somalis due to marginalisation; (ix) extremist Kenyan Muslims; (x) The Youth, whose members are susceptible to radicalisation; (xi) Al-Shabaab and (xiii) the international community. The actions of some stakeholders have been seen to ignite conflicts. For instance, the government's response through politicisation of the counterterrorism and the scapegoating of certain ethnic and religious groups helped bolster Al-Shabaab's propaganda and resolve. Further, the proliferation of small arms provides more opportunities for violence and wider insecurity.

8.2 The Nature of Conflicts

According to Dowd and Raleigh (2013) Kenya was considered the seventh most violent country in the ACLED (*Armed Conflict Location and Event Data Project*) dataset, with just over 3500 political violent events recorded between 1997 and September 2013. Parts of the country witnessed low levels of persistent violence which have become unremarkable to many citizens (Scott-Villers et al. 2014: 2; Gibbons 2014). Ombaka (2015) posits that many Kenyans now treat rampant theft and robbery as a normal burden of citizenship. The incidences of violence peaked between 1997 and September 2013, with 341 cases recorded during January–March 2008 (Dowd/Raleigh 2013).

Kenya's continued experience of cyclical violent insecurity and resultant terrorist attacks have not only destabilised large parts of Kenya's peripheral counties but also instilled fear. Scholars agree that the increasing and spreading insecurity has fomented fear and stoked ethnic and regional divisions, precipitating security crack-downs and affecting the country's infamously tumultuous politics (Ombaka 2015; Lind et al. 2015). The *Kenya National Commission on Human Rights* (KNCHR) 2014 presents a damning report on the increase in the incidences, gravity and intensity of violence since 2010 that included "persistent terror attacks, inter-community conflicts and violence targeting law enforcement officers as well as a big number of extra-judicial executions. The consequences of these flare-ups have been the death of people, displacement of large population, and disruption of the provision of education and health services in several counties including Baringo, Turkana, Wajir, Mandera, and Isiolo" (KNCHR 2014). Sources of income for people in Turkana, Mandera, Marsabit, Wajir Baringo, Mombasa and Lamu have been adversely affected due to insecurity. This has persisted despite the constitutional reforms implemented to address and prevent violence in Kenya in the wake of the 2007–2008 post-election violence (Lind et al. 2015).

Scholars agree that the multiple and overlapping conflicts sometimes coincide with electoral cycles that act as triggers for politically motivated violence (Dowd/Raleigh 2013; Elder et al. 2014; Halakhe 2013). The conflicts come in the form of militia activities in urban areas; communal violence in the Rift Valley and elsewhere; spill-over conflict from neighbouring Somalia with implications for unrest in north-eastern Kenya; and the confluence of separatism and Islamist mobilisation on the Kenyan coast. Notable cases of conflict include the 2007–2008 post-election violence, when ethnic identity-based clashes caused over 1,100 fatalities and widespread population displacement, especially in the Kenyan Rift Valley (Cox et al. 2014; Halakhe 2013). The central Isiolo region, a pastoralist dryland, has been a hotspot for violence for many years. There have been ethnic clashes since 2012 in the Tana River. There have also been clashes in Marsabit County in northern Kenya and ethno-religious riots in Mombasa. Scot-Villers et al. (2014) pointed out that violence in areas like Marsabit County is persistent and ranges from economic boycotts and hate speech to unexplained deaths and small wars. The causes of the various conflicts are related to Kenya's deep and persistent fault lines, relating to factors such as socio-political marginalisation and elite manipulation of ethnic identities for political mobilisation. The multiplicity and overlap of conflicts shape the nature of conflict and resultant security risks posed to the populations, thus requiring distinct responses. The persistence of these conflicts is partly attributable to unresolved past conflicts and lack of healing and reconciliation, which consequently lead to cycles of revenge attacks.

The *Constitution and Reform Education Consortium* (CRECO) 2012 baseline report, based on the mapping of conflict in the forty-seven counties in Kenya, established an interesting trend. According to the report, higher incidences of violence that were noted in most of its Kenya's counties were consequences of ethnic intolerance, border conflicts, political party zoning, competition over land and other resources, proliferation of small arms, weak security and poverty. The Rift Valley region experienced the highest levels of violence between 1997 and September 2013, followed by Nairobi. The other areas have also experienced high levels of violence and continue to experience the highest levels of poverty and underdevelopment. The gaps in security, governance and development in these peripheral areas occasion armed groups, weapons and jihadist ideology to gain ground and pose a threat to Kenya as a whole. Another significant cause of conflict is Kenya's porous borders compounded by insurgency and counter-insurgency forces supported by neighbouring states that add a regional security dimension to Kenya's conflict situation (Sharamo 2014: 3; TSA 2014: 21–26). The Kenyan state has nominal control of hotspots or regions considered the bedrock of insecurity. These regions cover more than half of Kenya's territory and straddle twelve of Kenya's forty-seven counties. They include West Pokot, Elgeyo-Marakwet, Baringo, Turkana, Samburu, Isiolo, Marsabit, Mandera, Wajir, Garissa, Lamu and Tana River. The major characteristic of the aforementioned is that they form the block of the marginalised counties in the country.

Kenya has the highest levels of inter-communal violence in Africa, as observed by Dowd and Raleigh (2013) and TSA (2014). The violence is perpetrated by ethnically identity based communal militias and often involves cycles of attacks and counter attacks. Competition over land-based resources often drives local conflicts, which are sometimes triggered by the migration of herders in search of water and pasture compounded by the lack of security and the proliferation of small arms and light weapons. Political mobilisation and attendant competition breed animosity between communities, stoking violence between ethnic and regional groups that compete over access to resources and power. This is a pressing danger, given that the decentralisation initiatives in Kenya, though promising to bring government and power closer to marginalised populations, may exacerbate tensions between communities with competing claims on ethnic homelands, rights to land, and political representation. In some cases, elite sponsorship is suspected to have been a factor in communal violence, especially in 2013 in Mandera and Moyale, and in the Tana River region in August 2012.

The multi-layered and complex conflicts in the pastoralist communities of Kenya have tended to take the form of cattle rustling, ethnic violence, displacements, massacres and revenge attacks. Cattle raids are undertaken for prestige and the acquisition of livestock for payment of bride prices, as well as competition over scarce and diminishing water and pasture resources. Another dimension is the commercialisation of cattle raiding, which involves excessive violence with the aim of mass sales in towns. Other drivers of conflicts include climate change; environmental degradation; drought, famine and other natural catastrophes; land-related conflicts (some relating to administrative and electoral boundaries); the politicisation of communal relations; the proliferation of small arms and light weapons; tensions with agricultural communities; and human-wildlife conflicts aggravated by competing uses of land for commercial ranching and wildlife conservation. These conflicts have become increasingly intractable as a result of weakened traditional governance systems; the breakdown of inter-communal social contracts; Elders' loss of control over youths; the persistence of *moranism* (warrior) culture; and politicisation of peacemaking processes (Sharamo 2014: 3; Okumu 2013: 1, 5). Some researchers suggest that the conflicts have intensified partly as a result of the weakness or unwillingness of the state to protect its citizens in the respective regions of the country (Okumu 2013). There has also been a shift in conflicts manifestation in pastoralist communities from the traditional resource-based incidents to clashes driven by economic and political gain (Gibbons 2014: 3; Sharamo 2014).

Further, conflicts are driven by institutional, political-economic and social spheres operating at national, regional and even global levels. The institutional factors driving conflicts include contested borders, unresolved land tenure issues, and failures of policing and justice. The political-economic factors include extractive commercial enterprises without adequate benefit-sharing with community members – like the current situation in Turkana after the discovery of oil – land alienation, divisive politics and corrupt local administrations; whilst social factors relate to historical marginalisation and exclusion, as well as issues of ethnic identity and gender resulting in development projects being accompanied by violence and militarism. The perpetrators and war financiers have used ethnicity to mobilise young people's engagement in violence as a means of providing their economic benefits.

Violent Islamist extremists have been prevalent in the north-eastern region of the country which borders Somalia, the coastal province in the south east, and Nairobi. This is exacerbated by the terrain in northern Kenya, which makes it difficult to secure and favourable for Al-Shabaab to infiltrate, and by Somalia's long-running state collapse and conflict spill-over act as 'external stresses' on Kenya's peace and stability (Lind et al. 2015: 4). The conflicts have further been fuelled by Kenya's domestic policies, politics and practices, especially in relation to uneven development programmes and the treatment of the country's Muslim populations. These internal domestic causes, together with external factors (i.e. Al-Shabaab, regional arms trade, drug trafficking), have affected eight counties, namely Mandera, Garissa, Wajir, Mombasa, Tana River, Kilifi, Lamu and Nairobi.

There have also been suggestions that Kenya's military operations in Somalia have a commercial objective, as the military is a beneficiary of the illicit trade in charcoal from Kismayo port in southern Somalia as well as the trade in contraband sugar from Somalia into Kenya (Lind et al. 2015: 16; Anderson/McKnight 2014: 9–10; Mc Evoy 2013: 6). Other researchers have held the view that Kenya is attempting to lay claim to gas and oil deposits in a contested maritime zone off the coasts of Kenya and Somalia

(Lind et al. 2015: 16; Anderson/McKnight 2014: 7; Mc Evoy 2013: 6). Since 2011, when Kenya stepped up its military involvement in Somalia (dubbed 'Operation Linda Nchi'), ostensibly to wipe out Al-Shabaab, there has been an upsurge in attacks across the north and the east coast of Kenya, as well as the September 2013 siege of Nairobi's Westgate shopping Centre, the June 2014 attack on hotels and bars on Mpeketoni, and village massacres, together with the targeted killings of police and religious figures. In April 2015, Al-Shabaab attacked Garissa University College, killing 148 people, 142 of which were students; three were soldiers and three police officers (African Union 2015: 1). During the invasion by Al-Shabaab insurgents, Muslim students were separated from non-Muslim students and released, while all the others were executed.

Between 2009 and September 2013, 63% of violent activities attributed to Al-Shabaab have taken place in the north-eastern part of the country (Dowd/Raleigh 2013: 7). The coastal region has had secession elements, with grievances based on distinct ethno-regional identities and claims of socio-economic marginalisation. However, the calls for secession have been largely non-violent. The marginalisation claims are of concern, especially when they overlap with religious narratives used by Muslim extremists in the volatile region. The violence in the coastal region manifests in relatively high volatility, reflected in sporadic spikes in violence, followed by relative quietness.

In the past, religious tensions have served as key flashpoints for violence, with rioting following the violent deaths of prominent Muslim clerics in the region. In 2012 there were reported cases of killings of Muslim businessmen, traders, clerics and activists (Lind et al. 2015: 27–28; Anderson/McKnight 2014: 18). The assassinations further radicalised the locals who were sympathetic to their cause. This exacerbated the breakdown in trust in the country's institutions by the coastal people as they blamed assassinations of Muslims on the country's security forces (Dowd/Raleigh 2013: 7; Lind et al. 2015: 28; Anderson/McKnight 2014: 18).

Another manifestation of conflicts is maritime crime and terrorism. Kenya has a major port serving Central and Eastern Africa, with significant sea traffic into Mombasa. Kenya's counter-terrorism efforts have further exposed her as a terrorism target. This kind of crime affects revenues due to reduced dockings and use of the port. Al-Shabaab, an Islamist terrorist group mainly active in Somalia, skilfully manipulated historical injustices, socio-economic inequalities, and the treatment of minority groups to recruit members. Al-Shabaab's tactics include hitting 'soft targets' such as "open public places where a large number of civilians congregate and where the attacks can instil maximum terror, extreme fear, horror, panic and chaos, cause the greatest number of casualties and provoke a high sense of insecurity, uncertainty and doubt about national security forces' ability to stop the carnage" (African Union, 2015: 4). The deliberate targeting of civilians has increased sharply since 2010 and early 2011.

8.3 Drivers of Conflict in Kenya

The conflicts in Kenya can be attributed to various sources. They are ignited by causes from institutional, political-economic and social spheres operating at national, regional and even global levels (Gibbons 2014). The institutional causes include, but are not limited to, contested borders, unresolved land tenure issues, and failures of the policing and justice systems. Politico-economic factors include mining business enterprises without adequate revenue-sharing, land alienation, negative politics and corrupt public administrations. Social factors relate to marginalisation and exclusion, as well as issues of identity, gender and ethnicity.

8.3.1 Natural Resources

Natural resources are the predominant cause of conflict between pastoral communities and along the international borders. The Turkana are sporadically involved in separate clashes with the Pokot, the Karamojong and communities living to the east (such as the Rendile and Boran). The conflicts are fuelled by competition over commonly shared resources – mainly pastoral land and water – the cultural practice of livestock raiding and the associated desire for revenge. Furthermore, valued economic activities and resources – such as oil, coal, gypsum, metals, sand, stone, wood, water and other minerals – continue to drive conflict. The governance of these regions is distinct and varies from community to community. Issues like favouritism, exploitation and discrimination against other communities are rife.

8.3.2 Land and Development Projects

Inequity in land ownership and access cause dissatisfaction, as do development projects that are not delivered in a conflict-sensitive manner. This happens to be the dominant driver of conflicts in the coastal, the Rift Valley and the Mount Elgon regions. It is also gaining currency as a flashpoint in regions with important resources like minerals (for instance, oil in Turkana and Merti) and valued commercial centres like the Lamu port. The management of land resources is fraught with vested political interests, with local communities rarely involved in the decision-making process on how land is allocated. For instance, in re-settling landless communities in times of humanitarian emergencies or resource exploitation, the locals are rarely consulted or involved, making them perceive the allocations as unjust and a means of depriving them of their ancestral lands.

Besides, land is a highly emotive issue in Kenya and therefore a driver in a wide variety of conflicts (Mbugua 2013). Roughly 80% of Kenyans depend on farming for their survival, but only 20% of Kenya's land mass is viable for agricultural production, leading to intense competition over productive land. Further, the deteriorating value of land due to climate change is also emerging as a major driver of conflicts in the country (Mbugua 2013: 25).

Similarly, grievances over inequity in land ownership and access have not been settled. Development projects, large-scale land acquisition, and internal migration by other ethnic groups have created tensions over land in places like Isiolo County and on the coast. For example, some mega development projects implemented by the government at times cause conflict because they are delivered in a very insensitive manner. Lack of creation of awareness on the creation of wildlife conservancies also causes tensions among pastoralist communities, as has been the case in Narok and Samburu Counties.

Further, the economic development has not only alleviated poverty but also changed the relationship between generations, which has impacted negatively on inter-communal relations and shaped conflict dynamics, especially in northern Kenya.

8.3.3 Political Activities

Political events such as elections, political electioneering processes, or any community and large group-orientated elections (such as large group farms like Mbo-I-Kamiti, savings and credit societies, labour unions) are key drivers of conflict. This is mainly because of the involvement of unemployed youths hired by the political class as 'goons' to create chaos and harass opponents. Political parties tend to have vested interests (favouring a particular clique or ethnic group), with narrow perspectives and strong elements of rivalry.

In this fashion, politicians and elites fuel conflicts by hiring militias to mete out violence and in other instances by engaging in polarising discourses against certain groups and failing to represent Kenyans on issue-based concerns. Simply put, violence has been a tool used by political elites to improve their chances of election victory, either by driving away opposition voters or intimidating them from voting. Elite behaviour is deeply tied to favouritism in which state resources, jobs and contracts are allocated on an ethnic basis, which continues to feed inter-ethnic competition and stereotyping.

The militia groups include private armies for elites and politicians and militant groups with the objectives of overthrowing, replacing or seceding from the state. The groups are normally short-lived militant units, constituted for the purpose of pursuing elite interests for a short period of time. They are used by elites to carry out certain acts of violence on their behalf, without appearing to be affiliated to the political violence. For instance, between 1997 and September 2013, political militias engaged in the violence against civilians in Kenya (Dowd and Raleigh 2013). Political entrepreneurs also use ethnic affiliation to manipulate ethnic grievances as the basis for political mobilisation to gain power and control over resources. The roots of some of the violence in Kenya lie in the struggle to influence the balance of power and the distribution of economic resources in the country (Scott-Villers et al. 2014: 2). While the intensity of violence and persistence appear to differ according to place or period, the underlying motive in all cases is to remain in politics. Kenyan politics, which are partisan in nature, appear to be the major trigger of conflicts. The different political affiliations in the country have been characterised by widespread political violence along ethnic and class lines.

Al-Shabaab has recognised how effectively it can exploit the local politics by using social and economic grievances to deepen political divides and further its own cause (Anderson 2014). Al-Shabaab's violence has thus interwoven with local Kenyan violence. For instance, in Mpeketoni, Al-Shabaab has opportunistically linked its message to local Kenyan politics (Anderson/McKnight 2014). Its successful attacks have been facilitated by the failure of the state apparatus and the instrumentalisation of ethnic violence.

8.3.4 The Bulging Unemployed Youth

Young people who are susceptible to radicalisation are a serious threat to Kenya's peace and security. Poverty, unemployment, unfulfilled promises, and relative deprivation remain the key factors that frustrate the youth and thus increase the likelihood of its involvement in militias and other groupings (Kituku 2012). Violence becomes an avenue to vent and presents a lucrative enterprise for the unemployed youth, which contributes to its perpetuation. The narrative presented to Muslim youths is that the Muslims in Kenya suffer humiliation, especially from the Christian occupation of coastal land; they are told they can obtain revenge by killing prominent preachers, and fed stories about the liberating potential of violence (Lind et al. 2015).

Thus, youths join extremist groups as a counter-response to what they perceive as government-imposed collective punishment on Somalis and Muslims in the country (Botha 2014). The other significant factors in youth radicalisation are said to be uneven socio-economic development and historic marginalisation of some parts of Kenya (Lind et al. 2015: 19; Botha 2013: 13).

8.3.5 Community and Cultural Rivalry

Rivalry is a major driver of conflicts among the most prominent communities, even though it has also affected some minority communities. It is mainly attributed to the country's political history, where in the majority of instances force and not dialogue has been used to solve issues. In addition, minority communities have often not been given space for expression, and where their voice has been expressed, it has not been listened to. It is also important to point out that there are some minority communities with issues that have not been addressed. For instance, unresolved massacres, such as the Wagalla Massacre and the Shifta war among others, are examples of the use of government force, the assassination of certain luminary politicians, political expedience in sharing resources and opportunities, and the political loyalties and interests that drive the development of different regions, all of which are indicators of this type of rivalry, thereby breeding disgruntlement amongst certain communities.

Pastoralist communities have a long history of competition for pasture and grazing land, which has sometimes led to violence in the form of cattle-rustling, ethnic violence, displacements and massacres (Sharamo 2014: 3). Proliferation of small arms, devolution and development have changed the nature of this violence. The traditional pastoralist systems have been unable to cope with these changing conflict dynamics.

8.3.6 Proliferation of Small Arms

The proliferation of small arms provides more opportunities for violence and wider insecurity. It has been pointed out that there are over six hundred small arms in Kenya, the majority being automatic weapons. In most cases, the regions with the most frequent occurrence of politically-related violence seem to have more individuals with access to small arms, whilst the use of explosive devices (such as ordinary grenades and improvised explosive devices) has become a feature in several locations in the country.

8.3.7 Proximity to Training Grounds in Yemen and Somalia

Kenya's geopolitical position vis-à-vis Al Qaeda and Al Shabaab impacts regional and internal security. It is an open secret that Al Qaeda – and, to a lesser degree, Al Shabaab – use Yemen for training and recruitment. As a result – and due to internal socio-economic factors – some elements from Kenya could be involved. Whilst there is no evidence that this is currently happening, the proximity and temptation for young people to join militias and violent groups is clear, as the Sabaot case best illustrates.

8.3.8 Corruption and Impunity

Corruption has severely compromised the security sector, while impunity and lack of justice legitimise violence and lead to revenge attacks. There is a perception that police corruption, impunity, and judicial tractability, partly as a result of the institutions being underfunded and undermined, are sustaining violence (Scott-Villers et al. 2014). The failure of central government to "resolve small wars indicates, in the minds of the citizens, a form of approval" when people see political leaders being rewarded for their belligerence. As a result of their economic vulnerability,

Kenyans feel they are offered little choice but to line up behind political leaders who encourage violence. In the absence of justice, people's ways of coping with the intermittent violence, such as through revenge attacks, often adds fuel to the fire, which reinforces the likelihood of further violence (Scott-Villers et al. 2014).

Kenya's instability is compounded by long-standing inequitable policies and endemic corruption which has "severely compromised the security apparatus like the police and the armed forces" (Ombaka 2015; Parrin 2015). Corruption compromises security in terms of weak security intelligence: a poor anti-terror strategy; poor and inadequate equipment; incompetent/botched investigations; poor local and regional co-operation and a cycle of impunity and disregard for the rule of law (Ombaka 2015; Sharamo 2014).

Internal violent events, such as inter-ethnic clashes, livestock theft, boundary clashes and inter-clan rivalries, have assumed an ominous character in that the combatants no longer fear the interference of the state security apparatus in their activities, resulting in the murders of the members of police personnel who have responded to the violence (Ombaka 2015; Mbugua 2013).

8.3.9 Discrimination and Marginalisation

Long-term discrimination and marginalisation of certain groups and regions have been exploited by violent extremists. Regarding this, the Muslims in Kenya have the perception that they are marginalised within the Kenyan state and feel that they are treated as second-rate citizens (Anderson/McKnight 2014). This has led most Muslim youth to increasingly believe that the discrimination they face is part of the wider systematic discrimination against Muslims in the world (Mwakimako/Willis 2014; Botha 2013). Consequently, lack of trust of Kenya's government is widespread among the country's Muslim population, making it harder to tackle the vice of radicalisation.

Many researchers agree that the major underlying drivers of conflict in Kenya are unemployment, horizontal inequality, and highly centralised ethnopolitics (Elder et al. 2014; Sharamo 2014). This has resulted in the marginalisation of certain groups and regions. Kenya's violent Islamist mobilisation and militancy are greatly shaped by the local environment. The alienation, disaffection, and dissent of Kenya's Muslim community play a major role. The Christian values which dominate Kenyan politics have "taken on a Pentecostal and stridently evangelical tone which at best excludes Muslims and at worst is openly hostile to them" (Botha 2013). The Kenyan government is perceived as lacking the will or capacity to seek reconciliation with its Muslim citizens, which means it is missing the opportunity to find a panacea for the conflicts.

8.4 Attempts and Measures to Solve Conflicts

The international community has invested heavily in ensuring Kenya's stability. Some of the initiatives related to this include investing in programmes to avert election-related violence, financial crisis, climate and desertification, youth unemployment and human insecurity (Cox et al. 2014). Kenya is considered by the international community to be an important player in efforts to maintain regional stability in the Horn of Africa and Great Lakes regions. This explains why there has been heavy investment in safeguarding its much prized stability. The resultant international support has enabled the Kenyan government to put in place structures such as the *Provincial Peace Forum* (PPF), *District Peace Committees* (DPC), *Divisional Peace Committees* (DvPCs), and *Location Peace Committees* (LPCs), to better manage conflict-inducing social cleavages (Halakhe 2013). The conflict prevention efforts are coordinated by the *National Steering Committee* (NSC) on peacebuilding and conflict management. Large investments in new technology, early warning systems, and capacity-building programmes for the country's peace infrastructure have been procured. Further, the *National Cohesion and Integration Commission* (NCIC) was established to lead national efforts to change inter-ethnic group attitudes in Kenya and construct a more cohesive, peaceful national identity.

The promulgation of a new constitution was in response to the 2007–2008 election violence. Reforms in the governance and security institutions are meant to improve service delivery and make governance much more accountable and equitable across the country, with power being devolved to the forty-seven counties. The constitution guarantees parliamentary seats to marginalised groups and prevents discrimination (Cox et al. 2014). The reforms of the governance and security institutions are designed to improve service delivery and make governance much more accountable and equitable across the country. Despite such positive progress, political institutions "remain weak and predatory in nature, as they have often been captured and manipulated by elites to protect ethnic-based, political and economic interests" since actors and institutions have not yet embraced the constitutional values and principles.

There have been disarmament campaigns, increased security presence, and peacebuilding efforts to counter communal violence. The government has at times responded to the high levels of communal violence by "proposing a disarmament campaign in rural areas, alongside deploying the paramilitary *General Service Unit* (GSU) to areas affected by sharp spikes in communal violence" (Dowd/Raleigh 2013). These efforts may not be successful, given that the political dimensions of the problem are not addressed and that the poor infrastructure and poorly paid police and security forces make policing and security hard.

In October 2011, Kenya deployed troops in Somalia as a military campaign christened Operation *Linda Nchi*, or Protect the Country, ostensibly meant to "create a 'friendly' buffer-zone state in Juba land that would work in Kenya's interests". Kenyan military forces were accused of involvement in the charcoal trade, despite a UN ban on its export. This has served to boost Al-Shabaab's resources, which are heavily dependent on taxes raised from the Somali charcoal trade.

Another initiative was Operation *Usalama* Watch, launched in 2014, which saw police carry out massive swoops in majority-Somali neighbourhoods. The sense of 'collective punishment' increased Muslims' sense of alienation. Threats to deport Somali refugees and plans to build a barrier on the border with Kenya also contributed to this sense of alienation. These counter-terrorism responses are said to play into the hands of Al-Shabaab. This is due to politicisation of the counter-terrorism response, resulting in revenge attacks against different ethnic groups.

These conflict mitigation and peacebuilding efforts have been supported by society, the government and the international community. Local civil society organisations and interreligious groups' peace efforts have been negatively affected by divisive politics at local level and lack of donor support (Cox et al. 2014). Despite being flawed and limited, these agreements have often been more successful at creating peace and a sense of justice than modern state law (Scott-Villers et al. 2014). For instance, the 2008 Maikona-Walda Declaration effectively ended active hostilities between the Borana and Gabra. Kenya has also initiated the erection of the wall along the Somalia-Kenya common border to seal off the porous border. This was meant to prevent incursions by Al-Shabaab. These responses by the government have caused increased widespread fear among Muslim and Somali communities, while doing little to stop the rise of Islamic extremism in Kenya (Parrin 2015). Experts and researchers point at poor governance and stalled police reforms as the real enemy within (Parrin 2015).

8.5 Conclusion and Recommendation

In some instances, political authorities are to blame for some of the conflicts. Evidence presented in the cases discussed in this chapter shows that there have been open clashes between the local people and the state, as in the north-eastern province. In other cases, conflicts have existed between potentially antagonistic ethnic groups, as in the case of the Rift Valley and Mombasa clashes.

The failure to address various problems embedded in Kenya's social structures makes peace a mirage. The problems which need to be fixed include economic inequalities, provision of basic essentials, the culture of violence as manifested in the formation of gangs used by politicians to settle political scores, gender inequalities and unequal power relationships between women and men, and the creation of an effective and accessible justice system that has made women and girls in the IDP camp susceptible to multiple levels and various forms of violence (Njiru 2014: 62).

This chapter makes the following recommendations to strengthen Kenya's security situation (Lind et al. 2015). It is necessary to: (i) address regional inequalities and historic marginalisation by following the moral intent of Kenya's 2010 Constitution and sincerely implementing its provisions to devolve powers and resources to new sub-national county governments; (ii) remove the institutionalized discrimination of Kenya's Somalis and Muslims through systematic reforms to undo "hierarchies in citizenship", thereby overcoming the inherent limits of the existing discrete counter-radicalisation efforts; and (iii) mobilise political support for the security sector and policing reforms designed to rein in abusive, predatory and corrupt practices and to promote accountability to all the citizens in need of protection.

Other important recommendations include (i) fast tracking security sector reforms; (ii) making sure that police are properly equipped and their welfare is looked after; (iii) addressing corruption; (iv) taking personal responsibility for security lapses; (v) adding human rights to the police curriculum; (vi) sensitising the public on national cohesion; (vii) addressing youth unemployment; (viii) investigating and prosecuting human rights violations; (ix) facilitating the movement of police vehicles; and (x) making public the findings of the Tana River Inquiry and conducting inquiries into other security lapses (KNCHR 2014: 11–12).

Further, organisations like the African Union recommend combating the threat of Al-Shabaab by: (i) strengthening international and regional co-operation and coordination; (ii) enhancing the capacity of defence, intelligence and security services and law enforcement; and (iii) heavy investment in prevention and building a strong partnership between the state and different national stakeholders, including the communities living in border areas, to win commitment and genuine support for the government's counter-terrorism efforts (African Union 2015: 5).

References

Anderson, David, 2014: *Why Mpeketoni Matters: Al-Shabaab and Violence in Kenya.* NOREF Policy Brief (Oslo: Norwegian Peacebuilding Resource Centre).

Anderson, David; McKnight, Jacob, 2014: "Kenya at War: Al-Shabaab and its Enemies in Eastern Africa", in: *African Affairs,* 114,454: 1–27,

African Union, 2015: *Incident Analysis: Terrorist Attack, Garissa University, Kenya.* ACSRT/Incident-Analysis-009-2015 (Alger: African Union).

Botha, Anneli, 2013: *Assessing the Vulnerability of Kenyan Youths to Radicalisation and Extremism.* ISS Paper 245 (Pretoria: Institute for Security Studies).

Botha, Anneli, 2014: *Radicalisation in Kenya: Recruitment to al-Shabaab and the Mombasa Republican Council.* ISS Paper 265 (Pretoria: Institute for Security Studies).

Cox, Fletcher D.; Orsborn, Catherine R.; Sisk, Timothy D., 2014: *Religion, Peacebuilding, and Social Cohesion in Conflict-Affected Countries* (Boulder, Co.: University of Denver); at: https://www.du.edu/korbel/sie/media/documents/faculty_pubs/sisk/religion-and-social-cohesion-reports/rsc-researchreport.pdf.

Dowd, Caitriona; Raleigh Clionadh, 2013: *ACLED Country Report: Kenya* (Brighton: ACLED).

Elder, Claire; Stigant, Susan; Claes, Jonas, 2014: *Elections and Violent Conflict in Kenya: Making Prevention Stick* (Washington, DC: United States Institute of Peace).

Gibbons, Sarah, 2014: *Draft Discussion Brief: Towards Peace and Security in Dryland Kenya: The Demand for a New Approach* (Pastoralist Parliamentary Group and Drylands Learning and Capacity Building Initiative for Improved Policy and Practice in the Horn of Africa).

Halakhe, Abdullahi Boru, 2013: *'R2P in Practice': Ethnic Violence, Elections and Atrocity Prevention in Kenya.* Occasional Paper Series No. 4 (Nairobi: Global Centre for the Responsibility to Protect).

Idris, Ahmed; Odoyo, Nicholas; Maruga, Peter; Kariuki, Nahashon; Maxwell, Daniel; Marshak, Anastacia; Richards; Simon; 2013: *Conflict Management and Disaster Risk Reduction: A Case Study of Kenya* (Medford, MA: Feinstein International Center).

ICG, 2014: *Kenya: Al-Shabaab – Closer to Home.* Update Briefing: Africa Briefing No. 102 (Brussels: ICG).

IRIN, 2015a: "Repatriation Threat Alarms Somali Refugees in Kenya", in: *IRIN* (Geneva: Refworld); at: https://www.refworld.org/docid/552cda914.html.

IRIN, 2015b: "Kenya's Anti-Terror Border Wall Sparks Heated Debate", in: *IRIN* (Geneva: Refworld, 17 April); at: https://www.refworld.org/docid/5535f9f44.html.

Kituku, Wambua, 2012: *Building a Culture of Peace in Kenya: Baseline Report on Conflict-Mapping and Profiles of 47 Counties in Kenya* (Nairobi: CRECO).

KNCHR, 2012: *29 Days of Terror in the Delta: KNCHR Account into the Atrocities at Tana Delta* (Nairobi: KNCHR).

KNCHR, 2014: *Are We Under Siege? The State of Security in Kenya: An Occasional Report (2010–2014)* (Nairobi: KNCHR).

Lind, Jeremy; Mutahi, Patrick; Oosterom, Maryjok, 2015: *Tangled Ties: Al-Shabaab and Political Volatility in Kenya.* IDS Evidence Report No. 130: Addressing and Mitigating Violence (Brighton: IDS).

Lynch, Gabrielle, 2014: "Electing the 'Alliance of the Accused': The Success of the Jubilee Alliance in Kenya's Rift Valley", in: *Journal of Eastern African Studies*, 8,1: 93–114; at: https://doi.org/10.1080/17531055.2013.844438.

Mbugua, Joseph, 2013: *Inter-Communal Conflicts in Kenya: The Real Issues at Stake in the Tana Delta.* Issues Briefs No.1 (Nairobi: International Peace Support Training Centre).

Mc Evoy, Claire, 2013: *Shifting Priorities: Kenya's Changing Approach to Peacebuilding and Peacemaking.* NOREF Report (Oslo: NOREF).

Ministry of State for Planning, National Development and Vision 2030, 2010: *2009 Population & Housing Census Results* (Nairobi: Ministry of State for Planning, National Development and Vision, 2030).

Mwakimako, Hassan; Willis, Justin, 2014: *Islam, Politics and Violence on the Kenya Coast*, Note 4 (Bordeaux: Observatoire des Enjeux Politiques et Securitaires dans la Corne de l'Afrique).

Njeri, Mary; Ogola, S., 2014: *My Action Counts: An Assessment of Gender-Based Violence Responses in Nine Counties of Kenya* (New York: International Rescue Committee).

Njiru, Roseanne, 2014: "Political Battles on Women's Bodies: Post-Election Conflicts and Violence Against Women in Internally Displaced Persons' Camps in Kenya", in: *Societies Without Borders*, 9,1: 48–68.

Okumu, Willis, 2013: *Trans-Local Peace Building among Pastoralist Communities in Kenya: The Case of Laikipia Peace Caravan.* Culture and Environment in Africa Series 3 (Cologne: African Studies Centre).

Ombaka, Dickson, 2015: "Explaining Kenya's Insecurity: The Weak State, Corruption, Banditry and Terrorism", in: *International Journal of Liberal Arts and Social Science,* 3,3: 11–26.

Owuor, Victor; Wisor, Scott, 2014: *The Role of Kenya's Private Sector in Peacebuilding: The Case of the 2013 Election Cycle* (Broomfield, Co: One Earth Future Foundation).

Parrin, Anjili, 2015: "Is Kenya's Security Policy the Real Enemy Within?", in: *IRIN* (Geneva: Refworld, 21 April); at: https://www.refworld.org/docid/553a3526a.html.

Scott-Villers, Patta; Ondicho, Tom; Lubaale, Grace; Ndung'u, Diana; Kabala, Nathaniel; Oosterom, Marjoke, 2014: *Roots and Routes of Political Violence in Kenya's Civil and Political Society: A Case Study of Marsabit County.* IDS Evidence Report No. 71: Addressing and Mitigating Violence (Brighton: IDS).

Sharamo, Roba, 2014: *The Politics of Pastoral Violence: A Case Study of Isiolo County, Northern Kenya.* Working Paper 095 (Brighton: Future Agricultures).

Thomas, Kylie; Masinjila, Masheti; Bere, Eunice, 2013: "Political Transition and Sexual and Gender-Based Violence in South Africa, Kenya, and Zimbabwe: A Comparative Analysis", in: *Gender & Development*, 21,3: 519–532.

TSA, 2014: *National Security Profile for the Republic of Kenya* (Port Louis: TSA).
UNICEF, 2012: *Violence Against Children in Kenya: Findings from a 2010 National Survey* (New York: UNICEF).
Warah, Rasna, 2015: "With Graft's Tentacles Reaching into Military, Insecurity is Guaranteed", in: *Daily Nation* (11 May).

Chapter 9
Human Rights and Sexual Abuse of the Girl-Child in Nigeria: Implications for Development

Janet Monisola Oluwaleye

Abstract The prevalence of sexual abuse of the girl-child in Nigeria in recent times has made it a subject of discourse and headlines in many newspapers. Despite the declaration in Nigeria's Sexual Offences Act that it is an offence to attempt intercourse with a child below eighteen years of age, the rising rate of the destructive offence right across the country is alarming. The question is: what are the reasons for the rising rate of sexual abuse of the girl-child in Nigeria? What is the nature and what are the implications of the abuse? This study, using secondary sources, analytically investigates sexual abuse of the girl-child in Nigeria, including the nature, scope, causes, and consequences of such a disheartening practice that has become rampant in Nigeria. The data gathered are descriptively analysed. The study identifies poverty, hawking, lack of values or parental control, pornographic pictures and the abuse of social media as culpable. Other causes include the get-rich-quick syndrome, unemployment, bad governance, and weak enforcement of law and culture. Sexual abuse of the girl-child is perpetrated by the rich and the poor, teachers and coaches, religious leaders and laymen, educated and illiterate people, and more especially by family members and neighbours. The study further argues that the attendant socio-economic effects of girl-child abuse in Nigeria include backwardness, cheating, rape, abortion, unwanted pregnancy, broken homes, single parenting, health hazards, population growth, an increase in unemployment, and underdevelopment. To curb the act, the study recommends realistic and effective government policies, the provision of jobs, a strong security base, intervention by community and religious leaders, and penalties for offenders.

Keywords Abuse · Assault · Girl-child · Rape · Sexual violence

Janet Monisola Oluwaleye, Lecturer, Ekiti State University, Ado-Ekiti, Nigeria; *Email*: jmoluwaleye@gmail.com and monisola.oluwaleye@eksu.edu.ng.

Ú. Oswald Spring and H. G. Brauch (eds.), *Decolonising Conflicts, Security, Peace, Gender, Environment and Development in the Anthropocene*, The Anthropocene: Politik—Economics—Society—Science 30,
https://doi.org/10.1007/978-3-030-62316-6_9

9.1 Introduction

Human rights are universal and belong to all human beings, including children. According to Grant (2017), human rights apply to people of all ages. It doesn't matter whether you are seven or seventy-two; we can all expect the same basic protection. These rights encompass civic, political, economic, social and cultural rights. Children and youths also enjoy certain human rights specifically linked to their status as minors and to their need for special care and protection (PDHRE 2018). The need to extend care to the child has been stated in the *Geneva Declaration of the Rights of the Child* (1924) and in the *Declaration of the Rights of the Child* adopted by the General Assembly on 20 November 1959 and recognised in the *Universal Declaration of Human Rights*, in the *International Covenant on Civic and Political Rights* (in particular in Articles 23 and 24), in the *International Covenant on Economic, Social and Cultural Rights* (in particular Article 10) and in the statutes and relevant instruments of specialised agencies and international organisations concerned with the welfare of children (OHCHR 2018; Humanium 2018). A child, according to Part 1 Article 1 of the *United Nations Universal Declaration of Human Rights* (UDHR), means every human being below the age of eighteen years (OHCHR 2018).

Among these rights are: the "human right to a standard of living adequate for a child's intellectual, physical, moral and spiritual development"; the "human right to freedom from cultural practices, customs and traditions harmful to the child", including female genital mutilation; "the human right to life and to freedom to education" – to free and compulsory elementary education, to equal access to readily available forms of secondary and higher education, to freedom from all types of discrimination at all levels of education; "the human rights to information about health, sexuality and reproduction" and "the human right to protection from all physical or mental abuse" (PDHRE 2018). The rate of sexual abuse of the child-girl is more rampant in recent times, all over the country, in spite of the declaration in the Sexual Offences Act that it is an offence to attempt intercourse with a child below eighteen years of age. The rights of the girl-child to a decent life have overall effects on her psychological, mental, social, economic and physical development.

According to Akpoghome/Nwano (2016), the sexual defilement of children is a global pandemic affecting children of various classes and both the educated and uneducated, regardless of their race, ethnic background or religion. Child sexual abuse has been described as any sexual act between an adult and a minor, or between two minors when one exerts power over the other, forcing, coercing or persuading a child to engage in any type of sexual act, including a non-contact act such as exhibitionism, exposure to pornography, voyeurism, or communicating in a sexual manner by phone or internet (Darkness to Light 2018). More than before, sexual abuse of minors by adults has become rampant in Nigeria in the last two decades.

A meta-analysis by the University of Barcelona of sixty-five research studies across twenty-two countries to estimate an 'overall international figure' of sexual abuse of children discovered that, geographically, the highest prevalence of child sexual abuse was found in Africa (Wihbey 2011). UNICEF estimates that 120 million girls under the age of twenty have been subjected to forced sexual intercourse or other involuntary sexual acts (Falade/Fasuan 2017). One in ten children is reported to have been abused before they are eighteen years old (Darkness to Light 2018). Research conducted by the *Centre for Disease Control* (CDC) estimated that approximately one in six boys and one in four girls are sexually abused before the age of eighteen. UNICEF (2017) reported that one of four girls and 10% of the boys in Nigeria have been victims of sexual violence. According to the US Department of Justice, only 10 per cent of perpetrators were strangers (Daniles 2018). Most perpetrators of sexual abuse are known to the victims. Flowing from the above are the following research questions: what are the reasons for the rising rate of sexual abuse of the girl-child in Nigeria? What is the nature and what are the implications of the abuse?

9.2 Theoretical and Conceptual Clarification of Sexual Abuse and the Human Rights of Children

Human rights, according to the Equality and Human Rights Commission (2018), are basic rights and freedoms that belong to every person in the world, from birth until death. These rights are extended to all categories of human beings in spite of age, from infant till death. This indicates that children are not exempted from basic rights and freedoms. In the same vein, the United Nations (2018) described human rights as:

> basic rights inherent to all human beings, regardless of race, sex, nationality, ethnicity, language, religion, or any other status. Human rights include the rights to life and liberty, freedom from slavery and torture, freedom of opinion and oppression, the right to work and education, and many more. Everyone is entitled to these rights, without discrimination.

The 'all' coverage includes children without discrimination. Children as part of the 'all' should not be denied liberty and should possess freedom from torture. UNICEF Canada (2018) refers to human rights as basic standards to which every person is entitled in order to survive and develop in dignity. All children's human rights apply to children at all times, without exception (UNICEF Canada 2018). Children should not be deprived of the basic standard rights because they are under age. They should be given full privilege to enjoy their rights at every stage of their development.

According to Weston (2018), human rights are rights that belong to an individual simply for being human, or as a consequence of inherent human vulnerability, or because they are requisite to the possibility of a just society. To him, human rights refer to a wide continuum of values or capabilities thought to enhance human agency or protect human interests and declared to be universal in character, in some sense claimed for all human beings, present and future, no matter their age.

Sexual abuse, on the other hand, for Ituma et al. (2013: 498), is unwanted sexual activity, with the perpetrators using force, making threats or taking advantage of the person or persons concerned, who are not able to give approval or consent. It is not as a result of mutual agreement between equals.

The definition of child sexual abuse formulated by the 1999 WHO Consultation on Child Abuse Prevention (62) stated that

> child sexual abuse is the involvement of a child in sexual activity that he or she does not fully comprehend, is unable to give informed consent to, or for which the child is not developmentally prepared and cannot give consent, or that violates the laws or social taboos of society. Child sexual abuse is evidenced by this activity between a child and an adult or another child who by age or development is in a relationship of responsibility, trust or power, the activity being intended to gratify or satisfy the needs of the person. This may include but is not limited to: the inducement or coercion of a child to engage in any unlawful sexual activity; the exploitative use of a child in prostitution or other unlawful sexual practices; the exploitative use of children in pornographic performance and materials. (WHO 2018)

From the above definition, the child is ignorant of the sexual activity, she or he is not mature enough to give informed consent to the involvement, and/or she or he is either coerced or induced (through petting or giving gifts to lure the innocent child). It also indicates that the act is to gratify or satisfy the needs of the perpetrator. It further reveals the exploitative use of child in prostitution or other unlawful sexual activity.

Child sexual abuse is a form of child abuse that includes sexual activity with a minor. A child cannot consent to any form of sexual activity, period. When a perpetrator engages with a child this way, they are committing a crime that can have lasting effects on the victim for years. Child sexual abuse does not need to include physical contact between a perpetrator and a child. Some forms of child sexual abuse include:

- Exhibitionism, or exposing oneself to a minor;
- Fondling;
- Intercourse;
- Masturbation in the presence of a minor or forcing the minor to masturbate;
- Obscene phone calls, text messages, or digital interaction;
- Producing, owning, or sharing pornographic images or movies of children;
- Sex of any kind with a minor, including vaginal, oral, or anal;
- Sex trafficking;
- Any other sexual conduct that is harmful to a child's mental, emotional, or physical welfare (RAINN 2018).

From the above, various forms of sexual abuse were identified, including exposing oneself to a minor, fondling, intercourse, sharing pornographic images to children, vaginal, oral or anal sex or sex trafficking. It is an abuse in whatever form it takes when the perpetrator engages with a child.

Two types of child sexual abuse have been identified. These include contact (touching) and non-contact (non-touching) abuse (NSPCC 2018; Stop it Now 2018). Contact abuse involves touching activities where an abuser makes physical contact with a child. It includes:

- Sexual touching of any part of the body whether the child is wearing clothes or not;

- Rape or penetration by putting an object or body parts inside a child's mouth, vagina or anus;
- Forcing or encouraging a child to take part in sexual activity;
- Making a child take their clothes off, touch someone else's genitals or masturbate.

Non-contact abuse involves non-touching activities, such as grooming, exploitation, persuading children to perform sexual acts over the internet and flashing. It includes:

- Encouraging a child to watch or hear sexual acts;
- Not taking proper measures to prevent a child being exposed to sexual activities by others;
- Meeting a child following sexual grooming with the intent of abusing them;
- Online abuse, including making, viewing or distributing child abuse images;
- Allowing someone else to make, view or distribute child abuse images;
- Showing pornography to a child;
- Sexually exploiting a child for money, power or status (child exploitation).

From the definition of the National Society for the Prevention of Cruelty to Children (NSPCC), it can be deducted that the sexual abuse of children does not just involve physical contact but extends to other such activities that can arouse or influence sexual sensitivity in children.

Sexual abuse of a child is an express abuse of the human rights of that child. A sexually abused child is subjected to physical and psychological torture which may last throughout the lifetime of the victim. Some victims are subjected to sexually transmitted diseases, thereby jeopardizing the child's right to the enjoyment of the highest standard of health. The psychological effects on the brains of abused children has been reported to have negative impacts on their educational performance, consequently determining the future status of victims (Darkness to Light 2018) The status in the long run has implications for nation-building and development, such as the nation's poverty level, mortality rate, fertility rate, standard of living, health standard, inequality and proportion of participation in the work force.

9.3 Theoretical Clarification

This study employs the *Four Pre-Conditions Model*. This model was postulated by Finkelhor in 1984 (SECASA 2017). He grouped all factors contributing to child sexual abuse into four pre-conditions:

i. *Motivation.* This includes Emotional Congruence, in which sexual contact with a child satisfies emotional needs; Sexual Arousal, in which the child represents the source of sexual gratification for the abuser; and Blockage, when alternative sources of sexual gratification are either not available or are less satisfactory. Only one of the three is needed to be present for sexual abuse to occur.

ii. *Internal inhibitions*. He posits that no matter how strong the sexual interest might be, if the abuser is inhibited by taboos then he will not abuse.
iii. *External inhibitors*. The potential abusers must overcome external inhibitors prior to sexual abuse. External inhibitors that may restrain the abuser's action include the family constellation, neighbours, peers and societal sanctions as well as the level of supervision a child receives. Lack of supervision has been found to be a contributing factor to sexual abuse. External inhibitors are easily overcome if the potential abuser is left alone with an unsupervised child.
iv. *The potential abuser* has to overcome the child's possible resistance to being sexually abused. Abusers may sense which children are good potential targets, who can be intimidated or coerced to keep secret or otherwise manipulated (SECASA 2018).

The proposition above specified some of the causes of sexual abuse, such as motivation from the abused which may be in the form of indecent dressing and exposing sensitive parts of the body. External inhibitors such as the carelessness of parents or care-givers in supervising the child were also identified. Finkelhor further noted possible resistance from the child as another condition. The second proposition is in form of restrictions on the abuser, who may be deterred by the fear of sanctions or the taboo. This model is relevant to explain some of the causes of and remedies for the sexual abuse of the girl-child in Nigeria.

The main limitation of the model is that it is essentially a descriptive framework which incorporates a range of dissonant theories and observed clinical data. As such, it cannot be viewed as a theory until it is tested empirically, in particular, in its application to treatment and prevention. However, it presents a comprehensive, multi-causal, hierarchical model with both psychological and sociological explanatory power for understanding why and how sexual abuse occurs, which is more than can be said of other theoretical approaches (SECASA 2018).

9.4 The Nature and Scope of the Sexual Abuse of the Girl-Child in Nigeria

9.4.1 The Spread

There is consensus among scholars that the sexual abuse of the girl-child is a universal problem (Drezett et al. 2001: 413; Singh et al. 2014; Akin-Odanye 2018; UNICEF 2018). The 2014 UNICEF study estimated that around 120 million girls under the age of twenty (about one in ten) have been subjected to forced sexual intercourse or other forced sexual acts at some point in their lives (UNICEF 2018). Chiedu – cited in Falade/Fasuan (2017) – affirmed that the perplexing prevalence of various forms of sexual abuse in Nigeria has gone up from 12.5 to 80%. Cases of sexual abuse of the child-girl in Nigeria have increased during the last two decades. For instance,

Ms Ajie estimated that at least two million Nigerian girls experience sexual abuse annually (Akinbobola 2020).

9.4.2 *The Perpetrators*

Most studies on the sexual abuse of the girl-child revealed that the majority of perpetrators were people familiar to the victims (Barajas 2017; Choji 2017; Advocates for Youths 2018; Darkness to Light 2018; Hargrove 2014; Stop it Now 2018). Advocates for Youths (2018) reported that perpetrators are relatives or close acquaintances of those they target. According to a paediatrician, Dr. Biola Ayodele, "What used to be a taboo and sacrilege gradually became a national spread, with everyday occurrences of horrifying cases of fathers abusing their young daughters, unabating cases of incest and rape of minors in their care" (Choji 2017). Advocates for Youths (2018) reported that in 93% of cases of a child's sexual abuse, the child knows the person that commits the crime. A similar body in the fight against child sexual abuse, Stop it Now (2018), noted that most perpetrators of child sexual abuse are relatives or close acquaintances of the youths they target. It is notable that most perpetrators were not strangers but people who were familiar to the abused, such as step-fathers, care-givers, parents, neighbours, teachers and other close acquaintance.

9.4.3 *The Culture of Silence*

In spite of the fact that disclosure by the victim is one of the ways in which sexual abuse of a child can be known, most victims of child sexual abuse often delay speaking of it or keep the secret till adulthood (Fontes/Plummer 2010: 493; Hargrove 2014; Choji 2017; Badam 2017). Child sexual abuse is believed to be uniformly shrouded in secrecy and silence in all cultures. Many scholars have posited that the victim's delay in disclosing sexual abuse may be associated with the feeling of guilt for not telling sooner, especially at the stage of 'grooming' (Fontes/Plummer 2010: 493). Some victims failed to disclose abuse for fear of stigma or violence. For instance, a victim referred to as Wanda from Benue State in Nigeria, who experienced sexual abuse from her nanny (who is also a close relative), revealed that, "I didn't want to vanish, so I had to keep quiet" (Edeh 2018). She could not tell her parents because of fear and the threat of being killed by the nanny. Others kept the abuse they suffered a secret because of shame or the cost of justice (Fontes/Plumme 2010; Racebaitr 2016). Hargrove (2014) submitted that many did not report because they were not abused by strangers but by a neighbour, a teacher, a family friend or even a parent. The culture of silence hinders perpetrators being penalised or punished to serve as a deterrent to others.

9.4.4 Abusers' Strategies

Child sexual abuse is carried out against the child using force, trickery, bribes, and threats or pressure (Howe 2013: 16). Darkness to Light (2018) refers to grooming as a process by which an offender gradually draws a victim into a sexual relationship and maintains that relationship in secrecy. At the same time, the offender may also fill roles within the victim's family that make the offender trusted and valued. The behaviour associated with grooming includes:

- Offering special interest, treats, gifts and attention to the child;
- Isolating the child from others;
- Fulfilling the child's unmet needs;
- Treating the child as if she is older;
- Gradually crossing physical boundaries and becoming increasingly intimate/sexual;
- Using secrecy, blame and other threats to maintain control (Darkness to Light 2018; Fradkin 2018).

Children should be enlightened and given sexuality education covering the above strategies to avoid falling prey to the abusers.

9.4.5 Coercion

Coercion is the intimidation of a victim to compel the individual to do some act against his or her will by the use of psychological pressure, physical force or threats (*The Free Dictionary* by Farlex 2018). According to MNCASA (2018), victims coerced have been badgered, tricked, threatened with being 'outed', or kept from eating, sleeping, using the bathroom or otherwise held hostage until they quit resisting. According to Minnesota Coalition Against Sexual Assault (MNCASA 2018), consent is not present when either partner:

- Fears the consequence of not consenting (including use of force);
- Feels threatened or intimidated;
- Fears being 'outed';
- Is coerced;
- Says no, either verbally or physically (e.g. crying, kicking or pushing away);
- Has a communication barrier that prevents the person from understanding what is being said;
- Has differing abilities that prevent the person from making an informed choice;
- Is incapacitated by alcohol or drugs;
- Lacks full knowledge of or information about what is happening;
- Is not an active participant in the activity;
- Is under the legal age of consent.

9.5 Causes of Sexual Abuse of the Girl-Child in Nigeria

A lot of reasons have been put forward by scholars on the causes of sexual abuse of girl-child in Nigeria. These include the following.

9.5.1 *Environmental Factors*

It was discovered that some situations make girl-children vulnerable to sexual abuse, such as living with a stepfather or a male adult who is not their biological father. Also, the WHO identified unaccomplished children, children in foster care, adopted children, stepchildren, and physically or mentally handicapped children to be at greater risk of being abused. Olusegun/Idowu (2016: 17) noted that children without proper parental care and supervision are easily abused. They opined that parents who send their daughters on errands at night, give them goods to sell, leave them for several hours or days in the care of servants, or allow them freedom of choice in lifestyle and clothing put them at greater risk of being abused. Abusers usually take advantage of a conducive environment to perpetrate their evil activities.

9.5.2 *Indecent Dressing*

Indecent dressing, especially by teenagers, has been discovered to be a motivating factor for sexual abuse. According to a survey carried out by Ayogu, cited in Hillary/Rimamsikwe (2014: 35), some clothes that parents buy for their children with inscriptions like, "I am a sexy girl", "Hug me tight", "Lovely babe-sexy babe", "I am for u", "Kiss me fast", "Touch me", "Deep kiss", "Romance me", "Devil's advocate", present a strong pull to rapists to carry out their exploits. Ituma et al. (2013: 506), in the same line, noted that provocative dressing sends out signals which stimulate lecherous thoughts and actions. Indecent dressing is like an invitation to sexual arousal which has landed many in a situation they did not originally intend.

9.5.3 *Poverty*

Poverty has been identified as another factor in sexual abuse. because of its educational disadvantage which causes children to be out of school and on the streets, with the aim of earning a living or combining the money they earn with their parents' income (Olusegun/Idowu 2016). This situation is believed to expose children to bad people who may abuse them sexually. For instance, Abubakar, an Islamic scholar, pointed out that girls selling kolanuts, packets of fried groundnuts and detergents on their heads are a common sight in Kano, and the girls are lured by rapists who pretend they want to buy their wares (IRIN News 2008). Some girls engaged in street trading who enter the houses of customers in an attempt to sell their wares and sell late into the night have fallen prey to abusers.

9.5.4 *Superstitious Beliefs*

Another factor for sexual abuse of the girl-child in Nigeria is the superstitious belief held by some men that having sex with a virgin is advantageous. Hilary/Rimamsike (2014: 35) and Abubakar (IRIN News 2008) revealed that some men believed that they could be cured of AIDS/HIV by having sexual intercourse with a virgin. This led men living with HIV/AIDS to resort to raping innocent girls in their quest for a cure (Hilary/Rimamsike 2014: 35). Another category of men believe that they can become rich by having sex with a virgin. This too has motivated rapists to commit the wanton act of child sexual abuse (IRIN News 2008). Such superstitious beliefs that are dangerous to the well-being of others should be made taboo, with relevant sanctions to instil fear and also serve as a deterrent to desperate abusers.

9.5.5 *Ignorance and Lack of Sex Education*

A survey by Ohia (2016: 138) discloses that many mothers themselves do not know the teachings that constitute sexual education, probably because of the nature of silence on sexuality and sex-related matters. She discovered through a survey study that many mothers are not in favour of giving sexuality education, while those who were in favour of giving sexuality education felt it should be after the first menstruation or when the child is about to get married. The Heal Project revealed that the culture of silence and shame around sex and sexuality, in a hyper-sexual society that exposes sexual imagery but does not talk about it, creates a breeding ground for child sexual abuse (Racebaitr 2016).

9.5.6 *Failure to Bring Perpetrators to Book and the Challenge of the Justice System in Nigeria*

Failure to bring perpetrators of child sexual abuse to book has been revealed as one of the factors perpetuating sexual violence against children. Barajas (2017) revealed that child sexual abuse is spurred by silence encouraged by the media, the legal system and culture. In the same vein, Awosusi/Ogundana (2015) opined that the non-reporting of sexual violence is a clog in the wheel of delivering the necessary judgement to assailants. They submitted that failure to disclose can leave the victims without support and thereby trapped in an unbroken cycle of sexual violence. The idea of settling such cases at home should be discouraged, while victims and their parents should be given the necessary support to ensure that perpetrators are punished.

9.5.7 Consequences of the Sexual Abuse of a Girl-Child in Nigeria

The sexual abuse of a girl-child has both short-term and long-term socio-economic implications for development and nation-building. According to UNICEF Canada (2018), policy-making that fails to take children into account has a negative impact on the future of all members of society. The organisation also noted that the inter-connected and equal importance of all rights should guide the government to act and make decisions which respect all human rights.

Article 34 of the United Nations Office of the High Commissioner committed States Parties to protect the child from all forms of sexual exploitation and sexual abuse. The States are to take appropriate measures to prevent the inducement or coercion of a child to engage in any unlawful sexual activity (OHCHR 2018). The rampant cases of sexual abuse of girl-children in Nigeria is a flagrant abuse of these rights, exploitation of the victims and denial of the rights to personal dignity. Moreover, it is an abuse of the right of every child to a standard of living adequate for the child's physical, mental, spiritual, moral and social development, as stated in Article 27 of the United Nations Office of the High Commissioner on Human Rights (OHCHR 2018). These, also, have long-term effects on society and the nation at large.

Medically, sexual abuse increases the victim's chances of infection due to lack of sufficient vaginal discharge. It may cause internal lacerations that result in bleeding, sexually transmitted diseases or permanent vaginal fluid (Choji 2017). These developments are against the rights of the child, as stated in Article 24 of the United Nations Human Rights: "the right of the child to the enjoyment of highest standard of health …" (OHCHR 2018). Victims of child sexual abuse are highly exposed to the danger of sexually transmitted diseases such as gonorrhoea, syphilis, herpes genitalis, chancoid, HIV/AIDS, and monolia, among others (Ituma et al. 2013: 507). According to Ndukwe, cited in Ituma et al. (2013: 507), sexually transmitted diseases reduce and banish their victims to the status of presently absent human beings. Thus, they are stigmatised and discriminated against, and consequently unable to contribute their quota to societal development.

In spite of Article 29 of United Nations Office of the High Commissioner, which supports "The development of the child's personality, talents and mental and physical abilities to their fullest potential", research studies have revealed that traumatic stress, such as that caused by sexual abuse, may cause notable changes in brain functioning and development (Mullen/Fleming 2014; Akoloh et al. 2016; Choji 2017).

Akoloh et al. (2016) reported differences in reading habits, mastery of the subjects, class participation, submission of assignments and passing grades. Consequently, the potential and opportunities of sexually abused children are obstructed.

Findings of various scholars also revealed that children who are sexually abused are at greater risk of post-traumatic stress and other similar symptoms, such as depression and suicide attempts (Briere/Elliot 1994: 57–58; Hilary/Rimamsikwe 2014: 37; Darkness to Light 2018). Such psychological problems can disrupt normal development, leading to far greater physical aggression, non-compliance and oppositionality than in non-abused children (Briere/Elliot 1994: 58, 62; Mulen/Flemming 2014; Darkness to Light 2018).

Additionally, some of the victims are introduced to erotic sexual behaviour which may lead to unplanned pregnancy, dropping out of school and becoming a single parent. These can hinder the capacity-building of such victims and their offspring, thus perpetuating cycles of poverty and obstructing national economic development.

Omede/Agahiu (2016), who revealed the importance of girl-children in nation-building, identified sexual violence as one of the factors hindering effective girl-child education. The economic impacts of sexually abused girl-children on nation-building and development include perpetuating the poverty cycle, deterioration in health, increased inequality, increased birth rates, increased infant and mortality rates, a reduction in participation in the female labour force and reduced earnings.

Thus, the influence of a girl-child on the socio-economic development of any country cannot be over-emphasised. According to This day (2018),

> She is the channel through which the balance of human existence is built upon. She is a life giver and sole connection between the idea of life and life itself. Girls who are educated, healthy and free can transform their communities when they become mothers and pass on the benefits to their children and posterity yet unborn.

The sexual abuse of girl-children will negatively impact the socio-economic as well as the psychological development of such a child, which will invariably have future consequences for nation-building and development.

9.6 Conclusion

The sexual abuse of a girl-child is against the child's human rights and it has short- and long-terms effects on the child and on national development. Factors which increase the risk of child sexual abuse include environmental factors, superstitious beliefs, poverty, ignorance, lack of education and society's failure to bring perpetrators to book. It has been discovered that the sexual abuse of girl-children deprives the victims of their rights to a good standard of living, quality education, freedom from exploitation and freedom from sexual abuse. Furthermore, the right to physical, mental, social, moral and spiritual development is also reduced. On nation-building and development, the effects of sexual abuse may perpetuate the poverty cycle, increase inequality, increase the birth rate, increase infant and mortality rates and reduce women's participation in the labour force and women's earnings. In order to end the cases of sexual abuse of girl-children in Nigeria, families, the government, non-governmental organisations and society as a whole all have roles to play.

9.6.1 Recommendations

The following recommendations are germane to ending the sexual abuse of the girl-child in Nigeria:

Enlightenment/Awareness. There is a need for sexuality education in the family, at school, by government institutions and non-governmental organisations. Sexuality education should be taught the way children are taught to avoid dangers – for instance, instruction not to touch a hot stove, to look right and left before crossing roads etc. The ten proposals put forward by the Child Mind Institute (2018) will be of immense help:

i. Talk about body parts early and use proper names.
ii. Tell them some are private and not for everyone to see.
iii. Teach your children body boundaries, that nobody should touch their private parts and no one should ask them to touch his private parts.
iv. Tell your child not to keep body secrets no matter what anyone tells them.
v. Tell your children that no one should take pictures of their private parts.
vi. Tell your child how to get out of scary and uncomfortable situation; she should learn to say 'no' and when to leave.
vii. Have a code word your child can use when feeling unsafe or wanting to be fetched.
viii. Tell your child she will not be in trouble for telling her body secret
ix. Tell her about secret touches instead of just 'good and bad' touches'.
x. Tell your child that these rules apply with people they know and even other children. Additionally, parents should monitor their children and carefully choose who they relate with.

The government should adopt policies that will prevent the sexual abuse of children, such as the provision of compulsory, free and quality education in a safe environment. This will reduce vulnerability, and consequently less condition for child abuse. Furthermore, governments should enlighten people using various strategies, radio, television and social media to break the silence and thereby prevent perpetrators. The government should be fully committed to implementing the Child Rights Acts.

Non-governmental organisations (NGOs) should commit funds to help the victims and their family in the course of seeking justice against the sexual abuse of children. They should also provide victims with moral and technical support in the course of seeking justice. They can also embark on an enlightenment programme to educate people about the occurrence and the means of prevention.

The community should unite to break the culture of silence and reprove the perpetrators instead of stigmatising the victims and their families. Perpetrators should be made uncomfortable in the community to demonstrate the gravity of the evil and to serve as a deterrent to others. Communities should support the victims and have zero tolerance for the sexual abuse of children.

Religious organisations should sensitise their members against acts of sexual abuse and debunk the beliefs that make people engage in it for financial breakthroughs and healing of HIV. Religious leaders should let people know that sexual abuse is an abomination which places the perpetrators under the anger and judgement of God. They should sponsor programmes in the media to create awareness and warn people against sexual abuse.

References

Advocates for Youth, 2008: "Sexual Abuse 1: An overview".

Akinbobola, Yemisi, 2020: "Social Media Stimulates Nigerian Debates on Sexual Violence", in: *UN African Renewal*; at: https://www.un.org/africarenewal/web-features/social-media-stimulates-nigerian-debate-sexual-violence.

Akin-Odanye, Elizabeth Oluwatoyin, 2018: "Prevalence and Management of Child Sexual Abuse Cases Presented at Nigerian Hospitals: A Systematic Review", in: *Journal of Health and Social Sciences*; 3,2: 109–124.

Akoloh, Laural; Okenjom, Godian Patrick; Obiahu, Chinyere Lynda, 2016: "The Effect of Child Abuse on Youths and their Academic Performance in Secondary Schools in Balyesa State, Nigeria", in: *Greener Journal of Education Research,* 6,4 (June): 170–176.

Akpoghome, Theresa. U. and Nwano, Theophilus. C., 2016. "Examining the Incidences of Sexual Abuse of Children in Nigeria", in: *Donnish Journal of Law and Conflict Resolution,* 1,2 (November): 1–9.

Awosusi, Ajoke Olukemi; Ogundana, C. Folakemi, 2015: "Culture of Silence and Wave of Sexual Violence in Nigeria", in: *AASCIT Journal of Education,* 1,3: 31–37.

Badam, Ramola Talwar, 2017: "Culture of Silence Shrouds Child Abuse, Experts Warns during Dubai Conference", in: *The National*, 22 November; at: https://www.thenational.ae/uae/culture-of-silence-shrouds-child-abuse-experts-warn-during-dubai-conference-1.677970.

Barajas, Sharae, 2017: "We Can't End the Cycle of Familiar Sexual Abuse in Silence", in: *Ms Magazine* (30 October).

Briere, John N.; Elliot, Diana M., 1994: "Immediate and Long-Term Impacts of Child Sexual Abuse", in: *The Future of Children, Sexual Abuse of Children,* 4,2 (Summer/Fall): 54–69.

Daniels, Natasha, 2018: "10 Ways to Teach Your Children the Skills to Prevent Child Sexual Abuse", in: *Child Mind Institute.*

Choji, Ruth, 2017: "Rising Tides of Sexual Abuse: Expert's Analysis", in: *Leadership* (3 June).

Darkness to Light, 2018: "Child Sexual Abuse Statistics".

Darkness to Light, 2018: "Step 1: Learn the Facts"; at: https://www.d2l.org/education/5-steps/step-1/.

Drezett, Jefferson; Caballero, Marcelo; Juliano, Yara; Prieto, Elizabeth T.; Marques, Jose A.; Fernandes, Cesar E., 2001: "Study of Mechanism and Factors Related to Sexual Abuse in Female Children and Adolescents", in: *Journal de Pediatria,* 77,5: 413–9.

Edeh, Oyinye, 2018: "Sexual Assaults and the Culture of Silence in Nigeria", in: *The Institute of Current World Affairs*; at: https://www.icwa.org/sexual-violence-in-nigeria/.

Equality and Human Rights Commission, 2018: "What are Human Rights?"; at: https://www.equalityhumanrights.com/en/human-rights/what-are-human-rights.

Falade, Bolanle Kemisola; Fasuan, E. Olawale, 2017: "Rape and Sexual Violence against the Girl-Child: Securing the Future Through Good Governance in Nigeria", in: *International Journal of Public Administraion and Management Research* (IJMPAMR), 4,2 (Oct/Nov).

Fontes, Lisa Aronson; Plummer, Carol, 2010: "Cultural Issues in Disclosure of Child Sexual Abuse", in: *Journal of Child Sexual Abuse*, 19,5 (September): 491–518.

Fradkin, H., 2018: "What Causes Sexual Abuse of a Child?", in: *Sharecare*; at: https://www.sharecare.com/health/sexual-abuse/what-causes-sexual-abuse-child.

Grant, Sam, 2017: "How Do Human Rights Protect Children?", in: *Rights Info* (20 November).

Hargrove, Stephanie, 2014: "What's Hidden in Plain Sight: A Look at Child Sexual Abuse", in: *American Psychological Association* (November).

Hilary, Chukwuka Achunike; Rimamsikwe, Habila Kitause, 2014: "Rape Epidemic in Nigeria: Cases, Causes, Consequences and Responses to the Epidemic", in: *IMPACT: International Journal of Applied, Natural and Social Sciences* (IMPACT: IJRANSS), 2,1 (January): 31–44.

Howe, Glenford, 2013: "Sexual Violence Against Children in the Caribbean, Report 2012", in: *UNICEF, United Nations Children's Fund Office for the Eastern Caribbean Area* (May).

IRIN News, 2008: "Child Rape in Kano on the Increase" (3 January).

Ituma, Ezichi A.; Uroko, Favour Chukwuemeka; Eskay, M., 2013: "The Problem of Sexual Abuse in Nigeria Socio-Religious Society", in: *International Journal of Research in Arts and Social Sciences*, 5: 498–512.

Humanium, 2018: "Rights of the Child", at: https://www.humanium.org/en/child-rights/.

Minnesota Coalition Against Sexual Assault (MNCASA), 2018: "What is Sexual Violence?"; at: https://www.mncasa.org/what-is-sexual-violence/.

Mullen, Paul E.; Fleming, Jillian, 2014: "Long Term Effects of Child Sexual Abuse, The America Academy of Experts in Traumatic Stress"; at: https://www.aaets.org/article176.htm.

OHCHR, 2018: "Convention on the Rights of the Child", in: *United Nations Human Rights*; at: https://www.ohchr.org/en/professionalinterest/pages/crc.aspx.

Ohia, Nkiru, 2018: "Mothers' Attitude to Giving Sexuality Education as a Check to Sexual Abuse of Primary School Girls in South-East Nigeria", in: *Journal of Educational and Social Research*, 6,1 (January).

Olusegun, Olaitan O.; Idowu, Amos A., 2016: "Child Abuse in Nigeria: Dimension, Reasons for its Persistence and Probable Solutions", in: *Child and Family Law Journal*, 4,1: 1–25 (Article 2).

Omede, Andrew A.; Agahiu, Grace Etumabo, 2016: "Implications of Girl-Child Education to Nation Building in the 21st Century in Nigeria", in: *Global Journal of Human-Social Science: G Linguistics & Education*, 16,3 (Version 1.0): 1–5.

PDHRE, 2018: "Human Rights and the Girl-Child", The People's Movement for Human Rights Education.

Racebaitr, 2016: "Fighting a Culture of Silence Around Sexuality that Breeds Child Sexual Abuse: Interview with The Heal Project Founder Ignacio Rivera" (September); at: https://racebaitr.com/2016/09/29/fighting-a-culture-of-silence-around-sexuality-that-breeds-child-sexual-abuse-interview-with-the-heal-project-founder-ignacio-rivera/#.

RAINN, 2018: "Child Sexual Abuse"; at: https://www.rainn.org/articles/child-sexual-abuse.

SECASA (South Eastern Center Against Sexual Assault and Family Violence), 2015: "Long-Term Effects".

Singh, Mannat Mohanjeet; Parsekar, Shradha S.; Nair, Sreekumaran N., 2014: "An Epidemiological Overview of Child Sexual Abuse", in: *Journal of Family Medicine and Primary Care*, 3,4 (Oct–Nov): 430–435.

Stop it now, 2018: "The Scope of Sexual Child Abuse Definition and Fact Sheet", at: https://www.stopitnow.org/faq/the-scope-of-child-sexual-abuse-definition-and-fact-sheet.

The Free Dictionary by Farlex, 2018: "Coercion"; at: https://legal-dictionary.thefreedictionary.com.

This day, 2018: "The Girl Child and Nation-Building" (13 October).

UNICEF, 2017: "Child Protection"; at: https://www.unicef.org/nigeria/child-protection.

UNICEF, 2018: "Child Protection from Violence, Exploitation and Abuse".
UNICEF Canada, 2018: "About the Convention on the Rights of the Child".
UN, 2018: "Human Rights".
Weston, Burns H., 2018: "Human Rights", in: *Encyclopedia Britannica*; at: https://www.britannica.com/topic/human-rights.
Wihbey, John, 2011: "Global Prevalence of Child Sexual Abuse", in: *Journalist's Resource*, (Harvard Kennedy School Shorenstein Center on Media, Politics and Public Policy); at: https://journalistsresource.org/studies/government/criminal-justice/global-prevalence-child-sexual-abuse.
World Health Organization (WHO), 2012: "Child Abuse and Neglect".
World Health Organization (WHO), 2018: "Child Sexual Abuse".

Chapter 10
The Farmer-Herder Conflicts in Nigeria's Open Space: Taming the Tide

Aminu Bakari Buba

Abstract With the wave of Nigeria's evolving farmer-herder conflicts, the mobile Fulbe group appears to be the centre of discourse among any other pastoral group in the country. Its nomadic lifestyle has constantly led to contact with farmers. This contact takes various forms, from mutual co-existence and co-operation to competition and conflicts over shared natural resources, such as green and lush pasture, vegetation, fresh water and land for livelihood. One of the drivers of this competition over scarce resources is the ecological change caused by global warming, which is already being felt in the northern part of Nigeria. It has been forecast that these environmental changes will significantly increase, with more irregular precipitation and rising temperatures. These changes are aggravating land degradation and increasing the frequency of droughts, and consequently, lead to declining food production and a decline in the availability of water, which is a major problem for food security. The farmer-herder conflicts might increase in frequency and intensity in the coming years. The Nigerian government is searching for an alternative to the situation. The government needs to consider the historical antecedent of the mobility and migration of the Fulbe as a lifestyle when formulating policies. The government should enhance security in the country, engage on an aggressive scale in combating the desertification which has been pushing people further south, recognise the fundamental features of Fulbe mobility, formulate long-lasting ranching policies, and ensure that the grazing routes and reserves that have been overtaken by development and farming are maintained or new ones established.

Keywords Conflict · Ecology · Farmer · Fulbe · Global warming · Grazing routes · Herders · Land resources

---

Aminu Bakari Buba, Lecturer, Department of History and Diplomatic Studies, Federal University of Kashere, P.M.B 0182, Gombe State, Nigeria; Email: bakariamin@gmail.com.

Ú. Oswald Spring and H. G. Brauch (eds.), *Decolonising Conflicts, Security, Peace, Gender, Environment and Development in the Anthropocene*,
The Anthropocene: Politik—Economics—Society—Science 30,
https://doi.org/10.1007/978-3-030-62316-6_10

10.1 Introduction

Ongoing conflicts in Nigeria between nomadic herders and farming communities in northern Nigeria have intensified in contemporary times and are moving southwards at an accelerated pace, menacing the country's safety and stability. The farmer-herder conflict is a clash between communities whose main occupation is farming and communities whose livelihood depends on raising cattle (Cabot 2017: 11). The economy of Nigeria is known for its oil and gas, but agricultural production occupies about 70% of its labour force. Subsistence farmers in the central and southern regions cultivate and harvest most of the country's root crops, palm produce and vegetable crops, while farmers (Manoma) and herders (Makiyaya) in the north produce grains and livestock. Over 90% of pastoralist nomads in Nigeria are classified as the Fulbe, a large ethnic group spanning West and Central Africa. In Nigeria, pastoralists own approximately 90% of the national herd (International Crisis Group 2017: 1) .

Farmer-herder conflicts have been in existence since the commencement of agriculture, either amplified or lessened in force and occurrence subject to environmental, economic and other factors. A good example of this is a rise in the magnitude of herds. As a result of the enhanced state of herds, pastoral herders have been forced to seek additional pastures beyond the limited grazing routes and reserves allocated by the government. The climatic factor has, by and large, played a major role in the ways of life of people and the patterns of economic activities (Ikimi 2004: 10). Climate change has constituted a great threat by putting great pressures on the land and thus provoking conflicts between groups. Since the 1980s there has been a striking expansion of Agricultural Development Programmes, such as *fadama* (wetland) cultivation. This means that both the farmers and the pastoralists are increasingly faced with fierce struggles for access to such valuable vegetation which, more often than not, result in increased conflicts and violence. The Fulbe pastoralists move around an extensive area, travelling between their major base in the semi-arid north during the rainy season and the wetter south during the dry season. Over the years, intricate systems of managing pastoral nomadism have evolved, resulting in the establishment of cattle routes in the northern parts of the country. Nomadic pastoralists and sedentary crop cultivators have developed beneficial exchange relationships as well as conflict management mechanisms (Shettima 2018: 9).

However, global warming and demographic, socio-political and environmental changes have seriously altered the customary ways of managing pastoral nomadism. Emerging social trends such as the *Boko Haram* conflict, cattle-rustling and climate change have further changed the seasonal pattern of the Fulbe nomadic movement in northern Nigeria. This chapter reviews the historical trajectories of pastoral nomadism in Nigeria, and the changes in land use practices between nomadic pastoralists and sedentary farmers. It further elaborates on the drivers of farmer-herder conflicts in northern Nigeria which impede the effective organisation of pastoral nomadism in modern times. The contest for survival between pastoralists and farmers hinges on environmental concerns which propel the dependent, independent and intervening variables (Abbass 2018: 3). This chapter suggests possible

Table 10.1 Livestock in Nigeria

S/N	Animal	Number in Millions	Value in NGN	Value in USD
1.	Cattle	20,560,933	₦2,056,093,300,000	$6,741,289,508
2.	Camels	279,802	₦85,339,610	$279,802
3.	Goats	73,879,561	₦2,615,033,266,105	$8,573,879,561
4.	Sheep	42,091,042	₦12,837,767,810	$42,091,042

Source FAOSTAT (2016): *Statistics of Live Animals*. The values in NGN and USD are the author's own computation

policy options to handle the incessantly evolving conflicts. Rather than seeing the herders as drivers of conflict, in taming the tide, their role should be seen from an economic point of view.

Herders contribute significantly to Nigeria's economy, with an estimated 20 million cattle, 73 million goats, 42 million sheep (FAOSTAT 2016) and about 975,000 donkeys (International Crisis Group 2017: 1). Livestock represents between 20 and 30% of total agricultural production and about 6–8% of overall *Gross Domestic Product* (GDP) (International Crisis Group 2017: 1). See the details of animal herds in Nigeria in Table 10.1.

10.2 Evolution of Pastoral Nomadism

Pastoral nomadism has historically been part and parcel of human evolution. Mankind is said to have evolved through stages of relative advancement in arts, culture and science which collectively constitute civilisation (Bashir 2018: 1). This advancement in civilisation gave birth to the Agricultural Revolution, which generated other sweeping transformations in the life of humankind and gave birth in turn to a sedentary lifestyle and animal husbandry. In an attempt to cope with the challenges posed by the terrain and climate, life further pushed humankind up the scale of human civilisation. This was the era of Neolithic revolution, when human beings continued to perfect a transhumance lifestyle, searching for good pasture for their herds, which, through the natural reproductive process, multiplied in number. Pastoral production appears clearly in the archaeological record in both East and West Africa from between 4,500 and 4,000 BCE (Attah 2018: 7). This advancement in civilisation gave birth to the Agricultural Revolution, which compelled other sweeping transformations in the life of humankind. It gave birth to a sedentary lifestyle and animal husbandry for sustainable livelihood. In an attempt to cope with the challenges of the environment, life further pushed humankind up the scale of civilisation.

This transformation in civilisation can be regarded as the precursor or the genesis of the modern-day "pastoral nomadic civilisation" (Bashir 2018: 1). In Nigeria, pastoral nomads include the Fulbe, Kanuri, Koyan, Shuwa Arabs, Tuaregs, Uled Suleiman and Yedina, and related peoples have remained in the semi-arid area of

Lake Chad (Agbegbedia 2013: 67) and the Sudan Savannah vegetation. However, changes in ecology and pressure on grazing have produced some adaptations among the Uled Suleiman (camel herders of Libyan origin), who now migrate between Niger and north-east Nigeria. The desiccation of wetland areas, coupled with pressure on grazing resources, has obliged them to explore further south, and the high levels of water extraction are making the Hadejia-Jama'are and Komadugu-Yobe wetlands dry up (Blench 2003). Of all the pastoral nomadic groups, the Fulbe appear to be the most widespread in transhumance across Nigeria.

The Fulbe group in Nigeria, also termed Fulani, Fulata, Bororo, and Sullubawa, is characterised by a common culture, ancestry, language and religious belief. The earliest centre of the Fulbe in Western Sudan was within the Senegal Basin (Abubakar 2008, 48). Nowadays fourteen million Fulbes live in a region stretching from Senegal to northern Ethiopia, and this geographical dispersion explains intra-group variations in their lifestyles (Cabot 2017: 13). The Fulbe in Nigeria live mainly in the northern states. They predominate in these states: Adamawa, Bauchi, Borno, Gombe, Jigawa, Kaduna, Kano, Katsina, Nasarawa Taraba and Sokoto, but only in Adamawa is their language – Fulfulde – widely spoken (Abubakar 2008: 48). However, across the whole region, the Fulbe generally have a collective pastoralist livelihood strategy. Due to the arid and semi-arid features of this region, the Fulbe generally practise nomadic or semi-nomadic lifestyles.

The greatest drivers of competition over scarce resources are global warming and ecological change, which is already being felt in the northern part of Nigeria. It has been forecast that these environmental changes will significantly increase, with more irregular precipitation and rising temperatures. This conflict claimed the lives of more than 2,500 people in 2016 (International Crisis Group 2017: 3) . Despite the precarious situation, governments' reactions at all levels have not been very effective, which is becoming a threat to peace and harmony. Hatred and mistrust are increasing between the farmers and herders, who have lived side by side for several decades.

10.3 The Fulbe Transhumance and Livelihood Transformation in Nigeria

The itinerant transhumance of the Fulbe began in the Senegal Basin in the eleventh century. By the sixteenth century the Fulbe had reached Hausaland (Abubakar 1972: 53). The Fulbe pastoral economy ultimately became an avenue of revenue sought after by the Hausa rulers and taxed intensely. This brought the Fulbe firmly into the orbit of the northern Nigerian political economy and contact with farmers (Shettima 2018: 11). The Fulbe pastoralists received protection from the Sokoto Caliphate founded after the 1804 Islamic revolution in Hausaland. In 1855, the Caliphate's economic fortune led to the development of trade, agriculture and pastoralism. In particular, cattle routes were introduced and protected, and wells were dug along routes (Shettima 2018: 11). Over time, the farmers and herders developed an intricate system of exchange relations by sharing the land, water, lush vegetation and other resources. Pastoralists required the calories produced by farmers, and the farmers required the protein and dairy products produced by the pastoralists.

Looking at the complex nature of the Fulbe transhumance, the British colonial power introduced grazing reserves in the early years of the 1900s. In 1940, seeing the evolving nature of the herders and conflict emanating as a result of inter-group relations, the British attempted to separate the grazing land from the farmland, but this attempt failed because they imposed land use controls (Frantz 1981). Lord Lugard forcibly settled between 2,000 and 3,000 Tuaregs in northern Sokoto (Thom 1972: 35; cited in Watts 1983: 294). In 1942, the idea of sedentarisation of the Fulbe nomads was put into practice in the Jos area, each pastoral household being allocated four hectares of pastureland to settle and engage in crop production (Shettima 2018: 14). This scheme eventually malformed because the allocated land was taken over by mining. Furthermore, the pastoralists regarded the sedentarisation policy of the colonial government as an imposition.

The idea of grazing reserves in Nigeria commenced in the 1950s when Hamisu Kano, working with pastoralists on livestock vaccination, anticipated the shortages of grazing land in northern Nigeria. Supported by the government and the *International Bank for Reconstruction and Development* (IBRD), 'grazing reserves' were recommended in major pastoral areas of Nigeria, on the premise that the reserves would make it easier to provide facilities such as marketing channels, watering points and veterinary pots as well as control bovine diseases. After independence, the IBRD report was reviewed and adopted as a consolidated programme named 'Fulani Amenities Programme', which provided the second impetus for the creation of grazing reserves. In 1965 the Nigerian Federal Government integrated the Northern Nigeria Government Fulani Amenities plan into the law regulating the 'grazing reserve'. The law empowered the Ministry of Animal and Forest Resources and the Native Authorities to acquire any given land and convert it into a grazing reserve (Shettima 2018: 16).

Thereafter, the Federal Government accepted the IBRD endorsements of grazing reserves. With the technical assistance of the *United States Agency for International Development* (USAID), various grazing reserves were established, including the Kachia, Rumka-Kukar-Jangarai, Gujba, Udubo, Garkida and Sorau grazing reserves (Awogbade 1987: 20, in Shettima 2018: 16). To date, the federal and state governments have acquired some 2.84 million hectares of grazing reserves. At present, Nigeria has a total of 417 grazing reserves all over the country, out of which only about 113 have been gazetted (Leo 2014: 17; cited in Kuna/Ibrahim 2015: 200). It was estimated that the 2.84 million hectares of grazing reserve represent only some 6.2% of the total need. In addition, farmers have seriously encroached on the existing grazing reserves (Shettima 2018: 14). Furthermore, most of the grazing reserves were poorly developed, with little or no traces of grass on them. They lack watering facilities, veterinary services, cattle dips or spray races and feed stores (FACU 1998: 11; cited in: Shettima 2018: 16). In addition to the establishment of grazing reserves, cattle routes otherwise known as *burtali* were demarcated across the major livestock corridors in the country. Unfortunately, such routes have mostly been taken over by crop production (Shettima 2018: 16).

10.4 Drivers of the Farmer-Herder Conflicts

The expansion of agricultural cultivation, and environmental pressures including climate change, are increasingly pushing farmers and pastoralists into competition and conflict. Deforestation, changes in land use, and overgrazing have led to a critical reduction in the natural vegetation cover in many parts of the country, to the extent that soils are exposed to rain and the effects result in wind and water erosion and the gullying of slopes. This could be seen as a major driver of the farmer-herder conflicts.Additionally, an increase in rural population has led to the encroachment of agricultural land on traditional pastures and is reducing the nomads' traditional economic basis (Attah 2018: 14). In the north east, the livestock corridor has been massively overtaken by farming.

Other immediate factors explain the intensification of the conflict. First is the rapid increase in ethnic militias, such as the Bachama (*bwatiye*) in the Numan area and that of the Fulani in Adamawa state, with illegally acquired weapons. The second aspect is the inability of the federal government to prosecute perpetrators of previous attacks. The final factor is the introduction of anti-grazing laws in November 2017, which were fervidly opposed by herders in Benue and Taraba states (International Crisis Group 2018: 3), and which have resulted in the mass migration of herders and cattle to neighbouring states, sparking clashes between farmers and herders. This was particularly evident in Adamawa State, which saw the death of more than one hundred people in December 2017.

Recently there has been the rise of divisive sentiments among the various groups who have hitherto been living side by side for several decades in the northern parts of Nigeria. Nigeria's independence was characterised by regional sentiments, ethnic nationalism, religious bigotry and identity politics. The combination of all these negative elements has a tremendous effect on intergroup relations in the country. This was what prompted the first coup of 1966 that saw the death of Nigeria's first and only Prime Minister. Decades of military rule in Nigeria after the civil war have seen the rise of a new wave of ethnic politics in the country with the coming of the Fourth Republic in 1999. Because of ethnic and religious sentiment fuelled by the political class to its own advantage, the Fulbe pastoralist groups, which are mainly Muslim and from the core north, have constantly been rejected by the sedentary communities with whom they were initially living cordially. The gradual southwards expansion of the Fulbe is responsible for ongoing ethno-religious conflicts in the central states of Nigeria, but to a greater extent in Plateau, Nasarawa and Benue States (Edward 2014: 71). Since the violent farmer-herder conflicts accelerated in January 2018, it was projected that 300,000 people have abandoned their homes for safety. Significant dislodgment and deteriorating security situations in areas of Adamawa, Benue, Nasarawa, Plateau and Taraba states have impeded farming activities as well as herding, thereby driving up the prices of food (International Crisis Group 2018: i) .

Lake Chad, a major wetland on which the people depend, is the largest inland drainage basin in Africa. It is situated in West/Central Africa between 6° to 24° N and 8° to 24° E and occupies a vast expanse of land. The basin covers an area of about 2,500,000 km^2. This is about 8% of the surface area of the African continent (UNEP 2006). The lake shares a boundary with Cameroon, Niger, Nigeria, Chad and the *Central African Republic* (CAR) as riparian states. Other states that have a stake within the greater basin include Sudan, the Republic of the Congo, Algeria and Libya (Bdlya 2018: 3). Over time the Lake has decreased by about 90% since the 1960s, from about 23,000 km^2 to 1,350 by 2001 (Bdlya 2018: 3). The drying of the Lake has led to declining resources on which farmers, pastoralists and fishermen depend. As a result of the decrease in the level of the Lake, the aquatic vegetation along the shores of the lake disappeared by 1973 (Bdlya 2018: 3). The Lake water has also continued to recede. Added to this is population increase. The current population within the region is estimated to be approximately 37.2 million people (UNESCO 2007), which has put pressure on resources, leading to conflicts between different groups. Added to the pressure on resources is the impact of dam construction, which altered water flow into the lake within and from outside Nigeria. Within Nigeria, two dams have been constructed along the sub-basins that flow into Lake Chad, the Yedseram and Ngadda dams, and the Hadeija-Juma'are-Komadougou and Yobe dams (Kuna/Ibrahim 2015: 38). They have had adverse effects on the water inflow into the lake, causing it to recede and negatively affecting irrigation for both fishing and pastoral activities.

Nigeria's extreme north is arid and semi-arid, with a long dry season spanning from October to May and low rainfall which historically stood at around 600–900 mm from June to September. In 2008, the National Meteorological Agency reported that over the preceding thirty years the annual rainy season dropped from an average of 150–120 days (International Crisis Group 2017: 3) . This has created ecological destabilisation and altered the pattern of the vegetation belt, especially in the northern parts of Nigeria (Odugbo 2006). In the last six decades, over 350,000 km^2 of the already arid region has turned to desert or desert-like conditions, a phenomenon progressing southwards at the rate of 0.6 Km per year (International Crisis Group 2017: 3). In Bauchi, Borno, Gombe, Jigawa, Kano, Katsina, Kebbi, Sokoto, Yobe and Zamfara states, estimates suggest that 50–75% of the land area is becoming desert (FAO 2017: 6). These environmental changes have wrecked agriculture and human livelihoods, forcing millions of pastoralists and others to migrate south in search of productive land (International Crisis Group 2017: 3). In recent times, some herders have preferred to permanently graze their livestock in the new abode they found for themselves. This situation has elicited rising disputes over lush vegetation, land and water use with central Nigeria's increasing populations of sedentary farmers.

The negative effect of climate change leads to encroachment on nomadic herders' corridors and grazing reserves by farmers (Kuna/Ibrahim 2015: 227), which in turn often leads to the blocking of cattle routes and watering points to the extent that pastoralists are severely handicapped in manoeuvring around planted fields. Herds are therefore forced to trample on crops (Gefu 2018: 7). This problem is compounded by the inability of the government to provide a suitable environment to settle the ever-increasing population of pastoralists. This has resulted in a decline in the productivity of the national herds, leading to massive importation of dairy products, beef and live animals, thereby draining the scarce financial resources on importation (Gefu 2018: 8)

The Grazing Reserves Law introduced by the government in 1965 was an attempt to settle nomadic pastoralists. Animal Grazing reserves are patches of land acquired by the government for the purposes of pastoral use. The law postulated procedures for land acquisition, development and administration. The law is now over fifty years old, but there are only a few grazing reserves which have been actually been gazetted, developed and become fully operational. Moreover, of all the grazing reserves earmarked for development, only fifty (13.7%) have been so far gazetted (Gefu 2018: 8). Overgrazing, fuelwood extraction, and degradation of land with heavy vehicles exploiting firewood have compounded the situation, leading to erosion and destruction of the ecosystem. Wildlife has also been hunted to extinction in these reserves.

Moreover, the intensification of Agricultural Development Programmes promoted by the government and supported by the World Bank, such as the *Fadama* agricultural programmes, has created a new conflict ground. *Fadama* is a Hausa name for irrigable lands in low-lying plains underlaid by shallow aquifers near major river systems (World Bank 2010). This programme, which is now in its third stage (*Fadama* III), has impacted the lives of rural farmers, raising their incomes by 63% (World Bank 2010). The intensification of these projects specifically across the northern parts of Nigeria has led to herders being deprived of access to lush vegetation and water by *Fadama* farming communities, who are in the driving seat. The problem has recently been complicated by the opening up of dry season irrigated crop production. Crop expansion programmes and projects under the World-Bank-assisted *Fadama* projects seem to have contributed immensely to the conflict situation in many parts of the ruminant-producing zones of northern Nigeria. This conflict situation could only be averted if development projects were thoughtfully planned and implemented with the interests of all stakeholders adequately considered and taken into account in project documentation and implementation (Gefu 2018: 39). Furthermore, high-value crops promoted by the National *Fadama* Development Projects, notably tomatoes and onions, produce little residue for livestock feeding, which is further diminishing available fodder (International Crisis Group 2017: 4).

Added to all this is the emergence of conflicts in the two major livestock corridors of the country, namely the *Boko Haram*. In the year 2009, Boko Haram became notorious for its fatal mutiny in the area of the Lake Chad Basin and for its abduction of 276 Chibok girls in 2014 (CTC Sentinel 2018: 23–24). In March 2015, the Boko Haram group increased its violence, especially its suicide bombings, particularly those conducted by women and children. The increase in magnitude of the Boko Haram insurgency resulted in the April 2014 Chibok Secondary School Girls abductions. Assessments of their fighting force increased ludicrously: based on a report by the Government of Cameroon, Ministry of Defence estimates in July 2014 revealed that the insurgent group possessed 15,000–20,000 fighters; on the other hand, a Nigerian-based journalist put forward that it had around 50,000 combatants by the close of 2014. Boko Haram insurgency had by this period exceeded the Islamic State to become the world's most lethal terrorist group in February 2015 (CTC Sentinel 2018: 23).

This conflict has pushed nomadic pastoralists further south, especially into the lush environment of central Nigeria. Nomadic pastoralists have been seasonally moving into the central zone for decades. However, in the past, pastoralists spent the dry season from December to May in the central zone and returned to the north by the onset of the rainy season. Over the last two decades or so, this pattern of movement has dramatically changed due to the depletion of pasture resources in the north and competition with farmers over scarce resources, in addition to the more recent challenges such as the *Boko Haram* conflict and rural banditry. As a result, pastoralists have been staying in the central zone much longer than usual. In recent times, many have actually chosen to graze their herds permanently "and this has triggered increasing disputes over land and water use with central Nigeria's growing population of sedentary farmers" (International Crisis Group 2017: 3) .

The proliferation of small arms flowing from North Africa to West Africa as a result of conflicts has intensified the farmer-herder conflicts, with lots of casualties (UNEP 2011: 25). It is estimated that seven to eight million small arms and light weapons are in use in West Africa today. However, simple machetes were notoriously the most-used weapons (Cabot 2017: 13). The proliferation of arms has led to an upsurge in criminality, with cattle-rustling the most prevalent illegal activity apart from Boko Haram. Various thefts and cattle-rustling incidents which have been going on are hard to estimate: many thefts, especially those occurring in remote villages or forests with a limited state security presence, go unreported. One report estimated that in 2013 more than 64,750 cattle were stolen and at least 2,991 herders killed in states across the north-central zone (International Crisis Group 2017: 5). Between 2011 and 2015, cattle-rustlers, bandits, and other gangs of criminals killed and maimed more than 1,135 people in the State of Zamfara, as revealed to the public by the Civil Defence Corps in Nigeria. Vigilante groups (variously known as *Yan Banga*, *Yan Sa Kai* and *Kato da Gora*) formed to combat bandits have compounded insecurity in some areas where the arrest and summary killing of cattle-rustlers has activated considerable vengeful conflicts most of the time. In other places, the vigilantes have evolved into raiders themselves, on some occasions collecting cash and livestock from local herders as "protection levy" (International Crisis Group 2017: 5). This is the most deadly conflict in Nigeria, accounting for over 2,000 deaths, in addition to tens of thousands of displaced people in Benue and Kaduna states in 2016 alone.

In the early decades of the twentieth century, herders and community chiefs agreed on herding routes (*burtol*). Disputes over wandering stock or damaged crops were typically resolved by village chiefs and herders' leaders. Those who disregarded the decisions of these community-level leaders were referred to local authorities. This system of conflict resolution has faded away since the 1970s (International Crisis Group 2017: 6), weakened by the intensification of government institutions such as the Nigerian police and civil courts. The pastoral herders hate these new institutions, which are perceived to be corrupt and prone to extorting money from them. Therefore, pastoral herders feel ever more disregarded and are mainly sceptical of political leaders as conflict mediators and peace-brokers. The absence of effective mediation mechanisms (*tsarin sulhu*), including sustained community level dialogues, has exacerbated violence.

Politics have played a major role in intensifying farmer-herder conflicts via societies that do not like the identity of the Fulbe group. These societies have accused the Fulbe of receiving government support, especially if the government in a particular dispensation is from the Hausa or Fulbe group. President Muhammadu Buhari's administration has been accused by groups of not implementing measures to solve the conflict simply because he is of Fulbe extraction. Generally, Fulbe herders are unwelcome in different states of northern Nigeria – for example, Benue, Plateau, Taraba, some parts of Nasarawa and Kogi State – where competition over land resources has been increasing. Conflicts with indigenous groups of these areas are very common. Herders are usually accused of destroying agricultural crops and encroaching on protected forests. These reports, coupled with the Fulbes' non-indigenous status in destination areas such as Benue, Kogi and Nasarawa, have produced a movement of indigenous people aiming to expel herders from their land on the grounds of livelihood deprivation and societal misconduct. Herders, on the other hand, resist expulsion because the areas offer the best conditions for their livestock, thus leading to conflict intensification (Rita 2012, 21). Herders have used a number of covert and overt strategies to stay, in spite of opposition. These strategies include paying compensation for crop damage and bragging about access to influential people as a result of the gifts and bribes they offer to traditional rulers in host communities, thus making friendships. Being physically aggressive was also used as leverage.

In Benue State the conflict in the Guma, Ado and Katsina-Ala local government areas (Agbegbedia 2013: 22) was exacerbated by the refusal of the Benue State Governor to accommodate the Fulbe herders or provide land for cattle colonies designed to resolve the incessant conflicts. According to the Governor, "The State has no land for the cattle colony proposed by the Federal Government" (sunnewsonline.com 2018). Senator Akume of Benue State accused Governor Ortom of using the insecurity as an excuse for the ineptitude of his government's performance.

According to a report by the *Coalition of Civil Society Groups for Peace and Security* (CCGPS), Benue and Taraba State Governments fuelled the crisis in the two states by sponsoring militias. The National President of CCGPS, Kola Salawu, condemned the killings in these states perpetrated under the banner of livestock guards set up after the passage of the Anti-Open Grazing Bill into law in Benue State (*Daily Trust* 2018). The bill was passed to stop the activities of open grazing by herders, which have been a source of concern and bloodshed in the two states. It has been reported on several occasions that hooligans and agents of anti-Fulbe herders dress in Fulbe attire and perpetrate crimes to indict the herders as agents of criminality in their host communities. However, there is evidence of genuine Fulbe herders attacking farming communities as a result of the seizure of their cattle, livestock poisoning, identity hate, discrimination and deteriorating co-operation. Government and community responses are urgently needed to stop this intractable situation.

Based on a report submitted to the National Executive Council in 2014 (when it was chaired by former Benue State Governor Gabriel Suswam) for the establishment of grazing reserves, President Goodluck Jonathan directed the Central Bank of Nigeria to release a total of ₦100 billion ($317 million) to the country's thirty-six state governments for ranch construction (International Crisis Group 2017: 10). Although the Central Bank released ₦100 billion to state governments, they failed to construct any ranches. As a result of this failure, in January 2017 the lower House of the National Assembly appointed a committee to scrutinise allegations that the funds provided for ranching facilities had been looted and to report back within four weeks. The findings of the committee remained unpublished at the time of writing this chapter (International Crisis Group 2017: 10).

There has been a spate of killings in Nigeria as a result of the farmer-herder conflicts. It was estimated that about 2,500 people were killed countrywide in 2016, a figure much higher than the Boko Haram insurgency over the same time frame, which stood at 422 (*Washington Post* 2016). In Benue, the worst affected state, Governor Samuel Ortom's accounts reveal that more than 1,878 people lost their lives between 2014 and 2016. From January 2015 to February 2017, thousands of people were displaced in Kaduna, Benue and Plateau states; just the fear of farmer-herder conflicts has driven residents to more secure areas. Authorities and donor agencies often ignore these conflicts, and affected communities receive far less support from the *National Emergency Management Agency* (NEMA) and international agencies than those affected by the Boko Haram insurgency. For women and girls, the impact is worrisome. The relatives of men killed in the conflicts are often evicted from family compounds and farmlands. Post-conflict economic and social disenfranchisement renders women more susceptible to sexual and economic exploitation. The economic effect has been colossal. Statistics revealed in 2015 indicated that the Apex government of Nigeria was losing the sum of 13.7 billion US dollars in revenue yearly as a result of herder-farmer clashes in states like Benue, Kaduna, Nasarawa and Plateau states. It was also found out that, on average, these four listed states lost 47% of their *internally-generated revenues* (IGR). In March 2017, Governor Samuel Ortom of Benue state declared that frequent attacks by herders migrating from far northern states, and perhaps also from neighbouring countries like Cameroon and Niger, had cost the Benue State Government a total sum of ₦95 billion (about $634 million at that time) between 2012 and 2014 (International Crisis Group 2017: 7).

On the legislative arm, the National Assembly has not been responding adequately in addressing the ever-increasing conflict. In 2011, a bill was sponsored by a Senator from Niger state to create a National Grazing Reserves Commission and livestock routes, but it was not passed and ultimately expired when the Seventh Senate lapsed in May 2015. From 2015 to 2016, three bills were sponsored by Senate members to create grazing reserves, livestock routes and ranches across the country. After much debate, all three were dropped in November 2016 on the grounds that land use was exclusively a state government duty and power. This is a clear setback that the 1978 Land Use Act is affecting development. All these constraints have limited the legislative Chamber to an avenue of holding public hearings and passing resolutions without impacts. The hearings have not been followed by any policy recommendations or actions towards ending the violence (International Crisis Group 2017: 8).

There is no judicial or legal framework for reconciling the conflicts and competition of sedentary and transhumance production systems over natural resources. It is indeed this legal lacuna that is largely responsible for the conflicts between farmers and herders, with crop farmers claiming lawless use of rangelands and pastoralists parading their livestock on to cultivated lands. As a result of the colonial and post-colonial state of the crisis, 'farmer-bias' in development policy blames the pastoral herders for intensifying the conflict, whereas, in actual fact, it makes sense to think of fields straying into grazing lands (rather than animals straying into fields) to describe a situation in which the agriculturalists – not the herders – are causing the conflicts (Traore 1996: cited in Shettima 2018: 18).

10.5 Solutions and Way Forward

Both farmers and herders would gain if they agreed to co-operate within some parameters. An immediate step to address the conflict is to improve security across the country. Intensification of security operations will help curtail cattle rustling. The government should arrest and prosecute individuals who carry illegal firearms (International Crisis Group 2017: 15). There are several security measures to be adopted by the government to improve security. First is to increase the number of security personnel, particularly in the most susceptible areas like Benue, Nasarawa and Taraba states. In the short term, security personnel currently engaged in duties in the south-eastern and south-western parts of the country should be posted to the distressed spots; the large numbers of police personnel attached to politicians and other privileged elites in Abuja and state capitals should also be reassigned to these areas (International Crisis Group 2018: 27) .

Environmental change is considered to be a major driver of this conflict. It has fuelled this crisis to a large degree, increasing competition among farmers and herders over scarce resources. To mitigate this, there should be an aggressive scale of combating the desertification which has been pushing people further south. Climate experts estimate that eleven states of the far north could become desert or semi-desert regions if considerable efforts are not put in place. Therefore, in the long term, the federal and state governments should strengthen the execution of the Great Green Wall Initiative for the Sahara and the Sahel region of the country. Initially, the project entailed planting a 15 km-wide belt of trees, spanning 7,775 km and traversing nine African countries from the Senegal basin to Djibouti. It was later extended to include the construction of water-retention ponds and other basic infrastructures, establishing agricultural production systems and promoting other income-generating activities. Nigeria's National Agency for the Great Green Wall aims to rehabilitate 22,500 km^2 of degraded land by 2030 (International Crisis Group 2017: 19). In another development, the government drafted a *National Adaptation Strategy and Plan of Action on Climate Change* (NASPA); in November 2012, it adopted a National Policy on

Climate Change. The country's official development policy, called *Vision 2020*, also contains climate change initiatives. These policies and plans should be implemented properly to enhance a long-lasting solution of the evolving environmental situations.

Governments should ensure that the grazing routes and reserves that have been overtaken by development and farming are maintained or new ones established. Concerted efforts should be made to gazette and develop additional grazing reserves to ease the pressure currently put on the existing grazing resources (Gefu 2018: 35). In March 2016, the federal government declared its intention to establish grazing areas across the country, but opposition from farming communities forced it to concede. The Federal Government, working with states, should survey, demarcate and officially document existing grazing reserves that have not been overrun by human settlements and infrastructural developments. The Federal Government also should follow through on its plan to establish new grazing reserves in the ten northern states that have already provided 55,000 hectares to that effect. It should help state governments develop these areas following the prototypical model provided by the *International Fund for Agricultural Development* (IFAD), which is funding three reserves in Sokoto State. Moreover, the Federal Government should take steps to encourage ranching.

The Buhari administration's *Agriculture Promotion Policy* (APP) 2016–2020 recognises "the cattle value chain which has become a security dilemma, as roaming of cattle is a source of friction between farmers and herdsmen". Ultimately, a shift is required through domestication of livestock in confined ranches. In April 2017 a northern leaders' summit suggested "a concerted development of ranches" as an option towards ending the clashes. Some initial steps have been taken, and ranching has been advocated by the Federal Government of Nigeria as the requisite solution to the incessant conflict. In April 2017, a policy dialogue initiated by the federal agriculture ministry, facilitated by the *UN Food and Agriculture Organization* (FAO) recommended that governments should formulate and implement a ten-year *National Ranch Development Plan.* The plan is geared towards obtaining support from conventional livestock producers by helping them to establish livestock co-operatives and networking them with monetary institutions like the *Bank of Agriculture of Nigeria* (BOA) and *Nigeria Incentive-Based Risk Sharing System for Agricultural Lending* (NIRSAL). Implementation of this policy will facilitate the productivity of herds.

Security has been a major issue in Nigeria for decades, which is apparent from dozens of conflicts and social vices unfolding in different dimensions. Putting this into perspective, in its report on *Harvest of Death: Three Years of Bloody Clashes Between Farmers and Herders* Amnesty International revealed that the inability of Nigeria's government to act fast on some security situations resulted in the death of 3,641 persons between 2016–2018. In some cases Nigeria's security organisations had information of such upcoming attacks but were mostly slow to act accordingly (Amnesty International 2018). This situation has been seen in the attack of Mubi, in northern Adamawa State in 2014. Similarly, warning of an impending attack on Gombe was received during the eve of the 2015 general election, yet Boko Haram attacked Gombe Metropolis for hours and fled without serious response on the part of Nigeria's security forces.

Moreover, this conflict is not only limited to Nigeria but other neighbouring countries in the West African sub-region which require regional co-operation. This will oblige Abuja to work with neighbouring countries to find a lasting security solution to the transnational border conflicts. Following disclosures that foreign herders were involved in attacks on farming communities in Nigeria, the Government is presenting proposals to the African Union "to compel member countries to take steps towards preventing their herdsmen from grazing into neighbouring countries, which is a major catastrophe if not addressed" (International Crisis Group 2017: 20). To that end, the Government should engage the governments of Cameroon, Chad and Niger, as well as the ECOWAS commission, to reach an agreement on how to collectively monitor and standardise continental pastoralism, in accordance with applicable international instruments including ECOWAS Protocols. With the help of international communities like ECOWAS, the African Union and the United States of America, the Government can get the required support and hardware to curtail the situation.

A national policy is indeed needed to address the situation. In line with this, a National Livestock Transformation Plan was masterminded by the government. It is aimed at making a gradual change from open pastoral grazing to a ranching system. "Open grazing is no longer viable; that's why we're switching to ranching," as stated by the Minister of Agriculture in 2018. The plan, running from 2018 to 2027, is a multifaceted intervention intended to modernise livestock management, improve productivity and enhance security. Under the new plan, ten states – Adamawa, Benue, Ebonyi, Edo, Kaduna, Nasarawa, Oyo, Plateau, Taraba and Zamfara – have been selected as the pilot states, with ninety-four branches to be modelled in clusters of four at twenty sites spread across those states. To participate in the plan, cattle herders are expected to organise and register as co-operatives that will then be able to rent land from state governments and also benefit from loans, grants and subsidies. The federal and state governments are expected to provide a total of ₦70 billion (about US $195 million) for the pilot phase, spanning three years, while private interests are expected to invest in excess of ₦100 billion (about US $278 million) between the fourth and tenth years (International Crisis Group 2018: 25) . The plan is a laudable effort to resolve the farmer-herder conflicts, but like earlier initiatives, it has drawn mixed responses. Instead of selecting some pilot states, legislation needs to be put in place to encourage more states to participate and offer land that will be used for such a purpose.

10.6 Conclusion

Fuelled by environmental change, the accelerating conflicts between farmers and herders are among Nigeria's most unrelenting security challenges of the twenty-first century. They could hypothetically engender bloodshed on an even wider scale unless the Government intervenes earnestly and makes ending the conflicts a national priority. In taming the tide of these conflicts, state governments need to formulate and implement unique strategies to solve ecological problems and dissolve the discontent among groups without taking sides. There should be the reinforcement of law. Instituting and defending grazing reserves would strike a change. In the long run, the grander contest will be addressing decadence in failed states across Africa so as to reduce the influx of arms. Moreover, environmental and climate change has been pushing people south, leading to increasing pressure on land resources like water, lush vegetation and fertile cultivable lands. Therefore the laxity in addressing these issues would result in further menace in Nigeria, which is already combating several security challenges. To sum up, the farmer-herder skirmishes have arguably overtaken the Boko Haram insurgency as the greatest danger to Nigeria's stability between 2016–2018. The Fulbe herders should not be seen as societal invaders or interlopers but as a sacrosanct subset of the agrarian economy.

References

Abdullahi, Suleiman, 2018: Personal Interview on Incessant Farmer-Herder Conflict in Adamawa State, Nigeria.

Abbas, Isah Mohammed, 2018: "Contest for Survival between Fulani Pastoralists and Farmers", Paper for the 1st National Conference on the Dynamics of Pastoral Nomadism in Contemporary Nigeria: Problems, Policies and Prospects, 16–17 April, Federal University of Kashere, Gombe State, Nigeria.

Abubakar, Sa'ad, 2008: *Lamibe Fombina: A History of Adamawa Emirate 1809–2008*. (Ibadan: Book Wright Nigeria Publishers, New Edition).

Abubakar, Sa'ad, 1972: 'The Establishment of Fulbe Authority in the Upper Benue Basin, *1809–47*' in: *Savanna,* 1,1: 67–80.

Agbegbedia, Anthony Oghenevwoke, 2013: 'An Assessment of the Methods of Managing Conflict Between Pastoralists and Farmers in Benue State, Nigeria' (PhD Thesis, University of Ibadan, Department of African Studies).

Amnesty International, 2018: "Nigeria: Government Failures fuel Escalating Conflict between Farmers and Herders as Death Toll Nears 4,000".

Attah, Noah Echa, 2018: "Global and Continental Perspectives on Nomadic Pastoralism", Paper for the 1st National Conference, on the Dynamics of Pastoral Nomadism in Contemporary Nigeria: Problems, Policies and Prospects, 16th–17th April, Federal University of Kashere, Gombe State, Nigeria.

Awogbade, Moses, 1987: "Grazing Reserves in Nigeria", in: *Nomadic Peoples*, 23: 19–29.

Azad, Abul Kalam; Crawford, Emily; Kaila, Heidi Kristina, 2018: "Conflict and Violence in Nigeria Result from the North East, North Central, and South-South zones". Preliminary Draft Report (Abuja, Nigeria: The World Bank – National Bureau of Statistics Nigeria).

Bdliya, Hassan Haruna, 2018: "Climate Change, Receding Lake Chad and the Challenges of Farmers/Herders in Nigeria", Paper for the 1st National Conference on the Dynamics of Pastoral Nomadism in Contemporary Nigeria: Problems, Policies and Prospects, 16–17 April, Federal University of Kashere, Gombe State.

Bello, Usman Alhassan, 2013: "Herdsmen and Farmer Conflicts in North-Eastern Nigeria: Causes, Repercussions and Resolutions", in: *Academic Journal of Interdisciplinary Studies,* 2,5: 129–133.

Blench, Roger, 2003: "The Transformation of Conflict between Pastoralists and Farmers", in: *African Journal,* 2,3; cited in: Agbegbedia, Anthony Oghenevwoke, 2013: "An Assessment of the Methods of Managing Conflict Between Pastoralists and Farmers in Benue State, Nigeria" (PhD Thesis, University of Ibadan, Department of African Studies).

Cabot, Charlene, 2017: *Climate Change, Security Risks and Conflict Reduction in Africa: A Case Study of Farmer-Herder Conflicts over Natural Resources in Côte d'Ivoire, Ghana and Burkina Faso 1960–2000* (Berlin-Heidelberg: Springer).

Daily Trust Newspaper, 2018: "Coalition Flays Sponsorship of State Militias, Wants Emergency in Benue and Taraba", in: *Daily Trust Newspaper* (11 January 2018); at: http://www.dailytrust.com.ng/coalition-flays-sponsorship-of-state-militias-wants-emergency-in-Benue-taraba.html (20 August 2018).

Edward, Okeke Okechukwu, 2014: "Conflicts between Fulani Herders and Farmers in Central and Southern Nigeria: Discourse on Proposed Establishment of Grazing Routes and Reserves", in: *International Journal of Arts and Humanities*, Bahir Dar, Ethiopia, 3,1: 9, 66–84.

Food and Agricultural Organisation, 2016: "Live Animal Data in Nigeria".

Food and Agricultural Organisation, 2017: Country Programming Framework, Federal Republic of Nigeria 2013–2017.

Gefu, Jerome, 2018: "An Overview of the Management of Pastoral Nomadism in Nigeria: Towards an Effective Livestock Management Framework, Paper for the 1st National Conference on the Dynamics of Pastoral Nomadism in Contemporary Nigeria: Problems, Policies and Prospects, 16–17 April, Federal University of Kashere, Gombe State, Nigeria.

Kuna, Mohammed; Ibrahim, Jibrin, (Eds.), 2015: "Rural Banditry and Conflicts in Northern Nigeria" (Abuja: Centre of Democracy and Development).

Ikimi, Obaro, (Ed.), 2004: *Groundwork of Nigeria History* (Ibadan: Heinemann Educational Books Nigeria Plc),.

International Crisis Group, 2017: "Herders against Farmers: Nigeria's Expanding Deadly Conflict", *Africa* No. 252 (Brussels: ICG).

Odugbo, Petèr, 2006: "An Analysis of Rainfall Patterns in Nigeria", in: *Global Journal of Environmental Sciences,* 4,2; 139–145.

Shettima, Abba Gana, 2018: "Managing Pastoral Nomadism in Nigeria: A Historical Review", Paper for the 1st National Conference on the Dynamics of Pastoral Nomadism in Contemporary Nigeria: Problems, Policies and Prospects, 16–17 April, Federal University of Kashere, Gombe State, Nigeria.

Singh, Ashbindu., et al., 1999: *Early Warning of Selected Emerging Environmental Issues in Africa: Change and Correlation from a Geographic Perspective* (Nairobi, Kenya: United Nations Environment Programme).

The Sun Newspaper, "No Land for Cattle Colony in Benue-Ortom", in: *The Sun Newspaper*, 16 January 2018; at: http://www.sunnewsonline.com/no-land-for-cattle-colony-in-benue-ortom (20 August 2018).

Traore, Samba, 1996: "Problems in Pastoral Land Management Related to Tenure: Policies versus Basic Practice", in: IIED; GRET: *Managing Land Tenure and Resource Access in West Africa* (Eds.): *Proceedings of a workshop held by IIED and GRET* (Goree: GRET–IIED).

UNESCO 2007: "Water Portal Weekly Updates No. 178 Lake Chad Basin".

Warner, Jason; Hulme, Charlotte, 2018: "The Islamic State in Africa: Estimating Fighter Numbers in Cells Across the Continent", in: *Combating Terrorism Centre CTC Sentinel*, 11,7: 21–27.

Washington Post, 2016: "The Brutal Toll of Boko Haram's Attacks on Civilians"; at: https://www.washingtonpost.com/graphics/world/nigeria-boko-haram/ (12 May 2018).

Watts, Michael, 1983: *Silent Violence: Food, Famine and Peasantry in Northern Nigeria.* (Berkeley, Los Angeles: University of California Press).

World Bank, 2002: "Reversal of Land and Water Degradation Trends in the Lake Chad Basin Ecosystem", Project Brief (Washington, DC: World Bank).

World Bank, 2010: "Rural Agricultural Project Fast Becoming a Household Name in Nigeria"; at: http://worldbank.org/en/news/feature/2010/07/28/fadama-iii-rural-agriculture-project-fast-becoming-a-household-name-in-nigeria (19 August 2019).

Yembilah Barre, Rita, 2012: "The Theoretical Milieu: Fulani herder-indigene conflicts in Tallensi-Nabdam, Ghana", in: *Ghana Journal of Geography*, 4: 18–41.

Chapter 11
Climate Rituals: Cultural Response for Climate Change Adaptations in Africa

Mokua Ombati

Abstract The threat posed by climate change to Africa's development and growth is real. To move along a climate-resilient-development pathway, Africa must respond and implement effective climate change adaptation measures, so that climate-related risks and opportunities support development objectives. This can partly be achieved by recognising the rich interactions between climate and culture, and harnessing cultural science, knowledge and technologies. Informed by the anthropological theory of cultural ecology, this chapter, framed on the prism of weather and climate-related modification rituals, sets an understanding of how communities across Africa have, over the centuries, characteristically, utilised this form of culture to respond and adapt to categories of climate change. Drawn from evidence from a swathe of extant data, this study interprets the ritual practices and explains the logical basis, wisdom, understanding, thought-patterns and imagination underlying and informing their foundation from purposively selected cases representing various regional zones across Africa. In considering the principal tenets structuring this body of knowledge across a multiplicity of socio-cultural, economic, political, geographical and anthropological settings, civilisations and histories, this chapter highlights insightful understanding of the existence of a pattern of unity and diversity in the 'climate rituals' which have developed as a human adaptation to climate change across Africa, and emphasises the value of such rituals as a prototype for cultural response adaptations to climate change. The study recommends incorporating the insights derived from this form of knowledge and science into the design of creative multipronged climate change response adaptations in contemporary Africa.

Keywords African indigenous knowledge · Climate change · Culture · Rainmaking rituals

Mr. Mokua Ombati is a research fellow affiliated to the Department of Anthropology and Human Ecology at Moi University, Eldoret, Kenya; Email: keombe@gmail.com.

Ú. Oswald Spring and H. G. Brauch (eds.), *Decolonising Conflicts, Security, Peace, Gender, Environment and Development in the Anthropocene*,
The Anthropocene: Politik—Economics—Society—Science 30,
https://doi.org/10.1007/978-3-030-62316-6_11

11.1 Climate and the Climate Change System

Climate describes the accumulated patterns of weather in specific locations. Climate change, understood as any systematic change in the long-term statistics of climate elements, that is, temperature, precipitation, or winds, whether as a result of natural processes of variability or anthropogenic (human-induced) activity, is probably one of the greatest and most complex challenges facing human, animal and plant species across the globe (UNFCCC 2006). Climate change is experienced when climatic conditions permanently shift either upwards or downwards. Shifts in the start, length and end of the rainfall season, the number, length and intensity of dry spells, or changes in total weather seasons, among others, all signify climate change. Climate change can thus be a shift in the mean climatic conditions, but can also exhibit itself as a change in the intensity and frequency of extreme climate events. Accordingly, climate change is commonly exemplified through rising atmospheric temperature, precipitation, strong winds, cloud cover, rising sea level, flooding, storms, the recession of glaciers and melting of sea ice, longer severe droughts, the extension of subtropical deserts, loss of biodiversity, and acidification of oceans, among other phenomena (IPCC 2014).

Of particular note for this study, climate change alters the quantity, intensity, frequency and distribution of rainfall. The increased intensity and frequency of higher temperatures, solar radiation, winds and prolonged periods without adequate rainfall cause droughts, which then result in water shortages. Drought triggers cascading negative consequences, which then undermine people's modes of livelihood. And yet, the future is predicted to bring even greater climatic changes (Kumari et al. 2018).

11.1.1 Threat of Climate Change to Africa's Growth and Development

Climate change represents a growing, urgent, serious and "potentially irreversible threat to human societies and the planet" (UNFCCC 2015: 1). Although climate change is projected on a global scale and is a borderless phenomenon that affects the entire planet, different regions and different ecologies are affected differently. People, too, face different aspects of climate change depending on which ecological zone they live in. Broad ecological zones have relatively similar climate change commonalities but also many local and regional variations.

The climate change reality threatens Africa's growth and development, and undermines its capacities, abilities and potentials. Countries in Africa are disproportionally vulnerable to the devastating challenges of climate change and the corresponding socio-economic, political, security and cultural losses, because of their limited social, financial and technological resources for effective adaptation and mitigation. High dependence on climate-sensitive natural resources for livelihoods and economic sustenance, poor governance, gender inequality, environmental and economic challenges, complex disasters and conflicts, and high levels of poverty further add to the uncertainty of daily life, and inherently increase the vulnerability to climate change in Africa.

Climate change not only affects people's economies, socio-political and religious organisation, health, safety and security, and other dimensions of life, such as settlement decisions, patterns, ritual practices, and agricultural schedules, but also their relationship to the natural environment (IPCC 2014). For Africa, the concern about climate change is not so much about projections of carbon emission and future environmental damage. It is more about the links between the manifestations of climate change, be they in the form of droughts, desertification, floods, storms or soil erosion, and the realities of "contemporary disaster events that threaten lives and livelihoods, and hinder the continent's economic growth and social progress" (Lisk 2009: 8).

Climate change in the form of drought is particularly pertinent and chronic in Africa, wreaking havoc, and creating enormous developmental challenges. Incidences of prolonged drought and unseasonal rains are reportedly getting more frequent (Ojwang et al. 2010). The complex nature of the onset and termination of drought has led it to acquire the title the 'creeping disaster', and, unlike other disasters, the effects of drought span all aspects of human life and continue to be felt long after its termination in the form of unending food insecurity (Masinde/Bagula 2011: 274).

Agriculture and farming (AGRA 2014), the backbone of Africa's subsistence and cash economy, experiences the most devastating impacts of climate change. This phenomenon is compounded by the fact that great proportions of livelihoods (over 70%) depend on vulnerable rain-fed small-scale farming operating in fragile environments that are ecologically, geographically and economically marginal. Fluctuating seasonal rainfall patterns, severe and prolonged droughts, desertification, flooding and other manifestations of climate change contribute to loss of arable land, soil erosion, land degradation and fertility loss, and subsequently to high pre- and post-harvest agricultural losses, increased incidences of pests and diseases, and invasive weeds. All these amount to crop failure and reduced agricultural yields. Decline in agricultural productivity leads to food shortages, and subsequent varying and high food prices that most households cannot afford. National economies are also immensely affected, as the impacts of climate change affect local and subsequently international prices, in turn affecting food demand, calorie availability, and ultimately human well-being (AGRA 2014; Nyasimi et al. 2013).

In the livestock sector, climate change contributes to inadequate pasture and water, increased incidences of livestock diseases and pests, conflicts over water and grazing resources, and livestock deaths (Nyasimi et al. 2013). For marine ecosystems, change in climate leads to acidification and fluctuating water quantities, in addition to increasing disease vector transmission and influencing marine pathogens (IFAD 2009; FAO 2007). In effect, all these cause disruptions to the reproductive patterns of fish and aquaculture, and the corresponding depletion in species and stocks (AGRA 2014).

The negative effects of climate change encompass long-lasting cultural disturbances and losses and susceptibility to diseases, as epidemiologies are affected by environmental factors, such as rising temperature and increasing precipitation. In terms of human health, climate change exacerbates the occurrence, intensity and spread of disease outbreaks (IPCC 2014). For instance, climate change phenomenon is indirectly linked to negative effects on human health through unclean water, poor sanitation and ecosystems degradation, thereby contributing to malnutrition, diarrhoea, amoebic dysentery, malaria, cholera and other disease outbreaks. Indeed, poor sanitation and water quality are linked to climate-induced droughts and floods, which account for over 20% of disease outbreaks in Africa. Further, among other factors, dryness, low humidity and dusty conditions are climatic factors that could cause meningitis (Nche 2014: 2).

More broadly, forest ecosystems contribute immensely to the natural, human, and animal species through a well-known range of services. Most importantly, forests represent important resources for millions of people and animal species who depend on them and their products for timber, shelter, food, fodder, fuelwood, recreation, medicines and ecosystem services. In addition, forests influence climate change, as sources of greenhouse gases when they are destroyed and as sinks for carbon when they grow or expand. Besides, climate change affects local revenues from tourism and recreational services when vast areas of dead or dying forests reduce scenic and recreational appeal (Bernier/Schoene 2009).

Forest ecosystems frequently respond sensitively to climatic changes, severely escalating the negative impacts to the people, societies, animal life, physical environment and socio-economic activities that depend on them. For example, water shortages and unpredictable rainfall, in combination with population growth and land degradation, increase pressure on forest ecosystems and their capacity to meet immediate livelihood needs. As drought reduces the productivity of adjoining agricultural lands, for instance, communities with limited economic and livelihood alternatives are likely to turn to forests for crop cultivation, grazing and illicit harvesting of forest products, aggravating the local loss of forest cover. The loss of forest cover increases the likelihood, frequency, extent or severity, and greater incidences of intense droughts, heatwaves, fires, floods, landslides and storms, leading to degradation of vital ecosystems, outbreaks of diseases and infestations of insects and other pests (Bernier/Schoene 2009).

Biodiversity is equally threatened as plants and animal species reduce their resilience to climate change (Nyasimi et al. 2013). In Africa, where the penetration of modern biomedicine is rather limited, a greater percentage of the people depend on indigenous medicine and herbal therapies as their sources of medicine, health, income and knowledge. Climate change reconfigures the geospatial positioning of biodiversity, compromising and leading to a dramatic decline and corresponding shift in distributional availability and access to medicinal flora and fauna. Increasingly, environmental quality impinges on ecological medicinal resources. This, in turn, determines the quality of traditional healing and medicine, since these overwhelmingly rely on existing mystical sites, a wide variety of herbs, roots, leaves, tree barks, and a multiplicity of animal parts and minerals for therapeutic potions and remedies (Jangu 2012). The loss of biodiversity, therefore, significantly affects the resource base of traditional medicine, health and knowledge.

Additionally, historically, African people gained their knowledge and practical experiences through continued observation, close attachment to nature, and direct dependence on natural resources. However, climate change negatively affects the ability of the younger generation to learn traditional medicine, health and knowledge, as there is reduced interaction between humans and the natural environment. Inevitably, all these further weaken people's vulnerabilities and adaptive responses. Indeed, as Fatorić/Seekamp (2017: 228) assert, "climate change poses serious threats to the protection, preservation, and transmission" of cultural "heritage and resources". Systematic access to scientific-based knowledge of cultural heritage and resources can inform current and future preservation and/or adaptation research, which in turn assists and supports decision-making that requires innovative ways of synthesizing fragmented knowledge.

Further, the consequences of climate change profoundly impact African traditional religion and practices. African traditional religion is inextricably interwoven with the culture of the people, all aspects of their daily lives, from what they "eat (or cannot eat), the way they farm, do everyday chores, hunt, make tools and clothes, arrange themselves in families, marry, divide work among family members, educate their children, treat illness, and bury the dead" (Nche 2014: 1–2). The idea is that African traditional religion finds "expression in nature, which goes to establish that anything that affects nature affects the religion, and anything that distorts the natural course of events, affects the traditional religious practices of Africans" (Nche 2014: 1). The consequences of climate change distort the weather, which was traditionally predictable, affecting the daily sequence of activities and patterning of life, and therefore altering the seasons of rituals, sacrifices and festivities.

Climate change also influences mobility and migration. As climate change affects the quality of the environment, reducing its capacity to sustain human livelihood, it may thereby create forms of displacement, causing people to move from their abodes in search of more favourable locations and sources of livelihood due to the inability of existing resources to sustain the demands of the human population which depends on its supply. This may trigger conflicts and violent contestations over resources in the receiving communities, and increase the potential for conflicts when migrants return to their areas of origin and issues such as resource ownership or rights of use arise. Climate change may also lead to further increases in rural-urban migration because of the degradation of resources in rural areas and people searching for better livelihoods in urban centres. The consequent results may be increased growth of informal settlements and competition for resources in Africa's cities. Compounding the climate change and migration predicament, inadequate living conditions and huge population numbers in urban settlements, greatly increase the risk of epidemic disease outbreaks.

The migration predicament caused by climate change triggers cyclical detrimental human security concerns. When people no longer have access to basic necessities critical for survival, they may adapt by migrating, consequently generating (in)security challenges when the capacity of the receiving communities is stretched to tipping point, making interactions and relations become conflictual rather than co-operative.

Focusing on climate change impacts, internal migration patterns and development in Sub-Saharan Africa, South Asia, and Latin America, a World Bank report (Kumari et al. 2018) projects that the number of internal climate migrants could reach more than 143 million by 2050, 86 million of whom from in Sub-Saharan Africa. Across East Africa, climate change could see more than 10 million people forced to move over the same period. Unless concerted action is taken at national and global levels, countries within the affected regions will have to find ways of coping with the stress of internal climate migrants.

The devastating consequences of climate change threaten to consign millions of people in Africa to extreme poverty. Other stressors, including high population growth rates, which double the demand for food, water, and livestock forage (Nyasimi et al. 2013), also compound climate change's negative impacts. These manifold stressors limit the multiple resources and diversity of livelihoods, intensifying specialisation in single resources and monoculture with high capital investment that renders 'modern' systems particularly vulnerable to climate change. Accordingly, the manifested negative impacts of climate variability on people's civilisations and resources, livelihoods, health, food production, economic opportunities, *gross domestic product* (GDP) and the continued survival of animal and natural systems are quite devastating, to say the least.

In the ensuing circumstances, the enormity of the climate change crisis requires concerted efforts from all stakeholders – governments, experts, educationalists, scientists, environmentalists, activists, communities, individuals and international partners – to join together in pulling down the barriers. Without this, the direct costs of climate change damage in Africa would potentially be insurmountable and considerably greater if indirect costs are included. The cumulative impacts have the potential to reverse much of the progress African countries have already made and hinder the attainment of any further development goals and objectives. For those reasons and many others, the gravity and the deleterious consequences posed by climate change call for increased capacities and the (re)formulation of viable response adaptations.

11.2 The Cultural Ecology of Climate

Cultural Ecology, the core theoretical premise whose insights underpin this study, is "a heuristic device for understanding the effect of environment upon culture", that is, how environmental forces influence humans and how human activities affect the biosphere and the Earth itself (Steward 1955: 30–42). The theory focuses its interpretation on how cultural beliefs and practices help human populations to respond and adapt to their environments and live within the means of their ecosystem. Coined by Julian Steward (1955), the father of ecological studies in anthropology, *Cultural Ecology* is a theory and methodology concerned with the complex multifaceted cultural aspects of reciprocity and interdependent relationships between humans and nature, and that explains how and why cultures adapt in one way and not another. Steward combined four approaches to frame the interaction between nature and culture: (1) an explanation of culture in terms of the environment where it exists; (2) the relationship between culture and environment as a process; (3) a consideration of small-scale environments; and (4) the connection of ecology and multilinear cultural evolution.

Cultural Ecology, then, focuses on how human populations, through their culturally-patterned behaviours and associated beliefs, shape their environments, and the subsequent manners in which these relations are expressed and maintained through linkages and feedbacks that make the population an active part of those environments through the organisational structures which cultures use to interpret their experiences of the natural world. In essence, cultural ecology attempts to provide a materialist explanation of human society and culture as the primary and most potent product of human adaptation to the environment (Moore 2009). As the mechanism which allows human societies and populations to adjust and adapt to their environments through cultural means, Cultural Ecology recognises that each environment requires different adaptations and that not every culture works towards a similar 'norm' of adaptation (Mokua 2019). As such, each culture has a distinct, even unique, pattern of ecological adaptation. To ensure effective adaptation, cultures employ practices designed to exert at least some level of control, or domesticate the environment and the resources within them. This is achieved through a variety of techniques, including the manipulation of landscapes, the management of individual resources, and attempts to influence the supernatural. Ritual activities designed to maintain the environment in its domesticated state are considered passive environmental manipulation (Sutton/Anderson 2010).

11.2.1 Culture for Adaptation

As such, viable response adaptations to the most existential crisis for humanity, that is climate change, require creative multipronged responses, as climate is a physical phenomenon with cultural aspects and implications. This means climate change has a socio-cultural, security, econo-political and human rights imperative, "something that cannot be solved through a reliance on science and technology alone" (Carter/Charles 2009: 15). In what Gaillard (2012: 262) terms the "climate gap", in most times, dominant modes, discourses and interventions of climate change science, policy, and planning "are disconnected from local realities, including people's needs, the cultural fabric and the traditional system[s]." They tend to elide the concerns and complexities of everyday life and to disregard the role of culture in producing knowledge about climate change and finding possible solutions.

Climate gap manifests itself in sentiments that acknowledge only physical scientific knowledge, material aspects, and discourses as the true answers to climate change and that, conversely, assume that cultural considerations and the knowledge of local communities cannot be considered valid, valued, or welcome. Such a perspective is irrelevant, ignorant of people's everyday lived experiences, and incapable of addressing the full range of challenges caused by climate change. It therefore follows that any interventions which ignore local knowledge and culture are subjective, bound to fail, or lead to maladaptive outcomes.

Cultural knowledge, then, is no less beneficial than scientific knowledge in strategies to address climate change. The ways in which people respond to climate bear culturally specific connotations. Culture frames the ways people identify, perceive, understand, experience, and respond to the world around them (Mokua 2019). Culture is therefore central in framing understandings, responses and giving meaning to climate change as a phenomenon of concern to society. Culture mediates every dimension of climate change, including the identification of opportunities and risks, decisions about adaptation and responses, and the means of implementation. Likewise, since culture is embedded in dominant modes of production, reproduction, consumption, lifestyles, political and social organisation, the impacts of climate change are given meaning through cultural interpretations (Hulme 2015; Adger et al. 2013). That is, in their day-to-day activities, people frame meanings and perceive climate risks, adversities and opportunities in ways that correspond to their cultural values and beliefs, and therefore shape and adapt their attitudes and behaviours accordingly. Therefore, climate needs to be understood, first and foremost, culturally.

Consequently, in order to develop and implement effective adaptation response measures, it is crucial to recognise and incorporate cultural science, data, knowledge, technologies and innovations, and recognise the rich interactions between climate and culture in a systematic way. Indeed, the premier global institutions on climate change, the *United Nations Framework Convention on Climate Change* (UNFCCC 2006) and its implementing instruments (protocols and conventions), and the *Intergovernmental Panel on Climate Change* (IPCC) recognise and advance indigenous knowledge (herein cultural knowledge) as a distinct guiding principle for coping with climate change. In actual fact, the IPCC (2007) (re)affirms indigenous knowledge as "an invaluable basis for developing adaptation and natural resource management strategies in response to environmental and other forms of change" (AR4). In addition, as stated in Mokua (2017: 78), African communities have a wealth of cultural "knowledge, science and innovations" they have "long used to sustain themselves in the face of climatic change".

Cultural knowledge, as used in this study, refers to African indigenous knowledge which denotes, as Mokua (2019: 126) indicates, "an African community's totality of knowledge" spanning all spheres of life and existence, built, "accumulated and handed down through generations". Such knowledge is place-based and rooted in local cultures, and is the result of cumulative experience, interaction and observation, tested in the laboratory of everyday life. Common terminologies used to refer to this body of knowledge include, but are not limited to, indigenous knowledge, traditional knowledge, cultural knowledge, traditional ecological knowledge, local knowledge, folk knowledge, indigenous traditional knowledge, ethnoscience, indigenous technical knowledge, indigenous technology, traditional climate knowledge and indigenous science (Mokua 2017, 2019). Although each term may have somewhat different connotations and reference groups, they all share sufficient meaning to be utilised interchangeably throughout this study.

11.2.2 Research Objectives

In view of that, this study, framed on the prism of weather and climate-related manipulation rituals, investigates how cultures across Africa have, over the centuries, typically utilised rituals to respond and adapt to climate change in their localities. The 'climate rituals', more prominently referred to as 'rainmaking rituals', are herein studied as a prototype of African cultural knowledge and response to climate change. Specifically, the study makes an explanatory interpretation, analysing the characteristic features, properties and areas which are symbolic or representative of rainmaking rituals and belief systems, presenting elements such as the logic, understanding, thought-patterns and imagination, behaviour, attitude, conduct, norms and value requirements underlying and informing the ritual foundations and performance.

In considering the principal tenets structuring this body of knowledge across a multiplicity of socio-cultural, economic, political, geographical and anthropological settings, civilisations and histories, the study sets insightful understanding of the unity and/or diversity of 'climate rituals' in the African anthropological space and universe. Finally, the study advocates the hybridisation of insights of this form of knowledge and science into the design of creative multipronged climate change response adaptations, planning, policy and research in contemporary Africa.

11.2.3 Research Design

This study, based on evidence from a wave of extant cultural data from purposively selected ethnographies representing various regional zones across Africa, assumes an ethnographic research design, providing detailed descriptions and interpretations of the climate rituals within the settings and socio-cultural contexts in which they take place from both interpretive *emics* (native point of view) and explanatory *etics* (external observer/'science') perspectives. The complementary balance between 'emic' and 'etic' validities, or 'insider-outsider' perspectives, ensures the aspect of 'interpretive validity', while contributing and maintaining the reliability of the study. 'Emic validity' is learning and understanding the rituals from the perspectives of the host communities. 'Emic validity' in this study is further enhanced because, as an African native, the researcher more easily and quickly understands and appreciates the subtleties of language, behaviour, thought systems, world views, and other finer cultural properties of African society. On the other hand, *etic* perspective is important because attributes of culture include dichotomies such as the ideal versus the real and the tacit versus the explicit, and therefore, throughout the study, the researcher maintains a sense of external, 'objective' framework as a trained research scientist, making interpretations and drawing conclusions solely on the strength of available data and evidence.

11.3 Climate Rituals in Africa

'Climate rituals' are climate manipulation and "weather modification ritual[s] that attempts to invoke rain", otherwise popularly known as 'rainmaking rituals' (Mokua 2017: 81). Although the most common form of contemporary weather modification is cloud seeding to increase rain or snow, usually for the purpose of increasing the local water supply, in this study the term 'rainmaking' is restricted to "the rites which accompany prayers for rain" (Mbiti 1975: 111). In the African anthropological thought and mindset "only God can make or produce rain" (Mbiti 1969: 180); and since 'making' rain is entirely God's mandate, it is the responsibility of men and women to summon God's rain. Consequently, in this study the term 'rainmaking' is used to refer to those traditional practices, rituals and ceremonies thought capable of controlling and manipulating the weather, and 'rainmakers' are the people bestowed with the power and ability to lead, intercede and conduct those practices.

Climate change adaptation, on the other hand, is, as Mokua (2019: 129) indicates, patterned "adjustments in ecological, socio-cultural" and/or "economic systems in response to actual or expected climatic stimuli and their effects" which enable coping and harmonise "livelihoods [...and the community's] own environment". It involves a corpus of patterned actions which adjust to particular weather patterns.

11.3.1 Rainmaking Rituals in East Africa

Comparing the rainmaking rituals of two Kenyan communities, the Kamba and the Bunyore, Akong'a (1987) explains that rainmaking can either be a public event in which all members of the community participate freely, or a private function led and conducted by known rainmakers who are able to diagnose the causes of either drought or too much rain and who have the power to reverse the situation. Akong'a cites studies in which Frazer (1942) explains that rainmakers were so influential that they could climb up the social ranks to be chiefs or kings in their communities. Were (1972) concurs, and cites a case in which a Luhyia ruler in Western Kenya was replaced by an alien – a Maasai – because he could not successfully intercede and make rain in response to a devastating drought and the resultant famine and loss of life. Jangu (2012: 100) reports a similar incident in Tanga, Tanzania, where locals "challenged the British authority for selecting a chief who failed to tame weather and bring rain". They claimed the chief lacked indigenous medicinal knowledge, was "deemed unfit to govern, and was asked to resign" (Jangu 2012: 100).

Further, Akong'a (1987) notes that the tradition of rainmaking may have emerged from communities living in arid and semi-arid regions as a means of coping with and mitigating unreliable and inadequate rainfall. Nonetheless, he acknowledges that rainmaking is also practised in some communities living in areas experiencing adequate rainfall. He observes that in the latter case, rainmaking is a prestigious profession practised by elites keen on helping or exercising control over their communities. He concludes that "rainfall rituals are based on faith and through faith the participants can make it rain" (Akong'a 1987: 83).

Writing on communal sacrifices and prayers performed to arrest drought and barrenness of land among the Iteso of both Kenya and Uganda, Ekeya (1984) notes that humanity's interference with the ecosystem withheld rain from falling. Cutting down trees – especially those regarded as sacred, such as the *ekure* tree – and killing wildlife such as the ground hornbill (*eutut*), believed to be a sacred bird, the diviner of rain, was tantamount to interfering with rain and was considered as detrimental to society as sorcery and witchcraft. As long as they lived in perfect harmony with their surrounding ecology, the Iteso felt assured of rain.

Despite the possibility of interference with the normal season of rain, the Iteso nonetheless believed that it was very necessary to pray to *Akuj* (Creator) for rain. Draped in full ceremonial dresses and regalia, and employing incantations, song and dance, adult Iteso men and women gathered in select spaces for the ritual prayer after being summoned by the diviner-prophet. Ekeya (1984) reckons that the ritual prayer seldom failed to result in rain falling, noting how practising rainmakers successfully conducted the prayer ceremony a few months before she arrived in the field and interviewed them. Whenever the people gathered and performed the prayer ritual, the rain fell either the following day or as soon as the ceremony was completed, giving the worshippers barely enough time to return to their homes. Giving a full description of a white colonial government administrator's account (Ekeya 1984: 83) of the ritual prayer performed in Teso, Uganda in April 1940, A.C.A. Wright, the colonial administrator, uncomfortably acknowledges the success of the ceremony saying:

> The service apparently proved disconcertingly successful as the next day an extremely heavy storm took place with heavy rain and such hail that many Abdim storks were killed and a lot of damage to cultivation done. This was followed by a week's steady rain.

In an inaugural lecture at the University of Nairobi, Kenya, Obasi (1976), a Professor of Meteorology, examined rainfall failure for prolonged periods of time in various regions of the world, giving rise to the occurrence of drought, and accompanying negative social-economic effects. The droughts devastated the world from the Sahel of north-west Africa to East Africa, Asia, the Americas, and Britain in Europe, decimating populations, animal herds and natural systems. The drought experienced in England was the worst in 249 years – so bad that Britons were prompted to "pray for rain as water crisis worsens" (Obasi 1976: 6). For the prayers, officials of a British Water Authority had a "prayer mat from Botswana" bearing the word *Pula* – meaning "let there be rain" in Setswana (Obasi 1976: 6).

Obasi (1976: 10) compares accounts of climate change instances in the Bible and the corresponding response of "rainmaking by the biblical God" with those of African "raindoctors" (as he refers to African rainmakers), and observes that the methods in all cases are somewhat obscure and mysterious. Nonetheless, the art can "qualify as science", although it would appropriately be categorised in the realm of metaphysics (Obasi 1976: 10). He acknowledges the African "Gods of rain" and emphasises that African rainmakers can only be regarded as "rain prophets" who bear no blame for any prediction which fails, since they do not promise rainfall but only report what the "Gods of rain" pass to them – they act as media (Obasi 1976: 10). He warns against being sceptical of their abilities, and appeals for the acceptance of their contributions, for we can bank on their ability to forecast with accurate precision in the "time scale of a few hours to days" (Obasi 1976: 9).

Obasi (1976) equates the grounding of this African knowledge with the maxim of signs given by the ancient philosopher, Theophrastus (373–286), and Chinese folk meteorology in which the appearance of the sky, animal and insect behaviour, the blooming and wilting of plants, and other signs are used as sensors for forecasting the occurrence of rain. Considering mechanisms involved in both nature's art of rainmaking as revealed by science and the art of rainmaking as practised by meteorologists (cloud seeding), Obasi concludes that given the shortfalls of each and the inability of scientists to conclusively endorse any one of them, there is a need to perfect the art of rainmaking to help alleviate world drought.

Sanders (1997, 2002, 2008) explores local ideas, ideals and models, as well as the history of rainmaking and rain-breaking practices, amongst the Ihanzu (or Isanzu), a Bantu-speaking agricultural group of north-central Tanzania. He points out that rain rites were divided into two distinct types: those that occur annually and those that occur following the failure of the annual rains. He documents a range of Ihanzu rain rites in detail, and simultaneously reproduces the local cosmological underpinnings that broadly inform and give meaning to those rites. Central to the Ihanzu cultural imagination lies the notion of gender complementarity or in a "variety of contexts how Ihanzu rain rites, as locally understood, are in the main 'about' the cultural construction of gender, […] and the relationship between these two categories" (1997: 247).

Sanders (1997) demonstrates how masculine and feminine principles of a gendered universe are seen as a site of cosmic and divine powers when combined. Thus, to "join the genders is to transform, to invigorate and to unleash the cosmic powers of the Ihanzu human, natural and divine worlds" (Sanders 1997: 247). To bridge the genders is to create, to rejuvenate. In addition, the people's understandings of magical and royal power and legitimacy remain a constant gender complementarity, and therefore control over rains is control over reigns. When "masculine and feminine elements of a gender universe combine, things happen […] the rains fall" (Sanders 1997: 247). Importantly, however, rainmaking serves as an efficacious process without which the Ihanzu would have "no rain […] no harvest, food, or beer […] no animals […] no villages" and thus no existence (Sanders 2008:101). In conclusion, Sanders argues that the gendered complementarity of the Ihanzu might be equally useful in explaining rain rites in the rest of Africa.

11.3.2 Rainmaking Rituals in the Sahel of North Africa

Abu-Zahra (1988) combines ethnographic fieldwork data with Qur'an and Arabic commentaries on Prophet Mohammad's traditions to trace local understandings on the cause of drought and consequent responses in Muslim society in the Sahel of Tunisia. The Sahel is part of the arid zone that extends from Morocco through Algeria, Tunisia, and Libya to Egypt, and indeed the whole of North Africa and the Middle East. The local people profess Islamic religion, and therefore their outlooks on the links between divine power, human actions and rain are based on the Qur'an and are expressed in the rain prayers prescribed by the Prophet. Abu-Zahra discusses Islamic beliefs relevant to rain rituals and provides the corresponding popular interpretation and symbolism as found and clearly explained in Islamic sources. The rain rituals are interpreted as rites of spiritual passage.

Abu-Zahra (1988) also presents data on the adaptation of agricultural methods to the scarcity of rain, the social means of combating the risks of dry farming, and people's images of the fertility of the land. Being agriculturists, local communities depend on the intensity and frequency of rainfall to ensure the quality and quantity of crop yields. However, the unreliability of rainfall threatens their livelihoods. Droughts are widely considered to be a result of sexual impurity, arrogance, or "injustice and ingratitude to God", but if people repent their sins in faith, "God may answer their prayers, spread His mercy, and send rain; thus pain may be followed by joy" (Abu-Zahra 1988: 519). This is because rain is a mystery only known to God and the Prophet, and only God has the keys to unlock the rains.

The rain prayers take place in open areas full of sacred landmarks. These are areas charged with spiritual powers. Sometimes people choose a place for prayers close to a source of water, with a similar act of optimism to express the hope that rain will be drawn to them. The performance of the rain prayers outside settlements is a symbolic rite of transition to free themselves from the 'boundaries of their sins' as they appeal to God in the uninhabited open space surrounded by sacred caves and shrines of saints, a place of vows, sacrifices and invocations for the success of the 'growing' children.

To indicate the need for immediate rain, the people supplicate through repentance of their sins, expression of their suffering, and demonstrations of their piety and humility to God, who brings them 'divine mercy' in the form of rain. To ask God to send them rain is to ask, "God have mercy on us" (Abu-Zahra 1988: 508). Thus, the concept of rain as divine mercy is at the heart of a nexus of numerous ideas that connect rain, human behaviour, and divine power. Obedience to God brings bounty and blessings, just as disobedience begets suffering in the form of drought.

Accordingly, rain prayers are rites of purification, entwined with fertility rites and rites to presage an auspicious near future. When they 'turnaround' their garments (wear clothes inside-out) in prayer, this symbolically acts to indicate that the sterile land will be 'turned' into fertile land, in the same fashion as the turning around of their garments. The rites performed symbolise the alternation of fortune and exchanging bad luck for good luck. Their purpose is to influence the world of the unknown, that is, the infinite, divine realm. Rain rituals are rites of spiritual passage from disobedience to reverence for God; from the known world, the world we sense and perceive (which, in the context of the rain rituals, is the present drought), to the world of God, that is, rain in the near future.

The rain rituals are figurative of rites of passage in which the people, through acts of penance and almsgiving, are symbolically separated from their former 'impure self' and transported or transition into 'becoming purified' in the rain prayers, and the rites of reversal presage a similar reversal in their worldly fortunes. Finally, the rites of incorporation presage integration back into community life and reconciliation between its members. Consequently, it is hoped that the spiritual passage will be accompanied by a passage in worldly affairs from scarcity to riches. The rain rites in the Sahel express the desire for spiritual rejuvenation and new beginnings.

11.3.3 Rainmaking Rituals in West Africa

Underscoring the intricate web labyrinth between water, power and socio-politics in traditional West African society, Wade (1997) identifies water as not politically neutral among the communities of the Mandara Mountains of Nigeria. In their autonomous realms, rulers powerfully demonstrate the political significance of water. Each chief, king or prince is firstly a rain-chief, as power over rain, over water, translates into power over citizens. A chief's authority is dependent upon bringing rain to the community, and the onset of the annual rains re-affirms this authority. The loss of rainmaking stones undermines political power, authority and control.

However, this is not necessarily so in part of the southern Mandara Mountains, where the Chadic-speaking Fali have rainmakers but not rain-chiefs. For the Fali chiefdoms, rain is a perennial uncertainty on when it falls and by how much (intensity). Droughts are such major threats of historic proportion that some are embedded in oral tradition. Rain – the symbolic precursor of fecundity – traditionally affects the pivotal and most notable biannual Fali fertility and initiation rite, during which it must rain to ensure the future potency and fecundity of the youth. Wade acknowledges having seen it rain frequently during these fertility and initiation ceremonies and rites.

Wade (1997) notes the rainmaking complex in Fali society and concludes that, though distinct from chiefs and wielding little in the way of enforcing the chiefs' coercive political authority, rainmakers are considered transformers par excellence. Their 'power' lies in their capacity to transform. The Fali rainmakers unite "heaven and earth in the life-giving act of rain", and *re-create* a meaningful time – that of sowing, growing and harvesting. With an act of rainmaking, the "old, dry, dead, hot, stifling, impotent year is transformed into the new one with all its rain-given promise of fecundity" (Wade 1997: 280). Rain is the product of potency, the power which renders the world fertile.

11.3.4 Rainmaking Rituals in Central and Southern Africa

Writing about Ronga rain rituals in Mozambique, Shaffer (2017: 7) recounts details of a public celebration she participated in during February 2008, when she "observed the collection of a beer tithe to offer the ancestors during the ritual, and [participants] were soaked by the rains that fell before the ritual ended". The Ronga rain rituals are a biannual activity in which the community petitions the ancestors to intercede with God and ask for rain prior to the growing season, and conducts a second ritual to give thanks for the rain that has fallen during the growing season and ask that the rains continue for a good yield. Shaffer opines that rain rituals could be understood as community-level adaptation interventions to reduce the uncertainties of rainfall variability and to control what is seen as a sacred phenomenon.

Schoeman (2006) uses the archaeological signature of rain-control sites to indicate the archaeological articulation and spatial manifestation of rain-control as a materialised ideology of farming communities in the ritualised *Shashe-Limpopo Confluence Area* (SLCA) of southern Africa. Specifically, she explores the archaeological manifestation of rain-control, its relationship with the ideologies of farming communities, and the relationship of the local communities with the elite Mapungubwe state formation, a relationship partially grounded in ritual and rain-control. She contends that after Mapungubwe state formation, rain-control was removed from nature to the farming communities, nationalised and centrally relocated to Mapungubwe hill.

Equally, Murimbika (2006) makes an ethno-archaeological analysis of centralised rainmaking in the Mapungubwe region, which cuts across the Shashe-Limpopo basin of the Sotho-Tswana in South Africa and Botswana, and the Shona in Zimbabwe. He explored Venda, Sotho-Tswana and Shona traditional belief systems through their oral histories, cosmologies and practices, and identified three systems of rainmaking practices. First is the rainmaking practice associated with kin-based chiefdoms, and this relates to rank-based societies (like the Pedi, Hananwa and Tswana). Second is the rainmaking practice among class-based societies with sacred leadership (common among the Venda and, to some extent, the Lovedu, sometimes pronounced Lobedu). Third is the rainmaking practice related to a bifurcated structure in which political and religious powers are split (for example, among the Shona). The third rainmaking practice represents the devolution of complexity after the disintegration of the Zimbabwe culture.

Similarly, Huffman (2009) applies archaeological data to ethnographic evidence to show that the traditional Bantu-speaking societies of the Mapungubwe region believed drought to be caused by serious breaches in the laws of pollution. The rules surrounding ground beans are principally important because the beans are particularly sensitive to pollution. A new variety of cowpea seed, for example, coming from somewhere else, was automatically considered to be ritually polluted. To overcome the pollution, it needed to be blessed or cleansed before it was planted. Because of this sensitivity, local farmers dug up and destroyed the plants during a severe drought. Additionally, strange men and menstruating women were not allowed to walk across the fields. If farmers ignored these rules about planting, harvesting and entering fields and severe drought followed shortly afterwards, the particular farmers were to blame.

As a rule, droughts should not occur if rainmakers perform their job, and if every citizen obeys the rules of the land. The ancestors safeguard these rules and only allow misfortune when there is a serious breach. Being farming communities, rainmaking was mainly tied to the agricultural cycle and, normally, rainmakers performed their duties in special sacred rain locations, or 'rain kraals', at the back of or behind their homesteads. On breach of the rules, the misfortune of drought followed and rituals were offered to appease the gods. However, when normal rituals and medicines consistently failed, and droughts persisted, rainmakers moved out of the 'polluted settlements' with their implements to special, steep-sided hills – the terrain of supernatural forces – to perform special rituals. People thought to have caused the drought burned their grain bins, and sometimes their houses, as a ritual of cleansing and built new ones on top.

Examining legends and myths related to rainmaking rituals in the Mutema chieftaincy of Chipinge region in Zimbabwe, Vijfhuizen (1997) indicates the central theme to be how people attribute drought to conflicts between and amongst traditional political leaders at various levels. Rain is associated with political authority, power and control. Local leaders are believed to have control over the rains, and therefore any conflicts amongst them will make the ancestors keep the rains away. Control over the rains is control over the people. During periods of drought all leaders must unite, co-operate, avoid conflicts and power struggles, and guide the people in rainmaking rites, requesting the ancestors for rain, otherwise the drought will continue. The leaders are thus regarded as spiritual leaders in addition to their chieftaincy. Hence, to ensure good rains, leaders must resolve their conflicts and come together for ritual worship of their ancestors. Drought is also thought to be caused by changing climate and human activities, including the cutting down of trees.

Vijfhuizen (1997) provides detailed information about procedures, equipment, people and the gender of the participants, and their exact roles in the consummation of the rainmaking rituals. Of particular importance is the role of women in rituals and rainmaking. For instance, beer for the ancestral spirits, a fundamental ingredient in rainmaking rituals, is only brewed by old post-menopausal women. This is because ancestors want hygiene and cleanliness. If something goes wrong in the beer processing, the ancestors will not communicate with the living. To consummate rainmaking rituals, the co-operation and participation of women is therefore, critical. In addition, women act as spirit mediums of the royal ancestral spirits. The royal ancestral spirits never fail to bring the rains after the rituals.

In what can be described as hybridity or blending of two religious cultures and belief systems, that of Christianity and African traditional religion, Müller (2008) provides a detailed witness description of a contemporary Christian cross-border pilgrimage from south Africa to Botswana with the purpose of praying for rain when Botswana experienced a severe drought. The entourage and the public in Botswana had high hopes of the leader of the pilgrims, the Bishop, as the "rain man" or "the man who has the key to the rain" (Müller 2008: 821) because of memories of a previous pilgrimage in which good sustaining rains had indeed been recorded subsequent to the visit. Müller reports witnessing the fall of "hard driving rain […] right through until dawn" in the night before the scheduled event (Müller 2008: 823). Nonetheless, the public was convinced that the Bishop was not a god nor did he have "any special powers over the rain" but was "merely praying for rain" (Müller 2008: 824). Equally, the Bishop openly voiced belief in the "power of prayer" (Müller 2008: 525).

The belief in the rainmaking powers of the Bishop can be interpreted as a transposition of older, pre-Christian ideas regarding the rainmaking powers of traditional rulers. "Traditional societies did not separate the secular and the sacred in the public sphere" (Müller 2008: 825), and consequently, traditional rulers' powers included mediation between the people and their ancestors. Accordingly, the Bishop's power of 'rainmaking' was understood from the perspective of the cultural female rulers, bearing the dynastic title *Modjaji*, the legendary 'Rain Queens', believed to have rain-bringing powers in southern Africa. This occurrence confirms "neither European contact nor Christianity has succeeded in breaking down the implicit belief […] in the power […] to make rain" (Müller 2008: 824). This study and many others before it show that "rainmaking and sacral leadership tended to converge in the figure of the traditional ruler" (Müller 2008: 826). Thus, the Bishop's divinity and status as a rainmaker appear inseparably intertwined, as he fulfils a similar role to that of a traditional ruler.

Responding to the rainmaking and rain prevention rituals which are practised amongst the Pedi tribes – also called the Northern-Sotho tribes – of southern Africa around the Limpopo region, Semenya (2013) presents a pastoral perspective. The rituals of rainmaking and rain prevention are an ancient science practised by the Pedi, and continue to be practised to this day. The rituals are practised in times of drought and also as rain prevention when activities such as weddings or traditional gatherings are organised to ensure that the rain does not disturb the gathering and that everything goes as planned. The *Moroka* (rainmaker) leads the community in performing the rituals.

Whilst the Sotho-speaking groups of Southern Africa believe in a Supreme Being who is ultimately responsible for the weather, rain rituals are predominantly directed at the ancestors who intercede on the people's behalf. Drought is believed to occur when they are negligent. Often the rainmaking rituals are performed using special instruments and tools at special sacred sites considered mysterious and sometimes even dangerous, "inhabited by superior powers" of "spirits or deities" (Semenya 2013: 2). In discussing the rain rituals and selected scriptural passages, Semenya offers the biblical view and, in conclusion, formulates principles that clarify the Bible's perspective on the rituals.

Rafapa (2008) sought to investigate rainmaking among the Mamaala ethnic community of South Africa in order to find out which intuitive knowledge about this art is objective and accurate, and which is subjective and distorted. Rafapa opines that the traditional rainmaking of the Mamaala had both idiosyncratic features and aspects in common with the way this religious-cultural art was practised among many African communities, and therefore considers the Mamaala accounts to be representative of the many variants in Africa. Rafapa acknowledges the Africans' ability to ritualistically bring about rain.

Offering a detailed Afrocentric account of the African religious-cultural practice of bringing rain, Rafapa (2008) dispels Eurocentric tendencies to see African rainmaking practices and rainmakers as inscrutable exotica. To preserve African indigenous knowledge as well as purge it of Eurocentric distortions, Rafapa submits for research into the disappearing practice of rainmaking, a phenomenon that has characterised many African communities over the centuries. Otherwise, forms of African indigenous knowledge are threatened by the intensification of foreign cultural elements and conflicts. These developments justify the preservation of African rainmaking history, which would otherwise be lost after these practices have vanished completely.

Gewald (2001) chronicles the absence of rain in the second half of the year 2000 and early 2001, and the development of the 'El Negro' thought and outlook in Botswana. 'El Negro', the remains of a Tswana man who, as a stuffed specimen, had been on public display in Europe for years, was flown to Botswana in October 2000 and given a state funeral. Gewald situates thoughts on 'El Negro' in the context of popular consciousness of rainmaking, magic and witchcraft within Tswana cultural thinking and discourse. As with all aspects of life, death in Tswana society is inextricably linked to rain. The spirits of the dead, *badimo*, live in a parallel world, similar to that of the living, and take a keen interest in the world of the living, influencing and determining events. They reward with good prosperity people who treat them with becoming respect and obedience, but punish with misfortune those who neglect them or who offend against the prevailing social code, of which they are the guardians. Hence, in order to retain the favour of the spirits, people expressly propitiate or domesticate them, in the belief that otherwise they face all manner of misfortunes, the worst of which is the absence of rain.

It is on the strength of this mindset that when, in early 2001, the rains failed, a thought emerged and developed that linked the absence of rain with the buried remains of 'El Negro'. This perspective was reinforced by the fact that in the first half of the year 2000 Botswana was blessed with extensive and long-lasting rains. It was only in the second half of the year and early 2001, after the arrival of the remains of 'El Negro', that the rains failed. So, in the Tswana anthropological thought, the arrival of 'El Negro' was to blame for the failure of the rains to fall in due season.

This outlook argued that the return of the body was witchcraft that was stopping the rain from falling. In essence, the Batswana thought that whatever they had buried was something that prevented the rain from falling. Underlying all these fears was the common issue of not knowing exactly who or where these buried remains (bones) originally came from, and the inability to gain an insight into why these bones came to be buried in Botswana in the first place. The gist of the 'El Negro' outlook was that the bones could be those of a person who did not come from Botswana (probably a Bushman), and who had either angered the ancestors, or was an extremely upset and troubled Tswana spirit ancestor inflicting wrath on Batswana for maltreatment.

Because of the 'El Negro' perspective, connotations of rain and weather associated with the term El Niño were transferred in the public consciousness to the term 'El Negro'. The two terms were unconsciously interchanged. In terms of popular consciousness, 'El Negro' came to be linked to a multitude of ideas relating to the natural weather phenomenon known as El Niño. In addition, 'El Negro', and the events surrounding the return of what were alleged to be the remains, came to be linked to a multitude of moral national concerns bedevilling the country at the time, ranging from Bushmen, the sale of human body parts, witchcraft, and upset ancestors to corrupt politicians. In popular conversations, the return of 'El Negro', in combination with one or more of the issues, came to be regarded as responsible for the absence of rain in Botswana.

In other words, popular discourse in Botswana implicitly accepted and acknowledged the important influence of the metaphysical and supernatural on their everyday life, and served as an explanation as to why things occur in the way they do. Like their Western contemporaries who watch daily weather forecasts, the Batswana know about high-pressure zones, and believe they know *how* the weather works. Though they cannot directly measure the weather variables, they can directly observe the bio-physical manifestations of climate with their senses. However, they do not know *why*, and it is on this fundamental issue that they seek an answer – in this instance: 'Why is it not raining in Botswana?' In attempting to answer this question, they linked the coming of 'El Negro' with current national concerns about the absence of rain.

11.3.5 African Anthropological Wisdom and Thought Processes on Rainmaking Rituals

Rain has always been a most valuable commodity occupying a central theme in the African anthropological thought and universe. Africans regard rain as the most sacred phenomenon on which their "physical life" as well as their "prosperity and wellbeing" depend (Mbiti 1969: 180). African people "observe days of rest to mark the start of the rain season", and greet the seasons with "ceremonies of thanksgiving and prayer for sufficient rain" (Mbiti 1969: 181). Failure of rain curses the land and renders it unproductive and infertile, its people barren and their animal herds perish.

Rain is therefore the most "explicit expression of the goodness and providence of God" (Mbiti 1969: 181), and as it comes from above, it maintains an "eternal mystical" link with the African people, their past, present and future genealogies, the earth and the divine in heaven. In addition, rain is so intimately associated with God that in some communities the same word is used for both, or when it rains people say that "God is falling" (Mbiti 1969: 181). In many other communities, the name for God means "Rain Giver" (Mbiti 1969: 181). Yet many personify rain or regard it as being controlled by a divinity (subject to God). In Botswana, for example, it is symbolised in the official term for the national unit of currency, the *Pula*, which means 'rain'. The word *Pula* is also used to indicate blessings and good fortune in the local Setswana language (Gewald 2001; Obasi 1976).

It is therefore not surprising that the mode of weather manipulation by means of ritual prayers, as a practical matrix of response adaptation to climate change, has occupied, dominated and reigned in the African anthropological outlook, understanding, thought and imagination from antiquity to contemporary civilisations. This mode of thinking and outlook was brazenly proven in the contemporary kingdom of Swaziland (now called Eswatini), when climate change in the first quarter of 2016 brought a severe drought that particularly hit subsistence farmers who rely on erratic rainfall as their source of water. The drought caused an extreme lack of water, subsequent falling crop yields and the death of thousands of livestock. In response to the increasingly precarious situation, Swaziland's monarch underscored the importance of prayer, saying that the country did not need an irrigation system, but had an irrigation system coming directly from God, who would bring down the rains to water their fields (Kenworthy 2016). In view of that, the country's political leadership prescribed prayer for rain as the solution to the crisis of drought the country was experiencing at the time.

A similar systematic pattern of thought, imagination and outlook was witnessed and discharged in Kenya when severe drought ravaged the country in the last quarter of 2016 and early 2017. While on a political campaign trail in one region of the country in January 2017, the populace unrelentingly implored President Uhuru Kenyatta to pray. Before addressing the crowd, the President reluctantly said a short prayer in the local Kikuyu dialect, asking God to "give us rains so that our animals have something to eat. We pray for a good harvest in our farms and growth in our country" and immediately afterwards a heavy downpour pounded the area (Wangui/Gita 21/01/2017). The act of the President making an effective prayer can be viewed, understood and interpreted in terms of leaders assuming responsibility for and control over not only the people, but also nature – the rains. Therefore, control, power and authority over the rains were symbolically representative of influence, authority and command over the people.

On the same mode of thinking, Gewald (2001) reiterates the central importance of rain and rainmaking ceremonies in the past and contemporary Tswana society past and present, noting that a chief is a chief by the grace of the people but, in return, a chief must ensure adequate rainfall. It was thus a sign of good luck when it rained on the day that Festus Mogae was formally inaugurated as President of Botswana in early 2000. Both in the past and the present, the failure of rain reflects the failure of a chief, king or ruler and indicates that the land is impure or bewitched. Indeed, ceremonies connected with the bringing of rain were and are considered the most important rituals over which a chief/ruler can preside.

Rainmaking specialists are accorded a high status in the African anthropological universe. In essence, the person who controls rain also determines the course of life and society. Indeed, by making an effective prayer for rain, the Kenyan President, Uhuru Kenyatta, was equated with respectable and distinguished Kikuyu forefathers and elders who, during dry spells, led the community in prayer, which God answered by bestowing the blessing of plentiful rainfall. In actual fact, his father, the late Jomo Kenyatta, who later became the first president of post-colonial Kenya, was, on his own admission, one of the pre-adolescents entrusted with leading rainmaking sacrificial processions in the Kikuyu community. The older Kenyatta (1938: 137) authoritatively writes of his own personal experience:

> In the ceremony in which I took part, I well remember that our prayers were quickly answered, for even before the sacred fires had ceased to burn, torrential rain came upon us. We were soaked, and it will not be easy for me to forget the walk home in the downpour.

Akong'a (1987) recounts a similar strand of thought and reasoning when elders of the Meru community in Kenya successfully re-enacted a rainmaking ritual on 28 April 1984. This was after the long-expected rains had failed to arrive in due season, sending ripples of fear of drought and famine across the horn of Africa. Referring to highly publicised media reports on the details and effective outcome of the ritual, Akong'a attests that, "hardly had the smoke from the sacrificial black lamb swirled into the evening sky, when a heavy downpour started that lasted the whole night" (Akong'a 1987: 83–84).

The formulaic notion of African anthropological leadership conceived as embedded with mystical powers is, perhaps, best articulated by Turner (1969), who stresses that the position of chiefs, kings and rulers in many African societies represents both the "apex of the structured politico-legal hierarchy and the total community as an unstructured unit" (Turner 1969: 98). They are, "symbolically, also the tribal territory itself and all its resources. Its fertility and freedom from drought, famine, disease, and insect plagues" (Turner 1969: 98) are bound up with their office and their physical and moral condition. Their daily invocations are for "fertility and continued health and strength of the land, its animals and vegetable resources, and of the people – in short, for the commonweal and public good" (Turner 1969: 99), the failure of which would render the land unproductive, infertile and barren. In that respect, leaders' legitimacies, authorities, abilities and influences are determined by their success in controlling and intervening in agricultural productivity, people's health, family fertility, and other social, political, religious, environmental, and security issues.

By this mode of thought, rulers are expected to have relevant indigenous and customary knowledge and skill in each sphere. In cases where they lacked particular knowledge and skill, they employed or invited experts who had such knowledge. It is on the strength of this anthropological mindset and thought system that chiefs among the Sukuma community of north-west Tanzania, around Lake Victoria, worked with diviners and specialised rainmakers famously known as *bagemi ba mbula* (rain harvesters) to ensure conditions conducive for agricultural production, farming and livestock-keeping. When conditions were bad, chiefs instructed diviners to consult the spirits to determine the reasons for those conditions and the actions that should be taken to change those conditions. Such actions often included the performance of rituals to appease spirits in order to bring rain. In such events, rainmaking rituals officiated by chiefs were solemnised by the *bagemi ba mbula* (Jangu 2012).

Accordingly, to the African people, the power of kings, chiefs and all rulers was historically associated with charisma, knowledge and skill, as well as religio-magical capabilities. They were not only authorities over their territory but also intermediaries and mediators between their people and God. The health and prosperity of their nation, therefore, depended on how the king accessed, controlled and manipulated the energies and powers God gifts to nature. That is why, in many other parts of Africa, a chief had overall concern for adequate and favourable distribution of rain on which the "survival of his people and, indeed his own political and social survival depended". They consequently carried out rituals to aid the agricultural cycle, consulted rainmakers about the prospects of the weather, and ritually prepared the seeds and soils for cultivation, planting and harvesting. They also ritually shaved their heads so that their hair grew with the crops, performed the ritual protection of the growing crops from birds and animal pests, and performed harvest celebrations with the ritual eating of food from the new crops (Jangu 2012: 100–101).

Obasi (1976) and Akong'a (1987) observe the continuation of the anthropological frame of mind and thinking on the apparent belief in the power of traditional rainmakers by contemporary post-colonial African leadership. During Kenya's independence celebrations in 1963, the Government invited renowned Nganyi rainmakers of the Bunyore sub-tribe of the Luhyia community to the capital, Nairobi, to ward off any rainfall during the historic occasion. Similarly, during the drought that affected most parts of Kenya in 1974/1975, an appeal was made to these rainmakers and the Government summoned them to Nairobi where they appeared on television.

Evidently, belief in the mystical powers of rainmakers to control and manipulate the weather dominates the African anthropological thinking, understanding and universe on climate change. Accordingly, rainmaking rituals are reported among many ethnic communities in the African continent, extending from the Sahel in the north all the way through the west, centre and east, to the bottom end of southern Africa. The rituals closely mirror what is reported in other regions of the world including the biblical story of the prophet Elijah, who caused rain to fall in Israel after a devastating three-year drought (Obasi 1976; Akong'a 1987). The accounts confirm the factual truth, reality and efficacy of rainmaking rituals not only in Africa but across the world.

11.3.6 Rainmaking Rituals in Perspective

The observations to be drawn from these data are quite clear. Rainmaking rituals are an ancient science and reality for most people and ethnic communities across the African continent, and are closely related to the experiences of their everyday lives. The rainmaking rites denote the African peoples' pragmatic adaptation to nature by the application of the empirical knowledge at their disposal, and the way they relate to their political, economic, legal, and socio-cultural environment and surroundings. They also systematically unpack the complex relationships between African humanity, anthropology, cosmology, ritual and power.

Although there are variations in the details and processes, the rituals are similar in several respects and consist of invoking the deity to bring rain during periods of drought or to prevent detrimental rainstorms or rainfall altogether. Thus, the purpose of the rain rituals is to influence the weather conditions in order to cause rain or drought, either for good or for destruction.

A major observational component of the rainmaking rituals is the association of immorality and apostasy with drought, on the one hand, and divine pardon or religious fidelity and rain on the other. Thus the utmost observation of purity and holiness is required when carrying out the rituals. For instance, although in most cases all community members are involved in various ways in the ritual activities, however, ultimate consummation is often preserved for persons considered most pure and holy, such as pre-adolescents and women past the menopause. They are also the only ones allowed to enter sacred sites and places. In addition, to observe purity, sexual intercourse is often strictly prohibited in the days preceding and during the rituals.

In most times, rainmaking rituals are solemnised by expert rainmakers. Rainmakers are specialised people, with a language, symbolism, knowledge, skills and practices of their own which are not easily revealed or accessible to ordinary people. They intercede with the gods of rain on behalf of the people. They have an intercessory inspiration and power to perform the art and science of rainmaking. They are the living embodiment of the rain gods, guarantors of the yearly season cycle, transformers of the clouds and weather.

Rainmakers perform the rituals using special instruments and tools at special sacred sites considered mysterious and sometimes even dangerous, charged with superior powers of spirits and deities. Sometimes people choose a place for rain prayers close to a source of water, with a similar act of optimism to express the hope that rain will be drawn to them. The performance of the rain prayers outside normal settlements is symbolic of freeing people from the 'boundaries of their sins', as they appeal to God in the uninhabited open space surrounded by mystical powers, a place of vows, sacrifices and invocations for the successful fortune of their land and the prosperity of their genealogies.

In the African anthropological view, mystical power pervades all manifestations of the physical world and is concentrated in certain people (such as rainmakers, seers, chiefs and kings), places (such as swamps, valleys, hills and mountains) and objects (such as rain stones) to higher degrees than in others. Therefore, humans engage in rituals because they enable them to mediate and balance interactions with multifarious preternatural forces that inhabit the human, physical and spiritual worlds. Rituals allow them to harness such power. In most cases, there is evidence that it rains immediately or shortly after the rituals are performed. It is therefore prudent that this disappearing strand of ethnoscience be reinvigorated, summoned and incorporated into the design of creative multipronged climate change response adaptations, planning, policy and research in contemporary Africa.

References

Abu-Zahra, Nadia, 1988: "The Rain Rituals as Rites of Spiritual Passage", in: *International Journal of Middle East Studies*, 20,4: 507–529.

Adger, W. Neill; Barnett, Jon; Brown, Katrina; Marshall, Nadine; O'Brien, Karen, 2013: "Cultural Dimensions of Climate Change Impacts and Adaptation", in: *Nature Climate Change*, 3: 112–117.

AGRA (Alliance for a Green Revolution in Africa), 2014: *Africa Agriculture Status Report: Climate Change and Smallholder Agriculture in Sub-Saharan Africa* (Nairobi, Kenya: AGRA); at: https://ccafs.cgiar.org/publications/africa-agriculture-status-report-2014-climate-change-and-smallholderagriculture-sub (15 September 2017).

Akong'a, Jason Joshua, 1987: "Rainmaking Rituals: A Comparative Study of Two Kenyan Societies", in: *African Study Monographs*, 8,2: 71–85.

Bernier, Pierre; Schoene, Dieter, 2009: "Adapting Forests and Their Management to Climate Change: An Overview", in: "FAO (Food and Agriculture Organization of the United Nations): Adapting to Climate Change", in: *Unasylva*, 60,231/232: 5–11.

Carter, Bob; Charles, Nickie, 2009: "Society, Nature and Sociology", in: *The Sociological Review*, 57 (supplement 2): 1–20.

Ekeya, Bette J.M., 1984: "The Emurwon – Diviner/Prophet – In the Religion of the Iteso" (unpublished PhD thesis, University of Nairobi, Kenya).

FAO (Food and Agriculture Organization), 2007: *Adaptation to Climate Change in Agriculture, Forestry, and Fisheries: Perspective, Framework and Priorities* (Rome: FAO).

Fatorić, Sandra; Seekamp, Erin, 2017: "Are Cultural Heritage and Resources Threatened by Climate Change? A Systematic Literature Review", in: *Climatic Change*, 142: 227–254.

Frazer, James George, 1942: *The Golden Bough: A Study in Magic and Religion* (New York: Macmillan).

Gaillard, Jean-Christophe, 2012: "The Climate Gap", in: *Climate and Development*, 4,4: 261–64; https://doi.org/10.1080/17565529.2012.742846.

Gewald, Jan-Bart, 2001: "El Negro, El Niño, Witchcraft and the Absence of Rain in Botswana", in: *African Affairs*, 100: 555–580.

Huffman, Thomas N., 2009: "A Cultural Proxy for Drought: Ritual Burning in the Iron Age of Southern Africa", in: *Journal of Archaeological Science*, 36: 991–1005.

Hulme, Mike, 2015: "Climate and its Changes: A Cultural Appraisal", in: *Geo: Geography and Environment*, 1–11.

IFAD (International Fund for Agricultural Development), 2009: *Livestock and Climate Change*.

IPCC (Intergovernmental Panel on Climate Change), 2007: *Fourth Assessment Report (AR4): Summary for Policymakers* (New York, Cambridge University Press).

IPCC (Intergovernmental Panel on Climate Change), 2014: *Climate Change (2014): Impacts, Adaptation, and Vulnerability*. Part B: Regional Aspects: Contribution of Working Group II to the Fifth Assessment Report of the Intergovernmental Panel on Climate Change Cambridge University Press (Cambridge & New York: Cambridge University Press).

Jangu, Menan Hungwe, 2012: "Healing Environmental Harms: Social Change and Sukuma Traditional Medicine on Tanzania's Extractive Frontier" (PhD thesis, University of Michigan, USA).

Kenworthy, Peter, 2016: "God Will Bring Down the Rains", in: *Pambazuka News*, 757; at: http://pambazuka.org/en/category/comment/96328 (2 April 2017).

Kenyatta, Jomo, 1938: *Facing Mount Kenya: The Tribal Life of the Kikuyu* (Nairobi: East African Educational Publishers).

Kumari, Rigaud Kanta; Sherbinin, Alex de; Jones, Bryan; Bergmann, Jonas; Clement, Viviane; Ober, Kayly; Schewe, Jacob; Adamo, Susana; McCusker, Brent; Heuser, Silke; Midgley, Amelia, 2018: *Groundswell: Preparing for Internal Climate Migration* (Washington, D.C.: The World Bank).

Lisk, Franklyn, 2009: "Overview: The Current Climate Change Situation in Africa", in: Besada, Hany Gamil; Sewankambo, Nelson K., (Eds.): *Climate Change in Africa: Adaptation, Mitigation and Governance Challenges*, in: Centre for International Governance Innovation (CIGI) Special Report (Waterloo, Ontario: CIGI).

Masinde, Euphraith Muthoni; Bagula, Antoine, 2011: "ITIKI: Bridge between African Indigenous Knowledge and Modern Science of Drought Prediction", in: *Knowledge Management for Development Journal*, 7,3: 274–290.

Mbiti, John Samuel, 1969: *African Religions and Philosophy* (Nairobi: East African Educational Publishers,).

Mbiti, John Samuel, 1975: *The Prayers of African Religion* (London: SPCK).

Mokua, Ombati, 2017: "Rainmaking Rituals: Song and Dance for Climate Change in the Making of Livelihoods in Africa", in: *International Journal of Modern Anthropology*, 10: 74–96.

Mokua, Ombati, 2019: "Ethnology of Select Indigenous Cultural Resources for Climate Change Adaptation: Responses of the Abagusii of Kenya", in: Brauch, Hans Günter; Oswald Spring, Úrsula; Collins, Andrew; Serrano Oswald, Serena Eréndira (Eds.) *Climate Change, Disasters, Sustainability Transition and Peace in the Anthropocene* (Cham, Switzerland: Springer International Publishing): 125–151.

Moore, Jerry. D., 2009: *Visions of Culture: An Introduction to Anthropological Theories and Theorists*, 3rd edn. (Plymouth, UK: AltaMira Press).

Müller, Retief, 2008: "Rain Rituals and Hybridity in Southern Africa", in: *Verbum et Ecclesia* 29,3: 819–831.

Murimbika, McEdward, 2006: "Sacred Powers and Rituals of Transformation: An Ethno Archaeological Study of Rainmaking Rituals and Agricultural Productivity during the Evolution of the Mapungubwe State, AD 1000 to AD 1300" (unpublished PhD thesis, University of the Witwatersrand, Johannesburg, South Africa).

Nche, George Christian, 2014: "The Impact of Climate Change on African Traditional Religious Practices", in: *Journal of Earth Science and Climatic Change*, 5,7: 1–5.

Nyasimi, Mary; Radeny, Maren; Kinyangi, James, 2013: *Climate Change Adaptation and Mitigation Initiatives for Agriculture in East Africa*, CCAFS Working Paper No. 60 (Copenhagen: CGIAR Research Program on Climate Change, Agriculture and Food Security [CCAFS]); at: www.ccafs.cgiar.org (2 April 2017).

Obasi, Godwin Olu Patrick, 1976: *The Art of Rainmakers*. 10th Inaugural Lecture (Nairobi: University of Nairobi, Kenya).

Ojwang', Gordon O.; Agatsiva, Jaspat; Situma, Charles, 2010: *Analysis of Climate Change and Variability Risks in the Smallholder Sector*; at: http://www.fao.org/docrep/013/i1785e/i1785e00.pdf (2 April 2017).

Rafapa, Lesibana Jacobus, 2008: "At the Heart of African Rainmaking", in: *Southern African Journal for Folklore Studies*, 18,1: 51–62.

Sanders, Todd Darrell, 1997: "Rainmaking, Gender and Power in Ihanzu, Tanzania, 1885–1995" (unpublished PhD thesis, London School of Economics and Political Science, London).

Sanders, Todd Darrell, 2002: "Reflections on Two Sticks: Gender, Sexuality and Rainmaking", in: *Cahiers d'Études africaines*, 166, XLII-2: 285–313.

Sanders, Todd Darrell, 2008: *Beyond Bodies: Rainmaking and Sense Making in Tanzania* (Toronto: University of Toronto Press).

Schoeman, Maria Hendrieka, 2006: Clouding Power? Rain-Control, Space, Landscape and Ideology in Shashe-Limpopo State Formation (unpublished PhD thesis, University of the Witwatersrand, Johannesburg, South Africa).

Semenya, David K., 2013: "The Making and Prevention of Rain amongst the Pedi Tribe of South Africa: A Pastoral Response", in: *Theological Studies*, 69,1: 1–5.

Shaffer, L. Jen, 2017: "Rain Rituals as a Barometer of Vulnerability in an Uncertain Climate", in: *Journal of Ecological Anthropology*, 19,1: 1–17.

Steward, Julian, 1955: *Theory of Culture Change: The Methodology of Multilinear Evolution* (Urbana: University of Illinois Press).

Sutton, Mark Q.; Anderson, E.N., 2010: *Introduction to Cultural Ecology*, 2nd edn. (Plymouth, UK: AltaMira Press).

Turner, Witter Victor, 1969: *The Ritual Process: Structure and Anti-Structure* (NY: Cornell University Press).

UNFCCC (United Nations Framework Convention on Climate Change), 2006: *United Nations Framework Convention on Climate Change* (UNFCCC): Consolidated Version with Amendments to Annexes (FCCC/INFORMAL/84-GE.05-62220 (E) 200705); at: https://unfccc.int/resource/docs/publications/handbook.pdf (2 April 2017).

UNFCCC (United Nations Framework Convention on Climate Change), 2015: *Adoption of the Paris Agreement*; at: https://unfccc.int/resource/docs/2015/cop21/eng/l09r01.pdf (6 May 2018).

Vijfhuizen, Carin, 1997: "Rain-Making, Political Conflicts and Gender Images: A Case from Mutema Chieftaincy in Zimbabwe", in: *Zambezia*, 24,1: 31–49.

Wade, James H., 1997: "Rainmakers and the Problematics of Power in Fali Society", in: Jungraithmayr, H.; Barreteau, D.; Seibert, U. (Eds.): *The Man and the Water in the Lake Chad Basin = Man and Water in the Lake Chad Basin* (Paris: IRD Editions): 271–284 (publication arising from Colloquia and seminars of the Mega-Chad Network, J. W. Goethe-Universität, Frankfurt, 13–14 May 1993).

Wangui, Joseph; Gita, Grace, 2017: "Did President Uhuru Kenyatta Make it Rain in Nyeri?", in: *The Daily Nation* (21 January).

Were, Gideon Saulo, 1972: "The Maasai and Kalenjin Factor in the Settlement of Western Kenya: A Study of Ethnic Interaction and Evolution", in: *Journal of Eastern African Research and Development*, 2,1: 1–11.

Chapter 12
Ethnically-Charged Wartime Sexual Violence: The Agony of the South Sudanese Refugees in Uganda

Robert Senath Esuruku

Abstract Since the outbreak of civil war in South Sudan in December 2013, thousands of South Sudanese have been subjected to sexual violence, including rape, gang rape, sexual mutilation, castration, and forced nudity. The outrageous and brutal sexual violence has resulted in undesirable physical, psychological, and social impacts on the survivors, who are largely living in refugee settlements in the West Nile region of Uganda. This paper provides a critical understanding of the disturbing narratives and experiences of the survivors of the ethnically-charged sexual attacks in the ongoing civil war in South Sudan. It examines how extreme acts of sexual violence were part of a strategy to terrorise, degrade, shame and humiliate both the victims and their ethnic or political group. This paper also examines the impacts of the horrendous sexual violence on the survivors, their families and society, and provides policy recommendations to address the wartime sexual violence.

Keywords Civil war · Ethnicity · South sudan refugees · Wartime sexual violence

12.1 Introduction

In early 2018, I conducted a political economy analysis study of refugees and host community relations for the *Aga Khan Foundation* (AKF) and the *World University Service of Canada* (WUSC) in Adjumani Districts. During the study, I interacted with a number of respondents who told me that they were brutally assaulted sexually because of their different ethnicities in the ongoing civil war in South Sudan. These individuals were mainly from the Dinka and the Nuer ethnic groups. My interest in conducting this study was augmented when I learned that the two ethnic groups were put in different refugee settlements and were not allowed to mix freely due to the animosities resulting from the civil war in South Sudan, which has pitched the two ethnic groups against each other.

Dr. Robert Senath Esuruku, Head, Department of Development Studies, Makerere University, Uganda; Email: esuruku@chuss.mak.ac.ug or robert.esuruku@gmail.com.

Ú. Oswald Spring and H. G. Brauch (eds.), *Decolonising Conflicts, Security, Peace, Gender, Environment and Development in the Anthropocene*,
The Anthropocene: Politik—Economics—Society—Science 30,
https://doi.org/10.1007/978-3-030-62316-6_12

The respondents told me that sexual violence is part of the systemic debasement of perceived enemies in the civil war which began in December 2013. The civil war has divided communities in South Sudan along ethnic lines, specifically the Dinka and the Nuer ethnic groups. The rape and mutilation of both male and female civilians by the military forces is well documented (Berger 2018; Cumming-Bruce 2016; Hove 2019; Murphy 2019). The focus of this chapter is on conflict-related sexual violence against women, men, girls and boys in South Sudan who now live in refugee settlements in the West Nile region of Uganda (see Fig. 12.1).

I conducted field research for this publication between September and December 2018 in Ayilo, Boroli, Mireyi, Nyumanzi, and Pagirinya refugee settlements in the Adjumani district in Uganda. Ayilo, Mireyi, and Nyumanzi are mainly settled by the Dinka ethnic group, while Boroli is settled by the Nuer ethnic groups. A total of thirty-eight survivors were interviewed from the Dinka and the Nuer ethnic groups. The survivors interviewed were mostly referred to the researcher by community members, other survivors and by the Refugee Welfare Councils which are the elected leaders in refugee settlements. The researcher only interviewed female survivors because the male survivors identified by the community members declined to be interviewed. In-depth interviews were conducted in Juba Arabic, and in three cases an interpreter was used. Pseudonyms are used in all cases for the protection and privacy of the survivors.

The researcher examined how extreme acts of sexual violence were used as a tactic to terrorise, destroy and humiliate both the victims and their ethnic or political group. The researcher examined the impacts of the horrendous sexual violence on the survivors, their families and society. The survivors also recounted the killings, looting, and destruction of property that they witnessed. These acts of sexual violence have left scores of women and men who have since taken refuge in the West Nile and other regions in Uganda with physical, psychological, and social dislocations.

The researcher ensured that survivors understood the purpose of the study and how the information would be used. All survivors provided their informed consent to participate in the research and for the information they provided to be published for academic purposes. The interviews were mainly conducted in the homes of the survivors or, in a few instances, in meeting spaces provided by the Refugee Welfare Councils. All the survivors interviewed are receiving medical, psychological, and livelihood support services from experts in the refugee settlement.

12.2 Background to the Civil Conflict

The Republic of South Sudan got its independence from neighbouring Sudan on 9 July 2011 amidst widespread optimism from both within and beyond its borders (See Fig. 12.2). Two years later, the country plunged into brutal civil war when forces loyal to President Salva Kiir Mayardit, a Dinka, started battling those loyal to Riek Machar Teny Dhurgon, his former Vice President, who is a Nuer. Since the outbreak of the civil war, more than 400,000 people have been killed, and up to two million

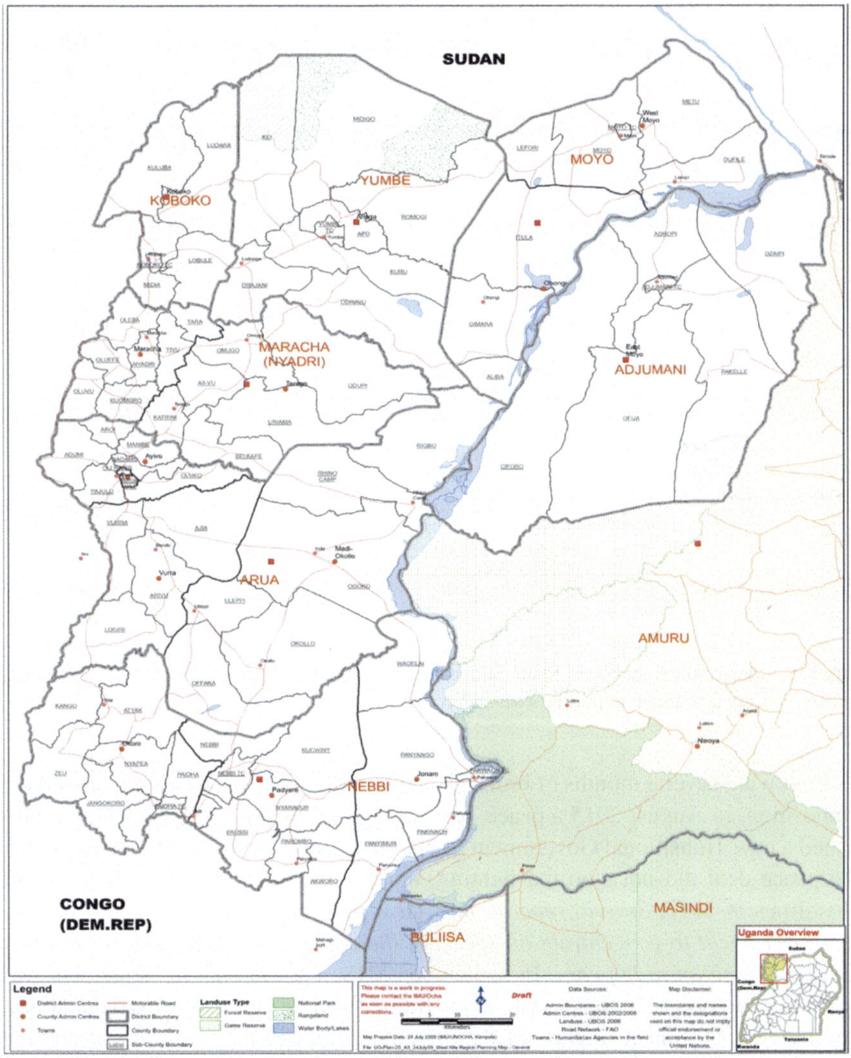

Fig. 12.1 Map of the West Nile region. *Source* OCHA (2009): https://reliefweb.int/sites/reliefweb.int/files/resources/221315C4ECA7DE72C125760400352369-map.pdf

people have left South Sudan in what has come one of the largest refugee crises in the world. More than 200,000 people now live in the UN-protected camps inside South Sudan. Millions more are internally displaced. The state has waged a brutal war in civilian areas, burned huts, destroyed foodstuffs, and raped and killed civilians. The fighting has seen unprecedented levels of mass killings and abhorrent violent sexual assaults, including the rape of women, girls and men, and the violation of bodies.

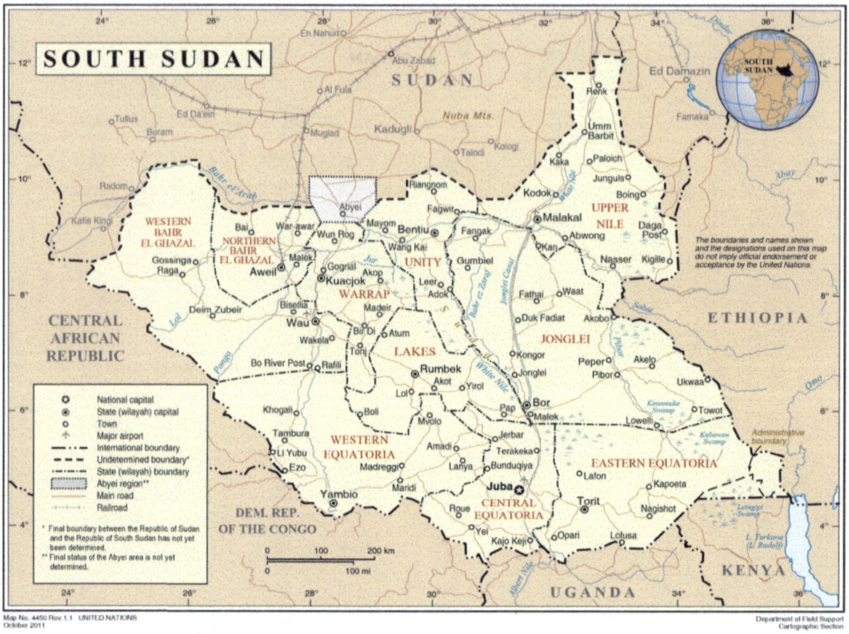

Fig. 12.2 Map of the Republic of South Sudan. *Source* United Nations (2011); at: https://www.un.org/Depts/Cartographic/map/profile/southsudan.pdf

Following several months of brutal armed fight between the Government and the Opposition, in August 2015 a peace deal was signed in Addis Ababa which established a new Transitional Government of National Unity for South Sudan. However, the peace deal did not stop the fighting, nor did the signing of the Cessation of Hostilities *in 2017. Several rebel groups have proliferated in the country, forcing the government to concentrate on securing the capital city, Juba, before securing sustainable nationwide peace.* In September 2018, a power-sharing agreement was signed in neighbouring Sudan, but it was fraught with delays and missed deadlines, and fighting continued in parts of the country. Although the Government and the Opposition have claimed to be committed to implementing the peace deal, both sides blame each other for abuses and for violating the agreement. In July 2018, the UN Security Council approved a US-sponsored resolution which imposed an arms embargo on the entire territory of South Sudan. It has also imposed sanctions on key figures in the conflict. However, the enforcement of the arms embargo has remained a challenge, as noted by international experts (United Nations Security Council 2018).

Following the signing of the Cessation of Hostilities Agreement on 21 December 2017, studies conducted by the *Armed Conflict Location and Event Data Project* (ACLEDP) indicated that conflict activity in South Sudan has reduced. The totals of monthly fatalities since the agreement have remained significantly lower than in the same period of the previous year (see Fig. 12.3).

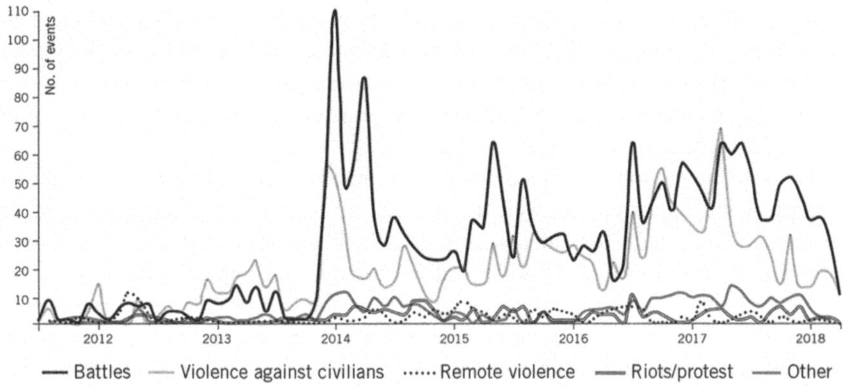

Fig. 12.3 Graph showing the number of conflict incidents from 2011 to 2018. *Source* Armed Conflict Location and Event Data Project [ACLEDP] 2018; at: https://acleddata.com/dashboard/#/dashboard

The relative calm can be attributed to the signing of the agreement by twelve of the parties engaged in the South Sudan civil war. According to the United Nations, the agreement is not stable because it was broken only hours after coming into effect, and there are continued clashes throughout the country. The groups that engage in conflict try to frame the attacks as criminal violence. However, battles between the government forces and the *Sudan People's Liberation Army-In Opposition* (SPLA-IO) have continued in Unity, Jonglei and Central Equatoria provinces. *T*he main signatories to the agreement have failed to meet the key ceasefire milestone of naming a government of national unity, missing a May 2019 deadline, as well as the second extension of 12 November 2019. The parties have agreed on a further 100 days to name the national unity government by the February 2020 deadline. Just days before the deadline, the Government and the Opposition reached an agreement entailing significant concessions on both sides and formed a unity government, lessening fears that failure to reach a deal by the 22 February deadline would trigger yet more violence. The unity government is led by President Salva Kiir Mayardit, Riek Machar Teny Dhurgon as First Vice President, and four other Vice Presidents.

Days before the 12 November 2019 deadline to form a unity government, the guarantors of the peace agreement, the Ugandan President Yoweri Kaguta Museveni and the Sudanese Sovereign Council head Lieutenant General Abdel Fattah Abdelrahman Al-Burhan, convened on 7 November 2019 with President Salva Kiir Mayardit and the main rebel leader, Riek Machar Teny Dhurgon, in Entebbe, Uganda, where they agreed to a 100-day extension in order to resolve the outstanding issues, and to review progress after fifty days. *The negotiated 100-day extension for naming a unity government has averted a crisis imperilling a ceasefire between South Sudan's main belligerents.* They also agreed to establish a mechanism to supervise the implementation of critical tasks, and asked the *Intergovernmental Authority on Development* (IGAD) to address the status of Riek Machar Teny Dhurgon. On 9 November 2019, IGAD convened a meeting of the signatories to the Revitalised Agreement in Addis

Ababa, at which all parties endorsed the 100-day extension. On 10 November, at an extraordinary session, the IGAD Council of Ministers welcomed the decision and directed its Special Envoy to South Sudan to facilitate the resolution of the issue of the number of states and their boundaries, undertake trust- and confidence-building measures and engage with the non-signatories.

The implementation of critical tasks continued to be delayed by the deficit of political will and trust, as well as financial constraints. The parties and other stakeholders repeatedly called upon the South Sudanese Government to honour its pledge to fully disburse US$100 million to fund the implementation process. While the South Sudanese Government has requested financial and material contributions from the international community, a lack of transparency and accountability continues to inhibit such support. On 19 November 2019, the South Sudanese Government announced its decision to allocate an additional US$40 million towards implementation.

12.2.1 Ethnic Divisions Within the SPLM

The *Sudan People's Liberation Movement* (SPLM) was established in 1983 under the leadership of John Garang de Mobior. The SPLM led a rebellion against Khartoum in a bid to realise a more secularised state (Warner 2016). The SPLM drew its initial members from the south, but as the liberation quest gained momentum, it incorporated some members from the north under the banner of liberating marginalised groups in the north (Barltrop 2010). Ethnically, the SPLM was, from its inception, a diverse organisation, but within that diversity, the Nuer and the Dinka were the majority by virtue of the sizes of their populations. They occupied polar positions within the organisation's hierarchy, something that is still visible today (Kiranda 2016). Other major ethnic groups are the Shilluk, the Toposa, the Otuho, the Bari and the Azande (see Fig. 12.4 with the map of ethnic groups in South Sudan below).

As the liberation pursuit was on its course, the SPLM grappled with various challenges, ranging from organisational, internal and leadership to financial and ideological challenges. Finding solutions to these challenges became an uphill task for the SPLM leadership, since these challenges were ethnicised as attempts by the dominant ethnic groups to find solutions that favoured their side. In the absence of a functioning united leadership, cracks emerged within the SPLM, and signs of splits began show right from its inception (Janssen 2017). Along similar lines, Mamdani (2014) argues that cracks within SPLM provided a fertile ground for the continued conflict, as the antagonised parties were confronted by the questions of equal ethnic representation in the struggle for power, and the path in which the power would follow. These divisions paved the way for the subsequent rivalries that rocked SPLM from within.

The first split that occurred at the nascent stages of the liberation struggle (1984–1985) was more ideological and was anchored on the determination of the path the liberation struggle was to take. On one side, Akuot Atem Mayem and Gai Tut Yang

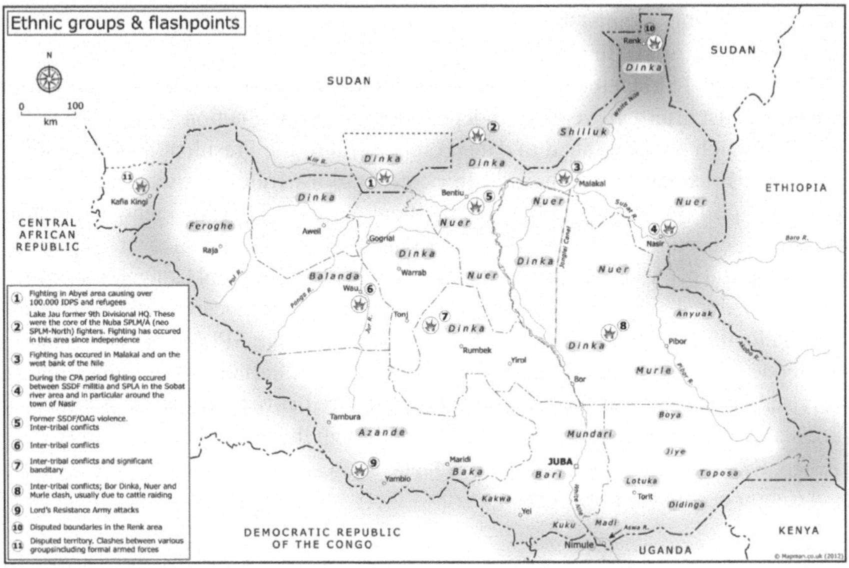

Fig. 12.4 Map of ethnic groups in South Sudan. *Source* LeRiche et al. (2013): *South Sudan: From Revolution to Independence* (Oxford: Oxford University Press: 2012)

called for an independent South Sudan, and on the other John Garang de Mobior, William Nyuon Banyi, and Kerubino Kuanyin Bol led the side that advocated for what they termed a New Sudan, which would be a more democratic, secular and pluralistic country. Both sides received support from diverse ethnic groups, but there were undertones that the quest for an independent South Sudan was an idea of the Nuer, while calls for a New Sudan resonated well with the Dinka (Kiranda 2016). Although the claims and insinuations were muted, they triggered an unending slugfest between the two dominant ethnic groups and dimmed the possibilities of a peaceful South Sudan.

The second split, which served as a litmus test on the leadership of the SPLM, occurred in 1991 after Riek Machar Teny Dhurgon joined forces with Lam Akol, a senior commander in the *Sudan People's Liberation Army* (SPLA), to trigger a change of the SPLM leadership. The two, together with others, called for the replacement of John Garang de Mobior as the leader of the SPLM (Sørbø 2014). They accused John Garang de Mobior of establishing close ties with the Ethiopian government, which they regarded as a move that would stymie internal reforms within the SPLM (Johnson 2014). This attempt did not come to fruition, and Riek Machar Teny Dhurgon led a splinter group in the form of the Sudan People's Liberation Movement/Army-Nasir, which continued to support the independence of the south from the north, even though it received military and financial support from the Government in Khartoum. Noteworthy, the confrontation between Riek Machar Teny Dhurgon and John Garang de Mobior has been viewed through an ethnic lens that pitted the Nuer against the Dinka in a contest that has transformed South Sudan into a crucible of war (Berger 2018).

The night of 15 December 2013 witnessed the third split, which originated from the SPLM. Just after two years of independence the young nation was embroiled in a conflict which has continued to date. Forces loyal to the President and those loyal to the Vice-President were engaged in confrontations following weeks of intense succession politics within the SPLM Political Bureau. The President, Salva Kiir Mayardit, accused Riek Machar Teny Dhurgon of plotting a coup against his government just as the party was preparing its May 2013 SPLM National Convention, which was supposed to discuss, among other issues, the party's flag-bearer in the 2015 presidential elections, the term limits of the chairperson of the SPLM, the Constitution and a code of conduct. An order to disarm members of certain communities within the presidential guard led to a mutiny that triggered revenge attacks of Dinka in Akobo and of Nuer at Bor (Johnson 2014). Although the alleged coup-plotters were arrested, Riek Machar Teny Dhurgon managed to escape from the country, but troops loyal to him continued the conflict.

12.2.2 Ethnic Division Between the Dinka and the Nuer

Ethnicity has remained an important variable in South Sudan's politics. The tyranny of numbers enjoyed by dominant ethnic groups has become an important instrument of ascending to power. Ethnic mobilisations based on historical rivalries and attachments explain the composition of the warring parties in South Sudan. Strong ethnic loyalty combined with a political system that allows winners to dominate government positions and get a larger share of the national cake cause political stakes to be heightened to the extent of violence. It is also important to note that other factors, such as availability of arms amongst civilians, competition for available resources and the role of Sudan, underscored features in contemporary discourses that have aided the continuation of the conflict, but have not explained why it must always be a Dinka aiming a cannon at the Nuer and vice versa, as occurred in the Bor Massacre and other subsequent confrontations (Berger 2018).

The seeds of the Dinka elite's plan were sown during the latter years of the North-South civil war. The starting point was the 1991 split in the SPLA, following the attempt by Nuer leader Riek Machar Teny Dhurgon to replace John Garang de Mobior, a Dinka, as commander of the rebel movement. Despite the later detente and Riek Machar Teny Dhurgon's re-joining of the SPLA in 1997, the wounds and suspicions have remained. Scores of Dinka officers were executed or beaten to death by Machar's loyalists in the immediate aftermath of his declaration that he was assuming control of the movement (Berger 2018). Relatives of many of the Dinka who were murdered are today among the most senior members of the Kiir government. But there were also retaliatory killings of Nuer by Dinka which involved the wholesale massacre of villages (Wild/Jok/Ronak 2018).

The Dinka and the Nuer have a shared history of extreme violence. Over the past five years of the war, these histories of violence have been revisited and reframed to fuel and legitimise the targeting of civilians and militarised groups. Similarly, efforts by South Sudanese in the diaspora and sympathetic Westerners to document atrocities in South Sudan committed by northern Sudanese during the long North-South civil war are feeding into the ruling elite's unfounded claims that the Dinka were the main victims of the war and also the ones who fought and won the independence of South Sudan.

12.2.3 The Bor Massacre and Its Implications for the Civil War

According to Wild et al. (2018), in 1991 a massacre with devastating consequences occurred in Bor, the capital of the Jonglei state. Located on the east of the River Bahl al Jabal, Bor was predominantly inhabited by pastoralist Nuer with pockets of Dinka communities. Years before the massacre there had been a series of inter-ethnic cattle raiding episodes between the Nuer and the Dinka in a bid to increase their herds. These raids were initially conducted by means of spears and well-orchestrated ambushes, but later, with an increase in the number of guns, firearms became the common tools of the trade. Regardless of the raiding methods, it is important to note that cattle are historical symbols of social status, and their products, which are of high nutritional value, are important sources of livelihood in the South Sudanese communities.

Prior to the massacre, there was a proliferation of arms among the civilians who had formed well-organised groups. While the Dinka had a local militia called the *Titweng*, the Nuer had the *White Army* that was originally formed to protect cattle but upon gaining widespread success in their raids became an important asset in the political sphere (Young 2016). This came against the backdrop of visible rifts in the SPLM leadership that provided the avenue for Riek Machar Teny Dhurgon to incorporate the Nuer White Army members into SPLM-Nasir, and with the support of the Khartoum government in the north, SPLM-Nasir orchestrated one of the deadliest massacres in the history of South Sudan. According to Wild et al. (2018), Riek Machar Teny Dhurgon, who was then entangled in ideological differences with John Garang de Mobior, mobilised over 20,000 members of the SPLM-Nasir to carry out an attack against the Dinka in Bor in what came to be known as the Bor Massacre. It saw the death of over 2,000 people of Dinka origin and the destruction of properties as well as other atrocities (Wild et al. 2018). Even though Riek Machar Teny Dhurgon offered an apology to the Dinka in 2011 when he was the Vice President, it is beyond doubt that the massacre left an indelible footprint of loss on the lives of the Dinka, and it has become a political tool that has been used against Riek Machar Teny Dhurgon in his quest to ascend to the highest office in the land (Chol 2011).

It would be easy to argue that a focus on the historical rivalry between the dominant ethnic groups is simplistic and superficial and that this would reduce the ongoing feud solely to an ethnic conflict, as it is had already been labelled by segments of the international media. However, efforts to sustain an ethnic conflict narrative have been quickly countered by the argument that the South Sudan government was a representation of diverse ethnic groups and that even after the December 2013 crisis, which saw a number of people arrested on the allegations of an attempted coup, the President did not spare those from his tribe (Pinaud 2014). Indeed, the ousted and the current Vice President belong to the Nuer and the President is from the Dinka, but the presence of people of diverse ethnic origins in the Government cannot be construed to mean a representation of ethnic interests, since African societies have the propensity to bestow ethnic responsibilities on particular individuals whose voices not only become the voice of the ethnic groups but also symbols of ethnic unity. Therefore, any kind of humiliation that targets these ethnic kingpins becomes an outright humiliation to the ethnic groups they represent, who then, on behalf of their leaders, may endeavour to seek revenge (Glowacki/Wrangham 2015).

12.3 Strategic Rape Theory

Strategic rape theory is currently the most influential theory of mass wartime rape. It is widely credited by activists and scholars. Since the Yugoslavian and Rwandan mass rapes, a consensus has been building that wholesale rape represents just another ordnance – like bombs, bullets, or propaganda – that the military can use to accomplish its strategic objectives (Brownmiller 1993). Rape is a tactic executed by soldiers in the service of larger strategic objectives. While supporters of this position do not always claim that military planners explicitly instruct soldiers to rape, the implication is clear. Wartime rape is a coherent, coordinated, logical, and brutally effective means of prosecuting warfare (Allen 1996; Thomas/Regan 1994).

Variations on the theory that wartime rape is strategic rape are based on the deleterious effects that mass rape has on enemy populations. It is credited with spreading debilitating terror, diminishing the resistance of civilians, and demoralising, humiliating, and emasculating enemy soldiers who are thereby shown to have failed in their most elemental protective duties. Further, mass rape is said to cast blight on the very roots of the afflicted culture, affecting its capacity to remain coherent and to reproduce itself. By raping women, soldiers split the familial atoms of which every society is composed. Raped women may become pregnant by the enemy, they may suffer grievous physical and psychological injuries, they may die, they may be abandoned or disavowed by shamed families and husbands, all of which degrade the ability of a culture to replenish itself through sexual reproduction. For these reasons, advocates of strategic rape theory often refer to it as *genocidal rape* designed, whether with full consciousness or not, to annihilate a people and a culture (Allen 1996; MacKinnon 1994).

While mass wartime rape can assuredly result in the damage discussed above, it remains possible that the supporters of strategic rape theory may be confusing the consequences of wartime rape with the motives for it. Just because these consequences may include demoralised populaces or fractured families does not mean that these were the goals for which the rapes were perpetrated in the first place. However, in the case of South Sudan, the strategic rape theory is an appropriate explanation for state-managed and ethically motivated rape. Although it may not be the sole reason for wartime rape, in the perspective of the Dinka and the Nuer in South Sudan, where the mass rape has occurred in the State-led civil war, the strategic rape theory remains apt.

12.4 Wartime Sexual Violence

The term *conflict-related sexual violence*, as used in this text, refers to rape, sexual slavery, forced pregnancy, enforced castration, forced marriage, and any other form of sexual violence of comparable gravity perpetrated against women, men, girls or boys that is directly or indirectly linked to a conflict. The victims are frequently the actual or perceived members of a persecuted political, ethnic or religious minority, or targeted on the basis of actual or perceived sexual orientation and gender identity. The climate of impunity is generally associated with State collapse and cross-border consequences, such as displacement or trafficking in persons, and/or violations of the provisions of a ceasefire agreement. According to Legal Action Worldwide (2016), the term also encompasses trafficking in persons when committed in situations of conflict for the purpose of sexual violence or exploitation.

According to Siefert (1996), rape has nothing to do with sexuality but with the exertion of sexual violence directed against women. Stiglmayer (1994) further described rape as an aggressive and humiliating act, as even a soldier knows, or at least suspects. He rapes because he wants to engage in violence. He rapes because he wants to demonstrate his power. He rapes because he is the victor. He rapes because the woman is the enemy's woman, and he wants to humiliate and annihilate the enemy. He rapes because the woman is herself the enemy whom he wishes to humiliate and annihilate. He rapes because he despises women. He rapes to prove his virility. He rapes because the acquisition of the female body means a piece of territory conquered. He rapes to take out on someone else the humiliation he has suffered in the war. He rapes to work off his fears. He rapes because it's really only some 'fun' with the guys. He rapes because war, a man's business, has awakened his aggressiveness, and he directs it at those who play a subordinate role in the world of war.

According to Swiss/Giller (1993), in war, rape is an assault on both the individual and her family and her community. As well as an attempt to dominate, humiliate, and control behaviour, rape in war can also be intended to disable an enemy by destroying the bonds of family and society. Rape can be both a military strategy and a nationalist policy. For sociocultural theorists, military rape is often the result of a complex combination of causal factors. However, as stressed above, the sexual impulses of individual soldiers are almost never allowed a significant place in the causal mix.

The armed conflict that erupted in December 2013, which is the main focus of this chapter, is only the most recent episode of violence in South Sudan's history. During the first (1956–1972) and second (1983–2005) Sudanese civil wars, the government of Sudan and pro-government militias fought against armed groups who sought autonomy for the southern regions of Sudan. Both periods of civil war were characterised by extreme violence against civilians, gross human rights abuse, and massive forced displacement. While there is little documentation of incidents of sexual violence during the first civil war, available literature indicates that abduction, rape, sexual slavery, and other forms of sexual violence became increasingly common during the second civil war (Bubenzer/Stern 2011). Southern Sudanese women were victims of sexual violence, committed by all parties to the fighting, including the Sudan Armed Forces, the SPLA and other armed groups.

Sudanese militia allied to the Sudan Armed Forces, for example, abducted thousands of Dinka women and children, particularly from the Northern Bahr el Ghazal region. Following their abduction, many of these women and children were subjected to sexual violence (Human Rights Watch 1995). The 1991 split in the SPLA resulted in heavy fighting between Southern Sudanese, including an increase in the targeting of women and girls through rape and abduction. For example, during the 1991 attack on Bor by the SPLA-Nasir faction, women were raped and mutilated. A 2004 survey of male and female South Sudanese refugees in Uganda found that approximately 10% had been raped, 8% had experienced forced prostitution or sexual slavery, and 33% had witnessed the rape of a woman (Unni 2004).

There is also some anecdotal information that Southern Sudanese men and boys were subjected to sexual violence during the war. According to a former administrator of Bentiu, during the 1980s, members of the Sudan Armed Forces brought Southern Sudanese displaced boys to their barracks in Bentiu and sexually assaulted them in exchange for food and clothing. Boys who were abducted and taken to northern Sudan were also, in some cases, subjected to sexual violence (Human Rights Watch 1995) .

During the second civil war (1983–2005), there was also an increase in sexual and gender-based violence within Southern Sudanese communities and homes, including rape, early and forced marriage, and domestic violence. According to Jok (2015), war introduced the tenet that it is the duty of men to defend the homeland while women should ensure 'cultural survival' in the face of high mortality rates due to war and famine by reproducing at an accelerated rate. As a result, soldiers returning from the front lines felt a sense of urgency and entitlement to sex, both within and outside marriage, and used violence if women did not yield to their demands. When transfer to another location or to the frontlines was imminent, soldiers had an additional incentive to commit rape, as they knew they would likely escape punishment. The increase in domestic violence has also been attributed to men's suspicions that their wives had committed adultery while they were away fighting, and to alcohol consumption (Berger 2018).

According to Anyieth (2011), during the second civil war, family and societal structures and norms, which may have helped to curb sexual and gender-based violence, were weakened, and this also contributed to the increase in sexual violence.

Contextual factors that disrupted these structures include the separation and displacement of families and communities, the weakened authority and influence of elders and chiefs, the general militarisation of society, the increased availability of weapons, the difficulty in securing redress for violent crimes, and the resulting culture of lawlessness.

The signing in 2005 of the *Comprehensive Peace Agreement* (CPA) between the Sudanese government and the rebel *Sudan People's Liberation Movement/Army* (SPLM/A) did not bring an end to violence within South Sudan, or to sexual violence specifically. According to Berger (2018), from 2005 to 2013, fighting often occurred between government forces and armed insurgent militias in the context of intercommunal violence often linked to land and cattle. During such fighting, sexual violence was perpetrated against women and girls. For example, abduction of women and children often aimed at or resulting in forced marriage and rape was a feature of back and forth attacks between Lou Nuer and Murle communities in Jonglei state in 2011 and 2012 (UNMISS 2012). Security forces also committed acts of sexual violence during government-led civilian disarmament campaigns, particularly in Jonglei state (United Nations Security Council 2012). Even outside theatres of conflict, there were numerous cases of rape and sexual assault committed by government officers during this period (United Nations Human Rights Council 2018) .

12.5 Rape as a Weapon of War

Rape as weapon of war is well studied and documented (Brownmiller 1993; Gottschall 2004). Rape was long considered an unfortunate but inevitable accompaniment of war. This was considered to be the result of the prolonged sexual deprivation of soldiers and insufficient military discipline (Nandy 1983). Its use as a weapon of war was gruesomely demonstrated during World War II, when both Allied and Axis armies committed rape as a means of terrorising enemy civilian populations and demoralising enemy troops (Brownmiller 1993). The 'Rape of Nanjing' by the Japanese army and the brutal mass rapes committed against German women by Soviet soldiers in Berlin were given as the worst examples of rape as a weapon of war (Eykholt 2000). Rape was regarded by men in the Soviet army as a well-deserved form of punishment, whether the civilians had anything to do with the war or not.

According to Gottschall (2004), the frequency, savagery, and systematic organisation of wartime rape increased in late 20th-century conflict. For example, in the former Yugoslavia, rape was used as an instrument of ethnic cleansing, whereby women belonging to dominated ethnic groups were deliberately impregnated through rape by enemy soldiers. In Rwanda, women belonging to the Tutsi ethnic group were systematically raped by HIV-infected men who were recruited and organised by the Hutu-led government.

In 1993, the United Nations Commission on Human Rights, which later became the United Nations Human Rights Council, declared systematic rape and military sexual slavery as crimes against humanity and violations of women's human rights

(Sellers 2006). In 1995 the United Nation's Fourth World Conference on Women declared that rape by armed groups during wartime is a war crime. In the former Yugoslavia and Rwanda, international tribunals were established to prosecute crimes committed in wars, including sexual violence as a war crime (Sellers 2006). In 1998, the Rwandan tribunal ruled that rape and sexual violence constitute genocide. The International Criminal Court, established in 1998, was subsequently granted jurisdiction over a range of women's issues, including rape and forced pregnancy.

In 2000, the United Nations Security Council Resolution 1325, on women, peace, and security, was adopted unanimously by the United Nations Security Council, which acknowledged the disproportionate and unique impact of armed conflict on women and girls. In a resolution adopted in 2008, the UN Security Council Resolution 1820 affirmed that rape and other forms of sexual violence can constitute war crimes, crimes against humanity or a constitutive act with respect to genocide (Sellers 2006).

The United Nations Human Rights Council established the Commission on Human Rights in South Sudan for a one-year term on 23 March 2016 and subsequently extended its mandate for an additional year in 2017, 2018 and 2019, with its current term at the time of writing due to expire in March 2020. On 6 February 2020, national, regional, and international non-governmental organisations wrote to the UN Human Rights Council to support the extension of the mandate of the *United Nations* (UN) Commission on Human Rights in South Sudan ("the CoHR") during the 43rd session of ("the Council" or "the HRC"), which was planned to take place from 24 February-20 March 2020. However, at the time of finalising this paper, it was not possible to independently verify whether its mandate was extended. The mandate of the Commission on Human Rights in South Sudan specifically includes investigation of *sexual and gender-based violence* (SGBV). The mandate is unique in South Sudan in that it extends beyond traditional human rights monitoring and investigation to include identifying alleged perpetrators and collecting and preserving evidence for the purpose of future prosecutions, truth-seeking, and reparations (United Nations Human Rights Council 2019) .

In March 2018, the Commission on Human Rights in South Sudan reported to the Human Rights Council that sexual and gender-based violence remained the central feature of the conflict in South Sudan. The Commission documented many accounts of rapes and gang rapes, forced stripping and nudity, forced sexual acts, castration and mutilation of genitalia, perpetrated by all parties to the conflict. The Commission identified different characteristics of SGBV perpetrated by the warring factions, including sexual violence as torture and the use of sexual violence as a weapon of war and method of punishment or humiliation. Sexual and Gender-Based Violence in South Sudan has also been characterised by its repetitiveness and gruesomeness, with many victims being assaulted multiple times and/or with the use of various objects (United Nations Human Rights Council 2019) .

Impunity for these crimes remains high in South Sudan as government and armed groups deny committing sexual violence. According to a 2017 study by the Global Women's Institute and the International Rescue Committee, more than 65% of women and girls in South Sudan have reportedly experienced some form of sexual violence at least once in their lives. UNICEF reports also revealed that 25% of those targeted by sexual violence are children, including the rape of girls as young as seven years. Elderly and pregnant women have also been raped. The Commission also received reports of male victims of sexual violence. Sexual and gender-based violence against men and boys is even more under-reported than that against women and girls as there is a greater level of stigma.

Sexual violence has continued to be employed as a tactic of war, with widespread and strategic rapes, including mass rapes, allegedly committed by several parties to armed conflict, mostly in conjunction with other crimes, such as killing, looting, pillage, forced displacement and arbitrary detention. The strategic nature of the violence was evident in the selective targeting of victims from opposing ethnic, religious or political groups, mirroring the fault lines of the wider conflict or crisis. Patterns of sexual violence have also been seen in the context of urban warfare, during house searches, operations in residential areas and at checkpoints. The systemic use of rape against civilian populations in South Sudan as a weapon of war is the direct outcome of the ethnic nationalism driving the now four-year civil war. Both male and female victims are being targeted by soldiers of the Dinka-led SPLA, and their allied forces. While some prominent human rights groups have referred to a "breakdown in the rule of law", the use of rape is a strategy employed by the SPLA commanders as part of their scorched-earth policies in the home territories of minority ethnic groups (Mednick 2017).

Among the Nuer, the main targets of the current war, rape has also been used by armed groups against civilian populations and, like the SPLA-led violence, has included grievous acts committed against victims. But the state-led offensives against non-Dinka regions in the Equatorias, western Bahr el Ghazal and Greater Upper Nile are responsible for the vast majority of sexualised violence. According to Berger (2018), the use of sexual violence in conflict areas of South Sudan is consistent with the strategy of a total war, in which the economic foundation of pastoralist communities in which the labour provided by women and girls and the wealth accrued by male relatives upon their marriage is considered a legitimate and important target for debasement and assault by combatants. According to Cumming-Bruce (2016), army-affiliated militias made up mainly of youths rape and abduct women and girls as a form of payment under an agreement that allowed them "to do what you can and to take what you can". The militias stole cattle and other property under the same understanding.

12.6 Narratives of South Sudanese Refugees

Now living in refugee settlements in the West Nile region of Uganda, several women and men are battling with the physical, psychological, and social impacts of brutal mass rape which have left some of them disabled, rejected by their family members, infected with HIV/AIDS, caring for children born as the result of rape, and fistula. All the survivors interviewed believed that they suffered because of their ethnicity. Studies conducted by institutions reveal aggravated acts of sexual violence against thousands of people across the country since hostilities began in December 2013 (Mercy Corps 2018; United Nations Human Rights Council 2017). Perpetrators come from both sides of the conflict, pitting the government forces of President Salva Kiir Mayardit, a Dinka, against the opposition forces of Riek Machar Teny Dhurgon, a Nuer, and their respective allied armed groups. According to Amnesty International (2017a, b), this is premeditated sexual violence on a massive scale. Women have been gang-raped, sexually assaulted with sticks and mutilated with knives.

Survivors interviewed during this study described the situation of sexual violence in South Sudan as endemic. Sexual violence takes place any time civilians come into close proximity with armed actors. This includes during military attacks on villages, during searches of residential areas by armed actors, along roads and at checkpoints, and during abduction or detention. Most incidents happened alongside other forms of violence, including killing, looting and the destruction of homes.

Anastasia, a Nuer by tribe, was walking back to the UNMISS base when she encountered several armed government soldiers of the Tiger Battalion near the Giyada military barracks.

> The soldiers were of the Dinka tribe. They told me, "Come here, you Nuer woman. You look so beautiful and you are the woman we have been looking for." They were able to identify me as a Nuer because of the marks on my face. One of the solders immediately pushed me down and ripped my clothes open. I tried to resist him and he slapped me hard on the face while the other laughed at me. They raped me one by one until I blacked out and found myself in the hospital. I was raped because I am a Nuer woman.

This story is not so different from the one of Rosetta, who recounted how she was raped by eight Dinka solders in turn and left unconscious.

> The first one looked at me and asked, "Do you want to be raped or do you want to die?" I kept quiet and he was moving closer to me. He ripped my clothes off, pushed me down and he started raping me. He gave me a hot slap when I resisted and ordered me to keep quiet. I told him, "It's painful," and he slapped me again. I was taken to the hospital and when I recovered, I came to Uganda to seek refuge. This boy you see here with me is the son of the soldiers who raped me and he resembles them. Sometimes he reminds of the suffering I am going through, but what can I do?"

These painful narratives of South Sudanese women represent acts of sexual violence they were subjected to because of their ethnicity. In addition, the long-term physical effects of sexual violence and the psychological burden borne by survivors is enormous, and can persist for years. Survivors interviewed described having nightmares, loss of memory, lack of concentration, and thoughts of revenge or suicide, which are common symptoms of psychological distress associated with post-traumatic stress disorder. Some expressed feelings of loneliness and despair. For many, their emotional distress was debilitating, and impacted their ability to carry out their day-to-day activities. While some organisations in refugee settlements do provide basic psychosocial support services for sexual violence survivors, these services are basic and do not comprehensively address the needs of the survivors.

Lucy, a Dinka by tribe, narrated the day when she and her friends were arrested, dragged into the bush and raped by four Nuer armed men.

> We were crying and screaming loudly, but they were relentless and they kept cheering themselves up and singing war songs. They raped us closed to three hours and we bled so much. I got pregnant and I carried the child of the men who raped me. Sometimes I don't like him because his presence brings back the pain of being raped and rejected by my family. Tell me, is this boy my son just because I gave birth to him?

According to Veronika, another victim of sexual violence, some of the sexual attacks were designed to terrorise, degrade and shame the victims.

> I was raped before my husband and when he attempted to defend me, they raped him too. I have been raped five times and at one point, I was gang raped by armed Dinka men almost to death. It's by God's grace that I am still alive and I can walk. I was paralysed, bedridden and could not control urine. At one point I wanted to commit suicide because I was rejected by my family, but I was counselled and encouraged to remain strong. But why should I be raped because I am a Nuer woman? I am not a soldier, I am not involved in the senseless wars, I am a woman, I am a mother and I am a South Sudanese.

While psychological distress is to a large extent a predictable consequence of sexual violence, survivors experience stigma and rejection by spouses and family members. Victims are blamed for what occurred to them and are sometimes stigmatised more than the perpetrators themselves. Survivors and families weigh their desire to seek medical and psychological support with the social cost of disclosure, and in the end often opt to conceal their distress. The silence often exacerbates both the medical and emotional problems. Beyond the traumatic effect on individual survivors, acts of sexual violence have created fractures within and between communities, contributed to inter-ethnic animosity, and fuelled continued cycles of abuse.

> Justina was abducted and raped by two Dinka soldiers for six weeks in Juba in 2016.
>
> They forced me to cook for them and wash their clothes. They could not let me go anywhere. Even when I wanted to go to toilet, I was escorted and brought back to the room. One Saturday after lunch, they played very loud music and drunk beer and they raped me in turns for close to five hours. When they realised that I was not able to hold urine and I was becoming weak, they let me to go. Sometimes when I go to bed, I see them in my dreams and I cry so much. They said they raped me because I am a Nuer and Nuer men raped their (Dinka) women too.

These horrendous wartime sexual acts have left the victims with devastating and life-changing consequences, including physical injuries and psychological distress. Many survivors have also been shunned by not only their family members, but stigmatised by the wider community. Survivors' feelings of shame, experiences of stigma, and rejection by spouses and family members added to their distress. Some survivors described being blamed for what happened to them, sometimes more than the perpetrators themselves. In some cases, their husbands beat or abandoned them after they disclosed their experiences. The shame, stigma and rejection contribute to long-term societal oppression of survivors of sexual violence, including by creating an obstacle for them to report crimes and seek medical and psychological support.

The psychological burden carried by survivors of sexual violence is enormously traumatic. For some, their emotional distress was debilitating, making it difficult to carry out their day-to-day activities. Problems with sleeping were among the most commonly described signs of psychological distress. Unfortunately, specialised mental health services are limited in the refugee settlements in the West Nile. While various organisations provide psychosocial support services for sexual violence survivors, they have limited capacity. Livelihood initiatives exist, but are inadequate compared to the needs.

The shame, stigma and rejection that survivors experience are often the results of discrimination and contribute to the long-term societal oppression of survivors of sexual violence. As noted previously, stigma deters survivors from reporting crimes and from seeking medical and psychological support. Addressing the stigma associated with sexual violence and its impacts is therefore critical to ensuring that survivors live with dignity and respect, and that they can access justice and reparations.

12.7 Conclusion and Recommendations

Since December 2013, South Sudan has been riven by virulent civil conflict. Widespread and systematic sexual violence has been used as a tactic of war to terrorise and persecute populations in a manner that indicates that it is ethnic as well as political in nature. The latest wave of unrest resulted in over two million people being displaced internally and another two million fleeing across borders, aggravating the risk of sexual violence by militias, armed youths and elements of the security forces. Fear spread by sexual violence has resulted in women restricting their movements, thus impeding livelihood activities. The situation is exacerbated by the proliferation of small arms and light weapons. Pervasive impunity has normalised extreme patterns of violence, as national institutions are unable to prosecute rape effectively or provide remedies. However, such incidents are challenging to document, as sexual violence in general is vastly under-reported owing to shame, stigma and fear of retaliation, lack of service coverage, and the requirement that service-providers refer cases to the police. The risk of 'honour crimes' or tribal solutions to rape perpetuates the silence and under-reporting.

For peacebuilding and reconciliation to take root, justice must be done for sexual violence survivors. Otherwise, the trauma, disease, perceived dishonour, and desire for vengeance will fester within communities. Those who have suffered sexual violence and the children born of rape should be reintegrated into their societies to avoid remaining susceptible to exploitation and recruitment. Women and girls who were excluded following sexual violence should be rehabilitated and supported economically to enable them to rebuild their livelihoods.

Sexual violence has profound and life-changing consequences for victims as well as their families and communities. While criminal accountability can serve to acknowledge the harm suffered by victims of sexual violence and ensure perpetrators are identified, held responsible and punished, reparative measures go farther, towards addressing their suffering and restoring their well-being. Victims of sexual violence have a right to prompt, effective and adequate reparation, including restitution, compensation, rehabilitation, satisfaction, and guarantees of non-repetition for the crimes committed against them.

To fulfil the right to reparation, authorities should ensure victims can access adequate medical and psychological services to help address the physical and mental harm caused by sexual violence. Upon return to South Sudan, the Government should publicly acknowledge the role of government actors in perpetrating sexual violence and make public apologies to victims. It should undertake efforts, such as educational and training programmes, to improve the status of women in South Sudanese society and to address the societal acceptance of certain forms of violence against women, and the stigmatisation of and discrimination against victims of sexual violence. The South Sudan government should also address the political and inter-ethnic fissures that continuously drive the cycle of sexual violence, as a way of preventing their repetition.

The physical consequences of sexual violence continue long after the abuse, often for a lifetime. Common medical problems experienced because of rape include sexually transmitted diseases including HIV/AIDS and fistula – tearing of the walls between the vagina, bladder, and anus. Some survivors have suffered serious internal wounds and damage to reproductive organs. Alongside sexual violence, many survivors interviewed were subjected to severe beatings which have left some of them disfigured. The international donors through UNHRC should ensure that survivors of sexual violence can access comprehensive sexual and reproductive health information and services. There is a need to improve the physical accessibility of services, and to combat the stigma that sometimes prevents survivors from seeking out services.

In the longer term, the Government of South Sudan must address the societal fractures that are both a consequence and cause of sexual violence as well as many other crimes. Acts of sexual violence have created fractures within and between communities, contributed to inter-ethnic animosity, and fuelled continued cycles of abuse. The sentiments of hatred, desire to dominate, or humiliate that drive perpetrators of sexual violence have inspired similar feelings in some survivors. As part of South Sudan's obligations to prevent the continued recurrence of sexual violence, particularly to the extent that it is motivated by and targeted based on ethnicity or perceived political allegiance, the Government of South Sudan must undertake efforts to address these social fractures.

References

Allen, Beverly, 1996: *Rape Warfare: The Hidden Genocide in Herzegovina and Croatia* (Minneapolis: University of Minnesota).

Amnesty International, 2017: *South Sudan: 'Do Not Remain Silent': Survivors of Sexual Violence in South Sudan Call for Justice and Reparations* (London: Amnesty International).

Amnesty International, 2017: "South Sudan: Sexual violence 'on a massive scale' leaves thousands in mental distress amid raging conflict"; at: https://www.amnesty.org/en/latest/news/2017/07/south-sudan-sexual-violence-on-a-massive-scale-leaves-thousands-in-mental-distress-amid-raging-conflict/ (13 December 2019).

Anyieth, D'Awol, 2011: "'Sibu Ana, Sibu Ana', 'Leave Me, Leave Me': Survivors of Sexual Violence in South Sudan", in: Bubenzer, Friederike; Stern, Orly (Eds.): *Hope, Pain and Patience: The Lives of Women in South Sudan* (Wynberg: The Institute for Justice and Reconciliation): 53–78.

Armed Conflict Location and Event Data Project, 2018: "South Sudan analysis".

Baker, Aryn, 2016: "War Crimes: Slaughter and Starvation Haunt a Broken South Sudan", in: *Time*, 187,20 (30 May): 20–27.

Barltrop, Richard, 2010: "Leadership, Trust and Legitimacy in Southern Sudan's Transition After 2005"; at: http://www.undp.org/content/dam/aplaws/publication/en/publications/capacity-development/-of-change/leadership/leadership-trust-and-legitimacy-in-southern-sudan-transition-after-2005/Leadershiptrust%20and%20legitimacy%20in%20Southern%20Sudan%20transition%20after%202005.pdf (23 September 2018).

Berger, Carol, 2018: *Ethnocide as a Tool of State-Building: South Sudan and the Never Ending War*. Working Paper; at: http://updm-rss.org/UPDM/Ethnocide_Dinka_strategy_dominance.pdf (11 March 2020).

Brownmiller, Susan, 1993: *Against Our Will: Men, Women, and Rape* (New York: Simon and Schuster).

Chol, Agereb Leek, 2011: "20 Years' Apology: A Political Campaign for Dr. Riek Machar Teny – A Case of 'Why Garang Must Go Now' is Now Recanted"; at: https://paanluelwel2011.files.wordpress.com/2011/10/dr-riek-political-apology.pdf (16 September 2018).

Cumming-Bruce, Nick, 2016: "Mass Rape, a Weapon of War, Traumatises South Sudan", in: *The New York Times* (11 March 2016).

Eykholt, Mark, 2000: "Aggression, Victimization, and Chinese Historiography of the Nanjing Massacre", in: Fogel, Joshua A. (Ed.): *The Nanjing Massacre in History and Historio-graphy* (Berkeley: University of California Press).

Gottschall, Jonathan, 2004: "Explaining Wartime Rape", in: *The Journal of Sex Research*, 41,2: 129–136.

Hove, Mediel; Ndawana, Enock, 2019: "Women's Rights in Jeopardy: The Case of War-Torn South Sudan"; in: *Sage Open* (October-December 2017): 1–13; at: DOI: 10.1177/ 2158244017737355 journals.sagepub.com/home/sgo (16 December 2019).

Human Rights Watch, 1995: "Children in Sudan: Slaves, Street Children and Child Soldiers"; at: www.hrw.org/reports/1995/Sudan.htm (27 March 2019).

Janssen, Jordy, 2017: "South Sudan: Why Power-Sharing Did not Lead to Political Stability" (MA dissertation, University of Leiden).

Johnson, Douglas H., 2014: "The Political Crisis in South Sudan", in: *African Studies Review*, 57,3: 167–174.

Jok, Jok Madut, 1999: "Militarisation and Gender Violence in Southern Sudan", in: *Journal of Asian and African Studies*, 34,4: 427–442.

Jok, Jok Madut, 2015: "The Paradox of Peace in Sudan and South Sudan: Why Political Settlements Failed to Endure", in: *Inclusive Political Settlements Paper 15* (Berlin: Berghof Foundation).

Kiranda, Yusuf; Mathias, Kamp; Mugisha, Michael B.; Donnas, Ojok, 2016: "Conflict and State Formation in South Sudan: The Logic of Oil Revenues in Influencing the Dynamics of Elite Bargains", in: *Journal on Perspectives of African Democracy and Development*, 1,1: 30–40.

Knopf, Katherine A., 2016: *Ending South Sudan's Civil War:* Council Special Report No. 77 (New York: Council on Foreign Relations, Centre for Preventive Action).

Legal Action Worldwide, 2016: "Accountability for Sexual Violence committed by Armed Men in South Sudan"; at: https://reliefweb.int/sites/reliefweb.int/files/resources/Legal-Action-Worldwide-Report-on-Accountability-for-Sexual-Violence-Committed-by-Armed-Men-in-South-Sudan.pdf (23 April 2019).

LeRiche, Matthew; Matthew, Arnold, 2013: *South Sudan: From Revolution to Independence* (Oxford: Oxford University Press).

MacKinnon, Catharine Alice, 1994: "Rape, Genocide, and Human Rights", in: Stiglmayer, Alexandra (Ed.): *Mass Rape: The War Against Women in Bosnia-Herzegovina* (Lincoln: University of Nebraska Press).

Mamdani, Mahmood, 2014: "South Sudan: No Power-Sharing without Political Reform", in: *New Vision* (18 February).

Mednick, Sam, 2017: "South Sudan Violence Against Women is Twice Global Average", in: *Associated Press* (29 November 29).

Mercy Corps, 2018: "Quick Facts: What You Need to Know About the South Sudan Crisis"; at: https://www.mercycorps.org/articles/south-sudan/quick-facts-what-you-need-know-about-south-sudan-crisis (9 August 2018).

Murphy, Maureen; Bingenheimer, Jeffrey B.; Ovince, Junior; Ellsberg, Mary; Contreras-Urbina, Manuel. 2019: "The Effects of Conflict and Displacement on Violence Against Adolescent Girls in South Sudan: The Case of Adolescent Girls in the Protection of Civilian Sites in Juba", in: *Sexual and Reproductive Health Matters*, 27,1: 181–191.

Nandy, Ashis, 1983: *The Intimate Enemy: Loss and Recovery of Self under Colonialism* (Delhi: Oxford University Press).

Pinaud, Clemence, 2014: "South Sudan: Civil War, Predation and the Making of a Military Aristocracy", in: *African Affairs*, 113,451: 192–211.

Seifert, Ruth, 1994: "War and Rape: A Preliminary Analysis", in: Stiglmayer, Alexandra (Ed.): *Mass Rape: The War Against Women in Bosnia-Herzegovina* (Lincoln: University of Nebraska Press): 54–72.

Seifert, Ruth, 1996: "The Second Front: The Logic of Sexual Violence in Wars", in: *Women's Studies International Forum*, 19: 35– 43.

Sellers, Patricia Viseur, 2006: "The Prosecution of Sexual Violence in Conflict: The Importance of Human Rights as Means of Interpretation"; at: https://www.ohchr.org/Documents/Issues/Women/WRGS/Paper_Prosecution_of_Sexual_Violence.pdf(13 May 2019).

Swiss, Shana; Giller, Joan, 1993: "Rape as a Crime of War: A Medical Perspective", in: *JAMA*, 270: 612–615.

The Global Women's Institute; The International Rescue Committee; CARE International UK; What Works; UK Department for International Development, 2018: "Intersections of Violence Against Women and Girls with State-Building and Peace-Building: Lessons from Nepal, Sierra Leone and South Sudan"; at: https://reliefweb.int/sites/reliefweb.int/files/resources/What-Works_Intersections-of-VAWG-with-SBPB-report_2018.pdf (16 December 2019).

Thomas, Dorothy; Regan, Ralph, 1994: "Rape in War: Challenging the Tradition of Impunity", in: *SAIS Review*, 14: 81–99.

UNICEF, 2016: "Rise in Child Recruitment as Conflict in South Sudan Enters Fourth Year"; at: www.unicef.org/media/media_94185.html (13 September 2018).

United Nations Human Rights Council, 2017: *Report of the Commission on Human Rights in South Sudan* (6 March 2017), UN Doc. A/HRC/34/63, para. 43.

United Nations Human Rights Council, 2018: *Report of the Commission on Human Rights in South Sudan, 26 February-23 March 2018* (Geneva: United Nations Human Rights Council); at: https://www.ohchr.org/EN/HRBodies/HRC/CoHSouthSudan/Pages/Index.aspx

United Nations Human Rights Council, 2019: *Prosecuting Sexual and Gender-Based Crimes in South Sudan: UN Commission on Human Rights in South Sudan Working Session on Strategies for Evidence Collection* (Geneva: United Nations Human Rights Council). https://www.ohchr.org/en/hrbodies/hrc/cohsouthsudan/pages/index.aspx (16 December 2019).

United Nations Security Council, 2018: "Letter dated 26 November 2018 from the Panel of Experts on South Sudan addressed to the President of the Security Council"; at: https://undocs.org/pdf?symbol=en/S/2018/1049 (16 December 2019).

United Nations, 2011: "Map of South Sudan: Department of Field Support Cartographic Section"; at: https://www.un.org/Depts/Cartographic/map/profile/southsudan.pdf (13 September 2018).

United Nations, 2016: "Report of the Secretary-General on Conflict-Related Sexual Violence".

UNMISS, 2012: "Incidents of Inter-Communal Violence in Jonglei State"; at: https://unmiss.unmissions.org/sites/default/files/june_2012_jonglei_report.pdf (13 September 2018).

Unni, Krishnan Karunakara; Neuner, Frank; Schauer, Margarete; Singh, Margarete; Hill, Kenneth; Elbert, Thomas; Burnha, Gilbert, 2004: "Traumatic Events and Symptoms of Post-Traumatic Stress Disorder Amongst Sudanese Nationals, Refugees and Ugandans in the West Nile, African Health Sciences"; at: www.ncbi.nlm.nih.gov/pmc/articles/PMC2141616/ (23 April 2019).

Warner, Lesley Anne, 2016: "The Disintegration of the Military Integration Process in South Sudan (2006–2013)", in: *Stability: International Journal of Security and Development*, 5,1: 1-20.; at: http://dx.doi.org/10.5334/sta.460 (13 April 2019).

Wild, Hannah; Jok, Jok Madut; Patel, Ronak, 2018: "The militarization of cattle raiding in South Sudan: How a Traditional Practice became a Tool for Political Violence", in: *Journal of International Humanitarian Action*, 3,2: 1–11.

Young, John, 2016: *Popular Struggles and Elite Co-optation: The Nuer White Army in South Sudan's Civil War* (Geneva: Graduate Institute of International and Development Studies).

Chapter 13
Traditional Conflict and Peacemaking Processes: The Case of Kurdish Tribes in Mardin, Turkey

Safiye Ateş Burç

Abstract Like contemporary Western approaches to peacemaking, peacebuilding and conflict, the traditional ones are also very effective and important. Traditional and local loyalties determine both political and social, even everday life among the Kurdish tribes which live in the south-eastern part of Turkey. These tribes generally need friends and enemies to continue their existence. So, for them conflict and reconciliation are common features. *Xwin* (blood feud) is the unique traditional conflict of this community. Also *béş/béj* (blood money) is the most common form of traditional peacemaking and *nané aşitiyé* (a dinner of peace) is accepted as a traditional peacebuilding ritual. In this study, the traditional conflict and peacemaking process will be examined in the context of Kurdish tribes which live in Mardin in Turkey and the basic conflict and reconciliation practices of the Kurdish tribes are discussed with the argument that traditional practices have an effective role in peacemaking and conflict.

Keywords Traditional peacemaking · Conflict resolution · Peacebuilding · *Xwin* · *Béş/béj* · *Nane aşitiye* · Mediation · Kurdish tribe · Turkey

13.1 Introduction

In 2010, peace was declared between the Türk Tribe and the Temelli Clan of the Metinan Tribe with a magnificent dinner. The two main agents of this peace declaration were Ahmet Türk, who both led the Türk Tribe and chaired the Democratic Society Congress which was in the position of the pro-Kurdish party at the time, and Şerif Temelli, who was one of the three leaders of the Metinan Tribe and who at the same time had been a chief village guard[1] in Derik (a county of Mardin)

Dr Safiye Ateş Burç, Research Assistant, Department of Political Science and International Relations, Mardin Artuklu University, Turkey. She has a PhD from the Political Sciences Faculty of Ankara University.

[1] This is a paramilitary system which was established among Kurdish tribes in 1985, when Turgut Özal was President.

Ú. Oswald Spring and H. G. Brauch (eds.), *Decolonising Conflicts, Security, Peace, Gender, Environment and Development in the Anthropocene*,
The Anthropocene: Politik—Economics—Society—Science 30,
https://doi.org/10.1007/978-3-030-62316-6_13

and an executive of a right-wing party. *Nané aşitiye* (a peace dinner) was held in the village of each leader. With this ritualistic activity (dinner), which immediately comes to mind when peacemaking is mentioned in Kurdish tribes, it is declared to everyone that the two sides have made peace, and since both parties enter each other's living space and sit at each other's table, they pay respect to each other in front of everyone and declare that the hostility between them is completely over. Around 15,000 people attended the second *nané aşitiye* (the dinner which was held in Ahmet Türk's village) between Türks and Temellis. Among them were many politicians. Women of the respective tribe cooked about a thousand goats and sheep that had been butchered in big cauldrons and served them to the incoming guests in thirty-four separate tents. Two leaders, accompanied by other notable tribesmen, flew the dove of peace and made a statement to the press that the accord between them was definite.[2] This reconciliation put an end to a twenty-six-year-old conflict that caused people from both parties to die, and de facto prevented the Türk tribe from entering Derik. This long-lasting conflict between the two tribes played a role in the daily lives of both tribes, determining their allies and enemies and deeply influencing their practices, especially marriage (wedding) and mourning (death), as well as basic political activities such as party administration, parliamentary membership, and village guarding. Their political loyalty has also been shaped by this conflict.

Tribes, which are a kind of a social and political organisation, are units in which local loyalties are very strong and blood-based kinship relations are also strong. Thus, definitions of a tribe offered by many anthropologists and sociologists have terms that have these points in common. However, a tribe is a phenomenon that must first be defined by a *spirit of unity*. This spirit refers to a mix of blood relations, place attachment and political preference, as well as emotional bonds. Especially in Kurdish tribes, the spirit of unity has a power which surpasses political preferences and emotional bonds. Therefore, local conflict and peace mechanisms, which make co-existence possible or have a role in breaking it in some cases, are of critical importance because these practices play an active role in keeping the spirit of unity alive. This spirit is not only a social phenomenon like conflict and peacemaking, but also a political one that requires political activity and appeals to strong collective emotions. As can be seen from the above, the peacemaking between the Türk and Temelli Tribes was equal to both a meso and macro conflict transformation: on one side, there was a tribe and a supporter of Turkish state power; on the other side there was another tribe but a supporter of Kurdish organisations. Already in all other Kurdish tribes in Mardin[3], and even in the south east, such intersectional conflicts and

---

[2] Aslan, Ferit; Bulut, Bayram; Karakaş, Emrullah, "Ahmet Türk'ten 15 bin kişilik barış yemeği", in: *Hürriyet*, 10 October 2010; at: http://www.hurriyet.com.tr/gundem/ahmet-turkten-15-bin-kisilik-baris-yemegi-16007135 (12 September 2018).

[3] The Kurdish tribes in Mardin (in the whole county) are: Alika, Aliki, Arbani, Arnas, Barava, Basıkil, Batuvan, Bubilan, Butikan, Çemkari, Çomeri, Dakori, Dasikan, Daşi, Derhavi, Dermemikan, Dımıli, Direveri, Dorika, Durıran, Dumanan, Elıki, Erolyan, Habızbıni (habesbeni), Hama, Harunan, Helecan, Hemmikan, Herikan, Hesar, Hessinan, Hevırki, Hıdıran, Kasıkan, Kiçan, Kiçani Hasenan, Kiçani Ömeran, Kiçani Şemsan, Kiki, Koçekan, Köse, Lef, Memmikan, Mendilkani, Metinan, Mhalmi, Milli, Mirsiniyan, Mişkin, Musareşan, Ömerka, Ömeryan, Poran, Pınarali, Raman, Rişmil,

peace usually coincide. However, the most important point is that in this region, these types of complex conflicts are resolved by the traditional methods and local culture. In this study, the traditional conflict and peacemaking process will be examined in the context of Kurdish tribes which are located in Mardin in Turkey, and the basic conflict and reconciliation practices (such as blood feud and blood money) of the Kurdish tribes are discussed with the argument that inter or intra-tribal conflict and reconciliation practices, which are regarded as micro issues, both determine the tribal macro-political stances and behaviours and provide an exemplar from the cultural sphere for macro-conflict and peace in the entire country, because traditional methods are so important and effective. First a conceptualisation of traditional peacemaking and conflict resolution approaches will be offered based on local evidence, then an analysis will follow of the *xwin*, *bêş* and *nané aşitiyé* as forms of traditional conflict, conflict resolution and peacebuilding rituals.

13.1.1 Traditional and Indigenous Peacemaking and Conflict Resolution

Conflict resolution is not a technical matter that is considered as a zero-sum game and that consists of strategic steps. Just as cultural elements legitimise direct and structural violence and make the conflict continue (Galtung 1990), they can also provide a solution to conflicts (Mac Ginty 2008a: 141). Therefore, the methods of conflict resolution brought about by the interest in culture and traditions can be effective in ending many micro, meso and even macro and mega conflicts. Nowadays, Western contemporary conflict resolutions have rediscovered traditional and indigenous peacebuilding methods. "Conflict is often culturally located" (Mac Ginty 2008a: 141), therefore it is necessary to have extensive knowledge of the culture in order to analyse the conflict properly and get closer to an effective solution. Today, traditional approaches that encompass knowledge of the local and location-specific solutions also play an important role in the transformation of contemporary conflicts (Boege 2009: 432; Liden 2009: 616). The combination of traditional approaches to conflict resolution and the view of modern conflict resolution can create an environment that reconciles cultures and makes peace in the local area permanent. However, this acceptance may be met with suspicion in tribal organisations where local loyalties are strong and the leader's power struggle is of critical importance. This is because in tribal organisations reconciliation – like conflict – is a power demonstration, and the dominant power-holders that maintain the conflict can transform even the peace itself into a means to consolidate their power. Therefore, in tribal organisations,

---

Rutan, Salihan, Sorkan, Spirti, Sürgücü, Şeb (şib), Şemsıkan, Şeyhan, Tat, Temikan, Teyyanı Kervan, Teyyanı Reşit, Zahuran, and Zivikan (Aydın/Emiroğlu/Özel/Ünsan 2019: 783; Sarı 2010: 347–348; Ateş Durç 2009: 48–50). Since Mardin is a city which has different ethnic and religious structures, different ethnic or religious identities can find a place within the same tribe. Although rarely seen at tribal level, there are sometimes Arabs turning into Kurdish or Kurdish turning into Arabs in the same tribe. This refers to an intersectional identity in Mardin.

reconciliation seems condemned to be influenced by the dominant power-holders. Paradoxically, both peace and social peace are ensured through the reconciliation of conflicting parties, thereby strengthening the power of the tribal leaders who provide peace. As a matter of fact, it is known that those who prolong conflict within a tribe or initiate conflict with another group are often tribal leaders. Although this paradox holds the conflict potentially alive, local reconciliation traditions (especially in the Kurdish tribes) always make the imagination of peace possible in society. In addition, the most important providers of social cohesion in this community are also local reconciliation methods. For example, among the Kurdish tribes, the rituals of reconciliation include the religious beliefs of the Kurds, and the peace ceremony begins with prayers in front of everyone who is at *nané aşitiyé*, as everyone who represents the parties to the conflict passes under the Qu'ran as a kind of oath. Conflicts cannot be transformed into peace by imposing it from the outside. Consequently, even though there are question marks, traditional reconciliation and conflict practices in tribes are very important for both social and political cohesion.

Instead of Western peacebuilding approaches that are accepted as top-down methods, traditional approaches, known as bottom-up (Mac Ginty 2010: 353) and with local legitimacy, are very effective in the transformation of conflict. Bottom-up approaches are participatory, culture-compliant, low-cost and sustainable (Mac Ginty 2008b: 120). Especially in Asian and African societies, some traditional methods kept alive by the local culture are tremendously effective in the peaceful transformation of conflict. Even though they adapt to modern times and have gone through some changes, these methods have the potential to make serious contributions to conflict resolution efforts (Boege 2011: 432). When their conflict-improving effect is added to the realistic perspective of Classical liberal conflict resolutions, an environment in which the cultures are reconciled and the peace is perpetuated at local level can be formed. When local-level or traditional methods in micro conflicts are modified to help solve macro conflicts, sustainable peace can be ensured. In fact, tradition doesn't always stand at a fixed point or place, and it, too, moves with the times and improves (Boege 2009: 437).

The only difference that distinguishes traditional methods of peacebuilding from Western-style methods is that these methods are specific to the area where the conflict occurs. Instead of standardised methods used by international actors, locally constructed methods are "participative, culturally intuitive and locally based" (Mac Ginty 2010: 354). As argued above, traditional approaches to peacebuilding are "sustainability, independence from expensive and artificial additives in the form of external peace support interventions and originating locally" (Mac Ginty 2010: 348). Furthermore, while Western approaches have universal practicality and are comprised of generally standardised practices, traditional methods are applicable only to the specific area they belong to (Boege 2009: 438). Notably in Africa, classical liberal peacebuilding is not effective in making sustainable peace. Local conflict resolution methods are preferred to contemporary methods which are viewed as being imported from outside, because "indigenous approaches to peacebuilding including mediation, accommodation, reconciliation and negotiation are rooted in the knowledge, customs and history of Africans" (Isifi 2015: 62). Thus, in the case of a conflict involving Africa and even Asia and the Middle East, many international peacemakers are trying to learn the methods of reconciliation which are found in the culture of the related country and are important to everyone. More precisely, international actors are rediscovering traditional and indigenous approaches to conflict resolution and even encourage traditional and indigenous peacebuilding and peacemaking approaches (Mac Ginty 2010: 355). For example, traditions such as *Gacaca* courts in Rwanda, the *Nahe Biti Process* in Timor-Leste, *Loya Jirga* local peacemaking in Afghanistan, the *Ubuntu* philosophy in South Africa, the *Mato Upot* in Northern Uganda, the *Omoluwabi* amongst the Yoruba natives in West Africa, *Kanye Ndu Bowi* in Ghana and Togo (Mac Ginty 2008a, 2008b; Isifi 2015; Olowu 2018) offer peacemakers serious assistance in building macro peace and transforming past conflicts and potential conflicts in the future in the related countries. Archbishop Desmond Tutu's philosophy of Ubuntu, which he included in the South African peace process, shows how social peace can be facilitated with traditional approaches. Tutu (1999) argues that *Ubuntu* is a soul and the person with this spirit intends to live in peace with others because, in *Ubuntu* philosophy, everyone is part of a great whole. *Ubuntu* is briefly defined as: "A person is a person through other persons, and I am because we are; we are because I am" (Isifi 2015: 65). Knowing the culture of every country and the traditional methods of reconciliation of the people residing there, and even witnessing the mechanisms of traditional conflict, are also very important when formulating a suitable roadmap for sustainable peace.

Traditional peacebuilding methods make it possible not only to include the traditions of the area of conflict in peacebuilding, but also to incorporate into the process common feelings from a more accurate point. What people who have grown up in the same culture get angry about and what they can tolerate, how these wounds can be healed, and which pains can be transformed into positives by which tools are revealed by means of these methods. Therefore, in order to be able to forgive and empathise, these methods give more important hints to the mediators or negotiators than universal methods. However, because of the emotional aspects of conflict and reconciliation, it is necessary to organise and manage the symbolic space well. It is necessary to know the culture in places where conflict is experienced and a solution is sought, and to know which method is truly sustainable, legitimate, applicable and acceptable. In this sense, peacemakers have serious responsibilities, especially in the resolution of micro and meso conflicts.

Tribes are at the centre of social life in geographies where Kurds live, as cultural and political societies in which localness is important and tradition is dominant. Local conflict and peacemaking practices, meaning traditional methods that are applied, which enable these structures to be divided, unite, break up or grow, can say important things in terms of conflict resolution. The effect of traditional methods of peacemaking and the contribution of the cultural to the determination of politics have not been deeply investigated (Boege 2011: 434). Yet, the conflict motivation of the local communities and their efforts to transform this conflict into peace have the potential to guide contemporary peacemakers. Surveys done by Oxfam in February 2008 in Afghanistan show that local mechanisms are much more effective than official interventions at resolving local conflicts and disagreements in that region (Waldman 2008: 3).

In tribes where traditionality prevails, the factor that ensures that tribe members are in solidarity is their blood ties. Hence, this tie is the common point in the definitions of a tribe. In Kurdish tribes in the south-eastern region of Turkey the bond to the tribe and hence loyalty to the kinship is above everything else. Here, the tribe is not only a phenomenon which determines social status, but also a political and martial phenomenon that determines the relationship with the central authority and provides security for its members. Kurdish tribes, which are one of the symbols of deep kinship bonds and powerful hierarchal structure, have always kept their traditional structure. Many researchers, from Ziya Gokalp to Friedrich Barth, have studied these traditional units.

In the Kurdish tribes, *blood feud* is the basic form of conflict; *blood money* and *peace dinners* are the most common peacebuilding and peacemaking methods. Revenge and the reciprocity of revenge (Boege 2009: 441) are the main characteristics of traditional conflicts. Blood relations and place attachment are very important in Kurdish tribes in which strategies of enmity are present. However, many of studies on these tribes (Van Bruinessen 1992; Yalçin-Heckmann 2006; Beşikçi 1992; Şerefxan, n.d.; Gökalp 1992; Özer 1988; Barth 2001; Leach 2001; Türkdoğan 2006; Ateş Durç 2009) have made it mandatory to define these units in a different way. Accordingly, a tribe is a unit that is not only built on blood relations and place attachment, but also on a spirit of unity, an obligation to stick together against enemies and to be safe against power-holding actors such as the State. If we set off from the definition which this study takes as a basis, we can say that solidarity, loyalty, peace, and power are the main points that determine the relationships of Kurdish tribes with macro actors. Conflict between or within tribes and traditional peacemaking practices deeply affect the social structure and daily life of the tribal members. However, traditional methods of peace within Kurdish tribes are quite different from the peacemaking methods in Africa and Asia. For example, the *Ubuntu* philosophy in South Africa or the unifying methods of other African native societies are methods that emphasise social peace and human rights that stem from being human. Mediation mechanisms are the strongest method that turns conflict into peace among Kurdish tribes. The way in which power is used and negotiation are the two main impulses that determine the local methods of peacemaking of Kurdish tribes. In conflicts among Kurds, if these conflicts are between local groups (*eşir* [tribe], *taif* [clan] and *malbat* [dynasty]), the conflict is more destructive, but the reconciliation is more procedural and prone to resolution. Mediators can be very effective in intra or inter tribal conflicts because they are chosen from that culture. In addition, since it is possible to see the honour and dignity of the tribe in the behaviour of the tribal leader, it is almost an obligation for them to keep their pledge. This pledge (promise) is the only factor that prevents the key power leaders in the tribe from using the peace process as a continuation of the conflict. Also, this word itself is almost binding for both parties of a peace agreement because that pledge (oath) is given in the presence of the *Qur'an*, which is one of the indispensable religious holy items of all parties. Nobody opposes the *Qur'an* in Sunni society, and all Kurdish tribes in Mardin are Sunni.

13.2 Traditional Conflict, the Peacemaking Process and Mediation in Kurdish Tribes

13.2.1 From Xwin *(Blood Feud) as a Tribal Conflict to* Béj *(Blood Money) as a Peacemaking Process*

Fights between or within tribes can be based on a wide range of economic, social, or political causes, from the grazing of one tribe's animals in the other tribe's pasture to arguments about land; from kidnapping women to killing men. Many of these quarrels are big conflicts that are known as *xwin*, in which people from both sides use violence of any nature against each other. The *xwin* between the Türk and Metinan tribes in Mardin is an example of such a tribal conflict. The hostility between the Türk tribe and Temelli clan of the Metinan tribe began in 1985. Şerif Temelli and his brother, Cebrail Temelli, the most influential men of the tribe, were gunned down by unidentified persons with automatic weapons when they drove to Bozbayır Village in the Derik District. In this incident, Cebrail Temelli died and four of Şerif Temelli's fingers were broken. The Kurdish Worker Party admitted responsibility for the incident. However, the Temelli family suspected Ahmet Türk and his tribe – the Türk tribe – of being the instigators and hated them. Thus, the *xwin* began. Because of this speculative bond, some members of the Temelli family joined a village guard system (a paramilitary organisation in Turkey) after 1985[4] (Ateş Durç 2009: 127). Other conflicts outside the blood feud are small-scale meso conflicts that affect a small range, which can be resolved by a senior member of the family intervening or by a money payment.

In general, the reasons why conflicts start in Kurdish tribes revolve around matters that will insult a tribe in front of the other tribes, thereby harming its fame, glory and reputation. The most important factor that makes a conflict harsher and makes it last for some time is that in these conflicts between or within tribes it is not the *reason* itself but the symbolic dimension that it creates. Thus, the reason that starts the conflict itself[5] does not matter very much; what matters is the harm that it brings to the tribe's symbolic value. At times, there are certain news items in the media, such as "Blood feud starts because of a cock" or "They killed each other for a trivial

[4] Yavuz, Ramazan; Avuka, Adnan; Sunar, Serdar, Karakaş, Emrullah, 2010: "Mardin'de 5 bin kişilik barış töreni", in: *Milliyet*, 29 September; at: http://www.milliyet.com.tr/mardin-de-5-bin-kisilik-baris-toreni-gundem-1295256/. Aslan, Ferit; Bulut, Bayram; Karakaş, Emrullah, 2010: "Ahmet Türk'ten 15 bin kişilik barış yemeği", in: *Hürriyet*, 10 October; at: http://www.hurriyet.com.tr/gundem/ahmet-turkten-15-bin-kisilik-baris-yemegi-16007135.

[5] The tribal system, which changed shape with migration from the country to towns, seems to have affected the conflict in cities. In fact, in some tribes that dominate the Tori region, the reason that started the conflict is now considered important and the price that is going to be paid is argued on this basis. In such instances, it is significant that the cause of the argument is considered important. This is not yet common to all Kurdish tribes. Further discussion of this is beyond the scope of this study, but since there appear to be no existing studies on this subject, it is could form a rich topic for other researchers.

reason". The main reason such news items appear is that the concrete reasons are of critical importance when it comes to the micro or meso conflicts in social life. However, what makes quarrels between tribes close to political conflicts is that the concrete reason is virtually turned into an excuse or that it has harmed the prestige of the tribe.

Xwin,[6] which is the first thing that comes to Kurdish minds with the word 'conflict' and which is experienced at a different intensity from region to region, is the biggest cause of tribal conflicts and creates the most destructive factions in Kurdish tribes (especially among those who settled in south-eastern Turkey). These conflicts, which have a sense of vengeance as their origin, and which are the new form of a *tit for tat* tradition (Özer 2003: 163) in present times, are well described in the following narrative by Friedrich Barth: "When one of us was killed by someone from another tribe his close relatives went after that tribe and killed the first man they saw. Sometimes they killed not one man but four or five. Then of course the other tribe came to take revenge, killing some of us. That could go on years and years; sometimes 50 to 100 men got killed before peace was finally made" (Barth 2001: 72–77). In an interview with Şerif Temelli, a list was compiled of those killed and injured as a result of attacks on the Metinan Tribe. This list is actually the list of those who were killed and injured in *xwin* to which the tribe was a party. Indeed, one of the items on this list is the death put forward as the cause of the *xwin* with Ahmet Türk's tribe, as mentioned previously (Ates Durc 2009: 127).

Xwins "traditionally, are intertribal affairs. When a Kurd is murdered by someone from another tribe, not only the lineage of the dead man, but the whole tribe comes together for an extra-juridical form of punishment, usually provoking countermeasures that lead to escalated tribal warfare."[7] These conflicts, which have a devastating impact on tribe members, are also "the continuation of inter-tribal policy by other means" (Van Bruinessen 1992: 65). This is because in these traditional conflicts where the most crude form of revenge is seen, we encounter not only a criminal situation but also a power struggle. If one party kills one person in the blood feuds caused by inter-tribal conflicts, the opposing side will surely match the offence and kill a member of other group. If a tribe whose members are killed does not kill someone from the other tribe and there is no attempt at reconciliation, the tribe will lose its social dignity and power and become open to all kinds of pressure. "If the tribe member does not follow the blood feud, if he does not avenge the death of his relative, if he does not respond the attack with a counter attack, he is considered dishonourable in the public opinion. His nose becomes down, his name becomes small, he has no value, no honour" (Gökalp 1992: 40). For example, the blood feud

[6]"Iraq: Blood Feud Tradition in Kurdistan Region", *Country of Origin Information Portal*, November 2017, para. 1.

[7]"Kurdish Families – Kurdish Marriage Patterns", in: J. Rank: *Marriage and Family Encyclopaedias*, n.d., para. 10; at: http://family.jrank.org/pages/1026/Kurdish-Families-Kurdish-Marriage-Patterns.html.

between the Arbani and Belayi tribes in the Kızıltepe district of Mardin, which reconciled with a long mediation process in the summer of 1998, lasted fifty years and eleven people lost their lives during this period.[8]

In these conflicts, which have an element of interpersonal struggle, the timescale for taking revenge can be extended to a very wide period (revenge for a murdered person can be taken three years later); the aim is to take revenge at the right moment. In addition, the person performing the act of killing in the blood feud is not the only target of the attack (Özer 2003: 164). All of the tribe's perpetrators are held responsible for the act of killing, which is often seen as punishing another tribe member instead of the offender. This is because such conflicts concerning the whole tribe are not individual, but between groups (Van Bruinessen 1992: 65) and are even related to the lineage. The only thing that matters in killing is that revenge is taken. In 1909, Ziya Gökalp (1992: 112) stated in his article *Firari Mahzunlar, Gıyabi Mahkumlar*: "Blood stain is only washed with blood. But whose blood? Directly of the murderer? No! It is not the individual responsible for the murder, it is the leader of the tribe."

In the city centre of Mardin, an armed conflict took place between two Arabic tribes: Sincar and Kabala. A respectful young member of the Kabala tribe was killed by Abdülkadir Sincar (the leader of the tribe) on 16 April 2015. A *xwin* began after that death. But there weren't any visible disputes between the two tribes for a long time. In fact, this silence was a preparation for revenge. After nearly four years, when two members of the Sincar tribe (including the son of the murderer) waited at a red light, a car approached them and a gun was fired. A young member of the Sincar tribe was killed and two members of the Kabala tribe went to prison.[9]

Vanlı (1997), who reported the observations of Western travellers visiting the areas where the Kurds lived in the seventeenth century, cited the traveller Niebuhr's words: "If it is correct to point out the character of a nation with a single example, the Kurds are extremely vindicatory" (Vanlı 1997: 57). In such conflicts today, understanding continues in the same way as in the previous century. However, in the killings, the target is not chosen by random selection of the killer's tribe. Normal targets are the killer himself and close relatives (Van Bruinessen 1992: 65). In addition, women and children are not among those who will be killed in a blood feud; objects of the vengeance are only men. In fact, children and adolescents, who are generally considered to be less culpable in legal terms and who are assigned to take revenge, and women are the ones who experience the greatest harm. However, tribal femininity, socially established, also plays an important role in the continuation and reproduction of the tribal system structured with a dominant masculine mentality. As a matter of fact, the mother and sister of the person who will take revenge in many blood feuds are also aware of the situation. It is additionally the woman who has socialised the child

[8]"50 Yıllık kan davası bitti!", *Milliyet*, 28 July 1998, para 1., at: http://www.milliyet.com.tr/the-others/50-yillik-kan-davasi-bitti-5350948.

[9]Güneş, Nezir, "Kırmızı ışıkta bekleyen araca silahlı saldırı: 1 ölü 4 yaralı", in: *Hürriyet*, 22 February 2019; at: http://www.hurriyet.com.tr/yerel-haberler/mardin/kirmizi-isikta-bekleyen-araca-silahli-saldiri-41126682.

as a tribal member since his/her birth and conveys the rules that he/she must obey. In tribal organisations, women are lower than men in the social hierarchy. Therefore, the ability of the woman, who is most affected by the conflict produced by this system and who makes serious contributions to the reproduction of the system, to challenge and transform this system is an issue that should be questioned in another detailed article. However, it is possible to say here that the phenomena of femininity, masculinity and conflict are intertwined and can be expressed as both feeding and consuming each other.

The segmentary lineage system detailed by Dale F. Eickelmann and Rudi Paul Lindner, which is appropriate for tribal organisations, shows that the basic principle of the tribal and inter-tribal conflicts is the unity of action. In this system, which is based on paternity, division is in the form of a dual division and this duality continues from the bottom up. Each unit (a, b, c, etc.) corresponds to a separate local community. These communities live in the same land and expect support from each other when they become a party to a conflict. All conflicts based on different reasons, especially murder, are considered to be directed at the entire tribe and the honour of the whole tribe (Lindner 1982: 693; Eikelmann 1989: 132). Therefore, the conflict does not only take place against the person who started it, but once it has started it concerns the entire tribe. In most cases, one of the branching groups may need the help of another during the conflict. So, for example, if d has made an attack on the dignity and honour of b, then c is expected to support b. As seen in Fig. 13.1, a + b = 1, c + d = 2, e + f = 3, g + h = 4; 1 + 2 = A, 3 + 4 = B; A + B = I. In this system, the groups consist of I's family. If an outside attack is carried out on one of I's family, then all local communities from the smallest to the biggest mentioned above become a party to the conflict. According to Yalçın-Heckmann (2002: 157) a, b, c,

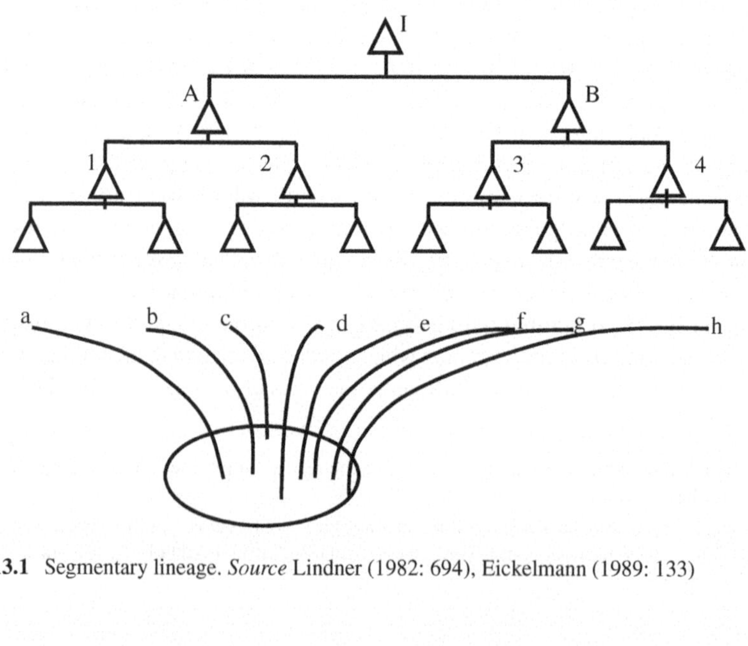

Fig. 13.1 Segmentary lineage. *Source* Lindner (1982: 694), Eickelmann (1989: 133)

d, e, f, g and h represent dynasties; 1, 2, 3 and 4 represent clans; A and B represent clans; and I represents a tribe.[10] In this model, the most important element that unites groups is the unity of action. According to this, "I am against my brother; me and my brother against my cousin; me, my brother and my cousin are against the World" (Lindner 1982: 693). Yalçın-Heckmann (2002: 147) adapts this unity of action to Kurdish tribes. As she said, "brothers, father and sons against uncle and cousins; all of these against patrilineal other cousins; again all of them against another tribe." It is stated that this system is activated especially in conflicts. "If one member of the a family is in conflict with a member of the b family, the conflict is between a + b. If it is between the a family and the c family, the conflict becomes a + b's clash against c + d. If one member of the a family clashes with someone from the h family, the conflict is seen as the opposition of all A and B clans" (Ateş Durç 2009: 26).

Blood revenge or revenge "is regarded as being morally acceptable and honorable" (Taş 2014: 141). The most important factor that directs and re-creates the traditional conflicts of Kurdish tribes is the loss of honour, dignity and prestige, pointing to a moral and symbolic dimension. These symbolic forms are very valuable because they are the elements that reinforce the power of the tribes. If the economic capital of a tribe is not enough to keep it alive, without the symbolic capital the tribe cannot continue as a strong social organisation. Therefore, a tribe which does not want to lose prestige can pursue a blood feud for many years. The cost of the conflict increases as the two parties continue to kill each other. As the cost increases, the conflict deepens. This is a psychological trap which is difficult to escape from, like the story above quoted by Barth. Revenge and triumph motivate the parties, but, as the years go by, the conflict becomes challenging for all parties.

Revenge seems to donate dignity and honour to the tribes, but in such communities where revenge is traditionalised, the whole group (in the tribe) continuously feels threatened.[11] For the tribe leaders, this threat perception of the community is

[10]Indeed a, b, c, d, e, f, g and h refer to minimal segments; 1, 2, 3 and 4 to minor segments; A and B to major segments; and I to the maximal lineage.

[11]In fact, in the areas where tribalism is visible and effective, not being from a tribe (bé eşir) or walking away from an attack on his tribe places the tribe member in an insecure situation. Perhaps because of this, "mutual obedience, unconditional submission, sacrifice, renunciation, and we are feeling" (Özer 2003: 101) are the basic traditional values that determine tribal relations. The tribe member fulfils every task and duty that falls on him to enable his tribe to be strong because he thinks that the tribe will ensure his security before the State (Ateş Durç 2009: 79). When an offence is committed, he first takes refuge in the family elders and the tribal leader. Thus, before the State's law enforcement power, the family or tribal leader intervenes, calculates what needs to be done, and fixes the situation if there is something that can be fixed. In other words, the law of tribe rules. If there is a serious crime like murder, then the tribesmen of the tribe come together and discuss how to proceed. Neither the attacker nor the person who has been attacked inform the gendarmerie or the police. If there is no official record of the attack, the family of the murdered person will not report the murder because social pressure and tribal law require revenge. Therefore, the courts of the State and law enforcement are used for a much simpler and more limited number of issues. Justice is distributed by tribal forces. The belief that justice is not fulfilled by state institutions (Özer 2003: 167) also affects this. However, in this situation, the tribal system shifts the punishment from an objective area to a subjective area. The idea that punishment is not enough is one of the key factors that nurtures and increases conflicts.

preferred because it interlocks group members. Indeed, clan revenge clarifies and hardens the boundaries of the group and unites them under the protection and power of a leader. As Edmund Leach (2001) emphasises, hostility benefits both leaders and tribesmen. Because of this hostility, tribal members can establish unity and solidarity, and this increases social cohesion (Leach 2001: 92). However, in addition to this threat perception, in the tribes where the culture of revenge is continuous "individuals, families or even the entire community are affected economically, socially and geographically. Some people may have to give up their wealth, land and property and leave where they have lived for hundreds of years to move to a different city or country in order to escape revenge acts" (Taş 2014: 141).

In major social conflicts such as blood feud, participation in the conflict, the size of the population to be attacked, and the traditional methods of conflict resolution to be applied vary according to whether the conflict is in a tribal or inter-tribal area.

Blood feuds are generally settled with the aid of mediators. These need to be people who are respected by both sides and whose authority all partipants in the feud are prepared to obey (Karadeniz 2012: 67). If there is a conflict between different clans within the same tribe, it is usually resolved by the tribe leader and the *Ruspi* (old wise man). Hence in such cases the peacemakers belong to the same tribe as the clashing clans. Also, since the two clans are part of a single tribe, the conflict does not require the whole tribe to take sides in the conflict. The sides are composed only of the actual clans who are in conflict and the families who regard themselves as close to those clans. Therefore, in this kind of meso conflict, the people to attack are also limited. Moreover, since they are in the same tribe, it does not take much for the conflict to deepen. Generally, after the first hot conflict, the dispute lasts for many years, but the killings will not be as high as in other blood feuds. Moreover, such conflicts, which would make the entire tribe uneasy, can be transformed into peace by the powerful people in the tribe calling on others to mediate. For example, the blood feud that began in 2013 among the big families of the same tribe in Ballıbaba (Gırésor) and Sisan villages of the Derik district of Mardin resulted in peace in 2018 based on the efforts of the leaders of that tribe and the corporation representing that tribe. The feud was five years old, but mediation continued for four years. Parties were to remain faithful in peace: "multiple imams took the lead in the reception by reading hymns. Members of the blood-feuding families passed under the Holy *Qur'an*, held in the air. Thus, with the oath made on the *Qur'an*, the blood defendants agreed to leave their animosity behind. Blood feud families shook hands in the presence of Qur'an and made peace."[12]

If the conflict is inter-tribal, then the entire tribe community is considered to be a party to the feud and every member (except women and children) is the potential target of the attack. It also takes a lot of effort to solve such cases. Peacemakers range from traditional leaders and civil rulers to the leaders of political parties. In inter-tribal conflicts traditional peacemaking rituals are more common and care is taken to implement them in the same way. Such experiences of conflict and peace are

---

[12]Kamer, Hatice: "Mardinde kavgalı aileler 5000 kişinin katıldığı yemekle barıştı", in: *BBC*, 14 August 2018, para. 18; at: https://www.bbc.com/turkce/haberler-turkiye-45171916.

very common in Mardin. It is known that there are conflicts between Kurdish tribes themselves, Arab tribes themselves and between the Kurdish tribes on one side and Arabs on the other, which have lasted for a long time and deeply affect the whole political and social structure in the region. For example, the Kikan tribe, one of the well-known tribes of Mardin, had a seventy-two-year feud which lasted from 1926 to 1998. Because this case was a blood feud which included the tribe members in Syria but not in Turkey, it influenced people's behaviour and their social life; therefore, the peace dinner was also held in Syria. The peace meals, which normally lasted one day, lasted four days in this case.[13] The hostility between the Dashi and Asiti tribes in the Kiziltepe district lasted twenty-eight years. Through the mediation of a political party executive of the period, the conflict ended with a peace dinner attended by 5,000 people on 29 September 2019.[14] However, there are numerous unsolved blood feuds in addition to those resolved. These cases not only disrupt social cohesion, but also allow for political polarisation and take a decisive role in determining the political attitude of tribe members and even their daily behaviour.

Social and traditional conflicts with increased costs, such as blood feuds, do not usually result in spontaneous peace. The hostility created by these disagreements does not end with the imposition or intervention of the state authority. In most cases, courts and prisons are transformed into vengeance places instead of solving inter-tribal problems (Ateş Durç 2009: 84). As emphasised earlier, the blinding feelings of vengeance prevent the parties from agreeing with each other. But with the inclusion of a strong and respected person in the region, the process evolves into peace-building. The important thing here is that the mediator must be someone respected and welcomed by both parties. A certain amount of blood money agreed between the parties in the blood feud of the same tribe is agreed and the case ends with the payment of the price (Van Bruinessen 1992: 68). However, the process of solving inter-tribal conflicts is more complex. The United Nations High Commissioner for Refugees' report on the Iraqi tribes in November 2016 emphasised the ability of tribes to solve their feuds by means of their traditions, and the blood money was presented as a traditional peacemaking mechanism (2018: 1).

Tribes commonly resolve disputes in line with tribal customs. Typically, tribes first seek a resolution through arbitration and the payment of financial compensation – "blood money" – to the family of the victim (*fasl* or *diyya*) in cases of murder, physical harm and damage or loss of property. The injured party in turn gives up the right to retribution. Only where tribes fail to resolve disputes between them by peaceful means do such conflicts turn into blood feuds, which may give rise to long cycles of retaliatory violence and revenge.

A long time may be needed to resolve the feud. Sometimes the parties are able to come together as a result of months or even years of negotiations. However,

[13]"Barış Yemeği", in: *Hürriyet*, 18 September 1998; at: http://www.hurriyet.com.tr/gundem/baris-yemegi-39038828.

[14]"Mardin'de 28 yıldır husumetli aileler barıştırıldı", in: *Habertürk*, 29 September 2019; at: https://www.haberturk.com/mardin-haberleri/72220285-mardinde-28-yildir-husumetli-asiretler-baristirildi (18 December 2019); "Mardin'de 30 yıllık husumet barış yemeği ile sona erdi", in: *Artukluhaber*, 29 September 2019.

these conflicts must also reach ripeness in order for the parties to be able to gather together and start making peace. So it is necessary to wait for the right time for reconciliation to arrive. Interveners to solve such conflicts do not have much difficulty in predicting the appropriate time because they know the history, parties and course of the conflict well. In addition, the demand for inter-tribal *dawa* (the name given to solving conflicts, etc.) usually comes from the party that initiated the conflict. Sometimes, both sides announce that they want to make peace with the powerful people or institutions they feel close to, and the struggle for reconciliation begins. The most effective traditional method of conflict resolution is *béş/béj,* which is also called blood money, as in tribal conflicts. This traditional method is the first stage in peacebuilding. When an agreement is reached and payment is made,[15] it can be said that the start of peacebuilding was successful. In the second stage, peace will be announced to everyone and peacebuilding will take place. The first phase of traditional peacemaking in the Kurdish tribes is called *diyet* or *kanlık* (Özer 2003: 170) by some researchers. Usually cash money is paid as *diyet* but sometimes instead of money, animal(s), a vineyard, garden and even women can be given.[16] All of these *diyets* are paid with the participation of the entire tribe. The sharing of *béş/béj* in the lineage is often unequal. However, every lineage member must make at least a symbolic contribution (Van Bruinessen 1992: 67). In other words, just like conflict, peace is social. Since the whole tribe is one side of the conflict, the payment demanded as blood money is charged not only to the person who initiated the blood feud or his close relatives, but also to the entire tribe.

Conflicts that can be considered small can be resolved by notable tribesmen or the leaders of other tribes via relatively short negotiations. However, blood feuds or the abduction of girls, which are both large-scale conflicts, require a much more serious and longer negotiation and strife period. The blood feud between the Türk and Metinan families is among the conflicts in which violence has been experienced in its purest form and therefore the intermediary person and institutions in the resolution of the conflict had to have lengthy meetings. Selahattin Demirtaş, one of the Peace and Democratisation Party (HDP- the pro-Kurdish party of that period) chairmen of the era, has uttered the following on the subject:[17]

> A complaint between families is being mentioned. This peace concerns more than 50-60 villages that are geographically dispersed. If such difficult matters can be resolved, this is an important example for the peace of the country. Here, the parties, peace parliament, associations, and persons who really made an effort for peace, had *meetings again and again*, and both parties had meetings with each other. As a result of these meetings, it

[15] "Kurdish Families – Kurdish Marriage Patterns", in: J. Rank: *Marriage and Family Encyclopaedias*, n.d., para. 10, http://family.jrank.org/pages/1026/Kurdish-Families-Kurdish-Marriage-Patterns.html.

[16] "Iraq: Blood Feud Tradition in Kurdistan Region", Country of Origin Information Portal, November 2017, para. 1.

[17] Aslan, Ferit; Bulut, Bayram; Karakaş, Emrullah, "Ahmet Türk'ten 15 bin kişilik barış yemeği", in: *Hürriyet*, 10 October 2010, para. 6; at: http://www.hurriyet.com.tr/gundem/ahmet-turkten-15-bin-kisilik-baris-yemegi-16007135.

became obvious that ailment and animosities benefit no one, and that living together like brothers is the appropriate solution.

Undoubtedly, the most important institution that keeps the mechanism working in local conflicts is tribe leadership. That the leader is active in the conflict and is able to detect who is an ally and who is the enemy determines the relationships of the tribe with the State and other political institutions. Indeed, what made it possible for the Metinan tribe leader Şerif Temelli to be the chief village guard for many years and assume the role of a delegate for right-wing parties (Ateş Durç 2009) is that such enmities are actively continued. In this way, leaders are influential actors in the peace ceremony. Ahmet Türk, as a powerful leader, said this about the peace between the Türk and Metinan tribes as well as the macro peacemaking between Kurds and Turks:

> We will endeavour to improve our brotherhood with the Temelli family. We want the Kurdish and Turkish peoples to take a sample of this peace and strive for lasting peace. We will continue our efforts in reconciling the Turkish and Kurdish peoples. We believe that there will definitely be peace between the Kurds and the Turks. We always wanted the brotherhood of the people, the unity, and we will insist on that from now on.[18]

13.2.2 Nané Aşitiyé *as a Ritualistic Peacebuilding Approach*

In addition to the reasons that provide a basis for peacemaking in Kurdish tribes, and the mediation mechanism, rituals and practices that must be fulfilled in every peace process hold a very important place. The most important ritual of peacebuilding is a peace ceremony which has mediators, *nané aşitiyé* and other practices. This dinner, with its attendees, those in the kitchen, the food that is cooked, and the way the dining area is organised, is virtually a spectacle. Many practices in these dinners, such as the foods that are supposed to be on the table, the number of animals that are butchered, how many guests there are, seating arrangements at the table and so on, have become traditions that have certain rules. The splendour with which these practices are carried out affects that tribe's prestige positively. Thus, peacemaking, just like conflict, is another show of strength.

Nané aşitiyé (Dost 2012: 89) not only serves as an announcement that the two parties have made peace, but also demonstrates the power of those who have made peace. The more and fancier the food, and the more people that attend, the bigger the glory and reputation of the parties that made peace. While this ritual is being carried out, both parties are anxious. This anxiety is reflected in the seating arrangement and preparation of the meals. Firstly, as shown in Fig. 13.2, there is usually rice (which is preferred because it is more expensive than bulgur), cubes of braised lamb browned in their own fat, and sheep's head for dinner. Pilaf is poured into different

[18]Yavuz, Ramazan; Avuka, Adnan; Sunar, Serdar; Karakaş, Emrullah: "Mardin'de 5 bin kişilik barış töreni", Milliyet, 29 September 2010, para. 11; at: http://www.milliyet.com.tr/mardin-de-5-bin-kisilik-baris-toreni-gundem-1295256/ (22 August 2018).

Fig. 13.2 Female members of a tribe preparing a dinner for peace after a blood feud in Mardin. *Source* Photographer Amar Kılıç (10 October 2010), who granted permission for its publication

sized trays, and big pieces of braised meat are put on top in such a way that one cannot see the pilaf because of the meat. The size of the chosen trays represents the prestige of the guests, which is why big trays are usually preferred for peace dinners. The dinner is not served to everyone on separate plates; the parties that made peace eat from the same tray so that there is no suspicion of poisoning. The tables are usually rectangular, and the host never sits at the shorter side of the table. There, the honourable mediators sit. The dinner hosts sit in the middle and at the edges of the longer side of the rectangle. Thus, they can attend to everyone. In addition, since sitting on the shorter side would constitute a demonstration of power, which would destroy the modesty which is needed for peace. The parties try to stay hospitable, modest, but powerful at the same time on the day of the dinner.[19]

When the parties come together on the day at the place of the peace dinner, they do not go to the dinner table immediately. There are various rituals which take place at almost every peace dinner before eating. First of all, a large group of mediators, parties and notables of the region are invited to this dinner. Among the guests there are necessarily *melle* (religious scholars). These scholars begin the ritual with prayers when everyone is gathered. A representative of the parties will then make speeches with statements of gratitude and loyalty of peace. Mediators also make short speeches emphasising the importance of peace. Then all parties, and often mediators, pass

[19]See: "Mardin'de barış yemeğine 21 bin kişi katıldı", 10 October 2010; at: https://www.internethaber.com/mardinde-baris-yemegine-21-bin-kisi-katildi-300554h.htm.

Fig. 13.3 Ahmet Türk, as both a host and the representative of the political party of which he is a member, leads partipants at the *nané aşitiyé*. Second on the right is Selahattin Demirtaş, a mediator of the feud. *Source* Photographer Amar Kılıç (10 October 2010), who granted permission for its publication

under the *Qur'an* while it is held in the hands of two religious scholars[20] and swear to the *Qur'an*, that they are loyal to peace. After all these practices, they go to dinner. In addition, just like the conflict, the peace ritual is an inter-male activity. If there is a woman at a peace dinner, she is most likely the representative, deputy or mayor of a political party. Women of the conflictual parties do not attend the dinner. In other words, the unequal relationship between men and women is maintained during reconciliation (Figs. 13.3 and 13.4).

In Kurdish tribes, local peace is not only social but also political. Sometimes the cause of conflict is political, sometimes the mediators are political actors, and sometimes the results will affect political loyalties, so these peacemakings are also political. As mentioned before, Kurdish tribes determine their allies, enemies, political preferences, and even marriages based on ongoing conflicts and the stance of the parties to these conflicts. For instance, because the Türk tribe seemed to be against the State, the Temelli clan began to strengthen its statist side after the 1980s and made it more apparent. The Temelli family has also been far (*dur*) from tribes which are close (*nézik*) to the Türks, and close to ones that are far. Namely, the most important

[20]Yavuz, Ramazan; Avuka, Adnan; Sunar, Serdar, Karakaş, Emrullah, "Mardin'de 5 bin kişilik barış töreni", in: *Milliyet*, 29 September 2010, at: http://www.milliyet.com.tr/mardin-de-5-bin-kisilik-baris-toreni-gundem-1295256/.

Fig. 13.4 Before the dinner, the conflictual parties pass under the *Qur'an* and promise to remain loyal to peace. *Source* Photographer Amar Kılıç (12 August 2018), who granted permission for its publication

factors that determine the political attitudes and behaviour of tribes are their allies and their enemies, but especially their enemies.

13.2.3 Mediators in the Kurdish Tribal System

Once the conflict starts in Kurdish tribes, reasoning is put on hold; a show of strength, accompanied by violence, starts, and the conflict dominates every aspect of the daily lives of those involved. The institution which determines the direction, severity and duration of such local conflicts is certainly tribal leadership. It is possible to see how strong the relationship between leadership and conflict is in Ayse Yildirim's study on the cultural and physical conflict between two Kurdish tribes in Ankara-Haymana. "Conflict is of critical importance in sustaining the hegemony of the leaders" (Yıldırım 2006: 195). The leader needs a state of conflict so as to build and sustain his authority. "At this point, an environment of perpetual conflict, composed by creating a symbolic enemy, will reinforce the loyalty of the tribe members to the leader" (Yıldırım 2006: 195). Bruinessen (1999) too argues with a similar point that the state of conflict makes group integrity possible, whereas the state of peace reduces this function and shatters the power of the leader. The leader can transform the tribe into a mass that is ready to be mobilised at any moment in wartime. He can order executions in family councils so that the clan can get its revenge or can persuade the tribe members to be on full alert at all times. However, Bruinessen leaves out the relationship of the leader and peacetime. Leaders need an enemy in their tribes even if not a concrete one, so that group members can cling together and revolve around him. Yet, the same leader increases his power by ending other tribes' conflicts or by resolving conflicts within his own tribe. If the tribe leader can organise the moment of peace well and make himself known as a mediator, his social dignity increases, therefore he can manage to consolidate his power (Fig. 13.5).

Leaders both reconcile tribe members in quarrels within the tribe and gain more power by taking part in the reconciliation of intertribal conflicts at the same time. As a matter of fact, intermediacy is a show of strength for Kurds, just like conflict and peace. For instance, the most influential person who takes part in many conflicts of all sizes in Mardin is Ahmet Türk. The fact that he is valued as an intermediary increases his political power, social dignity and prestige even more. Yet, what makes him the intermediary is ultimately this power that he has. One does not become an intermediary because one wants it; one can be the intermediary if one has (political, religious, or social) power.

In traditional societies, mediatorships are very important mechanisms for transforming long-lasting violence or weaponless conflicts into peace. In many traditional societies, the clan or tribal leaders, opinion leaders of the region, and governors (Özer

Fig. 13.5 At a peace dinner in Mardin, mediators and representatives of the two sides are holding hands, peace (in Turkish) is written on children's T-shirts, and they are flying a white dove, a symbol of peace. Those who are standing back are the guests who come to this event. *Source* Photographer Amar Kılıç (12 August 2018), who granted permission for its publication

1988) are common mediators. However, in Kurdish tribes the intermediaries who take part in conflict resolution vary. For example, marriage known as blood price marriage is one of the mediatorship forms that tribes have produced. This kind of marriage takes place between the children of two enemy families or tribes. If the injured tribe agrees, the family or tribe which has shot or seriously harmed someone from the other party marries the killer's sister to a male who is a close relative of the diseased. The woman who is forced into marriage cannot communicate with her family again; but she has been an efficient mediator in diminishing or putting an end to the hostility. In addition, political parties and organisations in which Kurdish people are active are other important forms of mediatorship. However, in such cases, based on the power of the party or organisation, the mediators can become dominant at times. In fact, in these situations, whatever that party or organisation says influences the decision.

13.3 Conclusion

Peace and conflict take up lots of time in the daily lives of the members of Kurdish tribes because these practices concern not only political and economic life but also greatly affect daily life. Tribe leaders have to consider the conflicts they are in as they build relationships with the State and other tribes. As a matter of fact, it is more accurate to do a reverse reading and state this: Kurdish tribes determine their relationships, attitudes, and behaviour according to the conflicts they are in. Specifically, they take into account enemy families and tribes as they make new friends and enemies, or when they embark on a political life, even when they vote. The conflict between tribes and the experiences which ensue demonstrate precisely this point.

Since conflicts within the tribe and intertribal conflicts and peacemaking affairs follow a local pattern, they may be called localised peace or conflict. Micro political affairs such as these local conflicts and peacemaking matters in tribes, which are traditional units, have a structure that can set an example for bigger peace efforts and conflicts which are macro political issues that concern the whole country. In particular, politico-symbolic activities such as the dinner of peace and mediation mechanisms which do not omit cultural traditions are two things which can serve as examples. Recognising and knowing the mechanisms of local reconciliation and conflict give clues for the permanence of macro-peace. As a matter of fact, technical procedures alone are insufficient for the maintenance of peace; there must also be the peace of cultures. In a possible solution to the Kurdish conflict in Turkey, the tips from traditional reconciliation mechanisms can be utilised. This is because the tribal model of organisation is still a dominant structure of Kurdish society. In fact, in these meso conflicts, violent conflict is transformed into peace, but the concerns of the parties to protect their forces and the very clear social hierarchy (between both men and women and the strong and the weak) are threatening social cohesion. After the conflict ends with a dinner of peace, the parties cannot apply physical violence to each other because moral and religious ties connect them. However, the desire to be strong enables them to continue symbolic violence. Therefore, the disappearence of these structures in social life occupies a very important place. Even if these structures are lost, the traditional conflict resolution and reconciliation methods which have developed in the culture of tribal type organisations will continue to exist. Although nowadays tribes cannot survive in rural areas, the mediation and reconciliation practices brought from there continue in almost the same form.

Contrary to popular belief, making use of traditional methods of peacemaking can be very effective in sustaining social peace. This is because while universal methods focus on primary disagreement issues, which can be considered technical issues, traditional methods approach peace as a cultural matter. Thus, the first places where peace is realised are not negotiation tables but symbolic places which are part of the social sphere.

References

Ateş Durç, Safiye, 2009: "Türkiye'de Aşiret ve Siyaset İlişkisi: Metinan Aşireti Örneği" ("The Relationship Between Tribes and Politics in Turkey: The Metinan Tribes Cases") (Masters dissertation, Ankara University, Social Sciences Institute).

Aydın, Suavi; Emiroğlu, Kudret; Özel, Oktay; Ünsal, Süha, 2019: *Mardin: Aşiret, Cemaat, Devlet* (Istanbul: Tarih Vakfı Yurt Yayınları).

Barth, Fredrik, 2001: *Güney Kürdistan'da Toplumsal Örgütlenmenin İlkeleri* [transl.: S.R. Şengül, H. Özsoy] (Istanbul: Avesta Yayınları).

Beşikçi, İsmail, 1992: *Doğuda Değişim ve Yapısal Sorunlar: Göçebe Arikan Aşireti* (Ankara: Doğan Kitabevi).

Boege, Volker, 2011: "Potential and Limits of Traditional Approaches in Peacebuilding", in: Austin, Beatrix; Fischer, Martina; Giessmann, Hans J. (Eds.): *Advancing Conflict Transformation. The Berghof Handbook II* (Opladen/Framington Hills: Barbara Budrich Publishers): 431–457.

Dost, Jan, 2012: *Adat u Rusumatnameé Ekradiye: Mela Mahmudé Beyazidi* (Istanbul: Nubihar).

Eickelmann, Dale F., 1989: *The Middle East: An Anthropological Approach* (Englewood Cliffs, New Jersey: Prentica Hall).

Galtung, Johan, 1990: "Cultural Violence", in: *Journal of Peace Research*, 27,3: 291–305.

Gökalp, Ziya, 1992: *Kürt Aşiretleri Hakkında Sosyolojik Tetkikler* (Istanbul: Sosyal Yayınları).

Isıfi, Abdul Karim, 2015: "Exploring Indigenous Approaches to Peacebuilding: The Case of Ubuntu in South Africa", in: *Peace Studies Journal*, 8,2: 59–72.

Karadeniz, Sıtkı, 2012: "*Aşiret Sisteminde Dönüşüm: Aşiretin Kentte Aldığı Yeni Biçimler, Batman Örneği*" (PhD dissertation, İnönü University, Social Sciences Institute).

Leach, Edmund, 2001: *Rewanduz Kürtleri: Toplumsal ve İktisadi Örgütlenme* [transl.: S. Birwan, H. Özsoy and S.R. Şengül] (Istanbul: Aram Yayıncılık).

Liden, Kristoffer, 2009: "Building Peace between Global and Local Politics: The Cosmopolitical Ethics of Liberal Peacebuilding", in: *International Peacekeeping*, 15,5: 616–634.

Lindner, Rudi Paul, 1982: "What was a Nomadic Tribe?", in: *Comparative Studies in Society and History*, 24,4: 689–711.

Mac Ginty, Roger, 2008a: "Indigenous Peace-Making Versus the Liberal Peace", in: *Cooperation and Conflict: Journal of the Nordic International Studies Association*, 43,2: 139–163.

Mac Ginty, Roger, 2008b: "Traditional and Indigenous Approaches to Peacemaking", in: Darby J.; Mac Ginty, R. (Eds.): *Contemporary Peacemaking: Conflict, Peace Processes and Post-War Reconstruction* (London: Palgrave Macmillian): 120–130.

Mac Ginty, Roger, 2010: "Gilding and Lily? International Support for Indigenous and Traditional Peacebuilding", in: Richmand, Oliver P. (Ed.), *Palgrave Advances in Peacebuilding* (London: Palgrave Macmillian): 347–366.

Olowu, Dejo, 2018: "Indigenous Approaches to Conflict Resolution in Africa: A Study of the Barolong People of the North-West Province, South Africa", in: *Journal of Law and Judicial System*, 1,1: 10–16.

Özer, Ahmet, 1988: "*Brukan Aşiretinin Sosyo Kültürel ve Ekonomik Yapısı*" (Masters thesis, Hacettepe University, Social Sciences Institute).

Özer, Ahmet, 2003: *Doğu'da Aşiret Düzeni ve Brukanlar* (Ankara: Elips Kitap).

Sarı, Engin, 2010: *Kültür, Kimlik, Politika: Mardin'de Kültürlerarasılık* (Istanbul: İletişim Yayınları).

Şerefxan, 2006: *Şerefname: Kürt Tarihi* [transl. M. Emin Bozarslan] (Istanbul: Deng Publications).

Taş, Latif 2014: *Legal Pluralism in Action: Dispute Resolution and the Kurdish Peace Committee* (London & New York: Routledge).

Tutu, Desmond, 1999: *No Future without Forgiveness* (London: Rider).

Türkdoğan, Orhan, 2006: *Doğu ve Güneydoğu: Kabile-Aşiret Yapısı* (Istanbul: IQ Yayıncılık).

Van Bruinessen, Martin, 1992: *Agha, Shaika and State: The Social and Political Structures of Kurdistan* (London: Zed Books Ltd).

Van Bruinnessen, Martin, 1999: "The Nature and Uses of Violence in the Kurdish Conflict", paper for the international colloquium, *Ethnic Construction and Political Violence*, Fondazione Giangianomo Feltrinelli, Cortona, Italy, 2–3 July.

Vanlı, İsmet Şerif, 1997: *Batılı Seyyahların Gözüyle Kürtler ve Kürdistan* (Istanbul: Avesta).

Waldman, Matt, 2008: *Community Peacebuilding in Afghanistan: The Case for a National Strategy* (Oxford: Oxfam International Research Report).

Yalçın-Heckmann, Lale, 2002: *Kürtlerde Aşiret ve Akrabalık İlişkileri* (Istanbul: İletişim).

Yıldırım, Ayşe, 2006; "İki Aşiret Arasında Kültürel Rekabet ve Çatışma", in: Suvari, Çakır Ceyhan; Yıldırım, Ayşe; Bozkurt, Tülin; İşoğlu, İlker M. (Eds.): *Anadolu'dan Etnik Manzaralar: Artakalanlar* (Istanbul: E. Publications).

Other Literature

"50 Yıllık kan davası bitti!", in: *Milliyet*, 28 July 1998; at: http://www.milliyet.com.tr/the-others/50-yillik-kan-davasi-bitti-5350948 (18 December 2019).

Aslan, Ferit; Bulut, Bayram; Karakaş, Emrullah, 2010: "Ahmet Türk'ten 15 bin kişilik barış yemeği", in: *Hürriyet*, 10 October; at: http://www.hurriyet.com.tr/gundem/ahmet-turkten-15-bin-kisilik-baris-yemegi-16007135 (12 September 2018).

"Barış Yemeği", in: *Hürriyet*, 18 September 1998; at: http://www.hurriyet.com.tr/gundem/baris-yemegi-39038828 (19 December 2018).

Güneş, Nezir, 2019: "Kırmızı ışıkta bekleyen araca silahlı saldırı: 1 ölü 4 yaralı", in: *Hürriyet*, 22 February; at: http://www.hurriyet.com.tr/yerel-haberler/mardin/kirmizi-isikta-bekleyen-araca-silahli-saldiri-41126682 (20 August 2018).

"Iraq: Blood Feud Tradition in Kurdistan Region", in: *Country of Origin Information Portal*, November 2017.

Kamer, Hatice: "Mardinde kavgalı aileler 5000 kişinin katıldığı yemekle barıştı", 14 August 2018; at: https://www.bbc.com/turkce/haberler-turkiye-45171916 (19 December 2019).

"Kurdish Families – Kurdish Marriage Patterns", in: J. Rank: *Marriage and Family Encyclopaedias*; at: http://family.jrank.org/pages/1026/Kurdish-Families-Kurdish-Marriage-Patterns.html (28 November 2018).

"Mardin'de 28 yıldır husumetli aileler barıştırıldı", in: *Habertürk*, 29 September 2019; at: https://www.haberturk.com/mardin-haberleri/72220285-mardinde-28-yildir-husumetli-asiretler-baristirildi (18 December 2019).

"Mardin'de 30 yıllık husumet barış yemeği ile sona erdi", in: *Artukluhaber,* 29 September 2019.

"Mardin'de barış yemeğine 21 bin kişi katıldı", in: Internet Haber, 10 October 2010; at: https://www.internethaber.com/mardinde-baris-yemegine-21-bin-kisi-katildi-300554h.htm (18 December 2019).

"Tribal Conflict Resolution in Iraq", 2018, in: *United Nations High Commissioner for Refugees* (Geneva: UNHCR).

United Kingdom: Home Office: "Country Policy and Information Note – Iraq: Blood Feuds", 10 August 2017; at: https://www.refworld.org/docid/59ae92424.html (27 January 2019).

Yavuz, Ramazan; Avuka, Adnan; Sunar, Serdar, Karakaş, Emrullah, 2010: "Mardin'de 5 bin kişilik barış töreni", in: *Milliyet*, 29 September; at: http://www.milliyet.com.tr/mardin-de-5-bin-kisilik-baris-toreni-gundem-1295256/ (22 August 2018).

Chapter 14
Family Shame and Eloping Couples: A Hindustani Warp in Time. Steps in Progress Towards Non-violence

David Evans

Abstract This chapter presents a qualitative study of social emergence in relation to forbidden marriage and its consequences (*Romeo and Juliet* style romance tragedy), predominantly in India and Nepal but also Pakistan. Despite Indian citizenship incorporating the right to free choice of marital partner, ongoing violations of this right have attracted worldwide publicity since 1993. Media awareness and increased presentation of outdated, patriarchal social rules and taboos around marriage, and their tragic consequences, has spawned a human rights movement, led mostly by women, which calls for legislative change to implement capital punishment for those murdering or persecuting couples in contentious marriages, and the provision of safe houses for eloping couples. The researcher uses philosophical value analysis to examine media reports, personal communications and Hindustani love legends to scrutinise local and national cultural obstacles to making these changes, and to establish evidence of social progress from lethal violence to non-violence, and finally to positive acceptance in family and community responses to love and marriage. The researcher concludes that within the communities of India, Nepal and Pakistan, enforceable, non-patriarchal law and order related to non-violence and human rights, and a shift in sociocultural thinking and behaviour rather than unquestioning adherence to centuries-old marriage rules, are needed to overcome ongoing *Romeo and Juliet* style romance tragedies.

Keywords Honour killings · India · Nepal · Pakistan · Romance tragedy · Social emergence

14.1 Introduction

This chapter is an updating of material from the researcher's PhD thesis entitled *Romance Tragedy of 'Romeo and Juliet' Style: A Study in Qualitative Research with*

Dr David Evans, Med. Dr, PhD, University of New England, Armidale, Australia; Email: ddeevans1@ gmail.com.

Ú. Oswald Spring and H. G. Brauch (eds.), *Decolonising Conflicts, Security, Peace, Gender, Environment and Development in the Anthropocene*, The Anthropocene: Politik—Economics—Society—Science 30,
https://doi.org/10.1007/978-3-030-62316-6_14

Special Focus on India and Nepal in the Time Period 1993–2013 (Evans 2014) – a title used to emphasise the phenomenon of forbidden love and its consequences that can be found in almost every culture with a recorded history.[1] The researcher critically analyses an extraordinary sequence of events involving local communities, the general public and the Government that began in 1993 when global media reported the illegal murder of young lovers in northern India. This report led the researcher to examine many media reports and personal communications from trusted friends in India and Nepal over a ten-year time period, and Hindustani legends to seek answers to two questions:

Why would Indian parents kill young people who have fallen in love, and make a public showing so their community can see what they have done? With the adoption of a Western-style democratic constitution and a police force available, why does the murder of newly married couples still happen in the present era?

The resultant findings led him to hypothesise about what progress looks like in relation to attitudes and reactions to 'forbidden' marriage, and to consider what the steps forward might be in social progress from violence to non-violence in relation to marriage for love. In this chapter, based on the thesis findings and updated media reports and personal communications (2014–2019), the researcher highlights generational steps in social change in India and Nepal, from violence to separation, from reconciliation to acceptance, and finally to celebrating marriage on the basis of love, not tradition. However, most reports on this subject focus on honour killings and related criminal acts that still go unpunished despite laws intended to gain justice for wronged couples, and evidence of personal change within families in favour of supporting their children in choosing their own marital partners.

This introduction provides a brief background to the patriarchal social, cultural, community, legal and non-legal powers at play that perpetuate honour killing to this day. It is followed by a statement of the study's aims, objective, methodology and conceptual framework, examples of media reports, personal communications and legends illustrating the ingrained practice of honour killing, with a few focusing on peaceful resolution of 'shameful' marriage and/or elopement, a discussion of this information with a view to seeking a path of progress towards non-violent family reconciliation, and a conclusion in which the researcher exhorts us all to continue to tell these stories to open the eyes of the world to these ongoing injustices as a step towards progressive social change, enforced and just laws, and an end to honour killings worldwide.

14.1.1 Background

Lethal violence towards lovers is criminal in Western countries and has also been illegal in India since 1950, when the country adopted a democratic constitution

[1] At: https://rune.une.edu.au/web/handle/1959.11/16964.

(Article 21). Prior to 1993, village authorities (*Khap Panchayats*) were free to administer their own laws relating to murders ('honour killings') arising from forbidden romances. However, these authorities took no action against the perpetrators. Even today, many *Khap Panchayats* still disregard national laws, and the phenomena of honour killing and lack of punishment for perpetrators continue to this day. For example, when a young Indian couple married for love in August 2019, the bride's father hired assassins to kill her husband.[2]

Family shame accounts for much of the killing. Families place great importance on social status, and breaking rules leads to public shame, with the added fear of social ostracism. Even as cases are investigated, there can be real doubt as to who performed the murder. The case of Sunita and Jasbir from Balla is a succinct example of this.[3] Sunita's father, Om Prakash, confessed to murdering his pregnant daughter and her boyfriend. An uncle and two cousins were among four others arrested. In Balla, many people believed the father had confessed to emphasise his support for his daughter's killing; to satisfy honour and to protect the real culprits among his family or village members.

Even after leaving home, many eloping couples have a desire to return to the family to receive their blessing. Individuals in good standing within the family may expect that it will all work out. If they are told that everything is all right at home and they return, they may be murdered. Either one or both families may be involved in the public denouncement and killing. Such deception and betrayal by family members is common. Young people can neither trust their own families nor rely on the authorities for help.[4] While safe houses are being established for runaway couples, from which they will have to re-emerge eventually and find a new life, many will have nowhere safe to go.[5] Some will choose suicide rather than renounce their relationship.

Media coverage of honour killings since 1993 has led to a community response that demands action by authorities including the *Khap Panchayats*, and has resulted in the conviction of some perpetrators. Reports have increased to include stories from Pakistan. In addition, repeating tragedies continue worldwide. It is reported from Australia and elsewhere that families continue to send young women overseas to meet other family members and face a forced marriage. In the United Kingdom, this has led to the formation of the 'Love Commandos', who attempt to rescue such

---

[2]See at: https://www.washingtonpost.com/world/asia_pacific/a-young-indian-couple-married-for-love-then-the-brides-father-hired-assassins/2019/08/19/3d1ce9a0-a1d0-11e9-a767-d7ab84aef3e9_story.html (18 December 2019).

[3]Reuters, 2008: "Indian Village Proud after Double 'Honor Killing'". (Simon Denyer, updated 16 May 2008); at: https://www.reuters.com/article/us-india-honourkilling/indian-village-proud-after-double-honor-killing-idUSDEL29449420080516 (11 September 2012).

[4]"Kin, Mob Assault Couple, Hubby Dies", in: *The Times of India*, 4 February 2006 [TNN]. See also: "Will Khap Defy the Constitution?" in: *The Times of India*, 13 April 2010.

[5]"Outsider Runaway Couples Find Sirsa Safe", in: *The Times of India*, 7 June 2011 [Bhaskar Mukherjee, TNN], at: https://timesofindia.indiatimes.com/india/Outsider-runaway-couples-find-Sirsa-safe/articleshow/8755718.cms (13 September 2012).

women.[6] As Prem Chowdhry, in her 2007 treatise *Contentious Marriages, Eloping Couples: Gender, Caste, and Patriarchy in Northern India*, concludes:

> Caste and its perpetuation through the observance of caste endogamy in marriage has not yet come under the scanner of public debate to be roundly condemned. The need is to understand the political economy of marriage, overlaid by cultural and ideological norms, in order to counter it effectively.

An updated review of reports of honour killings in India since the researcher completed his thesis, undertaken to ensure this chapter's currency, revealed little change in the situation. For example, in December 2016, *Huffpost* pointed out that a 2006 Supreme Court judgement on honour killings stated: "There is nothing honourable in such killings, and in fact they are nothing but barbaric and shameful acts of murder committed by brutal, feudal-minded persons who deserve harsh punishment.[7]

Yet the same post claimed that many years after this statement there were 28 reported honour killings in 2014 and 251 killings in 2015 (an increase of 796% in one year). Although suggesting "this spike could be a result of more vigilant reporting of these crimes, which largely go unreported", it recognised that despite efforts to "create a separate law dealing with honour killings".

Progressing to May 2018, The Oxford Human Rights Hub reported the Indian Supreme Court's stand against honour crimes, re-emphasising their statement of 2006 relating to Writ Petition (crl.) 208, submitted in 2002, vindicating the legality of inter-caste marriage.[8] The Supreme Court ruled that the most parents who "do not approve of such inter-caste or inter-religious marriage can do is … cut off social relations with the son or daughter." The report details preventive steps, remedial measures and punitive measures for the "police protection of couples in danger from their family and community".

And in June 2018, Sandeep Gopalan, Professor of Law at Deakin University, Australia, drew attention to the ongoing patriarchal disempowerment of women in India. The ABC News Australia reported an article he had written highlighting the honour killing of lower caste man Kevin Joseph as a symptom of women's disempowerment in India; in this case the lack of a woman's right to choose her own marital partner.[9] Professor Gopalan, with a view to empowering women, expressed

[6]BBC News, 2012: "The 'Love Commandos' Protecting Young Indian Couples." [Anu Anand, New Delhi, 5 June 2012], at: https://www.bbc.com/news/magazine-18280997 (13 September 2012).

[7]"Honour Killings Reported in India have Increased by 796% in a Year – Uttar Pradesh Tops the List", in: *Huffpost*, 2016 [Indriani Bassu, 7 December 2016].

[8]Oxford Human Rights Hub, 2018: "The Indian Supreme Court Takes a Stand against Honour Killings" [Karan Tripathy, 23 May 2018], at: https://ohrh.law.ox.ac.uk/the-indian-supreme-court-takes-a-stand-against-honour-crimes/ (4 April 2019).

[9]ABC News, 2018: "Kevin Joseph's 'Honour Killing' a Symptom of Women's Disempowerment in Kerala, India.[Sandeep Gopalan, updated 10 June 2018], at: https://www.abc.net.au/news/2018-06-10/kevin-joseph-honour-killing-womens-rights/9840020 (4 April 2019).

the opinion that marriage registrars need to certify that brides have stated consent to marry "of their own free will" without the presence of their parents. He claimed:

… to legally enhance women's rights, there is need for a campaign to educate parents that their daughters are not property, and to empower women. The marriage registration process must specifically ask women whether they consented to the union of their own free will.

Three months after Gopalan's article, *The Times of India* claimed there had been more than 300 honour killings in the past three years and listed eight cases that occurred in 2018, adding:

> … the irony is that there exists no separate law to punish those found guilty of such murders, and prosecutions are usually among various sections of the Indian Penal Code for homicide and culpable homicide not amounting to murder.[10]

These updated reports reinforce the researcher's belief that India's present need is to create a criminal code for 'honour killings' (as has been done in Pakistan) so that actual incidence statistics can be assessed. Police reports of each honour killing need to include coroners' reports and documentation of a statement regarding the killing from the village *Khap Panchayat* where the killing occurred. Such statements, which must be made freely available to the public, may lead to assessment of what is needed for genuine social change.

Nevertheless, in many Indian communities, the 'shameful' conduct of rebellious young people marrying across cultural norms has now been accepted. The social emergence of human rights (see Sawyer in Trajkovski 2009: 3), especially for women, is a work in progress in which documentation by the media, federal law enforcement and social protest are playing essential roles.

While India and Pakistan may appear to be leading the way for Arab countries who have not yet legalised a woman's right to choose her own marital partner, until India creates a criminal code specifically for 'honour killings', the woman's right to choose seems tokenistic and honour killing perpetrators may not be charged with murder.

14.2 The Study's Aims and Objective

The first aim of the study was to seek out and review case reports documenting the reality of illegal persecution and homicide in relation to marriage and choice of marital partner. The second aim was to attempt to show a positive axiological attitudinal progression from violent intervention, to disownment, to negative acceptance, to reconciliation, and ideally to positive acceptance and celebration of marriage. The

[10]"Honour Killings: More than 300 Cases in Last Three Years", in: *The Times of India*, 22 September 2018, at: https://timesofindia.indiatimes.com/india/honour-killings-more-than-300-cases-in-last-three-years/articleshow/65908947.cms (4 April 2019).

major objective was to work towards proposing a path of progress from violence to non-violence based on an appropriate balance between freedom of choice, the acceptance or rejection of advice, and family and community considerations; an ethical evolution of community thought and practice in regard to relationships fraught with difficulty.

14.3 Research Methodology

The researcher used qualitative research methodology because the study fits with Denzin/Lincoln's (2005: 3) second stream of future research: "a line of value analysis within society and the world at large, relying on philosophical values acceptable to both writer and reader." Philosophical issues such as equality, universalism and pragmatism are relevant, as is the suggestion of a 'sacred' element to be acknowledged as researchers look at, and reflect on, a world they can only partially understand (Denzin/Lincoln 2000). The researcher acknowledges spiritual issues but regards these as beyond objective scrutiny or definition. Unresolvable questions remain about the relevance of nature and nurture: questions pertinent to the study when lovers reject family-centred socialisation in favour of inner compulsive drives. There are also questions of jurisdiction in relation to individual and social justice, including the basic question of whose opinion matters (see Fig. 14.1, *Methodological Framework*).

Faced with the challenge of not having experienced the social issues under examination, the researcher co-opted trusted personal contacts from India and Nepal to initiate a collection of real-life stories of gross personal and community tragedy related to intermarriage. These stories involve mainly the Hindu and Muslim religions. Using these perspectives, the researcher took up Shakespeare's challenge – "our toil shall strive to mend" (Massachusetts Institute of Technology 1993) – and looked to grounded theory to show positive axiological progression from murder to non-violence based on the interplay of significant social and personal factors.

Real life and mythological stories that exemplify *Romeo and Juliet* style tragedy (Moore 1950) make up the grounded theory's codes. The concepts that emerge are causal reasons for the tragedies and the participants' responses to their situations, while the categories consist of the stories' outcomes. The theory is the road of progress from violence to good relationships.

Factual and philosophical starting points for the study were the time period from 1993 to 2013, with a focus on northern India and Nepal (factual), and choosing to accept the *Romeo and Juliet* concept as a phenomenon in its own right without further analysis (philosophical). The researcher chose this approach in accordance with Maurice Merleau-Ponty's (1991) claim that phenomenology is a philosophy that identifies essences of human existence and does not expect to arrive at an understanding of these from any starting point other than that of their 'facticity'. (For the purposes of this chapter, the time frame has been updated to include media reports from 2014 to 2019, as per the examples in the introduction.) Romance tragedy

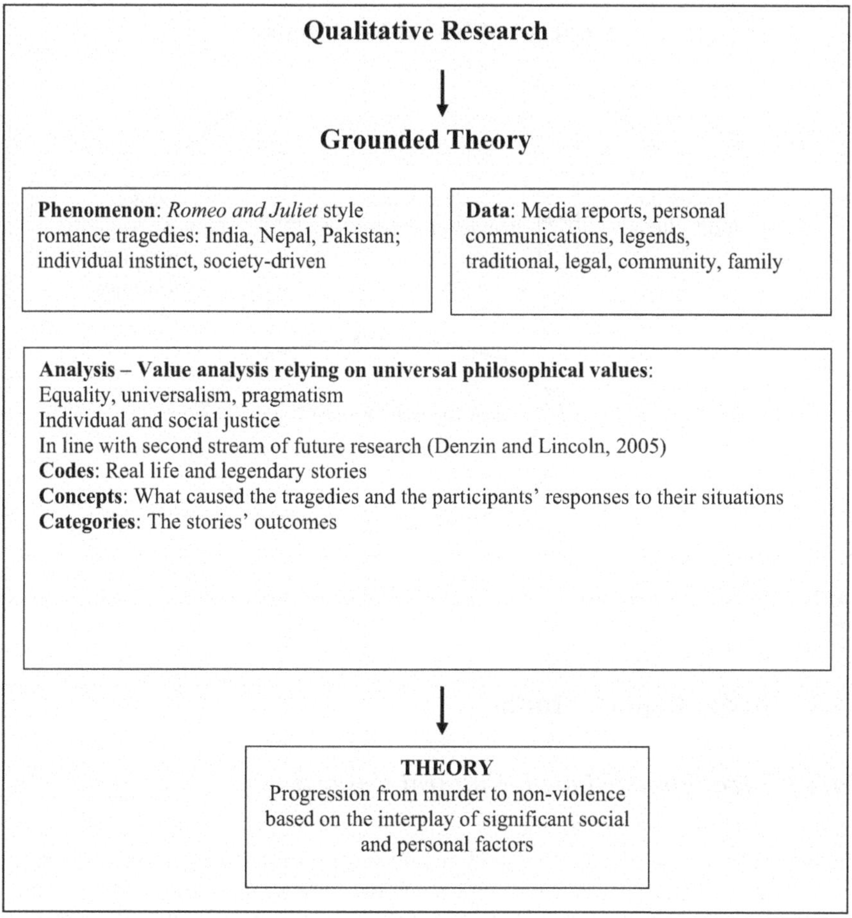

Fig. 14.1 Methodological Framework. *Source* The author

is presented as a two-sided phenomenon – basically instinctive but also society-orientated – which may or may not lead to tragedy, depending on the social milieu. Material for the study, which was sourced from media reports, Hindustani legends and personal communications, was selected because it fits within the study's conceptual framework (see Fig. 1.2) of social emergence arising from a "simultaneous focus on three levels of analysis: individuals, their interactional dynamics, and the socially emergent macro properties of the group" (Sawyer in Trajkovski et al. 2009: 3). This material is presented in the following sections to illustrate forces preventing or facilitating change (Fig. 14.2).

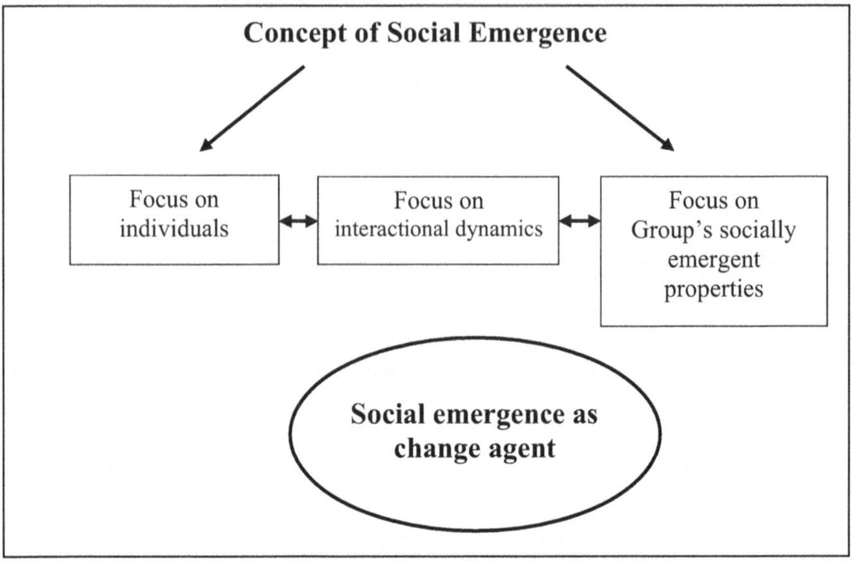

Fig. 14.2 Conceptual Framework. *Source* The author, developed from Sawyer (in Trajkovski et al. 2009: 3)

14.4 Media Reports: India

14.4.1 Traditional, Local, Cultural Context

Reported romance tragedies frequently come from rural northern India, especially from the state of Haryana. Most examples involve patrilineal families and communities which execute terrifying penalties for breaking accepted rules. Chowdhry (2007), in her treatise *Contentious Marriages and Eloping Couples,* emphasises the importance of village exogamy and caste endogamy in northern Indian villages, where land ownership and inheritance rights remain a prime consideration in the evolution of the rules of marriage. The *Hindu Succession Act 1956*, first amended in 2004 (*The Hindu Succession [Amendment] Bill 2004*) to give females equal property rights, emphasises that inheritance, especially of land, is paramount in choice of marital partner in the parts of rural India under consideration. Caste endogamy requires marriage within the same caste (eg, a Brahmin should marry a Brahmin). Village exogamy is more difficult to understand. It is like applying the Western brother-sister relationship to all peer group associates. School fellows and work associates from the same locality are considered consanguineous. Therefore, marriage between them is forbidden. The partner must be sought from a different village.

Traditional village leaders (*Khap Panchayats*) continue to hand down 'edicts' related to marriage to preserve traditional practices.[11] In many instances, they have responded to social change, which challenges their authority, by administering severe penalties (personal communication, Dilawar Chetsingh). These leaders link a 'brotherhood' of villages that work together in times of difficulty, such as disputes with neighbouring communities. They have a great bond with their land and will fight for it. Village society and social practices, including inheritance, have evolved to fit in with this type of society.

14.4.1.1 *Khap Panchayat* Edicts

Elders and village people usually accept annulment of marriage under a 'divorce or die' edict. All may be forgiven if the lovers renounce their commitment to each other and toe the line in the future. Gudiya and Mahesh Singh were faced with a 'divorce or die' edict from their village *Panchayat* but the couple refused to acknowledge it. Consequently, they were hacked to death and their body parts were burnt in a drain near their village.[12] Another couple, Amreen and Lokesh, who were given a 'divorce or die' edict in June 2009, chose the ultimate self-inflicted sacrifice of suicide in defiance of attempts to annul their inter-religious marriage. The BBC News reported their story of suicide in preference to earthly separation as India's *Romeo and Juliet* tragedy.[13]

'Expulsion from the village' is another commonly used edict, illustrated by the case of Ravinder Gehlawat and Shilpa Kadiyan in July 2009. Ravinder's entire clan was told to leave their native village of Dhrana because their son had married twenty-year-old Shilpa of Siwah village near Panipat. He was accused of violating a 500-year-old custom that the *Panchayat* felt was sacrosanct: "There can be no matrimony between the sects of Gehlawat and Kadiyan as they have a 'brotherhood' akin to consanguinity".[14] The Gehlawat family was given an ultimatum to leave the village within seventy-two hours. A penalty of Rs 21,000 was imposed on any villager who dared talk to them or help them.

In a similar situation in June 2000, the *Khap Panchayat* of village Jondhi determined that the marriage between Darshana, from the Gehlot got, and Ashish, from the Dagar got, should be annulled because it was against the rule of exogamy that prohibited intermarriage between members of these two gots (Chowdhry 2007). In

[11] See livemint, 2013: "Panchayats under the shadow of the 'khaps'". [Updated 1 May 2013]; at: https://www.livemint.com/Politics/nW03eT80k8lmHDfAUGh33J/Panchayats-under-the-shadow-of-the-khaps.html (4 April 2019).

[12] "Getting Away with Murder", in: *The Telegraph*, 6 February 2007 [Uddalak Mukherjee, Calcutta]; at: https://www.telegraphindia.com/opinion/getting-away-with-murder/cid/1027277 (5 July 2012).

[13] BBC News, 2009: "India: India's Romeo and Juliet Tragedy", 19 June 2009 [Sanjoy Majumder, Phaphunda]; at: https://news.bbc.co.uk/2/hi/8109805.stm (11 June 2012).

[14] "Haryana Panchayat Takes on Govt over Same-gotra Marriage", in: *The Indian Express*, 20 July 2009 [Dinker Vashisht Rohtak]; at: https://indianexpress.com/article/news-archive/web/haryana-panchayat-takes-on-govt-over-samegotra-marriage/ (4 April 2019).

this case, Darshana was subjected to an edict of 'forced divorce and forced remarriage'. Darshana's status reverted from that of a married woman to a daughter. Her father-in-law became her father and was responsible for finding her a new husband. Her son had to remain with her ex-husband Ashish, whose entire family had only one week to leave the village.

Attempting to retain social control in the face of modern technology, the *Khap Panchayat* in the village of Lank imposed "new restrictive rules for young women". In November 2010, Associated Press reported what appears to be an unsustainable edict – the banning of mobile phones for young women. Young men, however, could use mobile phones but only under parental supervision. The village leaders feared that young men and women were secretly calling one another to arrange to elope, breaking the rules of village exogamy and caste endogamy. The local women's rights group criticised the measure as backward and unfair,[15] but more recent reports indicate continuation of this practice. Reasons for the ban include mobile phones "distract [girls and unmarried women] from their studies"[16] and "cause disturbance in society".[17] The latter reason is due to village exogamy and caste endogamy rules being a root cause of violence towards eloping couples.

Ongoing reports of edicts such as this against girls and women show that *Khap Panchayats* remain powerful despite slow improvements regarding social freedom for women in India, as attested by Ki Awaaz (2012) in his article "Status of Women in India: From the 1950s To the 21st Century Youth".[18]

14.4.2 Federal Legal and Government Context: Progress Towards Non-violence

Caste endogamy, village exogamy and capital punishment are key issues on which the Indian Supreme Court rules. Legal challenges are now being made to family-orientated impositions on eloping couples. The following cases illustrate the steps being taken to address the issue of honour killing in India at a national legal level, but also highlight the strength of tradition and the slow pace of change.

---

[15]"Indian Village Bans Unmarried Women from Using Mobiles", *The Guardian*, 24 November 2010; at: https://www.theguardian.com/world/2010/nov/24/unmarried-women-mobile-ban (5 September 2012).

[16]Aljazeera, 2016: "India: Banning Women from Owning Mobile Phones" [News: International Women's Day, 26 February 2016]; at: https://www.aljazeera.com/news/2016/02/india-banning-women-owning-mobile-phones-160226120014162.html (18 April 2019).

[17]Independent, 2016: "Girls and Unmarried Women in India Banned from Using Mobile Phones to Prevent 'Disturbance in Society'" [Kayleigh Lewis, 22 February 2016]; at: https://www.independent.co.uk/news/world/asia/girls-and-unmarried-women-in-india-forbidden-from-using-mobile-phones-to-prevent-disturbance-in-a6888911.html (18 April 2019).

[18]See at: https://www.youthkiawaaz.com/2012/03/heres-how-the-status-of-women-has-changed-in-india-since-1950-till-date/.

14.4.2.1 Caste Endogamy Petition Upheld

A recent Supreme Court Judgement in the Case of Lata Singh (and Bramha Nand Gupta) illustrates the rule of caste endogamy. Federal law relating to inter-caste marriage states that there is no bar to an inter-caste marriage under the *Hindu Marriage Act* or any other law.[19] However, as illustrated in the preceding stories and the police inaction described later in this section, the law is still not fully implemented in accordance with the democratic Indian Constitution of 1950.

The court upheld Lata Singh's petition for protection from her brothers after she married Bramha Nand Gupta of her own free will. It also directed that criminal proceedings "shall be instituted forthwith by the concerned authorities against the petitioner's brothers and others involved in accordance with law". Although Gupta was from a different caste, Lata Singh, under federal law, had every right to choose whom she married because she was a 'major' at the time of her marriage. Her brothers had victimised members of Gupta's family, beating them and throwing them out of their house along with their belongings; holding Gupta prisoner and torturing him (withholding food and water for almost one week); destroying Gupta's crops and taking forcible possession of his field and a shop he owned; and lodging a false police report alleging that Gupta and his relatives had kidnapped their sister. This latter action resulted in the arrest and jailing of Gupta's sisters and one of his brothers-in-law. Lata Singh also alleged that her brothers were threatening to kill Gupta and his relatives, and kidnap and kill her. The judgement comments clearly indicate the need for legal change:

> We cannot see what offence was committed by the petitioner, her husband or her husband's relatives. The caste system is a curse on the nation and the sooner it is destroyed the better … inter-caste marriages are in the national interest as they will result in destroying the caste system.[20]

This case highlights the anger of the victimisers, which is likely to be multifactorial, especially if the eloping female has ignored the authority of the responsible male head of the family, bringing public shame to the family and questions about property inheritance. It also highlights the length of time taken for the judgement and the damage done in the interim. The incident and its continuing consequences began in the year 2000. Lata Singh's (and Bramha Nand Gupta's) writ petition for protection was submitted in 2002. The judgement in favour of the petition was not given until 2006. While there may be no available recourse for action in other instances, this example provides some hope of the law working as it should, albeit very slowly. However, there was no hope of family reconciliation in this case.

---

[19]"Lata Singh vs State of U.P. & Another on 7 July 2006", in: https://kanoonsangrah.com/lata-singh-v-state-of-u-p-another (4 May 2012).

[20]"Lata Singh vs State of U.P. & Another on 7 July 2006", in: *Indian Kanoon*, 2012 [Markandey Katju]; at:https://indiankanoon.org/doc/1364215/ (4 May 2012).

14.4.2.2 Legal Support for Village Exogamy Case

The case of Darshana and Ashish, whose marriage was unlawfully annulled by the village *Jondhi Panchayat* in 2000 (see Sect. 14.4.1.1), gives hope for social and legal justice. In 2004, Darshana and Ashish sought rehabilitation in Jondhi after favourable directives were received by the district administration from the Punjab and Haryana High Court. The Deputy Commissioner "said the administration was bound to rehabilitate the couple in the village and action could be taken against those who tried to derail the process", which the police superintendent would oversee.[21]

This appears to be an enormously progressive step considering that four years earlier police had stood by and allowed the *Panchayat* and villagers to act unlawfully without any repercussions. There had been no protection for Darshana and Ashish.

14.4.2.3 Capital Punishment Awarded and Commuted

Although human rights organisations may urge that the death penalty be abolished, the High Court in India recommends capital punishment for the 'rarest of rare' category of 'intolerable murder', which now includes honour killings.[22] However, the implementation of a capital punishment sentence for honour killing seems unlikely. While the first judgement of this kind happened in 2010, with more in 2011 and 2012, the 2010 capital punishment sentence for the 'community honour' murder of Manoj and Babli in Haryana has been commuted to life imprisonment.[23]

The murder of Manoj and Babli, who were considered siblings because both belonged to the Banwala gotra, a Jat community, has become a prime case for public scrutiny. According to *The Dharmastra*, any union between Manoj and Babli would be invalid and incestuous, despite them not being directly related. The ruling forced Babli to consume poison whilst four other family members strangled her husband with a rope in front of her. The brutality of the double murder shocked the court, resulting in the original capital punishment sentence. The fact that this ruling has been overturned leads to the expectation that other capital punishment sentences of this kind will be commuted as appeals are brought to court, with the likelihood that new judges will be more sympathetic to traditional values and actions. This illustrates how the path of progress in these matters does not run forward in a straight line, but dances around the edges of a fire in which competing flames of legal justice and ingrained, traditional, unlawful practices fight for oxygen.

[21] "Another Couple Seeks Relief from Khap", *The Tribune*, 5 November 2004 [Deepender, Jhajjar]; at: https://www.tribuneindia.com/2004/20041106/haryana.htm#6 (4 April 2019).

[22] Rediff.com, 2011: "S.C.: Award Death Penalty for Honor Killings".

[23] ABC News, 2010: "Five to be Executed for Honour Killings", updated 31 March 2010 [Sally Sara]; at: https://www.abc.net.au/news/2010-03-31/five-to-be-executed-for-honour-killings/386540 (18 April 2019). CBC News, 2012: "Honor Crimes the Scourge of South Asia", 30 January 2012 [Faiz Jamal]; at: https://www.cbc.ca/news/world/honour-crimes-the-scourge-of-south-asia-1.1159113 (5 September 2012). "Death Sentence Commuted in Manoj-Babli Case", *The Hindu*, 12 March 2011.

14.4.2.4 Political Complacency and Law Reform

The fact that the major political parties remain silent in relation to the Manoj-Babli case reflects the importance that parties attach to caste support and gotra (clan) politics, and does not bode well for reform. However, the Law Commission has raised questions in Parliament as to whether a new law for perpetrators of honour killings is necessary or whether perpetrators can be charged under laws related to murder.[24]

14.4.2.5 Local Police Protection

Media reports suggest that an increasing number of couples are obtaining police protection but also indicate that the protection is precarious. It may be impossible for the police to prevent kidnapping of, and violence towards, an eloping couple. One case in particular, however, gives cause for hope: the case of Mumtaz Khan and Mohsin Khan.[25]

This case is interesting for its counselling component and the awarding of protection by the police. Acting on Mumtaz and Mohsin's petition, the Punjab and Haryana High Court provided the couple with security cover. Mumtaz was apprehending threats from her family for marrying Mohsin against their wishes. Around forty youngsters and family members, including Mumtaz Khan's father, were present in court. In an unprecedented ruling, which took a huge step on the path of progress towards non-violence, the court asked the father and daughter to "talk it out" in the judge's chamber.

14.4.2.6 Local Police Inaction

Misplaced trust in rulings for police protection have put many lives in danger, as illustrated by the story of Ved Pal. Just as "the police stood by silently" during the public announcement of severe penalties to the affected families (Chowdhry 2004: 10) in the case of the village Jondhi incident referred to earlier, the police again stood by, unable to exert their authority whether they wished to or not, while Ved Pal was lynched by an angry mob as he went looking for his bride. He had trusted court protection, so, thinking that right was on his side and accompanied by police, he felt he would have sufficient support. Tragically for him, his trust was misplaced.[26]

---

[24] "Law Commission for Making Honor Killings Non-Bailable Offence", in: *The Indian Express*, 20 January 2012; at: https://indianexpress.com/article/news-archive/latest-news/law-commission-for-making-honour-killings-nonbailable-offence/ (4 April 2019).

[25] "Couple Gets Protection", in: *The Tribune*, 6 June 2011 [Tribune News Service, Chandigarh]; at: https://www.tribuneindia.com/2011/20110607/cth2.htm#7 (6 September 2012).

[26] "Honor killing: Ved Pal's In-laws among 12 Held Guilty", in: *The Times of India*, 25 September 2011 [Deepender, Deswal, TNN]; at: https://timesofindia.indiatimes.com/city/chandigarh/Honour-killing-Ved-Pals-in-laws-among-12-held-guilty/articleshow/10109897.cms (6 September 2012).

Yet again this is an incident that illustrates that court judgements and police protection mean little to members of communities steeped in traditional violence against 'wrong' marriages. And yet again, one step forwards and several steps back; another life lost and another ditch in the path of progress towards non-violence.

14.4.2.7 Community Pressure on Police

No story explains the obstacles to progress towards non-violence better than the report of the murders of Sunita and Jasbir from Balla.[27] The village elders (*Khap Panchayats*) remain committed to maintaining community honour and opposing threats to the established way of life in terms of religious affiliation, ownership of land, power and authority, thus preserving their public image. While democracy has brought new government laws to India, these laws threaten local authority. Democracy has raised awareness that change is coming and social backlash, viewed as a threat to village honour, may lead to an increase in restrictive punishments. The villages in which honour killings continue have social support for their injunctions. In 2008, for example, Sunita and Jasbir, from the same local community and therefore forbidden to become lovers or to marry each other, were murdered. Their bodies, half-stripped, were laid out on the dirt next to Sunita's father's house in Balla village as a warning to others who might try to emulate their 'immoral act' and as a sign that the family's honour had been restored. The village of Balla, just two hours' drive from India's capital New Delhi, was united behind this act. A farmer said, "We have removed the blot. We would not have had a face to show if we had not done this. It was the act of real men." Sunita's mother Roshni reportedly said, "Nobody would drink water in our house. … My daughter's action made us aliens in our own land. But we have managed to redeem our honour." At the small police post in Balla, a constable admitted the case was unlikely to reach prosecution, with the village putting enormous pressure on the police, and especially on Jasbir's family, to quietly drop the case.

This example of community power, while disheartening in the honour killing context and total disregard for national law, is the type of power that needs to be harnessed for positive social change.

Given that the culture of forced marriage, forced divorce and murder continues, well-known ancient Hindustani love legends depicting romance tragedies remain highly relevant, being played out every day in real life. These folkloric legends help explain the deep-seated traditions that underpin honour killings.

---

[27]Reuters, 2008: "Indian Village Proud after Double 'Honor Killing'", updated 16 May 2008 [Simon Denyer]; at: https://www.reuters.com/article/us-india-honourkilling/indian-village-proud-after-double-honor-killing-idUSDEL29449420080516 (11 September 2012).

14.5 Hindustani Love Legends

Hindustani love legends vividly illustrate the tragedy and depth of forbidden love. Four popular legends in the Punjabi tradition of India and Pakistan have a special place in a country of arranged marriages, where love may flower after marriage, but it is sometimes said that "love is a dream not to be experienced in this life" (personal communication 2012). These legends are *Heer Ranjha, Mirza Sahiban, Sohni Mahiwal* and *Sassi Punnun.*[28] Another popular legend, *Anarkali,* is possibly a true story from the lineage of Mughal Emperors.[29]

Even in modern-day cosmopolitan Australia, there are people who have not heard of Shakespeare and *Romeo and Juliet*, but if you ask an immigrant Indian taxi driver about these Punjabi romance tragedies, they usually respond immediately with a smile and ask how you know about them.

Given the media reports of romance tragedies preceding this section, it is apparent that the legends have a strong influence on romantic feelings and behaviour today. The researcher presents them here to illustrate the links between legend and everyday reality, played out over centuries.

14.5.1 Heer Ranjha

Heer Ranjha, written in 1766 by Waris Shah, is believed to be based on the true account of two sixteenth-century lovers, and it is thought that Waris Shah, in writing it, sublimated his unreturned love for a girl (Bhag Bhari).[30]

Heer Saleti was an extremely beautiful woman born into a wealthy Jatt family. Ranjha was a Jatt, too. Unlike his older brothers who toiled in the fields, Ranjha was his father's favourite son and led a life of ease, playing the flute. Ranjha left home after a quarrel with his brothers. He arrived in Heer's village and fell deeply in love with her at first sight. Heer offered Ranjha a job as caretaker of her father's cattle. Mesmerised by the way Ranjha played his flute, Heer eventually fell in love with him. They met each other secretly for many years until Heer's jealous Uncle Kaido and her parents, Chuchak and Malki, caught them. To save her family's honour, Heer's family and the local priest forced her to marry a man who lived in a distant village.

The heart-broken Ranjha decided to become a Jogi, piercing his ears and renouncing the material world. Reciting the name of the Lord, 'Alakh Niranjan', he travelled around the Punjab, eventually finding Heer. The lovers escaped and returned to Heer's village, where her parents agreed to their marriage. However, on the wedding day, Heer's Uncle Kaido arranged for one of his servants to lace some

---

[28] Punjabi World, 2007: "Love Legends in History of Punjab", updated 20 April 2007.

[29] "Anarkali's tomb to get back grandeur", in: *The Tribune*, 6 October 2008 [Afzal Khan, Islamabad]; at: https://www.tribuneindia.com/2008/20081006/world.htm#7 (18 April 2019).

[30] SikhiWiki Encyclomedia of the Sikhs, 2010: "Heer Ranjha"; at: https://www.sikhiwiki.org/index.php/Heer_Ranjha (26 July 26, 2012).

laddu (a sweet) Ranjha had sent to Heer with a deadly poison. Such was his jealousy that he would rather have her die than see her happily married to Ranjha. When told what Kaido had done, Ranjha rushed to save Heer but was too late; she breathed her last just as he arrived. Broken-hearted a second time, Ranjha finished the poisoned *laddu* and lay down to die beside Heer.

14.5.2 Mirza Sahiban

The *Mirza Sahiban* story (Ayesha Word Press 2006) reportedly started during the Mughal Emperor Shah Jehan's era when a child was born in the clan of Kharral Jatts in the region of Danabad, Nanakana Sahib. The child, popularly known as Mirza Jatt, became the "greatest legendary character among the race of his people known as Jatts". Mirza and his cousin Sahiban fell in love. Sahiban, about to be forcibly wed by her parents, sent Mirza a taunting message:

> You must come and decorate Sahiban's hand with the marriage henna. This is the time you have to protect your self-respect and love, keep your promises, and sacrifice your life for truth.

Mirza, a young full-blooded man and famed archer, lifted Sahiban up on to his horse and rode away with her, followed by her brothers. As Mirza lay under a tree to rest, Sahiban's brothers, their swords drawn, caught up with the lovers. Sahiban, a virtuous and beautiful soul, did not desire any bloodshed; she did not want her hands drenched in blood instead of henna. She thought Mirza could not miss his target and if he shot his arrows her brothers would surely die. Before waking Mirza, Sahiban broke his arrows and put his quiver away in the tree. She presumed that when her brothers saw her, they would forgive Mirza and accept him. Instead, the brothers killed Mirza. Sahiban, distraught, took a sword and killed herself.

People living in villages surrounding Sahiban's home town are reported to forbid their women to visit Mirza and Sahiban's mausoleum, fearing they may follow in those lovers' footsteps and elope, as the mausoleum caretaker's daughter did. On hearing the scandalous news, the caretaker died. His grave is near the mausoleum.[31]

[31] "They Hate Love Legend Mirza Sahiban", in: *Daily Times*, 8 May 2006 [Wasim Ahmad Qadri, pp. 7–27]. Originally available at: https://www.dailytimes.com.pk/default.asp?page=2006%5C05%5C08%5Cstory_8-5-2006_pg7_27 (29 August 2012). No longer available online.

14.5.3 Sohni Mahiwal

Sohni Mahiwal is the most famous work of poet Fazal Shah Sayyad (1827–1890) from Lahore.[32] This *Quissa* (non-religious traditional Persian-style prose narrative) tells the story of Sohni's anguish (*Encyclopaedia of Indian Literature* 1988). Sohni's tomb is in Shahdapur City, Sindh. According to the legend, the bodies of Sohni and Mahiwal were recovered from the river Indus near Shahdapur and buried there.

Sohni was the daughter of a potter named Tula, who lived in Punjab near the banks of the Chenab River. As soon as the Surahis (water pitchers) came off the potting wheels, she drew floral designs on them and transformed them into artistic masterpieces. Izzat Baig, a rich trader from Balakh Bukhara, came to Hindustan on business. When he saw the beautiful Sohni he was completely enchanted. He bought water pitchers and mugs every day to get a glimpse of her. Sohni lost her heart to Izzat Baig and started building love castles in her dreams. Izzat Baig sent his companions away and, calling himself 'Mahiwal', took a job as a servant in Tula's house. When people started spreading rumours about Sohni and Mahiwal's love, Sohni's parents arranged her marriage to another potter without her consent. Suddenly, a *barat* (marriage party) arrived at her house. Sohni was helpless. Her parents bundled her off in the *doli* (marriage cart) but they could not pack off her love.

Mahiwal renounced the world and started living like a *fakir* (hermit) in a small hut across the river. The earth of Sohni's land was like a *dargah* (shrine) for him. He had forgotten his land, his people and his world. Taking refuge in the darkness of night when the world was asleep, Sohni began going to the riverside. Mahiwal would swim across the river to meet her and regularly roast a fish to give her. Once, due to a high tide, it is said he could not catch a fish so he cut a piece of his thigh and roasted it instead. Seeing the bandage on Mahiwal's thigh, Sohni realised Mahiwal could not swim so she started swimming across the river with the help of an earthenware pitcher.

Rumours of their romantic meeting spread. One day, Sohni's sister-in-law followed her and saw the hiding place where Sohni kept her pitcher. The next day, her sister-in-law removed the hard-baked pitcher and replaced it with an unbaked one. When Sohni tried to cross the river with the help of the pitcher, it dissolved and she drowned. Mahiwal saw Sohni drowning. He tried to save her but he also died in the river. Sohni's courage is applauded in song by Punjabi women, which includes the words, "Sohni was drowned but her soul still swims in water."

---

[32]Folk Tales of Pakistan, 2007: "Sohni Mahiwal," in: *Folk Love Stories of Pakistan*; at: https://sites.google.com/site/folkstalesofpakistan/sohni-mahiwal(2 August 2012).

14.5.4 *Sassi Punnun*

There is evidence of Punnun's existence in the ruins of Punnun Fort near Turbat, believed to be between 6,000 and 8,000 years old. Much of the fort has disintegrated. Sufi saint Shah Abdul Latif Bhittai (1689–1752) preserved the story.[33] Countless folk songs sung by women narrate this love story even now.

Sassi was the daughter of the King of Bhambour (SCRIBD 2014). At birth, astrologers predicted she was a curse for the royal family's prestige and the Queen ordered the child be put in a wooden box and thrown in the river Indus. A washerman of the Bhambour village found the wooden box with the child inside. Having no child of his own and believing the child was a blessing from God, he took her home. As Sassi grew,.stories of her beauty reached Punnun, the handsome young Prince of Makran. He became desperate to meet her, so travelled to Bhambour and sent his clothes to her father for washing, just so he could catch a glimpse of her. They fell in love at first sight. Sassi's father was dispirited, hoping that Sassi would marry a washerman. However, he asked Punnun to prove he was worthy of Sassi by passing a test as a washerman. Punnun agreed to prove his love but tore the clothes and failed to fulfil the agreement. Before returning the clothes, he hid gold coins in the pockets, hoping this would keep the villagers quiet. His trick succeeded and Sassi's father agreed to the marriage, but Punnun's father and brothers were against it. His brothers threatened Punnun but he refused to relent. Pretending to enjoy the marriage celebrations, they enticed him to drink different types of wines. When he was intoxicated, they carried him to Makram on a camel.

The next morning, when Sassi realised she had been cheated, she became mad with the grief of separation and ran barefoot towards Makram, crossing miles of hazardous desert and calling Punnun's name. Sassi became thirsty and asked a shepherd for some water, which he gave her, but seeing her incredible beauty, he tried to force himself on her. Sassi ran away and prayed to God to hide her. God listened to her prayers and she found herself lost in a valley surrounded by mountains, where she died.

When Punnun woke and found himself in Makram, he began running back to Bhambour, calling "Sassi! Sassi!" He met the shepherd who told him the whole story. Punnun also prayed and died. He was buried in the same mountain valley as Sassi, where the legendary grave still exists.

14.5.5 *Anarkali*

The gravestone on *Anarkali's* tomb bears the inscription, "Could I behold the face of my beloved once more, I would thank God until the day of resurrection" (University

[33] "Hidden in Sassi and Pannu's Tale Lies Bhambore and Makran's Affair", in: *Pakistan Today*, 23 June 2011 [Bilal Farouqi]; at: https://www.pakistantoday.com.pk/2011/06/20/hidden-in-sassi-and-pannu%e2%80%99s-tale-lies-bhambore-and-makran%e2%80%99s-affair/ (23 October 2012).

of Alberta 2012). The legend has it that the Great Mughal Emperor Akbar (1556–1605) and his wife, Mariam-uz-Zamani, had a son, Prince Saleem (later Emperor Jahangir), who was sent away to the Army to learn the discipline required to rule the Empire (Wordpress 2016). The day of his return was one of great celebration. Akbar's harem decided to hold a great Mujra (dance performance), led by a beautiful girl named Nadeera, daughter of Noor Khan Argun. She was of exceptional beauty, 'like a blossoming flower', and a favourite of the Emperor, who called her *Anarkali* (blossoming pomegranate). Prince Saleem and Anarkali fell in love. They began seeing each other secretly. Saleem informed his father of his intention to marry Anarkali and make her the Empress. The problem was that *Anarkali*, despite her fame in Lahore as a dancer, was a maid and not of noble blood. Akbar banned Saleem from seeing Anarkali. Grave arguments followed and Akbar ordered Anarkali's arrest. He placed her in a jail dungeon in Lahore. The furious Saleem and one of his friends helped Anarkali escape and hid her near the outskirts of town. Saleem organised an army from those loyal to him during his fourteen years in Lahore and attacked the city. Akbar had a much larger army and quickly defeated Saleem's force, giving his son two choices: surrender Anarkali or face the death penalty. Prince Saleem, truly in love with Anarkali, chose the death penalty.

Anarkali, unwilling to allow Prince Saleem to die, came out of hiding and approached Akbar. She asked him if she could give up her life to save Saleem. After Akbar agreed, she asked for just one wish; to spend one night with Saleem. After her night with Saleem, Anarkali drugged him with pomegranate blossom, said a tearful goodbye and left the royal palace with Akbar's guards. They took her to the area near present-day Anarkali Bazaar in Lahore, where a large ditch had been made for her. Akbar's soldiers strapped her to a wooden board and lowered it into the ditch, then enclosed the top of the ditch with a brick wall, burying Anarkali alive.

These Hindustani legends make it clear that forbidden love equals death, but that lovers are willing to die for their chance of utmost happiness. Themes common in the media reports – murder, suicide, misplaced trust, family treachery and parents' right to choose who their children marry – echo the themes common to the legends, raising the question: is there any hope of change so that love equals life?

The researcher's visits to Nepal, a predominantly Hindu country with similar attitudes to inter-caste marriage as those in India and Pakistan, provided some hope for future change, as described in the next section.

14.6 Nepal – Gospel in a Nutshell

Like media reports from India, the *Kathmandu Post* report in February 2004 of a love affair between Jagadish Khadki, a Dalit youth of Bishanpur district, and Parbati Raut, a girl from the upper caste in the same *Village District Council* (VDC), which culminated in marriage on 31 December 2003, clearly illustrates the tensions created by inter-caste marriage. The couple were hurled into a quagmire that dragged twelve Dalit families into the mess with them, due to the ire of the upper caste people. The girl's relatives kidnapped the couple and two hundred upper caste villagers attacked the community of eighty Dalit people, taking and destroying their property, and forcing them to leave the village. However, a small ray of hope for change appeared when the Asian Human Rights Commission (AHRC) took up their case (SCRIBD 2014).

Upholding family honour remains paramount in the next story, but it is balanced with genuine parental sympathy for their daughter's choice. Forbidden romance, elopement, marriage, estrangement... all the hallmarks of a tragedy unfolding, ended with reconciliation after two years and the birth of a baby boy. The bride's parents told the researcher their story in a way that implied estrangement was a necessary imposition to demonstrate to family and friends how much they disapproved of their daughter's action. Inwardly it was a different story; a move towards acceptance.

Finally, illustrating some families' determination to create change by inviting newspaper reporters to an inter-caste wedding, comes a report from Nepal's national newspaper, *The Kathmandu Post* (2004):

> I didn't know that even the parents have started taking it normally, and, maybe the parents need more than enough examples of successful inter-caste marriages or maybe the people should themselves move ahead to change the face of society.[34]

14.7 Discussion

The preceding material clearly demonstrates the realities of ongoing persecution and homicide in relation to 'forbidden' marriage, and that only the stories from Nepal show the emergence of new but still family-based social mores in this regard, and examples of positive axiological attitudinal progression to positive acceptance and celebration of marriage for love.

Murder and suicide; forced divorce; misplaced trust; family treachery; jealousy; desperation; personal sacrifice; social disownment; total commitment to personal, community and religious honour; patriarchy; rejection of national laws; and lovers' determination to be together in life and/or death despite the sanctions imposed on them are played out many times over in this very small sample of the material available. Individuals' choices impact not only themselves but their families and communities; they are entangled in family, custom, culture, religion, law and community sanctions.

[34]*The Kathmandu Post*, 2004: "'New Faces': Invitation to an Inter-Caste Marriage" [Dilliram Khatiwada]: 67–68; at: www.ddeevans.org/assets/files/FREE%20Book%202.docx (Original online access, 4 September 2012, no longer available).

Table 14.1 Media reports 2006–May 2013

Year	No. reports
2013 (Jan–May)	4
2012	37
2011	3
2010	5
2009	1
2008	4
2007	3
2006	2

Source The author

The *anti-caste* blog site, which comments on caste, women's oppression, communalism and class struggle in South Asia from a Marxist perspective, gives an incomplete but highly useful list of media reports of horrific violence that has been threatened or carried out against inter-caste lovers in the years from 2006 to May 2013 (see Table 14.1).[35]

The likely interpretation of the data in Table 14.1 is that media reports were infrequent until 2012. As evidenced from the more recent reports from 2018 and 2019 presented earlier in this chapter, particularly the *Times of India* report from 2018, these atrocities continue and provoke continuing public protest.[36] At the time of writing his thesis, the researcher hoped that the statistics in Table 14.1 indicated a peak in 2012 and incidents would lessen in the future due to stronger application of federal laws. The lessening of the sentence for those responsible for the brutal murder of Manoj and Babli implies that the law is not yet ready to completely overturn the power of the *Khap Panchayats*.

Adherence to community law and the ongoing power of village elders (*Khap Panchayats*) to uphold their centuries-old traditions remains, and even when national law finds in favour of persecuted couples, there is no guarantee those couples, or indeed members of their families, will be safe.

The real-life story of Lata Singh and Bramha Nand Gupta, and the legal upholding of their petition against Lata Singh's brothers, challenges the patriarchal ideology of her brothers' guardianship and the authority of the *Khap Panchayats*, giving some hope for movement towards just outcomes in keeping with the fact of national law. The Hindustani legends symbolise the imperative of romantic love that is unobtainable due to family interference to maintain family honour. In *Heer Ranjha*, the apparent disparity of social standing might have been resolved happily had it not been for Heer's jealous uncle. Mirza and Sahiban's elopement might have been successful had Sahiban not faced the dilemma of loyalties to both her lover and her brothers.

---

[35] *Anti-caste*, 2013: "inter-caste couples, anti-caste".

[36] "Honour Killings: More than 300 Cases in Last Three Years", in: *The Times of India*, 22 September 2018.

Sohni and Mahiwal also died due to family interference in the form of Sohni's sister-in-law. It is not clear whether the sister-in-law was jealous or attempting to preserve the family's honour. The tragic legend of *Sassi Punnun* centres on family betrayal, the lovers' unwavering dedication to each other, and death as a means of regaining lost love. Finally, a Moghul prince wanting to marry a peasant girl (*Anarkali*) and make her Queen simply could not be allowed, the story again ending in death through devotion.

14.7.1 Challenging Community Values

The stories from Nepal point to the path of progress, with reconciliation and acceptance in place of violence and disownment. The work of this study was to look for, and present, a road of progress, and the Nepalese stories lead the way forward. Each time forbidden love challenges family and community traditions, a small step is taken towards change, albeit at the price of tragedy – often death. As challenges are made to family and community traditions, acknowledgement must be given to the sense of loyalty to tradition and intensity of emotion felt by responsible family heads and community leaders. The readiness for lovers to die and become martyrs rather than be separated leads to irreconcilable situations. However, the demand for better options comes from the young who, in Shakespeare's words, "will not endure" (Shakespeare 1980). In India today, there is growing general support for legal protection for eloping couples' safety, even though they may have to leave their home, family and community. The concept of honour is evolving, with a need to retain what is good and acceptable as globalisation of "rights to options" is fostered, albeit in certain circumstances at the sacrifice of "rights to roots" (Santos 2002: 26).

Danger to young lovers who challenge social boundaries remains real today. Community awareness led by the rebellious young, supportive families, media reports, human rights organisations and film-makers is leading the Indian government, through its legal system, to provide protection to impassioned lovers. Currently, the price for those without family support is social separation and the need to find a new life elsewhere. However, a 2012 study reported that in villages where more women held local council and/or chairperson positions, the incidence of reporting crimes against women rose dramatically.[37] This newfound confidence to report crimes came not only from women having female political representatives but also from "an increase in the responsiveness of the police under women representatives". This finding points to the need for more women in positions of political power across India to redress ongoing adherence to honour killing traditions.

Achieving reconciliation and acceptance remains an unrealised goal. Social progress related to romance tragedy in India and Nepal shows facets of the changing

[37] Ideas for India, 2013: "The Power of Women's Political Voice", 17 June 2013 [Lakshmi Iyer; Anandi Mani]; at: https://www.ideasforindia.in/topics/social-identity/the-power-of-womens-political-voice.html (18 April 2019).

balance between social regulation and social emancipation. A fair and just legal system can end violence and legalise options, but it cannot eliminate ostracism or enforce family and community honour. Heroic couples willing to sacrifice all for the sake of love are initiating the need for change, as exemplified in a recent report of increasing inter-caste marriages in Tamil Nadu.[38]

Despite most parents refusing to attend the weddings, the report states that some accept the situation later. However, effective change comes from leaders within the family or community who have the support of their family and social community.

Caste and its perpetuation through the observance of caste endogamy in marriage has not yet come under the scanner of public debate to be roundly condemned. The need is to understand the political economy of marriage, overlaid by cultural and ideological norms, in order to counter it effectively. Meanwhile, continued media reporting of honour killings, although problematic in its coverage, has helped draw attention to the shocking continued violation of constitutional and human rights (Chowdhry 2010).

14.8 Conclusion – A Path of Progress

This study provides some evidence that things are changing slowly. The imperative of love is emphasised by those who choose death rather than separation, becoming martyrs for their cause. This study has observed wider media attention to 'honour killings' since 1993, leading to public protest, the provision of safe houses and a call for implementing capital punishment. While this call was realised in 2010, it has not yet been implemented. There is hope that the violence against inter-caste lovers will lessen in the future. Those brave enough to influence social change have many tools at their fingertips (e.g. blog sites, social media) with which to inform global masses of the ongoing injustices in the name of family honour, and the ineffectiveness of national laws.

The researcher concludes that until a specific criminal code for honour killing is introduced to those countries that do not have one, and the law is implemented fully, the rate of progress away from violence will remain slow.

If 'peace' is the enjoyment of good relationships, non-violence must be the first important step. Effectively-practised federal law is imperative. Family estrangement remains a legal option but the path of progress, as Shakespeare might have desired, is from non-violence, to temporary separation, to reconciliation and acceptance, and then to celebration. Our responsibility is to tell the stories and spread the word.

[38]"Inter-caste Marriages on the Rise Despite Odds", in: *The Hindu*, 16 March 2016 [M.K. Ananth]; at: https://www.thehindu.com/news/cities/Coimbatore/intercaste-marriages-on-rise-despite-odds/article8358579.ece (18 April 2019).

References

Ayesha Word Press, 2006: "Mirza-Sahiban, Another Love Legend"; at: https://ayesha.wordpress.com/2006/05/08/mirza-sahiban-another-folk-legend/ (29 August 2012).

Chowdhry, Prem, 2004: "Caste Panchayats and the Policing of Marriage in Haryana: Enforcing Kinship and Territorial Exogamy", in: Gupta, Dipankar (Ed.): *Caste in Question: Identity or Hierarchy?* (New Delhi: Sage): 1–42.

Chowdhry, Prem, 2007: *Contentious Marriages, Eloping Couples: Gender, Caste, and Patriarchy in Northern India* (London: Oxford University Press).

Chowdhry, Prem, 2010: "Redeeming 'Honour' through Violence: Unravelling the Concept and its Application" (Cequin); at: https://cequinindia.org/wp-content/uploads/2018/01/Honour-killings-by-Prem-Choudhury.pdf (14 August 2012).

Datta, Amresh; Lal, Mohan (Eds.), 1988: *Encyclopaedia of Indian Literature,* vol. 2: *Sohni Mahiwal* (New Delhi: Sahitya Akademi): 1,266.

Denzin, Norman; Lincoln, Yvonna, 2000: *Handbook of Qualitative Research,* 2nd edn. (Thousand Oaks, California: Sage).

Denzin, Norman; Lincoln, Yvonna, 2005: *The SAGE Handbook of Qualitative Research,* 3rd edn. (Thousand Oaks, California: Sage).

Hindu Succession Act 1956: Act No. 30 of Year 1956, 17 June 1956; at: https://revenue.tripura.gov.in/sites/default/files/hindu-succession-act-1956.pdf (4 April 2019).

Massachusetts Institute of Technology, 1993: *The Complete Works of William Shakespeare: Romeo and Juliet. Entire Play in One Page, Act 1, Prologue,* Line 14; at: https://shakespeare.mit.edu/romeo_juliet/full.html (27 September 2013)

Merleau-Ponty, Maurice [1991], 2005: *Phenomenology of Perception* [translated by Colin Smith] (London: Routledge).

Moore, Olin H, 1950: *The Legend of Romeo and Juliet,* Ch. VI (Ohio: The Ohio State University Press): 43.

Santos, Boaventura de Sousa, 2002: *Toward a New Legal Common Sense: Law, Globalization, and Emancipation,* 2nd edn. (London: Butterworths LexisNexis).

Sawyer, R. Keith, 2009: "The Science of Social Emergence", in Trajkovski, Goran; Collins, Samuel G. (Eds): *Handbook of Research on Agent-Based Societies: Social and Cultural Interactions* (IGI Global), Ch. I: 3; at: https://www.igi-global.com/book/handbook-research-agent-based-societies/442#table-of-contents (18 April 2019); doi:https://doi.org/10.4018/978-1-60566-236-7.ch001.

SCRIBD, 2004: "Inter Caste Marriage: Problems Attached to Inter Caste Marriage and Ways to Handle" [uploaded by Annupriyanga]; at: https://www.scribd.com/doc/64946776/Inter-Caste-Marriage (4 April 2019).

SCRIBD, 2014: "Historical Love Stories – Sassi Punnu" [uploaded by Brij Singh, 18 May 2014], at: https://www.scribd.com/document/225426552/Historical-Love-Stories-Sassi-Punnu (4 April 2019). [Original 2012 reference no longer available.]

Shakespeare, William, 1980: *Twelfth Night,* Act II, Scene 3. Arden Shakespeare, 2nd series (London: Arden).

The Hindu Succession (Amendment) Bill 2004; at: https://164.100.47.5/rs/book2/reports/personnel/7threport.htm (18 April 2019). [Original 2013 online site no longer available.]

University of Alberta, n.d.: *Anarkali's Tomb.*

Wordpress, 2016: "Salim and Anarkali (Legendary Romance)", in: *Stories from All around the World*; at: https://mythologystories.wordpress.com/2013/12/16/salim-anarkali/ (4 April 2019).

Chapter 15
We Are not Victims: The Roma, An Outdoor Art Gallery and the Same Old Story – Critical Thinking in Communication for Development

Maria Subert

Abstract Based on three months of ethnographic research, this paper analyses the *Bódvalenke Fresco Village Project*, an open-air art gallery created to help villagers with Roma roots in a Hungarian settlement reaffirm their identity and develop tourism. Examining how the project (as help provider) and the locals (as help recipients) understand help similarly or differently, the author concludes that differences in concept, conduct, and action regarding social, political and economic empowerment have led to an intractable conflict between the Fresco Village Project and villagers. The emphasis is that helping disadvantaged Roma in Western societies should not repeat the basic error of developmental communication – telling a single story of poverty and misery while positioning helpers as *saviours* and recipients as *victims*. This concept of help becomes problematic when applied to Roma/Gypsies, who have lived within this society for hundreds of years. Ultimately, the author concludes that those who want to help Roma communities need to use a language that will reflect on their own help practices, one that calls for more advanced principles for analysing help and empowerment practices and offers critical metrics for doing so.

Keywords Effective help · Empowerment · Fresco village · Help · Help narrative · Hungarian roma · Roma · Roma development · Romungro gypsies

15.1 Introduction

Sustainable peace exists in a society "where the probability of using destructive conflict, oppression and violence to solve problems is so low that it does not enter into any party's strategy"; at the same time, "the probability of using co-operation, dialogue and collaborative problem-solving to promote social justice and well-being is so high that it governs social organisation and life" (Coleman 2012: 355). In this respect, providing 'help' to historically disadvantageous communities (such as the

Dr. Maria Subert, Assistant Professor, City University of New York, USA; *Email*: subert.ma@gmail.com.

Ú. Oswald Spring and H. G. Brauch (eds.), *Decolonising Conflicts, Security, Peace, Gender, Environment and Development in the Anthropocene*,
The Anthropocene: Politik—Economics—Society—Science 30,
https://doi.org/10.1007/978-3-030-62316-6_15

Roma) is also a peacebuilding mission; in its best form, 'helping' embraces co-operation, dialogue, collaborative problem-solving, is aware of social justice and communal well-being, and avoids conflict and all forms of oppression. However, we rarely see this version actualised.

In fact, hardly any other generation in modern history has communicated more than ours about helping and empowering the disadvantaged. With the end of welfare societies, the gap between the wealthy and the poor grew exponentially in Western societies. As NGOs, missionaries, and volunteers took over the responsibility from their governments to aid the growing masses in poverty, communicating about help became more regular and substantial. However, the growth of vulnerable masses in the West points to the uncomfortable truth that helpers, who often seek to demonstrate they are knowledgeable, socially, and advanced technically, and in particular, moral when it comes to helping the needy in the developing world, lack the skills to communicate with and provide help to the needy in their own societies (Warraq 2011). Therefore, we need to critically examine the rhetoric of help.

The concept of *helping the needy* is frequently applied to the Roma/Gypsy[1] (Timmer 2010). This study analyses the *Bódvalenke Fresco Village Project* (BFVP).[2] Created to aid a Roma/Gypsy village in Hungary, it functions as an open-air gallery of large murals painted by artists with Roma roots. Murals cover almost all the back walls of old Hungarian peasant houses or small houses built in the so-called *Gypsy row*[3] at the end of the village on plots that were useless to the peasant villagers (Csalog 1997). Besides managing the gallery, the BFVP collects and distributes donations and promises to create jobs.

The purpose of this chapter is to examine existing help narratives in the village to determine what kind of support people think is suitable for the community. Because the great power differential between the Roma and their frequently failing helpers makes standing up for themselves difficult, critically analysing help communication and calling attention to the controversies inherent in *help* and *empowerment* is vitally important to disadvantaged peoples and their communities – and other minorities, migrants, refugees, and impoverished majorities – in seeking aid and empowerment.

The BFVP was initiated by a Hungarian woman (called here the *Project Leader*), in the hope that murals by contemporary Roma artists would bring development to the Roma village. The Project Leader's goals were "to create something that our Roma peers could be proud of and that would give them the majorities' unconditional appreciation," and "to better the living conditions of this people" (Gábor 2001: 1). EU forums and the media praise the project as a good example of helping disadvantaged

---

[1]In this writing I use both names with the same non-pejorative intent. I differentiate between the two according to the word used in the original context.

[2]In Hungarian: *Bódvalenke Freskófalu* Project.

[3]In 1961, the Communist Party declared that the party had to end Gypsy poverty. In 1965 they started to demolish Gypsy ghettoes. Gypsies who worked full time could apply for subsidised loans for the 'CS' houses (houses with reduced value). 'CS' houses were typically built at the edge of the villages, so the segregation of the Gypsies continued, but housing conditions improved greatly. Csalog (1997) reports that in 1971, 75% of the Gypsies lived in Gypsy rows without water and electricity, yet most of them (85%) had steady jobs, and most of them were employed full-time.

Roma/Gypsies (Blake 2013; European Parliament's ASP5G-2 Conference 2011[4]; Field 2013). However, this chapter will show that the villagers question whether the BFVP accomplished this (Doros 2015).

I seek to answer two questions: "How do villagers narrate the BFVP's help?" and "How can we figure out what kind of help best fits with the Roma villagers' lifeway and lifestyles?" While many criticise the dominant narrative of paternalistic help, among them Marushiakova and Popov (2001), Raatzsch and Kállai (2006), and Rövid (2009), the forms of help that would be effective for the Roma communities are rarely discussed. Recommending how to avoid paternalistic help makes this study significant.

The term *community* is used here not only to signify a "geographically defined population sharing similar goals, values, and daily routines" but a more complex sense of belonging, where the role of the community is to provide justice, resolve conflicts, and help one another (Gal 2016: 290).

15.2 Contextual Background

In 2005, about 600,000–700,000 Roma lived in Hungary (Ringold et al. 2005). They are formerly nomadic people who arrived in Europe from eastern India in the fourteenth century.[5] Most Gypsies who immigrated to Austria-Hungary in this period were assimilated by the middle of the nineteenth century as Romungros (Hungarian Gypsies), but fully assimilating the Roma culturally and economically was unsuccessful (Marushiakova/Popov 2001; Subert 2016).[6]

Under socialism, the previously semi-sedentary Roma in this village were employed in the agricultural co-operative and in local heavy industry, and families moved into their own government-subsidised houses. However, with the fall of communism in 1989, people suddenly lost their jobs and remained unemployed (Subert 2019).

The village's population consists of one extended family of Romungro Gypsies.[7] In the village, there are about eighty Roma adults (thirty-eight couples) and approximately 105 children (two of which are non-Roma). Most of the villagers receive unemployment benefits or are employed in the community-service program (Bősz 2013; Laborczi 2012). While the official minimum wage per month in Hungary is 98,000 Ft (approx. 490 US$), in 2014, the villagers' estimated average monthly income was less than 8,000 Ft (about 40 US$) (Kovács 2014).

[4]European Parliament's ASP5G-2 Conference poster, 2011; at: https://www.socialistsanddemocrats.eu/events/transforming-reality-roma-people-films-photos-and-music-conference-which-will-take-place.

[5]Okely hypothesises that Gypsy culture might have an indigenous origin (Okely 1989; Stewart 2013).

[6]Vlach and Beas Gypsies, who arrived in Hungary at the end of the nineteenth century from Romania, ended up experiencing similar marginalisation.

[7]Romungro Gypsies make up 70% of the total Roma population in Hungary.

When the Project Leader contacted Roma civil rights leaders with her idea of the Roma outdoor gallery, they recommended that she work with Roma activists or social workers who knew the Roma's circumstances. It was then that the Project Leader started working with the head of the European Workshop on Culture and Community Education Society.

Through the European Workshop, an autodidact painter and social worker – to preserve his anonymity, I call him Painter G (PG) – and others were actively involved in finding a place for the BFVP project. The search included talking with the mayor and the villagers to discover their preferences. The gallery was the Project Leader's idea, and her role was to find sponsors. In summary, in the beginning three people were in leadership positions: the Project Leader was the fundraiser, the European Workshop's founder was the intellectual leader, and PG was the inventor of the technique, the expert in social work and the artistic director in the BFVP. PG was also active in painting five frescoes.

The online application process was started by the European Workshop and the first painters were selected by the Project Leader and PG together. As artistic director, PG suggested that the painters remain for a longer period of time (about thirty days) and recommended involving the village's grown-ups and children in the creation of the murals. As a general rule, before they started painting, the artists took time to talk with the villagers, sitting with them, playing the guitar and singing for them. So in the beginning, the BFVP was a community-building programme: the painters stayed in the village for weeks, made acquaintances and friends with the villagers, used real people as models and informants, painted images of village everyday life – animals, objects and events that had happened in the community – and used children as apprentices. This drastically changed in 2010 and 2012.

After having a disagreement with the leader of the European Workshop on how to spend the money that had been raised, in September 2010 the Project Leader took the BFVP to the Hungarian Reformed Church Aid Foundation, which agreed to sponsor the project. After having a quarrel about whether or not painting a Formula 1 car on one of the central walls would be a proper subject for the project, PG (the last Roma advisor) also left the project in 2012.

When the Project Leader separated herself from her Roma helpers, the initial concept of building the community was dropped and the concept of the outdoor gallery was emphasised and further developed. As manager of the BFVP, the Project Leader is curator of the gallery. She exclusively represents not only the frescoes but also the Bódvalenke people in various Hungarian and European Union forums.

The Project Leader has no expertise in Roma issues or social work, but she has good relationships in the world of finance, speaks English, and has the flexibility to travel from Budapest to Bódvalenke since she ran her own translation agency.

In 2016, she officially broke all ties with the Hungarian Reformed Church Aid Foundation. The last Dragon Festival (a cultural event in the village financed by various donors each year) was organised in 2017 with only few visitors. According to the Hungarian Reformed Church Charity Service's employee, in 2019 the Project

Leader withdrew from the village (though she did not officially resign). Disagreements with the community, her conflicts with the Roma grass-roots artists, and her ailing health led her to stop managing the project.

15.3 Theoretical Background

To make sense of and create a broader application for my critical inquiry, I used Harré's (2012) positioning theory, in which storyline, social positions, and actions/acts are interrelated. Drawing on different lived and communicated storylines, the same set of words or actions, such as 'help', may have very different meanings. The narrators' social position, and its power, affect what they say, do, and accomplish. Their position, in other words, has epistemological and moral consequences. However, Harré (2012) asserts, different agents in an interaction may position themselves and others in the interaction differently; this makes rights, duties and obligations a continually changing system. This is the meaning of local moral order (Barnes 2004). The theory lets us see help narratives as "shifting multiple relations in a community of practice" that inherently include issues of power (Linehan/McCarthy 2000: 441).

15.3.1 Field Site

Although Roma are minorities in Hungary, in this dead-end village in eastern Hungary they are a majority. Out of about 200 inhabitants, 95%nare Roma ("Bódvalenke Freskófalu," n. d.). I lived in a building housing the BFVP office. The building is in the centre of the village. I lived alone in a separate room that opened from the kitchen. I was self-supporting. I had no car. I spent my time either speaking with the villagers or just being present and listening. During my time there, I lived in similar material circumstances as the villagers, which means I ate once or twice a day only basic food such as milk, bread, butter, cucumbers, tomatoes, potatoes and apples. I spent all my time in the village.

The villagers speak Hungarian – they do not speak Romani any more – and dress like Hungarians; many of them have adopted a lifestyle similar to the Hungarians. Since I am of Hungarian origin, I had no language issues. I realised, however, that Bódvalenke people speak Hungarian with the vocabulary of the 1960s to 1970s. Their use of the language is closer to poetry than to everyday language. They use sounds that do not form words, but have exact meanings, like 'yay, yuuy, eee, huu, haa' which I try to preserve in the text. For example, here is a local woman, Ica's story:

> I don't have a father or mother. They died. Yaaa, when I was a child. When I came here, I was very young when I came here, twelve years old. I was a child. My husband he is from

> here, I am from Szendrő. I had nobody, so I stayed with him. He was twenty-two and I was twelve. I was a child yet. Haaa.

When Ica adopted her niece, she (Ica) had ten children of her own. As she notes:

> Haaa. I am older, I am her aunt. Haaa, her mom and dad put her out. She brought nothing, she came by herself [empty-handed]. Haaa, I gave her a place. At that time I had ten kids. I felt sorry for her.

But I cannot recreate the aura, the feeling – what it means, for example, to witness Lilla's speech: her authority, her energy, her emotions expressed with her large variety of inflections:

> We can mow the mountain. We can do this and that. And to do work for those who got money. This village is locked from the outside world. This place doesn't belong to the outside world. This is not a village. This is not a small settlement. This is a jungle.

This was most remarkable in the cases of the oldest generation and the poorest families.

15.3.2 Method

This study is the result of a larger research project[8] based on a three-month sample of ethnographic data collected between May and December 2014, plus three follow-up interviews in 2019. In 2014, I conducted seventy-eight semi-structured interviews with thirty-five Roma villagers out of eighty adults and with ten of the eighteen artists, engaging in informal conversation. Ultimately, I included only villagers in this analysis. In 2019, I interviewed three more individuals, two women (one is a social worker who works in the village) and one man.

To find my participants, I started talking with people living in the fresco gallery houses. I used convenience sampling and snowball sampling. Convenience sampling means talking with "basically everyone you can find who will cooperate" (Lindlof/Taylor 2011: 117). In snowball sampling, participants recruit other participants through a process of referrals (Warren 2002). For example, I went to Ica's house first and asked her about her 'picture' (mural). She told me her life story, including taking in Réka when Réka was twelve years old. I asked Ica to help me speak with Réka. Réka then told me her life story and mentioned that her two sisters also married into the village. I asked her to introduce me to her sisters. After a week, the villagers came to me to ask for food, to speak with me, or just to check on me.

I prepared an interview protocol that listed the key topics for the interview, adding questions or changing their order as necessary (Creswell 2009: 129). I structured the interview questions into five major groups. The first group asked the informants' age and length of residence in the village. The second group was open-ended: "Tell me

[8]The original work was the author's doctoral dissertation at Ohio University in 2015: *Storying Dreams, Habits and the Past: Contemporary Roma/Gypsy Narratives.*

something about yourself", "Tell me about your family", and "Tell me about your home". These questions allowed me to probe memories that people find important. The third group consisted of more socially orientated queries, like "Do you remember any changes in the village after the BFVP started?"; "Do you remember any conflict in the community?"; and "Can you tell me about the making of the frescoes?" These questions encouraged the respondents to recall stories of collective life and how the BFVP affected it.

The fourth group of questions was orientated towards understanding the villagers' experiences with the wall-paintings, such as "What do the frescoes mean to you?"; "What stories do the frescoes tell?"; and "What is the role of the frescoes in the village?" The fifth group closed the interview and allowed participants to ask me questions. This way the villagers understood that I was interested in hearing their story, as a person, a villager, and a community member concerned about the project, and that I would like to know their personal responses to the wall paintings and their relationship with the BFVP. They often equated the wall-paintings' presence with the BFVP.

According to Riessman's ()*Dialogic Performance Analysis* (DPA), the research interview must be dialogical and not dominated by the researcher's questions or by the structure of the interview. Investigators need to listen to the narrator in an "emotionally attentive" and "engaged" way to understand the narrator's experiences. I often had to minimise my words and change them to "uuu", "aha", "yuyy", "eee" vocalisations or be silent and listen while the villagers communicated in this way.

My interviews were frequently interrupted by their young children or other unexpected circumstances; so sometimes I conducted multiple interviews with the same person. I also engaged in informal conversations before, during, and after the semi-structured interviews. Therefore, while I used semi-structured interviews to obtain answers to my questions as much as possible, I also applied informal conversation techniques. I did not keep to this protocol with many people who came to me to speak freely. Also, I did not follow this protocol with the painters. I asked them to speak about their artwork and experiences with the BFVP. Here, I rarely refer to the painters because this article is focused on the villagers.

During my three-month stay, my research participation changed over time. My initial goal was to examine existing narratives and learn about the BFVP's success. Based on the media, I thought it was an exemplary model of helping the Roma (Blake 2013; Field 2013). After more people spoke about their struggle with the BFVP, I engaged in more informal conversations and spent more time with them without speaking to avoid the mistaken expectation many journalists brought to the village, viz., that they should chronicle the BFVP's success.

The difference between the BFVP's and the villagers' understanding of help emerged as dominant issue in my thematic analysis.

My 2019 interviews revealed some significant organisational change that might affect the project's future. To paraphrase my informants, the Project Leader withdrew from the village. Her duties were taken over by the Hungarian Reformed Church Aid Foundation's social services in 2016. She stopped her frequent visitations and gave up personal control over the fresco project in 2019.

15.3.3 Analysis

To understand the fresco project and its effect on the village, I used Riessman's ()*Dialogic Performance Analysis* (DPA) to analyze my data. As a method, DPA is a hybrid of thematic and structural analyses with special attention given to the dialogue between the narrative, the narrator, and their social, historical and cultural contexts.

Thematic analysis helped me identify emerging themes. As Riessman (2008) suggests, the weakness of this approach is that the macro context is highlighted while the local context is neglected. Therefore, I also conducted structural analysis. Structural analysis pays attention to the narrative form, the details of speech, linguistic cues, local meanings, composition, and organisation. All these add to the context and, therefore, the meaning (e.g., the form might be hypothetical narratives that express unfulfilled needs and make-believe events). Riessman notes that the weakness of this approach is that the macro context is often overlooked. However, I believe that selective use of both thematic and structural analysis with a dialogical approach balances the weaknesses.

DPA assumes that investigators need to listen to the narrator in an emotionally attentive and engaged way to understand the narrator's experiences. I was empathetic and reflective in every step of this research and careful to avoid any bias due to my non-Roma Hungarian origin by frequently asking interviewees to clarify their comments if I did not fully understand. I was dependent on the villagers' words, advice and corrections to double-check my understandings. During the analysis and writing phase, I frequently consulted Roma activists, scholars, artists and other citizens involved in the project.

I should also mention that, in order to protect my informants, I didn't use real names. I used upper case letters in alphabetical order instead of the artists' real names, so the first painter interviewed appears as *Painter A* (PA), and the second as *Painter B* (PB), etc.

15.3.4 Previous Research

Previous scholarship is divided on the concepts of *help* and *empowerment*: some understand help as patronage (pro-aid), while others critique it (anti-aid). To avoid the inherent bias of neoliberal economics – like an updated *laissez-faire* – and its concept of patronage, Aiello et al. (2013) advise using a communicative methodology; in it, including the Roma in enquiries about them would make their empowerment less biased. However, emphasising the construction of democratic knowledge does not prevent ideological biases (and hence ethical pitfalls) inherent in how we use concepts such as *empowerment* and *help* and how we position Roma in help narratives that recount their abilities or disabilities. Rövid (2009) speaks out against essentialising the Roma,[9] viz. reducing them to a single or unitary concept, such as victims, and

[9]Rövid (2009) describes the open stigmatization of an entire ethnic group.

denying them their full humanity, turning them into a caricature. In keeping with this, Marushiakova and Popov (2011) criticise patronage that creates and keeps a Roma help industry running. They call attention to the ethical questions that go unanswered in the dominant help practice.

15.3.5 Results and Contribution

As will be demonstrated in this text, the BFVP destroys an ethic or methods of self-help in the Roma community and denies villagers' their psychological and spiritual needs, including basic human rights. I demonstrate that the BFVP's help is unacceptable to the villagers and offer ways to figure out the community's concept of effective help. One unique contribution of this study to the greater body of research is pointing out that unacceptable help can be one result of the help process, providing means to find out what kind of help would fit such a community, and calling for more ethical help for the Roma.

I believe an analysis of the BFVP's failure to help the Roma villagers is morally and intellectually critical to our understanding of social justice and empowerment. We need to appreciate what constitutes effective help for this misunderstood people.

15.4 Discussion

15.4.1 The Roma Help Project's Help Narrative

The BFVP's agenda of help for Roma emphasises the villagers' material needs. Having no jobs and being tricked extensively by usurious lenders, many families have no food, sometimes for as long as two weeks. Villagers are open about their material needs. One of the village women, Réka, a mother of ten, recalls this period:

> [Those days] we … [would] cry more than … eat. And then I blame myself – I should blame the world, not myself – that we have so little income and we cannot keep the family alive and have food, drink and clothing.

Still, even though she does not deny a lack of resources, she says she doesn't like the way that the BFVP publicises the community's poverty in the media by exposing the depth of their needs. Instead, many locals compare the BFVP's asking for money for the villagers to badly executed begging. We can understand it as badly executed because, while begging, Roma women never subordinate themselves to the potential givers. They customise their communication to get what they want, but they never see themselves as lesser, as lower in the social hierarchy than their donors. Also, they do not share intimate information about themselves (about their starvation) that would place their children at risk or that would prompt child-protection agencies to take their children away (Tauber 2008).

Older members in the community take responsibility for managing extended families (Gal 2016). The villagers recall that they could manage their lives when they had jobs under socialism, when they were employed in heavy industry, the stone quarries, the forestry industry, agricultural co-operatives, and state-owned construction companies. Before socialism, they worked as herdsmen and collected mushrooms, berries, and flowers and sold them; earlier, they were semi-sedentary seasonal day labourers and musicians. But, as they say, they do not remember another time when, as they are now, they were collectively put in the spotlight as pitiable people unable to manage their own lives.

According to a village woman, Vali, people do not mind being helped by individual donors to acquire things, but they do not want to be dependent on the BFVP. Vali explains:

When people [visitors] came here … and they brought something, it felt much better when they gave [their donated] things to the people themselves, and not to the Project [BFVP] .

We can understand the main difference as follows: Their relationship with the BFVP thrusts them into an established hierarchy where the Roma are fixed in a subordinate position, while their relationship with individual donors does not (Piasere 2000; Tauber 2008). Vali continues:

> In my opinion it was better when the project was not here. Since even when they were not here, the gypsies were okay. Everybody was talking with everybody. But once the Project [BFVP] office opened, one-two people, three-four people were getting in, they cooked their soup [a *Gypsy meal*, often part of the tourist attraction], and there was always [a] big quarrel. … And if [in the future] there is no project office here, we might achieve something.

From the BFVP's point of view, the villagers were not very well off until the project arrived. Yet, as is clear from Vali's comment, she apparently believes that the BFVP destroys the brotherly and sisterly relationships with which they helped each other. She says they dislike the fact that they are never allowed to have access to resources available for the BFVP and that their relationships with the BFVP are governed by domination.

It became clear to me that the villagers are not ashamed of their simple life but just want to live with dignity. They expressed in multiple ways that they want to earn money other than through welfare or workfare, communicate directly with non-profit or governmental organisations without relying on somebody to mediate, and remain independent, maintaining their equality in their community. People expressed this as an unfulfilled wish that is the opposite of what they experience now. The BFVP had claimed ownership over something the villagers could not previously even imagine: the Gypsy houses – though only one wall – and through them, by implication, the villagers' biographies, struggles, stories and performances. People complain that their private, often intimate, stories are 'sold out' from under them to outsiders and the media but they don't benefit from it. The villagers suffer what it means to be called a Roma, while the BFVP makes this into a brand and invites a colonising gaze. The neo-colonialist or Eurocentric gaze sees the Roma either as 'savages' or the 'Oriental other'. The non-Roma's romanticising, exoticising and othering makes the Roma fresco village and villagers more marketable. This colonising gaze implies that the Roma people are essentially different from other humans.

There is a striking similarity between the Roma gallery and the Roma arts programme that the BFVP strives to establish as a brand and another event that Moghaddam and Ginsburg (2003: 247) describe: a case where an American graduate student brought back to the United States a plant from his trip to Ecuador, applied for a US patent, and gained an intellectual property right over the plant's medical effect. While the student had been legally positioned as the owner, the indigenous people who had used this plant for hundreds of years for the very same healing effect had been positioned as users, and 'their newly found duties [role] as *users* are increasingly (and unfairly) being enforced'. Likewise, the villagers, who had lived there for generations, had been positioned as users of the gallery and objectified as a background necessary to the murals.

With this premise, the gallery is a visual colonisation that attracts other types of colonisation to the village, such as objectifying the villagers' lifestyles – which still involved relating, celebrating, representing, and helping one another. As the first sign of rebellion, in December 2015, I witnessed that one family had built a huge fence around their house, so that from the street, neither the house nor the mural are visible.

Accordingly, while the BFVP is believed to be telling a new Roma story, it painfully repeats old practices that were demonstrated to be unsuccessful in helping Roma. Despite the large body of research that followed the early folklorists, the same one-way cultural influence returns that was criticised widely in the 1980s. This framework locates Gypsies in an inactive, inferior position, in which they only receive from their social and cultural surroundings and never give back (Görög ; Kríza). The BFVP believed that the villagers represented a culture of poverty; that the only source of Roma pride is art imported from the outside – with the Project's assistance; and that the Roma are fundamentally different from Hungarians.

15.4.2 The Roma Villager's Help Narrative

Evidently, if the community's help narratives conflict with the helper's, aid will be rejected. In the following section, I focus more on the villagers' need narratives. As we will see, these are constantly offset by pride.

The villagers reject the assumption that material needs – food and clothing – are more important than non-material needs – honour, equality and independence. Vali says that when the Project Leader or other non-Roma 'helpers' publicise their needs, they degrade and demoralise them by denying their capacity to fulfil their role as Gypsy women:

> The Project Leader says every bad thing about the village on the internet. That we always starve. That the children are starving. She does this not because she wants to hurt us. But it hurts that [she says] we starve. Because it is [something] that we don't have, but we go until we find something and then the kids have something. But she does this because then she gets support. But … we don't get anything from it.

For her, hunger is an intimate and private issue to be fully shared only with insiders. To avoid hunger, a few families collect mushrooms, and one of them weaves baskets to make extra money; others exchange with other family members wood that they collect in the forest for potatoes, flour, oil, and spices. Horváth (2006) describes exchange and begging as common practice for Roma women, whose responsibility is to feed their families even after they have run out of money. Horváth recalls: "If on these pennyless days the Roma woman says she has 'itchy feet', she thinks, 'Today I am going to walk a lot to get some food'" (Horváth 2006: 115). So food is something that women 'walk out' while they bargain in stores, beg for money or food from non-Roma, and visit relatives living nearby or in neighbouring settlements for food ingredients.

The villagers univocally expressed that they want to live independently and safeguard equality among themselves as much as possible. They need non-mediated access to resources and freedom to manage their lives independently. Among their needs, some villagers listed dynamic and empathic teachers who would to teach them useful skills, such as farming, livestock breeding,[10] and/or skills that enable them to enter the (global) job market. Accordingly, to find out what kind of help would better fit a community, we must pay attention to narratives of both material needs and non-material needs.

Even though these people have many needs, there are various things they are proud of. In the following excerpt, Lilla contrasts the Roma villagers' situation with the situation of "others", which signifies whites, Hungarians, peasants and the rich – in short, everybody who has never experienced life in a Roma community:

> We are better than others in that we are able to survive each day. … [O]ur kids are better [off… because they] experience what others do not experience, not even once in their lives. Because our kids cannot see the colour of bread, or the colour of meat, vitamins, medicines, like them. … And we cannot tell them that we don't have them. We must figure out what to eat. Then our men have to go for a piece of wood, [and so] they need to go and sell it. So we could give something to the children. They have to beg [others] to make them buy it. They have to beg very much. Very, very much. The others don't know this. They don't feel the heaviness of need that we have here. But they are not better than us, by no chance. They are not. We can feel what they can't.

In Lilla's passionate justification of the villagers' life, she explains that Roma pride in their empathy and solidarity demonstrates a shift in narratives of pride and need.

As the narratives show, Vali and Lilla are proud of their strong families, their loyalty, and their endurance of suffering; these fit into larger categories of cultural and spiritual pride. While Roma/Gypsies have adopted many of the social and cultural practices of their nation state, they have managed to save their own culture as well. This culture recognises that people must balance family and community and value a simple life based on cultural and spiritual principles.

Contrary to this, the BFVP assumed that there is little to be proud of in the village and that its sustenance and cultural significance, as with the Project's assistance, must

[10] In some neighbouring villages, mayors organise this with success.

be imported from the outside to create something the villagers could be proud of. The BFVP consequently overlooks the villagers' pride narratives, which balance the suffering of their individual and collective identities. The BFVP's agenda capitalises upon the Roma needs without realising the communal assignments (the community's self-sufficiency, self-support) they perform faultlessly and powerfully. Finally, the Project treats only material needs, overlooking the community's non-material needs. This degrades the notion of *need* to simple shortages of material resources and misery, which positions the Roma villagers at the bottom of the social hierarchy. Contrary to the BFVP's agenda, the villagers are very cautious about what and how they tell outsiders about their material needs, and they make a point of emphasising that they have the wisdom and strength to deal with these needs if they can live as equal citizens and balance family, community and their unsatisfied needs. This places them above others in their needs, not below. Even though, in this neo-colonial power distribution, Roma are stripped of their rights, as Vali and Lilla express, they regard themselves as morally superior in this situation.

Consequently, to listen to what kind of help would fit a community's needs, it is fundamental to consider pride and need narratives as dialectical counterparts that draw the boundary of a cultural and ethical community. Need and pride are not binaries; they are representative forms of saving dignity and assist the creation of balance and stability in the community. However, inattentive outside helpers tend to edit out pride narratives, since they are often emancipatory and reject neoliberal interests, including paternalistic help. In this village, pride narratives are juxtaposed to the BFVP's dominant need narrative to restore the dignity of the people. On the other hand, contrary to the BFVP's slogan about telling a different story, about the Roma searching for something they can be proud of, the BFVP refutes this pride instead of building on it.

In our case, the BFVP's dominant storyline is 'helping the Roma'. Yet the BFVP has failed to figure out who the Roma are, what their real needs are, and what would empower them. In the words of a local woman, Ica: "The Project Leader doesn't understand the language of the Gypsies. She hears what we say, but she doesn't understand us." While the BFVP clings to the colonising concept of patronage, this contradicts the villagers' non-material needs and pride that safeguard identity and cultural borders while also raising ethical issues. In the following, I discuss some questions connected to the Project Leader's attempt to stand up for the BFVP's approach to helping the Roma.

15.4.2.1 The Ethical Dimensions of Help

Who can speak for others is not only a communications but an ethical issue. Deleuze and Foucault (1977) oppose speaking for others because this speaking has epistemological and moral consequences, because it includes whether someone knows, or fails to know something, and whether that something is good or bad. Alkoff (1995: 99) holds that speaking for others is wrong because it does not disrupt (often reinforces) oppression and maintains discursive hierarchies; yet, she highlights that it is our political responsibility to speak out against oppression, which often involves speaking for others.[11] In this case we must be transparent and reflexive about our motivation, social location, and our privilege, accountability, and responsibility for what we say. Alkoff (1995: 111) realises that, most importantly, those involved in helping and empowerment – such as welfare agencies and institutions for development and aid – "are almost impossible to engage in dialogic encounters because under capitalism, dialogue is not advantageous enough for most development teams."

The fact that developers should expose development can result in a non-dialogical relationship with the people. Having a non-dialogical relationship with the villagers, the story told by the BFVP is fundamentally different from those of the villagers. Yet, since Roma empowerment funds require neither expertise in Roma issues nor the villagers' feedback, as time passes, the BFVP grows more powerful and the villagers more powerless. This not only makes the BFVP's help ineffective but renders it unethical (Harré/Moghaddam 2003).

Monitoring the social actors' positioning in this specific help project revealed two different models of help in the village – the Project Leader's individualistic and the Roma people's communitarian concepts. As previously discussed, the individualistic version places the helper in a higher fixed position, since he/she has the power to help vulnerable people who are allegedly unable to discharge their duties (Harré 2012).

Contrary to this, from a communitarian perspective, helpers care about the equilibrium of the community and carry out each other's duties if one of them becomes unable to perform a task. This requires us to see help as a series of interactions in which nobody is entirely powerful or powerless, in which the position of helper to helped can constantly change. In the communitarian understanding of help, duties and responsibilities are distributed among members of the community at the same level (Harré 2012).

It is valid to ask whether the helper should adapt to the cultural and ethical norms of the locals. Most scholars agree that different cultures develop different value systems (Chen/Starosta 2003; Hall/Hall 1990; Hofstede 2003; Kale 2003; Martin et al. 1998). Kale (2003) says empathy works best among people with similar cultural backgrounds and experiences. Others believe that there are some ethical values that people share globally, such as *common human traits* (Bennett 1979; Christians/Traber 1997; Condon 1981; Valde/Whedbee 2008; Kidder 1994; Loges/Kidder 1997). Kale (2003) chooses empathy as the principal common value, while according to Kidder (1994: 222), common human ethical values are "love, truthfulness, fairness, freedom, unity, tolerance, responsibility, and respect for life." Similar common moral values are described by Bok (2002: 226) as "positive duties" – mutual support, care, and loyalty. Remarkably, these positive duties and other universal ethical values, such as empathy, fairness, freedom, unity, tolerance, responsibility etc., are identical to the villagers' pride narratives – which, in the absence of dialogue, are overlooked by the BFVP.

---

[11] In terms of positioning theory, this kind of speaking is one of our duties that one owes another by reference to the other's vulnerabilities (Harré/Moggadam 2003).

Empowerment is referred to by many villagers as a *missing* outcome of the BFVP's presence. While many locals talked about this, I want to quote Gyöngyi's opinion as emblematic: "There are lots of people who decide about your life [but] you have no voice." Leo explains his version of *empowerment* as follows:

> Even if we dress in golden clothes, even then, we will be called *gypsy*. The letter *G* [signifying *gypsy*] is on their back. This could be improved only with more jobs, more money and more honor – but we do not get these.

He states this in a small dead-end village in Eastern Hungary, surrounded by a Roma help project infused with a donation of US$35,000 from a large Austrian Bank for the murals, US$29,000 from the state sub-region and a Hungarian ministry for connected projects, about US$30,000 from the charity organisation of the bank's former CEO for three years, and possible governmental support and EU grants. Yet Leo and his extended family still lack jobs, money, and honour.

One could argue that cultural events organised in the village (such as the Dragon Festival) have a community-development, strengthening function and balance the weaknesses rooted in the top-down 'help' provided by the BFVP. To evaluate this, we need to know the context.

At the beginning of the project, the Project Leader renamed the customary village day in Bódvalenke and started organising an event called the 'Dragon Festival'. Financed by various smaller donors each year, in 2012, 2013, 2015, and 2017, this had been the major event to promote the village to the outside world. In 2012 the event cost a million Ft (US$24,200) paid by the *Open Society Institute* (OSI), the Erste Foundation, and the National Cultural Fund.

The festival is organised exclusively by the Project Leader, who is in a strong competition with the mayor, who never participates in any Project-organised events. Villagers talk about the village festivals with disappointment. They feel that it is a waste of money to pay famous musicians if the villagers have unfulfilled basic needs. More wealthy villagers (typically the usurious lenders) host guests during these events, but the poor families are left out (Hernád 2015). Thus, these events do not improve the living conditions of the villagers but increase the inequality between them. In Vali's words, "one-two people, three-four people were getting in, they cooked their soup, and there was always [a] big quarrel. Even though we were brothers or sisters, we were angry with each other." A village man, Zalán, says that previously they were able to solve everyday problems, but now things have become rearranged and the desire for wealth, greed, and jealousy divides the extended family:

> One pulls here, the other pulls there. One is more jealous than the other. They would kill each other. Two gypsies, I mean, as if we were brother and sister and you pulled that way and I pulled this way. …Before it was different. Now Lenke [Bódvalenke] is pulled out so much, how can I say for you, as if it doesn't exist. Things in the past were more valuable than now. We lived better. I went to my brothers and everything was made. But now it is not like this. Now they kill each other.

This is why they want to be independent of the BFVP.

All villagers said that if money is available, it should be spent wisely. They said that the BFVP should spend money appropriately on things the community really needs. But since the village days are organised top-down, without the BFVP co-operating with the locals and local minority government, the villagers have no voice with which to change things connected to these festivals. Most of the artists who painted the murals don't participate.

Accordingly, the Dragon Festivals were never an ethnic/cultural festivity organised by the villagers; they were designed to promote a romanticised Roma image created by the majority, in which the Project Leader spoke for the villagers from a paternalistic position.

According to the Reformed Church Charity Service's employee, "after the Project Leader broke with the Hungarian Reformed Church Aid Foundation and withdrew from the village, there will be no Dragon Festivals" (the last post on BFVP's website was on 22 March 2016). Since the events had been organised exclusively by the Project Leader, in her absence the villagers do not plan to organise the festival in the future.

Community-building was an important concept at the beginning of the project in 2012–2013 but was later dropped after the Project Leader separated from her Roma helpers who started the programme with her. As one of the artists involved in this initial phase, PG said:

> The story later was only about money. It was not about building a community or about art. It was about rushing through something and picking up money…The project lost the salt, the flavour of the whole thing. I think something got lost.

Another painter, PA, recognises that the BFVP's move into the village was not preceded by careful and benevolent work:

> When I went down to Bódvalenke I asked myself, "Where is the place of these people?" And I have come to the realisation that we cannot force our ideas on to [them]. If we enter their life territory, we should first listen to them, deal with them. We should first wash up the dishes and set the table later. I am sure the villagers are not [just] irritated by the big sums of money flowing by and not being shared with them; they are irritated by the attitude those people show towards them. They want to be accepted. But this is not happening there.

PA critiques a system that makes this kind of 'help' a norm.

15.5 Conclusion

To answer the first question, "How do villagers narrate the BFVP's help?", I conclude that for the Roma people, the BFVP's assistance does not provide effective help. Drawing on different lived and communicated storylines, and acting from different positions, help has very different meanings, including different moral values for the villagers and the BFVP. However, the villagers have the agency to reject the BFVP's help – and they do.

The villagers narrate the BFVP's help as unacceptable (because ineffective) based on the following premises: the BFVP limits villagers to their ethnic/cultural identity, represents a romanticised Roma identity created by the majority, builds a need industry around them, and demonstrates a belief that Roma are passive, unable to manage their lives, and are dependent on their non-Roma surroundings. Furthermore, the BFVP focuses exclusively on material needs (food, clothes), overlooking non-material needs (equality, independence) and narratives of cultural pride that restore people's dignity. *The Project transforms the villagers into objects of aid instead of subjects of action*; *it claims ownership over the community, one wall of the Roma houses, and the villagers' everyday performances and stories.* The BFVP's help agenda invites a colonizing gaze to capitalise upon these people as exotic. It builds on the individualistic idea of help while ignoring the communitarian concept important to the villagers. Finally, the Project Leader speaks for the villagers, assuming the position of the powerful helper (allegedly) overseeing (but not really executing) critical responsibilities and universal ethical values such as empathy, fairness, freedom, unity, tolerance, responsibility etc. Avoiding this sort of paternalism would seem to be the first step towards providing effective help.

Paternalistic help overlooks the likelihood that various participants in the helper-helped interaction may position themselves and key others differently and see their rights, duties and obligations as part of a flexible, continually changing system (Harré 2012; Barnes 2004). They keep their right to conform to the local moral order and reject the BFVP's help.

To answer the second question, "How can we figure out what kind of help best fits the Roma villagers?", we need to know whether the helper should adapt to the ethical norms of the locals. In this, larger-than-local ethical principles – such as empathy, love, truthfulness, fairness, accountability – should guide us.

As this study has demonstrated, three steps are fundamental to finding out what kind of help will fit a community, that is, examining: (1) the major concepts of identity in the helper's narrative that reveal the helper's position and that of those helped; (2) narratives of material and non-material needs; and (3) narratives of pride and need as dialectical counterparts.

These points demonstrate that social actors' positioning in the help process determines social and moral locations and affects their understanding of and need for help. Helping the disadvantaged Roma in Western societies should not repeat the basic error of developmental communication – telling a single and simple story of poverty while positioning the helpers as *saviours* and local recipients as *victims*. This concept of help becomes problematic when applied to Roma/Gypsies, who have lived within this society for hundreds of years.

Therefore, to provide effective help, those who attempt to assist Roma communities in Western societies should assume responsibility for a more ethical posture towards those they are trying assist. They need to use a language that will critique their own help practices, one that calls for more advanced principles for providing and analysing local help and empowerment practices.

Funk and Said (2010:1) warn us that "dynamics associated with interventionism may replace one set of problems with another"; externally driven efforts often tend to leave locals "feeling marginalised and disempowered, and unable to fulfil aspirations for … sustainable transformations in the quality of life" (1). They identify "localising peace" as "a central challenge for the twenty-first century." I hope this chapter has shown that "localising help" to the Roma among us is a critical related challenge. Without localised help practices that build on co-operation, dialogue, collaborative problem-solving, and social justice, Roma cannot thrive.

References

Aiello, Emilia; Mondejar, Eduard; Pulido, Miguel A., 2013: "Communicative Methodology of Research and Recognition of The Roma People", in: *International Review of Qualitative Research,* 6,2: 254–265; doi: https://doi.org/10.1525/irqr.2013.6.2.254.

Alcoff, Linda, 1995: "The Problem of Speaking for Others," in: Roof, Judith; Wiegman, Robyn (Eds.): *Who Can Speak? Authority and Critical Identity* (Urbana – Chicago: University of Illinois Press): 97–119.

Barnes, Mary, 2004: "The Use of Positioning Theory in Studying Student Participation in Collaborative Learning Activities". Session presented at the Annual Meeting of the Australian Association for Research in Education, 28 November – 2 December 2004, Melbourne, Australia; at: https://www.aare.edu.au/data/publications/2004/bar04684.pdf.

Bennett, Milton, 1979: "Overcoming the Golden Rule: Sympathy and Empathy", in: Nimmo, Dan. (Ed.): Communication Yearbook 3 (Washington, DC: International Communication Association).

Blake, Isaac, 2013: "International Romani Day in Bodvalenke, Hungary, with Commissioner Andor, on 5 April", in: *World News* (5 April 2013); at: www.romaniarts.co.uk/tag/international-romani-day-in-bodvalenke/.

Bódvalenke Freskófalu: Hogy látva lássanak, (n. d.). *Bódvalenke Freskófalu Project.*

Bok, Sissela, 2002: *Common Values* (Columbia: University of Missouri Press).

Bősz, Anett, 2013: "Hogy Látva Lássanak – Éhező Csodatevők Borsodban [That They Should See – Starving Miracle Workers in Borsod]"; at: https://boszanett.blog.hu/2013/12/22/_hogy_latva_lassanak_ehezo_csodatevok_borsodban.

Csalog, Zsolt, 1997: *"A Cigány*kérdés Magyarországon 1980 Előtt [The Gypsy Question in Hungary Before 1980]", in: Vajda, Imre (Ed.): *Periférián: Roma Szociologiai Tanulmányok* (Tárnok, Hungary: Adriadne Alapitvány): 194–224.

Chen, Guo-Ming; Starosta, William J., 2003: "Asian Approaches to Human Communication: A Dialogue", in: *Intercultural Communication Studies*, 12: 1–15.

Christians, Clifford G.; Traber, Michael (Eds.), 1997: "Introduction", in: *Communication Ethics and Universal Values* (Thousand Oaks, CA: Sage).

Coleman, Peter T., 2012: "Conclusion: The Essence of Peace? Toward a Comprehensive and Parsimonious Model of Sustainable Peace", in: Coleman, Peter T.; Deutsch, Morton (Eds.): *Psychological Components of Sustainable Peace* (New York: Springer).

Condon, John, 1981: "Values and Ethics in Communication Across Cultures: Notes on the North American Case", in: *Communication*, 6,2: 255–266.

Creswell, John. 2009: *Research Design: Qualitative, quantitative and mixed method Approaches*, 3rd edn. (Thousand Oaks, CA: Sage).

Doros, Judit, 2015: "Bódvalenke, Végállomás [Bódvalenke, Final Station]": in: *Népszabadság* (15 August 2015).

Field, Jim, 2013: "A Visit to Bodvalenke Hungary's 'Fresco Village'" (6 September 2013); at: www.chicagohomeless.org/a-visit-to-bodvalenke.

Funk, Nathan C.; Said, Abdul Aziz, 2010: "Localizing Peace: An Agenda for Sustainable Peacebuilding", in: *CAHSS Journals*, 17,1.

Deleuze, Gilles; Foucault, Michel, 1977: "Intellectuals and Power", in: Bouchard, Donald (Ed.): *Language, Counter-Memory, Practice: Selected Essays and Interviews* (Ithaca, N.Y.: Cornell University Press): 203–217.

Gal, Tali, 2016: "'The Conflict Is Ours': Community Involvement in Restorative Justice", in: *Contemporary Justice Review*, 19,3: 289–324.

Gábor, Gina, 2011: "Bódvalenke: Köszöntő [Bódvalenke: Welcome]", in: *Bódvalenke A Freskófalu.*

Görög, Veronika, 1994: "Cigány Folklór [Gypsy Folklore]", in: Bódi, Zs. (Ed.): *Cigány Néprajzi Tanumányok,* vol. 2 (Budapest: Magyar Néprajzi Társaság: Mikszáth Kiadó): 122–147.

Hall, Esward. T.; Hall, Mildred Reed, 1990: *Hidden Differences: Doing Business with the Japanese* (New York: Anchor Books).

Harré, Rom, 2012: "Positioning Theory: Moral Dimensions of Social-Cultural Psychology", in: Valsiner, J. (Ed.): *The Oxford Handbook of Culture and Psychology* (Oxford: Oxford Handbooks).

Harré, Rom; Moghaddam, Fathali, 2003: "Introduction", in: Harré, R. (Ed.): *The Self and Others: Positioning Individuals and Groups in Personal, Political, and Cultural Context* (Westport, Conn.: Praeger Publishers): 1–11.

Hernád, Géza, 2015: "Bódvalenke – virtuális álmok földje a világ végén nlc"; at: https://nlc.hu/foto/20150817/bodvalenke-sarkany-fesztival/.

Hofstede, Gert, 2003: *Culture's Consequences: Comparing Values, Behaviors, Institutions and Organizations Across Nations* (Thousand Oaks, CA: Sage Publications).

Horváth, Kata, 2006: "Éhség Szövegek: A Szociológia, a Média és egy Cigány Közösség Éhséginterpretációi" [Hunger Narratives: Interpretations of Hunger in Sociology, Media and a Gypsy Society], in: Prónai, Cs. (Ed.): *Cigányok Világa Európában* (Budapest, Hungary: Nyitott Könyvműhely Kiadó).

Johannesse, Richard; Valde, Kathleen; Whedbee, Karen, 2008: *Ethics in Human Communication*, 6th edn. (Long Grove, Ill.: Waveland).

Kale, David, 2003: "Peace as an Ethic for Intercultural Communication", in: Samovar, Larry. (Ed.): *Intercultural Communication: A Reader*, 10th edn. (Belmont, CA: Wadsworth): 466–470.

Kidder, Rushworth M., 1994: *Shared Values for a Troubled World: Conversations with Men and Women of Conscience* (San Francisco: Jossey-Bass).

Kovács, István, 2014: "'Hogy Ne Haljunk Éhen' – Bódvalenke Léket Kapott ['So We Should Not Die of Hunger' – Bódvalenke Has Hit the Rocks]", in: *Magyar Narancs* (3 June 2014); at: https://magyarnarancs.hu/kismagyarorszag/hogy-ne-haljunk-ehen-bodvalenke-leket-kapott-90343.

Kríza, Ildikó, 1993: "Narratív Hagyomány a Magyar Cigányság Körében [The Narrative Tradition Among Hungarian Gypsies]", in: Barna, Gábor (Ed.): *Cigány Népi Kultúra a Kárpát-Medencében a 18–19 Században* [*Gypsy Folk Culture of the Carpathian Basin in the 18th–19thCentury*] (Budapest, Hungary: MNT): 91–97.

Laborczi, Dóra, 2012: "Ha Választhatnék, Megfognám Ezt a Házat a Négy Sarkánál És Letenném Valami Rendes Helyen [*If I Could Choose, I Would Grab This House and Put It Down in a Normal Place*]", in: *HVG.hu*; at: https://hvg.hu/itthon/20120307_bodvalenke_freskofalu_turizmus.

Lindlof, Thomas R.; Taylor, Bryan C., 2011: *Qualitative Communication Research Methods*, 3rd edn. (Thousand Oaks, CA: Sage).

Linehan, Carol; McCarthy, John, 2000: "Positioning in Practice: Understanding Participation in the Social World", in: *Journal for the Theory of Social Behaviour*, 30: 435–453.

Loges, William E.; Kidder, Rushworth M., 1997: *Global Values, Moral Boundaries: A Pilot Study* (Camden, Maine: Institute for Global Ethic).

Martin, Judith N.; Flores, Lisa A.; Nakayama, Thomas K., 1998: "Ethical Issues in Intercultural Communication", in: Martin, Judith. (Ed.): *Readings in Cultural Communication* (Thousand Oaks, CA: Sage).

Marushiakova, Elena; Popov, Vesselin, 2001: "Historical and Ethnographic Background: Gypsies, Roma, Sinti", in: Guy, Will (Ed.): *Between Past and Future: The Roma of Central and Eastern Europe* (Hatfield: University of Hertfordshire Press): 33–53.

Marushiakova, Elena; Popov, Vesselin, 2011: "Between Exoticization and Marginalization: Current Problems of Gypsy Studies", in: *Behemoth: A Journal on Civilization,* 4,1: 86–105; at: doi: 10.515/behemoth.2011.006.

Okely, Judith, 1989: *Traveller Gypsies* (New York: Cambridge University Press).

Piasere, Leonardo, 2000: "Antropologia Sociale e Storica Della Mendicità Zingara" [Social and Historical Anthropology of the Gypsy Beggar], in: *Polis,* 14,3: 409–428.

Raatzsch, Jenő A.; Kállai, András, 2006: [Roma] "Szerződés az Etnikai Hovatartozás Eladásáról" [(Roma) The Contract to Sell the Ethnicity], in: *Transit* (1 October 2013–2 February 2 2014); at: https://hu.tranzit.org/en//0/2014-01-10/roma-the-contract-to-sell-the-ethnicity.

Riessman, Catherine K., 1993: *Narrative Analysis* (Newbury Park, Calif.: Sage).

Riessman, Catherine K., 2008: *Narrative Methods for the Human Sciences* (Thousand Oaks, CA: Sage).

Ringold, Dena; Orenstein, Mitchell A.; Wilkens, Erika, 2005: *Roma in an Expanding Europe: Breaking the Poverty Cycle* (Washington, DC: The World Bank).

Rövid, Márton, 2009: "One-Size-Fits-All Roma?", in: *Romani Studies,* 21,1: 1–22; at: doi: https://doi.org/10.3828/rs.2011.1.

Stage, Christina W.; Matterson, Marifran, 2003: "Ethnographic interviewing as contextualized conversation", in: Clair, Robin Patric (Ed.), *Expressions of Ethnography: Novel Approaches to Qualitative Methods* (Albany, NY: State University of New York Press): 97–106.

Subert, Mária, 2016: "Intercultural Communication in The Roma-Hungarian Conflict: The Roma Murders", in: *Journal of Communication, Media, Economics*, 8,1: 101–127.

Subert, Mária, 2019: "Motives and Legacies Behind 2008–2009 Hungarian Roma Murders and Apologies", in: *Contemporary Justice Review*, 22,1: 3–22. at: https://www.tandfonline.com/doi/full/10.1080/10282580.2019.1576127.

Tauber, Elizabeth, 2008: "'Do You Remember the Time We Went Begging and Selling': The Ethnography of Transformations in Female Economic Activities and Its Narrative in the Context of Memory and Respect Among the Sinti in North Italy", in: Jakobs, Fabian (Ed.): *Romani/Gypsy Cultures in New Perspective* (Leipzig, Germany: Leipziger Universitäts-verlag): 155–177.

Timmer, Andria D. 2010: "Constructing the 'Needy' Subject: NGO Discourses of Roma Need", in: *Political and Legal Anthropology Review*, 33,2: 264–281.

Warraq, Ibn, 2011: *Why the West is Best: A Muslim Apostate's Defense of Liberal Democracy* (New York/London: Encounter Books).

Part III
Peacekeeping, Peacebuilding, Peacemaking and Transitional Justice in the Anthropocene

Chapter 16
A Discourse on the Norms and Ideologies of Peacekeeping

Jude Cocodia

Abstract Much peacekeeping literature focuses on the need of the international community to prevent human catastrophes, project liberal values, provide security or prevent conflict areas from becoming terrorist enclaves, which has been a major goal in the global war on terror since the September 11 attacks (Ramsbotham/Woodhouse 1996; Hultman 2013). This approach to peace studies impedes reflections on what is already known and limits activities in this area to classroom lectures. So, how do we keep in touch with the past while searching for more effective approaches for the present and future? To foster this process, this paper adopts a traditional approach to peace studies and examines the challenges of peacekeeping approaches, neo-realist manipulations, the unanswered issues that persist, the challenges on the ground to peacekeepers and host populations, and the implications for troop-contributing countries. This paper encourages people to revisit the ideological foundations of peace studies and the development of peacekeeping theories.

Keywords Civilian protection · Conflict · Effectiveness · Motives · Peacekeeping · Violence

16.1 Introduction

nThe literature on peace operations grew significantly at the end of the Cold War, and this is attributable to three factors: the surge of new operations since 1999, the release of the United Nations (UN) Brahimi Report in 2000 and the events of 11 September 2001 (Johnstone 2005: 1). The failure of the UN in Rwanda (1990–1993) and Bosnia (1995) prompted the Brahimi Report, which advocated a shift from the traditional approach of peacekeeping, which rested on the principles of host state consent, neutrality and minimum force, to a more robust and forceful approach bordering on the use of force to protect civilians and enforce peace where necessary. The justification for this transition was forged on humanitarian precepts and had at

Dr. Jude Cocodia, Senior Lecturer and Acting Head of the Department of Political Science, Niger Delta University, Yenagoa, Bayelsa State, Nigeria; Email: jcocodia41@gmail.com.

Ú. Oswald Spring and H. G. Brauch (eds.), *Decolonising Conflicts, Security, Peace, Gender, Environment and Development in the Anthropocene*,
The Anthropocene: Politik—Economics—Society—Science 30,
https://doi.org/10.1007/978-3-030-62316-6_16

its foundation the protection of civilians who had increasingly become the victims of civil wars.

Much of the extant peacekeeping literature focuses on the need of the international community to prevent human catastrophes or to prevent conflict areas becoming terrorist enclaves, which has been a major goal in the global war on terror since the 11 September attacks. However, even when peace operations are about preserving territory and shoring up government authority in mission areas, there are certain issues that have remained at the core of peace operations because they determine how well peacekeeping mandates are carried out. Debate on these issues, in as much as they relate to ending conflict and building negative peace, are the focus of this chapter. These debates are grouped into two major themes and will be discussed in this order:

- Motives and aims of peacekeeping,
- Effectiveness of peacekeeping.

The discourse on motives and aims dwells on state behaviour and the justification for intervention. The debate on effectiveness examines peacekeeping's approaches and challenges, and evaluates their impact in the short and long term. This discourse is done within a theoretical analytical framework since it makes reference only to those works that are necessary for the discourses contained therein (Willison 2014).

16.2 Motives and Aims of Peacekeeping

Individual states participating in peacekeeping have done so for a variety of reasons, ranging from troop remuneration, national prestige and derivation of international aid to strengthening a regime's hold on power. These motives are usually connected to debates of national interests or the need to preserve the international system. However, there are authors who maintain that states undertake peacekeeping responsibilities for liberal and humanitarian objectives.

16.2.1 Promoting States' National Interests

Peacekeeping is argued to serve certain aspirations which favour the powerful states and, as such, intervening to promote an order to protect a status quo that suits them is quite common. Authors in this school hold that the major powers act on the maxim that they have the special responsibility of maintaining peace and security in a system where they occupy privileged positions (James 1996: 172; Finnemore 2003: 8). They maintain that peacekeeping as a multilateral initiative provides legitimacy for the pursuit of the interests of these states and changes in these interests determine the path peacekeeping takes (Pugh 2004; Bellamy et al. 2010). Since the interests of these states come first, peacekeeping has seen a sliding towards re-enacting earlier practices of imperial policing where the concerns of international actors are placed ahead of those of local communities in mission areas (Rubinstein 2010: 457). A supporting view notes that peace operations serve a narrow problem-solving purpose of doctoring the global political economy within a framework of liberal imperialism, which raises suspicions about whose interest peacekeeping really serves (Flint/Falah 2004: 1391; Pugh 2004: 39).

16.2.2 Preserving the International System

A fundamental role of peacekeeping is regulating the state-based international system. Preservation of the international system by the maintenance of its basic state-building blocks is the principal classical function of peacekeeping (MacQueen 2006: 235; Zanotti 2006: 152). This school of thought asserts that international interventions function as part of an international disciplinary security regime used to normalise states that are perceived as potential sources of instability to the international community as a whole or to a region in particular (Bellamy/Williams, 2009: 43). The attempt to normalise states hinges on the concept of mutual democratic pacifism, which holds that peace is better sustained between democratic nations (Russet 1995; Friedman 1999).

Consequently, peacekeeping is employed to ensure that states remain or are transformed into democracies, and there is a tendency to equate securing peace with universalising democracy (Barnett/Finnemore 2005: 172). Several authors, while sharing this view, object to this practice on the grounds that peacekeepers cannot be effective peacebuilders and that democracy can only be viable when it is established from within rather than from without. For this reason, external peacebuilding efforts which are shored up by institutions remain fragile (Hazen 2007: 325–326; Fortna/Howard 2008: 293).

16.2.3 Promoting Liberal Norms

By getting involved in peacekeeping, states sometimes place liberal values over national interests (Newland/Meyers 1998; Hampton 2005). Finnemore (2003: 55) contends that "intervention looks odd from the conventional perspectives of international politics because it does not conform to the conceptions of interest that they specify". She argues that it will be difficult to calculate state benefit or accuse states of benefiting when the reduction of human suffering as a result of their actions is apparent. Using the intervention of the US in Somalia as an example, she asserts that it was a clear case of intervention without obvious interests.

Peacekeeping practices and norms emerge from prevailing global norms that legitimise certain practices and de-legitimise others. Peacekeeping's support for democracy lends credence to the view that it is shaped by a global culture. Peacekeeping's hasty democratisation of mission areas is derived from its conformity to the ideological reorientation of the organisations involved (Paris 2003).

Another reason why the international community resorts to peacekeeping can be found in the liberal ideology of human rights. Authors of this school qualify the need for intervention in terms of rights and security, namely; religious toleration, minority rights, human rights, international stability. When these are threatened, the international community has the right to intervene to keep the peace (Krasner 1999: 111; Evans 2008: 284; Hultman 2013: 60). Peacekeeping is thus justified on humanitarian grounds and on the basis that states have a responsibility to protect their populations from genocide, war crimes, crimes against humanity and ethnic cleansing (Bellamy et al. 2010: 29–30: Glanville 2011: 233). When they manifestly fail to do so, the international community acquires a duty to protect vulnerable populations in what is known today as the *responsibility to protect* (R2P) (Bellamy 2008; Dannreuther 2008; Bellamy et al. 2010).

16.2.4 Compensating for State Failure and the Responsibility to Protect

With emphasis on the safety of individuals, the R2P focuses on that area of human security that is concerned with the alleviation of chronic threats, and stresses protection from downward disruption. States draw their legitimacy from their ability to protect their citizens, and when they fail at this task, the international community reserves the right to intervene to fulfil this function (Bellamy 2008; Diehl/PharaohKhan 2010).

With the increase in intrastate conflicts in the post Cold War era, belligerents usually target civilian populations, causing astronomic casualties, rather than other militia groups or government forces. Due to the impact of this phenomenon on civilians there is growing support for humanitarian intervention on the grounds that, even when establishing peace fails, peace operations are tasked with curbing the worst excesses and alleviating at least the most acute deprivations (Bellamy/Williams 2012: 541; Vircoulon/Lesueur 2014: 3).

One interpretation sees the R2P as the responsibility of a state to its citizens. This view contends that a state's sovereignty and legitimacy are dependent on its ability to respect and protect the rights of populations in its care (Bellamy 2008: 615; Bellamy et al. 2010: 29–30; Glanville 2011: 233). The other interpretation conceives it as an international response to avoid shocking episodes of inhumanity, such as genocide and ethnic cleansing (Diehl/PharaohKhan 2010: 111; Pattison 2008: 128). The position here is that international intervention becomes necessary when states fail in their responsibilities to protect their populations from such atrocities and violations (Ramsbotham/Woodhouse 1996: 73; Evans 2008: 284; Hultman 2013: 60).

Challenging the concept of international intervention on humanitarian grounds is the moral hazard theory which contends that such deployments challenge the Westphalian tradition of states exercising sovereignty over their territories (Pupavac 2006: 256; Seybolt 2008: 1; Robiliard 2011: 25). This school also argues that the R2P instigates rather than prevents genocide and mass atrocities (Kuperman 2009; Hehir 2010). Kuperman (2008: 219) notes:

> Rebellion is the most common trigger for genocidal violence, 'yet' some militants may rebel despite the risk of provoking state retaliation because they expect any resulting atrocities to attract intervention that facilitates their rebellion. Ultimately, intervention may help their rebellion succeed.

The contention is the R2P prolongs suffering by encouraging repressed groups to continue armed resistance and reject reasonable negotiated settlements (Kuperman 2009: 285). Moral hazard theorists hold up this view as one of the major reasons for the onset and prolongation of the conflict in Darfur (Kuperman 2009: 285–292; Robiliard 2011: 23). Another critique against the R2P holds that it has become too reliant on international rescue such that civil society has neglected decisive roles which citizens could play in protecting themselves (Mégret 2009: 580). There is also the contention that since the powerful states decide on intervention in weaker states, such intervention challenges the concept of equality of sovereign states (Pupavac 2006: 256).

These criticisms of humanitarian intervention seem rooted in uneasiness about violations of sovereignty. It is almost certain that if states adhere to their responsibilities to protect their populations, there would rarely be the need for violent insurgencies or peacekeeping interventions in the first place. When states fail, intervention becomes necessary to stem civilian casualties from the conflict that may ensue.

16.3 Effectiveness in Peace Operations

16.3.1 Approaches

It is well-known that peace operations are conducted according to the stipulations of either the traditional Chapter VI of the Charter of the United Nations, or the robust Chapter VII, but experts remain divided over which is more effective and which should dominate the intervention landscape. The debate has been whether organisations mandated to keep the peace should adhere to the principles of the soft Chapter VI approach or wield a stick, as demanded by Chapter VII operations, to enforce compliance with agreements when necessary. This section revisits these debates.

16.3.1.1 Traditional Peace Operations

Since it is the obligation of the intervener to be seen as impartial in order to assist in the maintenance of long-term settlements by the parties, authors adduce that traditional peacekeeping which upholds neutrality is a more effective approach (Donald 2002: 21; Macqueen 2006: 234; Levine 2010: 1). Under the traditional approach, peacekeepers are mandated to administer, monitor or patrol conflict areas, usually with the consent of the parties to the dispute, and nearly always under the provisions of Chapter VI of the UN Charter. Peacekeeping under this traditional approach acts as a buffer between conflict parties and is governed by the trinity tenets of consent, impartiality/neutrality and minimum force. Authors note that given the limited size of these missions, they are never meant to use force even if they want to (Diehl 1993: 4–5; Donald 2002: 21; Pugh 2004: 47, Bellamy et al. 2010: 13; Levine 2010: 1).

It is argued that due to its emphasis on confidence-building in mediation, the less forceful peacekeepers are, the more likely they are to be seen as partners, and the less likely they are to be targets (Sloan 2011: 21). Since credibility and authority emanating from moral stature significantly matter in mediating disputes, an intervener will almost certainly fail to gain the confidence and co-operation of disputants against whom it applies force (Princen 1992; Nathan 1999). Authors in this school of thought contend that though short-term stabilisation results of robust operations may be positive, the long-term repercussions could be negative (Wright 2005: 60; Hunt 2013: 4). They are of the view that the modern force posture of peacekeeping and humanitarianism are opposing tenets, hence humanitarians maintain a distance from robust peace operations to safeguard their credibility (Pugh 2004: 50).

16.3.1.2 Robust Peace Operations

Authors in support of the robust Chapter VII approach are of the view that the majority of violent conflicts since the end of the Cold War have been intrastate and are better settled with some degree of coercion (Walter 1997: 336, Donald 2002: 21; O'Hanlon 2003: 10; Williams 2010: 3). Regan (2002: 72) and Eisenkopf and Bächtiger (2013: 74) contend in support that neutral policies are largely incapable of contributing to the termination of a conflict, and neutrality during an on-going conflict may be insufficient to convince the actors that stopping the fighting is in their best interest. Consent does not guarantee success, which is why peacekeeping had to shift from its traditional orientation to more robust and vigorous efforts to effect peace (Doyle/Sambanis 2007: 495). Proponents contend that the changing landscape of conflict (from inter-state to intra-state) made preserving the principles of consent, neutrality and minimum force almost impossible as the desire to save lives and the need for force took precedence (Levine 2010). The restraint on force advocated by Chapter VI missions, beyond making peacekeeping operations inefficient, also increases peacekeepers' risk rate in volatile areas, while neutrality emboldens spoilers and increases civilian casualties (Ciechanski 1997; Jett 2001; Dallaire 2005).

This study consents to the appropriateness of the traditional approach in certain conflicts, and recognises that while robust operations are more widely utilised in view of the brutal nature of modern intrastate conflicts, the robust approach comes with its own set of challenges which impact negatively, albeit unintentionally on mission areas.

16.3.1.3 Timing Interventions

Having examined debates pertaining to the aims and motives of peace operations, this section examines the issue of the appropriate time, if any, to deploy peacekeepers. Mark Paterson and Chris Saunders (2012: 2) contend that the "launch of a peacekeeping mission depends on a convergence of international, regional, and national demands: the interests of key Security Council members; the willingness of local warring parties to co-operate; and the support of neighbouring powers – particularly hegemonies". The arguments about the criteria for intervention will be grouped under the following: preventive negotiations (proactive) and de-escalating conditions (reactive).

16.3.1.4 As a Preventive Measure

While the issue of preventing war before it starts is ideal and comes with moral approval since it prevents the loss of lives, it also comes with its own problems. These problems sometimes sway efforts from the use of this method of conflict resolution.

Some schools of thought hold that peacekeeping as a conflict prevention measure is bound to be more efficient and cost-effective and contribute positively to peacebuilding. Since it consists of stemming conflict before destruction and bloodshed set in, they contend that the parties are bound to be more amenable to settlement. At this level of open conflict, there are signs of confrontation as parties make known their grievances and efforts are made to deal with it, and militarisation has not been undertaken as an option. It is at this stage that conflict prevention is most desirable and should be undertaken. As William Zartman (2005: 2) warns, with the onset of hostilities, "new and worse situations are created making reconciliation and reconstruction even more difficult". This view is shared by Diehl and Frederking (2010), who aver that once military action has been engaged, the conflict becomes difficult to resolve and spills into further violence which is often costly in terms of lives lost.

One of the problems associated with conflict prevention is getting the opposing factions to come together to agree to a settlement (Zartman 2005; Diehl 1993). Other problems include the need for large-scale bureaucracy and proof that the danger averted was really worth the expenditure (Bellamy et al. 2010). Based on years of experience working in the UN peacekeeping department, Marrack Goulding (2002: 335) cautions that Inter Governmental Organisations (IGOs) should not rush into settling conflicts until enough initiative for resolution is shown by conflict parties.

This, he contends, assures a higher degree of success. What happens, then, when the initiative for resolution shown by the parties is deemed insufficient?

16.3.1.5 As Conflict De-escalation

This refers to the situation when certain conditions have been fulfilled to pave the way for intervention during conflict. Often within the traditional purview of peacekeeping, this consists of intervention taking place after negotiated cessation of hostilities by conflict groups.

It has been observed that while the majority of interstate wars have ended on the bargaining table, intrastate wars have often been fought to the death or ended either in genocide or expulsion of the losing side. This is often the case when there is no third party to intervene and guarantee a peace agreement (Walter 1997: 335–336). The point here is that combatants in intrastate conflicts often choose to fight to the finish if an intervener that can guarantee and enforce peace agreements cannot be found. Correspondingly, Sitkowski ()nnnsees peacekeepers as *enforcers-mediators* whose presence is indispensable in halting the prolongation of conflict and creating reassurances. With the majority of wars today being intrastate conflicts, the necessity for such an intervener is apparent.

Some authors take a different view and contend that parties seek outside intervention only when they have reached a hurting stalemate (Zartman 1985). Whatever the difference in the positions of Walter (1997), Sitkowski (2006) and Zartman (1985), they point to trust in a mediator. This trust in an intervener stems from the inability of the conflicting parties to arrange settlements themselves (Walter 1997: 342). Subsequently, only when an outside enforcer steps in to guarantee the terms do commitments to disarm and share political power become believable. Only then does co-operation become possible (Walter 1997: 336).

It is the norm to accuse the UN and regional peacekeeping organisations of not acting soon enough in conflict, in a bid to compel them to urgent action, but ceding to this demand would be an operational catastrophe if the warring parties are not ready to negotiate (Goulding 2002: 335). Intervention should therefore be based on the parties' level of seriousness to find a solution to the crisis (Diehl 1993). However, if groups at war never get to this point without first reaching a hurting stalemate, a lot of destruction and misery would have already been unleashed, which makes this reactive approach quite unpalatable. As such, proactive intervention would seem more admirable than reactive approaches.

16.3.1.6 Challenges in Peacekeeping

One major challenge for peacekeeping is the national interests of states. This often impedes international goodwill and prevents peacekeeping organisations from matching rhetoric with action (Patterson 2008; Diehl/Frederking 2010; Knopf 2011). States do not always act with the intention of conflict management, but rather base their decisions on strategic considerations other than the well-being of the country experiencing the conflict (Bellamy/Williams 2009; Aydin 2010; Cottey/Forster 2010). They reckon that the ensuing gap between the intention and practice of peacekeeping organisations denotes a weak political will which produces doctrinal lapses and porous mandates that often leave peacekeeping missions ineffective and prolonged (Ciechanski 1997; Lipson 2007). When intervention is prolonged, conflict intensifies and human suffering increases (Regan 2002; Peksen 2012).

Another challenge facing peacekeeping is its demand that peacekeepers should lay down their lives for foreigners. This is quite an unnatural expectation even for soldiers who are well trained. Thomas Jacobson (2012: 4) contends that peacekeeping is an aberration because of this demand. This factor has been at the core of the many failures of peacekeeping and is the reason why peacekeepers often cower in the face of armed opposition and do not protect women and children. Despite the deployment of the UN peacekeeping mission in 1999 in the *Democratic Republic of Congo* (DRC), and despite being the second-largest UN peace operation, the country remains rife with conflict. Armed groups roam freely, attacking villages and raping women, a phenomenon that has earned the DRC the title *rape centre of the world.* Going by such examples, Jacobson (2012) contends that it is difficult to justify peacekeeping as wise or in the safety and best interests of the people it is purportedly meant to protect. As such, the peace and safety attributed to peacekeeping are an illusion. Paterson and Saunders (2012: 2) share the same view in their observation of the lack of concern exhibited by peacekeepers who were accused of looting, raping and in rare cases murdering local people in the eastern Congo.

16.3.1.7 Unintended Consequences

The expansion of mandates from the strict supervision of ceasefires and separation of warring factions in accordance with traditional peacekeeping to the more complex roles of post-conflict reconstruction and peacebuilding in recent times has had unintended consequences. Despite these consequences, which have at times severely undermined the reputation of peacekeeping missions and their sponsoring IGOs, they still affirm peace operations to be indispensable in the management of conflict.

Supporting the humanitarian focus of peacekeeping is the argument that examining security with the state as the focal point is not reflective of the multifaceted concerns that have come to be conceptualised as human security (Hudson 2005: 791). Recognising this shift in emphasis to humanitarian needs, academics contend that the unintended impact of peace operations leaves populations of host communities exposed and vulnerable to varying forces (Aoi et al. 2007; Willet 2010). They acknowledge that though no mission has yet failed as a direct result of the unintended consequences generated, many have had their effectiveness questioned (Aoi et al. 2007: 168).

Where there is military presence there is demand for prostitutes, and authors have gone on to examine this phenomenon from the perspective of *sexual exploitation and abuse* (SEA) of the local women of host communities (Aoi et al. 2007: 171; Bellamy et al. 2010: 364; Smith/Smith 2011: 126). Some authors have taken the debate further and examined the impact of the zero-tolerance policy of peacekeeping organisations, which prohibits the exchange of money, goods and services between peacekeepers and the local populace for sex, on curbing SEA (Kent 2007; Jennings/Nikolić-Ristanović 2009). These scholars identify the constraints in imposing norms and implementing compliance in view of the cultural diversity of peacekeeping personnel and the fact, too, that peacekeeping is often conducted in societies where poverty prevails and respect for the rule of law is acutely low. They note that the zero-tolerance policy towards SEA has been caught between protecting local populations and protecting the image of peacekeeping operations from disrepute (Jennings 2008: 36, 45).

The sex industry is but a part of the wider peacekeeping economy. Another unintended phenomenon, according to scholars, is the creation of parallel economies – one which serves the local populace, and which is adversely affected by the other, which serves the needs of peacekeeping missions (Aoi et al. 2007: 170). They argue that the dependence of the local economy on the peacekeeping economy affects the capacity of the former to sustain itself in the aftermath of peacekeeping operations and prompts the elongation of the mission to prevent the collapse of the host economy and a rapid return to conflict (Carnahan et al. 2006: 9; Aoi et al. 2007: 170; Aning/Edu-Afful 2013: 18).

The 2010 cholera outbreak in Haiti that killed over 10,000 Haitians and made over 800,000 ill presents a new perspective to the issue of unintended consequences of peacekeeping. Prior to this period Haiti had been cholera-free, but peacekeeping worsened this poor country's health challenges. The cholera epidemic was the outcome of peacekeepers from Nepal who arrived in Haiti suffering from the virus, which was prevalent in Nepal at the time. Up until March 2016, six years after, an average of thirty-seven cholera deaths were still being recorded in Haiti each month (Agerholm 2016; Sengupta 2016).

While the experience in Haiti demonstrates the potential of peacekeepers to create health hazards in mission areas, they can just as well carry communicable diseases from their places of deployment to their home countries and so create or worsen health challenges. It was estimated that 11–12% of Nigeria's 5,000 peacekeepers who served in Liberia returned home HIV-positive (Afaha 2016: 14). Most of these officers and men were ignorant of their HIV status and ignorantly passed the virus on to their wives (Oyeniyi 2010: 9).

Contributing peacekeepers confers significant benefits on the *troop-contributing countries* (TCC's). These include improving the military hardware, and the combat and negotiating expertise of troops from these countries, and raising the state's international profile (Norden 1995: 436). Authors have, however, pointed out the downside to the gesture of contributing troops. They argue that by contributing troops to peacekeeping, despotic governments are given access to international aid (Aoi et al. 2007: 172; Sotomayor 2007: 176–178; Baledrokadroka 2012: 109). Also, the experience garnered by troops in combat and community relations during peacekeeping missions is sometimes used to perpetuate authoritarian regimes in their home countries when they return (Cunliffe 2014; Sotomayor 2014). This shows that unintended consequences go beyond peacekeeping mission areas.

16.3.1.8 Evaluating Effectiveness in Peacekeeping

Due to the variety of interpretations that have developed over the years, it is difficult to set a standard for peacekeeping's success (Druckman/Stern 1999; Johnstone 2005). Assessing the impact of intervention is not straightforward, since much depends on the purpose and size of the intervention and how far into the post-war era assessments should go (Pikering/Kisangani 2006: 363). In attempting to resolve this problem, academics have categorised assessments into short- and long-term variants (Diehl/Druckman 2013). This categorisation tallies with Francis Fukuyama's (2004) three distinct phases of peace operations, namely: (a) the initial stabilisation of a war-torn society, (b) the recreation of local institutions of governance, and (c) the strengthening of those institutions. Short-term evaluations are indicative of the first phase, which entails conflict cessation, protecting lives and keeping environments secure. The second and third phases are concerns of long-term evaluation, since they consist of peacebuilding, infrastructural and institutional reconstruction within the host community, and the viability of these institutions.

16.3.1.9 Short-Term Evaluations

Short-term evaluations assess the connection between actions taken by field operatives, conflict reduction and achieving stability. In evaluating the success of peacekeeping operations in this regard, authors have proffered several metrics. These include assessing whether the mandate was fulfilled, whether the specific accomplishments were met, and the immediate impact of the operation on the local population (Diehl 1993; Johansen 1994).

Some authors question these criteria, arguing that on the issue of mandate, there is the problem of clarity and interpretation (Druckman/Stern 1999: 2). The accomplishment criterion, they argue, leaves open the question of just how many accomplishments it would take. With reference to impact on the local population, they advocate that the criterion begs the question and ignores the impacts on the broader political processes that may have larger impacts on the lives of the local people. Supporting this critique, Chiyuki Aoi et al. (2007: 6) contend that while the intent of peacekeeping may have been achieved, it may propel other problematic issues which question the initial success. From this perspective, analysing peacekeeping's effectiveness based on immediate outcomes, such as providing security and enabling stability, is premature (Fortna 2004: 269).

16.3.1.10 Long-Term Evaluations

There is the contention that peacekeeping success should be rated on factors such as government obligations to uphold norms of peace, the functionality of democratic institutions, the prosecution of war crimes, and the professionalism of peacekeepers (Johansen 1994). This approach is adopted by Devon Curtis (2012), who assesses the success of peacebuilding in Burundi based on the vibrancy of its transitional justice, economic and democratic reforms ten years after the end of conflict. Other authors who favour the long term approach argue that success should be based on the duration of peace after the exit of peacekeepers (Fortna 2008; Hultman et al. 2015). One advantage of this approach, contends this school of thought, is that it accommodates peacekeeping externalities, a phrase which refers to the duties assigned peacekeepers and the unwarranted effects that come with them (Lipson 2007; Kent 2007; Jennings 2008).

Since it is unknown how long it would take for these long-term effects (such as respect for rights and firm democratic institutions) to take root, short-term assessments become a necessary stop-gap in the evaluation process. Not knowing how far into the post-conflict period researchers have to look in order to determine whether peacekeeping has been successful is a major problem with long-term assessment (Pickering/Kisangani 2006: 363). Supporting this assertion, David Quinn et al. (2013) contend that the immediate impact or short-term assessment is a more realistic metric of evaluation.

Evaluation studies on peace operations by organisations such as the *African Union* (AU) should adopt the short-term approach, since the primary aim of AU peace operations is to keep environments secure and protect lives. For such studies, it would be inappropriate to anchor success on long-term metrics. For studies focusing on the impact of peacebuilding which manifests over a long period, the long-term approach becomes a viable option. Studies focusing on UN peacebuilding can apply this yardstick.

16.4 Conclusion

This reflection focuses on peacekeeping's aims, motives and effectiveness. The objective of this study is to ensure that the issues that really matter are not forgotten amid all the complex theories, strategic planning, bureaucracy and pursuit of interests that come with studying and preparing peace operations.

The major contention of proponents of traditional peacekeeping is that peacekeeper neutrality and the minimum use of force sustains mediator credibility. An intervener will fail to gain the co-operation of disputants against whom it applies force. Authors in support of robust peacekeeping contend that passivity, as demanded by neutrality, in modern-day conflict theatres serves the interest of parties resistant to peace, and increases human suffering. A robust approach is inevitable if atrocities are to be prevented by peacekeepers.

Robust missions are usually large, and authors have pointed out that this generates certain unintended effects within the target area and beyond. These include the increase in sex trafficking, the creation of a dual economy, and the manipulation of peacekeepers' expertise by their home governments to sustain autocratic regimes. Unfortunately, international institutions responsible for peacekeeping have little control over these consequences and it is difficult to plan for them.

Peacekeeping institutions have had to cope with the challenge of harmonising states' interests that impact on the provision of stability in conflict areas. Beyond promoting their national interests, certain states use peacekeeping to preserve a favourable *status quo* in the international system, spread liberal ideals and support humanitarian norms. The legitimacy and authority of international peacekeeping organisations derives partially from their ability to harmonise these different motives and aims into a cohesive whole.

This study shows that unintended consequences vary. While peacekeepers often facilitate these consequences, especially in areas of increased sexual activities, sometimes they are solely responsible for epidemics, as seen in the Haiti case, while at other times they become victims to epidemics in mission areas. This reflection also captures the argument that TCC's are prone to unintended (though sometimes calculated) consequences just like the mission areas.

There are many factors that influence peacekeeping deployments. Apart from happenings on the ground, the rational calculation of peacekeeping organisations and what they hope to achieve are also determinants. Peacekeepers are deployed to either prevent the escalation of conflict (proactive) or to help end the conflict and assist in rebuilding (reactive). While the former prevents the loss of life, but leaves everyone guessing at the level of destruction that was averted, the latter, after all the destruction, is the product of the will and consent of the conflicting parties to end the violence.

There are many more debates in the field of peacekeeping, but the themes discussed in this paper often constitute the core of the discourse. Paying attention to these debates will serve to deepen understanding of these problems and contribute to improving peacekeeping practice.

References

Afaha, Philip, 2016: "Nigeria: Counting the Cost of the ECOMOG Operations", in: *Conflict Studies Quarterly*, 16: 3–16.

Agerholm, Harriet, 2016: "UN Admits Playing Role in Cholera Outbreak that Killed almost 10,000 People in Haiti", in: *Independent*, 19 August; at: https://www.independent.co.uk/news/world/americas/un-cholera-role-haiti-outbreak-deaths-united-nations-peacekeepers-helped-spread-a7198861.html (9 August 2017).

Aning, Kofi; Edu-Afful, Fiifi, 2013: "Unintended Impacts and the Gendered Consequences of Peacekeeping Economies in Liberia", in: *International Peacekeeping*, 20, 1: 17–32.

Aoi, Chiyiki; De Coning, Cedric; Thakur, Ramesh, 2007: "Unintended Consequences, Complex Peace Operations and Peacebuilding Systems", in: Aoi, Chiyuki; De Coning, Cedric; Thakur

Ramesh (Eds.) *Unintended Consequences of Peacekeeping Operations* (Tokyo: United Nations University Press): 3–20.
Aydin, Aysegul, 2010: "Where do States Go? Strategy in Civil War Intervention", in: *Conflict Management and Peace Science*, 27,1: 47–66.
Baledrokadroka, Jone, 2012: "The Unintended Consequences of Fiji's International Peace-keeping", in: *Security Challenges*, 8,4: 105–116.
Barnett, Michael; Finnemore, Martha, 2005: "The Power of Liberal International Organisations", in: Barnett, Michael; Duvall, Raymond (Eds.): *Power in Global Governance* (Cambridge: Cambridge University Press): 161–185.
Bellamy, Alex; Williams, Paul, 2009: "The West and Contemporary Peace Operations", in: *Journal of Peace Research*, 46,1: 39–57.
Bellamy, Alex, 2008: "The Responsibility to Protect and the Problem of Military Inter-vention", in: *International Affairs*, 84,4: 615–639.
Bellamy, Alex; Williams, Paul; Griffin, Stuart, 2010: *Understanding Peacekeeping* (Cambridge: Polity Press).
Bellamy, Alex; Williams, Paul, 2012: "On the Limits of Moral Hazard: The 'Responsibility to Protect', Armed Conflict and Mass Atrocities", in: *European Journal of International Relations*, 18,3: 539–571.
Carnahan, Michael; Durch, William; Gilmore, Scott, 2006: *Economic Impact of Peacekeeping: Final Report* (New York: United Nations Peacekeeping)
Ciechanski, Jerzy, 1997: "Enforcement Measures under Chapter VII of the UN Charter: UN Practice After the Cold War", in: Pugh, Michael (Ed.): *The UN, Peace and Force* (London: Frank Cass): 82–104.
Cocodia, Jude, 2018: *Peacekeeping and the African Union: Building Negative Peace* (London: Routledge).
Cottey, Andrew; Forster, Anthony, 2010: "Enhancing Regional Peacekeeping Capabilities", in: *The Adelphi Papers*, 44,365: 51–68.
Cunliffe, Philip, 2014: *Legion of Peace: UN Peacekeepers from the Global South* (London: Hurst Publishers).
Curtis, Devon, 2012: "The Contested Politics of Peacebuilding in Africa", in: Curtis, Devon; Dzinesa, Gwinyayi, (Eds.): *Peacebuilding, Power and Politics in Africa* (Ohio: Ohio University Press): 1–28.
Dallaire, Romeo, 2005: *Shake Hands with the Devil: The Failure of Humanity in Rwanda* (London: Arrow Books).
Dannreuther, Roland, 2008: *International Security: The Contemporary Agenda* (Cambridge: Polity Press).
Diehl, Paul; Druckman, Daniel, 2013: "Peace Operation Success: The Evaluation Framework", in: Druckman, Daniel; Diehl, Paul F. (Eds): *Peace Operation Success: A Comparative Analysis* (Boston: Martinus Nijhoff Publishers): 11–27.
Diehl, Paul; Frederking, Brian, 2010: "Peace and Security Affairs", in: Diehl, Paul; Frederking, Brian (Eds.): *The Politics of Global Governance: International Organizations in an Interdependent World* (Boulder: Lynne Rienner Publishers): 139–141.
Diehl, Paul; PharaohKhan, Elijah, 2010: "Financing UN Peackeeping: A Review and Assessment of Proposals", in: Diehl, Paul F.; PharaohKhan, Elijah (Eds.): *The Politics of Global Governance: International Organizations in an Interdependent World* (Boulder: Lynne Rienner Publishers): 111–138.
Diehl, Paul, 1993: *International Peacekeeping* (Baltimore: The John Hopkins University Press).
Donald, Dominick, 2002: "Neutrality, Impartiality and UN Peacekeeping at the Beginning of the 21st Century", in: *International Peacekeeping*, 9,4: 21–38.
Dorn, Walter; Libben Joshua, 2018: "Preparing for Peace: Myths and Realities of Canadian Peacekeeping Training", in: *International Journal*, 73,2: 253.
Doyle, Michael; Sambanis, Nicholas, 2007: "The UN Record on Peacekeeping Operations", in: *International Journal: Canada's Journal of Global Policy Analysis*, 62,3: 495–518.

Druckman, Daniel; Stern, Paul, 1999: "Perspectives on Evaluating Peacekeeping Missions", in: *International Journal of Peace Studies*, 4,1.
Eisenkopf, Gerald; Bächtiger, André, 2012: "Mediation and Conflict Prevention", in: *Journal of Conflict Resolution*, 57,4: 58–86.
Evans, Gareth, 2008: "The Responsibility to Protect: An Idea Whose Time has Come ... and Gone"? in: *International Relations*, 22,3: 283–298.
Finnemore, Martha, 2003: *The Purpose of Intervention: Changing Beliefs About the Use of Force* (Ithaca, NY: Cornell University Press).
Flint, Colin; Falah, Ghazi-Walid, 2004: "How the United States Justified Its War on Terrorism: Prime Morality and the Construction of a Just War", in: *Third World Quarterly*, 25,8: 1379–1399.
Fortna, Virginia; Howard, Lise, 2008: "Pitfalls and Prospects in the Peacekeeping Literature", in: *Annual Review of Political Science*, 11: 283–301.
Fortna, Virginia P., 2004: "Does Peacekeeping Keep Peace? International Intervention and the Duration of Peace After Civil War", in: *International Studies Quarterly*, 48: 269–292.
Friedman, Thomas, 1999: *The Lexus and the Olive Tree* (New York, NY: Anchor Books).
Fukuyama, Francis, 2004: *State-Building: Governance and World Order in the 21st Century* (Ithaca, NY: Cornell University Press).
Glanville, Luke, 2011: "The Antecedents of 'Sovereignty as Responsibility'", in: *European Journal of International Relations*, 17,2: 233–255.
Goulding, Marrack, 2002: *Peacemonger* (London: John Murray).
Hazen, Jennifer, 2007: "Can Peacekeepers be Peacebuilders"? in: *International Peacekeeping*, 14,3: 323–338.
Hehir, Aidan, 2010: "The Responsibility to Protect: Sound and Fury Signifying Nothing"? in: *International Relations*, 24,2: 218–239.
Hudson, Natalie, 2005: "En-gendering UN Peacekeeping Operations", in: *International Journal*, 60: 785–806.
Hultman, Lisa, 2013: "UN Peace Operations and Protection of Civilians: Cheap Talk or Norm Implementation"? in: *Journal of Peace Research*, 50,1: 59–73.
Hultman, Lisa; Kathman, Jacob; Shannon, Megan, 2015: "United Nations Peacekeeping Dynamics and the Duration of Post-Civil Conflict Peace", in: *Conflict Management and Peace Science* (25 March): 1–19.
Hultman, Lisa; Kathman, Jacob; Shannon, Megan, 2013: "United Nations Peacekeeping and Civilian Protection in Civil War", in: *American Journal of Political Science*, 57,4: 875–891.
Hunt, Charles, 2013: "UN Peace Operations and 'All Necessary Means'", in: *Asia Pacific Centre for the Responsibility to Protect*, 3,3: 1–10.
Jacobson, Thomas, 2012: *UN Peacekeeping: Few Successes, Many Failures, Inherent Flaws* (Colorado Springs: International Diplomacy and Public Policy Centre, LLC).
James, Alan, 1996: "The Dual Nature of UN Peacekeeping", in: Bourantonis, Dimitris; Evrivi-ades, Marios (Eds.): *A United Nations for the Twenty-First Century* (The Hague: Kluwer Law International): 171–186.
Jennings, Kathleen, 2008: "Protecting Whom? Approaches to Sexual Exploitation and Abuse in UN Peacekeeping Operations", in: *GSDRC Fafo-Report No. 36* (Oslo: Governance, Social Development, Humanitarian, Conflict).
Jennings, Kathleen; Nikolić-Ristanović, Vesna, 2009: "UN Peacekeeping Economies and Local Sex Industries: Connections and Implications", in: Microcon Research Working Paper 17 (Brighton: Microcon).
Jett, Dennis, 2001: *Why Peacekeeping Fails* (New York: Palgrave Macmillan).
Johansen, Robert, 1994: "UN Peacekeeping: How Should We Measure Success?", in: *Mershon International Studies Review*, 38: 307–310.
Johnstone, Ian, 2005: *Peace Operations Literature Review* (New York: New York University Center on International Cooperation).
Kent, Vanessa, 2007: "Protecting Civilians from UN Peacekeepers and Humanitarian Workers: Sexual Exploitation and Abuse", in: Aoi, Chiyuki; De Coning, Cedric; Thakur, Ramesh (Eds.):

Unintended Consequences of Peacekeeping Operations (Tokyo: United Nations University Press): 44–66.
Knopf, Payton, 2011: "Rethinking Peacemaking in Darfur", in: Council on Foreign Relations (Ed.): *Policy Innovation Memorandum*, No. 3 (April).
Krasner, Stephen, 1999: *Sovereignty: Organised Hypocrisy* (Princeton, NJ: Princeton University Press).
Kuperman, Alan, 2009: "Darfur: Strategic Victimhood Strikes Again"? in: *Genocide Studies and Prevention: An International Journal*, 4,3: 281–303.
Kuperman, Alan, 2008: "Mitigating the Moral Hazard of Humanitarian Intervention: Lessons from Economics", in: *Global Governance*, 14: 219–240.
Levine, Daniel, 2010: *Peacekeeper Impartiality: Standards, Processes, and Operation* (University of Maryland: Centre for International and Security Studies).
Lipson, Michael, 2001: "Peacekeeping: Organized Hypocrisy"? in: *European Journal of International Relations*, 13,1: 5–34.
Lundgren, Magnus, 2018: "Backdoor Peacekeeping: Does Participation in UN Peacekeeping Reduce Coups at Home?", in: *Journal of Peace Research*, 55,4: 508–523.
MacQueen, Norrie, 2006: *Peacekeeping and the International System* (London: Routledge).
Mégret, Frederic, 2009: "Beyond the Salvation Paradigm: Responsibility to Protect (Others) vs the Power of Protecting Oneself", in: *Security Dialogue*, 40,6: 575–595.
Nagar, Shantanu, 2018: "UN Peacekeeping Challenges, Solutions and South Asia's Stance", in: *South Asian Survey*, 23,1: 54–68.
Nathan, Laurie, 1999: "When Push Comes to Shove. The Failure of International Mediation in African Civil Wars", in: *Track Two*, 8,2: 1–23.
Norden, Deborah, 1995: "Keeping the Peace, Outside and In: Argentina's UN Mission", in: *International Peacekeeping*, 2,3: 346–347.
O'Hanlon, Michael, 2003: *Expanding Global Military Capacity for Humanitarian Intervention* (Washington DC: Brookings Institution Press)
Oyeniyi, Bukola, 2010: "From Liberia with Love: Officer's Wives Confronting HIV/AIDS", in: *International Resource Network*, 3: 1–17.
Paris, Roland, 2003: "Peacekeeping and the Constraints of Global Culture", in: *European Journal of International Relations*, 9,3: 441–443.
Paterson, Mark; Saunders, Chris, 2012: "What Does UN Peacekeeping Cost African Countries?" (Cape Town: Centre for Conflict Resolution).
Patterson, Patrick, 2008: "Darfur and Peacekeeping Operations in Africa", in: *Military Review* (July-August).
Pattison, James, 2008: "Humanitarian Intervention and a Cosmopolitan UN Force", in: *Journal of International Political Theory*, 4, 1: 126–145.
Peksen, Dursun, 2012: "Does Foreign Military Intervention Help Human Rights?", in: *Political Research Quarterly*, 65,3: 558–571.
Pickering, Jeffery; Kisangani, Emizet, 2006: "Political, Economic, and Social Consequences of Foreign Military Intervention", in: *Political Research Quarterly*, 59, 3: 363–376.
Princen, Thomas, 1992: *Intermediaries in International Conflict* (Princeton, NJ: Princeton University Press).
Pugh, Michael, 2004: "Peacekeeping and Critical Theory", in: *International Peacekeeping*, 11,1: 39–58.
Pupavac, Vanessa, 2006: "The Politics of Emergency and the Demise of the Developing State: Problems for Humanitarian Advocacy", in: *Development in Practice*, 16,3: 255–269.
Quinn, David; Wilkenfeld, Jonathan; Eralp, Pelin, Asal, Victor; Mclauchlin, Theodore, 2013: "Crisis Managers but not Conflict Resolvers: Mediating Ethnic Intrastate Conflict in Africa", in: *Conflict Management and Peace Science*, 30,4: 387–406.
Ramsbotham, Oliver; Woodhouse, Tom, 1996: *Humanitarian Intervention in Contemporary Conflict* (Cambridge: Polity Press).

Regan, Patrick, 2002: "Third-Party Interventions and the Duration of Intrastate Conflicts", in: *Journal of Conflict Resolution*, 46,55: 55–73.
Robiliard, Valentin, 2011: "The Dafurian Rebellion and the Moral Hazard of Humanitarian Intervention", in: *McGill International Review*, 1,1: 18–26.
Rubinstein, Robert, 2010: "Peacekeeping and the Return of Imperial Policing", *International Peacekeeping*, 17,4: 457–470.
Russet, Bruce, 1995: *Grasping the Democratic Peace: Principles for a Post-Cold War World* (Princeton, N.J.: Princeton University Press).
Sengupta, Somini, 2016: "UN Apologises for Role in Haiti's 2010 Cholera Outbreak", *The New York Times* (1 December); at: https://www.nytimes.com/2016/12/01/world/americas/united-nations-apology-haiti-cholera.html (9 August 2017).
Seybolt, Taylor, 2008: *Humanitarian Military Intervention: The Conditions for Successes and Failure* (Oxford: Oxford University Press).
Sitkowski, Andrzej, 2006: *UN Peacekeeping: Myth and Reality* (Westport, CT: Praeger Security International).
Sloan, James, 2011: *The Militarisation of Peacekeeping in the Twenty-First Century* (Oxford: Hart Publishing).
Smith, Charles; Smith, Heather, 2011: "Human Trafficking: The Unintended Effects of United Nations Intervention", in: *International Political Science Review*, 32,2: 125–145.
Sotomayor, Arturo, 2014: *The Myth of the Democratic Peacekeeper: Civil-Military Relations and the United Nations* (Baltimore: John Hopkins University Press).
Vircoulon, Thierry; Lesueur, Thibaud, 2014: "The CAR Crisis: Thinking Beyond Traditional Peacekeeping", in: *International Crisis Group* (17 June).
Walter, Barbara, 1997: "The Critical Barrier to Civil War Settlement", in: *International Organisation*, 51,3: 335–364.
Willet, Susan, 2010: "Introduction: Security Council Resolution 1325: Assessing the Impact on Women, Peace and Security", in: *International Peacekeeping*, 17,2: 142–158.
Williams, Paul, 2010b: "Enhancing Civilian Protection in Peace Operations: Insights from Africa", in: *Africa Centre for Strategic Studies, Research Paper*, No. 1 (September).
Williams, Paul, 2016: "Lessons Learned from Peace Operations in Africa", in: *Africa Security Brief*, No. 3 (March).
Willison, Martin, 2014: "What is the Difference Between Literature Review, Theoretical Analysis and Conceptual Analysis?"; at: https://www.researchgate.net/post/What_is_the_difference_between_literature_review_theoretical_analysis_and_conceptual_analysis (24 August 2017).
Wright, Steve, 2005: "Violent Peacekeeping: The Rise and Rise of Repressive Techniques and Technologies", in: *Politics and Ethics Review*, 1,1: 60–69.
Zanotti, Laura, 2006: "Taming Chaos: A Foucauldian View of UN Peacekeeping, Democracy and Normalization", in: *International Peacekeeping*, 13,2: 150–167.
Zartman, William, 1985: "International Mediation: Conflict Resolution and Power Politics", in: *Journal of Social Issues*, 41,2: 27–45.
Zartman, William, 2005: *Cowardly Lions: Missed Opportunities to Prevent Deadly Conflict and State Collapse* (London: Lynne Rienner Publishers).
Ziegler, Charles, 2016: "Russia on the rebound: Using and misusing the Responsibility to Protect", in: *International Relations*, 30,3: 353.

Chapter 17
Governance by Violent Non-state Actors as a Challenge to Sustainable Peace in Brazil

Marcos Alan S. V. Ferreira

Abstract This chapter examines the role of *violent non-state actors* (VNSA) and their capacity to offer governance in competition with the Brazilian state, specifically in poor urban communities, where state security forces fail to provide the basic needs of their citizens in terms of physical protection, human dignity and social security – primarily in areas with a high level of structural violence. Scholars have pointed to the emergence of areas that are not governed by the State. In such cases, informal governance structures led by criminal organisations emerge and unbalance actors and processes aiming a sustainable peace. Using the case study method on two criminal organisations – *Primeiro Comando da Capital* (PCC) (First Capital Command) and *Família do Norte* (FDN) (Northern Family) – the chapter argues that the State's failure or negligence in the areas with high structural violence in the cities give these VNSAs legitimacy, promoting rules of conviviality, resolution of litigation and also promoting material goods for their criminal members.

Keywords Brazil · Criminal organisations · Governance · Peace · Violence

17.1 Introduction

Latin America is one of the most violent regions in the world, given the high index of homicides, crime and the presence of strong violent criminal organisations (Ferreira 2017). While the violence in this area has been extensively analysed by experts in the last few years (for example, by Imbush et al. 2011; Gledhill 2015; Sain/Games 2014), there is not much literature on these new forms of governance caused by the so-called *Violent Non-State Actors* (VNSA).

The explosion of violence through competition among VNSAs, clashes between them and the State, or a spillover effect on neighbouring societies has generated different forms of informal governance, which are causally linked to problems of limited statehood or the State's institutional weakness.

Marcos Alan S.V. Ferreira, Ph.D., Associate Professor, Department of International Relations, Federal University of Paraiba (UFPB), João Pessoa, Brazil; Email: marcosalan@gmail.com.

Ú. Oswald Spring and H. G. Brauch (eds.), *Decolonising Conflicts, Security, Peace, Gender, Environment and Development in the Anthropocene*,
The Anthropocene: Politik—Economics—Society—Science 30,
https://doi.org/10.1007/978-3-030-62316-6_17

Such a context exists in Brazil. In the South-East and Central regions, the criminal organisation *Primeiro Comando da Capital* (PCC) [First Capital Command] has controlled not only drug trafficking but also entire neighbourhoods and the prison system in at least three Brazilian states.[1] Its business targets Europe and North America as end consumers, and uses West Africa as a transit region. The same applies to *Família do Norte* (FDN) [Northern Family], a VNSA with branches in Colombia and Peru that is engaged in managing the drug route from the Amazon River to Europe, Africa and North America, and that also controls the lives of thousands of people living in poor areas in northern Brazil.

To understand this unique scenario, this case study[2] tries to understand the PCC and the FDN governance in prisons and suburban areas under the control of these criminal organisations. Both organisations were chosen because they are similar in terms of power, resources, and their informal governance capacity. This informal governance by the two VNSAs has limited statehood and also causes institutional weakness in Brazilian territory. The research focus is to analyse and interpret the following question:

> *How is the informal governance by these VNSAs in Brazilstructured, more specifically the agency of the PCC and the FDN in suburban areas under crime control?*

A combination of sources was examined for this study. Narratives of official reports on crime were compared with news from different sources because "news resources provide unrivalled coverage and amounts of information", although "it is always a good idea to combine news reports with other sources" (Öberg/Sollenberg 2011: 71). Among these other sources were selected publications by experts on organised crime in Brazil. The period studied extended from 2006 to 2018, and thus included the year of striking violence of PCC attacks in the state of São Paulo, as well as the year of the creation of the FDN, with the 27th IPRA Conference at which this paper was first presented marking the cut-off point.

This chapter is divided in five parts. After this introduction (Sect. 17.1), the conceptual framework (Sect. 17.2) presents the grounds for the analysis, more specifically, the concept of the violent non-state actor and its governance capabilities. In the subsequent Sect. (17.3), the historical background of the two most powerful VNSAs in Brazil, the PCC and the FDN, is explained. Next, the specificities of the PCC and FDN governance in poor neighbourhoods in several Brazilian cities is analysed (Sect. 17.4). The final section presents the conclusions and identifies open questions for future research (Sect. 17.5).

[1]It has branches in several South American countries and can be regarded as a transnational actor, as it operates in drug trafficking and violent actions not only in Brazil but also in Paraguay and Bolivia, and is allied with violent groups in Colombia (Brazil 2013).

[2]Case studies are appropriate for analysis of particular topics and place emphasis on interpretation, the most distinctive characteristic of qualitative inquiry (Stake 1995: 8). Using them for this particular issue is beneficial, given the uniqueness of the governance structure of the PCC and the FDN.

17.2 Violent Non-state Actors and Governance Capabilities

In the context of social disintegration and inequality as seen in South America, illicit criminal networks plant their foundations in a fertile ground where thousands of deaths and immense obstacles to regional development rapidly accumulate (Heinemann/Verne 2006: 2). Thus, as a result of this combination of social inequality and the use of this same condition by non-state actors and often by the State itself, in many of the corners of South American society even negative peace has not been achieved (Gledhill 2015). Therefore it is not surprising to note that it is on the unequal continent of South America, in particular, that violence is maximised and crime finds its strength and lethality (Declaration of Geneva 2015; UNODC 2014).

The alarming rates of violence are a picture showing that South America is far from being a peaceful region, which makes the debate on organised crime more urgent and relevant. In this context, as affirmed by Johan Galtung (1964: 2), "there are many borderlines cross-cutting mankind, creating steep gradients in degree of integration and willingness to use violence. Only some of these borders are national borderlines".

According to the assumptions of Peace Studies, in this chapter the unit of analysis must be the causes of violence (see Wallensteen 2001; Galtung 1996). Furthermore, it is here regarded that organised crime is a non-state actor that causes or disseminates violence in the societies here examined.

The simple definition of a non-state actor can be conceptually ambiguous if we include actors who apparently "only have in common that they are not the state, nor the government" (Peters et al. 2009: 4). In general, broader definitions of non-state actors may include multinational corporations, non-governmental organisations, intergovernmental organisations, and violent actors such as guerrillas, gangs and criminal groups.

To overcome the ambiguity of the non-state concept, this chapter focuses on a very specific typology of non-state actor: the one that uses violence as a tool to achieve its goals. Mandel (2013: 42) defines *violent non-state actors*, or VNSAs, "as relatively autonomous organistions (which are not under full and direct state control) with significant and persistent coercive capacities for organised violence". Although there is some disagreement about the types of VNSAs, there is also a consensus in the literature that criminal organisations clearly shape themselves as a type of VNSA (Ezrow 2017; Mandel 2013; Schneckener 2006). In addition, as indicated by Ulrich Schneckener (2006, 2009), these groups have two additional characteristics: non-integration into formalised state institutions, and a degree of autonomy in relation to politics, military operations and resources.

Phil Williams (2008) adds that VNSAs challenge the State's monopoly of force (Williams 2008: 4). Thus, when we have both a state that is negligent in controlling some social space[3] and a VNSA with enhanced capacities in a more interconnected

[3]Social spaces are here understood according to Pierre Bourdieu's definition:

The structure of social space manifests itself, in the most diverse contexts, in the form of spatial oppositions, appropriated physical space functioning as a spontaneous metaphor for the social order.

world, we can find non-state or private actors that can "play important roles in the authoritative allocation of values for societies by developing, monitoring and even enforcing standards, rules and practices that regulate some aspect of social life" (Breslin/Nesadurai 2018: 188). Thus, a VNSA has the capacity not only to use violence, but also to create norms and rules in the territories under its control, forming new settings of authority and domination for a given population. However, these VNSAs do not always challenge the State. In some cases, the state institutions support the crime through corruption and even material support, thereby endowing VNSA operations with more power (see also Snyder/Durán-Martinez 2009; Durán-Martinez 2015).

In short, where state monopoly of violence fails to provide the basic needs of citizens in terms of physical protection, human dignity and social security – mostly in areas with a high level of structural violence – this increases the possibility of the emergence of areas that are not governed by the State. In such cases, informal governance structures led by criminal organisations emerge and unbalance social peace actors and processes, as seen in parts of Brazil.

The governance of the PCC and the FDN in the prison system is extensively studied by prominent Brazilian social scientists (see, for example, Biondi 2017; Dias/Darke 2016; Dias/Salla 2013; Manso/Dias 2018; Feltran 2018; Siqueira/Paiva 2019; Candotti et al. 2017). Nevertheless, the governance in suburban areas is still sparsely systematised in the literature, particularly under the analytical lenses of the VNSA concept and peace research. The critical perspective outlined below focuses on two violent non-state actors that have a direct impact on the lives of millions of Brazilians.

17.3 Brief Historical Background of Brazilian Criminal Organisations: The PCC and the FDN

Since the 1970s, Brazilian society has experienced the emergence of criminal groups organised in prisons. During the military dictatorship (1964–1985), criminals took advantage of interaction with leftist organisations in prisons and shaped drug trafficking groups outside the cells, one example being the *Comando Vermelho* (CV)

There is no space that does not express social hierarchies and distances in a more or less distorted fashion, especially through the effect of naturalization associated with the durable inscription of social realities in the physical world. The structure of the spatial distribution of powers records the balance of social struggles over the profits of space, which are waged individually (as indicated by mobility) and collectively (through political contests over housing policy, for instance). The stake of these struggles is the construction of spatially based homogeneous groupings, that is, segregation that is both cause and effect of the exclusive usage of a space. These profits take the form of profits of localization, rents of situation, profits of rank and profits of occupation. The ability to dominate appropriated space depends on the capital possessed, which allows one to keep undesirable persons and things at a distance and to draw desirable ones closer. Yet one can physically occupy a location without inhabiting it properly if one does not dispose of the means tacitly required for that, beginning with the proper dispositions, for it is the habitus that makes the habitat (Bourdieu 2018: 106).

[Red Command] in Rio de Janeiro (see Oliveira Filho 2013: 15). These groups expanded and eventually split up into several new ones, such as *Amigos dos Amigos* (ADA) [Friends of Friends] and *Terceiro Comando* (TC) [Third Command].

Possessing strong connections with Colombian drug cartels, from the 1970s to the 1990s the CV and other groups based in Rio de Janeiro were hegemonic in the retail of drugs in Brazil. The context changes in the 1990s, when communications technology facilitated the emergence of new criminal organisations inside prisons. Two of these organisations have grown fast and now control streets and prisons in key Brazilian states: the *Primeiro Comando da Capital* [PCC] – hegemonic in São Paulo, Paraná and Mato Grosso states – and the *Família do Norte* [FDN] – the main VNSA in the key state of Amazonas, bordering Peru, Colombia and Venezuela.

17.3.1 The Primeiro Comando da Capital (PCC)

The PCC under the command of Marcos Herbas Camacho, aka Marcola, has become one of the main VNSAs in Brazil and neighbouring countries. Created in 1993 in Taubaté Penitentiary (São Paulo state) to advocate for human rights for prisoners, since 2001 the group has been transformed into a complex organisation. A turning point in this change was seen in the mega-rebellion organised simultaneously in twenty-nine penitentiaries that involved 28,000 prisoners, showing a unique organisation of a group neglected by authorities (Adorno/Salla 2007: 4; Brazil 2006: 193; Christino/Tognoli 2017).

Five years later, the PCC exceeded the boundaries of the prison through coordinated attacks against authorities and the civilian population. The 2006 attacks paralysed the São Paulo metropolitan area and its 18 million inhabitants: 51 prisons mutinied and 251 terror attacks resulted in 90 buses set on fire and 42 policemen and prison officers killed. The strong response of the police generated numerous criticisms of the public opinion and human rights institutions. The violence statistics of these attacks and the aftermath of events are diverse. However, until 21 May at least 493 deaths were registered – 451 of them civilians – and four disappearances (see Adorno/Salla 2007: 7–8; Biderman et al. 2019).

Between 2006 and 2012, an organisational change was seen in PCC. According to Biondi (cited by Feltran et al. 2010), "Marcola has promoted the insertion of 'equality' into the motto and practices of the PCC". Since then, the PCC has operated in cells under the leadership of five members called *sintonia*[4] *final geral* (final general tune, or general fine-tuning).[5] Even though PCC claims there is no central leadership, Marcola is the de facto top leader of the organisation (Marques 2010: 312; Brazil

[4]*Sintonia* is the slang used to refer to direction; in this case, it is the central management of the PCC.

[5]Marcos Herbas Camacho, aka Marcola; Abel Pacheco de Andrade, aka Vida Loka; Rogério Geremias de Simone, aka Gegê do Mangue (killed); Roberto Soriano, aka Tiriça; Daniel Vinicius Canônico, aka Cego; Fabiano Alves de Souza, aka Paca (killed), Edilson Borges Nogueira, aka Biroska (killed); Júlio Cesar Guedes de Moraes, aka Julinho Carambola (Adorno/Costa 2018).

2013:51; Adorno/Costa 2018). Moreover, according to PCC statute, the *sintonia final* is highest level of the organisation, in which several other *sintonias* operate below – as we will see in next section (Feltran 2018: 307).

Most of the PCC's members contribute workwise and financially to the PCC. It is estimated that 80% of PCC profits come from trafficking approximately forty tons of cocaine per year, totalling US$200 million/year (Hisayusu 2016). Nevertheless, the sociologist Gabriel Feltran indicates that the money earned by the PCC from drug trafficking is higher than stated by Hisayusu (2016). For him,

> If the value of a kilogram of cocaine sold in retail in Berlin is, for example, €100,000.00, we can imagine the value in each helicopter that brings 400 kg, like the ones seized in Brazil between 2015 and 2018, bringing the drugs from Pedro Juan Caballero [Paraguay] to São Paulo state countryside, or Bahia state south, or even Ceará state. There are potentially €40 million [...] transferred *in each* aircraft (Feltran 2018: 81).

While drug trafficking is the main PCC business, the remaining profits are the result of car and bank robberies, kidnappings, arms trafficking, raffles for detainees and the monthly fee paid by prisoners and families in exchange for protection (Hisayasu 2016). Thus, the organisation is not only producing substantive profits, but also provides collective goods for 'baptised' members and invests in the organisation's development and expansion. The PCC's financial power directly reflects its geographical extension and the intensity of its members' violence.

17.3.2 The *Família do Norte* (FDN)

The *Família do Norte* (FDN) [Northern Family] is the third-largest criminal organisation in the country and has a different context than the two largest, CV (Red Command) and the PCC. The group's existence was only recently discovered, thanks to an operation entitled 'La Muralla', begun on 20 May 2014 and completed on 19 January 2016 by the Federal Police in the state of Amazonas.

The origin of the FDN was unclear until 2016, when the investigation by the Federal Police discovered that this VNSA was created in 2006 by two prisoners – Gelson Lima Carnaúba (aka 'G') and José Roberto Fernandes Barbosa (aka Zé Roberto da Compensa, Messi, or Pertuba) – aiming to control drug trafficking outside penitentiaries through the use of lawyers and cell phones. The group was a transformation of the *Primeiro Comando do Norte* [First Northern Command] and adopted the new name after breaking off its relationships with the PCC and the CV in the struggle for control of drug trafficking in the Amazon region (Candotti et al. 2017).

One of the first steps of the FDN to construct a more successful operation was the creation of the group's statute, which sets out the hierarchical character of the organisation, since all decisions must pass through the filter of the founding members and second-level leaders (Brazil 2016). Camila Dias and Bruno Manso bring to attention the similarities between FDN and PCC. They affirm that.

> The new faction wanted to take advantage of the privileged channels of drug supply in the border region and organise the retail sale in the Amazon and other states. The drugs from the FDN arrive from Colombia via the Solimões river, which offers an advantage for traffickers who are able to get into the jungle. They created status, *salves* [criminal calls for action to its members], began to charge monthly, replicating the successful management model of the PCC (Manso/Dias 2018: 37).

Flávio Costa, an experienced and reputable journalist who has covered FDN's activities over the last decade, wrote that two top leaders have distinct roles in the organisation:

> (…) in the division of tasks, José Roberto is the operative face … He was commanding a large gang of traffickers in Manaus with the origin in the neighbourhood of Compensa, and Gelson is the guy… who begins to create contacts between the leaders of the Red Command … (Costa, cited in Framento 2018).

In 2010 the FDN formed a drug distribution network with a large reach in north and north-east Brazil, extending, according to the authorities, to 13,000 members (Manso/Paes 2018: 38). These members not only controlled the drug market but also generated new forms of governance in poor areas of Manaus and other cities in the north of Brazil. Vigilante groups and rules generated by the FDN were adopted in spaces where the state has little (or no) control (Siqueira/Paiva 2019) – so-called *quebradas* (poor urban communities) – in a similar way to PCC operations in the state of São Paulo.

In 2015, the year when the PCC intensified its expansion throughout the FDN-controlled regions, the violence between the two VNSAs has intensified. In October 2016, in the Penitentiary Monte Cristo in the state of Roraima (in northern Brazil), PCC members attacked FDN members, killing ten people. In January 2017, the war between the gangs experienced unprecedented violence with a brutal FDN response. The most vicious attack was led by FDN on the first day of 2017, when fifty-six detainees, probably linked to PCC, were killed and the corpses quartered and burned.

This invisible war between the gangs is related not only to illicit markets (like drug and arms trafficking, extortion, and others), but also to the control of power in two important social spaces for Brazilian VNSAs: prison systems and poor suburban areas (or *quebradas*). The following section analyses how these governance procedures by the PCC and the FDN operate in poor urban communities or *quebradas*.

17.4 Prison Actors Governing Outside the Cells: VNSAs Controlling Lives and Territories

As the PCC and the FDN became more powerful over time in terms of human and financial resources, their ability to generate norms and to govern social groups or even entire populations also increased. This governance by a VNSA is not necessarily exclusive in a given social space. It is common to see a hybrid governance, in which the State and VNSAs compete for the power to exert control over people's lives, as seen in the case of the regulation of life and death in suburban areas of the city of São Paulo (see Willis 2015) or even in the Brazilian prisons (see Feltran 2018; Manso/Dias 2018). It is in these spaces that we find more clearly the challenge to the authority of the State and, most worrisome, the emergence of a parallel power with the capacity to affect the people.

In the last three decades Brazil has experienced the growth of VNSAs linked to organised crime that operate by taking advantage of the failure of the Brazilian State in some social spaces. In particular, such groups are considered proponents of state instability, violent conflict and conditions of insecurity. To such traditionally attributed roles is added their capacity to harm human rights, public safety, and socio-economic development (Davis 2009; Krause/Milliken 2009; Rodgers/Muggah 2009). Here our focus will be specifically on VNSA governance in suburban neighbourhoods, given the aforementioned extensive literature on governance by the PCC and the FDN in the prison system.

17.4.1 Spillover of the PCC's Governance Structure from Prisons to the *Quebradas*

In the final years of the 2000s' first decade, the prisoners 'baptised' by PCC happened to identify the potential for the group's ideology to spill over from the penitentiaries. The PCC's *disciplina* (meaning both the PCC's ideology and its rules of coexistence) created methods for conflict resolution, with violence the last resort among the 'baptised'. Among these regulations are, for example, the 'trial' (*debates*) when some discussion or betrayal emerged among prisoners. Through this *disciplina*, the PCC-controlled prisons regulated aspects of the daily lives of the prisoners and condemned rape, the use of crack and homicide (Feltran 2018: 18–19). Through the regulation of the violence inside a chaotic environment, the PCC gradually gained recognition among prisoners, extended to all Brazilian prisons and eventually became hegemonic in at least three Brazilian states – São Paulo, Mato Grosso do Sul and Paraná (Manso/Dias 2018: 333).

The practice of PCC's regulations is labelled the *proceder* (to proceed, or to behave) by its members (see more in Marques 2010). However, if a prisoner fails to follow the *proceder* of PCC, "it may be forced to transfer to other prisons or may be punished with physical violence, even death" (Dias/Darke 2017: 23). In turn, the killing of someone only can happen with the approval of PCC representatives in each prison. Moreover, this VNSA "has become a way to organise the negotiations, licit and illicit, with officials and the direction of penitentiaries" (Feltran 2018: 19).

Being a member of the PCC made the individuals more respected in the slums and poor urban neighbourhoods (*quebradas*) when they were freed. Most of them continued to commit crimes after being released from prison – official statistics show that 70% of Brazilian prisoners return to crime – but now with a subtle difference: they had learned that the well-organised regulation and practice of governance can help to achieve objectives, in the same way that the PCC managed to pacify penitentiaries through its *disciplina*. Thus, little by little, the *disciplina* also became the moral regulation for crime business in the streets.

Organised in networks in the penitentiaries under its control, the PCC emulated this model for the *quebradas*. Below the aforementioned *sintonia final geral* there are two branches. The first is organised by geographical divisions, and the second is based on issue areas. There are two geographic divisions: the São Paulo state *sintonia* and the *sintonia* for other states and countries. In São Paulo state the cells are divided according to the telephone areas. For example, in areas where the telephone prefix is 013 (the coastal region of São Paulo state), at least two cells operate: one for the prisons and one for the poor neighbourhoods (Lessing/Willis 2019; Manso/Dias 2018: 335). In the *sintonia* for other states and countries, each one has its own cells for both the prisons and the *quebradas*. All those cells have specific missions and responsibilities. The PCC's human resources are estimated to be between 20,000 and 30,000 'baptised' members spread across Brazil and abroad (Lessing/Willis 2019; Adorno/Costa 2018; Feltran 2018: 91).

The above-mentioned PCC cells which are based on issue areas are organised according to the VNSA's activities. Most recent field research points to four key tasks:

(a) the *sintonia dos gravatas* (neckties management) is responsible for hiring and paying lawyers for PCC-enrolled members;
(b) the *sintonia da ajuda* (support management) is responsible for providing basic goods or emergency support for its members and families;
(c) the *sintonia do cadastro* (enrolment management) keeps general records of enrolment, punishment and expelled members;
(d) the *sintonia financeira* (financial management) organises finance, and this is divided into sub-cells: *progresso* (for drugs); *cebola* (monthly payments of PCC members); *rifa* (raffles); *cigarro* (for cigar-smuggling and commerce) (Manso/Dias 2018: 336–337).

Within this complex governance structure, the PCC's authority and command can be understood in terms of social space (Willis 2015: 58). Such governance is not only clear in prisons but also in those social spaces in which the PCC's cells are strong and full of human and material resources, like the *quebradas*. In poor urban areas, the group emulates the governance in the prisons, disseminating their ideology and violent practices.

This is seen, for example, in the instrumental violence used to enforce the group's norms in all areas under its control. The *debates* (i.e. trials conducted according to the PCC code) have the function of making corrective procedures against individuals who are an obstacle to the criminal organisation. The *debates* often serve to eliminate enemies like prison agents who no longer co-operate with the organisation, members who betray it, and individuals involved in crimes not allowed by the PCC's code of conduct, such as the rape of women, paedophilia and corruption/misconduct in financial management. However, the penalty can also be a warning in the form of physical punishment or public humiliation.

An example of a punishment and public humiliation was described in the research of a former resident in Capão Redondo (São Paulo city), one of the poorest areas in the city and mostly under the control of the PCC. A criminal had stolen a television from a small house in 2016; however, stealing in PCC territories is usually not permitted, especially if it involves taking something from a *pessoa de bem* (ordinary citizen). The thief was subjected to a *debate*. After the 'trial', the thief received the penalty of walking in the neighbourhood with a TV on his shoulders and giving the item back to the rightful owner. Additionally, he had to move to another neighbourhood or he would be killed if new robberies were reported. The culprit never has appeared again in the area.

The story shows that there was a spillover from one social space – prisons – to another in which acceptance of the PCC's verdicts became easier given the failure of the Brazilian state. As summarised by Graham Denyer Willis,

> Recognised for its relative fairness, it influences everything from the fairness of lineups on family visitation day at prisons to the rules in prisons where not a single baptised PCC member is housed. When I did research in a PCC-controlled community, residents claimed they could sleep worry free with their doors and windows unlocked. This was something entirely new [...].

The PCC's legitimacy is thus deeply woven into social relations. Those who live under this system engage in a form of self-governance. Even where PCC is not explicitly present, its norms are. The PCC notion of right and wrong is transcendent and diffuse both for those understood to be within the 'família' [family], meaning not necessarily the *irmãos batizados* [baptised brothers], but also for all of those people living under the PCC's code (Willis 2015: 58).

PCC's notion of the right or wrong, as well as its criminal governance structures, has not only influenced poor neighbourhoods outside the prisons of São Paulo. Strong VNSAs in other states emulated PCC's ideas to put in motion similar systems of governance that filled the gap of the State in deprived social spaces.

17.4.2 Emulating the Enemy: The FDN Governance

Like the PCC, the ascendency of the FDN in prisons in the state of Amazonas also provided new forms of governance in poor neighbourhoods. Despite a less organised structure and resources, the monopoly of the latter VNSA spilled over through the streets, bringing norms, rules and practices that regulate social spaces. With subtle differences, it is clear that the FDN's actions emule the PCC's practices (Manso/Dias 2018; Framento 2018; Candotti et al. 2017; Ferreira/Framento 2019).

First of all, it is important to emphasise that the FDN has emerged as an opposition to the PCC. After the worsening of conflict between rival gangs in Amazonas's state prisons and the increasing rebellions in 2013, the PCC members were progressively transferred to penitentiaries in other states. This change in the demography brought a monopoly of FDN, making it easy for the group to impose its own rules in the prison system.

Although the PCC has a stronger presence and power throughout Brazil, the FDN has a strategic advantage: it is geographically positioned in a state near the countries that produce most of the cocaine consumed in the world, i.e. Peru, Bolivia and Colombia. The VNSA takes advantage of this context to control the drug route from the Andean countries to other lands, making it stronger in terms of material resources and influence (Framento 2018; Ferreira/Framento 2019).

In an effort similar to the PCC's, the Amazonian organisation has been seeking to exercise control of its members through an enrolment composed of name, neighbourhood and crime specialism, with members then receiving a registration number. In addition, each member must be directly connected to one of the aforementioned leaders. As in the PCC, they should contribute monthly to the financing of the group. According to Framento (2018: 92), the profits from these contributions are used to finance the group's actions, mainly drug trafficking and alleviation of the precarious conditions of prisoners in the state of Amazonas (see also Brazil 2016).

Controlling the drug outlets in the main conurbation in northern Brazil, Manaus, FDN operates in small 'teams' or cells. These teams control the drug business and also perform violent acts. Analogous to its enemy, the FDN's practices include 'crime tribunals', at which a Council decides the punishment in cases of murder, violent robberies, kidnappings and torture in the spaces controlled by that VNSA. These courts regulate the crimes that may be committed, and may even endorse execution in the case of non-compliance. It is also common to execute individuals in a symbolic way, such as recording the acronym of the group on the victim's body (Brazil 2016). Thus, the FDN virtually assumes the role of the State, providing a *sui generis* kind of governance through the dissemination of violence and decisions about the lives of thousands of people.

Some members of these teams not only use and control violence in FDN-controlled areas, but also provide funding for social activities considered important in their communities, for instance the football contests in neighbourhoods like Compensa, regarded by FDN as the property of one of its leaders, José Roberto (Brazil 2016). Thus, as said by two experts, FDN goes beyond the notion of a prison gang. An example of this is the well-known mobilisation of *bondes* (death squads) and groups of assailants that reinforce the 'vision' in their considerations about crime, including the participation of women in the ethics propagated by the FDN (Siqueira/Paiva 2019).

Evidence from police investigations and academic research shows that women's participation in FDN's practices is more frequent than in other criminal organisations operating in Brazil (Bastos 2018; Siqueira/Paiva 2019). Luciane de Lima, wife of one of the FDN's key leaders, José Roberto, was responsible for accountability and the Amazonian VNSA's money laundering after the imprisonment of her husband. Similarly, Pamela Pereira, wife of Gelson Carnaúba, was responsible for FDN's finance (Bastos 2018).

Another example of the FDN's power and authority is its influence in regional elections. A scandal involving the criminal organisation was revealed by the press in the 2014 elections. The magazine *Veja* reported that Carliomar Brandão, the Under Secretary of Justice and Human Rights of the state of Amazonas, had met with

José Roberto at the COMPAJ Penitentiary to negotiate the support of the group for the re-election of the Governor José Melo. In the recording obtained by the magazine, the group leader promised more than 100,000 votes and, in turn, the Under Secretary promised that the government would not intervene within the FDN-dominated prisons. Later, Brandão stated that the visit was actually intended to avoid deaths inside the prison (Leitão 2017). The Under Secretary was eventually dismissed and the Governor eventually re-elected (Ramalho 2017; Prazeres 2017).

17.4.3 A Governance for or Against the Oppressed?

Although the governance of criminal groups in poor neighbourhoods in areas controlled by the PCC and FDN is eventually seen as positive by the local population due to the sense of security (Willis 2015: 57–58), it is not really apparent that such influence creates a more peaceful environment of social justice. It is noticeable that this governance has been much more characterised by monopolising violence, most often for the benefit of businesses run by the criminal organisations, than by improving the local conditions of the population that is not involved in crime. To summarise, the *sintonias* from the PCC's organogram, or the 'teams' in the FDN structure, do far more to serve the crime business than to promote positive peace in the *quebradas*.

The complex governance structure involving dozens of cells with specific aims, especially in the PCC but also in the FDN, does not aim to improve the poor conditions in Brazilian urban peripheries. Using the words of PCC's statute, the key here is "to participate in projects to create solutions to social and financial abandonment [of its members] to support the helpless *members*" (Feltran 2018: 306). In other words, the attention is directed at enrolled members, but occasionally this focus can also benefit other people linked to associates.

Nevertheless, much of the PCC's and the FDN's legitimacy arises from the unsatisfactory performance of the State in poor areas. In a setting of high structural violence and, in poor areas, a weak state presence embedded with an authoritarian approach that creates more tension than pacification (see Gledhill 2015), there is room for the triumph of VNSAs' legitimacy. Thus, in areas where the State is barely present – or, when present, inflicts severe violence on the poor – crime is the saviour that brings at least some hope of justice for citizens who are often the victims of structural and cultural violence.

Regarding the VNSAs' violence management through force or crime tribunals, the theme here is not governance for mitigating inequality or for the betterment of *quebradas*, but again to run activities for the advancement of the VNSA. The individual who is punished by the PCC or the FDN is often someone who challenged the VNSA's business or its management of its illicit (criminal) activities. Eventually the bandit who stole a TV from the ordinary citizen is severely punished, but this castigation aims to gain legitimacy from ordinary citizens to continue to carry out drug trafficking and other activities in the neighbourhood. Thus, often the victim is someone inside the poor areas, and not an agent of oppression of the population. Here a more just society is not at stake, but a justice that serves to legitimise and carry out illegal business, making the crime stronger.

There is also a paradox in the PCC and FDN governance slogans and practices which claim to defend the helpless or the oppressed (see paragraph 18 of the PCC statute cited by Feltran 2018: 309). The defence of the poor people from the urban areas of Brazil is made through the recruitment of this same public for their business. Young people without prospects in an unjust and unequal society like Brazil's become involved with criminal gangs by performing small activities to earn some money. These activities include simply giving alerts to other criminals when the police are approaching. Gradually these youths are engaged in more complex activities until they reach a position of respect in the organisation. Or, as is also common, these youths are arrested and learn in the penitentiary to perform more complex crimes with PCC and FDN top leaders.

The paradox here is that by engaging these young people in crime, the VNSAs are at the same time reproducing the conditions that they criticise, becoming also an agent responsible for both direct and structural violence. In other words, as summarised by Blackwell and Duarte (2014: 111–112):

> […] social exclusion becomes a form of structural and cultural violence which prevents thousands of people from achieving personal fulfillment. This structural violence disproportionately affects the society's most vulnerable members, such as women, youth, and ethnic minorities. Members of organised crime groups and gangs have often responded to tough-on-crime government policies by recruiting and involving youth to partake in their criminal activities.

Thus, the structural violence that prevents people from having opportunities also reproduces violence when the same people become more engaged in crime after being recruited by powerful VNSAs. Structural violence is then useful for criminal organisations in both directions: legitimising the discourse that the crime is defending the oppressed, but also providing the conditions to recruit more and more young people for their business.

17.5 Conclusions

The governance by VNSAs in Brazil has been transformed over time. At first this governance system focused on the prison system to advocate for human rights and better conditions for detainees. Progressively these demands spilled beyond the cells, reaching poor neighbourhoods and creating new practices and norms to regulate social spaces.

At the same time that the PCC and the FDN try to keep a pacific governance in the areas where they are hegemonic, their territorial expansion tends to disseminate more violence. Furthermore, the State's inability to act effectively within the most deprived areas in the cities, especially in the promotion of education and income, provides space for these VNSAs to achieve legitimacy by promoting rules of conviviality, resolving litigation and also promoting material goods to their criminal members that eventually spill over to populations living in poverty. In addition, the groups themselves are a source of income for many of the young people most impacted by the cultural, structural and direct violence that plagues the poorest neighbourhoods in urban areas.

Here it is clear that it is difficult to imagine a hybrid violence governance system generating peace (positive and negative). Pablo Policzer (2005: 3) argued that "creating sovereign states – with the right and the capacity to monopolise coercive force inside their territories – is the most effective way to achieve peace and order." Moreover, "it is far better to live under states that have the monopoly of force [...], than outside states and be prey to roving bands of gangsters, criminals, and rebels." Nevertheless, the scenario tends to be worrying if we consider the expansion of Brazilian VNSA domestically and abroad[6] (Manso/Dias 2018; Adorno/Costa 2018; Ferreira 2019), weakening the social fabric and generating more violent divisions in social spaces.

When we take into account that the influence of both the FDN and the PCC is expanding and that they are violent non-state actors with "sustained coercive capabilities for organised violence" (Mandel 2013: 42), some questions are raised for future research, such as: Will the continuous expansion of VNSAs generate more clashes among the groups and against the State? Will the State continue to neglect poor areas when the killings are concentrated in social spaces with high structural violence?

Whatever the response to those questions, the fact is that the violence will continue to plague the lives of millions of Brazilians in the following years. There is no clear public policy to undermine the power of VNSAs nowadays. Consequently, unfortunately those questions raised above show that the issue will remain an important subject for analysis by peace researchers and social scientists in general who aim at a more peaceful and ordered Brazilian society.

References

Adorno, Luis; Costa, Flavio, 2018: "PCC em Expansão", in: *UOL Notícias*; at: https://www.uol/noticias/especiais/pcc-o-partido-do-crime.htm#pcc-em-expansao.

Adorno, Sergio; Salla, Fernando, 2007: "Criminalidade Organizada nas Prisões e os Ataques do PCC", in: *Estudos Avançados,* 21,61: 07–29.

Bastos, Isabela, 2018: "De Donas de Casa a Chefes do Tráfico: um olhar sobre as Mulheres do Crime", in: *Portal Em Tempo*, 17 March; at: https://d.emtempo.com.br/policia/97768/de-donas-de-casa-a-chefes-do-trafico-um-olhar-sobre-mulheres-no-crime.

[6]In an interview with the *The Wall Street Journal* in 2017, the Colombian Defence Minister Luis Carlos Villegas expressed concern, saying that Brazilian criminal organisations are courting the FARC dissidents to obtain heavy weapons and military training, while also hoping to expand into the highly competitive Colombian criminal territory (Jelmayer et al. 2017).

Biderman, Ciro; Lima, Renato Sérgio de; Mello, João Manuel Pinto de; Schneider, Alexandre, 2019: "Pax Monopolista and Crime: The Case of the Emergence of the Primeiro Comando da Capital in São Paulo, in: *Journal of Quantitative Criminology*, 35: 573–605.

Biondi, Karina, 2017: "Movement Between and Beyond Walls: Micropolitics of Incitements and Variations among São Paulo's Prisoners' Movement the 'PCC' and the Prison System", in: *Prison Service Journal*, 229 (January): 19–23.

Blackwell, Adam; Duarte, Paulina, 2014: "Violence, Crime and Social Exclusion", in: OAS – Organization of American States (Ed.): *Inequality and Social Exclusion in the Americas: 14 Essays* (Washington: OAS): 111–134.

Bourdieu, Pierre, 2018: "Social Space and the Genesis of Appropriated Physical Space", in: *International Journal of Urban and Regional Research*, 42: 106–114; at: https://doi.org/10.1111/1468-2427.12534.

Brazil, The Chamber of Deputies, 2006 (November): *Relatório da Comissão Parlamentar de Inquérito destinada a Investigar as Organizações Criminosas do Tráfico De Armas* (Brazil: Câmara dos Deputados).

Brazil, Polícia Federal, 2016: "Superintendência Regional no Amazonas", in: *Operação La Muralla* (Manaus: Delegacia de Repressão a Entorpecentes).

Brazil, São Paulo State Public Prosecutors, 2013: *Procedimento Investigatório Criminal nº 336/10*, Presidente Venceslau (9 September).

Breslin, Shaun; Nesadurai, Helen E.S., 2018: "Who Governs and How? Non-State Actors and Transnational Governance in Southeast Asia", in: *Journal of Contemporary Asia*, 48,2: 187–203; at: DOI: https://doi.org/10.1080/00472336.2017.1416423.

Candotti, Fábio Magalhães; da Cunha, Flávia M.; Siqueira, Ítalo B. L., 2017: "Crime e Estado no Amazonas", in: *Le Monde Diplomatique Brasil*, 28 March; at: https://diplomatique.org.br/crime-e-estado-no-amazonas/.

Christino, Marcio; Tognoli, Claudio, 2017: *Laços De Sangue – A História Secreta do PCC* (São Paulo: Matriz).

Davis, Diane, 2009: "Non-State Armed Actors, New Imagined Communities, and Shifting Patterns of Sovereignty and Insecurity in the Modern World", in: *Contemporary Security Policy*, 30,2 (August): 221–245.

Dias, Camila; Darke, Sacha, 2016: "From Dispersed to Monopolized Violence: Expansion and Consolidation of the Primeiro Comando da Capital's Hegemony in São Paulo's Prisons", in: *Crime Law and Social Change*, 65: 213–225.

Dias, Camila; Salla, Fernando, 2013: "Organized Crime in Brazilian Prisons: The Example of the PCC", in: *International Journal of Criminology and Sociology*, 2: 397–408.

Durán-Martinez, Angelica, 2015: "To Kill and Tell? State Power, Criminal Competition, and Drug Violence", in: *Journal of Conflict Resolution*, 59,4 (June): 1–27; at: DOI: https://doi.org/10.1177/0022002715587047.

Ezrow, Natasha, 2017: *Global Politics and Violent Non-State Actors* (Thousand Oaks, CA: Sage).

Feltran, Gabriel, 2018: *Irmãos: Uma História do PCC* (São Paulo: Companhia das Letras).

Feltran, Gabriel; Dias, Camila N.; Biondi, Karina; Marques, Adalton, 2010: "16 Perguntas sobre o PCC – Crimes no Brasil", in: *Estado de São Paulo*, 23 January.

Ferreira, Marcos Alan S.V., 2017: "Criminality and Violence in South America: The Challenges for Peace and UNASUR's Response", in: *International Studies Perspectives*, 18,1: 64–80.

Ferreira, Marcos Alan S.V., 2019: "Brazilian Criminal Organizations as Transnational Violent Non-State Actors: a case study of the Primeiro Comando da Capital (PCC)", in: *Trends in Organized Crime*, 22,2: 148–165.

Ferreira, Marcos Alan S.V.; Framento, Rodrigo, 2019: "Degradação da Paz no Norte do Brasil: O Conflito Entre Primeiro Comando da Capital (PCC) e Família do Norte (FDN)", in: *Revista Brasileira de Políticas Públicas e Internacionais*, 4,2: 91–114.

Framento, Rodrigo, 2018: *A Degradação da Paz no Norte do Brasil: Um Exame a partir da Violência entre Primeiro Comando da Capital (PCC) e Família do Norte (FDN)* (Monograph – BA in International Relations, Universidade Federal da Paraíba, João Pessoa, June 2018).

Galtung, Johan, 1964: "Editorial", in: *Journal of Peace Research*, 1,1: 1–4.
Galtung, Johan, 1969: *Peace by Peaceful Means: Peace and Conflict, Development and Civilization* (London: Sage).
Geneva Declaration, 2015: *Global Burden of Armed Violence 2015: Every Body Counts* (Cambridge: Cambridge Univ. Press).
Gledhill, John, 2015: *The New War on the Poor: The Production of Insecurity in Latin America* (London: Zed).
Heinemann, Andrea; Verner, Dorte, 2006: "Crime and Violence in Development: A Literature Review of Latin America and the Caribbean", in: *World Bank Policy Research Working Paper 4041* (Washington, D.C.: the World Bank).
Hisayasu, Alexandre, 2016: "PCC 10 Anos: O Poder Financeiro", in: *Infográficos Estado*; at: https://infograficos.estadao.com.br/cidades/dominios-do-crime/poder-financeiro (10 February 2018).
Imbusch, Pierre; Misse, Michel; Carrión, Fernando, 2011: "Violence Research in Latin America and the Caribbean: A Literature Review", in: *International Journal of Conflict and Violence*, 5,1: 87-154.
Jelmayer, Rogerio; Vyas, Kejal; Pearson, Samantha, 2017: "Brazilian Gang Enlists FARC Rebels for Drug Trade", in: *The Wall Street Journal* (21 January).
Krause, Keith; Milliken, Jennifer, 2009: "Introduction: The Challenge of Non-State Armed Groups", in: *Contemporary Security Policy*, 30,2: 202–220.
Leitão, Leslie, 2017: "Governo do Amazonas Negocia Apoio de Traficantes para o 2º Turno", in: *Portal Veja*; at: https://veja.abril.com.br/blog/reinaldo/governo-do-amazonas-negocia-apoio-de-traficantes-para-o-2-turno/ (1 April 2018).
Lessing, Benjamin; Denyer Willis, Graham, 2019: "Legitimacy in Criminal Governance: Managing a Drug Empire from Behind Bars", in: *American Political Science Review*, 113,2, 1–23.
Mandel, Robert, 2013: *Global Security Upheaval: Armed Nonstate Groups Usurping State Stability Functions* (Redwood City, US: Stanford Security Studies).
Manso, Bruno; Dias, Camila, 2018: *A Guerra: A Ascensão do PCC e o Mundo do Crime no Brasil* (São Paulo: Todavia).
Marques, Adalton, 2010: "'Liderança', 'Proceder' e 'Igualdade': Uma Etnografia das Relações Políticas no Primeiro Comando da Capital", in: *Etnográfica: Revista do Centro em Rede de Investigação em Antropologia*, 14,2: 311–335.
Möller, Frida, 2011: "News Reports and Written Narratives: Collecting Information using Different Types of Empirical Sources", in: Höglund, Kristine; Öberg, Magnus: *Understanding Peace Research: Methods and Challenges* (London: Routledge).
Öberg, Magnus; Sollenberg, Margaretta, 2011: "Gathering Conflict Information Using News Resources", in: Höglund, Kristine; Öberg, Magnus: *Understanding Peace Research: Methods and Challenges* (London: Routledge).
Oliveira Filho, Roberto Gurgel, 2013: "O Tratamento Jurídico Penal das Organizações Criminosas no Brasil" (PhD thesis, Pontifical Catholic University of Rio de Janeiro); at: https://doi.org/10.17771/PUCRio.acad.21215.
Peters, Anne; Koechlin, Lucy; Zinkernagel, Gretta F., 2009: "Non-State Actors as Standard Setters: Framing the Issue of an Interdisciplinary Fashion", in: Peters, Anne; Koechlin, Lucy; Förster, Till (Eds.): *Non-State Actors as Standard Setters* (Cambridge: Cambridge University Press).
Policzer, Pablo, 2005: "Neither Terrorists nor Freedom Fighters", Paper presented at the annual meeting of the International Studies Association, Honolulu, Hawaii, 5 March 2005.
Prazeres, Leandro, 2017: "Investigado por Negociar Votos, Ex-Subsecretário da Justiça teve 3 Encontros com Líderes da FDN", in: *UOL Notícias;* at: https://goo.gl/bvibW1 (22 May 2018).
Ramalho, Renan, 2017: "TSE Decide Manter Cassação do Governador do Amazonas", in: *Portal G1 Amazonas*; at: https://goo.gl/i84w2a (3 May 2018).
Rodgers, Dennis; Muggah, Robert, 2009: "Gangs as Non-State Armed Groups: The Central American Case", in: *Contemporary Security Policy*, 30,20 (August): 301–317.

Sain, Marcelo; Games, Nicolas, 2014: "Tendências e Desafios do Crime Organizado na América Latina", in Reginaldo Nasser; Rodrigo Moraes (orgs.), *Brasil e a Segurança no seu Entorno Estratégico: América do Sul e o Atlântico Sul* (Brasília: IPEA): 119–144.
Schneckener, Ulrich, 2006: "Fragile Statehood, Armed Non-State Actors and Security Governance", in: Bryden, Alan; Caparini, Marina (Eds.): *Private Actors and Security Governance* (Berlin: LIT - DCAF): 23–41.
Schneckener, Ulrich, 2009: *Spoilers or Governance Actors? Engaging Armed Non-State Groups in Areas of Limited Statehood*, SFB-Governance Working Paper Series, No. 21 (October), Research Center (SFB) 700 (Berlin: Free University of Berlin).
Siqueira, Ítalo B. de L.; Paiva, Luiz F.S., 2019: "'No Norte tem Comando': As Contradições e os Efeitos Políticos do Encarceramento em Massa", in: *Revista Brasileira de Sociologia*, 7,17: 125–154.
Snyder, Robert; Durán-Martínez, Angelica, 2009: "Does Illegality Breed Violence? Drug Trafficking and State-Sponsored Protection Rackets", in: *Crime, Law and Social Change*, 52,3, 253–273.
Stake, Robert E., 1995: *The Art of Case Study Research* (London: Sage).
UNODC (United Nations Office on Drugs and Crime), 2014: *World Drugs Report 2014* (Vienna: UNODC).
Wallensteen, Peter, 2001: "The Growing Peace Research Agenda", in: *Kroc Institute Occasional Paper 21*, Op. 4 (South Bend: Univ. of Notre Dame).
Williams, Phil, 2008: "Violent Non-State Actors and National and International Security", in: *International Relations and Security Network*; at: https://www.files.ethz.ch/isn/93880/VNSAs.pdf (11 June 2017).
Willis, Graham Denyer, 2015: *The Killing Consensus: Police, Organized Crime, and the Regulation of Life and Death in Urban Brazil* (Oakland: University of California Press).

Chapter 18
Art of Peace: Cultural Practices and Peacebuilding in Mexico

Alfonso Hernández Gómez

Abstract The last decade has been a constant bloodshed in Mexico due to drug wars and as a consequence of the lack of an effective strategy against this critical problem. The government has focused on the deployment of the military and the police to control the drug cartels and local gangs, with poor results. The main question is how to mitigate violence and what kind of strategies are effective to solve this problem. This article explores different approaches of peacebuilding done by local artists and organisations in communities with high levels of violence. Using ethnographic research and direct fieldwork on some of the most violent neighbourhoods in Mexico, I want to show how people have organised themselves and resist violence using cultural practices like murals, hip-hop or artistic interventions in public space. Following peace theories, I use the concepts of structural, cultural and interpersonal violence to frame the cultural practices and understand their art as a contribution to peacebuilding that can transform their communities. This study contributes to the current debate on innovative strategies to mitigate violence, the possibility of cultural activities for peacebuilding, and the alternatives to security policies on violence prevention in Mexico.

Keywords Peacebuilding · Cultural practices · Urban violence · Street art · Youth · Resilience and engaged research · Visual anthropology

18.1 Introduction: Why We Should Stop Talking About Violence?

Mexican and Latin American scholars address the high levels of violence in Mexico (Reguillo 2010), and how dangerous some cities and neighbourhoods are. The only image that Mexico has abroad is the stigma of being dominated by drug cartels and criminals, and sadly this is partially true. Cultural industries like Hollywood, Netflix and TV shows, and musical trends, like gangster rap in the US (Chang 2003) and

---

Mr Alfonso Hernández Gómez, Ph.D. student, UAM-I, Mexico City;
Email: ahimsamusik@gmail.com.

Ú. Oswald Spring and H. G. Brauch (eds.), *Decolonising Conflicts, Security, Peace, Gender, Environment and Development in the Anthropocene*, The Anthropocene: Politik—Economics—Society—Science 30,
https://doi.org/10.1007/978-3-030-62316-6_18

the 'Narcocorridos' in Mexico (Valenzuela 2003), have attracted many viewers with their sensationalised treatment of violence (Wood 2018; Valkenburg/Taylor 2017).

There are many academic analyses of violence in Mexico (Giménez 2017; Salazar 2015; Hernández 2010; Valencia 2010; Reguillo 2010; Valenzuela 2015; Domínguez 2015; Barrios 2014), from journalism, the social sciences, criminology and literature, where intellectuals are looking for answers and alternatives to solve this key problem. These studies offer a broader understanding of its causes, consequences, and development, but there is still a lack of studies and testimonies that highlight efforts to react to violence which are not official, but provided by ordinary citizens.

This chapter focuses on the peacebuilding actions of unknown and extraordinary people in daily life and from marginalised communities, and gives 'voice' and 'visibility' to those who struggle against this current violence. In the social sciences there is great interest in studying the causes, the history and the narratives of violence, so we must discuss how academia could offer answers. This chapter does not discuss violence, but proposes to shift the perspective in academic work towards a culture of peace, to stop studying criminals and to focus on the people who are resilient and the communities that resist violence. This creative and vital force will be considered below.

This article shows the findings of an ongoing project that examines the role and usages of cultural practices for peacebuilding in violent contexts in Mexico. This study is complemented with a documentary about the projects I work with, and the impact they may have on mitigating violence in their communities. In this documentary I conveyed the stories of artists, cultural managers, organisers and activists who deal with violence in Mexico and exercise peacebuilding through arts. I will show some examples of these peacebuilding practices, to show why they are relevant, and how they contribute to the efforts of preventing, healing, and decreasing violence in Mexico.

Moreover, the documentary works as a methodological tool of research and communication, making use of images to complement the text. Photographs and videos enabled me to generate a dialogue and an understanding of the participants' perspective of their cultural practices. I could also communicate this to a broader audience, initiating a new exchange of ideas with the people portrayed, the real heroes of this story. With the use of video I developed a participatory and engaged methodology which gave me the opportunity to explore more deeply the meanings and symbolic potential of what I call the *art of peace*.

18.1.1 The Context of Violence in Mexico

Mexico has been suffering from a severe crisis of violence and insecurity for more than twelve years, with the beginning of a supposed drug war resulting in a bloodbath with thousands of lives lost for no reason. Since 2006 violence has increased constantly and there is no sign of the situation improving in the future. According to

Table 18.1 Homicide rates in Mexico between 2011 and 2017

Year	2011	2012	2013	2014	2015	2016	2017	Total
Homicides	27,213	25,967	23,063	20,010	20,762	24,559	31,174	172,748

Source INEGI, Índice de mortalidad por homicidios en México (September 2018)

the Global Peace Index's vision of humanity, the peace in Mexico is degrading and there is no guarantee of security for the people (GPI 2018).[1]

Mexico is currently living in the midst of a drug war, and the results are gruesome: more than 200,000 dreadful homicides since 2006 (HRW 2011), more than 40,000 missing people (IBERO 2017), nine femicides per day (El País 2019), and a feeling of insecurity everywhere. Another consequence of violence is at the cultural level, in the social feeling of fear, the psychological assimilation of danger and the symbolic normalisation of death in daily life. The cultural impact of violence is seen mainly in childhood and youth (World Bank 2012) by those who are raised in an environment of permanent risk and loss, deprived of social bonds and prey to marginalisation.

In recent years, Mexico has been among the top five countries (without a declared war) with the highest homicide rate in the world, and is one of the most unsafe and least peaceful countries in the world (GPI 2019), with the homicide rate rising from 9,000 in 2004 to 10,000 in 2006, 25,000 in 2010 and more than 30,000 in 2017 (INEGI 2018).

Table 18.1 shows the track of homicides in recent years in Mexico. It illustrates that the tendency is for the number of homicides to increase each year. Although the last government invested massive resources in militarisation, that did not reduce the criminal violence, but made it worse.

Along with this problem of the drug cartels and their fight in the drug trade, other forms of violence have spread across the whole country, such as gender violence, femicides (Segato 2016), poverty, youth gangs (Nateras 2013), forced disappearances, and political violence (WB 2012). In marginal and peripheral spaces, corruption and impunity also play a role in the replication and increase of violence. Under these many forms of violence, the interplay of structural factors and cultural settings reproduces this violent scenario, providing the infrastructural conditions for the existence of a culture of violence (Auyero 2012). This stands for the social reproduction of meanings and identities on people raised and shaped in violent environments.

In the present context, violence is systematically used for many different political and economic interests, impacting on the daily life of a large excluded population living on the outskirts of towns or in the abandoned countryside of Mexico. The work of Argentinian sociologist Javier Auyero, who explains the endemic violence and aggressive behaviour in the shanty towns of Buenos Aires, illustrates the way that structural violence is embodied and signified as part of the inhabitants' identity,

[1] See: "Mexico Peace Index Overall Score"; at: https://visionofhumanity.org/indexes/mexico-peace-index.

as a frame for understanding reality. Auyero considers violence a 'total social fact'[2] (Barrios 2014) that gives meaning to life. Without being so deterministic, I would understand violence in Mexico as a component of reality. While it is necessary to understand the cultural meanings of death and aggression in the current Mexican context, it is also important to understand the impact of everyday violence on the identity of the people who suffer it permanently.

In Mexico the State itself has become part of the criminal economy, and politicians are the business partners of the cartels (Hernández 2010, 2019), a topic that has been denounced and portrayed in journalism, literature and the cinema. Criminality has effectively become part of Mexico's national culture (Domínguez 2015).[3] Many political struggles make dealing with corruption the target of their movement because of this state of generalised impunity. The present government regards corruption as the country's biggest challenge (Beyerle).[4] The institutions lack transparency and no longer provide security to the citizens; this weakens the legitimacy of the State regarding its capacity to fight against this problem (Ayala 2013). In the case of Mexico, police corruption, military violence and the government representatives involved in organised crime have made citizens lose trust in the State. Now citizens are uniting and have armed themselves to provide security to their families and territories against criminal gangs, as seen in rural areas of Michoacán and the mountains of Guerrero (Gledhill 2013: 91).[5]

This is the present scenario in Mexico and that is the challenge for citizens, as part of the problem but also the solution. I want to explore the different approaches to transforming the problems of violence and creating the foundations for a lasting peace. Communities are coming together and looking for alternative paths, because violence only brings more violence, and we just want to live. From this perspective, the present research is a way to contribute to the efforts to achieve peace.

18.2 Researching Peacebuilding: A Methodological Approach

In the summer of 2015 in Ciudad Juarez, when the level of violence was supposed to be going down, and the city was recovering from a severe crisis that had positioned it as the most violent city in the world in 2011 and 2012 (INEGI 2012), I was with some of my local contacts, waiting for Lulu, a community and youth worker in the

[2]This is a reference to Marcel Mauss' concept that means: "an activity that has implications throughout society, in the economic, legal, political, and religious spheres" (Mauss 1971: 56).

[3]For some examples see: *El Infierno* (Luis Estrada 2010), and *Heli* (Amat Escalante 2013).

[4]The current President of Mexico, López Obrador, reached power with a discourse against corruption.

[5]For a closer look, there are documentaries that portray this situation, for instance: *Cartel Land* (Matthew Heineman 2015), *Ayotzinapa, el Paso de la Tortuga* (Enrique García Meza 2018), *Guerrero* (Ludovic Bonleux 2017), *La Libertad del Diablo* (Everardo González 2018), and *La Tempestad* (Tatiana Huezo 2016).

neighbourhood. I didn't know her yet; it was the very first time I would meet her, but I had heard much about her: she had worked for more than ten years with kids and teenagers, trying to rescue them from drug cartels, gangs, addictions, and other risks of street life. My contacts in the city recommended me to interview her, so we were there in her house, waiting with great expectation. Suddenly she arrived, but remained outside the house taking a phone call. It was already dark when she came into the house and sat in the middle of the room. We were there looking at her, and without even saying 'Hola', she just exploded:

> One of my boys was killed last night. His body was found just now, so the boys called me to let me know. He had disappeared two days ago, and now they've found him. Why? He is the second this week. The other boy was just seventeen. He was beaten badly, tortured, his fingers were cut off.[6] Why? Two in this week! How many more?

She held her face in her hands, trying to calm the imminent crying.

I was shocked, like all the people who were there listening to her, at the glimpse of what happens every day on the solitary hills of Anapra. While my contacts talked with her about the issue and tried to calm her, she turned her eyes to me, and asked: "Who is he?" They introduced me to her, explained what I was doing, and asked her about doing an interview later, if she could. She turned to me again and asked the question: "Are you a journalist? Because I don't want to talk about this to anyone. You are just interested in morbid news. I don't like journalists." I replied that I was an anthropology student who was collaborating with the organisation *Del barrio a la comunidad*.[7] So she said: "Okay, but do you think that you can do something useful? Have you heard what just happened? Would your interview be of any help? Can you do something to help stop this happening?" I didn't really know what to say. I was shocked by her testimony. I started going to Anapra and looking her work with the kids. After some time of visiting her, she agreed to give me the interview. It was a long conversation. She told me about her life, her childhood, the story before she came to Juarez, her pain of raising a daughter alone, the harsh life in Anapra, and the beginning of her work in the community, giving food to homeless kids. That long interview created a bond between us. I was inspired and also motivated to help her in some way, at least from a distance. I gave her a big hug when the interview ended, as the only way to give back, at least for the moment.

Many similar episodes with other 'informants' happened during my fieldwork. You create connections and empathise with the people you want to study and work with. As an anthropologist, when you study highly emotional realities and get closer to the people involved, it's very difficult to just have the position of the researcher, since you are part of the same reality and also have a relationship with the people. This is what Fals Borda, a Colombian sociologist, declared as the principles of research

[6]Later on someone explained me that it was the symbol of punishment for thieves of the narco, to cut off hands or fingers of those who steal from them.

[7]"From the neighbourhood to the community"; it is a municipal programme which aims to prevent violence in the city. Its organisers were my primary contacts when exploring the suburbs and neighbourhoods where violence is high and almost endemic.

that is also action, a means of gaining knowledge that requires both the mind and the heart (Falso-Borda 1987).

In his *participatory action research* (PAR) approach he described the different ways in which research can be part of transformative practices, and an aid to the communities we claim to study (Fals-Borda/Muhammad 1991). The PAR principles applied to this research mainly consist of having a subject-subject, horizontal relationship, that tries to make beneficial changes with the engagement of the same people who are affected by certain problems. The way I found to do this was by visual anthropology (Zirion/Dorotinsky 2017), using photography and video not only as tools to record evidence and keep track of the fieldwork, but also as a means of capturing the symbolic elements of reality (Rouch 1973) and as a channel of communication with the subjects of this study. I sometimes used photography as a method of intervention in the field, to build trust, create a bond with the people I was portraying, and also give me a role to play when I was with them. In this sense, the camera was more than just a tool; it became a key to access my subjects' point of view, understand their culture and obtain evidence of their activities, which was very useful and necessary in some cases (Paniagua 2007).

Mindful of this, I had to undertake the research in a way that was of some use against violence and of some help to the social actors I was working with. This approach was applied to the research into cultural projects as part of a process involving the community and trying to address its needs in some way. This was also the approach of the development agencies that are working hand in hand with the community and are making use of cultural projects to this end (Barrios/Chaves 2014). I talked to people, and asked about the best way my research could be of some help or of any interest to them. One way we agreed that I could be of some help and the research useful was to tell their stories, to let the world know that there are empowered people who create some positive changes, women and men who do not accept violence as an unavoidable evil, but understand it as a dynamic phenomenon that can change.

As part of my methodology to make contact with people, I have experience as a cultural promoter and activist, which helped me to forge links with many organisations in different parts of Mexico. I did a 'snowball' process of meeting people, and came to know social actors like Lulu while working in Ciudad Juarez. In every city a similar approach was employed. The criteria for choosing the cities were the official data regarding those cities (mainly based on the homicide rate), their history of connections with the criminal economy, and media reports. The goal was to explore some of the most violent communities in Mexico and, once in them, focus on the cultural projects in sensitive neighbourhoods. With my local contacts and participatory observation (Marshall/Rossman 1989), I selected the specific localities, cultural projects, single artists, and organisers to work with. That's why the research was done in five different localities in Mexico: in Tijuana (Baja California), Ciudad Juarez (Chihuahua), Acapulco (Guerrero), Ciudad de Mexico and Ecatepec (Mexico).

Once I had made the selection of ethnographic work in diverse social contexts of violence, I chose case studies based on the following criteria:

- Independent projects that do not depend on the Government;
- Community engagement and participatory processes;
- Use of cultural practices and creativity for peacebuilding;
- Working with marginalised communities and victims of violence.

My research methodology aims to foster collaboration with the communities and strives to be a problem-solving knowledge that focuses on the subjects and their practices. The PAR approach is used in Latin America by the postcolonial schools of sociology (de Sousa Santos 2009), who seek a non-colonising research method (Tuhiwai 2012: 8) and want to give more voice and reflexivity to the people forming part of the research (Kraniauskas 2015). So both approaches, the collaborative and the decolonising, which are characteristic of cultural studies (Hall 2019), guided my way of forging contacts and shaped my participation on the ground. Of course, the result was somewhat different from traditional research, based on the subject-object relationship and extractive techniques of traditional sciences. Instead, this research was a process of dialogue between two subjects (Fals Borda 1991) in need of each another.

My involvement with the participants and the relationship of the research with their projects triggered a process of dialogue with local actors, conducted not only by the researcher's interests but by the people as well. Because of this style of collaboration, I worked alongside the local people, in a process of mutual learning. The way I did it was mainly by the use of photography. I acted as a cameraman, documenting and following up different projects and giving photos and videos to the participants for personal use.

My goal in the field was to generate co-learning and collective action, both relevant aspects of PAR methodologies (Ander-Egg 1990). Through the use of the audiovisual component of the research, I had enough information to communicate the stories to the people, and the projects I study, and they had a visual testimony of their work, which is important to highlight their local efforts towards peace. In this sense, we both gained something: they now have a video to show to the world and to evidence the violence they experienced and their strategies to change it. Now they have more visibility in front of different audiences and their message continues to be transmitted and spread. In this sense, the research has been useful and meaningful for some of the participants. They have seen the documentary and were satisfied and grateful for being part of this project: in that sense I am creating a community with them and with those who watch it.

I realised the potential of researching with a camera, because by photographing and recording cultural activities, it is possible to have a closer look and examine subtler and deeper cultural meanings of the actions of the subjects. This allows us to see cultural patterns, explore languages and to express better what they feel and have experienced.

The audiovisual methodology in anthropology shares some ideas with peace journalism (Youngblood 2016): it can highlight and call attention to urgent social issues originated by violence and to potential threats to peace, analyse causes and consequences, suggest solutions for the conflicts, and propose alternatives to violence (Shaw et al. 2011).

How do we make our research useful? How can ethnography be meaningful to people who are outside the boundaries of academia? What is the way to make research helpful, useful and an aid for the people we work with? I found it necessary to think about the position we have in the field when we work with difficult realities, directly with people who are in a highly emotional state, and also exposed to violence. In this sense, this research is part of a broader personal commitment and academic engagement (Rylko-Bauer/Merril 2006), a sense of political action that responds to people's needs.

Personally, the reason I am interested in engaged anthropology is that I am committed to an anthropological practice that respects the dignity and rights of all humans and has a beneficent effect on the promotion of social justice (Low/Merry 2010: 204).

The research on violence is a conflictive arena where the personal emotions, the academic world, and the local realities, meet; where we are exposed to many and contradictory thoughts, feelings, and risks not always easy to deal with (Ghassem-Fachandi 2009). In my research, I have sometimes experienced depression, and been in danger and risky situations, that affect my perception of reality. Adaptation to violence is also something that has happened to me in places I have spent more time, so that I can deal better with the emotions and the constant violence surrounding me.

I started figuring out how to include my subjects more actively in my research, and how to make their efforts against violence visible. Doing such research requires the participation of others, and video was a direct way to collaborate with them. Because this is an engaged research, the question is: how to give back something to them, when the action is necessary to their communities? How can we use social research as a tool for empowering communities and supporting projects? Low and Merry identify a number of ways to achieve this: "sharing and support, teaching and public education, social critique, collaboration, advocacy, and activism" (Low/Merry 2010). The documentary *Arte de la Paz* (60 min, 2017) was the first outcome of the methodology applied, and a way to give back – as a methodological step in PAR methodology (Pineda 1987; Fals Borda/Rodríguez 1987) – and make a meaningful contribution to the people involved. The documentary was screened in every place it was filmed, and the guests of honour at each presentation were all of those who had shared their stories with me, the subjects and co-authors of this research. The best thing was that almost all of them attended the film, and all expressed gratitude for and acceptance of what and how they are portrayed in the documentary.

The documentary is a great way to give a voice to the people, and bring their stories closer to society, so their perspective can be understood, which is fundamental to peacebuilding. Proximity is a very important aspect of this kind of engaged research and emotional knowledge (*sentipensante*), so the opportunity to look and listen to the people can provide us with a more intimate and deeper look at their cultural reality. In many screenings of the documentary, some people in the audience cried, some others expressed their feelings of empathy and asked how they could take action, and most people showed their concern about this issue; it generated discussion, dialogues among people, the exchange of ideas and hopes. The research accomplished its purpose.

Since we look for profound engagement with the artists and their actions, and given the fact that they are spread all over Mexico (and the world), I am using digital technologies and social media to create interaction, network, and generate more data about the actions that the artist put into practice in their communities. I wanted to create a transmedia product, and the way to do this was with an online platform with several resources, such as the documentary, essays, and a global map with artists and projects which work for peacebuilding through cultural projects all over the world.

I'm developing this use of technologies further as another PAR technique and engaged methodology for research in anthropology, because it can be helpful for assessing social processes, self-generating relevant data, and creating virtual interactions and collective dialogues beyond physical borders.

18.3 Cultural Interventions and Peacebuilding

Some of the first studies in Latin America about the impact of cultural projects in marginalised territories were on the *favelas* of Brazil, in a context of conflict between the drug traffickers and the police that resulted in a huge number of civilian casualties (HRW 1993).[8] In that conflict one of the first experiments of the securitisation (Gledhill 2013) of the urban ghettos was undertaken, to be afterwards replicated across the whole continent (Calveiro 2002). But as violence was rising, community leaders started cultural projects in the *favelas*. One of the works that better systematises these cultural experiences is a project called Solos Culturais (Cultural Grounds) by the Observatorio do Favelas (Observatory of Shanty Towns)[9] (Barbosa/Gonçalves 2013), which included an evaluation of the impact of cultural interventions in the *favelas* with more violence, and also trained a group of young people living in five *favelas* of Brazil. They created an atlas and guide of cultural projects in Brazil, as well as a group of cultural promoters in the communities they work with. This example has shed light on the issue that motivates this research: whether the use of arts can deal with violence, if it is an effective element of social change, if it can really modify behaviour, transform the symbolism of violence, and change people's lives. What is the range of action of culture in violent contexts? What is the effect of cultural interventions on peacebuilding?

In Brazil, by the reinvention of their tradition and by different strategies of community building and negotiation inside communities with violent actors, people are

[8]One of the most notorious massacres in the *favelas* took place on 29 August 1993 in Vigario Geral, the biggest *favela* and one of the most dangerous ones. In one night, the military police shot and killed twenty-one innocent people; see at: https://www.amnesty.org/en/documents/amr19/015/2003/en/

[9]You can see their work at: https://observatoriodefavelas.org.br.

changing the violent patterns and improving their quality of life. These projects combine the need for social services, like education, health care and sanitation, with the reclamation of the community's cultural rights. The projects were started by members of the *favelas*, neighbours who could not bear violence, fear and death anymore, so they empowered themselves and became leaders of their community. How can these examples be useful for Mexico?

Why is the use of art relevant? As mentioned above, the cultural aspects can be drivers of emotions, collective belonging and making us aware of our humanity. Educators can use artistic workshops to transmit values related to non-violence, and develop in others a critical attitude to violence, trying to stop the normalising and acceptance of violence in daily life (OEA 2011). Art can be both pedagogical and performative because it impacts the material and symbolic levels of culture, inhabits the language that shapes our social reality, and ignites the values that create communities.

In this sense, what I have seen is that most of the cultural projects I have worked with practise emancipatory or popular education (Freire 1990), which develops the capacities of the marginalised and poorest sectors of society to use the knowledge to solve their problems, face their reality, and to improve their lives. In this sense, when working together, cultural development and peace education can be useful to prevent individuals from using violence as a means of conflict resolution (Cascón 2010).

There are some recent studies in Mexico and Latin America that are trying to discover how cultural strategies can be ways of tackling violence, and which could be the right approach (Covarrubias 2012; Barrios/Chaves 2014; Martínez Ruiz 2016; Jiménez 2016; Catzoli-Robles 2016). Some studies consider that these practices can help to reduce violence, mend the social fabric, engage youths, prevent domestic and gender violence, and promote values and different alternatives to communities (Covarrubias 2012: 76). Others want to show how these projects can help in the development of a community, strengthen the social fabric, break the dynamics of fear and social isolation, and empower communities to work for the improvement of their condition of insecurity (Barrios/Chaves 2014).

There are a number of cultural initiatives organised by local governments and civil society organisations that promote cultural access and skill development. These actions have been studied and some of their actors and specialists have shared experiences and research from all over Mexico to contribute to the analysis of the meanings of the artistic experience for the people. Lucina Jiménez, an anthropologist and current director of the *Instituto Nacional de Bellas Artes* (INBAL) (one of the most important cultural entities in Mexico), has gathered information about a range of governmental and citizen actions and projects. She explores the role of art in fostering collaboration and active citizenship, and in promoting peace education as an alternative to violence. In her opinion, there is a need for the democratisation of culture to bring people closer to new experiences, and for citizenship-building through the cultural experience (Jiménez 2016). This work, along with reports, different evaluations and the systematisation of cultural methodologies, currently forms some of the main material for understanding the social effects of violence and the ability of culture to offer solutions.

There are strategies that designed to trigger transformative experiences in people and their environment, starting with the recognition that art builds self-knowledge and generates experiences and is not just a means of expression, thought this last is crucial in terms of human rights and participatory democracy (Jiménez 2016).

Even though these studies are an approach to the problem, they still have some limitations. The first is that they have an institutionalised and top-down perspective of the problem. Most of the analyses focus on structural issues, or on explaining the impact of programmes or NGOs' initiatives at a higher level. They lack contact with the communities they work with, and have no links with the people who implemented them. The studies are only descriptive or theoretical, focusing on problems at a regional level and around political policies. This makes even more evident the other weakness of these studies, that they are not listening to the people, and do not explain the local realities and individual factors involved in violence and peacebuilding. These studies do not respond to their needs and ignore them.

To have a different approach, in this research I used the work of Mirian Bautista, a Mexican social scientist whose methodology is focused on the narratives of violence from an individual perspective in different regions of Mexico (Bautista 2017). She coined the term "social whisper", which stands for the "understanding of ordinary citizen's discourse, which is produced and circulates – different from the official one – in daily life, feeding itself from the subjective experiences and other discourses" (Bautista 2017: 12). My research makes use of the notion of *social whisper* in the analysis of cultural practices against violence, in order to have a broader and a more comprehensive perspective of the social actors involved.

This study aims to provide a closer look at the different actions and social realities of violence so that we can better understand them from a cultural and subjective perspective. The aim is to offer the intimate view that is missing in previous works – a goal which can be achieved with the methodological approach of engaged anthropology, and the kind of research that can also support action.

I assume in this research that arts and cultural practices can mobilise actions and senses in society, and have the potential to transform reality. Gandhi understood art as a way to work towards the development of the individual, giving to the world the best of each person: "Physical drill, handicrafts, drawing, and music should go hand in hand in order to draw the best out of the boys and girls" (Gandhi 2008: 200). He believed in the use of education to change individuals and make them part of the struggle for freedom, and arts, particularly poetry, could play an important role in this by changing people's minds and hearts (Martínez Ruiz 2016). In their correspondence, he and Rabindranath Tagore, winner of the 1913 Nobel Prize in Literature, exchanged ideas about the role of artistic education in securing India's independence. They were aware of the influence of the arts in the development of the individual, and also of the potential of the arts to be a powerful political weapon that could bring peace and liberty. Is it possible to translate this conception to the problem of contemporary violence?

First of all, we have to understand violence, its causes and consequences, the different levels that exist (e.g. the cultural), the intersection of structural and subjective violence, and how our conceptions of peace should reflect the complexity of the concept we face (Galtung 1969: 15). This essay accepts three concepts of peace: positive peace as conflict transformation, opposed to the causes of violence, and seeking reconciliation, reparation and restoration (Lederach 1997: 48); neutral peace, as the elimination of cultural and structural violence (Jiménez 2011); and imperfect peace, as the perpetual effort towards peace, as something never fully achievable, but as a progression over time (Ramos 2015).

Based on this theoretical diversity, I have devised my own definition of peacebuilding as the process of healing violence at the individual and social levels; and I use the term 'heal' because I want to focus on the subjective and collective levels of recovering from traumatic events and foster resilience (Cyrulnik 2011).

Considering violence an illness that propagates like epidemics (Slutkin 2016) but also understanding violence as a phenomenon with cultural and social causes as well as psychological roots (Raine 2013), the idea of the cure of a disease comes to my mind as the first step of the peacebuilding: in ailing communities and individuals no peace is possible. That's how the Department of Public Health of the University of Illinois developed the Cure Violence Programme,[10] which is "a public health approach to gun violence reduction that seeks to change individual and community attitudes and norms about gun violence. It considers gun violence to be analogous to a communicable disease that passes from person to person when left untreated" (Butts/Roman 2015). Art has been used on a large scale in different detention centres, community colleges, social centres and so on, in order to help deal with the traumas and violent shocks affecting the individuals concerned.

We need to understand the different kinds of violence we face: the border wall, poverty, lack of care, pollution and inequality are the kinds of violence that destroy individuals and communities – what social psychology calls "social suffering", the collective feeling that violence produces "suffering that originates in human relationships as a whole" (Gampel 2002: 17). So the impact of violence on individuals turns them into possible multiplicators of violence, or passive and fearful citizens. Therefore the first step to empower communities and work for peacebuilding is to heal individuals and social bonds in order to heal the fabric of society. Artistic expression and cultural practices are being used for this purpose.

How do we respond to violence? There are different reactions to violent trauma. Some people can become frightened without knowing what to do, some can just ignore it and continue with their lives, others can never get rid of the pain and consequently live in sorrow (Cyrulnik 2004). But in most cases, people respond positively, restore their lives and also fight and resist violence: this is resilience (Cyrulnik 2001).

I am interested in exploring the capacity of resilience in those who truly know what it is to experience violence in their own flesh and souls. I am talking about, for example, those with missing relatives, people who live in constant fear because

[10] See at: https://cvg.org/.

of the everyday insecurity, young people related to organised crime, mothers of the disappeared women, thousands of families losing some loved ones – they are the people who I am talking about, people who did not give up. I consider resilience to be a necessary step in the action against violence, and I have found during my research and ethnographic work how artistic actions can contribute to healing violence and also to this resilience at a personal and social level.

In this scenario, the peacebuilding approach I have researched is bottom-up, led mainly by the action of individuals at the community level, local organisations, or individuals who will bring about modifications at a micropolitical scale, displaying different strategies to deal with violence and propose alternatives. I want to show that when, as in Mexico, the State and governmental power cannot provide citizens with security, the capacity to resist and solve social problems is led by ordinary people at grass-roots level. In the next section I use case studies to illustrate this display of social agency and political engagement by the people. Studying the aesthetics of peace and the critical role they play in resisting and subverting the rhetoric of fear and silence imposed by violence is greatly facilitated by knowing the protagonists of this story.

How do cultural practices and peacebuilding relate to each other? In what ways can the arts help to promote non-violence and change behaviour? What potential do symbolic practices have to promote changes in society and attitudes towards violence? That is the goal of this chapter: to study different practices of peacebuilding, show how the arts are used as ways of dealing with conflicts, and display different methods of intervention in violent contexts. In many senses, these activities support or complement the principles of peace education (Fisas 2011). The main qualities of the cultural projects studied are that they are community-based, grass-roots supported, critical of violence, promote alternative education, and are self-organised.

Through experience and ethnographic work with these movements, I found that the use of the arts is widespread and in their discourse practitioners show faith in the capacity of culture to bring about changes in society regarding violence. People who are engaged in this kind of work trust the arts in different ways and for a variety of purposes: (1) as a peace education tool which promotes values in people (solidarity, harmony, non-violence), (2) as a way to activate citizenship in the outcast and offer opportunities, (3) as a communicative tool to denounce violence and advocate for justice, and (4) as a strategy for working with high-risk youths or sensitive communities.

Activists and organisers talk about culture in the sense of humanisation and something to help recover lost social bonds. They believe that symbolic practices can be ways to rebuild the social fabric in vulnerable communities (Vich). Culture is relevant in the development of individuals, helps to improve communication among people, empowers the outcast, helps to solve conflicts, transforms communities and gives security to public spaces. All of these factors promote values related to a culture of peace (Fisas 2011).

From my fieldwork experience with artistic collectives and different cultural promoters, I analyse how concepts like peacebuilding, community empowerment, strength of social networks, democratisation of culture and violence prevention are present in their actions. It is necessary to read from the ethnography the meanings of their performance and to interpret the concepts related to non-violence and peacebuilding in their work. For these cultural practices, which are created from outside the laws of the market, the process is more relevant than the results, and the goal is not the beauty or the business, but the development of the human being and the co-responsibility with the community (Rodríguez 1994: 366).

Mexico has recently experienced the rise of individuals, organisers and communities who deal with violence by the means they have at hand, relying on their cultural traditions as the source of their symbolic actions. This phenomenon has been growing, spreading the use of different artistic expressions and cultural strategies in the process of peacebuilding. Artists who work in sensitive communities with disadvantaged populations, who have the will to change structural and cultural violence, are the main subjects of this research. It is this kind of aesthetic and cultural practice that I consider to have the potential to create a new sense of social life and promote the practice of a new set of values for a non-violent and healthy society.

The art of peace has a double meaning. On the one hand, it is the art which aims to address violence, a kind of artistic expression that has non-violence as an inspiration and search, an art that pursues peace. On the other hand, and related to this, it is the idea that peace is a creative process in itself, which requires us to imagine a different world and a new way of living in it. The art of peace, like any aesthetic execution, needs to be crafted and created, most of the time from scratch, with sensitivity and as a mirror of the world. This art also has to be radical, disruptive, disobedient and stay with those at the bottom of society (Patella 2019).

18.4 Case Studies of Arts and Non-Violence in Mexico

Now, let me whisper to you some stories about ordinary people, like you and me, who are striving for change with no other resources than their imagination and their dreams: they are the artists of peace.

18.4.1 First Case: Anita Cuellar – Symbolic Actions as a Way to Face Loss and Be Resilient

I am walking under an extremely hot sun in a rough street on the outskirts of Ciudad Juarez to meet with Anita Cuellar, who is painting a mural of her daughter. When I arrive she explains to me the meanings of the different symbols in the mural, which depicts her daughter, Jessica Padilla, who was kidnapped back in 2011 and has been missing since then. The mural represents her dressed as a 'Quinceañera' (a girl celebrating her fifteenth birthday at a traditional celebration), with a Virgin of Guadalupe on one side and an angel on the other. Around her are the portraits of twenty-five other missing young women who disappeared in the same month as Jessy. Anita explains that as well as being a tribute to her daughter, this is her way of denouncing the injustice and trying to find her. The meanings are multiple, intersecting the personal with the political and combining the demand for justice with the exercise of collective memory and individual healing. As such, the mural is immensely powerful.

The whole process of making this mural lies completely on her shoulders: get the paint, summon the artists, find the wall, call the people, spread the message, and, above all, make this action a protest of the injustice and an expression of her personal drama. At the inauguration of the mural, which was also Jessy's birthday, the community gathered around Anita and her family to support her in the struggle to find her missing daughter. The surrounding community is very important for the resilience process (Melillo 2001) and also provides collective support – what Colombian anthropologist Myriam Jimeno calls "emotional communities" The bond created between those who have suffered violence is a relevant part of the healing process, and also the only chance of reparation after violent shocks and injustices (Jimeno 2011).

Gender violence is one of the biggest current problems in Mexico, where an average of nine femicides a day are reported – one of the highest ranks in Latin America (CNI 2020). In Ciudad Juarez, the statics are among the highest in Mexico and since the 1990s it has been a place of danger for women. Forced disappearances, rape, kidnapping and femicides have been documented and denounced for more than twenty years. This kind of violence has become naturalised and part of the culture of the city (Monárrez 2000).

Jessy's mural gives hope to her mother and feminist activists in the city, because it also has the political implication of denouncing femicides. Resilience in this case is not only the ability to recover after the loss; it is also the capacity to stand against violence, to join political action and to be disobedient, challenging authority in the demand for truth and justice (Patella 2019). Violence grows in the silence, hidden in the dark, and is fed by the loss of social memory. In this sense, unveiling injustice, recalling the victims, and reinserting those who are missing into social life represents a big challenge to the politics of fear, social oblivion, and institutional omissions. But even more important, Anita goes through a process of self-healing from trauma by the act of painting, of remembering and making it public in the political arena. My personal reaction and how much her stories touched me led me to put Anita's story at the centre of my documentary, as one of the strongest actors in the creation of peace.

With the execution of the mural, social memory is activated, and Anita displays her agency by the use of arts as a mean to recall and share her loss. In several talks we had and in some interviews she refers to her mural in these terms: "It has two intentions, one of life, just in case she's still alive, and another of death, in case she isn't still alive. She knows that she will be always in our hearts and in the hearts of the people" (interview with Anita Cuellar, September 2015).

Murals and symbolic actions, like performances, weaving and protesting, have become the way she can express her hopes, her pain, and denounce injustice. Regardless of the results, the symbolic actions have had the capacity to mobilise people and meanings, and to bring back to memory those who are no longer with us. This is one potential use of art to people who experience violence and have no other way to resist. Artistic expression allows Anita to denounce her situation and to enter into political defiance, and at the same time, to heal her own grief and transform it into action in daily life. In her testimonies, she mentioned many times that she wants this to stop happening: "If reparation is not possible, at least I hope this won't continue happening to other mothers like me." No repetition is all that she wants, and the truth is all that she deserves.

Anita Cuellar celebrates her missing daughter's birthday every year with a symbolic protest, trying to do something new every year. In the five years since I first met her in 2015 she has constantly been performing new symbolic actions, all of them with heavy emotional content and based on her own cultural capital and mindset: images of popular culture, religious symbols, simple words and drawings that impact the heart. Her struggle will never end, and she is aware of that, but the hope and will for justice is what makes her resist and empower herself, becoming a leader and reference in the movement against gender violence in Ciudad Juarez.

18.4.2 Second Case: Transforming the Border Wall Through Art

The border wall between Mexico and the US has been a controversial part of the Trump administration's political and electoral strategy. But in fact, the wall has been always there, in many ways. The US-Mexican border, like most borders, was established by violence – and its architecture is the architecture of violence (Chomsky/Cairns 2013). The US-Mexico border is one of the main causes of violence in the region, threatening human lives, dividing families and cultures, and even damaging the ecosystem.

So the idea of building up a higher and more fortified wall is just a discourse and a pretext to gain more control over it; it is simply a façade overlying more critical issues like racism, drug smuggling, and gun control. But like any wall, it is a perfect target for artists, who create interventions that aims to erase the boundaries, to connect the disconnected and to question the need and existence of that visible and almost permanent violence.

Tijuana is one of the busiest and most crowded borders in the world. It sits on the edge of both the First and the Third World, and at the far corner of Latin America (Herzog 1990). This borderland existence produces a variety of mixtures, exchanges and interactions that have created a hybrid culture that reinvents the categories of nationality and identity (Grimson 2011). The massive borderline that starts with a tall fence that dives into the sea is a visible injustice established against people and common sense, structural violence materialised in border patrols, helicopters and eternal customs controls. The configuration of the border has some emotional metaphors. Mexican anthropologist José Manuel Valenzuela defines the border as a scar, an open wound that still hurts Mexican history (Valenzuela 2003: 33). This is what makes it so conflictive. The inequality that it represents still makes the loss and the feeling of inferiority more evident. But it is also the privileged crossing point of goods, services, lives and deaths: the border is a rupture of two nations and one history.

An example of how art is dealing with this issue in Tijuana is the work of Mexican artist Enrique Chiu, who is painting a collective mural along the border wall, trying to change and redefine its meaning. He started an initiative called the Mural of Brotherhood. When interviewed, he told me: "Usually the border is something ugly, something we don't want to see, so we want to change the vision of this wall, transforming it into something beautiful, shifting from a wall of division into a wall of connection between the two countries and among the people" (Enrique Chiu, interview November 2017).

For this task, independent artists and citizens from both sides of the border gathered and painted the wall and the stripes of this horrible fence that claims the lives and consumes the dreams of the outcasts of Central America, who have fled from their countries, running away from violence, poverty, and oppression. Painting the wall has become a way to resist and try to transform the meaning of the wall from hatred to reconciliation, division to unity; Chiu transforms the wall into a bridge.

This initiative was started in 2017 and is still going on, with the goal becoming the largest border mural of the world. Chiu has organised painting workshops for the people, talks with immigrants and exhibitions of this project in several galleries, trying to call attention to the issues of the border. If the wall represents division, this action wants to unite people and promote social understanding among citizens on both sides, denouncing the violence against immigrants and the lack of policies to protect them.

He is trying to do this through his initiative, contributing to the creation of a new global citizenship by joining people through art. I was present a couple of times when he was painting the mural, and people were gathering around the art, kids, youngsters, old people, all painting together. People from San Diego were also present, and one man who lived there told me that for him the initiative was really positive for the community, because they could meet the immigrants and talk to them, so this could help to prevent racism and exclusion. I agree that when people come together and know each other, fear vanishes and violence is less likely to happen.

Another initiative designed with the intention of breaking the division was the installation of some seesaws on the border close to Ciudad Juarez (*El País* 2019). There are some villages along the border where families live on both sides due to displacement or immigration. The fence is a harsh obstacle to their reunion and the most visible sign of division and violence. Ronald Rael, a Californian architect, installed the seesaws through the gaps between the vertical slats in the fence, creating a playground for kids, who can now play together despite being on opposite sides of the border. The symbol of people playing together in front of the border is so powerful that it actually breaks the barrier and reunifies both nations. The intervention looks to reinforce the connections and to erase the border, transforming it into something fun. The ludic element is present and the message is very powerful: we need to focus on what unifies us rather than what divides us.[11]

Both examples show the potential of creativity to unify and to break metaphorical barriers. The violence in Mexico is mainly concentrated in the border area, where inequalities are high and the cartels have their territories (Solis/Morriconi 2018). So the actions being taken symbolise the desire for change and are critical of the message of violence that the border represents.

These actions need to send a clear message and also to have a projection over other borderland problems, like femicides, drug cartels and the gun traffic. We cannot assume that artistic interventions can solve the problematic violence or that by painting murals the border will disappear. Some interventions in the border appear to be just propagandistic,[12] while others can have a lasting effect and trigger a social response. Such examples present us with strategies of intervention and show how artistic practices have the potential to call attention to problems and change the imaginaries around social issues. The message is that the mere act of painting can trigger new social relationships and new ways of collaborating to address violence, offering the possibility of creating a community beyond borders.

We cannot lose the critical perspective and think that art by itself will solve social problems – that would be naïve and simplistic. What I interpret from these actions is the subversion of symbols and how creativity can foster new dialogues regarding structural violence and how art can have a cultural impact. These interventions, carried out by artists on both sides of the border, are instances of how transnational citizenships are shaped and work together to remove barriers between human beings who look for togetherness.

18.4.3 Third Case: Rap Artists and Cultural Managers Using Hip-Hop

In the border cities of Tijuana and Ciudad Juarez the hip-hop culture is very popular and is used effectively with gang members in extremely violent neighbourhoods. One example of this is *Colectivo Fundamental* (Fundamental Collective), a youth organisation born in the ghettos of Ciudad Juarez amidst the turmoil of the violence in the city.

It is an interesting project that has worked for more than ten years, motivating youths to quit habits like drug consumption, drug trafficking, robbery and violent clashes with other gangs. Those involved also try to conciliate and prevent retaliation,

---

[11] See at: https://elpais.com/elpais/2019/07/29/album/1564392988_733199.html#foto_gal_2.

[12] For instance, JR photo installations, which are an isolated intervention and a less critical way of talking about the problem of the border. See at: https://www.nytimes.com/2017/09/07/arts/design/jr-artist-mexico-border-wall.html.

working closely with youths, and mediating when there are conflicts between gang members. The primary forms of contact are painting graffiti, breakdancing, singing rap, and creating beats or DJ'ing: the four elements of the so-called hip-hop culture (Chang 2005: 238). The organisation has been working since 2009 within the most problematic neighbourhoods in town with young gang members, who find a meaning to life and an identity in these crews, the '*barrios*', as they call themselves.

One of the main reasons for the violence in Ciudad Juarez is the omnipresence of the drug economy in the life of the city. I heard many testimonies about the close relationship of the population with the drug traffic, how the 'narco' and daily life are tied together, and how this creates a culture around it, with characteristic outfits, music and language (Reguillo 2010). Everyone knows someone working for the cartel, and a significant proportion of the local economy is controlled by the drug economy. The Juarez and Sinaloa cartels are the main actors in the violence in the city. Since 2009, the quest for control of the routes and crossing points used for smuggling narcotics has generated a silent war that continues to the present, with varying levels of violence (Barrios 2014).

The youth of the suburbs and poor neighbourhoods are the main source of labourers for the cartel's armies. The gangs work for the drug cartels, and are the primary recruiters. Violence increases in communities when residents start working for a criminal gang. Drugs, guns and money are available to them, so it's easy for the *sicarios* (hitmen) to strike down rivals. Based on direct testimonies with gang members, adherence to a gang and working for a cartel are common destinies for youths. That is one of the reasons for the high youth homicide rate in Juarez. The lack of opportunities, the discrimination and addictions make it difficult for them to stop the cycles they live in and walk away from the culture of violence, so they continue the path of the street, which often leads to death or jail (Salazar 2015).

The fellows of the Fundamental Collective are former gang members who want to highlight and rescue the positive values of gangs: solidarity and sense of belonging, loyalty and service to the group are all strong components in shaping the identity of young kids living on the periphery in marginalised areas of their city. Hip-hop culture as a whole is attractive and gives meaning to this generation of excluded individuals, who, without intervention, are likely to die in the streets or become agents of violence, working as hitmen for the cartels (Short 2018).

For Tony Briones, founder and leader of *Colectivo Fundamental*, hip-hop gives them a different perspective and vision of the future. He has witnessed how these young boys and girls are changing some of the values of the gangs related to violence, and also how some of them are trying to change those patterns into something more positive, serving as role models to other young people and working for them.

The *Fundamental Collective* has gained trust among the gangs and inspired others to gain respect and social recognition through the abilities they have acquired in the different elements of hip-hop, or also through showing group leadership. Gang-related violence has been one of the main problems of this city, and gangs have their own culture: values, rituals, language, clothing, music, and a shared history (Nateras/Valenzuela 2013). It is the premise of this project to work within the gangs, to speak their language and to share their story.

In this sense, the cultural approach is one of the most effective ways to change the behaviour associated with violence and promote reconciliation between gangs. Let me offer an example of this. Once I was with one of the graffiti artists of Fundamental. Borus is his artistic name; he is a promoter of art and was with another young boy who wanted to paint some graffiti in the neighbourhood where he lived. We arrived at the place where other guys were waiting for us. Borus gave them spray paints and some advice on how to improve their graffiti. They were painting, I was recording, and suddenly a guy appeared, threatening the boys with a knife and yelling: "Who allowed you to paint this wall? This is my territory!"[13] I was taking pictures and didn't know what to do, since he was approaching the group with a very aggressive attitude. At that moment Borus intervened and started talking to the guy, explaining who they were and the purpose of the painting. The perspective of the guy was that he was also from a graffiti crew and they used to paint that area, so he didn't want another gang's graffiti on that wall. Instead of arguing, Borus convinced him to join them in painting and offered to give his gang spray paint as well. Finally, they were even laughing and the guy allowed the boys to finish their graffiti, and became very friendly. I've always thought that it could have ended up badly, but the mediation of Borus was effective in preventing violence. His main mediation skill was to understand the street codes and youth language.

I have noticed how artistic expressions like rap and graffiti are ways to reduce violence, offer opportunities, and intervene in the public space, transforming the perception that people have of themselves, and also motivating a change in their values and aspirations, a new sense of self and community responsibility. The work of *Fundamental Collective* has changed the lives of many young gang members, inspiring them with a different vision of reality, and it is still transforming the neighbourhoods and lives of one of the most important sectors of society and those left behind: youth.

In the words of Tony-Tocka Briones, the work they do is to decrease violence through art, bringing out the best of the people through cultural activities, and providing youths with alternatives. He told me once:

> I grew up in a hood with lots of fights among all of us. Then I realised that those who I was fighting were suffering from the same poverty, violence and same problems as me, so I thought it would be better to try to know each other and try to work together and move forwards. Youth in Ciudad Juarez has been left behind for decades, so it's no surprise that there are so many gang members and hitmen. We are trying to solve this problem of abandonment and lack of opportunities for youth.[14]

What they want to address, primarily, is the lack of space, access to education and legal jobs for the young people of this city. They realised that these structural problems were so great that they were the main source of violence among the city's youth, so they decided to face this problem by offering opportunities to young people. They also started a process of reconciliation between gangs, using hip-hop as a form of mediation and a way to bring together gang members. Fundamental Collective

---

[13]Let's point out that the fight for territory is one of the main causes of conflict between gangs.

[14]Interview with Tony Briones, April 2017.

uses hip-hop as a way to democratise culture, open spaces for youth and conciliate in conflicts between gangs and in personal disputes as well.

Instead of focusing on the violent content of the mainstream hip-hop artists of the massive rap industry, they want to rescue the original sense of hip-hop culture, which is that of solidarity, brotherhood and sisterhood, to denounce injustices and not fall into violence. All of these elements formed the basis of the original hip-hop culture in New York in the 1970s, with exponents like Afrika Bambata (Chang 2005). In the beginning hip-hop culture sought to bring an end to gang-related violence and abusive drug consumption, and promote the cultural identity of the subaltern classes of American society, like Afro-Americans and Latinos. *Fundamental* uses the original idea of hip-hop culture.

By the strategy of using hip-hop cultural expressions, *Colectivo Fundamental* has developed this network in sensitive neighbourhoods and had the chance to interrupt violent conflicts and mediate disputes between gangs, so they cannot spread and grow with fatal consequences. Hip-hop is an instrument for reconciliation and conflict-solving in some of the most problematic communities in Mexico. During the last few years, the organisation has been working with the "Cure Violence" methodology from Chicago (Butts/Gouvis 2015), a model of violence prevention and homicide reduction based on the epidemiological approach to violence, which considers it to be a disease and intervenes to interrupt the propagation of the epidemic (Slutkin/Ransford 2013). One of the members of *Fundamental*, Dani Mundo, who is a rap singer and youth worker, explained it to me in this way:

> Our duty is to think: okay, the weapons are there, drugs are there in their reach (the young boys of the neighbourhood), so how to stop them doing it, to interrupt the route between the young men and the gun, or the traffic. Because if they do it, probably there will be violence. So we offer them cultural activities, cultural workshops, inviting them to sing rap with us, to paint graffiti on the walls, to organise breakdance matches, or get involved in silk-screening T-shirts, so they can sell them and get money. In this way we stop the process of them becoming criminals or getting killed.[15]

This methodology, which requires strong involvement at local level, can be applied to this project since this artistic collective has gained trust and respect from the community, and has in-depth knowledge of the local social context. So we can see how a cultural collective has become one of the key actors in the peacebuilding process in Ciudad Juarez since 2009 through creating a movement within the *barrios* (neighbourhoods) and promoting alternatives to violence to those young people who do not have many options in the future (Salazar 2009).

[15]Interview with Daniel Mundo (MC), Grafico, September 2017. Daniel is a rap singer and cultural promoter in Ciudad Juarez, and has been working with Fundamental since 2009.

18.4.4 Fourth Case: Performance Against Gender Violence in the Centre of Mexico

They usually dress in white, with splashes of red and violet. Their make-up will also be white to highlight the wounds, and sometimes they will cover their faces to represent their invisibility within society. Sometimes they wear 'Quinceañera' dresses. They walk in the streets in demonstrations against femicides, stay with the victims, deal with the State and demand justice. The recent movement against femicide all over the country, with different mobilisations because of the daily occurrence of this problem, has widely used performance as a political resource (Taylor 2016). Many public rallies or activities have some component of creativity and incorporate innovative ideas; they are performative actions against violence in the public space, trying to call the attention of society to this terrible problem (Wright 2011). Let us now examine the role of performance art in addressing the critical problem of gender violence in Mexico.

The concept of performance has two different meanings. One is a form of dramatic representation in which the body and the use of symbols play a relevant part; it can also be considered action art or live art, without a fixed structure. Performance has always been implicated in radical politics and considered since the beginning as a critical art (Taylor 2012: 21). The second meaning comes from anthropology, and is defined as a symbolic practice that gives meaning to the different rituals of society (Turner 1999). These rituals can vary: from the celebration of a deity to a sporting event or a political parade, all are rituals in our society and have performative actions. What is relevant in the rituals is their capacity to give meaning to social life through performative actions and the circulation of symbols which impact on the perception of the people and their behaviour (Butler 2010: 96).

From both definitions, we can understand performance as a body or linguistic action that makes sense and gives significance to something relevant to society. The interpretation of the cultural and political implications of the performance aims to provide new meanings to society. For cultural practices and performances to bring about peace, they need an intention to break the naturalisation and the embodiment of violence. The possibility of resistance is the performative capacity to change the norms, and to provide agency to the subject when facing obstructions, which means creating a different kind of identity and a new set of values by prompting a shift in the cultural scripts we all live by (Butler 1997).

After the multiplication of cases, femicide has been normalised by society, so these performances represent, in front of the eyes of the people, the faces and bodies of those absent. It is a call for action; it is a public denunciation and a way to claim justice; it is also a political struggle for the women's movement, which demands safe environments, freedom to be and the right to live. In contexts where violence has become so naturalised that it is no longer perceived as an anomaly, performative actions can make the pathology of violence evident and be a powerful way to symbolise it and compell the people to reject it.

Anthropologist Nancy Scheper-Hughes conceptualised "everyday violence" as the normalisation of violence in marginal social spaces in which a range of violence (from direct to structural) intersects and produces the "social indifference to death" (Scheper-Hughes 1993). Symbolic and cultural violence is reproduced in the everyday, creating a normalisation of the anomaly, naturalising violence as a way of living, or, as Bourdieu defined it, the internalisation of oppression by the oppressed, with the best of its examples in gender violence (Bourdieu 2000) Through their different symbolic actions, the artists and cultural workers I am collaborating with look to subvert the social meaning of violence in the death zones, creating a sense of purpose for life in a new key.

Performativity on the aesthetics of non-violence is relevant because it can act as an emotional reliever, enabling people affected directly or indirectly by violence to express their feelings, which aids the process of sharing and communicating sorrow and traumatic experiences, creating supportive communities which people can lean on and rely on (Jimeno 2011).

In a town called Ecatepec, on the periphery of Mexico City, there was recently a high occurrence of femicides with total impunity, and they were increasing year by year.[16] In this context, Manuel Amador, a theatre actor and college professor, decided to organise a performance group at his school to make femicide more visible and denounce the impunity of these crimes to the community. He started this group in 2015, and declares: "The purpose is to recover the community, to build what is destroyed, in society and in the people."[17] He explained that it was a way to give a voice and a face to those who are gone, and have lost the capacity to be, a symbolic act that can recall the injustices and, at the same time, raise awareness in society about this issue. We can see the critical role performance has in this process of activating citizenship and strengthen resilience, responding to violence with dignity and meaningful actions, transforming their political identities from victims into social actors who defy the State and cultural norms (Ravelo 2010).

Participants march through the streets wearing white dresses, their faces covered with white masks, their bodies disfigured by 'wounds', divulging the details of the crimes against women, identifying themselves with the victims, and using their own bodies to tell the victims' stories. They suddenly stop at some place with a crowd, yelling: "My name was … and I used to live here," then tell those who approach: "I was taken in the street, close to my house. I never came back." Then they recount the full real case of a victim, concluding: "I was kidnapped, raped and killed." Several dramatic elements are displayed, along with a protest that aims to touch people and awaken society to do something against the violence and prevent more femicides. This method of creating a collective memory is a powerful way of encouraging people to rethink violence, and of changing attitudes towards it at the symbolic cultural level, where violence reproduces and is validated by social norms, beliefs and traditions.

---

[16]The last incident was of a couple who murdered more than twenty women in the span of few months.

[17]Interview with Manual Amador in July 2019.

These street performances have had a powerful impact on society, attracting media and making more visible a problem of violence that the institutions are not solving or responding to properly. The actions taken by a small group of young women who fear for their lives, and who are also victims of different kinds of aggression and violence, send powerful messages to society and demand justice for the victims. Every performance denounces and re-signifies the logic of the existing war against women (Segato 2016). The performers' actions condemn violence and symbolically restore the victims to collective memory.

The performativity of cultural practices provides individuals with the ability to subvert detrimental structures through the use of symbols and the production of meanings. This technique can be applied to violence as well, since it offers the opportunity to create alternative and relevant world-views, in which violence is transcended by a new signification of reality. Instead of fear, unity; from cruelty to hope, and from pain to life. Through cultural action, the different actors open up the possibility of ending symbolic violence, creating a space to reawaken social agency which momentarily slept and was brought down by the fear that violence infuses in people.

18.5 Conclusions

In this paper I have explained how cultural actions can engage citizens and make them resilient, and how cultural practices can be a strategy against violence in many different ways. There is no unique formula that can be used, so I affirm that if we want to know the potential of the use of arts for non-violence, we need to use a multilayered and intercultural approach that reveals subtle changes in the individuals concerned, and shows how arts and cultural practices can impact different issues and specific kinds of violence. As pointed out earlier, art is not an end in itself, and for the sake of peace it should be harnessed for social purposes, and adapted to communities and local actors, serving as an instrument for the social good. Moreover, research methodology also needs to be tailored to address this issue and collaborate in efforts to find solutions to practical problems by being a tool for communities and not just an instrument of academic knowledge (Fals Borda 1986: 27).

Cultural work has demonstrated its efficacy but cannot erase violence by itself or solely through murals, songs or performances; ways to address structural and cultural violence must also be taken into consideration. The limits of these practices depend upon the kind of violence involved and the concept of peace underlying it. How people understand peace is radically different for each context and actor, for Anita Cuellar is not the same as Enrique Chiu; even though both are painting murals, their purposes are different, which indicates that the idea of peacebuilding is not homogeneous. It also depends on the intersection of factors that create specific conditions of violence in each context and the cultural background of the actors.

As seen in the ethnographic examples, the way Anita Cuellar deals with violence is through symbolic actions, such as the mural of her daughter, which enabled her to denounce injustice while simultaneously performing a ritual for Jessy, in the sense of activating memory and paying homage to her. The performances are more direct and politically strong, but also have the same goal, of remembering the femicides in Ecatepec. Both actions against gender violence aim to deliver a message to society about the consequences of violence. Cultural practices are ways of healing violence, whether for the mother of a missing woman or a young boy in a gang: both need to expunge the pain of trauma and the effects of violence on their own flesh.

As I have seen in the boys and girls from *Fundamental*, many came from a very difficult background involving criminal records, drug consumption and domestic violence. To many of them, life has changed since they joined the group, as the very moment that they started doing something they enjoy, in an environment of mutual respect and solidarity, they found a fresh meaning to life. Cultural activities restore the sense of community and rebuild social fabric for those who suffer violence. It is not that a graffiti workshop can reduce violence on its own; what can change lives is the social process of organising and delivering it, the fact that people gather around a mural. Opening the opportunity for a marginal individual to paint a wall simultaneously opens the door to social inclusion, cohesion, resilience and active citizenship.

Cultural practices can ignite social agency by the mobilisation of new meanings to social life, and by the display of cultural practices in vulnerable communities acts of citizenship can contribute to resilience and peacebuilding as the best actions against violence (Balibar 2015). On the other hand, the practice of this kind of artistic expression can be a way of treating trauma, emotional pain and social suffering in ways that help to heal the victims of violence from an emotional perspective.

References

Ander-Egg, Ezequiel, 1990: "Repensando la Investigación – Acción Participative".

Balibar, Étienne, 2015: *Violence and Civility: On the Limits of Political Philosophy* (New York: Columbia University Press).

"Balancines sobre el muro de México", in: *El País*, 30 July 2019; at: https://elpais.com/elpais/2019/07/29/album/1564392988_733199.html#foto_gal_1.

Barbosa, Jorge Luiz; Gonçalves Dias, Caio, 2013: *Solos Culturais* (Rio de Janeiro: Observatório de Favelas).

Barrios Nogueira, Andrea; Chaves Zaldumbide, Ángel P., 2014: *Transformar la Realidad Social desde la Cultura: Planeación de Proyectos Culturales para el Desarrollo* (México: CONACULTA.)

Barrios Rodríguez, David, 2014: Las Ciudades Imposibles: Violencia, Miedos y Formas de Militarización Contemporánea en Urbes Latinoamericanas: Medellín – Ciudad Juárez (México: UNAM).

Bautista Arias, Miriam, 2017: El Murmullo Social de la Violencia en México: La Experiencia de los Sujetos Afectados por la Guerra contra el Narcotráfico (Mexico: CESOP-UAM).

Beyerle, Shaazka, 2014: *Curtailing Corruption: People Power for Accountability and Justice* (Boulder, CO: Lynne Rienner).

Boal Augusto, 2002: *Teatro del Oprimido: Juegos para Actores y No Actores* (Barcelona: Alba editorial).

De Sousa Santos, Boaventura, 2009: *Una Epistemología del Sur: La Reinvención del Conocimiento y la Emancipación Social* (Mexico City: siglo. XII).
Bourdieu, Pierre, 2000: *La Dominación Masculina* (Madrid: Anagrama).
Butler, Judith, 1997: *Lenguaje, Poder e Identidad* (Madrid: Síntesis).
Butler, Judith, 2010: "Performative agency", in *Journal of Cultural Economy*, 3:2, 147–161 (London: Taylor and Francis).
Butts, Jeffrey A.; Gouvis Roman, Caterina; Bostwick, Lindsay; Porte, Jeremy R., 2015: Cure "Violence: A Public Health Model to Reduce Gun Violence", in: *Annual Review of Public Health*, 36: 39-53.
Buvinic, Mayra; Morrison, Andrew; Orlando, María Beatriz, 2005: "*Violencia, Crimen y Desarrollo Social en América Latina y el Caribe*", in: *Papeles de Población*, 11,43 (January-March): 167–214 (Toluca: Universidad Autónoma del Estado de México).
Chang, Jeff, 2005: *Can't Stop Won't Stop. A Hip-Hop History* (California: MacMillan Books).
Chomsky, Noam, 2013: "*How the US-Mexico Border Is Cruel by Design*", in: *Architecture MPS* (28 October); at: https://www.alternet.org/2013/10/noam-chomsky-americas-suburban-nightmare-and-how-us-mexico-border-cruel-design/.
Clarke, John; Coll, Kathleen; Dagnino, Evelina; Neveu, Catherine, 2014: *Disputing Citizenship* (London: Policy Press).
Covarrubias Valderrama, Gerardo (Ed), 2012: *Violencia y cultura en México*, Colección intersecciones, 28 (México : Consejo Nacional para la Cultura y las Artes)
Cyrulnik, Boris, 2001: *La Maravilla del Dolor* (Barcelona: Granica).
Cyrulnik, Boris, 2004: *Del Gesto a la Palabra* (Barcelona: Gedisa).
Domínguez Ruvalcaba, Héctor, 2015: *Nación criminal*. Narrativas del crimen organizado en México (México: Ariel).
Espinosa, Valeria; Rubin, Donald, 2014: "Did the Military Interventions in the Mexican Drug War Increase Violence?", in: *Journal The American Statistician*, 69,2: 65-98.
Fals Borda, Orlando, 1986: "La investigación-acción participativa: Política y epistemología", in: Álvaro Camacho G. (Ed.): *Colombia de hoy* (Bogotá: Cerec): 21–38.
Fals Borda, Orlando; Rodríguez, Brandao C., 1987: *Investigación Participativa* (Montevideo: Oriental).
Fals Borda, Orlando; Anisur Rahman, Muhammad (Eds.), 1991: *Action and Knowledge: Breaking the Monopoly with Participatory Action Research* (New York: The Apex Press).
"Feminicidios en México", 2019: in: *El País*; at: https://elpais.com/tag/c/63d2707764492e98c02d7e1abd8f9bc9.
Ferrándiz, Francisco, 2009: *La Etnografía como Campo de Minas: De las Violencias Cotidianas a los Paisajes Post Bélicos* (Madrid: Centro de Ciencias Humanas y Sociales del CSIC)
Ferrándiz, Francisco; Feixa, Carles, 2004: "Una Mirada Antropológica sobre las Violencias", in: *Alteridades*, 14,27: 159–174 (Mexico City: UAM).
Fisas, Vincent, 2011: *Educar para una Cultura de Paz*, Quaderns de construcció de pau nº 20, (May).
Freire, Paulo, 1990: *Pedagogía del Oprimido* (Mexico City: siglo XII).
Galtung, Johan, 1969: "Violence, Peace, and Peace Research", in: *Journal of Peace Research*, 6,3: 167–191.
Gampel, Yolanda, 2002: "El Dolor de lo Social", in: *Psicoanálisis*, Asociación Psicoanalítica de Buenos Aires, 24,1-2: 12-25.
Gandhi, M.K., 2008: *All Men are Brothers: Life and Thoughts of Mahatma Gandhi as Told in His Own Words* (Ahmedabad: Navajivan Publishing House).
Gledhill, John, 2013: "Límites de la autonomía y autodefensa indígena: experiencias mexicanas", in: *Oxímora*, 2: 1-2 (Barcelona, CRAI Universitat de Barcelona).
Global Peace Index, 2019: "Vision of Humanity: Mexico"; at: https://visionofhumanity.org/indexes/mexico-peace-index/.
Ghassem-Fachandi, Parvis (Ed.), 2009: *Violence: Ethnographic Encounters* (New York: Berg Ed.).

Giménez, Gilberto; Jiménez, René (Eds.), 2017: *La violencia en México a la luz de las ciencias sociales* (México: UNAM).

Grimson, Alejandro, 2011: *Los Límites de la Cultura: Crítica de las Teorías de la Identidad* (Mexico City: siglo XXI).

Hall, Stuart, 2019: "Cultural Studies and Its Theoretical Legacies", in: Morley, David (Ed.): *Stuart Hall: Essential Essays, Volume 1: Foundations of Cultural Studies* (Durham: Duke University Press)

Henderson, Nan; Milsen, Mike, 2003: *Resiliency in Schools: Making it Happen for Students and Educators* (Phoenix: Corwin Press).

Hernández, Anabel, 2010: *Los Señores del Narco* (Mexico City: Grijalbo).

Hernández, Anabel, 2019: *El Traidor: El Diario Secreto del Hijo del Mayo* (Mexico City: Grijalbo).

Human Rights Watch (HRW), 2011, "Neither Rights nor Security: Killings, Torture, and Disappearances in Mexico's 'War Drugs'"; at: https://www.hrw.org/sites/default/files/reports/mexico1111webwcover_0.pdf.

Jiménez Bautista, Francisco, 2011: *Racionalidad Pacífica: Una Introducción a los Estudios para la Paz* (Madrid: Dykinson).

Jiménez, Lucina, 2016: *Arte para la Convivencia y Educación para la Paz* (México: CFE).

Jimeno, Myriam, 2007: "Lenguaje, subjetividad y experiencias de violencia", in: *Antípoda. Revista de Antropología y Arqueología* (Colombia), No. 7 (July-December): 169–190.

Jimeno, Myriam; Castillo, Ángela; Varela, Daniel, 2011: "Experiencias de Violencia: Etnografía y Recomposición Social en Colombia". in: *Sociedade e Cultura* (Goiânia: Universidade Federal de Goiás) 14,2 (July-December): 275–285.

Jimeno, Myriam, 2004: *Crimen Pasional: Contribución a una Antropología de las Emociones* (Bogotá: Universidad Nacional de Colombia).

Kraniauskas, John, 2015: *Políticas Culturales: Acumulación, Desarrollo y Crítica Cultural* (México: FLACSO).

Lederach, John Paul, 1997: *Building Peace: Sustainable Reconciliation in Divided Societies* (Washington: U.S. Institute of Peace).

Low, Seetha; Engle Merry, Sally, 2010: "Engaged Anthropology: Diversity and Dilemmas: An Introduction to Supplement 2", in: *Current Anthropology*, 51,S2: *Engaged Anthropology: Diversity and Dilemmas* (October): S203-S226.

Marshall, Catherine; Rossman, Gretchen B., 1989: *Designing Qualitative Research* (Los Angeles: Sage).

Martínez Ruiz, Xicoténcatl, 2016: *Poética Educativa: Artes, Educación para la Paz y Atención Consciente* (México: IPN).

Mauss, Marcel, 1971: *Ensayo sobre el Don, Forma y Función del Intercambio en las Sociedades Arcaicas* (Madrid: Sociología y Antropología).

Melillo, Aldo; Suárez Ojeda, Néstor (Eds.), 2001: *Resiliencia – Descubriendo las Propias Fortalezas* (Buenos Aires: Paidós).

Mendieta Ramírez, Angélica, 2016: "El Feminicidio en México: Las Cifras de la Barbarie", in: Escalante López, Sonia (Ed.): *Derecho Procesal Convencional y la Inconvencionalidad Textos Jurídicos en Homenajea: Dr. Eduardo Ferrer Mac-Gregor* (México: UNAM).

Monárrez Fragoso, Julia E., 2000: "La Cultura del Feminicidio en Ciudad Juárez, 1993–1999", in: *Frontera Norte*, 12,23 (January-June).

Organización de Estados Americanos (OEA), 2011, "Informe sobre Seguridad Ciudadana en las Américas"; at: https://www.oas.org/dsp/alertamerica/Alertamerica_2011.pdf.

Paniagua Ramírez, Karla, 2007: *El Documental Como Crisol: Un Análisis de Tres Clásicos para una Antropología de la imagen* (México: CIESAS).

Patella, Pietro, 2019: "Paz Desobediente: No-Cooperación hacia las Órdenes Inhumanas", in: *Polisemia*, 14,26: 1-26.

Valkenburg, Patti M.; Taylor Piotrowski, Jessica, 2017: *Plugged In: How Media Attract and Affect Youth* (London: Yale University Press).

Raine, Adrian, 2013: *The Anatomy of Violence: The Biological Roots of Crime* (London: Allen Lane).
Ramos Mueslera, Esteban, 2015: *Paz Transformadora (y Participativa): Teoría y Método de la Paz y el Conflicto desde la Perspectiva Sociopráxica* (Tegucigalpa: UNAH).
Ravelo, Patricia, 2010: "*We Never Thought it Would Happen to Us: Approaches to the Study of the Subjectivities of the Mothers of the Murdered Women of Ciudad Juárez*", in: Domíguez-Ruvalcaba, Héctor; Ignacio, Corona (Ed.): *Gender Violence at the US-Mexico Border: Media Representation and Public Response* (Tucson: University of Arizona Press).
Rigal, Luis y Siervent, Maria Teresa, (2012)*: "Investigación Acción Participativa: Un Desafío de Nuestros Tiempos para la Construcción de una Sociedad Democrática*"; at: https://www.flacsoandes.edu.ec/libros/digital/56482.pdf.
Rodríguez, Martín, 1994: "Educar para la Paz y la Racionalidad Comunicativa", in: *Educando para la Paz: Nuevas Propuestas* (Granada: Universidad de Granada).
Rouch Jean, 1973: "Camera and Man", in: Hockings, Paul (Ed.), 1995: *Principles of Visual Anthropology* (Berlin: De Gruyter): 79–98; at: https://der.org/jean-rouch/content/index.php.
Rylko-Bauer, Barbara; Singer, Merrill; van Willigen, John, 2006: "Reclaiming Applied Anthropology: Its Past, Present, and Future", in: *American Anthropologist*, 108,1: 178-190.
Salazar Gutiérrez, Salvador, 2015: *La Cárcel es Mi Vida y Mi Destino: Producción Sociocultural del Castigo: La Vida del Joven en Prisión* (Cd. Juárez: FronterAbierta).
Salazar Gutiérrez, Salvador, 2009: *Espacios de Socialidad-Sociabilidad en Colectivos Juveniles Urbanos: Idealizar el Triunfo, Enfrentar la Sobrevivencia* (Cd. Juárez: UACJ).
Salazar Gutiérrez, Salvador; Curiel García, Martha Mónica, 2012: *Ciudad Abatida: Antropología de la(s) Fatalidad(es)* (Chihuahua: Universidad Autónoma de Ciudad Juárez).
Salazar, P. Robinson; Rojas, Ivonne Yenissey, 2011: *"La Securitización de la Seguridad Pública: Una Reflexión Necesaria"*, in: *El Cotidiano*, 166 (March-April): 33–43 (Mexico City: Universidad Autónoma Metropolitana, Unidad Azcapotzalco); at: https://www.redalyc.org/articulo.oa?id=32518423004.
Schensul, Le Compte, 1999: "La Observación Participante como Método de Recolección de Datos".
Scheper-Hughes, Nancy, 1993: *Death Without Weeping: The Violence of Everyday Life in Brazil* (Berkeley: University of California Press).
Seaga Shaw, Ibrahim; Lynch, Jake; Hackett, Robert A., 2011: *Expanding Peace Journalism: Comparative and Critical Approaches* (Sydney: Sydney University Press).
Sedgewick, Peter, 2002: *Cultural Theory: The Key Concepts* (London: Routledge).
Segato, Rita, 2016: *La Guerra contra las Mujeres* (Madrid: Traficantes de sueños).
Seifert, Kathryn (Ed.), 2012: *Youth Violence Theory, Prevention, and Intervention* (New York: Springer).
Short Jr., James F., 2018: *Poverty, Ethnicity, And Violent Crime* (New York: Routledge).
Slutkin, Gary; Ransford, Charles R.; Decker, Brent, 2015: "Cure Violence: Treating Violence As a Contagious Disease", in: Maltz, Michael D.; Rice, Stephen K. (Eds.), 2015: *Envisioning Criminology: Researchers on Research as a Process of Discovery* (New York: Springer).
Solís Delgadillo, Juan Mario; Morriconi Bezerra, Marcelo, 2018: *Atlas de la Violencia en América Latina* (San Luis Potosí: Universidad Autónoma de San Luis Potosí).
Taylor, Diana, 2012: *Estudios avanzados de performance* (México: CFE).
Taylor, Diana, 2016: *Performance* (New York: Duke University Press).
"The Killings in Candelária and Vigário Geral: The Urgent Need To Police the Brazilian Police", in: *Human Rights Watch*: 5,11 (November 1993).
Tuhiwai Smith, Linda, 2012: *Decolonizing Methodologies: Research and Indigenous Peoples* (London: Zed Books).
Turner, Víctor, 1999: *La Selva de los Símbolos*: *Aspectos del Ritual Ndembu* (Madrid: Siglo XXI).
UNESCO, 2005: *Informe de la Comisión Mundial de Cultura y Desarrollo* (Paris: UNESCO).
Valencia, Sayak, 2010: *Capitalismo Gore* (Madrid: Editorial Melusina).
Valenzuela Arce, José Manuel, 2003: *Por las Fronteras del Norte: Una Aproximación Cultural a la Frontera México-Estados Unidos* (México City: Fondo de Cultura Económica).

Valenzuela Arce, José M., 2003: *Jefe de Jefes: Corridos y Narcocultura en México* (Havana: Casa de las Américas).
Valenzuela Arce, Manuel José; Nateras Domínguez, Alfredo, 2013: *Las Maras: Identidades Juveniles al Límite* (México: UAM-I).
Vich, Victor, 2014: *Desculturizar la Cultura: La Gestión Cultural como Forma de Acción Política* (México: siglo XII).
Wood, Mark A., 2018: *Anti Social Media: Crime-watching in the Internet Age* (London: Palgrave-Macmillan).
World Bank, 2012: *Special Report on Violence in Mexico* (Washington, D.C.: World Bank).
Wright, Melissa, 2011: "Necropolitics, Narcopolitics, and Femicide: Gendered Violence on the Mexico-US Border", in: *Signs*, 36,3 (March): 7–31.
Youngblood, Steve, 2016: *Peace Journalism Principles and Practices: Responsibly Reporting Conflicts, Reconciliation, and Solutions* (London: Routledge).
Zirión, Antonio; Dorotinsky, Déborah; Levin, Danna; Vázquez, Álvaro (Eds.), 2017: *Variaciones sobre Cine Etnográfico: Entre la Documentación Antropológica y la Experimentación Estética* (Mexico City: UNAM, UAM).

Films

Ayotzinapa, el paso de la Tortuga (Enrique García Meza, 2018, Mexico)
Cartel Land, (Matthew Heineman, 2015, Mexico-USA)
El Infierno (Luis Estrada, 2010, Mexico)
Guerrero (Ludovic Bonleux, 2017, Mexico)
Heli (Amat Escalante, 2013, Mexico).
La Libertad del Diablo (Everardo González, 2018, Mexico).
La Tempestad (Tatiana Huezo, 2016, Mexico).

Chapter 19
Grass-Roots Post-conflict Peacebuilding: A Case Study of Mosintuwu Women's School in Poso District, Central Sulawesi, Indonesia

Putri Ariza Kristimanta

Abstract Many works of conflict resolution and peacebuilding focus on the Government or top-level leadership efforts. However, peace efforts require a comprehensive approach that is simultaneously conducted at the grass-roots level. This study examined a case study of a grass-roots peacebuilding movement in a post-conflict society, in the Poso District, where a communal conflict occurred (1998–2001). The Mosintuwu Women's School, scattered in fourteen sub-districts and forty-two villages of the Poso District, in the Central Sulawesi Province, Indonesia, implemented the peace education, trauma healing, and social training for grass-roots women. This study aims to explore the contribution of this Women's School and the participation of grass-roots women in peacebuilding. The theories used in this study are grass-roots peacebuilding, women in peacebuilding, and peace education. This study is a case study with a qualitative approach. The sources of data are from qualitative interviews, observation, and documents. The study proves that grass-roots movements are essential to peacebuilding efforts because of their contribution to creating sustainable peace, restoring social conditions, implementing peace treaties, disseminating peace issues, establishing a culture of peace, teaching the understanding of conflict and peace, and facilitating societal independence. The study also proves that grass-roots women play a critical role in peacebuilding because they prioritise transformation, female empowerment, healing and forgiveness, and also uphold gender justice and equality. The suggestion put forward is that the Government should integrate grass-roots movements as a part of peacebuilding efforts.

Keywords Central Sulawesi · Grass-roots post-conflict peacebuilding · Indonesia · Peace education · Women in peacebuilding · Women school · Poso district

---

Putri Ariza Kristimanta, Researcher, Pusat Penelitian Politik, Lembaga Ilmu Pengetahuan Indonesia (Research Center for Politics, Indonesia Institute of Sciences); *Email*: putri.ariza@gmail.com.

Ú. Oswald Spring and H. G. Brauch (eds.), *Decolonising Conflicts, Security, Peace, Gender, Environment and Development in the Anthropocene*,
The Anthropocene: Politik—Economics—Society—Science 30,
https://doi.org/10.1007/978-3-030-62316-6_19

19.1 Introduction

The democratic transition process of Indonesia in 1998 was marked by a series of violent conflicts in Jakarta, Maluku, Central Sulawesi, Kalimantan, Lombok, Aceh, and Papua. As a country that has a diversity of religions, ethnicities, races, and beliefs, Indonesia has a high potential for conflict. Violence often occurs during the fall of an authoritarian regime and drastic economic downturns. From 1998 to 2003, during the transition from authoritarianism to democracy and centralisation to decentralisation, Indonesia experienced many changes in the political, economic, and social fields at national level. Times of massive change are the critical point when institutions and interests arise that take advantage of the situation. In this context, local governments often use violence to strengthen and expand control of the local political stage (Barron et al. 2014). Furthermore, there are tensions between the intertwined religious, political, and economic issues which underlie society and create complicated roots of conflict.

The Poso Conflict (1998–2001) was a time of communal violence between Muslims and Christians that occurred during the democratic transition period. A combination of religious, political, criminal, and economic factors made the Poso Conflict far too complicated to resolve quickly. There were five phases during its four years: the *first phase* (December 1998), the *second phase* (16 April–3 May 2000), the *third phase* (23 May 2000), the *fourth phase* (July-December 2001), and the *fifth phase* (after the peace accord of 2001). This conflict had been de-escalated since the peace accord, the Malino Declaration, was signed by both conflicting parties. The Government estimated that 577 people were killed, 384 were injured, 7932 houses were destroyed, 510 public facilities were burnt or damaged, and there were 86,000 displaced persons (HRW 2002).

The efforts of the national government to deal with communal violence through means such as the Malino Declaration peace accord failed to establish a sustainable reconciliation. The provincial and district governments encouraged a series of dialogues between traditional leaders, but they too proved ineffective. This failure was marked by a series of small outbursts of violence that occurred after the signing of the peace accord. In 2002, a year after the Malino peace accord, a series of cases of home-made bomb explosions occurred in settlements scattered throughout the Poso District: Ratulene Village, Poso City, Toini Village, Tagolu Village, and Ronoruncu Village. In the same year, there were six cases of mysterious shootings scattered across several villages, such as Masani Village, Matako Village, Mayoa Village, Sepe Silanca Village, and the Batu Gencu Village. Instances of home-made explosions and mysterious shootings also occurred throughout 2004 and 2005, spread across villages in the Poso District. In addition to explosions and shootings, there was also one case of burning houses of worship, one case of village-burning, and several clashes between people that harmed lives and property (Komnas HAM 2005).

After more than ten years the Poso Conflict has ended. This conclusion is inferred from the absence of open violence involving large numbers of people in recent years (Noorsalim 2010). However, some issues remain, for example, minority persecution cases, the existence of several extremist violent groups, and several targeted shootings. In *Indonesia's Yearly Report on Religious and Belief Freedom* between 2008 and 2018 only one case needs to be addressed carefully, namely the *Fajar Nusantara Movement* (Gafatar), a minority group that considered committing blasphemy in Poso. The *Indonesian Council of Ulama* (MUI) of Poso District mentioned that Gafatar is a misleading and dangerous group because it is not following peaceful Islamic teachings (The Wahid Institute 2018). In addition to the Gafatar group, the threat of terrorism from the Santoso group living around Poso is alarming enough to disrupt the peaceful situation in the region. This extremist violent group claimed to be affiliated with Al-Qaeda. Indonesian military and security forces have been carrying out military operations to immobilise the group since 2015.

Even though Poso has reached a relatively safe situation, social life is still in question. The first problem is that the displaced people were uncomfortable about returning to their place of origin. Most Christian internally displaced persons (IDPs) chose to live in Tentena City, whereas Muslim IDPs centralised in Poso City. The idea of returning home triggered fears and worries because of their traumatic experiences and grievances. The second problem is that some victims continued to resent the perpetrators of violence. This sense of resentment prevented the reconciliation working smoothly because neither of the conflicting parties engaged in social interaction (Noorsalim 2010).

Post-Malino occurrences transformed the initial horizontal communal conflict into a vertical conflict between the extremist groups and the Government. As a former conflict area, the groups took advantage of the massive arsenal of weapons and ammunition being left in the region. For instance, the case of police-targeted shootings in front of the Bank Central Asia Palu (the capital of Central Sulawesi) in 2011 by Santoso and his group used the remaining weapons of conflict (Ismail 2016). The peace declaration has not ended these unresolved threats of violence, intolerance, and extremism. Johan Galtung calls this condition negative peace, namely the absence of physical or direct violence (Galtung 1969). It is undeniable that positive peace is an endless task for the stakeholders and society building sustainable peace.

Achieving sustainable peace in a post-conflict area requires more than a peace accord and a dialogue at national level. It needs comprehensive, integrated, and collaborative peacebuilding efforts involving grass-roots actors. After reviewing the brief information about the conflict reconciliation, it is clear that the peace declaration which was formulated by the national government was not sufficient to establish sustainable peace for Poso. This issue is important because government policies, programmes, and efforts often do not reach grass-roots society, which prevents the post-conflict reconciliation process going smoothly. In addition to government efforts, individuals and communities need to be directly involved in grass-roots movements. Peace movements at grass-roots level are also acts of reconciliation that support the establishment of sustainable peace.

Having noted that grass-roots peacebuilding is crucial to achieve sustainable peace in a post-conflict setting, this study examines (1) how a grass-roots movement and grass-roots leadership support post-conflict peacebuilding and (2) how grass-roots women engage in post-conflict peacebuilding. Using a case study approach, this chapter elaborates further on their contribution in a deeply-divided society. It is hoped that this positive example from the small community will inspire others, particularly in regions that have similar characteristics and roots of conflict.

19.2 Methods

The method used in this research is a case study approach, which is limited in place and time. The case study approach is particularly useful when there is a need to obtain a holistic and in-depth exploration of an issue, event or phenomenon of interest, in its natural, real-life context. The data in this study was obtained from qualitative interviews, qualitative observations, and a literature study of related documents. The researcher conducted a semi-structured interview with eight respondents from various backgrounds to obtain a comprehensive understanding of the grass-roots women's movement in Poso (Table 19.1).

The researcher has selected respondents on the basis that the reflections of peace-building actors should come from top, middle, and grass-roots leaders. As representatives of the top leader, the researcher has selected three respondents from various backgrounds – the district government (Respondent W), the district police office (Respondent A), and armed forces in the district (Respondent R). Representatives of the middle leaders were Respondents L and D. Finally, Respondents L, E, and M represented grass-roots actors. In addition to qualitative interviews, the researcher conducted three semi-structured qualitative observations of Women's School classes and events during October 2016. In every interview and observation, the researcher used guidelines to assist the fieldwork.

Table 19.1 List of interviewees. *Source* the author

No	Interviewees	Affiliations	Date of interview
1	Respondent W	National Unity, Politics, and Community Protection Agency (Bakesbangpol); Poso District	3 October 2016
2	Respondent A	Departmental Police (Polres), Poso	6 October 2016
3	Respondent R	Military District Command (Kodim) 1307, Poso	7 October 2016
4	Respondent G	Founder and director of Mosintuwu Institute	13, 17 October 2016
5	Respondent E	Women's School 1st batch (2010–2011)	11 October 2016
6	Respondent M	Women's School 1st batch (2010–2011)	12 October 2016
7	Respondent L	Religious leader, Christian representative in Malino I Declaration, gender activist, rector	19 October 2016
8	Respondent D	Religious leader, coordinator of a church crisis centre, book author	15 October 2016

19.3 Poso District: A Brief Context

Indonesia has thirty-four provinces, one of which is Central Sulawesi. The province is divided into thirteen districts, one being Poso. It covers an area of 7112.25 km^2, and 251,185 people live in the nineteen sub-districts, which have a population growth rate of 19.5%. The capital of Poso is Poso City, which is located at the heart of Tomini Gulf, a six-hour hilly drive from Palu, the provincial capital of Central Sulawesi. Poso City was once the capital of the province until this status was transferred to Palu. Its geographically strategic location has made Poso a preferred destination for migrants from both South and North Sulawesi (Fig. 19.1).

Fig. 19.1 Location of the Poso District in Central Sulawesi and Indonesia by Bagas Chrisara; at: https://commons.wikimedia.org/wiki/File:Locator_Kabupaten_Poso.svg. Licensed under CC SA 4.0 https://creativecommons.org/licenses/by-sa/4.0/legalcode

The population in Poso varies widely, and includes indigenous people and migrants; 46.7% of the district's population is Protestant, 31.2% Muslim, and 1.1% Catholic. Most Muslim residents live in the cities and coastal villages, whereas the indigenous Protestant majority (Pamona, Mori, Napu, Besoa, Bada tribes) live in the uplands. Built by Dutch missionaries, *Gereja Kristen Sulawesi Tengah* (GKST; 'Central Sulawesi Christian Church') has been the dominant church in Poso since the twentieth century. Migrants from South and North Sulawesi (Bugis, Buton, Gorontalo) also make up the Muslim population in the district. The tradition of Arabic traders has played an essential role in the religious and educational environment.

Furthermore, in the districts there are also transmigration residents who came from densely populated areas, such as Java, Bali, and Lombok. In the 1990s, migrants dominated the economic sector of the districts. Chinese and Muslim Bugis traders own the local cultivation of products, including cacao, coffee, cloves, and copra. In early 1997, when an economic crisis began, more Muslim migrants arrived and dominated the land and soil, which had become significant resources due to the devaluation of the Rupiah.

The population projections in the Poso district for 2018 were 251,185 people consisting of 129,914 male inhabitants and 121,271 female inhabitants. Based on the projection in 2018, the population growth of Poso was 19.504%. In 2018, the ratio of male to female population was 107:100. The population density in the Poso district in 2018 reached 35 people/km^2, with the average number of residents per household being four persons. The population density in the nineteen subdistricts varied greatly. The highest population density of a subdistrict occurred in Poso Kota with a density of 2155 people/km^2, and the lowest was in Lore Peore with just seven people/km^2.

With regard to economic activities, the population aged fifteen years and above could be divided into two groups: those who were in the labour force and those who were not. In Poso district the number of people in the labour force population – based on the National Labour Force Survey in August 2018 – amounted to 140,392, with 60.77% being male and 39.23% being female. The total workforce for 2018 was 136,930. Of these, 3.6% were unemployed. For 2018, the unemployment rate in Poso was 2.47%. The poverty rate of the Labour Force Participation is one of the primary determinants of community welfare. In 2018 in the Poso district, 449,834 of the people were poor. The percentage of poor people in the Poso Regency in 2018 was 16.71% of the population (BPS Poso 2019).

Poso has a local motto, *sintuwu saroso*, which means "strong when united". This local wisdom has three meanings, *tuwu mombetuwunaka* (respect each other), *tuwu membepatuwu* (love each other), and *tuwu mombensunka* (help each other). *Sintuwu maroso* is a social life based in the community, which has been an essential social capital for many years, even during the conflict and post-conflict setting. For example, even if a non-Christian IDP moved to a Christian-dominated area, the *mosintuwu* spirit – whereby neighbours help each other at wedding and funeral ceremonies – would prevail regardless of different religions. *Sintuwu maroso* is not a slogan or a doctrine. It is a characteristic manifestation of Poso's people and a way of life (Ruagadi et al. 2007).

19.4 Theoretical Framework

19.4.1 Grass-Roots Post-conflict Peacebuilding

John Paul Lederach states that peacebuilding is an overarching concept that encompasses, produces, and sustains the processes, approaches, and stages required for the transformation of conflicts towards a peaceful and sustainable relationship. This term includes efforts made before and after the peace agreement. According to Lederach, peace is a dynamic social construct, and he does not see it as a stage (Lederach 1997: 20).

Lederach's pyramid formulates three approaches to peacebuilding: top-level, intermediate, and grass-roots. According to him, each approach has its own role and influence, and there is urgent need for co-operation between all actors in the reconciliation process. At the top level, highly visible leaders participate in high-level negotiations and ceasefire efforts. At the intermediate level, general training in problem-solving, conflict resolution, and peace commissions takes place. Finally, grass-roots approaches include local peacekeeping commissions, grass-roots training, peace education, prejudice reduction, and trauma healing. All three approaches are needed in the process of reconciliation, which means that individuals, communities, and governments must work together, coordinate, manage and implement such programmes (Khan 2009). In this study, I will focus on the grass-roots peacebuilding approach.

Peacebuilding at grass-roots level is the effort made by ordinary people who make extraordinary efforts to bridge the gap between conflicting groups and build sustainable peace (Qurtuby 2013). In the case of the Women's School, this activity is conducted through peace education by communities established after the conflict. The majority of the school founders and the participants are survivors of the Poso conflict who have experienced the physical and psychological trauma of conflict. This movement provides peace education in materials of tolerance, peace, gender, politics, economic rights, social rights, civil and political rights, and community management.

Peacebuilding at grass-roots level is essential because of the need for agreed implementation of the peace agreement in the form of real efforts and activities (Hermann 2004). Through grass-roots movements, such activities will be seen as more substantial and enduring because of the insistence on maintaining tolerance, solidarity and understanding, and the obligation to participate in upholding human rights, according to the amended peace agreement (Anasarias/Berliner 2009). Also, through grass-roots peacebuilding with a bottom-up approach (in accordance with the Lederach pyramid), interpersonal relationships in post-conflict societies will increase, which is the key to successful reconciliation (IDEA 2003: 25).

19.4.2 Women in Peacebuilding

Mazurana and McKay stated that women's participation in peace-building is essential to prevent conflict, end armed conflict, and rebuild communities. In the women's grass-roots movement, women are transformed from victims into peace actors in NGOs (McKay 2002: 124). Furthermore, peacebuilding is not easy because of the dominance of men in the public sphere. However, if women engage in grass-roots peacebuilding, the vision of peace, equity, and social justice can be achieved.

From a feminist perspective, Mazurana and McKay outline approaches to grass-roots peacebuilding made by women as a process that emphasises relationships, basic needs, non-violence, forgiveness, and the process of reconciliation (McKay 2002: 126). The movement also pays attention to healing psychological stress by campaigning for gender empowerment and social justice. However, funds and financial support are often limited. Women's participation in peacebuilding focuses on gender awareness and empowers women in the political, social, economic, and human rights fields. Women seek to reduce sexual violence; reduce the physical and psychological harm that occurs during and after the conflict; ensure gender justice; and improve the status of women in post-conflict societies. Since female victims are usually marginalised in post-conflict situations, grass-roots groups and NGOs become influential and bring changes. Through such groups, many women are involved in peacebuilding as well as increasing their capacity as peacebuilding actors. Because of this crucial female contribution, it is vital for post-conflict peacebuilding to involve them as essential actors (McKay 2002).

A role in peacebuilding has been found for women in several places in Indonesia, especially in places that have experienced communal conflicts. The first example is the *Gerakan Perempuan Peduli* (GPP) in Maluku, formed by Sister Brigitta Renyaan. Sister Brigitta, together with figures from other religions, gathered with the goal of stopping the violence and fighting for the rights of women and children. All activities in this movement led to reconciliation between people of different religions when the conflict in Maluku broke out. Various efforts are being made to build interfaith dialogues, during which it is important to open initiatives to communicate with stakeholders to urge men to stop the conflict. There are also religious-based women's movements such as *Aisyiyah* and *Fatayat NU* from the two largest Islamic organisations in Indonesia, *Muhammadiyah* and *Nahdlatul Ulama* (NU), and the Women's Commission of the Christian Church. These movements have the same goal of empowering women in all aspects of life – education, health, economy, social welfare, legal awareness, political education, and women's empowerment – so that women can increase their participation in the public world.

19.4.3 Peace Education

Peace education is transformative. It explores the knowledge, skills, attitudes, and values that led to the formation of thoughts or acts of violence. From that point, a transformation is needed to build understanding, develop awareness, and generate societal action that will construct an environment that is non-violent, fair, and environmentally friendly (Navarro-Castro/Nario-Galace 2010: 27).

Today, the need to build a culture of peace is a testament to our self-reflection in a world of greed, hatred, egoism, and ambition. Conflict and violence are often driven by political, economic, military and religious interests. The most powerful way of shaping a culture of peace is through peace education. This education needs to be accepted and implemented by everyone to form a culture of peace. A unique and different approach to education is required to enable young people to face the complexity of the challenge. Education needs to be orientated towards peace, non-violence, and international co-operation, and should not glorify war (Navarro-Castro/Nario-Galace 2010). Peace schools for women from *The Asian Muslim Action Network* (AMAN) already exist in Indonesia, and there is one in Poso.

19.5 About Mosintuwu Women's School, Poso

One of the grass-roots movements in post-conflict peacebuilding in Poso is the Mosintuwu Women's School founded by Respondent G in 2009. Respondent G is a Christian Poso conflict survivor who founded the Mosintuwu Institute after completing a thesis entitled *The Political Memory of Women and Children in the Poso Conflict*. The Women's School tries to realise the vision and mission of peace in post-conflict Poso through the intellectual fulfilment of economic, social, cultural, and civil rights, especially for women. Respondent G also wrote on her website that at this time,

> In addition to being a focal point in a similar movement in Central Sulawesi, the concept and model of this school has been known in several similar organisations in Indonesia to be used as a conventional model.

In addition to creating good relationships and trust between followers of Islam, Christianity, Hinduism, and other religious tribes, participants are actively involved in peace campaigns and conflict prevention through issue management. Beginning with only a few participants, now the Women School has conducted peace education in forty-two villages in the Poso District. There are eight curricula run by this school, (1) tolerance and peace, (2) gender, (3) women and culture, (4) speech and reasoning, (5) women and politics, (6) public service rights, (7) political and civil rights, and (8) community economic management. Respondent G's dream was to provide a space for communication for women of various tribes and religions to discuss and debate openly. Speakers for these curricula come from diverse backgrounds, and include academics, religious leaders, activists, humanists, and other practitioners.

The *Mosintuwu Women's School* was chosen as a case study because the school has carried out its education in many villages, and the learning model is comprehensive and is adopted by other communities. Also, this school makes women the main peace agents at grass-roots level. This school brings together mothers from vulnerable groups in Poso and provides them with information and insights through its eight school curricula. Most of these mothers are direct and indirect victims of conflict.

This school was also chosen because it had convened the first Women's Congress in Poso District that made recommendations to stakeholders. This congress took place on 25–27 March 2014 in Tentena. The Poso Women's Congress was attended by 450 grass-roots women from seventy villages from fourteen sub-districts in the Poso District. The attendees included housewives, farmers, fisherfolk and labourers from various tribes and religious backgrounds (Islam, Christianity, Hinduism). The Women's Congress produced 135 recommendations addressed to the Poso District Government, the Village Government, and civil society organisations. In addition to the congress, the Mosintuwu Institute is actively involved in counteracting domestic violence, and raised awareness of the issue over a 16-day commemoration of International Human Rights Day in 2019. Additionally, the Institute is concerned about the sovereignty of women and the well-being of villages, which it highlights by holding an annual Crops Festival involving grass-roots women.

19.6 Findings

19.6.1 Women's School and Post-conflict Peacebuilding

In the case of Poso, the conflict that occurred in 1998–2001 and continued sporadically until 2007 was far too complicated to be defined as a religious conflict (Respondents L and D, personal communication, October 2016). The intertwined religious, political, economic and criminal root causes made it even harder to resolve. Both conflicting parties fought even when they did not truly understand what caused it (Respondent G, personal communication, October 2016).

In addition to the intertwined roots of the conflict, media provocation played a crucial role in intensifying the escalation of the conflict by spreading fear (Respondent G, personal communication, October 2016). When internet technology was not accessible to all people, local newspapers and television were their only sources of news. National mainstream media generally framed the violence as pure criminality, while local Muslim and Christian media increasingly invoked national or global religious conspiracies. Both local media created narratives about 'masses' and malevolent 'masterminds' (Aragon 2005). The present situation of the media is no different. Mainstream media have been framing Poso as an unsafe battlefield area during a security operation to combat a violent extremist group (Respondents R and W, personal communication, October 2016).

Religious leaders from both the Muslim and Christian communities have also played a major role in disseminating messages of both peace and hate. Their teaching content and delivery has been crucial in both escalating and de-escalating the violence. This point was stressed during the peace process. Religious leaders should evaluate the content of their sermons before delivering them to the masses in mosques and churches (Respondent D, personal communication, October 2016).

Based on the research interviews and observation, in terms of the present situation, Poso has achieved a 'negative peace' as described by Galtung. According to the police, even though there were several random and scattered individual post-conflict acts of violence, i.e. juvenile delinquency, there was no open violence between community groups (Respondent A, personal communication, October 2016). The people had reached a point where they were annoyed by certain parties who wanted to divide them (Respondent W, personal communication, October 2016). Based on the fieldwork observation in several villages in the district over the course of one month, Poso is now generally very safe compared to Palu (the capital of Central Sulawesi) and Jakarta (the capital of Indonesia). The relatively peaceful social life is complemented with low crime rates.

Local wisdom and tradition have also had a positive impact on peacebuilding. For example, the people hold an annual harvest feast called *padungku* as a sign of gratitude to God. This lively tradition is celebrated by most people from every ethnic and religious group by dancing, eating, drinking, and hanging out. When the communal violence took place, *padungku* was no longer celebrated by the community. Today, this tradition has not only resumed, but continues to become a very much more anticipated celebration than *Eid* or *Christmas* Day. In addition to *padungku*, Poso has '*sintuwu maroso*' as local motto, which means 'strong when united'. This motto has been embedded for years in people's daily life. It is said to be the most effective way to establish a culture of peace and unite the people. *Sintuwu maroso* is a manifestation of the Poso social life of those who live side by side bearing joy and sorrow together.

However, it cannot be said that Poso has achieved a positive peace because there are still many vertical conflicts, such as communities versus governments, communities versus the police or the army, and communities versus corporations (Respondent D, personal communication, October 2016). An example of a vertical conflict case is between illegal gold miners versus the army. Miners demand the legalisation of smallholder mining in Dongi-Dongi, North Lore. Another case was the assault of a soldier, who was shot by armed civilian groups, the Santoso group, in a shoot-out at Kilo Village. These sporadic and vertical individual cases are numerous in Poso and other cities in Indonesia.

Positive peace is closely related to welfare, justice, and co-operation in society. In economic terms, Poso remains less affluent than the surrounding districts, for instance Banggai, Parigi Moutong, and Sigi. The level of poverty in Poso District is the second-highest in the Central Sulawesi province, affecting 16.62% of inhabitants in 2018 (BPS Poso 2019). A few respondents affected by hardship stated that the absence of conflict is not enough to achieve sustainable peace. Consistent economic growth and improvement in the district to create jobs and improve people's lives are also required (Respondent A, personal communications, October 2016). Improving

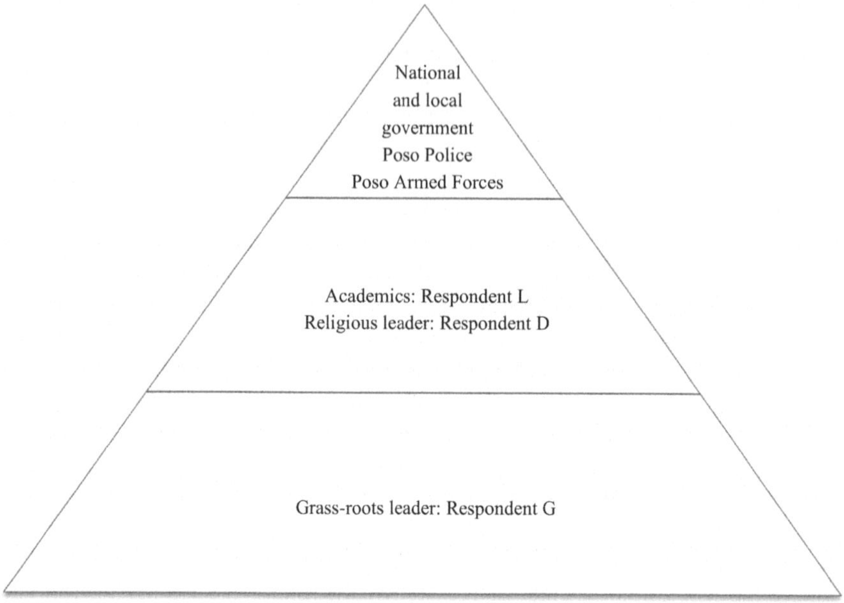

Fig. 19.2 Identification of peacebuilding actors according to the pyramid of Lederach. *Source* This figure was inspired by Lederach (1997: 39)

the economic life of the people is one way to establish a sustainable peace because if people are busy building their lives and working, they will not think of fighting and destroying each other (Respondent L, personal communication, October 2016).

Different actors can provide a series of definitions of peacebuilding, as shown above. The identification of peace-building actors relevant to this research, if linked to the Lederach pyramid, is as follows (Fig. 19.2).

In this study the top-level leadership was represented by the national government, the district government of Poso, the Poso Police, and the Armed Forces in Poso as the main stakeholders in the Poso District. According to Lederach, people at the top leadership level have a higher profile and more influence and power than other actors. However, they are exposed to the media, so they need to be careful about taking action and issuing statements if they want to maintain their position and avoid damaging their personal and institutional names (Lederach 1997: 40). Their movements are monitored by the local mass media, so their performance and activities are known to the people. Because people in this category often represent government agencies that depend on funding from the national and district budget, their peacebuilding activities may vary from year to year depending on the budget. This is what characterises top-level leadership; actors in this category have operational constraints because they are exposed to the public and are rarely able to personally control or implement long-term programmes or policies.

Mid-level leaders in peacebuilding have access to top-level figures, but are also close to the grass-roots community. These are people who have prominent positions in society or institutions, such as religious leaders, community leaders, NGO leaders, and academics. Some of these people have experienced war or conflict, and often take part in conflict and post-conflict situations (Lederach 1997: 41–42). The intermediate peace-building actors interviewed in this study were (1) Respondent L, an academic who served as a religious figure, women and peace activist; (2) Respondent D, a religious leader who became a Christian representative in the Malino Declaration.

Finally, at the bottom of the pyramid, there is the grass-roots level leadership that is the main focus of this research. At this level, people who are engaged in peacebuilding are often conflict survivors who experienced violence, trauma, and difficulties during and after the conflict, and have therefore been motived to become involved in grass-roots movements (Lederach 1997). Respondent G is a Poso woman who emerged as a peacebuilding actor at grass-roots level. Beginning with her research into the role and reconciliation of women and children's memories of the humanitarian tragedy in Poso, she started the Women's School with two of her friends. When the conflict was chaotic, she was taking Bachelor's and Master's degrees in Java, but her entire family still lived in Poso and experienced the situation of war. Staying in refugee camps for one year and hearing the stories of women inspired her to understand that Poso stories and narratives of peace during the conflict must be preserved and maintained, not merely obscured by the horror and the sadistic stories that are often remembered. She stressed that memory is an option; the mechanism of the human brain chooses what it will continue to remember (Gogali 2008: 38). If the memorable one is a hateful horror story of a particular group, the reconciliation process will not go smoothly because feelings of resentment and prejudice will continue to exist. She recruited grass-roots women to her school. The recruitment process was quite selective. She invited women who have never been included in community activities in their village, the grass-roots women who have never been consulted in the community decision-making (Respondent G, personal communications, October 2016). She is a central figure in the grass-roots movements of the *Mosintuwu Institute*, especially the Women's School, which is spread over fourteen sub-districts from a total of nineteen sub-districts in the Poso District.

The Lederach pyramid demonstrates that in addition to becoming grass-roots leaders, grass-roots women can also be actors in peacebuilding. Therefore, for this study, interviews were conducted with Respondent E and Respondent M, two great women who have grown stronger after being motivated by the Women's School. Respondent E, who used to be a housewife taking care of the family, has broadened her skill set and grown in confidence by becoming an activist against violence against women and children. She is now responsible for the Mosintuwu Women's and Children's Home Protection which is in charge of litigation and non-litigation of reported cases of violence (Respondent E, personal communication, October 2016). Meanwhile, Respondent M is now the organising coordinator of the Mosintuwu Institute, which oversees village reform, the Circle of Discussion and Action, and the Home for the Protection of Women and Children. Respondent M is also active in lobbying the Government for women to be more involved in the planning, supervision, and implementation of village activities based on Law No. 6, adopted in 2014, about village laws and village funds (Respondent M, personal communication, October 2016).

Synergic coordination is required between the three levels of peacebuilding actors in Lederach's pyramid. At the policy level, top-level leadership, such as the Malino Declaration, is effectively used as the momentum for the ceasefire. Middle leadership is needed to build a bridge between the top level and grass-roots level. Meanwhile, the grass-roots leaders have a direct impact and influence the community a lot because they have a common goal: sustainable peace. Once the peacebuilding actors have been identified on the basis of Lederach's pyramid, the discussion continues as to how grass-roots contributions can foster post-conflict peacebuilding. This study describes the contribution made by Mosintuwu Women's School to post-conflict peacebuilding in Poso.

First, the Women's School helps to create sustainable peace at grass-roots level by preventing conflict from resurfacing. This bottom-up approach can improve the interpersonal relationships between conflicting groups, which is critical for successful reconciliation (IDEA 2003: 25). The Women's School was formed because Respondent G's goal of reconciliation would not progress smoothly if the Poso community's memories, stories and narrations of the conflict were filled with fear and loathing of the opposing group. Respondent G argued that the people of Poso need to retain positive memories, not merely ones focusing on the horrors of the conflict.

This idea became the central element in the material taught at Poso's first-ever Women's School. With the reintroduction of narratives and stories of peace during the conflict by grass-roots women, it is hoped that the people of Poso can recover. The Women's School has also contributed to the establishment of sustainable peace in Poso by creating a culture of peace. Both directly and indirectly, its curriculum shapes women to become peace agents in their own communities. Sustained peace is partly characterised by the presence of justice in society, something that is also taught by the Women's School. Women are given insights, knowledge and briefing sessions on how to participate in village management, starting with the planning, implementation, and overseeing of Village Law, which is often dominated by men and patriarchal culture.

Second, the Women's School contributes to the recovery of post-conflict societies. The recovery done at grass-roots level is non-physical, because physical recovery has generally been arranged by the top-level leadership. In Poso, although the peace declaration was signed in 2001, sporadic and unstructured violence continued until 2007. The Women's School was first established in 2010 when prejudices and suspicions against opposing groups remained in society. This is acknowledged by Respondent M (a Christian) and Respondent E (a Muslim), who both reported that when first gathered in a class, they felt uncomfortable and disturbed when facing the opposing group of the other religion. They regarded each other as murderers because of the religion they were associated with.

Along with providing teaching materials, especially relating to tolerance and peace, the school has changed people's perspectives and way of thinking about the Poso conflict and reduced prejudice against the opposing group. Psychological recovery at the level of the individual mindset and perspective unconsciously restores the condition of society itself (Respondent L, personal communication, October 2016). Although the resulting impact cannot be measured numerically, it became evident that grass-roots women who succeeded in becoming peace agents started with their family.

Third, the Women's School contributed to the implementation of the peace agreement reached by the top-level leadership. Points in the Malino Declaration that were translated into action were: "to get rid of entire defamation and dishonesty against all parties and uphold mutual respect, and forgive one another, to create harmonious living together." As mentioned earlier, at the beginning of the establishment of the Women's School, Respondent G wanted to resurface the narratives, memories, and stories of peace that emerged from the grass roots sector during the conflict to prevent them always being overshadowed by horror stories, grim narrations and horrible memories of the opposing group. Respondent G believes that almost all Poso people are, directly or indirectly, victims as well as perpetrators. The points of the peace agreement were settled by the top level leaders and can chiefly be regarded as determining the moment of the ceasefire, while the people of Poso could only watch and had no say in the process. Respondent D consequently stressed that the peace agreement in the Poso reconciliation process was *de jure*, while the grass-roots peacebuilding was de facto. Therefore, it was important to maintain the mutual respect and forgiveness for post-conflict reconciliation to be sustainable.

19.6.2 Women and Transformative Peace Education

Peace education is a general term used to describe a series of formal and informal learning activities undertaken to disseminate peace in schools and communities through the planting of skills, characteristics, and values of non-violence with a view to reducing conflict and promoting tolerance and respect for differences (Buckland 2005: 60). This definition describes what the Women's School is doing. This informal school disseminates skills, insights, and peace values to establish a condition that upholds tolerance, resilience, and respect for diversity.

Transformative peace education is usually thought of in terms of curriculum i.e. what is taught, rather than how learning takes place and the environment in which it takes place. What the Women's School does is transformative peace education. As mentioned in the book *Peace Education: a Pathway to the Building Culture of Peace*, peace education explores the necessary knowledge, skills, attitudes, and values that lead to the formation of thoughts or acts of violence. From these matters, a transformation is needed to shape understanding, develop awareness, and challenge social actions that will construct an environment that is non-violent, fair, and environmentally friendly (Navarro-Castro/Nario-Galace 2010: 27).

Through their learning modules, this informal education is trying to empower women to become local peace agents. One of the earliest elements of the curriculum was *Religion, Tolerance, and Peace*, which became a central theme and the starting point for shifting mindsets (Respondent M, personal communication, October 2016). The material in the *Religion, Tolerance, and Peace* class creates an opportunity to open the lines of communication between two conflicting groups and enables them to understand each other. This open communication can change the way people think about other religions. This communication space is manifested in one way by visiting each other's places of worship and speaking to the *imam* at the mosque and the pastor at the church. At the mosque, Christian women asked about things they didn't understand, such as what *jihad* is, the obligation of *jihad*, and what form it takes (Respondent E, personal communication, October 2016). Step by step, this *Religion, Tolerance, and Peace* class can change the perspective of women and help them to understand the Poso conflicts and other groups. Respondent M, a Women's School member originating from the Bukit Bambu village where most of the people are Christian, said the class changed her way of looking at Muslims. This change of perspective is in line with the opinion of Salomon (as quoted in Blumberg et al. 2006: 33) linking peace education to peace psychology, namely how education can change the mindset that has been formed against other groups.

In addition to the *Religion, Tolerance, and Peace* curriculum, women are provided with a Talking and Reasoning curriculum, which is very useful for improving their confidence in communicating in the public forum, so that women are not trapped in the dominance of patriarchy in every aspect of social life. In addition to women being able to speak in public, this curriculum is expected to give women the courage to report any acts of violence they experience in the household (Respondent M, personal communication, October 2016).

The Community Service Rights curriculum provides women with knowledge and understanding of their rights as citizens, which must be met by national, provincial and district governments through policies, regulations, and legislation. This curriculum can help and encourage women to advocate for themselves if they do not receive their rights. By making women aware of their rights, the Community Service Rights curriculum enables them to help themselves by advocating for and demanding their rights if these are being denied to them by, for instance, authorities such as the village head (Respondent E, personal communication, October 2016).

Moreover, peace education is an effort to prevent systemic and long-lasting conflicts and is carried out from the bottom up (Keating/Knight 2004). This is in line with the Lederach peacekeeping pyramid, which emphasises that peacebuilding at grass-roots level has a broader and more sustained impact on the community as the efforts are based on a better general understanding of the needs, hopes, and dreams of conflict victims (Lederach 1997: 41).

Some essential things must be taken into consideration in peace education in Indonesia. *First*, all elements of society, from the Government to each individual, should understand and gain expertise in how to handle conflict (Panggabean 2006). Fostering understanding and expertise is closely related to the aims of Mosintuwu Women's School. The extent to which women understand the material given influences how the material is applied in everyday life, and the degree to which using the material becomes a habit in community life.

Second, in pursuing peace education, the Government and society must understand conflict as a problem to be solved, not as an attack that must be suppressed (Panggabean 2006). Concerning the Women's School, its third subject took the central issue of women's roles in the village because of Law No. 6 of 2014 about the village, whereby the participation of women is required in the planning, implementation, and supervision of government and village activities. There is a close relationship between building a village and peace. According to Respondent G, if the community can get together to build their village, surely they will not have time to think about conflict or join radical groups. A better situation certainly brings peace, prosperity, and the sovereignty of Poso, making people reluctant to sacrifice their property and lives for war (personal communication, October 2016).

Lastly, the most crucial aspect of peace education is that it should contribute towards community self-reliance by giving citizens the freedom to choose what is good for them (Panggabean 2006). It was mentioned earlier that peace alone is not enough in post-conflict areas. Justice and the sovereignty of the community also must also be achieved. Therefore social and economic independence must be pursued. Through the third strand of the curricula, Respondent G empowered a community-based economy at village level. She built community independence at village level – the smallest system of government that was closest to village inhabitants and easiest for them to influence – which immediately benefited the local society. In her opinion, women must actively participate in the management of their village, starting with knowing the potential of natural and human resources in their village, the challenges faced in managing the potential and how to overcome them, and having a strategy for achieving the sustainability of the village (Respondent G, personal communication, October 2016).

Sustainability is intertwined with economic independence, which is closely related to livelihoods or work in the village. If the potential of the village's natural resources is recognised and sustainably managed, the community does not need to leave the village to earn a living. An example of economic independence encouraged by Respondent G and the Women's School is *Kojo Coffee* in Bancea Village, one of the coffee-producing villages in Poso District. Respondent G helped encourage and promote the village's flagship product through the media. She also recognised the potential of soya beans in the village of Saojo to be used for *tempeh*, thereby enabling the village to produce a competitive and marketable product. Additionally, some handicrafts are done by the women in the village, for instance making products from plastic waste. Using village resources properly not only increases the income of those involved but helps them feel proud to be a villager (Respondent G, personal communication, October 2016).

19.6.3 Women's Participation in Post-conflict Peacebuilding

The four main aspects of women's participation in peace-building are prioritising forgiveness and healing, transformation, empowerment, and fighting for gender justice. This non-violence movement is characteristic of women's peacebuilding methods. It aligns with McKay's definition, which outlines the approach of peacebuilding made by women at grass-roots level as a process that emphasises relationships, basic needs, anti-violence (non-violence), remission (forgiveness), and the process of reconciliation (McKay 2002: 126). Moreover, the movement also pays attention to healing psychological stress by campaigning for gender and social justice.

First, the Women's School, the curricula succeeded in transforming individuals who had initially been victims of the prejudice and suspicion of opponents into agents of social change and peace in their environment. Women transform themselves as part of their participation in the peace-building process. A prominent transformation is the change of mind and of the way of thinking because it is these basic things that shape the behaviour and attitude of peaceful love, which is useful for reconciliation. This is in line with Mazurana and McKay's suggestion that in a grass-roots women's movement, women are transformed from victims into perpetrators of community redevelopment (McKay 2002: 124).

The agent of change and peace can start with the family, the smallest unit of a new society, then enter the public space where women can be more involved in community development. One of the dominant curricula in transforming women is the curriculum of *Religion, Tolerance, and Peace*, which provides an understanding of other groups and rectifies issues that have been misinterpreted, such as the issue of *jihad* in Islam. With this understanding, women can proudly have friends of a different religion and help to explain to their community issues that are not true. This is in stark contrast to the situation when the new Women's School was formed, and women from various religious and tribal backgrounds were gathered. The prejudices and the ice that once existed have turned into familiarity, and space has opened for communication. These results indicate that the transformation of women who participated in the Women's School has been successful. This is not only with regard to self-transformation; women can also transform society by conveying the values of peace to their community. Therefore, it is not excessive to refer to the women of the Women's School as a bridge of peace because the transformation process is thriving at the grass-roots women level. In addition to the women of the Women's School successfully becoming small agents of peace in the community, they have begun to enter the public sphere, something that is difficult because of the dominant patriarchal culture that traps women in the domestic realm. It is therefore essential for women to participate in peacebuilding at grass-roots level to gain a vision that peace, equality, and social justice can be achieved.

Second, women participate in the economic, social and political empowerment of women. Just like the transformation, their empowerment can start with themselves, then spread to the empowerment of other women. As stated by McKay, women's empowerment aims to reduce and prevent violence by shaping an environment in which women, men, and children can live in situations of non-violence, justice, equity and upholding human rights (McKay 2002). Regarding economic empowerment, women are actively involved in exploring the potential of natural resources in their villages. In addition to encouraging the utilisation of natural resources, they participate in promoting products and making loan applications, and they also held *Festival*

Hasil Bumi (Crops Festival) in November 2019. People from twelve villages in the Poso District and two villages of the Morowali District participated in this annual festival. Respondent G reported that this parade of crops is one way of demonstrating the potential of the village to build peace and justice in Poso because the products of the land managed by the village community are what support them every day.

Third, women are participating in the construction of peace through forgiveness and healing. In contrast to the dominant male approach to physical improvement and development, such as infrastructure, women pay more attention to psychological improvements. Forgiveness is essential for reconciliation because without forgiving one another, there will be an unresolved grudge in the future that could cause further violence. Respondent G, through the Women's School, tries to raise the profile of narratives and stories of peace during the conflict as part of a reconciliation of memories to make the process of forgiveness and healing work.

In the context of the Poso conflict, the Poso people are both perpetrators and victims because, in the definition of the victims, injury is not limited to physical damage, but is also psychological. An example of this psychological damage is experiencing fear, threats, and terror. Therefore, it is crucial to explore, raise, and discuss the narratives of peace at the same time, and to provide accurate information about religion (Respondent G, personal communication, October 2016). This reconciliation of memories was experienced by Respondent E and Respondent M, especially in *Religion, Tolerance, and Peace* materials, and proved effective in changing outlooks and ways of thinking, so it can be interpreted as motivating people to forgive each other over completed conflicts. Peace-loving women can remind their husbands to avoid potential violence (Respondent D, interview on 15 October 2016). It is acknowledged that women tend to avoid conversations that rake up past injuries that have the potential to create unresolved grudges. The forgiving nature of women is considered to balance the belligerent masculine egoism possessed by men, and can therefore foster the psychological process of forgiveness and healing.

Fourth, women are involved in peacebuilding by promoting gender equality and justice. Justice and peace are inseparable; it is impossible to achieve peace if justice is not enforced. However, these two things are not enough; both justice and gender equality are required because both during and after a conflict women are the most disadvantaged. They experience not just violence but multi-layered discrimination, and often bear an even more substantial burden because many of them are left behind to manage as best they can while their male relatives are at war. Some husbands never return. In post-conflict life bereaved women must bear the double burden of being both breadwinner and housekeeper. The influential patriarchal culture in Indonesia in general, and Poso in particular, increases their marginalisation and makes their situation more precarious.

Therefore, the movement undertaken by the women of the Women's School prioritises issues of equity and gender equality. Their struggle starts in the village. With the understanding that women and men are equal before the law, plus their knowledge of village laws, women have begun to fight for their civil rights at the village level. As people who often spend more time in the village than men who earn a living beyond its boundaries, women are perceived to better understand what is happening in the village and what is needed by the village for better development. Hundreds of women came to the Poso Women's Congress conducted by the Women's School to formulate policy recommendations for the Poso District Government. These recommendations covered various aspects, ranging from peace and equality to community service rights and community economics. With the involvement of the Women's School, grass-roots women have the confidence to contribute to the development of Poso and avoid the scarcity and marginalisation experienced by women who are trapped in the vortex of patriarchal culture and blinkered men.

19.7 Conclusion

Firstly, this study analysed the Mosintuwu Women's School's contribution to post-conflict peacebuilding in Poso. In addition to peace education, this grass-roots movement has provided healing and social training to women with the goal of facilitating peace, justice, and sovereignty. Respondent G founded this informal school with the initial and primary aim of raising the profile of grass-roots narratives and stories that have long existed during and after the Poso conflict so as not to obscure them with hateful horror stories and narratives. The re-appointment of the narrative of peace is important because the mass media never spread it while memories of the civil war still surrounded people. If hateful horror memories are not muted, silenced, and covered, the peace-building process will be hampered by prejudice, resentment, and distrust of the opposing group. Therefore, based on the results of data collection, the Women's School can contribute to post-conflict peacebuilding in ways that can prevent conflict from resurfacing and establish sustainable peace, which fosters the reconciliation process in Poso. The Women's School can restore the situation and condition of society; its participants can implement peace declarations; disseminate peace issues; establish a culture of peace; teach an understanding of conflict and peace; and contribute towards community independence.

Secondly, this research reviewed the participation of the women of Mosintuwu Women's School in peacebuilding in Poso. The peace process that is often highlighted in the media tends to be patriarchal and elitist, such as the dedication of houses of worship, official visits, or the inauguration of monuments, while the voices and peace stories that come from women tend to be quiet. In post-conflict Poso, women's role in peacebuilding can start in the family environment, then shift to the public sphere. A peaceful family environment and love are the beginning of a peaceful society, making children and adolescents less likely to engage in fights between students or youths. In the public sphere, women can be an inspiration and role model for their environment by advocating community rights, involving women in village development, sharing opinions in male-dominated discussion forums, and advocating in cases of violence against women and children at village level. All this has happened for the sake of peace and justice in *Tana Poso*.

References

Anasarias, Ernesto; Berliner, Peter, 2009: "Human Rights and Peacebuilding", in: River, Joseph De (Ed.): *Handbook on Building Cultures of Peace* (New York: Springer): 181–196.

Aragon, Lorraine V., 2005: "Mass Media Fragmentation and Narratives of Violent Action in Sulawesi's Poso Conflict", in *Indonesia,* 79: 1–55.

Barron, Patrick John; Jaffrey, Sana; Varshney, Ashutosh, 2014: *How Large Conflicts Subside: Evidence from Indonesia* (Jakarta: The World Bank).

Blumberg, Herbert H.; Hare, A. Paul; Costin, Anna, 2006: *Peace Psychology: A Comprehensive Introduction* (Cambridge: Cambridge University Press).

BPS Poso (Badan Pusat Statistik Kabupaten Poso), 2019: *Kabupaten Poso dalam Angka 2019* [Poso Years in Figures 2019] (Poso: BPS Kab. Poso).

Buckland, Peter, 2005: *Reshaping the Future: Education and Post-Conflict Reconstruction* (Washington DC: The World Bank).

Galtung, Johan, 1969: "Violence, Peace, and Peace Research", in *Journal of Peace Research,* 6,3: 167–191.

Gogali, Lian, 2008: *Tragedi Poso (Rekonsiliasi Ingatan): Gugatan Perempuan dan Anak-Anak Dalam Ingatan Konflik Poso* (Yogyakarta: Galang Press).

Hermann, Tamar S., 2004: "Reconciliation: Reflections on the Theoretical and Practical Utility of the Term", in: Bar-Siman-Tov, Yacoov (Ed.): *From Conflict Resolution to Reconciliation* (Oxford: Oxford University Press): 39–60.

HRW (Human Rights Watch), 2002: *Breakdown: Four Years of Communal Violence in Central Sulawesi* (New York: Human Rights Watch).

IDEA (International Institute for Democracy and Electoral Assistance), 2003: *Reconciliation After Violent Conflict: A Handbook* (Stockholm: IDEA).

Keating, Tom; Knight, W. Andy, 2004: *Building Sustainable Peace* (Canada: University of Alberta Press).

Khan, Waheeda, 2009: "Political Violence and Peacebuilding in Jammu and Kashmir", in: Montiel, Cristina Jayme; Noor, Noraini M. (Ed.): *Peace Psychology in Asia* (New York: Springer): 65–84.

Komnas HAM (Komisi Nasional Hak Asasi Manusia), 2005: *Poso, Kekerasan Yang Tak Kunjung Usai (Refleksi 7 Tahun Konflik Poso)* (Jakarta: Komnas HAM).

Lederach, John Paul, 1997: *Building Peace: Sustainable Reconciliation in Divided Societies* (Washington DC: United States Institute of Peace).

McKay, Susan, 2002: "Gender and Post-Conflict Reconstruction", in: Ho Won Jeong (Ed.): *Approaches to Peacebuilding* (New York: Palgrave Macmillan): 123–146.

Navarro-Castro, Loretta; Nario-Galace, Jasmin, 2nd edn., 2010: *Peace Education: a Pathway to a Culture of Peace* (Quezon City: Miriam College).

Panggabean, Samsu Rizal, 2006: "Educating to Handle Conflict and Avoid Violence", in: Coppel, Charles A. (Ed.): *Violent Conflicts in Indonesia: Analysis, Representation, Resolution* (London: Routledge): 217–228.

Qurtuby, Sumanto Al, 2013: "Reconciliation from Below: Indonesia's Religious Conflict and Grassroots Agency for Peace", in: *The Canadian Journal of Peace and Conflict Studies,* 44, 2: 135–162.

Ruagadi, Agustanty; Waru, Darwin; Lempadely, Sonny V.E.; Agus, Syamsul Alam, 2007: "Bersatu Kita Teguh di Tana Poso", in: Amirrachman, Alpha (Ed.): *Revitalisasi Kearifan Lokal: Studi Resolusi Konflik di Kalimantan Barat, Maluku, dan Poso* (Jakarta: International Center for Islam and Pluralism): 202–270.

The Wahid Institute, 2018: *Laporan Tahunan Kebebasan Beragama/Berkeyakinan dan Intoleransi 2018* (Jakarta: The Wahid Foundation).

Other Literature

Ismail, Noor Huda, 2016: "Memburu Santoso, Menyibak Konflik Poso"; at: Deutsche Welle https://www.dw.com/en/memburu-santoso-menyibak-conflik-poso/a-19178840 (10 October 2017).
Noorsalim, Mashudi, 2010: Catatan Lapangan: Sepuluh Tahun Pasca Konflik Komunal di Poso; at: https://referensi.elsam.or.id/wp-content/uploads/2014/12/Sepuluh-Tahun-Pasca-Konflik-Komunal-di-Poso.pdf (1 March 2016).

Chapter 20
Hydro-diplomacy Towards Peace Ecology: The Case of the Indus Water Treaty Between India and Pakistan

Saima Sabit Ali and Mansee Bal Bhargava

Abstract This chapter presents the institutional characteristics of the Indus Water Treaty (IWT Indus Waters Treaty, 1960: "The Indus Waters Treaty", signed in Karachi on 19 September 1960: 300–365, 1960) between India and Pakistan in the literature of peace ecology and discusses the complexities of including climate change discourse in the Treaty using the hydro-diplomacy approach in transboundary water management. The introduction to the chapter is followed by the conceptual framework of the study, the historical background of the Indus Water Treaty and the Indus River Basin, and the key aspects of the Indus Water Treaty. The second part includes institutional analysis of the Indus Water Treaty from the peace ecology and climate change perspectives. Towards the end this chapter notes that the challenges of the Indus Water Treaty justify the importance of hydro-diplomacy in the transboundary water management. The chapter concludes with a discussion on the convergence of hydro-diplomacy and peace ecology with climate change supplements for the rich and vulnerable Indus River Basin.

Keywords Climate change · Commons · Conflict · Hydro-diplomacy · Institution · Indus river basin · Indus water treaty · Peace ecology · Transboundary water management · Water war

20.1 Introduction

Conflicts over natural resources and the environment are among the greatest challenges and threats to human security in twenty-first-century geopolitics. However,

Saima Sabit Ali, M.Sc. Urban Management and Development, MAES Pvt. Limited (Engineering and Planning Consultants) Lahore-54660, Pakistan; Email: ar.saimaali@gmail.com.

Mansee Bal Bhargava, Prof. Dr, entrepreneur, researcher, educator; Consulting Partner at Environmental Design Consultants, Ahmedabad, Gujarat, India; Email: mansee.bal.bhargava@gmail.com

This chapter developed from two conference papers and incorporates feedback and suggestions.

Ú. Oswald Spring and H. G. Brauch (eds.), *Decolonising Conflicts, Security, Peace, Gender, Environment and Development in the Anthropocene*,
The Anthropocene: Politik—Economics—Society—Science 30,
https://doi.org/10.1007/978-3-030-62316-6_20

they can serve as a peacebuilding tool if managed in a sustainable manner. The *Indus Water Treaty* (IWT or Treaty 1960) between India and Pakistan is one such case of successful hydro-diplomacy (UN[1]; WB[2] websites) towards peace ecology. The IWT was signed between the governments of India and Pakistan on 19 September 1960 at Karachi with the aim of attaining workable water utilisation on the upstream and downstream of the Indus River basin, where India is in the upstream and Pakistan is in the downstream riparian zone. The Treaty covers six rivers of the Indus River Basin located in both countries, namely Indus, Jhelum, Chenab, Ravi, Beas and Sutlej.

The Treaty is of great importance in global hydro-diplomacy scenarios where transboundary conflicts arise and are aggravated when water matters snowball into geopolitics and diminish the peace and prosperity of the two countries. Regarded as one of the most successful transboundary treaties among the three hundred plus existing agreements globally, the Treaty has survived frequent tensions, including conflict, and has provided a framework for irrigation and hydropower development in the region for six decades. In addition, the impact of climate change is already observable on the Indus Rivers Basin, requiring strict compliance with close monitoring, and the sharing of information is emerging as a new yet strong connector for the two countries in the drive to uphold the Treaty with greater regard for environmental concerns and regional peace. The inter-dependency of the two countries due to the interconnected social-ecological systems (Ostrom 2009) is an undeniable strengthening feature, providing a strong reason for exploring the potentials of the connective capacity (Edelenbos et al. 2013) of the Indus water governance augmented by climate change for continued peace in the region, despite the growing geopolitical-cultural conflicts leading to several irreversible events and disturbances.

Many who study institutions assume that the six decades of sustained hydro-diplomacy reflect robust institutional arrangements (Ostrom 2005) of the Treaty and a resilient socio-ecological setting of the Indus River Basin. The Indus water governance has averted any damage to its collective institutional arrangements, despite several socio-ecological events and disturbances (Cox 2010), such as conflicts of geopolitics between the countries, communal conflicts within each country, and deeper order impacts of climate change. The Treaty is intensely studied from various perspectives (Alam 2002; Biswas 1992; Sahni 2006; Miner et al. 2009; Briscoe 2010; Kokab/Nawaz 2013; Kamran et al. 2017; Mahmood 2018). With the increasing impact of climate change, the Treaty and the Indus River Basin are becoming important research topics (Mirza 2007; Yang et al. 2014; Karki et al. 2011; Leos 2015; Mustafa/Wrathall 2011; Palmer et al. 2008). This study discusses whether climate change will add to the reasons to respect the sustenance of the Treaty, provide knowledge-driven solutions (Mahmood 2018), and guide actions towards building trust to restore peace in the region.

---

[1] See at: https://treaties.un.org/doc/Publication/UNTs/Volume%20419/volume-419-I-6032-English.pdf.

[2] See at: https://www.worldbank.org/en/region/sar/brief/fact-sheet-the-indus-waters-treaty-1960-and-the-world-bank.

20.2 Conceptual Framework

Environmental peacebuilding, or peace ecology, has emerged as a new frontier in interdisciplinary studies, offering a conceptual and operational framework for understanding the positive peacebuilding potential and process across the conflict lifecycle while mitigating potential risks (SDG Academy n.d.). With the rising climate change impacts, one may presume that transboundary hydro-diplomacy (Briscoe 2010; Cooley/Gleick 2011; Mahmood 2018) and peace ecology (Kyrou 2007) can be further strengthened within a climate change policy framework. To understand the Treaty from peace ecology and climate change perspectives, we investigate (1) the Indus hydro-diplomacy within the institutional agreement between the two countries to attain peace; and (2) whether, and if so, what institutional arrangements are possible and planned to address climate change. The philosophical part of the study is the normative interrelationship of climate change and peace ecology (Kyrou 2007), which is now a global concern and need. The conceptual framework is presented in Fig. 20.1.

This study highlights the characteristics of the Treaty exercised by India and Pakistan in managing the transboundary Indus River Basin as institutional arrangements (such as hydro-diplomacy) and as an agreement to attain peaceful relations (i.e. as peace ecology). This chapter is part of a desk study on the search for peace between India and Pakistan using the Treaty as a starting point of knowledge exchange and a tool to strengthen the peace process. Transboundary knowledge exchange/dissemination has a connective capacity just like the water flows that transcend boundaries. This chapter comprises a brief historical background (Sect. 20.3) and an introduction to the Indus River Basin and the Indus Water Treaty followed by discussions on hydro-diplomacy, climate change and the convergence of them into a peace ecology narrative. The chapter concludes by mooting the possibility of updating the Treaty to comply with the obligation to monitor and exchange data in

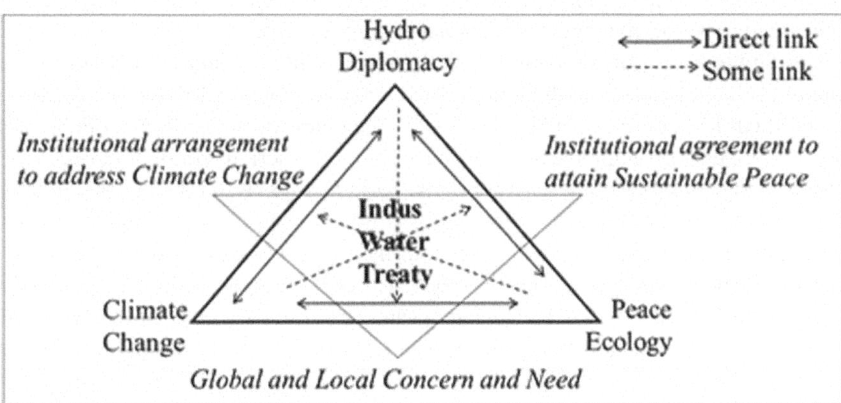

Fig. 20.1 Conceptual framework: *Source* The authors

response to the challenge posed by climate change, something that could also reduce the trust deficit – historically, a major problem between the Treaty parties. Climate change concerns are seen to protect the environment but also to have the potential to push collective thinking and efforts towards peaceful solutions.

20.3 Historical Background

International guidelines and water treaties are designed to deal with the complex issues of transboundary water management where political and jurisdictional lines are crossed by resources that know no boundaries (Cooley/Gleick 2011). More than 3,600 international water treaties on transboundary water issues have existed since antiquity and can be traced back from 805 to 1984 (FAO[3]). The free flow of rivers across the administrative boundaries of nation states presents challenges, and conflicts over their control have posed serious threats to human and environmental security.

There are over 300 transboundary river basins worldwide that cross the geographical boundaries of two or more nations, and they cover nearly 47.1% of the world's land surface (Global Water Partnership Technical Committee, in Wolf et al. 1999; Cooley/Gleick 2011). These rivers serve as the lifeline and spine for most nations in transboundary arrangements (McCracken 2017). To govern these vital resources 295 international water agreements have emerged since 1948 (Rao 2017). When the twentieth century's transboundary water agreements were established, the perceived water utilisation needs of the riparian stakeholder countries were deemed to be important, and the focus was on uses such as hydroelectricity (39%), water allocation (37%), flood control (9%), industrial uses (6%), navigation (4%), pollution (4%) and fishing (1%). With the rising and threatening impacts of climate change and with no direct provision for dealing with such threats in most agreements, there is a need for hydro-diplomatic routes to assess and address the climate change aspects while remaining within existing water treaties (Cooley et al. 2011) to respond to the challenges of transboundary water management in water-scarce regions.

The South Asian subcontinent has three major transboundary river basins – the Indus, the Ganges and the Brahmaputra – that influence the social, cultural, economic and political lives of over 1.6 billion people. The complex orientations of the rivers and the competition for their water share have resulted in serious conflicts between the neighbouring countries of these river basins. To address the water distribution among the basin countries and to resolve the nagging conflicts, Memoranda of Understanding and treaties were signed between neighbouring countries, such as the *Indus Water Treaty* – the oldest – that was signed in 1960 between India and Pakistan; the *Mahakali Treaty* that was signed in 1996 between India and Nepal; and the Ganges Treaty signed in 1996 between India and Bangladesh (Khan 2013). This study of the Indus Water

[3]See at: https://www.fao.org/land-water/water/water-management/transboundary-water-management/en/.

Treaty analyses the opportunity it presents for co-operation over shared resources and for addressing any challenges or conflicts arising from the share.

The nature and general characteristics of transboundary agreements globally deal with the complexity posed by two broad types of arrangements: (i) general principles of international behaviour and law; and (ii) specific bilateral or multilateral treaties negotiated for specific river basins (Cooley/Gleick 2011). The international agreements, treaties and charters devised for the protection of transboundary water commons and for peace as such are contained in the Helsinki Rules (1966), the UN Water Convention, the International Law Commission, and the Law for non-navigational waterfalls within the first arrangement. The second arrangement mostly addresses the general and specific features of the first arrangement. Together, they establish principles that are crucial binding criteria for the operation as well as the evaluation of transboundary water agreements: co-operation and information-sharing to ensure transparency; notification, consultation and negotiation geared towards peaceful conflict resolution – the aim of all rules and agreements. The laws endow all people with the right to water for reasonable personal use. To avoid conflicts between countries a significant harm principle is usually in place to protect vulnerable riparians from possible acts not spelled out specifically or through detailed definitions of broader terms used in an agreement and/or from unforeseen circumstances during the process of setting up institutional arrangements between co-riparian states that could harm a riparian party to an agreement (Cooley/Gleick 2011; Mahmood 2018). The Indus Water Treaty embeds these principles.

India and Pakistan, together known as a subcontinent until the mid-twentieth century, have shared history, land, resources, culture, and ecology inherited from the Indus Valley civilisation. It was during the British colonisation from the seventeenth century to the mid-twentieth century that the largest irrigation system in the world was built in the Indus River Basin. When British colonisation on the subcontinent ended, Pakistan and India became sovereign republics on 14–15 August 1947. The post-colonial Partition led to severe historic socio-ecological disasters in the immediate aftermath, due to the geopolitical division of two nations that had traditionally been living together with collective assets and resources. Soon after the Partition, the independent states realised the threats and opportunities which both nations had acquired on the division of the largest irrigational system in the world into a geographic downstream-upstream position, when India, being upstream, choked the water flow to Pakistan in 1948. The shared resources of the Indus River Basin thus emerged as a threat with increased risk of a war over resources which neither country was in a socio-political position to afford. Building trust since then was as complex as the Partition, a process which did not take place in an environment of friendliness and which contributed to hostility (Mahmood 2018). After nine years of negotiations, the governments of India and Pakistan came to a consensus on a treaty to resolve the resistance pertaining to the shared resources of the Indus River Basin. The Treaty was signed between the Government of India and the Government of Pakistan on 19 September 1960. The World Bank which was involved in the mediation during the negotiations is also a signatory to the Treaty, with its role procedural and limited to helping the two countries to protect the Treaty (Briscoe 2010).

20.4 The Indus River Basin

The Treaty is based on the Indus River, also known as Sindhu River, which is the longest river in Asia and twelfth largest river in the world (Ojeh 2006). The river originates near Lake Manasarovar in the north of the Himalayan range on the Kailash mountain in China at an elevation of 5,500 m. Flowing through the Himalayan mountain range and the Thar desert region of India and Pakistan, the river disgorges into the Arabian Sea near Karachi in Pakistan through the Indus Delta. The Indus River consists of six rivers located in the two countries, namely the Indus, Jhelum and Chenab in Pakistan, known as western rivers, and the Ravi, Beas and Sutlej in India, known as eastern rivers.

The data on the Indus River Basin spread and share between the countries vary in different sources, which is a cause for concern. According to the Water Research Institute (WRI 2003), the Indus River is 2,900 km in length (*Encyclopaedia Britannica* 2006) with a basin size of 1,081,718 km^2. The Indus River Basin is shared by four countries: Pakistan (63%), India (29%), China and Afghanistan (8%). According to the *FAO-AQUASTAT Survey*[4] of 2011, the Indusriverbasin has a total area of 1.12 million km^2 distributed between Pakistan (47%), India (39%), China (8%) and Afghanistan (6%). Nevertheless, it is significant that all four countries have built dams in the Indus basin in their own territories. Although the Indus River Basin is shared between four countries, the Treaty of 1960 was signed only by India and Pakistan due to their large percentage share of the river basin and the historical compulsion to come to an institutional agreement on water use and to establish peace after the partition of the subcontinent in 1947 that led to the division of land, assets and resources (Thakkar 2016).

The Indus River has been of great importance throughout history and is a vital resource. It offers an identity and a landmark as a socio-ecological reference as the home of one of the oldest civilisations in the world: the Indus Valley Civilisation. It is among the world's richest biodiversity zones, providing a variety of complex and essential ecosystem services for billions of people. It provides more than prosperity and stability for the region (Romshoo 2012). With the seventh largest mangrove forests globally, the Indus delta is the fifth largest delta in the world and is not only home to rare migratory birds and the Indus dolphin but forms a unique freshwater ecosystem. With its rich ecology, the region is also well-known for its art, culture and social structure. It is estimated that around 300 million people live in the Indus River Basin and are directly and indirectly influenced by the process and progress of the Treaty's activities on a regular basis (AQUASTAT 2011).

[4] AQUASTAT Survey (2011): *Irrigation in Southern and Eastern Asia in figures.*

20.5 The Indus Water Treaty of 1960

The Indus Water Treaty was signed by President Field Marshall Muhammad Ayub Khan for Pakistan and Prime Minister Jawaharlal Nehru for India. The World Bank that mediated and aided in the formulation of the Treaty was the main witness through its President, Mr Eugene R. Black. In safeguarding the Treaty, the collective agreement respects the intent and opinion of the involved governments. The Treaty includes twelve Articles, including the definitions and final provisions, with eight Annexes.

The four key components of the Treaty (Briscoe 2010) and the corresponding articles and annexes are simplified and explained by Ali and Bhargava (2019) as:

1. The division of the waters;
2. The financing plans;
3. The use of the hydroelectric potential of 'Pakistan's rivers' before they reach Pakistan.
4. The conflict resolution mechanism, through the Indus Commissioners or through external arbitration (Annexes F and G).

These four key features and corresponding articles and annexes are summarised in Table 20.1.

Articles II–IV deal with the use of the river, including the quantity of the water that may be used by both parties. The Treaty has allocated the rights of development to India for the three eastern tributaries, Sutlej, Beas and Ravi, which are exhausted; and to Pakistan for the three western tributaries: Chenab, Jhelum and Indus. The transboundary water flow has remained vulnerable to the continuous tense political, diplomatic and military relations since 1948, and is now further challenged by climate change impacts. Article V provides the financial settlement by both countries involved.

Articles VI-IX of the Treaty are to maintain co-operative arrangements and peacebuilding while governing the rights and obligations of either country without prejudice. The Treaty respects the opinions, rights and obligations of each party as essential elements of 'good faith' and 'co-operation' as crucial factors for the settlement of conflicts and for the peace process (Ahmed 2017; Mahmood 2018). While 'good faith' is an international norm of any treaty, it does not pressure either government to make any concessions; rather, it aims at effective negotiations. Hence as an effective negotiation tool, the declared intent of the Treaty in good faith identifies the Treaty as a successful model because of the elements of compliance for implementation and the mutual intention to co-operate in the agreement (O'Neill 2001; Mahmood 2018).

A *Permanent Indus Commission* was set up in article VIII as an institutional arrangement for sharing information, giving and responding to notices, for co-operation and dialogue, and it provides the means to resolve questions, differences and disputes (SANDRP 2017). The Commission consists of two Commissioners for Indus Waters appointed by each country as representatives of their governments. The commissioners meet once a year alternately in either country and prepare an annual report that updates the developments and data. Although the Commission is affected

Table 20.1 Summary of key features of the IWT (1960). *Source* Adapted from Briscoe (2010); IWT (1960)

Key features of the IWT (Briscoe 2010)	Corresponding articles in IWT (1960)	Corresponding annexes of the IWT (1960)
The Division of Waters	• ***Article II*** Provisions Regarding Eastern Rivers • ***Article III*** Provisions Regarding Western Rivers • ***Article IV*** Provisions Regarding Eastern Rivers and Western Rivers • ***Article XI*** General Provisions	**Annex B** Agricultural use by Pakistan from certain tributaries of The Ravi (Article II (3)) **Annex H** Transitional Arrangements (Article II (5)) Part 1: Preliminary Part 2: Distribution of the waters of Ravi Part 3: Distribution of the waters of the Sutlej and the Beas in Kharif in Phase I Part 4: Distribution of the waters of the Sutlej and the Beas In Kharif in Phase II Part 5: Distribution of the waters of the Sutlej and the Beas In Rabi Part 6: Water Accounts At Ferozepur Part 9: General Part 10: Special Provisions for 1960/1961
The Financing Plans	• ***Article V*** Financial Provisions • ***Article X*** Emergency Provision	**Annex H** Transitional Arrangements (Article II (5)) Part 1: Preliminary Part 7: Financial Provisions Part 8: Extension of Transition Period
Use of the hydroelectric potential of 'Pakistan's Rivers' before they reach Pakistan	• ***Article III*** Provisions Regarding Western Rivers • ***Article IV*** Provisions Regarding Eastern Rivers and Western Rivers • ***Article XI*** General Provisions	**Annex C** Agricultural use by India from the Western Rivers (Article III(2)(C)) **Annex D** Generation of hydro-electric power by India on The Western Rivers (Article III (2)(D)) Part 2 Hydro-Electric Plants in operation, or under construction, as on the effective date **Annex E** Storage of Waters by India on the Western Rivers (Article III (4))

(continued)

Table 20.1 (continued)

Key features of the IWT (Briscoe 2010)	Corresponding articles in IWT (1960)	Corresponding annexes of the IWT (1960)
The conflict resolution mechanism	• ***Article VI*** Exchange of Data • ***Article VII*** Future • Co-operation • ***Article VIII*** Permanent Indus Commission • ***Article IX*** Settlement of Differences and Disputes • ***Article XI*** General Provisions • ***Article XII*** Final Provisions	**Annex F** Neutral Expert (Article IX (2)) Part 1: Questions to Be Referred To A Neutral Expert Part 2: Appointment and Procedure Part 3: Expenses **Annex G** Court of Arbitration (Article IX (5)) **Annex H** Part 10: Special Provisions for 1960 and 1961

by the occasional geopolitical conflicts between the parties, the exchange of data and the settlement of differences and conflicts on complex issues of trust (Briscoe 2010; Mahmood 2018) have continued. Through the Commission the dialogues between both governments continues due to the need for developments and conflict resolution.

The conflict resolution mechanism is managed by the Indus Commissioners with the assistance of a neutral expert from the World Bank whenever required (Briscoe 2010). The World Bank delegation is charged with offering suggestions and speeding the dialogue between both countries through the Commission. A provision for establishing a Court of Arbitration is also available with assistance from the World Bank. In the case of differences and disputes on the development of hydro projects, the flow of water, or geopolitical post-conflict situations, the Court of Arbitration gets involved.

20.6 Institutional Agreement for a Peace Ecology

Water has been used and abused in wars throughout history[5] as the vulnerability of the resource becomes a target in war. The concept of 'water wars' was anticipated historically, which is why treaties were negotiated as preventive measures. A Transboundary Freshwater Dispute Database developed by scholars at Oregon State University (2005) provides a list of water-related events where co-operation prevailed over conflict. As many as 6,400 cases of water conflicts are listed for the years 1948 to 2005. The danger of an increase in future potential water conflicts related to transboundary waters has been declared a regional as well as an international peace and security threat (Council on EU water diplomacy in Foreign Affairs, Council meeting, Brussels, 22 July 2013). The Asian situation is alarming, with India having the highest number of 2,100 active socio-environmental conflicts in the

[5]A detailed chronology is available at: https://worldwater.org/chronology.html (Cooley/Gleick 2011).

world, followed by China, Bangladesh, Pakistan and Indonesia (EJAtlas[6] of 2017). The ambiguities in the conflict resolution provisions in the Treaty are due to both interpretation and technical issues that are inadequately addressed, such as the allocation of rights to use the resource, the definition of the nature of conflicts within the framework of the Treaty, and the speedy resolution of conflicts when they arise. In addition, unforeseen issues like climate change impacts were not provided for in the Treaty (Mahmood 2018). The differences are highlighted in the general and conflict resolution meetings by the Commission, the Neutral Expert and the Court of Arbitration.

Shared water resources and transboundary waters are also studied for their potential to foster peace rather than conflict. The literature containing optimistic views and suggestions for co-operative solutions is just as abundant as that written from a more cautious and troubled perspective. The United Nations Economic Commission for Europe (UNECE 2015) has published extensively and issued a *Policy Guidance Note* on the benefits of transboundary water co-operation with regard to identification, assessment and communication. The processes and perceptions leading to international water conflict resolutions emphasise similar key issues in advancing co-operation and robust implementation of international water treaties (Abukhater 2013).

The framework of a peace ecology (Kyrou 2007; Amster 2015; Oya Dursun-Ozkanca 2016) built on concepts of bioregionalism, place, sustainability and interconnectedness or, in other words, environmental peacemaking, provides a methodological approach for identifying and measuring resources for the transformation of conflicts and violence at the intersection of peace and ecology. The fundamental basis of the IWT adheres to the idea of a peace ecology as manifested through the preamble of the Treaty that refers to "…a spirit of goodwill and friendship, the rights and obligations of each…". The combined vision of peace and environmental justice is an inherent aspect of the IWT. The survival of the IWT for six decades can be considered an "effective environmental peacebuilding model" (Kyrou 2007) based on its efficacy in reaching narrow and specific institutional arrangements and agreements that safeguarded the environment and/or peace during several geopolitical events and disturbances.

Water security is of prime importance for sustainability. There are also occasions when water itself may not be the root cause of conflict but may become a catalyst, an excuse and a weapon for political unrest and gains (UNESCO 2013; Vink 2018). The Treaty has survived between both countries despite two major wars in 1965 and 1971; a limited war in 1999; and a series of geopolitical-military conflicts (1987, 1989–90, 2002, 2008) (Ranjan 2016).

It is crucial for both countries and the neutral expert of the IWT to acknowledge the opportunity costs involved in the peace process and the achieved outcomes in

---

[6] Environmental Justice Atlas (EJAtlas) is a map that was created at Universitat Autonoma de Barcelona's Institute of Environmental Science and Technology by Rubén Escobar (13 June 2017): "What are the socio-environmental conflicts in the world?"; at: https://nonprofit.xarxanet.org/news/what-are-socio-environmental-conflicts-world.

terms of both environment and peace, and, further, in making both more resilient with each passing event and disturbance. The Treaty is a vital sustainable peace ecology model for other transboundary economic-socio-cultural exchanges between the people of the basin, as its transparency helps to build bridges in the prevailing environment of distrust (Briscoe 2010; Mahmood 2018), with information-sharing and compliance with the Treaty being key to a sustainable peace process.

20.7 Institutional Arrangements to Address Climate Change

Among the physical effects of climate change are the variability of precipitation and the rise in the number and intensity of hydro-meteorological hazards that result in flooding and drought and thus create fierce competition between countries sharing transboundary waters. Thus, preventing water conflicts has become a crucial task for many nations. The South Asian region has experienced much devastation due to climate change. Archaeologists have suggested that the Indus Valley Civilisation may have perished thousands of years ago because of climate variability (Appalachian State University 2014; Schug et al. 2013; University of Cambridge 2014). Asia has 57 transboundary river basins that cover at least 40% of the area in international basins. About 70 transboundary aquifers require transboundary agreements to be revisited for climate change adaptation (Cooley/Gleick 2011).

Climate change is a regional, local and a global problem, and like all other rivers, the Indus River Basin is also affected. Such historical events are precedents and reminders for a rethinking of suitable institutional arrangements to address the challenges and complexity of climate change. The Indus River Basin is among the ten most vulnerable basins in the world due to the impacts of climate change (Wong et al. 2007). As the region is vulnerable to climate change impacts, addressing the challenges of both ecology and peace in the region is urgent, as without appropriate action the entire region may perish, just like the Indus Valley Civilisation did. The impact of climate change is already visible in the Indus River Basin, and is reflected in the rising water stress in the region, with frequent instances of floods and droughts. The Indus River Basin is expected to suffer a 50% water deficit by 2030, while the demand for water is expected to double due to the growing population that is making the lower riparian sub regions extremely vulnerable. The situation is further worsened by institutional interventions, for example, India restricting flows from the Ravi and Sutlej rivers to reduce the environmental degradation upstream; and Pakistan suffering from flow restrictions resulting in an accumulation of silt on the riverbeds downstream which is adversely affecting the fragile ecology of the area. Climate change and the upstream and downstream linkages are here very important and cannot be dealt with in isolation without understanding the inter-state linkages with climate change impacts. The effect of 1.5–2 °C temperature increase is a benchmark for forecasts on the effects and future planning strategies, focus and direction for the Indus Basin. The impacts of this global-warming on the cryosphere, which is a major contributor to the Indus River, would be severe: 33–49% of the water flow would disappear (Dilshad et al. 2018). The depletion of glacial melting will decrease the water flow in the Indus River Basin by 30–40% (Nax 2016).

Since the impacts of climate change transcend geographical boundaries, it is even more important to combat this common threat with collective action. However, this collective action requires sovereign states to have the political will and perception to take action to achieve and safeguard peace and security for both humankind and the environment (Oswald Spring 2007) and to understand that the concept of a peace ecology contributes to positive externalities.

Climate varies in the Indus River Basin from subtropical arid and semi-arid to temperate sub-humid in the plains of the Sindh and Punjab provinces and to Alpine in the mountainous highlands of the north. Climatic variability is influenced by altitude. The mean monthly temperatures in the lower plain vary in winter (December to February) from 14 to 20 °C and in summer (March to June) from 42 to 44 °C; and in the upper plain in winter vary from 2 to 23 °C and in summer from 23 to 49 °C. The average annual rainfall in the Indus River Basin is about 230 mm (WCD 2000). Annual precipitation ranges between 100 and 500 mm in the lowlands to a maximum of 2,000 mm on mountain slopes. Snowfall at higher altitudes (above 2,500 m) accounts for most of the river run-off (Ojeh 2006). Annual precipitation from 100–200 mm in the northern valley floors increases to 600 mm at 4,400 m altitude, and further increases to 1,500–2,000 mm at 5,500 m altitude. The precipitation in winter (October to March) is spatially also diverse. The flow of the Indus River decreases during the winter and floods the banks during the monsoons.

Climate change has been evident in the Indus River Basin for some time. For example, from 1961 to 1999 there were significant increases in the winter, summer and annual precipitation, and significant warming in winter, with more cooling trends in summer. These trends have an impact on water resource availability in the Indus River Basin (Fowler/Archer 2005). Because of the projected increasing water stress, some peace and conflict researchers anticipate as a possible outcome water conflicts over the allocation of water, particularly in transboundary settings like the IWT (Chellaney 2013; Akhtar 2010; Alam 2002; Cooley et al. 2012; Mahmood 2018).

The IWT is not a water-sharing transboundary water resource management system but a water allocation exercise in principle that requires co-operation (Kamran et al. 2017; Mahmood 2018), and with the new and continued climate change challenges, it becomes crucial that the new institutional arrangements address climate change concerns with co-operative arrangements. While the Treaty does not mention climate change impacts, institutional agreements are crucial for addressing the same and for building water resilience in the region. During the 113th Commission meeting in 2017, Pakistan proposed[7] deploying a monitoring mechanism to assess the adverse effects of climate change on the water flows of the Indus rivers system. It argued that because climate change will influence the flow of the rivers, it cannot be excluded

[7] See at: https://www.thenews.com.pk/print/193124-Pakistan-keen-to-set-up-climate-change-monitoring-system-for-Indus-rivers.

from the Treaty framework. Paragraph 29 of Annex G of the Treaty mentions the international conventions and customary international law through which the environment and climate change concerns can be addressed within the Treaty. There is an urgent need for joint monitoring of the Indus basin so that steps can be taken under the guiding principles of the Treaty to effectively mitigate the adverse effects of climate change. Both countries should develop a mechanism on watershed management as another tool to help make water flows smooth and ensure the real-time exchange of data related to river flows. Monitoring and sharing information will certainly benefit both countries and should be a welcome proposal leading to early action (SANDRP 2017).

With the growing impacts of climate change, the institutional framework of climate change concerns and adaptation becomes a strong tool for both countries for better collective decision-making and for designing collective strategies to address the water woes in the region, thereby indirectly contributing to the building blocks of peace ecology. There are huge opportunity costs involved in further energy development pertaining to climate change impacts which must be weighed against the benefits that the peaceful resolution of conflicts can bring (Thakkar 2016, 2017). It is in the interest of both countries to ensure compliance with the Treaty with respect to addressing issues of climate change.

20.8 Challenges of the Treaty

Although the IWT has survived during wars and border conflicts between both countries, there has been continuous unrest over the division and declaration in the Treaty. One of the unsaid crucial differences in the interpretation of the Treaty is that India considers the Treaty to be generous to Pakistan while Pakistan has perceived it as being discriminatory since its inception (Chandio 2014). The Treaty poses some socio-ecological challenges for both countries, including the new climate change impacts that shift the geopolitics of both countries to the back.

The Treaty deals with the allocation of water between India and Pakistan, but not with maintaining the environmental flows in the salient rivers in the case of the construction of dams, hydropower projects and barrages. It likewise has no provision for the collective assessment of projects on the rivers pertaining to carrying capacity, social and environmental impact assessments, or management plans, and no advance information on sharing any actions. There are not even any collective studies, monitoring or evaluations, let alone collective projects (Thakkar 2016). The rational choice of the countries may negatively affect the Treaty in the long run. For example, the impact of excessive groundwater extraction on the water flows in the rivers and river pollution across the boundaries right up to the delta is yet to be part of the collective discussion or decisions on resolving them. The annual report prepared by the Commissioners with updates on the developments, situation, data, and the agreement is first a country-specific activity and next little more than a scheduled duty with scant room for discussion or the evolution of the Treaty. It is important to understand the case of the IWT, which is not a regular case of high and lowland interactions but a case of two regions with a long history of conflict. It is also pertinent to note the design of the Treaty, which explains much about the nature of the Treaty and its parties. It is a water allocation treaty (Kamran et al. 2017) because sharing was not possible as the parties could not agree terms, hence no joint or co-management options were listed (Mahmood 2018). Amidst this, climate change has arrived, affecting the river basin significantly and requiring consideration in the functioning of the Treaty.

This refers to the important discussion on the heavy top-down nature of the Treaty. The Treaty undermines the importance and perception of the local community of the Indus River Basin. It was designed at a meta-constitutional level (Ostrom 2005), and the national – rather than local – governments had the upper hand. Among the 300 million people living in the Indus River Basin, several local community groups have stood up to their respective governments from time to time over issues related to the mismanagement of the basin water. It is estimated that 86% of over 300 transboundary water agreements do not include all riparians (Jagerskog/Phillips 2006). With new knowledge and developments some treaties now also address the SDGs and environmental concerns. Careful flexibility and adaptability in treaties are useful in managing new developments and unforeseen externalities by devising a response strategy for extreme events like climate change (Cooley et al. 2011) while complying with the agreement for sustainable and peaceful resource sharing.

A crucial part of the Treaty is that opting out of the Treaty hypothetically is not an option for either country, as an alternate 'better than this' is difficult to envisage. In this zero-sum water-sharing game (Arfanuzzaman/Syed 2018), one side cannot increase its share without diminishing that of the other. The Treaty is thus a classic case of a water common or common pool resource with high subtractability and low excludability where collective action through the institutional agreement is the only and best option available (Ostrom 1990). In critiquing the extensively used Hardin's (1968) model of *the Tragedy of the Commons* and the promotion of government and private interventions to control resources, Ostrom (1990) encouraged social groups to manage shared resources, arguing that, "common pool resources must be governed in the same fashion while they operate under different governance protocols". Over the years, the Treaty Commissioners may well have understood this fact and may thus have either held the meetings merely as part of the Treaty protocol or worked towards peaceful transboundary water management with a hydro-diplomatic process by default despite occasional geopolitical conflicts in the region. But leaving things as they are and hoping that, with improving political relations, a more reasonable and constructive spirit towards the operation of the Treaty will prevail on both sides in the future has also challenged the Treaty.

20.9 Opportunity for Hydro-Diplomacy

The South Asian context for hydro-diplomatic success is challenged by a substantial trust deficit and nationalistic sensitivities, coupled with socio-political complexities (Mahmood 2018), and has been a strong reason against the development of collaborative programmes for benefiting from shared resources (*Gulfnews*[8]). While the *South Asian Association for Regional Cooperation* (SAARC) is greatly important and influential in resolving regional issues and working for regional prosperity, it is ill-equipped to advocate hydro-diplomatic solutions in the region (Khan 2013) especially between countries with long estranged relations. Both India and Pakistan face geopolitical volatility where hydro-diplomacy is essential for peacebuilding. Hydro-diplomatic approaches that are well-designed are crucial for catalysing co-operation for sustainable development and peacebuilding (Vink 2018) in situations where influential global or regionally volatile politics and varying policy domains may trigger war on resources, specifically water wars.

Hydro-diplomacy (Cooley/Gleick 2011) places the focus on peace building and stability above water co-operation's technical focus on water allocation and availability. The water co-operation characteristics of sharing data, pooling information, and technical expertise exchange are embedded in hydro-diplomacy as basic requirements of working towards peace. Additionally, hydro-diplomacy is a tool designed to prevent or peacefully resolve conflicts and facilitate co-operation and enhanced mutual benefits between political entities (Pohl et al. 2018). Hydro-diplomacy is a powerful way to handle the problems of larger socio-ecological systems (Ostrom 2009) with effective political tools like negotiation processes and conflict resolution (Oswald Spring 2007; Pohl et al. 2018). A systematic hydro-diplomatic approach can resolve issues through changing an outlook on common-pool resources into a collective vision (Ostrom 1990) with a concern for and sensitivity towards the environment that encompasses both life (man, wildlife, flora and fauna) and natural resources in a larger envelope beyond the geographical bounds of sovereign states. It may not only change the focus of institutions to the protection of the shared resources but also trigger actions which are mutually beneficial and boost peace and security throughout the region (Oswald Spring 2007). Hydro-diplomacy is advocated as being instrumental in maintaining peace and providing solutions to avoid the tragedy of the commons envisaged by leading scholars and theorists (Oswald Spring 2007; Pohl et.al. 2018; Vink 2018; Khan 2013). Further, since the climate change impacts which lead to water distress are oblivious to boundaries, it is anticipated that hydro-diplomacy to advance the peace process is key to rethinking the focus of transboundary water management in the same spirit of goodwill and care for the greater good with which the Treaty was initiated.

When addressing a panel on water diplomacy in Stockholm on 28 August 2018, the Deputy UN Secretary-General, Amina Mohammed, argued:

---

[8]See at: https://gulfnews.com/opinion/op-eds/pragmatic-water-diplomacy-need-of-the-hour-in-south-asia-1.2002488.

> Water can represent a source of cooperation, shared growth and mutual support. Getting caught up in the 'water-war' rhetoric will be a mistake for the international community. When we examine history, we see that co-operation over water can prevail over conflict over water. Through water diplomacy, neighbouring states can be reminded of the benefits of co-operating around water resources. Water, if fairly shared, can become a confidence-building measure. Such confidence-building measures are urgent.[9]

20.10 Discussion

In the emerging water war paradigm (Akhtar 2010), although using water as a catalyst for cooperation and adaptive co-management may be challenging, it is nevertheless crucial for peaceful coexistence and enduring stability. International political leaders, humanitarians, environmentalists and climate experts from the Indus basin countries and globally warn of debates on water war and advocate water co-operation. The role of hydro-diplomacy as part of a peacebuilding process is crucial for countries like India and Pakistan that are susceptible to wars due to a long history of mistrust and unfinished conflicts in geopolitics that could result in water wars.

As oceans cannot be parcelled (McKean 1996:228), ecological problems associated with transboundary water flows are extremely challenging and complex to divide by geopolitical boundaries, so the fact that humanity is governed by a system of such boundaries is bound to place ecology in a stressful and vulnerable situation. Water has that connective capacity (Edelenbos et al. 2013) to bring people and politicians across borders to a converging dialogue, especially since it is a fundamental truth that billions of people on both sides are dependent on the ecological resources of the Indus River (Hussain 2016). There is a need for a collective culture of care for the environment by thinking beyond geopolitical boundaries and adopting an ecological as well as an anthropocentric approach (Amster 2009). One positive aspect of the advent of climate change is that it pushes us towards collective thinking when seeking peaceful solutions.

The impacts of climate change exacerbated by population growth affect the most water-stressed regions. The complications that lie deeper are the conflicting geopolitics that dominate water management. Hydro-diplomacy coupled with water co-operation offer hope for complex transboundary water situations, especially where one riparian is stronger politically, militarily, economically, or technically and within regions that comprise estranged neighbours and unstable border settings (Oswald Spring 2007).

With a framework for analysing key international water treaties using indicators such as water as an arena for future cooperation, employing a multi-disciplinary, transformative approach to conflict resolution offers researchers ideas for various geopolitical contexts, spatial and temporal diffusion, equity analysis, and how to mitigate the spatial and geographic patterns of water distribution which make even

[9]Speech by UN Deputy Secretary-General on 28 August 2018 in Stockholm.; at: https://economictimes.indiatimes.com/news/politics-and-nation/indus-waters-treaty-provided-framework-for-resolving-disputes-on-water-use-un-dy-secretary-general/articleshow/65580647.cms.

the most dynamic spatial models inefficient and uncertain (Abukhater 2013). For example, the cooperation and adaptive co-management of the Indus waters will remain critical for both countries and may be a way forward in the governance of water resources shared by India and Pakistan within the IWT. There are characteristics that are crucial for the successful governance and there is room for improvement of those (Rao 2017). Conflict resolution mechanisms are an element of the Treaty which have been exercised from time to time. The possibility for improvement and adaptation of the Treaty will enhance its effectiveness, while monitoring will increase transparency. Making institutions more robust eventually enables the facilitation of cooperation in new areas. While most international treaties include co-operation mechanisms for monitoring, information exchange and conflict resolution – which is always a vital objective of any treaty – a few overlooked elements, such as failure to include all stakeholders and riparians sharing the resource may constitute externalities that go unaccounted for or create bigger problems and add to institutional vulnerabilities over time. In the work following this study, we have organised an understanding of guiding principles of the Treaty into dual design principles or rules of successful resource management developed at the Ostrom workshop (Ostrom 1990–2005).

Managing diplomatic relations and the transboundary water between India and Pakistan has been a continuous challenge since the division of the subcontinent. Both countries have tried to manage both under extremely stressful geopolitics. The Treaty has been a link in the relationship between two countries where a trust deficit is the most critical issue (Mahmood 2018), as it brings them back to the negotiating table despite several raging conflicts between them (Thakkar 2017). The 'other' factors influencing the institutional arrangement that may have larger than perceived impacts tend to remain unspoken and unaddressed in institutional analyses. The threat of water wars and potential risks to the Treaty in the case of Pakistan and India are also critically linked to other unresolved issues that began with the partition of the subcontinent (Mahmood 2018). The disputed territory of Kashmir is an unspoken and unfinished war between both the countries, with a context that is still relevant and almost inseparable as all Western Rivers flow through the disputed territory (Schoefield 2000; Mahmood 2018). Such unspoken links have been observed to have stronger and much more devastating impacts in transboundary water management, and sustaining peace and solutions must be understood within the context of both countries and not through purely academic theories or hypothetical panaceas (Mahmood 2018). The ambiguities and the possibility of a suspension of data-sharing leave room for further manipulation by the upper riparian, being technologically and politically stronger, making the lower riparian vulnerable, insecure and dependent on the whims of the stronger state, as in the case of the Ganges Water Treaty (Oswald Spring 2007; Kamran et al. 2017; Thomas 2017: Akhtar 2010; Mirza 2008). This vulnerability of lower riparian states was also referred to by Briscoe (2010), Mahmood (2018), Chandio (2010) and Akhtar (2010), who claimed that "this ball is very much in India's court." The analysts have raised concerns and many possibilities (all arising from the choking of water in 1948 which had sown seeds of distrust and a realisation of the extreme vulnerability of the lower riparian) that India can choke Pakistan's flow of water for pressure or create significant harmful effects by drying up river-canals or flooding rivers (Chandio 2014). Trust-building is complex yet important and needs comprehensive solutions in this case.

The Treaty has larger socio-ecological impacts that transcend the shorter political motives of (changing) elected governments. What the Commissioners can do is continue exercising this peace process with more rigour using climate change as a stronger reason to remain in the Treaty, since, though the basin and the Treaty may have a charted boundary, climate change impacts transcend those boundaries and demand more action towards humanity. Although neither Commissioner seems to ask for any amendment to the Treaty, dealing with climate change, international treaty law and customary international law is insufficient as they do not offer any mandate. Since climate change has arrived and is urgent now! A review of the Treaty to include the concerns related to climate change is suggested after conducting a strategic socio-ecological impact assessment of the basin area. At this stage, omissions of any article are not suggested and may be a less helpful strategy for promoting transboundary water management than keeping the Treaty intact, even though it was compromised after a decade of negotiations.

However, additions or additional accords are suggested that may (i) help secure political commitments and will continue with the same spirit of leadership that prevailed at the time when the formulation of the Treaty built bridges, with a common goal of tackling the challenge of climate change; (ii) help enforce compliance with the Treaty to help build confidence and continue the transboundary resource-sharing; and (iii) define and describe articles to clear up confusions that give space to the type of manipulation which is not helpful in reducing the trust gap.

The history of transboundary water diplomacy in the Asian region provides lessons and models that are successful to achieve solutions in the increasingly complex institutional and contextual arrangements. An important lesson is that, in order to devise strategies, it is as vital to govern transboundary water resources as it is to understand the inter-disciplinary and trans-disciplinary nature of the subject. Ideas like "adaptive water governance" (Mirza 2007) and "joint and integrated development" (Bhatnagar 2008) are already being explored and applied in the Indus Basin studies. The idea of converging hydro-diplomacy and peace ecology with the crucial climate change component into the ongoing discourse on the rich and vulnerable Indus River Basin is certainly worth discussion.

Case studies of un/successful transboundary water diplomacy and peace processes further improve our understanding of the institutional arrangements that have worked, under which conditions, and when they collapsed. It is also important to note that although the IWT may be considered a successful treaty because it survived even in the midst of the conflicts between the two countries including wars, it should not be regarded as capable of preventing the chance of future differences erupting or of resolving disputes just because it has survived for more than sixty years. The survival can be attributed to other unspoken factors which may not apply now, and the future relevance of this past success is questionable (e.g. Pandey 2014; Thomas 2017; Mahmood 2018; Nax 2016). Scientific knowledge combined with knowledge from other disciplines as social sciences, environmental science, diplomacy and ethics may help connect the dots and build the bridges of cooperation for the resolution of issues (Mahmood 2018; Dilshad et al. 2018).

The collapse of societies (Tainter 1988) is attributed to socio-economic complexities pertaining to conflicts. Thinking humanity with the advent of climate change impact may eventually become a matter of prime importance and allow peace to prevail. While it is hoped that humans will not see the worst of the climate change impacts, at the same time there is a wish to see the day when climate change will become a peace messenger in the region.

References

Abukhater, Ahmed. 2013: *Water as a Catalyst for Peace: Transboundary Water Management and Conflict Resolution* (London: Routledge).

Ahmed, Imtiaz (Ed.), 2017: *South Asian Rivers: A Framework for Cooperation* (Cham: Springer International Publishing).

Akhtar, Shaheen, 2010: "Emerging Challenges to Indus Waters Treaty", in: *Regional Studies*, 28,4; 3-66.

Alam, Undala Z., 2002: "Questioning the Water Wars Rationale: A Case Study of the Indus Waters Treaty", in: *Geographical Journal*, 168,4: 341-353.

Ali, Saleem H., 2008: "Water Politics in South Asia: Technocratic Cooperation and Lasting Security in the Indus Basin and Beyond", in: *Journal of International Affairs*, 61,2: 167–182; at: https://www.jstor.org/stable/24358117.

Ali, Saima Sabit; Bhargava, Mansee Bal, 2019: "Institutional Arrangement of the Indus Water Treaty: A Transboundary Water Management between India-Pakistan", Paper for the 17th International Association for the Study of Commons Conference, Lima, Peru, 1–5 July.

Amster, Randal, 2015: *Peace Ecology* (Abingdon: Taylor & Francis).

Arfanuzzaman, M.; Syed, M.A., 2018: "Water demand and ecosystem nexus in the transboundary river basin: a zero-sum game", in: *Environment, Development and Sustainability*, 20,2: 963–974.

Bhatnagar, Manav, 2008: "Reconsidering the Indus Waters Treaty", in: *Tulane Environmental Law Journal*, 22,2: 271.

Black, Julia; Kauffmann, Celine, 2013: "Transboundary Water Management", in: *International Regulatory Co-operation: Case Studies,* 3: 59-102.

Biswas, Asit K., 1992: "Indus Water Treaty: The Negotiating Process", in: *Water International*, 17,4: 201-209.

Boyd, Robert; Richerson, Peter J.; Meinzen-Dick, Ruth; De Moor, Tine; Jackson, Matthew O.; Gjerde, Kristina M.; Harden-Davies, Harriet.; Frischmann, Brett M.; Madison, Michael J.; Strandburg, Katherine J.; McLean, Angela R.; Dye, Christopher, 2018: "Tragedy Revisited", in: *Science*, 362,6420: 1,236–1,241.

Briscoe, John, 2010: "Troubled Waters: Can a Bridge be Built over the Indus?", in: *Economic and Political Weekly*: 28–5.

Chandio, Khalid, 2014: *India Re-thinking Indus Water Treaty* (Islamabad: Policy Research Institute).

Chellaney, Brahma, 2011: *Water: Asia's New Battleground* (Washington, DC: Georgetown University Press).

Chellaney, Brahma, 2013: *Water, Peace, and War: Confronting the Global Water Crisis* (Lanham: Rowman & Littlefield).

Cooley, Heather; Gleick, Peter H., 2011: "Climate-proofing Transboundary Water Agreements", in: *Hydrol Sciences Journal*, 56,4: 711-718.

Cooley, Heather; Christian-Smith, Juliet; Gleick, Peter H.; Allen, Lucy; Cohen, Michael J., 2012: "Climate Change and Transboundary Waters", in: *The World's Water* (Washington, DC: Island Press): 1–22.

Dilshad, Tanzina; Syed, Abu; Hassan, Tanvir; Uddin, Nasir; Ferdous, Jannatul; Gupta, Sushanta; Hoque, Fahim; Al Faroque, Mohiuddin; Mallick, Dwijen; Rahman, Atiq, 2018: *Climate Change Adaptations at River Basins: Upstream and Downstream Linkages* (Kathmandu: HI-AWARE).

Dixit, Ajaya, 2018: "Trans-Boundary Water Governance in South Asia: The Beginning of a New Journey", in: Ahmed, Imtiaz (Ed.): *South Asian Rivers* (Cham: Springer International Publishing): 81–101.

Dursun-Ozkanca, Oya, 2016: "The Peacebuilding Assembly-Line Model: Towards a Theory of International Collaboration in Multidimensional Peacebuilding Operations", in: *International Journal of Peace Studies*, 21,2: 41-57.

Edelenbos, Jurian; Bressers, N.; Scholten, Peter (Eds.), 2013: *Water Governance as Connective Capacity* (London: Routledge).

Hussain, Bilal, 2016: "China, India, Pakistan: Moving Beyond Water Wars", in: *The Diplomat*, 6 October.

Indus Waters Treaty, 1960: "The Indus Waters Treaty", signed in Karachi on 19 September 1960: 300–365.

Kamran, M. Asif; Aijaz, A.; Shivakoti, Ganesh, 2017: "Institutions for Governance of Transboundary Water Commons: The Case of the Indus Basin", in: *Redefining Diversity & Dynamics of Natural Resources Management in Asia*, 2: 207–223 (Amsterdam: Elsevier).

Karki, Madhav Bahadur; Shrestha, Arun Bhakta; Winiger, Matthias, 2011: "Enhancing Knowledge Management and Adaptation Capacity for Integrated Management of Water Resources in the Indus River Basin", in: *Mountain Research and Development*, 31,3: 242-252.

Khan, Raja Nazakat Ali, 2013: *Indus Water Treaty: A Geo Political Study* (PhD dissertation, University of Kashmir, India); at: https://hdl.handle.net/10603/14384.

Kokab, Rizwan Ullah; Nawaz, Adnan, 2013: "Indus Water Treaty: Need for Review", in: *Asian Journal of Social Science*, 2: 210-218.

Kyrou, Christos N., 2007: "Peace Ecology: An Emerging Paradigm in Peace Studies", in: *International Journal of Peace Studies*, 12,1: 73-92.

Laghari, Abdul Nasir; Vanham, Davy; Rauch, Wolfgang, 2012: "The Indus Basin in the Framework of Current and Future Water Resources Management", in: *Hydrology and Earth System Sciences*, 16,4: 1,063–1,083; at: www.hydrol-earth-syst-sci.net/16/1063/2012/.

Lemos, Maria Carmen, 2015: "Usable Climate Knowledge for Adaptive and Co-Managed Water Governance", in: *Current Opinion in Environmental Sustainability*, 12: 48-52.

Mahmood, Ashfaq, 2018: *Hydro-Diplomacy: Preventing Water War Between Nuclear-Armed Pakistan and India* (Islamabad: Institute of Policy Studies, IPS Press).

McCracken, Melissa; Meyer, Chloe, 2018: "Monitoring of Transboundary Water Cooperation: Review of Sustainable Development Goal Indicator, 6.5.2 Methodology", in: *Journal of Hydrology*, 563: 1-12.

McCracken, Melissa, 2017 *Measuring Transboundary Water Cooperation: Options for Sustainable Development Goal Target 6.5* (Stockholm: Global Water Partnership [GWP]).

McKean, Margaret A, 1996: "Common-Property Regimes as a Solution to Problems of Scale and Linkage", in: Hanna, S. S.; Folke, C; Mäler, K.-G (Eds.): *Rights to Nature: Ecological, Economic, Cultural, and Political Principles of Institutions for the Environment* (Washington, DC: Island Press): 223–243.

Miner, Mary; Patankar, Gauri; Gamkhar, Shama; Eaton, David J., 2009: "Water Sharing between India and Pakistan: A Critical Evaluation of the Indus Water Treaty", in: *Water International*, 34,2: 204-216.

Mirza, M., 2007: "Climate Change, Adaptation and Adaptive governance in Water Sector in South Asia", in: *Physical Science Basis*: 1–19.

Mirza, Mohammad Nasrulla, 2011: *Water, War and Peace* (Saarbruecken: LAP Lambert Academy Publishers).

Mirza, Mohammad Nasrulla, 2008: "Water, War, and Peace: Linkages and Scenarios in India-Pakistan Relations", in: *Heidelberg Papers in South Asian and Comparative Politics*, 37 (February).

Mustafa, Daanish; Wrathall, David, 2011: "Indus Basin Floods of 2010: Souring of a Faustian Bargain?", in: *Water Alternatives*, 4,1: 72-85.

Najam, Adil; Yusuf, Moeed (Eds.), 2013: *South Asia 2060: Envisioning Regional Futures* (London: Anthem Press).

Nax, Natalie, 2016: *Looking to the Future: The Indus Waters Treaty and Climate Change* (Master's dissertation, University of Oregon, Conflict and Dispute Resolution Program).

Ndura-Ouédraogo, Elaive; Amster, Randall, 2009: *Building Cultures of Peace: Transdisciplinary Voices of Hope and Action* (Cambridge: Cambridge Scholars Publishing).

Nostbakken, David; Khan, A., 1946: "The Ecology of Peace", in: *Ecology Global Network* (17 September 2014), "Constitution of the United Nations Education Science and Cultural Organization".

O'Neill, John, 2001: "Representing People, Representing Nature, Representing the World", in; *Environment and Planning C: Government and Policy*, 19,4: 483-500.

Ostrom, Elinor; Shivakoti, Ganesh (Eds.), 2002: *Improving Irrigation Governance and Management in Nepal* (Oakland, CA: ICS Press).

Ostrom, Elinor, 1990: *Governing the Commons: The Evolution of Institutions for Collective Action* (New York: Cambridge University Press).

Ostrom, Elinor, 2005: *Understanding Institutional Diversity* (Princeton: Princeton University Press).

Ostrom, Elinor, 2009: "A General Framework for Analyzing Sustainability of Social-Ecological Systems", in; *Science*, 325,5939: 419-422.

Oswald Spring, Úrsula, 2007: "Hydro-Diplomacy: Opportunities for Learning from an Interregional Process", in: Lipchin, Clive et al (Ed.): *Integrated Water Resources Management and Security in the Middle East* (Dordrecht: Springer): 163–200.

Palmer, Margaret A.; Reidy Liermann, Catherine A.; Nilsson, Christer; Flörke, Martina; Alcamo, Joseph; Lake, P. Sam; Bond, Nick, 2008: "Climate Change and the World's River Basins: Anticipating Management Options", in: *Frontiers in Ecology and the Environment*, 6,2: 81-89.

Pandey, Punam, 2014: "Bangladesh, India, and fifteen years of peace: future directions of the Ganges Treaty", in: *Asian Survey*, 54,4: 651–673.

Philip, Diana, 2015: "Water Co-operation Quotient", Strategic Foresight Group (Mumbai: Mail Order Solutions India Pvt. Ltd.); at: https://www.strategicforesight.com/publication_pdf/28799WCQ-web.pdf.

Pratap, Ravindra, 2018: "Building Peace over Water in South Asia: The Watercourses Convention and SAARC", in: *Athens Journal of Law*, 4,1: 7–26; at: https://doi.org/https://doi.org/10.30958/ajl.4-1-1 doi=https://doi.org/10.30958/ajl.4-1-1.

Raman, Durgeshree, 2018: "Damming and Infrastructural Development of the Indus River Basin: Strengthening the Provisions of the Indus Waters Treaty", in: *Asian Journal of International Law*, 8,2: 372–402; at: doi: https://doi.org/10.1017/S2044251317000029.

Ranjan, Amit, 2016: "Disputed Waters: India, Pakistan and the Transboundary Rivers", in: *Studies in Indian Politics*, 4,2: 191-205.

Rao, S., 2017: "Governance of Water Resources Shared by India and Pakistan under the Indus Waters Treaty: Successful Elements and Room for Improvement", in: *NYU Environmental Law Journal*, 25: 108.

Riffat, Fatima; Iftikhar, Anum, 2016: "Water Issues and their Implications for India-Pakistan Relations", in: *Journal of the Punjab University Historical Society*, 28,2: 11-20.

Romshoo, A.S., 2012: *Indus River Basin Common Concerns and the Roadmap to Resolution* (New Delhi: Centre for Dialogue and Reconciliation).

Sahni, Hamir K., 2006: "The Politics of Water in South Asia: The Case of the Indus Waters Treaty", in: *SAIS Review of International Affairs*, 26,2: 153-165.

Sattar, Erum; Robison, Jason; McCool, Daniel, 2017: "Evolution of Water Institutions in the Indus River Basin: Reflections from the Law on the Colorado River", in: *University of Michigan Journal of Law Reform*, 51: 715.

Schoefield, Victoria, 2000: *Kashmir in Conflict: India, Pakistan and the Unfinished War* (London: IB Tauris).

Schug, Gwen Robbins; Blevins, K. Elaine.; Cox, Bret; Gray, Kslsey; Mushrif-Tripathy, V., 2013: "Infection, Disease, and Biosocial Processes at the End of the Indus Civilization", in: *PLOS One*, 8,12: p.e84814.

Strategic Foresight Group, 2017: "A Matter of Survival" (Geneva: Global High-Level Panel on Water and Peace); at: https://www.genevawaterhub.org/resource/matter-survival.

Tabassum, Shaista, 2004: *River Water Sharing Problem between India and Pakistan: Case Study of the Indus Waters Treaty*, vol 24 (Colombo: Regional Centre for Strategic Studies).

Tainter, Joseph. A., 1988: *The Collapse of Complex Societies* (Cambridge: Cambridge University Press).

Textbook for Massive Open Online Course on "Environmental Security and Sustaining Peace" (2 February 2019).

Thakkar, Himanshu, 2016: "So Who will Suffer in the Indus Water Imbroglio?", in: *South Asia Network on Dams, Rivers and People* [SANDRP] (4 October 2016).

Thakkar, Himanshu, 2017: "India Pakistan Resume Talks on Indus Treaty: Stakes are High", in: *South Asia Network on Dams, Rivers and People* [SANDRP] (21 March 2017).

Thomas, Kimberley Anh, 2017: "The Ganges water treaty: 20 years of cooperation, on India's terms", in: *Water Policy*, 19,4: 724–740.

UNESCO, 2013: *Free Flow – Reaching Water Security Through Cooperation* (Paris: United Nations Education Science and Cultural Organization).

University of Cambridge, 2014: "Decline of Bronze Age 'Megacities' Linked to Climate Change", in: *Science Daily* (26 February).

Vink, Maria, 2018: "The Role of Water Diplomacy in Peace Building", in: *Routledge Handbook of Environmental Conflict and Peacebuilding* (London: Routledge, GSE Research), 283,294: 283–294.

Wada, Yoshihide; Gain, Animesh K.; Giupponi, Carlo, 2016: "Measuring Global Water Security towards Sustainable Development Goals", in: *Environmental Research Letters*, 11,12: 2-13.

Wolf, Aaron T., 1999: "The Transboundary Freshwater Dispute Database Project", in: *Water International*, 24,2: 160-163.

Wong, C.M.; Williams, C.E.; Pittock, J.; Collier, U.; Schelle, P., 2007: *World's Top 10 Rivers at Risk* (Gland, Switzerland: WWF International).

Yang, Yi-Chen E.; Brown, Casey; Yu, Winston; Wescoat Jr, James; Ringler, Claudia, 2014: "Water Governance and Adaptation to Climate Change in the Indus River Basin", in: *Journal of Hydrology*, 519: 2,527-2,537.

Weblinks

https://wwf.panda.org/knowledge_hub/where_we_work/indus_delta/ (3 November 2018).

https://www.worldbank.org/en/news/press-release/2016/11/10/world-bank-urges-mediation-for-india-pakistan-over-indus.

https://dailytimes.com.pk/29778/indus-water-treaty-the-facts/(21 July 2019).

https://digital.library.unt.edu/ark:/67531/metadc30531/.

https://economictimes.indiatimes.com/news/politics-and-nation/indus-waters-treaty-provided-framework-for-resolving-disputes-on-water-use-un-dy-secretary-general/articleshow/65580647.cms.

https://ipripak.org/india-re-thinking-indus-water-treaty/.

https://mea.gov.in/bilateral-documents.htm?dtl/6439/Indus.
https://pangsau.com/2018/06/05/talking-water/.
https://sandrp.in/2016/10/04/so-who-will-suffer-in-the-indus-water-imbroglio/.
https://sandrp.in/2017/03/21/india-pakistan-resume-talks-on-indus-treaty-stakes-are-high/#_edn2.
https://timesofindia.indiatimes.com/blogs/toi-edit-page/shed-the-indus-albatross-indus-waters-treaty-offers-one-sided-benefits-to-pakistan-world-bank-too-is-partisan/.
https://www.indiatoday.in/india/story/nitin-gadkari-indus-water-treaty-pakistan-1520635-2019-05-09 (21 July 2019).
https://www.researchgate.net/publication/259563718_Conflictual_sustainability_the_Complexity_of_trans-boundary_water_conflicts.
https://www.un.org/sg/en/content/dsg/statement/2018-08-27/deputy-secretary-generals-remarks-high-level-panel-water-diplomacy.
https://www.webcitation.org/6GsC7TTei?url=http://www.unwater.org/downloads/worldstop10riversatriskfinalmarch13_1.pdf.
https://www.worldbank.org/en/region/sar/brief/fact-sheet-the-indus-waters-treaty-1960-and-the-world-bank.
Talk 11 on Opportunities and Challenges in Implementation of the Indus Water Treaty <talk11.htm.
Talk 25 Water Cooperation in Central Asia: History and Trends <talk25.htm>.
TNN| Sep 12, 2011; at: https://timesofindia.indiatimes.com/india/All-the-world-is-made-up-of-3600-water-treaties/articleshowprint/9950748.cms. Accessed 2/16/2019
www.sciencedaily.com/releases/2014/01/140116162019.htm.

Chapter 21
The Rohingya Refugee Crisis: Implications for Regional Security

Md Nurul Momem

Abstract Bangladesh now hosts more than one million stateless Rohingya people. Most of them have entered the country since August 2017, amid military crackdowns on them in the state of Rakhine in Myanmar. The stateless population is considered to be one of the biggest threats to regional peace and security in South Asia. This chapter explores the security challenges and concerns from three different perspectives: (i) local livelihood, (ii) political competition, and (iii) future risks of radicalisation that may emerge from the statelessness of the Rohingya. This paper highlights the current situation that requires particular actions on behalf of regional governments. Political will is needed and measures must be initiated with a view to successfully handling security implications.

Keywords Arakan · Bangladesh · Myanmar · Rohingya refugee · Security

21.1 The Nature of the Inquiry

With only 0.3% of the global *Gross Domestic Product* (GDP), Bangladesh now accounts for 4.7% of the world's refugees (Sattar 2019). This refugee crisis places burdens on local capacities; poses security threats to the host country; and also puts tremendous pressure on the international community and development partners to tackle a humanitarian crisis. These conditions may worsen in the region if there is no time frame or durable solution for the repatriation process. However, it is worth mentioning that forcibly displaced people have been increasing substantially at the global level, from 43.3 million in 2009 to 70.8 million in 2018, due to persecution, conflicts, violence or human rights violations (UNHCR 2018), and some regions suffer from the crisis of statelessness. The Rohingya refugees in Bangladesh are *forcibly displaced Myanmar nationals* (FDMN). Their situation constitutes one of the most complicated issues globally. It has deteriorated over several decades and has yet to be solved.

Prof. Dr. Md Nurul Momen, Department of Public Administration, University of Rajshahi, Rajshahi-6205, Bangladesh. Email: nurulmomen@ru.ac.bd (work).

Ú. Oswald Spring and H. G. Brauch (eds.), *Decolonising Conflicts, Security, Peace, Gender, Environment and Development in the Anthropocene*,
The Anthropocene: Politik—Economics—Society—Science 30,
https://doi.org/10.1007/978-3-030-62316-6_21

At least ten million people worldwide are stateless, and, more importantly, one third of them are children (OSCE n.d.). Stateless people are now considered a serious threat to long-term global stability and the peaceful development of society. For many reasons stateless people recently increased manifold in global politics due to violence against religious minorities or as specific groups or ethnic communities on the basis of religion or race, and these forcibly displaced people are unable to protect their civil and political rights. Even their national identity is questioned, regardless of the fact that everyone has the right to claim nationality and is entitled to receive state protection according to international law. This text explores the world's most persecuted and vulnerable FDMN group, the Rohingya refugees and their security implications for the host country and the region. In August 2017, Myanmar military forces started a violent repression campaign against the Rohingyas. Due to the large-scale indiscriminate killings within Arakan state, the Rohingyas were forced to flee from Myanmar to Bangladesh and some other neighbouring countries (Anupama 2017). This chapter reviews the Rohingya refugees as a case study to examine statelessness as a threat to regional peace and security. This chapter primarily relies on secondary literature for the theoretical understanding of security concerns emerging from the Rohingya refugee crisis. The author visited Cox's Bazar in July 2018 and conducted two *focus group discussions* (FGD) with the people who had arrived in August 2017.

21.2 Conceptualising and Defining Refugees and Stateless People

'Refugees' were discussed by the international community for the first time after World War II (Yesmin 2016), when the *International Refugee Organisat*ion (IRO) was formed in 1947 to deal with the refugees crisis in Europe. As a further initiative, the General Assembly of the *United Nations* (UN) adopted the Statutes of the *United Nations High Commissioner for Refugees* (UNHCR) on 14 December 1950, which replaced the IRO (Feller 2001). In 1951 the International Convention on the Status of Refugees and the 1967 Protocol on the Status of Refugees provided a universally accepted definition of a 'refugee' (Gunning 1989). The UNHCR Statutes and the 1951 Convention on the Status of Refugees did not perceive a 'refugee' as a displaced person, thus narrowing its definition (Kennedy 1986). According to Article 1.A. (2) of the 1951 Refugee Convention:

> a refugee is a person who, [o]wing to well-founded fear of being persecuted for reasons of race, religion, nationality, membership of a particular social group or political opinion, is outside the country of his nationality and is unable, or owing to such fear, is unwilling to avail himself of the protection of that country; or who, not having a nationality and being outside the country of his former habitual residence as a result of such events, is unable or, owing to such fear, is unwilling to return to it.

The Organisation of African Unity's 1969 Convention on Refugee Problems in Africa, namely the African Convention, operationalised the notion of forced refugees based on foreign aggression:

> … every person who, owing to external aggression, occupation, foreign domination or events seriously disturbing public order in either part of the whole of his country of origin or nationality, is compelled to leave his place of habitual residence in order to seek refuge in another place outside his country of origin or nationality (Gunning 1989).

At the global level there are two *United Nations* (UN) Conventions which deal with stateless people: the 1954 Convention regarding the Status of Stateless Persons (hereafter the 1954 Convention) and the 1961 Convention on the Reduction of Statelessness (hereafter the 1961 Convention). Article 1(1) of the 1954 Convention defines the concept of a 'stateless person' as a person who is not considered to be a national by any State under the legal framework (OSCE n.d.).

As can be seen, the Rohingyas fulfil every criterion of the above definitions: they have been denied citizenship by the state of Myanmar; they are persecuted as a minority religious and ethnic group; they experienced repression, killings and torture by the State security agencies; and finally they were expelled en masse from their homes and forced to flee to Bangladesh and other places in the region (Naushin et al. 2013). Given the circumstances, the UN sees the Rohingyas as the most persecuted ethnic Muslim community and 'humanitarian catastrophe' in the world.

21.3 Institutionalised Discrimination and Statelessness: The Rohingya People

Rohingyas are a Muslim community, who comprise one third of the population of Rakhine state in Myanmar. Rakhine state is the least developed region in Myanmar, and 78% of the people are poor, compared with 37.58% of the national average (Albert/Chatzky 2018). The Myanmar government has institutionalised the discrimination against the Rohingya by promulgating and enforcing many discriminatory laws (Anupama 2017). Discriminatory policies of the Myanmar government have forced several hundred thousand Rohingyas to flee their homeland and seek refuge in neighbouring Bangladesh since the late 1970s. Their rights of citizenship were denied by the military government of Myanmar in 1982 because they were not the original inhabitants of the land. To the Myanmar government, they were illegal migrants who basically came from Bangladesh a very long time ago, and therefore they were not entitled to receive citizenship rights, hence they were not considered among the 135 official ethnic groups in Myanmar (Alam 2017).

The Rohingya experienced decades of systematic discrimination under successive governments operating within the legal framework of Myanmar. Discrimination basically started with the Foreigners' Act of 1864 and continued until the Myanmar Citizenship Act of 1982. Even the Union Citizenship Act of 1948, enacted by the

British, refused to acknowledge their citizenship rights. Later on, after the independence of Burma (now Myanmar), conditions further worsened when military ruler General Ne Win, as the head of the Government, declared them to be illegal migrants from Bangladesh. To him, the Rohingyas were brought to Myanmar during British rule. In 1974 a new constitution was formulated by the military government, which introduced the Immigration Act of 1974 which denied the Rohingyas the right to citizenship (Hamzah et al. 2016). Most importantly, the citizenship law of 1982 did not regard them as any of the 'national races' (Mohsin 2019). The Rohingyas thus became stateless people, although many historians documented that they had been living peacefully in Arakan as early as 1430 during the reign of King Narameikhla of Arakan, who had about 30,000 Muslim soldiers (Hamzah et al. 2016).

However, at no point did the international community or development partners make sustained efforts to find a solution to the Rohingyas' citizenship or ethnic identity, despite the continuous non-recognition by the state of Myanmar (Mohsin 2019). However, the recent violent attacks by military forces and displacement after August 2017 have created a further crisis. Brutal attacks are justified by the Myanmar military as retaliation for a series of armed strikes by the *Arakan Rohingya Salvation Army* (ARSA) on civilians and security establishments in which more than a hundred people were killed. Violent clashes between the ARSA and the military compelled thousands of Rohingyas to flee to Bangladesh and some other neighbouring countries to seek political asylum (Albert 2018). A report by UNICEF mentions that among the total number of refugees about 608% were children (Reuters 1 September 2017).

21.4 The Refugee Crisis from a Security Perspective

The paradigms of security theories changed and international relations scholars and foreign policy advisors offered a new way of thinking after the end of the Cold War (McSweeney 1999). During the Cold War era security problems and concerns mostly dealt with territorial sovereignty, national interests, and military forces, but these issues are no longer considered to properly address the transnational issues of many new security challenges since the 1990s (Edwards 2009).

McGahan (2009: 3) applied securitisation to immigration issues, arguing that securitisation "offers a lens through which to highlight certain actors and processes in analysing immigration policies, particularly how societal threats are constructed and defended". He further argued that "ordinary citizens, apart from elites, are routinely involved in routine practices that shape the securitisation of migration and contribute to a general culture of fear". McGrahan's emphasis is on examining the securitisation policy of the host country in the context of refugees and migrants. Political, economic, and social dynamics of the host country may affect the way that security issues are handled. It is argued that the securitisation of refugees relates to how they are treated, particularly from a security perspective. Furthermore, the different views of politicians and the media have laid the foundation for refugee asylums and their security links, and influence whether refugees may be seen as a threat to the host country (Watson 2007).

On the other hand, Mogire (2011: 71–77) illustrated six issues for the security discourse on the refugee crisis. First of all, the author explored the stiff competition between refugees and the local community over basic needs and local resources, such as land, water, health services and firewood, for their livelihood. Secondly, environmental degradation in the form of deforestation and pollution caused by refugees may create conflicts between refugees and host communities. Thirdly, the author also identified competition in the labour market as a significant source of conflict. Fourthly, issues may occur when refugees are involved in the local politics or national issues of the host country and choose to take part in a conflict. The fifth source of conflict is observed when the host community perceives unequal treatment regarding the assistance that refugees receive from development partners or international organisations, as the host community considers itself equally eligible to receive similar support. And finally, conflict occurs when there is an absence of understanding between the host community and displaced persons with respect to each other's cultural differences.

The securitisation theory considers refugees to be a security threat based on the different perspectives of political and media analysis. To address this crisis, some host countries have introduced tough legislative frameworks, although this contradicts the provisions of international refugee law. For instance, in 2018 Hungary passed an anti-immigrant 'stop Soros' law that criminalises any individual or group that offers to help an illegal immigrant claim political asylum. There have already been some securitisation issues with regard to refugees in Bangladesh, such as the growth of religious extremism in refugee camps; refugees being used as carriers for arms and drug trafficking; stiff competition with the local community for jobs and control of resources; and high inflation rates on basic commodities in adjacent refugee camps (Driscoll 2017). Apart from the humanitarian crisis of refugees, there is a significant concern for Bangladesh and regional governments which must be examined from security perspectives. Bangladesh has already started to perceive the refugee crisis not as a solely humanitarian issue; rather, it is increasingly being considered as a future internal security threat, which may affect regional relations. Political analysts in Bangladesh consider the crisis to be a 'non-traditional security threat' (Wolf 2017).

21.5 The Rohingya Refugee Crisis: Three Perspectives

21.5.1 Local Livelihood

Local livelihoods are seriously affected as the Rohingya people are found responsible for deteriorating law and order and are increasingly involved in different illegal activities, such as human and drug trafficking and criminal offences, including robbery, sexual assault, domestic violence, and forced labour. The two main refugee camps in Nayapara (in the subdistrict Teknaf) and Kutupalang (in the subdistrict Ukhia) in Cox's Bazar District in Bangladesh have higher crime and criminal practices that may have caused negative spillover effects on local communities. The *International Organization for Migration* (IOM) identified 420 cases of trafficking between December 2018 and June 2019, which had increased fourfold during this period (Karim 2019).

After the large influx of people in 2017, two important studies (Wolf 2017; Tani/Rahman 2017) also concluded that refugees were involved in different criminal activities in the refugee camps. Wolf (2017) identified several underlying factors and issues that require a rethink on local security concerns. Both the existing literature and the field study report identified refugees as a significant challenge for some imminent economic issues and their effects on security and local livelihoods in the host country, such as:

(a) Refugees caused extra demographic pressure on food, shelter, healthcare services and education for children in a densely populated country with limited resources, such as Bangladesh.
(b) Refugees provided a source of cheap labour in the local market, which created socio-economic inequalities and caused distortions in the market system. The mean daily wage in the agriculture sector declined by more than 158% after the influx, and this decline prevailed adjacent to the camp area, where refugees were in competion with the local labour force when trying to find temporary jobs for their survival (Sattar 2019).
(c) There has been concern about the future potential ecological threat allegedly caused by refugees, due to the exploitation of firewood and deforestation for their better livelihood in their refugee camps. A loss of Tk 456.08 crore (Bangladeshi currency) has been reported due to the destruction of trees on 6,163.5 acres of hilly land to construct thirty makeshift camps, which have also caused Tk 1,400 crore in ecological damage (Kamol 2019). The exploitation of firewood and consequent deforestation by Rohingya refugees in their quest to create a better settlement negatively affects sustainable livelihoods (Tani/Rahman 2017).
(d) Refugee camps were built in Cox's Bazar District in Bangladesh, where the main seaport is located as an important strategic business area. It is predicted that the huge security measures in the seaport areas may have serious consequences for the business and tourism sectors. Given the circumstances, refugee issues are regarded as a security risk which could obstruct regional connectivity projects, such as the proposed *Kaladan Multi-Modal Transit Transport Project* (KMMTTP), which is designed to connect the Indian port of Kolkata with the port in Sittwe of Myanmar, the provincial capital of the Rakhine State (Wolf 2017).
(e) Refugees now outnumber the local population two to one in the *upazilas* of Ukhia and Teknaf in Cox's Bazar District (Sattar 2019). The level of frustration has risen between refugees and local communities in Bangladesh, as the influx of refugees has increased the demand for daily commodities, thus putting pressure on the cost of living. Further, refugees have made allegations against the host country and aid agencies, especially the UNHCR, claiming that

they are receiving inadequate basic support. On the other hand, tensions also exist among local communities who do not receive any support from development partners, even though they are also in desperate need of the very same necessary items given to refugees. Therefore, some local communities view the Rohingya people as illegal economic migrants. This perception of them may create future conflicts of interest between the locals and refugees (Wolf 2017). Refugees have often engaged in violent practices to safeguard their economic interests; their economic desires are likely to be connected with radical practices. Given the economic context, the refugee problem has created a 'vicious circle' that could lead to the further deterioration of the security establishment, which may threaten the quality of governance at the local level in the host country. Global governance institutions such as the UNHCR and refugees have praised the Government of Bangladesh for allowing refugees in, but this decision may have long-term negative effects on the local community and local resources.

The Bangladesh-Myanmar border has quickly become a hub for transnational crime. This is perceived in Bangladesh to be the fault of the Myanmar Government, which has so far not made a commitment to control the supply of Yaba, a dangerous meth pill which has particularly disastrous effects on the youth generation (Uddin 2014). Since the influx in August 2017, security forces in Bangladesh have seized more than ten million Yaba pills from refugee camps (Alsaafin 2018). The border between Bangladesh and Myanmar has been recognised by the Bangladesh Department of Narcotics Control as the key point for illegal drug trafficking. Some factories in the border area within Myanmar are producing illegal drugs, including Yaba, that are being brought to the Bangladesh market, despite the adoption of various measures by the Bangladesh security forces. It has been noticed that refugees are being used as carriers, or intermediaries, in the border area between Bangladesh and Myanmar due to the hilly terrain.

21.5.2 Political Competition

Refugees create an additional burden for the country's weak political institutions, and their camps are a breeding ground for corrupt practices. For instance, demanding, offering or accepting bribes for allowing refugees free movement, or for helping them to receive official documents and other benefits, foster an illicit camp economy and are deciding the status of documented refugees.

The refugee crisis has negative consequences for the political culture in Bangladesh. Since its independence, politics has been characterised as a zero-sum game between the ruling regime and their opposition political parties, which affects all spheres of public life (Momen 2020). The ruling political parties and their leaders and supporters depend on muscle power during the electoral process to limit the opportunities of potential parties to gain power. However, local politicians attempt to use the refugee issue for their political ends with the support of local government offices in order to have their names included in the electoral constituency. There has been some evidence that local political parties and their leaders are interested in capturing the vote banks of refugees. In doing so, a nexus between political and administrative offices has been developed that makes it easy for refugees to get false Bangladeshi birth certificates, national identity cards and sometimes even a 'Bangladesh Passport' (Wolf 2017). The enrolment of refugees in the list of voters is illegal and a criminal offence which has already been reported (Masud 2018). Although from time to time the *Election Commission* (EC) of Bangladesh has formed a Special Committee and initiatives to verify and update the voters list, there are still thousands of refugees enrolled on the list (Wolf 2017).

The crisis also has the potential to affect the domestic political landscape in Bangladesh. For example, the *Bangladesh Nationalist Party* (BNP) has been highly critical of the Awami League government for failing to draw the attention of the global community to the refugee crisis and to bring international bodies and powerful neighbouring countries like India and China into the process of resolving the crisis (Martin et al. 2017). Importantly, Sheikh Hasina's stance on welcoming the influx of refugees into Bangladesh has been praised by Western powers, and her welcoming attitude towards refugees will further boost the power of her authoritarian government and compromise the country's democratic process.

On the other hand, the decision of the Government of Bangladesh to allow the Rohingya influx may create unfair imbalances between the State's philosophy of secularism and religion (Joehnk Tom Felix 2017). However, the central office of the largest radical Islamist group in Bangladesh, *Hefazat-e-Islam*, is located in Chittagong, which is approximately 200 km from the refugee camps. It should be noted that *Hefazat-e-Islam* once called for *Jihad* for an independent Arakan State for Rohingya Muslims, and also threatened to wage war against Myanmar for killing and violating rights of the Rohingyas (Joehnk Tom Felix 2017). It is further pointed out that the current refugee crisis has given Hefazat-e-Islam an important role and benefits in the politics of Bangladesh, as the third-largest Muslim-populated country in the world, which may undermine the secular philosophy of the State (Joehnk Tom Felix 2017).

21.5.3 Future Risks of Radicalisation

As stated, there is the likelihood of the refugee crisis being hijacked by radicals. Hence, the crisis cannot be simply presented as solely a humanitarian crisis; rather it needs to be analysed from a broader security perspective, which may identify the possible risks of radical Islamist groups flourishing in the host country and spreading across the region in the future (Wolf 2017). It is noteworthy that organised Rohingya militancy started in the 1940s. The Rohingyas supported the then British colonial ruler, and fought against Japanese forces and were also involved in the war against Myanmar (former Burmese) guerilla forces who were struggling for liberation from British colonial rule. However, Myanmar became an independent state called 'Union of Burma' in 1948. In the 1950s, a new *mujahidin*-based militant group emerged, but was totally destroyed by the security forces of Myanmar (Singh 2017). After decades of clashes between Rohingya Muslims and Buddhist Rakhine at the local level, alongside Burmese military forces at the centre, the militant struggle received special momentum in the early 1980s, when the *Rohingya Solidarity Organisation* (RSO) was established. The RSO was formed to protect Muslims from future attacks of Buddhist Rakhine and military forces at the centre, and it also planned to create an independent Arakan state for Rohingya Muslims (Brennan/O'Hara 2015).

In order to fulfil the stated plan, the RSO conducted occasional low-profile attacks against Myanmar's security forces. On the other hand, the security forces in Bangladesh also organised a huge crackdown on the activities of the RSO in the border area. It is important to note that the RSO never received any noteworthy support or endorsement for its activities from ordinary Rohingya people. As a result, it became dysfunctional at the end of the 1990s. So the RSO's armed struggle for a separate Arakan state for Rohingya Muslims in the 1980s and 1990s failed to create a substantial threat to the internal security establishment in Myanmar (Wolf 2017). But a significant change in militancy was observed in October 2016, which was quite different from the earlier attacks made by the RSO. The new militants were trained in military operations and their attacks were fully planned and coordinated. This new form of well-organised militant operation aided the formation of the *Arakan Rohingya Salvation Army* (ARSA). Subsequently, the activities of ARSA, as a militant organisation in the Rakhine province, led to it being declared a terrorist organisation in 2017 by *Myanmar's Anti-Terrorism Central Committee* (MATCC). In August 2017, the Myanmar military forces conducted massive operations to uncover ARSA members, which led to the huge influx of Rohingya people, as the forces burnt civilian homes and killed ordinary people in the name of searching for ARSA members.

It is believed that ARSA maintains international links and enjoys the support of Islamic terrorist organisations. After the collapse of the Soviets in Afghanistan, Islamist radicals from Afghanistan started to morally support the armed struggle of the Rohingya to fight against the Government of Myanmar (Singh 2017). Although there is no concrete evidence, it is assumed that the *Pakistan Inter-Services Intelligence* (ISI) unit might be involved in the process of providing logistic support for the armed struggle of ARSA in order to maintain contact with international terror outfits (Wolf 2017). Once it was claimed by Myanmar officials that the top leadership of ARSA received training from the Taliban in Pakistan (Wolf 2017). A research study of the *International Crisis Group* has also noted that ARSA members may have received some weapons training from extremist groups (ICG 27 August 2017) in Pakistan and in Arab countries.

Furthermore, Abdus Qadoos Burmi, a well-known driver of the *Jihadist* movement, who is believed to be the head of the group *Harkat ul Jihad Islami-Arakan* (HUJI-A) based in Karachi in Pakistan, is of Rohingya descent. He supports the militant activities of ARSA by providing training in refugee camps in Bangladesh and recruiting new members in Pakistan. HUJI-A followers seek sympathy for their cause by posting images and videos of vulnerable Rohingya people on social media.

As part of its strategy, HUJI-A also maintains a large network of regional radical Islamist groups across the border (Roul 2017).

The global Jihadist organisation *al-Qaeda* has also shown immense interest in supporting the militant activities of ARSA by calling for a Jihadist movement in Myanmar. To do so, al-Qaeda set-up a new branch in South Asia called *Al Qaeda of the Indian Subcontinent* (AQIS), and after its establishment it named the Arakan part of Myanmar as one of its militant targets (Singh 2017). It is important to note that AQIS would like to extend its militant operations to the whole South Asian region, which is clear from the words of the AQIS leader Asim Umar, who referred to a 'Caravan of Jihad'. To him, AQIS might commence operations in Pakistan, move across the India border and subsequently reach both Bangladesh and Myanmar. That means that AQIS aims to reach and support oppressed Rohingya Muslims in Arakan state in Myanmar (Wolf 2017). Furthermore, the Rohingya influx into Bangladesh and the alleged crimes committed against Rohingyas has provided a space for previously apolitical Rohingya people to engage in radicalised activities, and these could result in a long-lasting civil war. Radical views maintained by religious extremist groups may benefit from the refugee crisis, and there is the likelihood that refugees might be recruited for radical practices. In short, there are some worries that refugee problems are fuelling militant activities which may spread to other countries in the region (Idris 2017).

In essence, having observed the frustration of refugees, there is the likelihood that oppressed and repressed refugees may choose to support militant organisations, such as ARSA or some other regional radical Islamist organisations (Martin et al. 2017). Additionally, the Islamist movement in Bangladesh led by Hefazat-e-Islam, which reportedly runs 25,000 Madrasa (religious schools) and whose headquarters are located in Chittagong, has called for a Jihadist movement to show solidarity with refugees for the establishment an independent Rakhine state (Martin et al. 2017). In September 2017 an estimated 20,000 Hefazat followers organised a mass rally to protest against the violence against Rohingya Muslims, while another radical Islamist group has appealed to the Government of Bangladesh to declare war on Myanmar, and asked it to provide arms and logistical support tos Rohingyas and liberate Rakhine state for them (Martin et al. 2017). However, it has been observed that refugees and their narratives sympathise with radical Muslims in Bangladesh, thus they are vulnerable to be recruited by radicals who can exploit refuges.

21.5.4 Security Issues Inside and Outside the Camps

Some security threats were observed even within the refugee camps. Refugees have also accused local security forces of not providing adequate security services in the camps. There is a possibility that refugees in Bangladesh will engage in terrorism. Ullah (2011) found that a radical extremist terrorist organisation, *Harkat-ul-Jihad-i-Islami* (HuJI), maintained links with the camps, and the largest Islamic political party in Bangladesh, *Jamaat-i-Islami*, had been a financier of the Rohingya militants' struggle (Ullah 2011).

Referring to the potential involvement of Rohingya refugees in militant activities, Rahman (2010: 235) noted: "The Rohingya camps in Cox's Bazar District are fertile grounds for the recruitment by Islamic militants. With little love for Myanmar, and alienated from Bangladesh, the stateless Rohingyas are vulnerable and desperate, and will likely become militant to uphold their interests." After two years of exploring the extent of security risks, a study by USAID (2012) warned that refugees staying in Bangladesh who were denied citizenship rights in Myanmar might potentially be recruited by extremist religious groups (Idris 2017). In the past, *Jamaat-e-Islami* was engaged in recruiting militant members among refugees (Fair/Oldmixon 2015), and they are still interested in the influx of new refugees. If there is no comprehensive and durable solution to address the plight of the refugee crisis, the camps may become a hub of Islamist militant groups seeking to recruit refugees (Fair/Oldmixon 2015).

For the prevention of a militarisation of the camps, strategies forrefugees are needed to improve their education. The introduction of a supportive refugee law in the host country is also emphasised that may provide temporal work permissions to refugees, and may also consider a short term citizenship status (Rahman 2010: 239). These proposals were made by different humanitarian organisations to allow refugees to integrate with the local community; however, it is unlikely whether the GoB will adopt this advice.

Complaints were made that refugees continually face harassment by the security forces as soon as they leave the camps. In general, undocumented or non-registered refugees often resort to desperate and violent measures outside their camps for their livelihood (Ticktin 2006). They always try to find work in the illegal market, and documented refugees without work permits also rely on the illegal market, which prevents them from becoming normal citizens and makes it difficult to integrate them into the host country (Ticktin 2006).

21.5.5 Security Concerns in the Region

Religious extremist groups in the region have been calling for *Jihadist* movements since the refugee influx into Bangladesh in 2017. Singh/Haziq (2016: 2) claimed that "a similar jihadist flare-up is now developing in the wake of the latest atrocities". In Indonesia, Malaysia and Pakistan, for example, bloggers and online radical activists have used social networking sites for sharing the violation of human rights with respect to the Rohingya people, such as stating opinions, posting unedited videos and pictures, and uploading a route map that shows how to reach Myanmar to join the militants' struggle (Singh/Haziq 2016).

The threats from religious extremist groups against the Government of Myanmar are described as "threatening to deepen sectarian tensions across the region.....as Muslim communities...grow increasingly angry over the treatment of Muslim Rohingya by Buddhist Rakhine" (*The Japan Times* 2017). Politicians in Islamic political parties and Muslim majority countries (including Malaysia, Indonesia, Bangladesh and Pakistan) share provocative comments, thereby increasing the risks of religious conflicts in the region. Similarly, the global terrorist organisation *al-Qaeda* has appealed to all Muslim people in the region to support the Rohingya people in Myanmar in all spheres of activity, and highlighted that: "The savage treatment meted out to our Muslim brothers…shall not pass without punishment. The government of Myanmar shall be made to taste what our Muslim brothers have tasted" (*The Japan Times* 2017).

The Government of Bangladesh has already informed India of the possible security threats that have emerged from Rohingya refugees. The Bangladesh High Commissioner in Delhi said that due to the huge influx of refugees into Bangladesh, the current security situation is a grave concern due to the fact that Islamist groups may look for new recruits among Rohingya refugees (Mitra 2017). It was also pointed out that "this has implications for India's own security. So what is happening there [in Bangladesh] will spill over into India" (Mitra 2017).

Transnational religious extremist groups, including *al-Qaeda* and *ISIS* (Momen/Hossain 2020), appear to have a connection with local militant groups in Bangladesh. Local radical groups may be used to carry out assassinations and small-scale bomb blasts and to target secular people and their voices. The large influx of Rohingya into different refugee camps could become a potential target for enrolment by transnational religious extremist groups. If the issue is not dealt with effectively, the whole region will face imminent threats to people's livelihoods and security establishments (Khan 2017).

There are reports that many Rohingya refugees have already entered the north-east and Kashmir areas in India through the Bangladesh border. In a report published by *Indian Express* (English Daily), approximately 5700 Rohingyas are self-settling in slum areas in Jammu (Khan 2017). It is important to note that a right-wing Hindu organisation, *Vishwa Hindu Parishad* (VHP), has urged the Government of India to deport Rohingyas from Jammu to avoid them destabilising the security establishment in India, although the claim behind VHP is related to religious sentiments that Rohingyas are Muslims. Defence and strategic analysts have also urged the Government to be aware of the future security risks in India. Among major security concerns is the likelihood of recruitment by religious extremist groups operating in the conflicted Jammu and Kashmir area, which is considered to be the most vulnerable Indian security establishment (Khan 2017). If local youths embrace the radical views of Rohingya militants, they could be pushed to implement their agenda, and the symbols and religious places of Buddhism across India may be future targets for radical Islamist groups (Khan 2017).

21.6 Conclusion

The Rohingya refugees represent an ethnic minority Muslim community in Myanmar which experienced state oppression and repression. They were forced to leave their homes and fled to the neighbouring country, Bangladesh. This crisis worsened when Bangladesh as the host country was not fully capable of providing basic necessities for all refugees. The current Rohingya refugee situation is characterised by the vulnerabilities of local livelihoods, political complexities and future risks of radicalisation. However, as *the* crisis has unfolded, global institutions, especially the *United Nations* and regional governments, have largely been sidelined and failed to play an effective role in finding a permanent solution to the issue. Bangladesh should also deeply reconsider the Rohingya refugee problem, treating it not solely as a humanitarian crisis; rather, it needs to comprehensively reconsider the activities of Rohingyas within and outside refugee camps, which may become a threat to the peace and security of the host country and pose future possible threats to regional security establishments.

References

Alam, Mayesha, 2017: "5 things you need to know about the Rohingya crisis and how it could roil Southeast Asia", in: *Washington Post*, 14 September 2017; at: https://www.washingtonpost.com/news/monkey-cage/wp/2017/09/14/5-things-you-need-to-know-about-rohingya-crisis-and-how-it-could-roil-southeast-asia/?utm_term=.b0a2e255524c..

Albert, Eleanor, 2018: "The Rohingya Crisis", Council on Foreign Relations, 20 April 2018.

Albert, Eleanor; Chatzky, Andrew, 2018: "The Rohingya Crisis", Council on Foreign Relations, USA, 5 December; at: https://www.cfr.org/backgrounder/rohingya-crisis.

Alsaafin, Linah, 2018: "'I carry yaba to survive': Rohingya and Bangladesh's meth trade", Al Jazeera, 20 August; at: https://www.aljazeera.com/indepth/features/carry-yaba-survive-rohingya-bangladesh-meth-trade-180818115319992.html.

Anupama, Kumari, 2017: "A Case Study of the Rohingya Crisis in Myanmar and India's Concern", in: *International Journal of Academic Research and Development*, 2,5: 477–481.

Brennan, Elliot; O'Hara, Christopher, 2015: "The Rohingya and Islamic Extremism: A Convenient Myth", in: *The Diplomat,* 29 June; at: https://thediplomat.com/2015/06/the-rohingya-and-islamic-extremism-a-convenient-myth/.

Driscoll, Dylan O., 2017: "Bangladesh Rohingya Crisis: Managing Risks in Securitisation of Refugees", University of Manchester, 20 October: 1–14.

Edwards, Alice, 2009: "Human Security and the Rights of Refugees: Transcending Territorial and Disciplinary Borders", in: *Michigan Journal of International Law*, 30,3; at: https://repository.law.umich.edu/mjil/vol30/iss3/8.

Feller, Erika, 2001: "The Evolution of the International Refugee Protection Regime", in: *Journal of Law and Policy*, 5,1: 129–139.

Gunning, Isabelle R., 1989: "Expanding the International Definition of Refugee: A Multicultural View", in: *Fordham International Law Journal*, 13,1: 35–85.

Hamzah, Intan Suria; Daud, Sity; Idris, Nor Azizan, 2016: "Migration and Security Threat in Malaysia: Analysis on Rohingya's Ethnic", in: *Global Journal of Human-Social Science*, 16,3: 1–9.

ICG (International Crisis Group), 2017: "Myanmar Tips into New Crisis after Rakhine State Attacks", in: Brussels, Belgium: International Crisis Group [ICG]), 27 August.

Idris, Iffat, 2017: "Drivers of Extremism, Violent Extremism and Terrorism (EVET)", in: *Bangladesh: K4D Helpdesk Report 139* (Brighton, UK: Institute of Development Studies).

Japan Times, 2017: "Security Concerns Build in South Asia as Rohingya crisis worsens", in: *The Japan Times*, 15 September.

Joehnk Tom Felix, 2017: "How the Rohingya Crisis is Changing Bangladesh", in: *New York Times*, 6 October; at: https://www.nytimes.com/2017/10/06/opinion/rohingya-bangladesh-myanmar.html.

Kamol, Ershad, 2019: "Bangladesh Loses Forests Worth Tk 1,800cr for Rohingya Settlements", in: *New Age Bangladesh*, 21 April; at: https://www.newagebd.net/article/70469/bangladesh-loses-forests-worth-tk1800cr-for-rohingya-settlements.

Karim, Naimul, 2019: "Rohingya Refugees in Bangladesh Warned to be Wary with Human Trafficking Rising", Thomson Reuters Foundation, 23 August; at: https://uk.reuters.com/article/bangladesh-myanmar-trafficking/feature-rohingya-refugees-in-bangladesh-warned-to-be-wary-with-human-trafficking-rising-idUKL8N2551NX.

Kennedy, D., 1986: "International Refugee Protection", in: *Human Rights Quarterly*, 8,1: 2.

Khan, Md. Intekhab A., 2017: "Myanmar's Rohingya Crisis: An Analysis of Security Threats for South Asia", in: *International Journal of Management and Social Science Research Review*, 1,32: 155–159.

Masud, Humayun, 2018: "Rohingyas in Voter List in Exchange for Money", in: *Bangla Tribune*, 24 August; at: https://en.banglatribune.com/country/news/11119/Rohingyas-in-voter-list-in-exchange-of-money.

Martin, Michael F.; Margesson, Rhoda; Vaughn, Bruce, 2017: "The Rohingya Crises in Bangladesh and Burma", in: *Congressional Research Service 7–5700*, 8 November: 1–34.

McGahan, Kevin, 2009: "The Securitization of Migration in Malaysia", in: *APSA 2009 Toronto Meeting Paper*; in: SSRN; at: https://ssrn.com/abstract=1449344.

McSweeney, Bill, 1999: *Security, Identity and Interests: A Sociology of International Relations* (Cambridge: Cambridge University Press).

Mogire, Edward, 2011: "Victims as Security Threats", in: *Global Security in a Changing World* (London: Routledge).

Mitra, D., 2017: "Public Anger Brews in Bangladesh over India's Stance on Rohingya Crisis", in: *The Wire*, 10 September; at: https://thewire.in/176013/bangaldesh-rohingya-crisis-myanmar-india/.

Mohsin, Amena, 2019: "Caught between the Nation and the State: Voices of Rohingya Refugee Women in Bangladesh", in: *Asian Journal of Comparative Politics*: 1–14; at: https://doi.org/10.1177/2057891119874882.

Momen M.N; Hossain E. 2020: "*Examining the Contemporary Conflict Regions in the World*". In: Leal Filho W., Azul A., Brandli L., Lange Salvia A., Özuyar P., Wall T. (eds) No Poverty. Encyclopedia of the UN Sustainable Development Goals. Springer, Cham. https://doi.org/10.1007/978-3-319-69625-6_41-2

Momen, M.N. 2020: "*Jamaat-ul-Mujahideen Bangladesh: analysis of organizational design and activities*", Dynamics of Asymmetric Conflict, https://doi.org/10.1080/17467586.2020.1821072

Naushin, Parnini Syeda; Redzuan Othman, Mohammad; Ghazali, Amer Saifude, 2013: "The Rohingya Refugee Crisis and Bangladesh-Myanmar Relations", in: *Asian and Pacific Migration Journal*, 22,1: 133–146.

OSCE, n.d.: *Handbook on Statelessness in the OSCE Area: International Standards and Good Practices* (Vienna: Organization for Security and Co-Operation in Europe (OSCE) – Warsaw: Office for Democratic Institutions and Human Rights (ODIHR) – Geneva: High Commissioner on National Minorities (HCNM) – Geneva: Office of the United Nations High Commissioner for Refugees (UNHCR) https://www.osce.org/handbook/statelessness-in-the-OSCE-area?download=true..

Rahman, Utpala, 2010: "The Rohingya Refugee: A Security Dilemma for Bangladesh", in: *Journal of Immigrant & Refugee Studies*, 8,2: 233–239.

Reuters, 2017: "UN Chief Calls for Restraint to Avoid 'Humanitarian Catastrophe' in Myanmar's Rakhine State", 1 September; at: https://www.reuters.com/article/us-myanmar-rohingya-uniniversity/u-n-chief-calls-forrestraint-to-avoid-humanitarian-catastrophe-in-myanmars-rakhine-state-idUSKCN1BC5L3.

Roul, Animesh, 2017: "Myanmar's Muslim Insurgency Gaining Prominence With Jihadist Groups", in: *Terrorism Monitor*, 15,5 (10 March), (Washingon, DC: The Jamestown Foundation).

Sattar, Zaidi, 2019: "Rohingya Crisis and the Host Community", in: *The Financial Express*, 30 July; at: https://thefinancialexpress.com.bd/views/rohingya-crisis-and-the-host-community-1564498784.

Singh, Jassmider, 2017: "Rohingya Crisis in Southeast Asia: The Jihadi Dimension", in: *RSIS Commentary* (3 April), (Singapore: S. Rajaratnam School of International Studies [RSIS]).

Singh, Jasminder; Haziq, Muhammad, 2016: "The Rohingya Crisis: Regional Security Implications", in: *RSIS Commentaries*, No. 293 (Singapore: Nanyang Technological University).

Tani, Masakazu; Rahman, Abiar Md., 2017: *Deforestation in the Teknaf Peninsula of Bangladesh: A Study of Political Ecology* (Heidelberg: Springer).

Ticktin, Miriam, 2006: "Where Ethics and Politics Meet: The Violence of Humanitarianism in France", in: *American Ethnologist*, 33,1: 33–49.

Uddin, Jashim M., 2014: "Prospects for Attaining a New Height in Bangladesh/Myanmar Relations", in: *Bangladesh Perspective*, 35,1: 1–22 (Dhaka: Bangladesh Institute of International and Strategic Studies).

Ullah, Ahsan A.K.M., 2011:"Rohingya Refugees to Bangladesh: Historical Exclusions and Contemporary Marginalisation", in: *Journal of Immigrant and Refugee Studies*, 9,2: 139–161.

UNHCR, n.d.: "The Refugee Convention, 1951"; at: https://www.unhcr.org/4ca34be29.pdf.

UNHCR, 2018: "Global Trends: Forced Displacement in 2018" (Geneva: UNHCR); at: https://www.unhcr.org/5d08d7ee7.pdf.

Watson, Scott D., 2007: "Manufacturing Threats: Asylum Seekers as Threats or Refugees", in: *Journal of International Law and International Relations*, 3,1: 95–118.

Wolf, Siegfried O. 2017: "Genocide, Exodus and Exploitation for Jihad: The Urgent Need to Address the Rohingya Crisis", in: *South Asia Democratic Forum* (SADF) *Working Paper 6: 1–42.*

Yesmin, Sultana, 2016: "Policy Towards Rohingya Refugees: A Comparative Analysis of Bangladesh, Malaysia and Thailand", in: *Journal of the Asiatic Society of Bangladesh* (Hum.), 61,1: 71–100.

Chapter 22
An Unsustainable Price: The Opportunity Costs of Transitional Justice

Helen Ware

Abstract For campaigners from Pope John Paul II to Angelina Jolie, and especially for those from legal backgrounds, it is axiomatic that there can be 'No Peace without Justice'. It is, of course, possible to define peace in such a way that this becomes a simple truism. But what if peace is defined much more basically as the prolonged absence of physical warfare within a country and the cessation of killing? This chapter proposes the argument that there is, indeed, peace without justice across the world and that economic justice should be given higher priority than forensic justice. The poor know that the world is an unjust place and, if given a chance to voice their choices, put food before truth or retribution.

Keywords Africa · Economic justice · Gender · Peacebuilding · Solomon islands · Timor leste · Transitional justice · Truth and reconciliation commission

22.1 Introduction

Speaking out against transitional justice is unpopular, especially amongst peacebuilding donors, practitioners and lawyers and others with legal training. The claim is often made that there can be no peace without justice, but this is only factually true when peace is defined in advance to include the delivery of legal justice. The reality is that peace may well be unattainable without the attainment of economic and social justice, but those arguing the need for transitional justice are thinking more along the lines of the United Nations' (2010: 2) definition: "transitional justice is the full range of processes and mechanisms associated with a society's attempt to come to terms with a legacy of large-scale past abuses, in order to ensure accountability, serve justice and achieve reconciliation".

Transitional justice is often the right answer to the wrong question. Less helpful questions in play include: 'How can the new, hopefully post-conflict, regime at least appear to distance itself from the crimes associated with the previous regime?';

Prof. Helen Ware, Ph.D., Peace Studies Centre, University of New England, Armidale, New South Wales, Australia 2351; Email: hware@une.edu.au.

Ú. Oswald Spring and H. G. Brauch (eds.), *Decolonising Conflicts, Security, Peace, Gender, Environment and Development in the Anthropocene*,
The Anthropocene: Politik—Economics—Society—Science 30,
https://doi.org/10.1007/978-3-030-62316-6_22

'How can the new government demonstrate to the international community that it is serious about turning over a new human rights leaf?'; and 'How to satisfy the people, relatively cheaply, that their primary priorities are being met?'. A more helpful question for those wishing to create positive peace could be much closer to: 'How can the government achieve a fair distribution of the nation's wealth?'.

UN Principle 10 requires those involved to "strive to ensure transitional justice processes and mechanisms [*whatever they may be*] take account of the root causes of conflict and repressive rule, and address violations of all rights". In El Salvador an oligarchy of fourteen families (*Las Catorce*) ruled the country, with just six families owning more land between them than the poorest 80% put together. Land maldistribution was the core cause of the conflict (Beverley 1982: 55). This could only be cured by nationwide land reform, not by any number of trials of abusive military leaders or truth and reconciliation commissions (TRC).

The purpose of this chapter is to explore a range of specific difficult issues associated with transitional justice, whilst always keeping in focus the need to consider the case for putting economic justice before the pursuit of the truth, legal retribution, or other paths to reconciliation. This becomes increasingly important as localised reconciliation from the grass roots up is increasingly promoted as the essential alternative to top-down initiatives (McEvoy/McGregor 2008). This is because it will be very difficult to create a nation-wide solution from a very large number of small local endeavours without great expenditure of human resources. One bridging solution in some contexts might be to design economic and infrastructure projects which involve collaboration between former enemies in building schools and clinics or upgrading physical markets. Replacing reconciliation meetings by such productive activities could bring doubly beneficial outcomes and might well gain donor support. Others have proposed using reparation schemes to improve access to microfinance on a broad scale (Bernstein 2009). In recent years a number of scholars and legal activists have proposed extending transitional justice to include righting socio-economic wrongs (see review by Waldorf 2012; McEvoy 2007; Uruena/Prada-Uribe 2018). This author believes this to be overloading a term which has been highly expanded already and bringing lawyers into an area where few have any expertise (Carranza 2008). The argument here is that righting socio-economic wrongs should remain a separate goal and should be given greater priority than legal justice issues in line with the views expressed by ordinary citizens of countries in transition (see below).

In this discussion of transitional justice the emphasis is upon trials and truth commissions because this is where most of the action has been, with limited numbers of cases of reparation and lustration and amnesties very much depending upon the local political context. The sections of this chapter that follow interweave discussions of a number of specific problems which may lead from transitional justice to transitional injustice with examinations of the transitional justice contexts and experiences of countries from the Ivory Coast to Rwanda and South Africa to the Solomon Islands and Timor Leste. The emphasis is also on African involvement because the reasons why transitional justice is needed and how it is implemented vary greatly on a regional basis (Dancy/Weibelhaus-Brahm 2011) and because, South Africa excepted, African experience, especially that of the women and men in the streets and markets, has been neglected. Nagy (2008: 275) has even asked "To what extent does transitional justice appear from on high as 'saviour' to the 'savagery' of ethnic, especially African 'tribal' conflict?" The aim here is to avoid the legal issues which have already been much debated in a dozen or more legal journals, and to focus wherever possible on the grass-roots experiences of ordinary villagers and slum-dwellers who often have little or no knowledge of these transitions which are supposed to be protecting them from future conflicts.

22.2 Theories of Transitional Justice

> Transitional justice is a term of art, an academic discipline, a career, a nuisance or simply jargon. (Weinstein et al. 2006: 9)

Despite the fact that there have been over 2000 publications on transitional justice (Olsen et al. 2008), "transitional justice is in general under-theorised" (Gready/Robins 2017: 956). Publications with promising titles are often concerned with the application of theories from other areas to issues of transitional justice or with very small-scale aspects of the topic. This is partially because donor funding is more likely to be available for studies of 'how' to do transitional justice than for whether to include what has become a standard item in the peacebuilding 'toolkit'. Dancy and Wiebelhaus-Brahm (2011: 1) argue that "In fact, 'transitional justice' has attained the status of orthodoxy among global peace-promoting actors and international institutions." Having extensively reviewed the literature, they choose Lie et al. (2007) and Meernik et al. (2010) as the best of the statistical studies. Lie and colleagues found that, for conflicts between 1946 and 2003, whether the conflict ended in a clear victory for one side or in a negotiated settlement was by far the most important determinant of the duration of peace, with regimes established through military victories being far more durable. In contrast, they found that the effect of transitional justice was weak and inconsistent. Meernik et al. explored the impact of international tribunals and domestic trials on the duration of peace and found no evidence that either mechanism made countries more or less prone to renewed conflict in the 1982–2007 period. For themselves, they "seek to address the lack of hypothesis-testing and dialogue by constructing an integrated model of peace transition, duration and conflict recurrence that incorporates the newest information" from the University of Minnesota data set covering 117 countries with internal wars or democratic transitions from autocracy. Their conclusions are that "Depending on the model chosen, we show that transitional justice mechanisms at worst have no effect, and at best slightly extend the duration of peace following violent conflict". As with all statistical studies in this area, queries can be raised as to some of the measures. Why should the number of amnesty laws be significant (the Philippines has had sixteen)? Does the fact that Senegal has had five periods of civil war followed by five periods of peace make the country the most at risk of more conflict? Do longer truth commissions have more impact or are they longer because they are receiving less government support? (The average timing of a truth commission is over eight years into peace spells, and the analysis showed that for every additional year that a truth commission operates the survival time for peace decreases by around 10.5%). There is an interesting definitional issue which has emerged with the increasing popularity of truth commissions: if countries such as Australia and Canada, which are not in transition, hold truth commissions to investigate and highlight issues relating to justice towards their indigenous populations, are they too a form of transitional justice?

One way of looking at different theoretical approaches to transitional justice is to examine the distinctions between the realists and the legalists. The realists tend to see social outcomes as the result of slowly developing changes in structural factors, whereas legalists tend to believe in the possibility of instituting progressive changes through new and better rule-based practices. Working to reduce economic imbalances may well take time, but given political will, new land reform legislation or major tax reform could have a quick impact (Pham et al. 2019). Even provisions for semi-autonomy to meet separatist grievances do not require years to implement. "For every theoretical or empirical claim concerning the beneficial social potential of a transitional justice accountability mechanism, there is a counter claim cautioning of the dangers of unintended consequences" (Dancy/Wiebelhaus-Brahm 2011: 11). For a government emerging from civil war there are many choices of definitive good advice. If amnesties and reparations may promote peace or war, depending on the local circumstances, improving the lot of the poorer citizens is a much less debatable goal.

Orentlicher (2007: 12 and 16), Professor of International Law at the American University's Washington College of Law and United Nations' Independent Expert on Combating Impunity 2004–5, is certainly a legalist, but one who "affirms the central importance of promoting the broad participation of victims and other citizens in the process of designing as well as implementing programmes of transitional justice". She also supports the quest for "the whole truth and as much justice as is possible". Thus she agrees that "no truth commission, however inspiring, can by itself set a country that has been deformed by widespread depredations right again. But an effective truth commission can engage society in an inclusive process of reckoning and repair. Truth commissions can, moreover, speak to the future through their recommendations for institutional reform". Sadly, however, such recommendations are almost invariably ignored. She acknowledges that: "trials, for example, may seem an outsized expenditure of precious national resources if a society's basic social and economic needs remain unmet", adding that far more needs to be done to ensure that trials and truth commissions are known about and their functions and purposes understood even in remote villages. Otherwise there can be situations as in Northern Uganda where the same villagers support both trials of and amnesties for the leaders of the Lord's Resistance Army (ICTJ/HRC University of Berkeley 2005: 30). A study in Nepal found that "Almost no one believed that a TRC would end the culture of impunity or violence or bring equality, and almost half appeared to have little idea about the mandate of a TRC" (Robins 2009: 324). "More than a third thought that trials would endanger peace, and a similar number that it (sic) could endanger reconciliation"; only a third understood what amnesty was and when asked who delivers justice 52% either did not know or thought that human rights agencies did. "This confirms that ordinary Nepalis understand very little of the justice system" (Robins 2009: 324). When victims were asked an open-ended question about their priorities, the overwhelming responses were in terms of basic needs, with only 3% prioritising judicial processes. Asked to choose between trials and compensation, the majority opted for compensation.

One element that emerges from Orentlicher's 2007 paper (although she does not discuss it directly) is that after an internal war there may be at least two categories of victims: one is those who were politically against the government and were singled out for abuse of their human rights for that reason, often even before the war; the other group is those who just became caught up in the violence through being in the wrong place at the wrong moment and may have been traumatised by either government or rebel soldiers until they became victims or were forced to become perpetrators in their turn. Orentlicher (2007) explains that she has been influenced by talking to many politically inspired victims, but appears to have encountered few of the semi-literate villagers in the second group.

Bell (2009: 5–6), who is Professor of Public International Law at the Transitional Justice Institute of the University of Ulster, Northern Ireland, has argued that, contrary to the claims of the opening editorial of *The International Journal of Transitional Justice* (2007: 1–5), transitional justice, is NOT "a praxis-based interdisciplinary field" but "rather is a label or a cloak that aims to rationalise a set of diverse bargains in relation to the past as an integrated endeavour, so as to obscure the quite different normative, moral and political implications of the bargains". It involves "an attempt by nonlaw disciplines to colonise the field". Given the lack of agreement as to what transitional justice actually is, the lack of grand theories to explain its nature and success or lack thereof is hardly surprising.

Equally, practitioners often refer to transitional justice as providing a 'toolkit' for countries in transition. One is not expected to have theories about a toolkit, just a core belief that the tools will work, albeit some better than others in particular contexts (Olsen et al. 2010a).

In 2010 Tricia Olsen, Leigh Payne and Andrew Reiter used evidence from their new Transitional Justice Database to examine which transitional justice mechanisms and combinations of mechanisms positively or negatively affect human rights and democracy. They did not start out from a theoretical base but rather analysed the data first and then provided reasons why their findings reflect how the mechanisms work in the real world. They found that specific combinations of mechanisms, that is trials and amnesties, and trials, amnesties and truth commissions, together generate improvements in human rights and democracy. However, all else being equal, truth commissions standing alone have a negative impact on the two political objectives of human rights and democracy, but do contribute positively when combined with trials and amnesties. They propose that their 'justice balance' approach works because trials provide accountability and amnesties provide the necessary "economic and political stability" (Olsen et al. 2010a: 977) – not otherwise defined. They even argue that regime collapse may result in economic crises which will likely limit widespread, expensive trial proceedings: "in these cases, for pragmatic economic reasons, new democracies may opt to use scarce economic resources to prosecute only a select number of perpetrators, while granting amnesty to the rest" (Olsen et al. 2010a: 1002). They do not suggest how governments should select those to be tried as opposed to those to receive amnesty in order to achieve a happy democratic balance, although they do suggest that timing may be a significant factor, with amnesties coming first, to be followed at some distance of time by truth commissions and trials. Their data does show that countries with greater economic growth are more likely to opt for trials rather than do nothing or provide amnesties (Olsen et al. 2010a: fn.101). They discuss a number of country case studies, including Uruguay, Greece and South Korea, but no cases involving immediate post-civil-war situations.

22.3 Assuming that the New Government is Sincere

The first crucial assumption associated with transitional justice is that the current government, which is usually a new government although it may often include abusive figures from the old regime who are too powerful to be omitted, is sincere in wishing for justice and national transformation. The 'victorious' armed forces also need to support, or at a minimum accept, the required changes. In reality what happened when *From Madness to Hope: The Report of the Truth Commission for El Salvador* (*Comision de la Verdad Para El Salvador* [CVEX]), was produced by three international commissioners, headed by the former Colombian President Belisario Betancur, was that the civilian government and the armed forces rejected the report and no follow-up organisation was established. Indeed just five days after the release of the report, amidst rumoured threats of a military coup, the legislature passed a general amnesty law covering all crimes related to the civil war (the Commission had received evidence of over 11,000 extrajudicial killings) (Beverley 1982: 55). That many truth and reconciliation commission reports have been rejected by the government of the day, as in Chile, El Salvador, Liberia, Nigeria, Solomon Islands and Uganda (three times), is in part a testimony to the potential power of such documents, but it also demonstrates that holding an investigatory commission is not enough; the commission has to make recommendations that will be followed up.

The Sri Lankan case is instructive. The President established the *Lessons Learned and Reconciliation Commission* in 2010. Around the world many, both within and beyond the Tamil diaspora communities, felt that this was simply designed to deflect attention from the arguments for trials of alleged war criminals after the President had promised that no 'war heroes' would be prosecuted. The reality, as subsequent events have demonstrated, was that the Government had no interest in truth or reconciliation on any terms other than its own (Jayakody 2017). Many transitional justice processes take place in messy post-conflict situations where there are no clear 'winners' and the losers are poised to take up the fight again if their demands are not satisfied. In

Sri Lanka the government was definitively victorious and had little need and even less desire to reconcile with its defeated foe.

22.4 Rwandan Insincerity: Using Transitional Justice to Build Control of the State

The use of the 'traditional' *gacaca* transitional justice process in Rwanda has been much debated. Loosely translated, 'gacaca' means 'justice on the grass', formerly most usually a judgement in a local minor property dispute delivered under a tree before the people by a local leader. Proponents argue that the *gacaca* courts represented a local, grass-roots response to overwhelming numbers of genocide participants needing to be brought to trial, estimates of whose numbers ranged from two hundred thousand to two million (Straus 2004). Opponents argue that the courts denied many standard legal protections, such as due process, any right to legal counsel, an unbiased jury and the use of judges with legal qualifications (Clark 2009). (Even before the genocide only 45 of the 708 magistrates in Rwanda held law degrees.) Perhaps because much of this debate has been between human rights lawyers, much less attention has been devoted to the use of the *gacaca* trials "as a policy tool by the Rwandan government to strengthen political order in the country and consolidate the control of a de facto single party state" (Loyle 2018: 664). As Oomen (2005) concludes: "This particular form of justice has been made subservient to the government's political mission." Proponents of the *gacaca* court solution often stress the potential of the grass-roots trials to lead to reconciliation without actually demonstrating just how this happens on the ground. It is equally – if not more – plausible to argue that a process which puts a minority, Tutsi-dominated government in charge of demonstrating (some would say demonising) across the entire country, over a ten-year period, the sins of the Hutu majority is most unlikely to lead to reconciliation between the prosecuted and the prosecutors (Des Forges/Longman 2004). In all, 1.5 million Rwandans or one in five of the adult population appeared before the *gacaca* courts (Kanyesigye 2011), but their mandate excluded any prosecutions of those Tutsi who attacked Hutu.

Rwandan genocide 'accountability' was deliberately entirely one-sided. This is a common problem with transitional justice: the victors in the conflict have very little motivation to expose their own people to prosecution or public ignominy; amnesty is a much more politically attractive option (Cronin-Furman 2015). In the Rwandan case, Tito Rutaremara, who worked with the 1994 *Rwanda Patriotic Front* (Tutsi) Human Rights Commission, argued: "We were finding people who committed genocide in '62, '64, then '72, '91. Finding people who committed genocide four, five times and each and every time, the former regimes would give amnesty" (Loyle 2018: 669). Understandably, neither equity in prosecutions nor reconciliation were priorities for the government. As Jean de Dieu Mucyo, a former Minister of Justice, argued: "Justice was the priority. You can't sit there and look at someone who killed your husband or your children and keep moving forward while still seeing him before you" (quoted by Loyle 2018: 669). (It should be noted that this was exactly what the majority of Cambodians were required to do for decades.) The government determined that "every single participant in the genocide, from the central planners, down to coerced peasant farmers and opportunistic looters, would be prosecuted and punished for their crimes" (Burnet 2010: 97; Rwandan Organic Law 40/2000). Once such a broad net had been cast, anyone who expressed any opposition to the *Rwandan Patriotic Front* (RPF) government could be put up for trial, in a process which was controlled by the government from the capital down to the local *gacaca* court sitting under a tree.

In many cases, post-conflict governments are relatively unstable, but where they have the power, they can and often will use transitional justice to attack their enemies and entrench their power (Nalepa 2010). "Transitional justice policies are particularly well suited for advancing domestic political order and consolidating power behind a given regime" (Loyle 2018: 671), and that regime does not have to be democratic. Historically, Smith (2012: 50) found that "the winner defined justice and constructed the framework of a [war crimes] trial for his political purposes". This is a very long way from Sikkink's 'justice cascade', in which global support for accountability has supported domestic demands for transitional justice (Sikkink 2011; Simmons 2017). It needs to be remembered that the demography of Rwanda, especially its post-genocide demography, means that it is the vast majority which is being required to be accountable by a numerically limited minority.

Indeed, the most significant weakness of Loyle's case against the RPF Rwandan government is the extent to which she stresses how the government is building on its Tutsi power base, without simultaneously acknowledging that any power base built on less than a tenth of the population is always, inevitably, going to be problematic. In part, President Kagame has tried to circumvent the issue by making the use of Tutsi/Hutu identifiers illegal. Everyone has to identify as 'Rwandan'. There is one exception to this: in 2009 the official designation of the genocide was changed from "the 1994 Genocide" to "the Genocide against the Tutsi", deliberately ignoring the fact that many politically moderate Hutu were massacred.

The Rwandan case is unique. The 800,000 genocide deaths in tiny Rwanda in 1994 represented more than 90% of all conflict deaths across the world (Uppsala 2018). There were three levels of justice in the decade that followed: an international criminal tribunal (the *International Criminal Tribunal for Rwanda* [ICTR]); national courts with genocide crimes prosecuted through a special chamber; and the *gacaca* courts. As noted, one in five of the adult population was put on trial. Rwandan governments are nothing if not thorough. The Kagame government has been in a unique position to use global genocide guilt to acquire economic benefits (hence its highly advantageous coffee deals with Costco and Starbucks) and its top-down authoritarian control has facilitated the implementation of the 2020 Economic Plan to the point where income from tourism now exceeds that from agriculture. Thus, to quote one of its supporters (sic): "Rwanda has thrived despite an authoritarian government that is devoid of respect for human and political rights and opposition" (Friedman 2012: 277). Peace without justice for the majority Hutu is possible.

It is to be hoped that more genuinely democratic post-conflict governments could achieve something close to Rwanda's 'economic miracle', but this can only be the case if governments do not become bogged down in issues relating to transitional justice, choosing instead to give first priority to economic restoration and development. Relatively rich countries, such as South Africa or the richer countries in Latin America, can try to pursue justice in the legal and economic spheres at the same time, but many post-conflict countries are desperately poor and have to make hard choices. Rwanda's grass-roots *gacaca* courts had few costs beyond the time of the local peasantry, who understandably became ever more reluctant to devote time which they could have spent farming to listening for a day each week to familiar sad tales of the greed of neighbour against neighbour. Little of the discussion or evidence on the effectiveness of the pursuit of truth has covered the significant opportunity costs of the pursuit of justice in the legal sense of the term to the detriment of the pursuit of economic equity. The common slogan has been "no peace without justice". Failing to define whether this essential justice includes economic justice has led to an undesirable prioritising of truth before livelihoods from which to feed the children by elites, diasporas and international lawyers who are not the ones who are going hungry.

The argument here is that working towards economic justice and combating poverty should be a higher priority than expensive transitional justice mechanisms. At a time when the entire country had less than ten fully functioning secondary schools, Liberia paid nine Truth and Reconciliation Commissioners for three years and held hearings with Liberian diaspora communities in eleven US cities as well as in Ghana, Nigeria and Sierra Leone. Often, international donors may help with the funding of transitional justice mechanisms, but beyond the expense there is the message of the expenditure. The *International Criminal Tribunal for Rwanda* (ICTR) employed sixteen full time and nine occasional judges and over one thousand staff in total at a cost of over US$1 billion (Wippman 2006). Taken together with the International Criminal Tribunal for Yugoslavia, their joint budgets exceeded one seventh of the total UN regular budgets. Justice is not beyond price, especially for Africans whose annual incomes could have been doubled for the cost of the Rwandan Tribunal. The Special Court for Sierra Leone cost US$300 million to prosecute a total of nine individuals at a time when the total annual budget of the Sierra Leone government for all of its activities was US$414 million. Justice is not beyond price and there is no way that such an imbalance can be justified. As a Sierra Leonian teacher whose daughter had had both of her hands cut off by the rebels said: "I would do anything to get a steady source of income for my daughter. Justice and truth, she cannot eat them" (Millar 2011: 530).

22.5 Victims, Perpetrators and Witnesses

The transitional justice model assumes that in internecine conflict there are three groups – victims, perpetrators and witnesses – and that these groups rarely overlap;

in short, that, as in a law court, there are good guys and bad guys. Thus the United Nations' Guiding Principle 6 requires transitional justice programme implementers to "ensure the centrality of victims in the design and implementation of transitional justice processes and mechanisms". The reality faced by the implementers is almost invariably much more complex, with, for example as in Northern Uganda, children who were victims when the rebels arrived and shot their parents, but were then forcibly recruited and coerced to shoot other children's parents in their turn, gradually coming to commit further atrocities for pleasure rather than because of duress. In countries which have experienced brutal civil wars, rebels, government soldiers, and members of previous and current governments are all likely to have committed multiple human rights abuses. This is why new governments appointing Truth and Justice Commissions take such extreme care in defining exactly which years' events are going to be covered so as to exclude their own crimes as far as possible.

22.6 Double Jeopardy

Transitional justice has tended to move from a choice between trials and a truth commission to the attempt to encompass both. This was the case in Sierra Leone, where the simultaneous existence of the *Truth and Reconciliation Commission* and the trials caused great confusion, especially for people who might be at risk of being tried for crimes committed in the course of the war (Tejun-Cole 2003). The Sierra Leone Working Group on Truth and Reconciliation (2006), Cockayne (2005) and Shaw (2005) all heavily criticised the extent to which transitional justice processes in Sierra Leone were designed more to meet international needs than to respond to the voices of local people, many of whom believed it would be better to forget than to remember and that the money could be better spent on schools and clinics. Anthropologist Millar (2011: 515) found that in Makeni, a town in northern Sierra Leone, the "the TRC's restorative approach was unable to generate a sense of post-war justice, and was, to many, experienced as a provocation". His conclusions "support an alternative *distributive* conception of justice and show the local conception of rights, experiences of infringement and needs for redress demand social, cultural, and economic considerations be taken seriously in transitional justice cases". One local leader asserted that the TRC should have brought a tractor to till untended fields as the best contribution to local welfare.

Often, as in East Timor, the aim is to send the most grave crimes to the courts whilst lesser abuses are more extensively covered by the truth commissions (Stahn 2001). Generally, it has not been explained how trials and truth commissions are supposed to work together, much less how they actually co-exist in complex, messy post-conflict situations. In the Solomon Islands the truth commission came long after former combatants, who had been imprisoned together in highly restricted spaces following their convictions, had worked out their own semi-formal reconciliation ceremonies (Vella 2014a).

22.7 South Africa Is Not a Useful Precedent

Whilst there can be no such thing as a typical transitional justice process, it is most unfortunate that the South African Truth and Reconciliation Commission has come to represent the example that most people think of when transitional justice is mentioned. Apartheid was a unique problem and there will never be another similar truth commission. For example, at present a truth commission in Myanmar with Rohingya witnesses free to testify is unimaginable. In the 1980s many feared that the transition to majority rule in South Africa could lead to civil war, but, dreadful and inhumane as Apartheid was, South Africa had not experienced a civil war since the Boer Wars of 1880–81 and 1899–1902. In the event, the transition was peaceful, but the South African murder rate, with over 20,000 murdered in 2018, remains at a level equivalent to a significant civil war. Some 48 of the 57 persons killed each day are men, reflecting the economic factors which have resulted in an unemployment rate of 28%. The black unemployment rate is 39%, compared with just 8% for whites; even for blacks with university degrees the unemployment rate is still 19% (Business Tech 27 August 2015). Under these conditions, very few blacks consider that transition has delivered any kind of economic justice, and without economic justice see reconciliation as a meaningless catchphrase.

22.8 The Solomon Islands' Experience: Timing Really Matters

The Australian government funded South African Archbishop Tutu to go to the Solomon Islands in 2008 to launch the first Truth and Reconciliation Commission in the Pacific. Its remit was to examine the 'tensions' between 1998 and 2003, which had resulted in some 200 deaths and the displacement of some 35,000 people, or one in twelve of the total population. The small number of Solomon Islanders beyond the capital who had enough contact with modern media to be aware of this visit were simply puzzled as to how the Commission was going to add to all the local efforts at reconciliation which had already been undertaken, from village chiefs arranging reconciliation meetings to the forgiveness prayer services by the Christian Melanesian Brotherhood (some of whose members had been spectacularly beheaded) and to ex-militants reconciling whilst imprisoned together (Vella [research officer with the Solomon Islands TRC] 2014a, b). To the Solomon Islanders, 'reconciliation' has a very specific meaning, usually involving specific ceremonies and the exchange of token wealth in shell or feather money, not the preparation of a bureaucratic report for a largely traditional population unfamiliar with paper documents. The preponderantly rural people expected the Commission to 'do' reconciliation through meaningful public ceremonies, not just write about it. The Archbishop might not be expected to understand local customs, but might have questioned the commonalities between multiracial South Africa with a population of 49 million and the distant Melanesian

Solomon Islands population of 400,000. The original idea behind truth commissions, as established in South America and continued in South Africa, involved finding out a truth that had been hidden, as in what had happened to opposition figures who had been 'disappeared' (Hayner 2011). In the Solomon Islands people knew what had happened and had even developed 'research fatigue' in recounting their experiences to yet one more foreign investigator (Vella 2014a). Many in the bureaucracy felt that the government had agreed to the appointment of the Commission largely to appease the aid donors and the United Nations system upon which the country was very heavily dependent, as 'the troubles' had bankrupted the Treasury. In the end, the final Report of the Solomon Islands' Commission was unofficially electronically published by its editor, Bishop Terry Brown, in 2013 because the government of the day, to whom the Report had been submitted, was unwilling to publish a document critical of some of its own members. They may also have been justifiably concerned that the Report could stir up grievances which had settled down over the decade since the events in question. As with many other truth commissions, the Solomon Islands faced two sets of issues: what had happened during the fighting and why the fighting had occurred. In this case, the 'why' was not going to be resolved by reconciliation, since it largely revolved around long-standing disputes over land ownership where outsiders had been welcomed at a time when there was plenty of unused land, but came to be rejected as available land became much scarcer and more in demand for commercial crop production.

Even within Africa, South Africa sets a poor precedent, being more of a developed than a developing country, with an ample supply of qualified lawyers and a strong tradition of adherence to the law which had survived throughout the Apartheid regime. Côte d'Ivoire represents a very different situation.

22.9 Côte D'Ivoire: Can Youth Lead the Way?

"In 2010, after ten years of intermittent conflict in which youth and the education sector were used to advance political ends of the elites in power, often through violent mobilisation, the post-conflict government of Côte d'Ivoire called for the creation of several transitional justice measures. In 2013, the Children and Youth Programme of the *International Centre for Transitional Justice* (ICTJ) and the *United Nations Children's Fund* (UNICEF) undertook a joint programme to engage children and youth in Côte d'Ivoire in their country's process of reckoning with its violent past and help them become responsible and valued actors in their own right. Young people played a highly visible role in the horrors of the war, provoking a social tendency to characterise Ivorian youth as reckless thrill-seekers taking out their frustration with life on the society around them by destroying it" (Ladisch/Rice 2014: 363).

In the event, the youths turned out to be far more responsible, peace-seeking and forward-looking than their government, which, having appointed a truth commission, refused to publish its report once submitted, despite the fact that it had performed the herculean task of summarising 60,000 submissions. Undeterred, the young people chose to establish their own education programme for peace, looking far more to the future than the past because they were only too well aware of major faults on all sides. Whereas truth commissions try very hard to remain apolitical, the Ivorian youths were keen to discuss the political origins of the fighting and debate future political solutions. In some respects, the search for truth and the search for retribution are both ageist projects, favouring the old over the young, many of whom, especially in high-birth-rate developing countries, were born since the tragic events in question occurred. In the case of trials, the old men who planned and led the war crimes escape and the young men who carried out the actual fighting are brought before the courts or are left standing on street corners with no civilian training or employment to go to. There is now a whole discussion of *demobilisation, disarmament, and reintegration* (DDR) and its relationship to transitional justice, but meanwhile the youths are still standing at the corners or looking for new employers for their military talents (Patel 2010).

22.10 Re-igniting the Flames

Apparently without irony, Alexander Karn writes that "the past can be a serious impediment to peace". He adds that this is the case "particularly where it has entailed large-scale violence and/or severe historical repression. Even where historical facts are not in dispute, rival groups often imbue them with divergent meanings, adding yet another layer to the conflict. For peacebuilders and conflict mediators, this makes the past a dangerous minefield of contested memories, but also a potentially useful tool for cultivating intergroup dialogue" (Karn 2015: 207). Although it is difficult to find statistical evidence, it would appear that local populations are often more concerned that transitional justice will reawaken the dogs of war best left asleep. Stover/Weinstein (2004) reviewed data from surveys of the general public on views about transitional justice in the Balkans, Rwanda, Uganda and Iraq. For Uganda they found highly complex relationships between justice, peace and reconciliation, in which attitudes were modified by ethnicity, geography, education and exposure to traumatic events.

In an unusual study in a country without a transitional justice programme, Bratton (2011) examined what ordinary Zimbabweans thought about transitional justice. The survey of 1200 randomly selected Zimbabweans across the country started by asking what people thought peace and justice meant to them. Some 36% defined peace by the absence of violence, 30% by unity and harmony, 16% by political freedom and 9% by stability and tranquillity. Justice meant fair treatment to 38%, truth and honesty to 17%, socio-economic equality to 14%, and rule of law to a mere 6%. In the Zimbabwe context some wanted public trials: "If they confess publicly it will be better because the people may end up feeling pity for them and understanding why they committed the crimes they did. But if we just forget about it [the memory] will always be painful. You will be thinking one thing and the perpetrator will be thinking another." Others felt "I think it is better for us to forget about [violence in the past]. Right now we do not know some of the offenders … If they are going to appear for a public trial all that pain will be revived. For now we comfort ourselves by saying that our brothers disappeared during a time of war … Digging those things up may be too painful to bear." In 2019, the bones of some of those who were killed during the Matabeleland massacres were indeed dug up, but it is too early to see whether this will contribute to peace or justice.

The eastern Democratic Republic of the Congo is one of the most dangerous places where a public opinion study of attitudes to transitional justice has been undertaken. Indeed, some sample sites had to be replaced because of renewed fighting in some of the areas originally selected (Vinck/Pham 2008: 401). The authors found that "basic survival needs and security must come before mechanisms that deal with justice issues and reparations. Respondents expressed fear of reprisal if they were to talk openly about their experience in the conflict, which poses an obstacle to any truth-seeking process". As in Uganda (Pham et al. 2007), "justice, reintegration and reconciliation issues were not frequent priorities among respondents when peace, security and basic needs were not yet met". Both education (according to 32% of respondents) and development (21%) were early priorities.

There have been a number of Afrobarometer studies which have asked questions relating to transitional justice, including in Mali. Much more use could be made of this data on the opinions of the general public to inform decisions on governmental programmes. Backer and Kulkarni (2016) have also compiled a useful study on the role of survey research in studying attitudes towards violent conflict and its aftermath, including views on transitional justice.

22.11 When the Enemy Has Left the Country: Timor Leste

The transitional justice model established in Timor Leste by the United Nations and the new post-independence government is often quoted as an ideal example to be followed, especially by other Asian countries. South Korea, which has held a whole series of truth commissions, is rarely quoted as an example, in part because its reports are not published in English. However, Timor Leste also had the unique good fortune that the Indonesian occupying forces had retreated and gone home to Indonesia. It is challenging to imagine what the South African Truth and Reconciliation Commission could have been like if all the white South Africans had left the country and gone to live in a still white-controlled Rhodesia. Thus, Timor Leste was in a very unusual position in which the country faced two quite distinct sets of issues in pursuing crimes and promoting reconciliation: dealing with the former occupier Indonesia with a 2005 population of 227 million as compared to Timor Leste's 1.03 million; and dealing with those Timorese who had sided with the Indonesian occupiers. All the major alleged instigators of the violence had gone home to Indonesia and were out of reach of the new Timor Leste government. Some of the lower level Timorese who had supported the Indonesian occupiers had usually committed minor crimes, but there was the *Serious Crimes Unit* to deal with the more significant offenders. Given Indonesia's massive geographic and population size just across the land border in West Timor, the new government had to be very pragmatic in how far they pursued the Indonesians responsible for war crimes. The *Timor Leste Commission for Reception* [of Timorese who had supported Indonesia], *Truth and Reconciliation* (usually known by its Portuguese acronym CAVR) with seven main commissioners and twenty-nine regional commissioners operated from 2002 to 2005 and then published its public report: 'Chega' or 'No More'. This named the Indonesian Minister of Defence, General Wiranto, as responsible for many deaths, but the Indonesian government continued to support him and he could not be touched. Eventually, in 2005, the two countries established the first modern bilateral truth commission (the Commission of Truth and Friendship) to examine the violence associated with the 1999 independence referendum. Acceptance of its findings by both countries represented the first time that Indonesia had accepted responsibility for the 1999 killings, but, contrary to much truth commission practice around the world, institutions rather than individuals were named as being responsible for the violence.

The Chega Report was unusual in addressing economic issues. The figure of 102,800 reported deaths included deaths from starvation and illness, which were said to have been caused by the Indonesian occupying authorities. This figure is now generally rounded up and reported as 200,000 by non-government organisations such as Amnesty International. As with many such estimates, including any guesstimates of deaths in conflicts in the Democratic Republic of the Congo, their accuracy depends on knowing what death rates were before the conflict, which involves choosing from a range of highly varied estimates, with this choice determining the final results. Timorese did not have to confront those they were accusing, and the Chega Report from Timor Leste could readily blame the departed Indonesians for deaths from starvation and a lack of medical resources; in the much more common situation, where the government establishing the truth commission has taken over from fellow nationals, allocating blame is much more problematic. Timor Leste also had the advantage of being a small country with small communities who knew when individuals disappeared. Timorese from the East who had committed serious crimes could usually choose to move to and stay in West Timor, which is still part of Indonesia.

22.12 Gender

Discussions of transitional justice which exclude gender issues, although common, are inevitably biased towards the views of the men who control almost all transitional justice negotiations. Given the chance, women may well see issues of transitional justice from a different perspective. Thus, rather than examining possible instruments of justice such as truth commissions and reparations, an ICTJ workshop in 2005 examined the nature of official truth, women's testimony, sexual violence, engaging with courts, victim identity, and TJ and nation building, even extending to issues for Aboriginal children in Australia (Columbia Journal of Gender and Law 2006).

Often without including female and male death rates, it is claimed that women suffer disproportionately from armed conflict (UNDP 1995: 45). Women rarely benefit significantly from transitional justice procedures. The tendency is to see women as victims and to ignore their agency and desires (Cockburn 2013). Women are appointed to courts and truth commissions because such appointments are easy gestures and help to mollify Western donors who are helping to pay the bills. The *International Centre for Transitional Justice* (ICTJ) (2006) has produced a manual on how truth commissions should handle gender issues, which may delight Western feminists but is far removed from the realities lived by poor women in poor countries. For example, it requires staff positions to be given to "gender activists and feminist scholars".

Bell and O'Rourke (2007) have examined the question "Does Feminism need a Theory of Transitional Justice?" After a literature review of "where are women, where is gender, and where is feminism in transitional justice?" they conclude that "feminist theory should focus on how transitional justice debates help or hinder broader projects of securing material gains for women through transition, rather than trying to fit a feminist notion of justice within transitional justice frameworks". They argue that transitional justice is supposed to move countries 'from' male-defined political violence 'to' liberal democratic frameworks, which do not necessarily include or assist women. They point out that in Afghanistan and Iraq part of the justification for external interference was allegedly to assist women, whose rights were generally being ignored. Afghanistan presents an excellent case where, in the event of an ultimate deal with the Taliban, women's rights may well be sacrificed in the interest of securing peace.

Efforts to add gender to transitional justice provisions have been most effective in extending the legal treatment of sexual violence in conflict, especially in including rape amongst the most serious war crimes. The Rome Statute of the International Criminal Court (Articles 7 and 8) explicitly recognises rape, sexual slavery, enforced prostitution, pregnancy, sterilisation and other forms of sexual violence as crimes against humanity and war crimes.

In the case of rape, women often suffer at least twice: once from the actual rape and twice from having to publicly recount the circumstances of the rape in cultures in which rape is inevitably considered to be the woman's fault and where the victim is shunned by her own family and former friends. Some have argued that truth commissions are better for women than trials because adversarial processes can provide a very negative experience for traumatised women (Mertus 2004; Campbell 2004; Bell/O'Rourke 2007). For the Solomon Islands' Truth and Reconciliation Commission the local women banded together and agreed that individual women would not testify. Instead, women would make a group submission as to what had happened, which would be signed off as being true by a respected group of women elders (Fangalasuu et al. 2011). This was especially important in the Melanesian context because a significant proportion of the rapes had involved women being attacked by their relatives in what the local culture defined as incestuous contacts, and no individual woman recounting this in public would have survived the experience.

In Nigeria, Boko Haram rebels continue to wreak havoc across the north east of the country. They have targeted women and especially school girls who go against their views of the necessary subservience of women. At this point in time, it is very difficult to envisage even a distant a future in which these abducted girls get to testify before a truth commission or a court. Boko Haram is radically Muslim and there have been very few truth commissions in Muslim countries. The priority must be to restore these girls and the older less educated women to a position where they can earn their own livings, because they are very often rejected by their home communities and even their families. Women and girls who have become pregnant during their abduction and informal imprisonment and slavery face terrible futures. Truth will bring more pain but nothing new to them, and may harm their children.

In terms of what women want, when they were asked in Afghanistan, Burundi and Timor Leste, they prioritised socio-economic rights before issues of legal justice (Chinkin 2004). One common demographic impact of violent conflicts is the creation of many female-headed households, yet little is done to acknowledge and meet their special needs. Overall, the increased participation of women does not necessarily "equate in any simple way with a feminist reshaping of either peace processes or transitional justice mechanisms" (Bell/O'Rourke 2007: 34). Women need to be present when processes are being chosen and designed, but they also need to be able to participate in contexts where the outcome of the process has not been largely predetermined and it is still possible to make significant changes to the male-created drafts on the table.

22.13 Conclusion

Canadian Judge and Former *UN High Commissioner for Human Rights* (UNHCR) Louise Arbour told New York University Law School in 2006 that:

> Without losing its *raison d'etre*, I believe that transitional justice is poised to make the gigantic leap that would allow justice, in its full sense, to make the contribution that it should to societies in transition. Transitional justice must have the ambition of assisting the transformation of oppressed societies into free ones by addressing the injustices of the past through measures that will procure an equitable future. It must reach to, but also beyond the crimes and abuses committed during the conflict which led to the transition, into human rights violations that pre-existed the conflict and caused, or contributed to it. When making that search, it is likely that one would expose a great number of violations of economic, social and cultural (ESC) rights and discriminatory practices. (Arbour 2006: 2)

This is splendid oratory but hardly points to a practical action plan. It claims too extensive a role for transitional justice. As Campbell (2004) has argued, post-conflict justice "requires a fundamental change to the social order which made possible the original trauma of crimes against humanity. In this sense justice remains the event yet to come." Often there is a strong need for major reforms in the security sector and in the shares of the economy which go to the benefit of the rich and the poor. Governments, which are responsible for levying taxes, generally have a very good idea of just how inequitable their system is. Equally, governments, which provide basic services, know just how limited those services are away from the capital and the major urban centres. Their citizens experience these inequities on a daily basis. Their need is for action, not yet more statistical reports. Their need is for a government with both the will and the resources to grow the national cake and to divide it more equitably. It is most unjust when citizens are not treated equitably, but that is a matter to be resolved by economic policy-makers and service delivery departments, not by lawyers.

It is a cliché to say that each country which has experienced so much violent conflict that transitional justice may be an appropriate response faces a different set of problems to begin to resolve in what is hopefully the post-conflict period. Like many clichés, it is at once true and untrue, helpful and unhelpful. This chapter argues that where choices must be made in the transition period, greater priority should be given to economic needs than to justice issues. Indeed, governments which are able to achieve economic progress by targeting the poorest half of the population will do significantly more for the well-being of their citizens than those who can deliver truth and forensic justice. Building schools is more important than building courts.

References

Akhavan, Payam, 2009: "Are International Criminal Tribunals a Disincentive to Peace? Reconciling Judicial Romanticism with Political Realism", in: *Human Rights Quarterly,* 31: 624–654.

Arbour, Louise, 2006: "Economic Justice for Societies in Transition", Second Annual Transitional Justice Lecture (New York: New York University Law School).

Backer, David; Kulkarni, Anupma, 2016: "Humanizing Transitional Justice: Reflections on the Role of Survey Research in Studying Violent Conflict and its Aftermath", in: *Transitional Justice Review,* 1,4: 187–232.

Bell, Christine, 2009: "Transitional Justice, Interdisciplinarity and the State of the 'Field' or 'Non-Field'", in: *International Journal of Transitional Justice,* 3: 5–27.

Bell, Christine; O'Rourke, Catherine, 2007: "Does Feminism Need a Theory of Transitional Justice?", in: *International Journal of Transitional Justice*: 23–44.

Bernstein, Anita, 2009 "Tort Theory, Microfinance and Gender Equality Converge in Pecuniary Reparations", in: Rubio-Marin, Ruth., (Ed.): *The Gender of Reparations: Unsettling Sexual Hierarchies while Redressing Human Rights Violations* (Cambridge: Cambridge University Press).

Beverley, John, 1982 "El Salvador", in: *Social Text,* 5: 55–72.

Bratton, Michael, 2011: "Violence, Partisanship and Transitional Justice in Zimbabwe", in: *Journal of Modern African Studies,* 49,3: 353–380.

Burnet, Jennie, 2010 "Truth, Reconciliation and Revenge in Rwanda's Gacaca", in: Hinton, Alexander. (Ed.): *Transitional Justice: Global Mechanisms and Local Realities after Genocide and Mass Violence* (New Brunswick: Rutgers University Press).

Campbell, Kirsten, 2004 "The Trauma of Justice: Sexual Violence, Crimes against Humanity, the International Criminal Tribunal for the Former Yugoslavia", in: *Social and Legal Studies,* 13: 329–350.

Carranza, Rubin, 2008: "Plunder and Pain: Should Transitional Justice Engage with Corruption and Economic Crimes?", in: *International Journal of Transitional Justice,* 2: 310–330.

Chinkin, Christine, 2004: "Peace Processes, Post-Conflict Security and Guarantees of Women's Rights: The International Context Considered", 9th Turkel Opsahl Memorial Lecture, Queen's University, Belfast.

Clark, Phil, 2009: "The Rules (and Politics) of Engagement: The Gacaca Courts and Post-Genocide Justice, Healing and Reconciliation in Rwanda", in: Clark, P.; Kaufman, Z. (Eds.): *After Genocide: Transitional Justice, Post-Conflict Reconciliation in Rwanda and Beyond* (New York: Columbia University Press).

Cockayne, James, 2005: "The Fraying Shoestring: Rethinking Hybrid War Crimes Tribunals", in: *Fordham International Law Journal,* 28: 616–637.

Cockburn, Cynthia, 2013: "War and Security, Women and Gender: An Overview of the Issues", in: *Gender and Development,* 21,3: 433–452.

Cronin-Furman, Kate, 2015: *Just Enough: The Politics of Accountability for Mass Atrocities* (PhD thesis, Columbia University, New York).

Dancy, Geoff; Wiebelhaus-Brahm, Eric, 2011: "Justice and the Peace: A Time Sensitive Empirical Evaluation", Paper presented at the Annual Meeting of the International Studies Association, Montreal.

Des Forges, Alison; Longman, Timothy, 2004: "Legal Responses to Genocide in Rwanda", in Stover, Eric; Weinstein, Harvey H. (Eds.): *My Neighbour, My Enemy: Justice and Community in the Aftermath of Mass Atrocity* (Cambridge: Cambridge University Press).

Fangalasuu, Judith et al., 2011: *Herem Kam Stori Blong Mifala Olketa Mere: Women's Submission to the Solomon Islands Truth and Reconciliation Commission,* written on behalf of Stori Blong Mere workshop participants (Honiara: Solomon Islands Truth and Reconciliation Commission).

Friedman, Andrew, 2012: "Kigame's Rwanda: Can an Authoritarian Development Model be Squared with Democratic and Human Rights", in: *Oregon Review of International Law:* 14: 253–277.

Gready, Paul; Robins, Simon, 2017: "Rethinking Civil Society and Transitional Justice: Lessons from Social Movements and 'New' Civil Society", in: *International Journal of Human Rights,* 31,7: 956–975.

Hayner, Patricia, 2nd edn., 2011: *Unspeakable Truths: Confronting State Terror and Atrocity* (New York: Routledge).

Hughes, James; Kostovicova, Denis, 2018 "Introduction: Reconciliation and Transitional Justice after Conflict", in: *Ethnic and Racial Studies,* 41,4: 617–623.

International Centre for Transitional Justice, 2006: *Truth Commissions and Gender: Principles, Policies and Procedures* (New York: ICTJ).

Jayakody, Aruni, 2017: "To Prosecute or Not to Prosecute: The Need for Justice in Post-Conflict Sri Lanka", in: *UNU Articles and Insights,* 13 December.

Kanyesigye, Frank, 2011: "Only 97 Cases Pending in Gacaca", in: *The New Times,* 10 June: 1.

Karn, Alexander, 2015: *Amending the Past: European Holocaust Commissions and the Right to History* (Madison: University of Wisconsin Press).

Ladisch, Virginie; Rice, Joanna, 2014: "Addressing the Recent Past in Schools: Reflections from Côte d'Ivoire", in: Clare Ramirez-Barat and Roger Duthie (Eds.): *Transitional Justice and Education: Learning Peace* (New York: Columbia University Press).

Lake, Milli, 2017: "Building the Rule of War: Postconflict Institutions and the Micro-Dynamics of Conflict in Eastern DR Congo", in: *International Organization,* 71,2: 281–315.

Lie, Tove; Binningsbo, Helga; Gates, Scott, 2007: "Post Conflict Justice and Sustainable Peace", in: *World Bank Policy Research Working Paper 4191* (Washington, DC: World Bank).

Loyle, Cyanne, 2016: "Transitional Injustice: Subverting Justice in Transition and Post Conflict Societies", in: *Journal of Human Rights,* 15,1: 1–35.

Loyle, Cyanne, 2018: "Transitional Justice and Political Order in Rwanda", in: *Ethnic and Racial Studies,* 41,4: 663–680.

McEvoy, Kieran, 2007: "Beyond Legalism: Towards a Thicker Understanding of Transitional Justice", in: *Journal of Law and Society,* 34,4: 411–440.

McEvoy, Kieran; McGregor, Lorna (Eds.), 2008: *Transitional Justice from Below: Grassroots Activism and the Struggle for Change* (London: Hart Publishing).
Meernik, James; Nichols, Angela; King, Kimi, 2010: "The Impact of International Tribunals and Domestic Trials on Peace and Human Rights after Civil War", in: *International Studies Perspectives,* 11: 309–334.
Mendeloff, David, 2004: "Truth-Seeking, Truth Telling, and Postconflict Peacebuilding: Curb the Enthusiasm?", in: *International Studies Review,* 6,3: 355–380.
Mertus, Julie, 2004: "'Shouting from the Bottom of a Well': The Impact of the International Trials for Wartime Rape on Women's Agency", in: *International Feminist Journal of Politics,* 6: 110–129.
Millar, Gearoid, 2011: "Local Evaluations of Justice through Truth Telling in Sierra Leone: Post-War Needs and Transitional Justice", in: *Human Rights Review,* 12: 515–535.
Nagy, Rosemary, 2008: "Transitional Justice as Global Project: Critical Reflections", in: *Third World Quarterly,* 29,2: 275–289.
Nalepa, Monika, 2010: *Skeletons in the Closet: Transitional Justice in Post-Communist Europe* (New York: Cambridge University Press).
Olsen, Tricia; Payne, Leigh; Reiter, Andrew, 2008: "Demand for Justice: Domestic Support for Transitional Justice Mechanisms", Paper presented at International Studies' Association 49th Annual Convention, San Francisco.
Olsen, Tricia; Payne, Leigh; Reiter, Andrew, 2010a: "The Justice Balance: When Transitional Justice Improves Human Rights and Democracy", in: *Human Rights Quarterly,* 32,4: 980–1007.
Olsen, Tricia; Payne, Leigh; Reiter, Andrew, 2010b: *Transitional Justice in Balance: Comparing Processes, Weighing Efficiency* (Washington DC: USIP Press:).
Oomen, Barbara, 2005: "Donor-Driven Justice and Its Discontents: The Case of Rwanda", in: *Development and Change,* 36,5: 887–910.
Patel, Ana, 2010 "Transitional Justice, Disarmament, Demobilisation and Reintegration and Security Sector Reform" (New York: International Centre for Transitional Justice).
Pham, Phuong, et al., 2007: *When the War Ends: A Population-Based Survey on Attitudes about Peace, Justice and Social Reconstruction in Northern Uganda,* Report by the HRC, the Payson Centre and ICTJ (New York: HRC, Payson Centre, ICTJ).
Pham, Phuong; Gibbons, Niamh; Vinck, Patrick, 2019; "A Framework for Assessing Political Will in Transitional Justice Contexts", in: *International Journal of Human Rights,* 23,6: 993–1009.
Ramirez-Barat, Clara; Duthie, Roger (Eds.), 2015: *Transitional Justice and Education: Learning Peace* (New York: Columbia University Press).
Robins, Simon, 2009: "Whose Voices? Understanding Victims' Needs in Transition: A Review of *Nepali Voices, Perceptions of Truth, Justice, Reconciliation, Reparations and the Transition in Nepal by the International Centre for Transitional Justice and the Advocacy Forum, March 2008*", in: *Journal of Human Rights Practice,* 1,2: 320–331.
Sikkink, Kathrine, 2011: *The Justice Cascade: How Human Rights Prosecutions are Changing the World* (New York: WW Norton).
Shaw, Rosalind, 2005: *Rethinking Truth and Reconciliation Commissions: Lessons from Sierra Leone,* USIP Special Report 130 (Washington, DC: USIP).
Sierra Leone Working Group on Truth and Reconciliation, 2006: *Searching for Truth and Reconciliation in Sierra Leone: An Initial Study of the Performance and Impact of the Truth and Reconciliation Commission* (Freetown: Transitional Justice Working Group).
Simmons, Beth, 2017: "Domestic Attitudes towards International Jurisdiction over Human Rights Violations", in: *Human Rights Review,* 18,3: 321–345.
Smith, Charles, 2012: *The Rise and Fall of War Crimes Trials: From Charles 1 to Bush 11* (New York: Cambridge University Press).
Snyder, Jack; Vinjamuri, Leslie, 2003/4: "Trials and Errors: Principle and Pragmatism in Strategies of International Justice", in: *International Security*, 28,3: 5–44.
Sriram, Chandia, 2007: "Justice as Peace? Liberal Peacebuilding and Strategies of Transitional Justice", in: *Global Society,* 21,4: 579–591.

Stahn, Carsten, 2001: "Accommodating Individual Criminal Responsibility and National Reconciliation: the UN Truth Commission for East Timor", in: *American Journal of International Law,* 96: 952–966.

Stan, Lavinia; Nedelsky, Nadya (Eds.), 2013: *Encyclopedia of Transitional Justice* (Cambridge: Cambridge University Press).

Stan, Lavinia; Nedelsky, Nadya (Eds.), 2015: *Post-Communist Transitional Justice: Lessons from Twenty-Five Years of Experience* (Cambridge: Cambridge University Press).

Stover, Eric; and Weinstein, Harvey H. (Eds.), 2004: *My Neighbour, My Enemy: Justice and Community in the Aftermath of Mass Atrocity* (Cambridge: Cambridge University Press).

Straus, Scott, 2004: "How Many Perpetrators were there in the Rwandan Genocide? An Estimate", in: *Journal of Genocide Research,* 6: 85–98.

Tejun-Cole, Abdul, 2003: "The Complementary and Conflicting Relationship between the Special Court for Sierra Leone and the Truth and Reconciliation Commission", in: *Yale Human Rights and Development Law Journal,* 6: 139–150.

UNDP, 1995: *Human Development Report* (New York: UNDP).

United Nations, 2010: *Guidance Note of the Secretary-General: UN Approach to Transitional Justice* (New York: United Nations).

Uruena, Rene; Prada-Uribe, Maria, 2018: "Transitional Justice and Economic Policy", in: *Annual Review of Law and Social Science,* 14: 397–410.

Vella, Louise, 2014a: "Translating Transitional Justice: The Solomon Islands Truth and Reconciliation Commission", in: *State, Society & Government in Melanesia Discussion Papers,* 2014/2.

Vella, Louise, 2014b: "'What Will You Do with Our stories?' Truth and Reconciliation in the Solomon Islands", in: *International Journal of Conflict and Violence,* 8,1: 91–103.

Vinck, Patrick; Pham, Phuong, 2008: "Ownership and Participation in Transitional Justice Mechanisms: A Sustainable Human Development Perspective from Eastern DRC", in: *International Journal of Justice,* 2,3: 398–411.

Walker, Margaret, 2013: "Nunca Mas: Truth Commissions, Prevention, and Human Rights Culture", in: May, Larry; Edenberg, Elizabeth (Eds.): *Jus Post Bellum and Transitional Justice* (New York: Cambridge University Press).

Wippman, David, 2006: "The Costs of International Justice", in: *American Journal of International Law,* 100,4: 861–881.

Part IV
Peace, Development and Education

Chapter 23
Simultaneous Intervention Strategies at Local Ecosystems for Sustainable Development Goals and Peace: Design and Systems Perspectives

Amar K. J. R. Nayak

Abstract The objective of this chapter is to revisualise all the issues and goals from a design and systems science perspective to be able to provide a common minimum and simple framework for interventions and fewer sets of indicators to work with and yet be able to achieve all the *sustainable development goals* (SDGs) faster at a lower cost and in a sustainable manner. Based on the author's community-based action research over the last fifteen years, extensive surveys and case studies across India for over two decades, the text outlines six major dimensions and Forty factors (15 embedded under Production dimension) towards building sustainable ecological systems. First, this paper lays down the major issues and challenges of the present times and how the SDGs provide a framework that encapsulates these issues and concerns, and then it highlights the challenges of the design and implementation of these SDGs. Second, it provides a linear analysis of the six dimensions – viz. relationships, institutions, production, organisations, governance and ecology – and the factors that could help all seventeen sustainable development goals to be achieved. Third, it provides a dynamic analysis by demonstrating the interconnectedness of the factors within each dimension and the interconnectedness of the factors of one dimension with the factors of the other dimensions. It then goes on to show the embedded nature of the six dimensions in an ecological system. Fourth, it discusses how simultaneous interventions in the six dimensions could easily help the SDGs to be achieved faster and at a lower cost. Fifth, it summarises the principles of simultaneous interventions in local ecosystems for SDGs and peace from sub-system design and systems perspectives. The text argues that simultaneous interventions rather than sequential interventions in each of the six dimensions in an ecosystem could help the seventeen *sustainable development goals* (SDGs) to be achieved multiple times in any given ecological system. By mapping the SDGs against the dimensions chosen for simultaneous interventions, the article shows that all SDGs can be achieved twice over, and as many as six SDGs can be achieved five times (500 per cent) more.

Amar K. J. R. Nayak, Prof. Dr., Professor of Strategic Management, Xavier Institute of Management Bhubaneswar, Xavier University Bhubaneswar. Email: amar@ximb.ac.in.

Ú. Oswald Spring and H. G. Brauch (eds.), *Decolonising Conflicts, Security, Peace, Gender, Environment and Development in the Anthropocene*, The Anthropocene: Politik—Economics—Society—Science 30,
https://doi.org/10.1007/978-3-030-62316-6_23

Keywords Ecology · Governance · Interactions · Interconnections · Interdependence · Institutions · Organisation · Peace · Production · Relationships

23.1 Issues and Challenges of Sustainability

Today, humankind faces several issues and challenges: social, cultural, economic, political and environmental. There has been growing social unrest and psychological turmoil, cultural intolerance, and economic insecurity arising from production and transaction slow-down in some regions and recession in other regions. Increasing inequality arising from systemic imbalances with poor and corrupt governance has led to increasing political unrest across the world. The growing fear, uncertainty, unrest, short-sightedness in planning and development has led to environmental destruction. In short, fuelled by social unrest and economic uncertainty, ecological systems are under great threat.

The seventeen SDGs developed and shared by the United Nations best describe the pressing challenges of countries around the world and the goals through which these issues and challenges can be resolved. The issues and challenges relate to poverty, hunger, health, education, gender, water-sanitation, energy, employment-economy, industry-innovation-infrastructure, inequalities, communities-cities, consumption-production, climate, life under water, life on land, peace-justice-institutions, and partnerships. The above issues are so many and varied that they encompass all the different bodies of knowledge systems.

Among the seventeen goals, the tenth goal, reducing inequalities, reflects one of the core issues of increasing inequality across the world. Piketty (2014) showed that global income inequality has increased in the last 190 years and that the *Gini coefficient* has been increasing during this period. Income inequality in the emerging countries has been growing much faster in recent decades. For instance, in India, income inequality was lower during the 1950s to the 1980s, when there was a social economy. Since the 1980s, when India started liberalising, inequality in terms of the *Gini coefficient* has been rising steadily. Further, wealth inequalities across countries have been much greater than income inequalities. We have often dismissed this phenomenon as a development paradox that no longer seems to be tenable in the context of growing unsustainability.

Inequalities are not only a phenomenon of the last 200 years. Rousseau (1762) wrote about the inequalities more than 250 years ago. In his book, *The Social Contract*, Rousseau argued: "Man is born free and everywhere he is in chains. One man thinks himself the master of others but remains more of a slave than they are." Income and material inequality has been a constant source of fear and loss of control. As far back as 4000 years, *Chakra Vyuha* and *Labyrinths* from across the world show the symbols of fear and loss of control by the common man. Inequality in control does not only concern physical or material resources or technology, but also psychological and mental control of the human person by market forces.

Further, it is unlikely that the SDGs can be realised by 2030, due to the inherent technical inconsistencies across sectors, dimensions, subsystems and systems across countries and geographies. The language, logic and values of the mainstream economy and industry do not seem to be compatible with the language, logic and values of sustainability.

The seventeen SDGs and the 304 draft indicators of the *United Nations Statistical Commission* (UNSC) are very well thought out; but the majority of them are complex and difficult to achieve. With the above background, realising the goals on the ground – i.e. within communities and ecological systems – seems to be a herculean task. There appear to be several issues that need to be reviewed for the SDGs and peace to be achieved.

First, the approach to identifying indicators to measure the sustainability of the goals has been primarily reductionist – the normal scientific approach. The apparent set of independent goals and indicators send out the signal that policy-makers and executives at state level should visualise the goals separately.

Second, extending the reductionist view, the goals have been perceived by policy-makers and administrators as independent of each other, and action plans are accordingly drawn up by the respective departments of different governments. However, as the above issues are interconnected, the lack of coordination, alignment and convergence between the different departments of a government lead to chaos and waste, and are counter-productive and often unsustainable.

Third, arising from the premise that these SDGs are independent of one another, these goals are intended to be achieved at national and global level, so they have not been designed to be achieved at the local or respective community level. However, the issues of hunger, water, health, education, sanitation, nutrition, etc. are all interconnected and often turn out be the problems of the same individual or the same community, hence bottom-up planning would be relevant.

Fourth, the number of indicators identified has therefore been so large, and only 16 per cent of these indicators rated as AAA are feasible, appropriate and very relevant (UNSC 2015); for the majority of the indicators, this is difficult to comprehend. The larger the number of indicators, and the more difficult they are for people on the ground – i.e. at community level – to understand, the greater the challenges for implementation.

Fifth, and most importantly, while the goals have been set, the means to achieve these goals have not been discussed. Any goal without a plan of action is of little value. It is in the path and processes of the action plans that the possibility of achieving the strategic goals lies (Teece et al. 1997). In other words, elaborating the common minimum areas of engagements or interventions that will help realise these goals is critical to achieving the SDGs in a small community or a country. Further, setting goals without creating an enabling environment, as we know, would take us nowhere. There is a need to create an enabling environment with simple and relevant activities, as understood by the people at ground level – i.e. at community level – for the various stakeholders to be able to participate and work towards achieving these interrelated goals.

The objective of this chapter is to revisualise all the issues and goals from a design and systems science perspective in order to provide a common minimum and simple framework for interventions and fewer sets of indicators to work with, and yet enable all SDGs to be achieved faster, at a lower cost and in a sustainable manner. Based on the author's community-based action research over the last fifteen years and extensive field surveys and case studies across India for over two decades, this text outlines six major dimensions and thirty factors towards building sustainable ecological systems.

This text, as described above, *first* lays down the major issues and challenges of the present times and how the SDGs provide a framework that encapsulates these issues and concerns and then highlights the challenges of the design and implementation of the SDGs. *Second*, it provides a linear analysis of the six dimensions and the factors that could help all seventeen SDGs to be achieved. *Third*, it provides a dynamic analysis by demonstrating the interconnected and embedded nature of the factors across the six dimensions. *Fourth*, it discusses how simultaneous interventions on the six dimensions could easily help the SDGs to be achieved faster and at a lower cost. *Fifth*, it summarises the principles of simultaneous interventions in local ecosystems for SDGs and peace from sub-system design and systems perspectives.

23.2 Key Dimensions and Factors for Sustainability: A Linear Analysis

I regard the following six dimensions – *relationships*, *institutions*, *production*, *organisations*, *governance* and *ecology* – as key components for interventions in any given community and ecological system. Each of the dimensions and critical factors within each dimension are discussed in this section.

Relationships: The key factors of study and interventions include a *sense of interdependence, notion of capital and wealth, mental construct, morals and values* and *faith and belief*. The dimension of 'relationships' is explored here from a systems science perspective and a spirituality perspective. The spirituality perspective has been drawn from theology (religious studies) and the study of philosophy. The five factors of the relationship dimension addresses relationships at different levels. *Faith and belief* are at the core of an individual. *Morals and values* are an outcome of relationships at a family level. *Mental construct* is an outcome of our education, training and experience. A *sense of interdependence* is an outcome of the relationship with neighbours and in the individual's own small community. The notion of *capital and wealth* is the overall societal orientation towards what is perceived as capital and wealth.

The nature and state of relationships often shape the sustainability of our endeavours in building institutions, production systems, organisational designs, community governance, and ecology where we live. Indeed, relationships seem to have been at the core of human engagements and our endeavour towards sustainability. It appears that unless the relationships are not in alignment with the ecological principles, inter-generational sustainability shall remain a distant dream.

The significance of relationships for the sustainability of communities and ecosystems is expressed in the working definition of sustainability:

> Sustainability is a dynamic state of deep relationships among the people and all the constituents both living and non-living within a micro ecological unit that strongly values the acts of sacrifice, reciprocity and love for each other; where the priority is to strengthen the weakest and the spirit of high internal competition with high external co-operation not only drives its own ecological unit to peace, joy and happiness but also inspires other micro ecological units for such deeper interrelationships. (Nayak 2014a: 111–118)

Institutions: This dimension includes both formal institutions and informal institutions. Accordingly, institutions are the norms, rules and conventions that regulate the functioning of an ecosystem. Deciphering the issue and factors of an institution has been one of the most challenging tasks. Increasingly there has been the realisation that without appropriate institutions, sustainability can only be a distant dream. The key factors of this dimension include *norms and conventions, rules and regulation, principles of justice, interaction intensity* and *institutional loading*.

Production: Production includes the primary, secondary and tertiary nature of production in any ecosystem. Agriculture, the primary production activity, is being greatly impacted by climate changes and is increasingly becoming unsustainable across the world. It appears that in the course of taking agriculture forward for greater productivity through intensive external inputs, we have made many small farmers unviable, increased the risks to food safety, and degraded the environment. Secondary (the manufacturing and processing sector) and tertiary (the service sector) production are being studied from the lens of an SDG compass. The key factors that are being discussed include *infrastructure* (farm forestry, support structures, local networks), *resource base* (moisture, raw materials, machinery), *competence base* (seeds, products, technology, process), *systems services* (soil-systems coordination), and *diversity* (production scope, actors and actants, cultures).

Organisation: Organisations have been the key engines of economic growth in human enterprise systems. However, today's organisational designs, especially in the secondary and tertiary sectors, seem to facilitate private financial capital creation far more than social wealth creation in a society. Both *Producer Organisations* (PO) and *Industrial Organisations* (IO) need to be designed such that they can evolve to be community enterprise systems rather than private enterprises.

The *farmer producer organisation* (FPO) has been a special case of study under the *organisation* dimension. FPOs, or the POs in this analysis, include different forms of collective – namely, primary co-operative, SHGs, farmers' clubs, producer organisations, and producer companies. Accordingly, organisational design factors that can facilitate greater frequency of interactions between the members and owners, a greater number of transactions throughout the year, and help members find greater value through these interactions and transactions need further exploration in the context of secondary and tertiary sectors.

The key design factors for studying FPOs and organisations in general include *size, scope, technology, ownership* and *management*. Size refers to the number of members and geographical extent. Scope refers to the number and type of activities that an organisation can engage in. Technology refers to the process and product technology suitable for an organisation. Ownership refers to the shareholding structure in the organisation. Management refers to management structure and the type of managerial skills appropriate for an organisation.

Governance: Governance has been an encompassing dimension in the subsystems and systems of our society. It has therefore been an important component with regard to the sustainability of our community systems. The focus of this study has been to identify the factors and principles of governance that can facilitate sustainable community systems at the lowest level of governance, i.e. the *Ward* or *Gram Panchayat* level, as well as be technically consistent for sustainability at higher levels of governance, namely, district, state-province and national, and global levels. The key factors of study under governance include *frequency of interactions, decision-making methods, problem-solving approach, resource dependency* and *governance architecture and responsibilities.*

Ecology: Ecology in our study represents a basic comprehensive unit (habitat-ecosystem) of our planet that is driven by the fundamental principles of nature. Ecology includes all living (biotic) and non-living (abiotic) objects in a given habitat. The natural principles of interconnection, interdependence, interactions, and caring for the weakest are the principal axioms in the study of ecology. The design and systems analysis of the other concentric and embedded layers of any ecosystem will also be determined by the above natural principles. The key factors that will be discussed include *changes* (atmospheric), *renewability* (of resources), *balance* (of species), *compatibility* (of human and natural systems*)*, and *openness* (in ecosystems).

Each of the factors described under different dimensions have been presented in Fig. 23.1, where the possible spectrum of each factor is outlined. The spectrum provides an extreme position of each factor that allows for it to be mapped as it is in the present and as it should be to achieve sustainability in each of the dimensions. On the whole, the left-hand side of Fig. 23.1 shows conditions of unsustainability, while the right-hand side shows conditions of sustainability. In the present reality, all the factors covered under the six dimensions of our society appear to be largely positioned or are moving towards the left-hand side of their respective spectrums, raising alarming concern over the sustainability of our ecological systems. The challenge has been to discover how communities and societies can transit or reverse the direction of this massive movement from the left side to the right side of the spectrum outlined in Fig. 23.1 and achieve a balance for sustainability.

←		→
	Relationships (Systems Science & Spirituality)	
Low (Distrust, Competition, Rivalry, & Contractual)	Sense of Interdependence	High (Trust, Co-operation, Friendly & Non-Contractual)
Financial Capital & Material Wealth	Notion of Capital & Wealth	Social Capital & Social Wealth
Scarcity, Fear & Inequity	Mental Construct	Plenty, Fearless & Equity
Regulating Others & Concern for Strongest	Morals & Values	Self-Regulation & Concern for Weakest
Hierarchical & Asymmetric	Faith & Belief	Non-Hierarchical & Symmetric
	Institutions (Social-Economic-political))	
Universal	Norms & Conventions	Contextual
Contextual	Rules & Regulation	Universal
Instrumental & Delayed	Principles of Justice	Fairness & Timely
Shallow	Interaction Intensity	Deep
Heavy	External Institutional Loading	Light
	Production (Primary-Secondary-Tertiary)	
Shallow	Infrastructure (Farm Forestry, Support Structures, Local Networks)	Deep
External & Non-Renewable	Resource Base (Moisture, Raw Material, Machinery)	In-situ & Renewable
Exotic	Competence Base (Seeds, Product Technology, Process Technology)	Indigenous
Disabling & Conflicting	Systems Services (Soil-Systems Coordination)	Enabling & Synergetic
Low	Diversity (Production Scope-Actors & Actants Cultures)	High
	Organisational Design (CES–FPOs at GP Level)	
Large	Size	Small
Few	Scope	Multiple
High Cost	Technology	Low cost
Private	Ownership	Common
Complex	Management	Simple
	Governance (GP-PS-ZP Governance)	
Low	Frequency of Interactions	High
Indirect Participation (of elite groups)	Decision-Making Method	Direct Participation (of diverse groups)
Positivism	Problem-Solving Approach	Constructivism
External	Resource Dependency	Internal
Unbounded	Governance Architecture & Responsibilities	Bounded
	Ecology (GP Ecology)	
Discomforting	Changes (Atmospheric)	Comforting
Non-Renewable	Renewability (of Resources)	Renewable
Imbalance	Balance (of Species)	Balanced
Low	Compatibility (of Human & Natural Systems)	High
Control	Openness (in Ecosystem)	Freedom

Fig. 23.1 Spectrum of dimensions and factors for optimizing internal designs and systems consistency in the six dimensions of intervention. *Source* The author

23.3 Interconnected and Embedded Nature of Factors Across Dimensions: A Dynamic Analysis

While the previous section shows that the factors within a dimension are interconnected, this section outlines the interconnection between factors from one dimension with factors from other dimensions and how the dimensions themselves seem to be embedded in a pattern to produce the basic unit of an ecological system.

23.3.1 The Interconnected Nature of Factors Within Dimensions

As shown in Fig. 23.1, the factors seem to show a dynamic relationship with each other within any given dimension. As an illustration, let us see how the five factors of *organisation* – size, scope, technology, ownership, and management – are dynamically interconnected with each other in the specific case of, say, a farmer-producer organisation (a co-operative or producer company).

As the organisational size increases in terms of membership and geographic spread; the scope of activities in the organisation tends to become limited. In other words; as the size increases, the organisation tends to become specialised. As it tends to become specialised, it has to rely more on technology, which leads to subsequent technology intensification in the organisation. As the process and product technology intensifies or is required, this doesn't come free to the organisation; but through huge investments. These investments are brought in by some investors who would like to take an ownership position in the organisation either directly or indirectly through greater management control. With more capital inflow to the firm, the ownership and management structure gradually gets modified as per the directions of the major investors. As all four factors shift towards one side of the spectrum (to the far left, as in Fig. 23.1) – i.e. size increases, specialisation increases or scope of activities reduces, technology intensifies, and ownership-management becomes concentrated – the producer organisation is obliged to adopt a very complex management structure to reduce transaction costs. Interestingly, these interconnected changes could be initiated by a change in any one of these five factors; a process or product technology intensification, for instance, can lead to a subsequent shift towards specialisation, and so on.

On the one hand, when the size grows, specialisation increases, technology intensifies, ownership-management gets concentrated and management becomes more complex, an organisation often evolves into a large multinational corporation. Unfortunately, large corporations seem to show signs of unsustainability in the future. Several historical research studies (Schumpeter 1943, 1977, 2009; Nayak 2009) on large multinational corporations around the world tend to make this point. Was the bankruptcy of Fagor, a flagship unit of Mondragon Cooperative Group, a result of such dynamics? Are the internal tensions in India's best cooperative, namely

Amul-GCMMF (Gujurat Cooperative Milk Marketing Cooperative Ltd.), due to its design flaws? The challenges of large corporations in the global economy could be understood from these basic organisational design factors.

On the other hand; when these factor positions are at the other end of the spectrum (to the far right, as in Fig. 23.1), we get small informal organisations, such as *self-help groups* (SHGs) or small primary co-operatives in India. While SHGs groups have been good social units for economically poor women, they do not seem to possess the technical viability to undertake more than a few limited functions. So it appears that one of the challenges of sustainability for producer organisations has been to find optimal positions of design factors. Small is beautiful (Schumacher 1973) indeed; however, what would be an optimal organisational or community size in the current challenging context to facilitate transition towards a sustainable community system?

The inter-industry dynamics across organisations in primary-secondary-tertiary sectors appears to be complex. Let's analyse organisations directly related to primary production. Greater volume of production of a crop in a given geography or collection by a producer organisation which may be initiated by itself or triggered from outside often requires a processing or manufacturing activity that gradually leads to more technology-dependency; first the dependency is on process technologies and then it is on product technologies. The increased transaction cost due to the larger operation is then dealt with further, scaling-up processes. With scaled-up production capacity, a higher price signalling mechanism is often used to source inputs by processing and manufacturing units in the secondary sector, which initially comes as a boon in terms of higher prices for farmers but gradually becomes a bane to producers in the primary sector, as observed in commercial farming across the world.

With greater demand to deal with complexities in the secondary sector, the tertiary sector flourishes in terms of global trade investments and all supporting services, including education and training in various fields – technical, management, economics, diplomacy, etc. – that are often path-dependent on the existing organisations. In the given context, the criteria for efficiency are different for individuals and organisations in different sectors. For instance, diversity is efficient for farmers in agriculture, whereas specialisation is efficient for processing/manufacturing units, and both scale and scope are required for retail organisations. With such technical contradictions across sectors, while pursuing their individual goals, they together perpetuate greater asymmetries across the primary-secondary-tertiary sectors in an economy (Nayak 2014a), the long-term impacts of which seem to play out more clearly in highly industrialised economies.

Similarly, the factors of *production* – infrastructure, resource base, competence base, systems services and diversity – in all three types of production system, namely the primary production (say, agriculture), secondary production (manufacturing and processing), and tertiary (services) sectors, are deeply interconnected. To simplify the matter, we can analyse the case of primary production (agriculture), where the factors include moisture, soil, seed, diversity, and farm forestry.

On the one hand, sustainable agricultural systems thrive and grow under deep ecology and high diversity of crops, horticulture and farm animals. On a farm with plenty of foliage, where the soil is not exposed to direct sunlight due to deeper ecology, the water requirement of the farm is reduced; the moisture requirements are met through in situ water conservation. Rainwater run-off is slowed down and soil erosion is reduced. Under these conditions, the soil is richer with organic life that enhances the soil health, and there is no need for inorganic fertilisers. Harmful pests in such a diversified farm have been observed to be low. When genetically stable indigenous seeds, carefully selected by the farmers, are planted in such a farm, they provide very high yields, effectively aided by bees, butterflies, and other species that flourish in a diversity farm system.

On the other hand, single crop production in a season with little farm diversification requires high-yielding seeds, such as the genetically modified seeds, large amounts inorganic fertilisers, and a large amount of water. Farm ecology is not much of a concern on such a farm. The system of operation for such a farm becomes complex over time and, more importantly, the net income of the farmer remains more uncertain. The choice on any one factor of agriculture will lead a farmer in a particular direction of the farming system: an integrated ecological sustainable agricultural system or an industrial mono-cropping system.

The factors of *Governance* are also deeply interconnected. Although it is desirable for local communities to self-govern themselves, the positivist problem-solving approach at the higher level of governance in a state drives the system to centralised governance. The positivist orientation typically relies on professionals over volunteers to achieve efficiency in public service delivery. While these orientations, principles and methods fit better with centralised decision-making, centralisation gradually reduces the direct participation of the people in the community. With lesser participation and ownership by the people in local governance, issues of transparency and accountability arise. The idea of a community, its culture, heritage, social wealth and common/pooled resources erodes gradually, leading to increasing costs and failures in the governance system.

The factors of *Institutions* are equally deeply interconnected among themselves. The large size of a state and its large budget would logically lead it to divide its responsibilities between separate departments and functions. Accordingly, the policy signals of the state institutions for regulatory and administrative mechanisms tend to become compartmental in nature.

The overall development approach of such a state is dispersed rather than focused in any cluster or community. With resource constraints to meet all requirements of several constituencies of the population the government is forced to design specialised programmes and schemes. Further, to be perceived well by the general public across constituencies, the government often implements these schemes in a dispersed manner rather than in the holistic manner that is technically required for resolving the various issues of people in a given small community.

Compatible institutional architecture to execute the policies of each Ministry or State department accordingly requires a top-down hierarchical institutional architecture that is largely external-market-determined rather than bottom-up community-driven. The institutional architecture arising from these compulsions serves more as a delivery agency of public services, where the institutional cost of providing services can only increase over time and put a greater fiscal burden on any government in power.

With the above policy signal, the developmental approach and the institutional architecture of the State, the institutional support will tend to become piecemeal rather than holistic. The input–output market network in a given community will also adopt a borderless loose market network that is in line with the overall competitive market economy framework.

In other words, the institutions evolve and consolidate with policy signals that are compartmental rather than convergent in nature, a development approach that is dispersed rather than focused, institutional architecture which is top-down and market-determined rather than bottom-up community-driven, institutional support which is piecemeal rather than holistic, and input–output networks which are borderless loose market networks rather than deep local market networks.

The factors of *Relationships* are deeply interconnected and often the most challenging to change. Depending on the faith and belief system, a person may be rooted in hierarchical and asymmetric beliefs at the core or otherwise. Accordingly, the morals and values may be towards regulating others and inherently focused on the strongest or may be otherwise. Mental construct that is not only developed through upbringing but more through education, training and work experience could shape thinking in terms of scarcity, fear and inequity versus feelings of plenty, fearlessness and equity. If the sense of interdependence is low for a person, he/she is likely to have a feeling of distrust, competition and hate, and if the sense of interdependence is high, he/she is likely to exhibit trust, co-operation and love in the local community. Depending on one's notion of capital and wealth, a person may see them in terms of financial capital and material wealth versus social capital and social wealth.

Similar to other dimensions, the key factors of *Ecology* are deeply interconnected. For instance, changes in *atmospheric conditions* affect the *renewability of resources* and the very existence and *balance of species* in an ecological system. The *compatibility of human and natural systems* has significant impacts on the above three factors of the ecology. Further, the ability of the human system to provide *openness* (in an ecosystem) impacts the capacity for renewability and natural balance. Interestingly, one could explore several bi-variate and multi-variate interrelationships between the five factors of the ecology dimension.

23.3.2 The Interconnectedness of Factors Across the Dimensions

Let us now analyse the interconnectedness of factors of one dimension with factors of other dimensions, as in Fig. 23.1. As an example, if the faith and belief (a factor of *relationship*) of people in a community is based on hierarchy and asymmetry, norms and convention (a factor of *institution*) will tend to be universal rather than contextual and institutional loading (another factor of *institution)* will be heavy rather than light.

Hierarchical perspective is likely to promote large size (a factor of *organisation*), a positivistic problem-solving approach (a factor of *governance*), and lower diversity (a factor of *production*) as well as encourage control over freedom of openness of the ecosystem (a factor of *ecology*).

Let's say if the mental construct (a factor of *relationships*) of people in a community were prima facie based on scarcity, fear and inequality, rules and regulation (factors of *institution*) are likely to be contextual rather than universal; the resource base (a factor of *production*) is likely to be non-renewable rather than renewable; ownership (a factor of *organisation*) is likely to be private than common; governance architecture and responsibility (a factor of *governance*) is likely to be unbounded, as a fearful mind would try to extend its limits of influence while minimising its responsibilities; further, the sense of fear and scarcity among people would invariably lead to economic activities that lead to the non-renewability of resources (a factor of *ecology*).

Let's take another example: if the sense of interdependence (a factor of *relationships*) is low, it is likely that there will be greater distrust, rivalry and competition between members of a community. Interaction intensity (a factor of *institution*) between members of that community is likely to be shallow. Interestingly, interaction intensity (a factor of *institution*) would be the opposite of this if the sense of interdependence were high. Further, interaction intensity could have a similar impact on the sense of interdependence (a factor of *relationship*). With a low sense of interdependence, the decision-making method (a factor under *governance*) is likely to be indirect participation. The frequency of interactions (a critical factor of *participation* under *governance*) impacts interaction intensity (a factor of *institution*) and sense of interdependence (a factor of *relationship*). When size (a factor of *organisation*) is large; the above three factors – a low sense of interdependence (a factor of *relationships*), shallow interaction intensity (a factor of *institution*), and indirect participation (a factor of *governance*) – are often observed. Further, with larger size systems, services (a factor of *production*) are likely to be disabled and conflicted. All these factors are likely to lead to an imbalance of species (a factor of ecology).

23.3.3 The Embedded Nature of Dimensions

The six dimensions – relationships, institutions, production, organisation, governance and ecology – seem to be deeply embedded within society. Relationships seem to be the base of all the other five dimensions. As we evolve in our relationships, we tend to develop informal and formal institutions to stabilise and strengthen our relationships. Accordingly, our institutions are fundamental to the other dimensions: production, organisation, governance. Our institutions greatly impact our ecology.

Organisation has been perceived to be the engine of growth in the present market economy and accordingly, governance and institutions; the other critical layers of a society increasingly appear to get shaped by the demands of the leading organisations. Inherent design deficiencies in such organisations can fuel further asymmetries, more unequal relationships, and deeper divisions among people that perpetuate a vicious cycle of unsustainability of the organisation.

Similarly, each of the other dimensions – production, organisation and governance – impacts all the other dimensions. While ecology gets impacted by the other dimensions, it encompasses and provides overall cover to all the other five dimensions. Indeed, all these six dimensions seem to form concentric circles and appear to be beautifully embedded within each other, providing a harmonious ecological system framework, as shown in Fig. 23.2.

The interconnection between the second hand, the minute hand and the hour hand of a clock or the interconnected gears in a machine demonstrate the nature of interconnections. The interconnectedness and embedded nature of these thirty factors of the six dimensions has been the reason for high complexity and lock-in problems in the systems and processes of our present times – a context where apparently no single individual, organisation or institution can possibly reverse the vicious cycle of

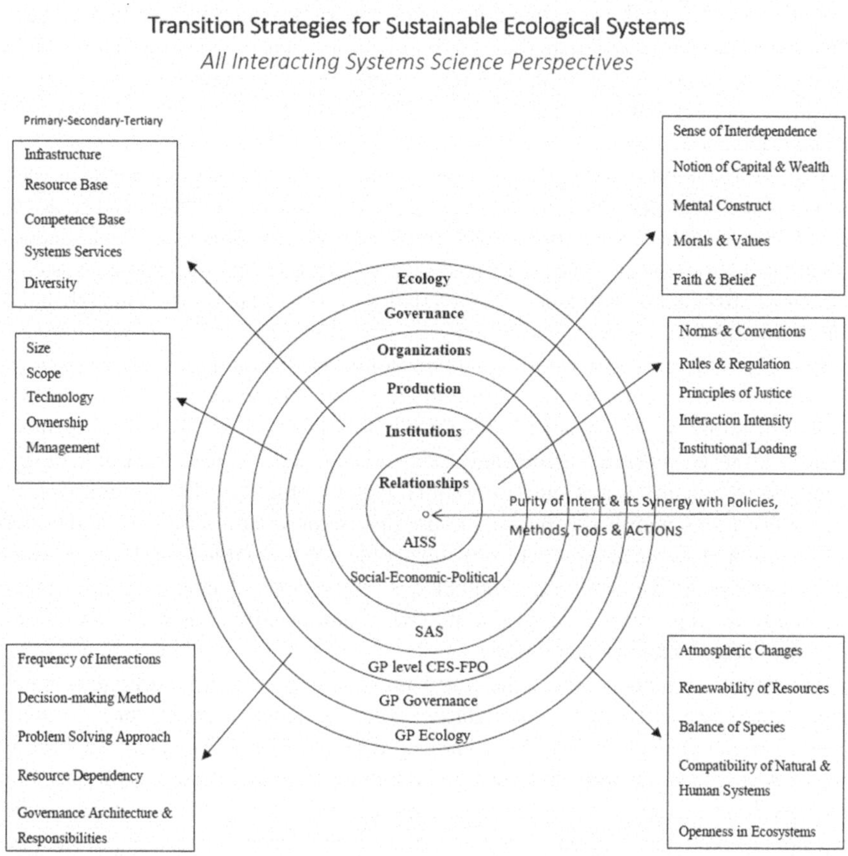

Fig.23.2 Six dimensions and factors for simultaneous interventions at a local community (Ward/GP) level. *Source* The author

unsustainability. Under the above highly interconnected complex dynamics where our own designs and systems at different levels appear to perpetuate lock-in effects and external controls, the chances of our becoming unsustainable seem to be far greater than becoming sustainable, unless the various stakeholders work together on multiple dimensions simultaneously.

23.4 Simultaneous Interventions on All Six Dimensions and Potential Outcomes of SDGs

Resolving interconnected issues in a community or a nation may draw the analogy of tightening the nuts and bolts while fixing a door. The interconnectedness of the various issues of the SDGS – such as poverty, health, education, water, energy – in the context of a family and community make it pertinent that necessary interventions to achieve SDGs need to be simultaneously undertaken. Based on the last two decades of a combination of research methodologies of surveys and case studies across India and engaging action research on multiple dimensions in six communities, I suggest simultaneous interventions on all of the six dimensions laid out earlier, though with different intensities.

Step 1 for scientific and systematic intervention is to review our designs at each level of production, organisation, governance, institutions, relationships and ecology. Figure 23.1 provides a spectrum to each of the thirty factors and shows that characteristics of factors towards the right-hand side suggest communities and ecosystems to be more sustainable than when characteristics of factors are towards the left-hand side of Fig. 23.1.

Step 2 is to identify a basic unit where the common minimum interventions are to be made in different communities. A community system not only consists of a certain number of families in a cluster of villages but also a certain extent of geographical spread with its ecosystem. Accordingly, micro-watershed, the basic unit of ecology, forms the technical base of our community system. However, for optimisation of size for economic viability, especially under diverse production systems, and easier identification of physical boundaries by the people, a *Gram Panchayat* (GP) or *Ward* can be taken as the basic unit for simultaneous *interventions*. Depending on whether it is in a hilly, plain or coastal region, this basic unit usually consist of 2–4 micro-watershed areas in the Indian context.

Interestingly, countries across the world are structured on similar primary boundaries of their respective local communities, such as *Gewong* in Bhutan, *Parishad Council* in Bangladesh, *GN* in Sri Lanka, and *Union Council* in Pakistan. Countries of Africa have a similar political and administrative governance structure. For instance, Tanzania has *ward* at the lowest level and *district* and *municipality* at the next upper level. The United Kingdom does not have such structures but has the traditional, informal institutional structure of a *parish* at the lowest level of community organisations. Advanced industrial economies usually have *Ward* at the lowest level and *District* at the higher level. In other words, as small community systems exist in countries around the world, holistic sustainability initiatives in such small communities in their respective nations could be undertaken for the gradual but overall sustainability of a nation.

The respective governments need to facilitate the elected representatives and local executives of a *Ward* or *GP* in systematic planning, develop convergence plans and execute schemes for health, education, essential public services, infrastructure, renewable energy systems and mobile communication within the Ward/GP.

Step 3 includes systematic and simultaneous intervention in production, organisation, governance and institutions.

Production, the core economic activity includes food production and other material production. Agriculture, the primary production activity, is being greatly impacted by climate change and is increasingly becoming unsustainable across the world. It appears that in the course of taking agriculture forward for greater productivity through intensive external inputs, we have made many small farmers unviable, further jeopardised food safety, and increased environmental degradation. Agricultural production science needs to follow the principles of agroecology that can enable a small farmer's agricultural field to become sustainable and the farmer viable in the short run and sustainable in the long run (Nayak 2018).

Strategically, a one-hectare farm model for, say, about 200 farmers in a GP/Ward could be initiated. Financial support for this intervention could include funding for in situ water conservation, soil health improvement, indigenous seeds, and an integrated farm with multiple crops, horticulture and livestock. It could also provide training to enable select farmers or groups of farmers to implement sustainable agricultural systems on their respective patches of land. The model would be designed to facilitate gradual development of a diversified integrated one-hectare sustainable farm capable of generating about INR 250,000 (approximately 3250 Euros.) per year within three years in the Indian context.

In the secondary and tertiary sectors, businesses are in dire stress and their sustainability is under question today. Business-as-usual models are not likely to survive. In some corporate social responsibility budgets the question of how companies earn has become more important than how we spend. It should also be recognised that businesses are frequently path-dependent on unsustainable routes, hence they need more time, creativity and commitment to transit from their current production and organisational design to a sustainable production and organisational design. Given these constraints, corporations need to phase their transition in two stages.

First, adopt the SDG compass (sdgcompass.org 2018) – i.e. comply, volunteer, partner, integrate and report. The SDG Compass provides details of how a company can integrate sustainable practices into its value chain that can either have positive impacts or reduce negative impacts on sustainability goals. Greening technologies, greening supply chains, and reducing energy consumption methods are part of integration.

Second, corporations can look out for partners in related and similar industries, so that the outputs and waste produced by one company can be the input for another company in a local industrial area and gradually develop a circular economy. This requires high trust and co-operation and not rivalry and competition between the companies of the circular economy.

Third, corporations need to gradually redesign their organisations in terms of their size, scope, technology, management and governance. The language, logic and values of corporations need to change from the paradigm of a competitive market economy to a co-operative social economy. To begin with, corporations need to move away from incentive intensity, very high hierarchical structures and high transaction cost operations to smaller and flat business units with higher interaction intensity within the business unit and minimal transaction costs.

Whether in primary, secondary or tertiary sectors, there is a need to facilitate *organisational design factors* that enable a higher frequency of interactions between the members and owners and a greater number of transactions throughout the year, and that help members find greater value through these interactions and transactions. The minimum support required for the community organisation or *community enterprise system* (CES) at a basic unit of governance includes providing basic common physical infrastructure at the cluster (*Ward/GP*) level, systematically facilitating the collation of surplus agricultural produce from the members of the *Ward/GP* level CES, and undertaking value addition and marketing of agricultural produce. In addition, retail services, community banking, and other essential services should be undertaken by the CES for members of the community, along with co-ordinating and undertaking all livelihood-related development programmes or employment opportunities at the *Ward/GP* level. Medium and small-scale enterprises also need to receive appropriate support to facilitate processing and value addition at the community level. The design needs to facilitate not only financial capital in the short term but also greater social capital formation in the long term to ensure a sustainable wealth creation process.

The nature and type of *institutions* at the district, state, national and global levels can either facilitate or destroy sustainable principles adopted in the other layers, namely production, organisation, governance and ecology, with a culture of relationships at the core. However, even if there are inconsistencies in the institutions at the higher levels, the coherence of institutions at district, block and *GP-Ward* level can greatly facilitate sustainability in the other five layers of dimensions.

Governments need to facilitate institutional linkages and collaborate with local institutions and district officials to pool resources and expertise at the *Ward/GP* level of community organisations. Further, it is necessary to collaborate with the district administration and state departments to invest in a three-tier value-added institutional structure at the *Ward/GP* to district level to help a district achieve all sustainable development goals and make the whole district sustainable. National governments, state governments and local governments need to carefully assess the design of their formal institutions to ensure that they are consistent with the informal institutions of local communities.

While the factors of interventions under different dimensions and the direction of the factors within the given spectrum have been laid down, the actual intervention is indeed simultaneous and not sequential. It is also an iterative process.

Undertaking these simultaneous and engaging interventions and engagements on multiple dimensions in collaboration with the community facilitates trust and co-operation among the people within the community – critical characteristics which lie at the core of a resilient and sustainable community system. The relationships between the actors and actants should be guided by the ecological principles of interconnectedness, interdependence, interactions and caring for the weakest.

23.4.1 Potential Outcomes of SDGs

Simultaneous interventions on the six proposed dimensions or components in any local ecological system or community system would yield a very rich output on the sustainable development goals. Figure 23.3 maps the various goals that could be achieved through the interventions in each of the six dimensions.

It is interesting to observe that most of the individual goals are achieved multiple times through intervention in different dimensions. Table 23.1 shows the dimensions (relationships, institutions, production, organisation, governance and ecology) of interventions and the shading shows the seventeen SDGs that can be achieved.

The last column shows the number of times that each SDG can be achieved through this approach of simultaneous intervention in any given ecological system. Each SDG can be achieved at least twice (200 per cent). Six of the seventeen SDGs can be achieved five times (500 per cent). Three of the SDGs can be achieved four times (400 per cent). Six of the SDGs can be achieved three times (300 per cent).

While interventions towards building relationships and institutions are quite tricky and time-consuming, they seem to play a major role in achieving all the SDGs. *Production* and *governance* each have an impact on as many as ten of the seventeen SDGs. Although interventions on the local ecology seem to have a direct impact on only five of the SDGs, ecology does indeed have an overarching role in sustaining the ecosystem. Only when the average conditions of ecology are not too far off track has the above impact been visualised. Interventions on organisation in the primary, secondary and tertiary sectors seem to contribute to achieving only four of the seventeen goals. However, when the other five dimensions are weak in an ecosystem, individual human actors play only a small role, whereas organisations of people in different sectors have a catalysing effect on other dimensions. For instance, it has increasingly been realised that *producer organisations* are a critical component in building rural agricultural communities in different developing countries of the world.

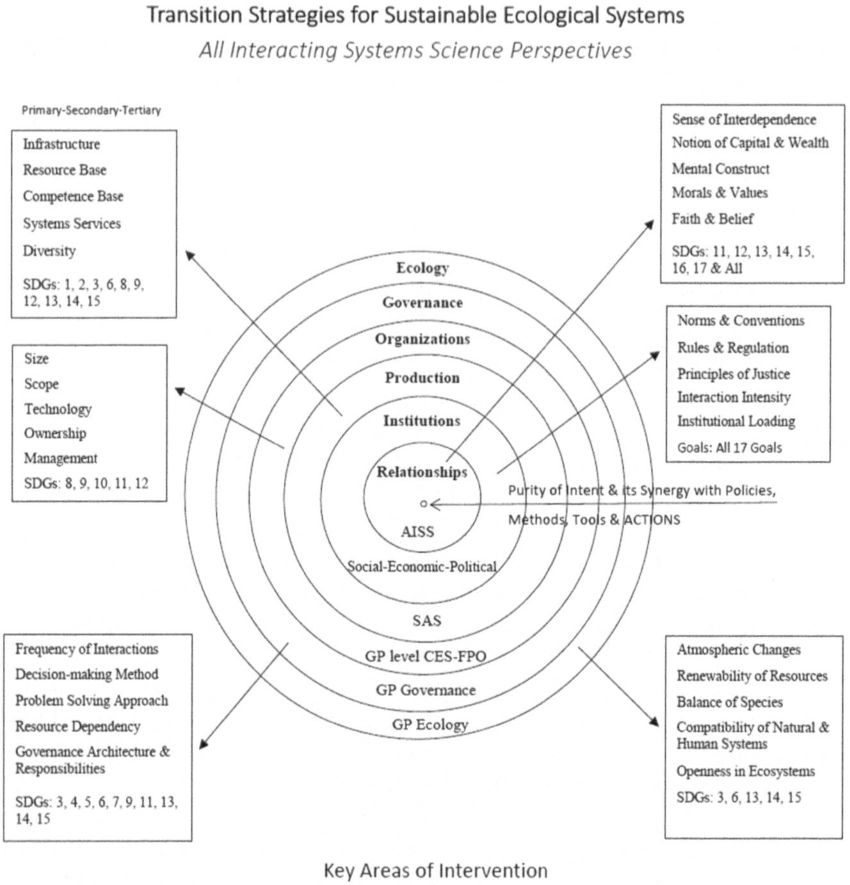

Fig. 23.3 SDGs that can be achieved by simultaneous intervention on the six dimensions in a local ecological system (Ward/GP) level. *Source* The author

23.5 Principles of Simultaneous Interventions at Local Ecosystems for SDGs and Peace

Strategies for the six key areas of intervention include:

(a) *Deepen relationships between members* within the community to foster greater social capital in the community.
(b) *Strengthen and balance the informal institutions* of local communities with formal institutions of the local and higher levels of governance.
(c) *Facilitate sustainable agricultural systems* on farmers' fields, taking the lowest level governance unit (Ward/GP) as a cluster, while, using the SDG Compass, enterprises in the secondary and tertiary sectors develop a sustainable design of value chains for their respective businesses.

Table 23.1 Number of times SDGs can be achieved with simultaneous interventions of the six dimensions if implemented in an ecosystem

Dimensions SDGs	Relationship	Institutions	Production	Organisation	Governance	Ecology	Times
SDG 1							3
SDG 2							3
SDG 3							5
SDG 4							3
SDG 5							3
SDG 6							5
SDG 7							3
SDG 8							4
SDG 9							5
SDG 10							3
SDG 11							4
SDG 12							3
SDG 13							5
SDG 14							5
SDG 15							5
SDG 16							2
SDG 17							2

Source: The author

(d) *Operationalise a GP/Ward level community organisation* as a community enterprise system and federate it at higher levels. A similar system of local enterprise networks of small and medium size can be conceived for the secondary and tertiary sectors. Further, collabate with the district administration and state departments for three-tier value-addition institutional architecture in respective districts.

(e) *Facilitate better planning, convergence and execution* for better health, education, management of commons, renewable energy and infrastructure in the respective GP/Ward (or the equivalent basic unit of governance system in other countries).

(f) *Nurture the local ecology with diverse flora and fauna* in the ecosystem of the basic unit of intervention, (in India, the *GP/Ward*).

The above interventions can be undertaken using the internal design and systems perspectives so that the internal design factors within each dimension are consistent with sustainability principles and those design factors across the dimensions are consistent with each other to ensure that the basic unit of the ecosystem under intervention is sustainable.

The core ecological principles that drive the above design and system's perspective to simultaneous interventions across the six dimensions include all actors and actants in an ecological system that are interconnected, interdependent and interact with each other. Growing evidence from evolutionary biology, game theory analysis, forest sciences, agro-ecology and sustainable agricultural systems shows the inherent interdependence and co-operation between species in an ecosystem for survival. Systems science, in particular, has increasingly been observing the significance of this natural or ecological phenomenon and has developed the logic that the strength of a chain is the strength of the weakest link.

Social sciences, including moral philosophy, theology and jurisprudence, have applied this very logic in their various arguments. Examples of this include the Christian exhortation to "Do unto others as you would have them do unto you"; the Kantian maxim of "Act as you would want all other people to act towards all other people"; Rawl's "theory of veiled ignorance"; the epitome of the Vedas, "Vasudhaiva Kutumbakam" ('the world is one family'); the Biblical instruction to "love your neighbour as yourself"; and "brotherhood", the basis of the Qur'an. The ecological principles of interconnectedness, interdependence and interactions and the follow-up logic to care for the weakest are being increasingly discovered, understood and appreciated by thuman beings as we transit from an anthropocentric view to an 'all interacting systems science' view of our planet.

With regard to research possibilities in the domain of sustainable ecological systems, the field appears to have been largely unexplored and neglected. Therefore, it presents a huge opportunity rather than a crying need to study and facilitate the transition process in both practice and policy. Indeed, there is no dearth of studies relating to individual dimensions –production, organisations, institutions, governance, relationships and ecology. However, most of these studies have broadly followed the reductionist approach of science and hence do not seem to have taken a systems view either within the dimensions or across these dimensions.

Further, past studies relating to factors in each of the above dimensions have often been limited to studying these factor positions at one end of the spectrum, leaving out the multitude of alternatives. In other words, the language, logic and values of analysis have largely been limited to the perspective of competition or a hybrid of competition and co-operation rather than exploring sustainability from the perspective of co-operation (Nayak 2014b).

This chapter has listed only forty core factors or topics for study and deliberation. However, there seem to be far greater numbers of sub-factors and possible combinations of these factors and sub-factors under different contexts that need to be studied to facilitate the process of transition to sustainability across communities.

The methodologies of research to explore alternatives for sustainability are likely to be different from the current methodologies of research and inquiry. Empiricism may not provide clues to sustainability, as in the present reality most practices and policies do not seem to be internally consistent with sustainability principles. Interdisciplinary research, systems thinking, action research and holistic implementation processes could be some of the potential ways to explore new alternatives towards successful transition process.

References

GRI, UN Global Compact, WBCSD, 2015:"SDG Compass, The Guide for Business Action on the SDGs"; at: https://sdgcompass.org/ (15 July 2018).

Nayak, A.K.J.R., 2009: "Optimizing Asymmetries for Sustainability: Design Issues of Producer Organizations", *XIMB Sustainability Seminar Series, Working Paper* 1.0 (Bhubaneswar: Xavier Institute of Management).

Nayak, A.K.J.R., 2014a: "Efficiency, Effectiveness and Sustainability: The Basis of Competition and Cooperation", in: *Vilakshan, XIMB Journal of Management*, XIMB, 14,1: 111–118.

Nayak, A.K.J.R., 2014b: "Logic, Language and Values of Cooperation versus Competition in the Context of Recreating Sustainable Community Systems", in: *International Review of Sociology*, 24,1: 13–26 (London: Routledge).

Nayak, A.K.J.R., 2018: "Economies of Scope: Context of Agriculture, Smallholder Farmers, and Sustainability", in: *Asian Journal of German and European Studies*, 3,2 [Open Access]; at: DOI https://doi.org/10.1186/s40856-018-0024-y.

Piketty, Thomas, 2014: *Capital in the Twenty-First Century* (Cambridge, MA: Harvard University Press).

Rousseau, Jean J., 1762: *The Social Contract* (Amsterdam: Marc Michel Rey).

Schumacher, Ernst F., 1973: *Small is Beautiful: A Study of Economics as if People Mattered* (London: Bonde and Briggs).

Schumpeter, Joseph A., 1943; *Capitalism, Socialism and Democracy* (London: Routledge).

Shah, Tushaar, 1996: *Catalysing Co-operation: Design of Self-Governing Organizations* (New Delhi: Sage Publications).

Teece, David J.; Pisano, Gary; Shuen, Amy, 1997: "Dynamic Capabilities and Strategic Management", in: *Strategic Management Journal*, 18,7: 509–533.

UNSC, 2015: *Technical Report by the Bureau of the United Nations for Statistical Commission* (New York: UNSC); at. https://www.sustainabledevelopment.un.org:content/documents/6754Technical%20report%20of%20the%20UNSC%20Bureau%20(final).pdf (15 July 2018).

Vernon, Raymond, 1971: *Sovereignty at Bay: The Multinationals Spread of US Enterprises* (New York: Basic Books).

Vernon, Raymond, 1977: *Storm over the Multinationals: The Real Issue* (London: Macmillan Press).

Vernon, Raymond, 2009: *In Hurricane's Eye: The Troubled Prospects of Multinational Enterprises* (Cambridge, MA: Harvard University Press).

Chapter 24
Citizen-Led Assessment and the Participatory Approach to Peace Education in Nepal

Rajib Timalsina

Abstract In 2015, in response to long-term instability, Nepal became a federal state with a number of powers devolved to 753 new local governments. These important powers include regulating education and health care. Education is one of the key responsibilities of local governance in such a context. This paper examines education governance from the perspectives of the participatory peace education approach in Nepal. This paper focuses on challenging two aspects of current education governance. The first is the problematic nature of global and national agendas in learning assessment, which don't match the realities and needs of local communities in Nepal, enforcing traditional 'pen and paper' based paper assessments. To this end, this chapter discusses an exemplary project which engaged in participatory pedagogic methods and mobilized hundreds of young undergraduate students to assess the reading and numeracy skills of all children aged between five and sixteen years in the sampled households, and in partnership generated new approaches to learning issues. Second, this addressed the problem of teachers' use of sometimes violent disciplinary techniques. In this second approach, the paper outlines how the project engaged teachers in participatory exercises that trained teachers to use non-violent communication. The researchers and teachers designed a scheme of work for the classes together, discussed solving the classroom challenges non-violently, and reflected on their classroom experiences.

Keywords Citizen-led assessments · Culture of peace in schools · Innovative peace education · Local governance · Non-violent communication · Participatory peace · Peace education

Rajib Timalsina, Assistant Professor, Department of Conflict, Peace and Development Studies, Tribhuvan University, Nepal; Council Member, International Peace Research Association (IPRA); Email: rajib.timalsina@gmail.com.

The author acknowledges the travel support from the IPRA Foundation to make it possible to present this paper at IPRA Conference 2018 in India. I am grateful to Úrsula Oswald Spring, Hans Günter Brauch, Unto Vesa, Sakar Pudasaini, Budanilkantha Municipality, Brihaspati Vidya Sadan School, and Citizen Volunteers for their support in preparing this paper.

Ú. Oswald Spring and H. G. Brauch (eds.), *Decolonising Conflicts, Security, Peace, Gender, Environment and Development in the Anthropocene*, The Anthropocene: Politik—Economics—Society—Science 30, https://doi.org/10.1007/978-3-030-62316-6_24

24.1 Context

The South Asian region, populated by almost two billion people (World Bank 2017), is full of stories related to conflicts and natural disasters. "There were seven countries with active armed conflicts in Asia and Oceania in 2018: Afghanistan, India, Indonesia, Myanmar, Pakistan, the Philippines and Thailand" (Stockholm International Peace Research Institute 2019). According to the *SIPRI Yearbook 2019*, the war in Afghanistan was the world's most lethal armed conflict in 2018, killing more than 43,000 combatants and civilians. The conflict between India and Pakistan over Kashmir is protracted and unresolved. For the period 2015–2018 a total of 274 cases of violence were recorded in the North Eastern part of India (Uppsala Conflict Data Project 2019).

Nepal also experienced a decade-long armed conflict until 2006, which ended with a comprehensive peace accord. The political instability and several notable civil unrests were major features of governance and politics in Nepal for almost a decade, even after the official end of the armed conflict. In 2015, in response to a long period of instability, Nepal became a federal state where some powers were delegated to 753 new local governments. Schedule 8 of the Constitution of Nepal 2015 listed significant responsibilities for local governments, which received several important powers, including regulating education up to twelfth grade, health care, raising revenue through land taxes and budgetary powers (Constituent Assembly Secretariat 2015: 156). All this will impact the lives of millions of Nepali citizens. To a greater or lesser degree, a similar devolution is found in federalised and decentralised countries across the Global South (Karki 2014) – Kenya, Pakistan, Nigeria and India are a few examples. Thus, studying Nepal as a case might provide some ideas for governance practice in recently federalised or decentralised countries in the Global South as a response to structural conflict and violence.

In Nepal, education, in particular, will amount to approximately one in every ten rupees that that local government spends (National Campaign for Education Nepal 2018). Thus, it is impossible to speak of better governance without speaking of better education. More specifically, parents and citizens are concerned with a quality education. For the question, 'why should we care if students are learning?', the reasons – from greater life expectancy to better citizenship, from increased lifetime earning to social justice – are manifold. Thus, it is important to know what type of decision-making process has been adopted by a newly established local body as a new development practice to make sure it achieves the target of the *Sustainable Development Goals* (SDGs), including but not limited to quality education. Similarly, today in a global context, millions of children are at risk of being left behind (UNICEF 2018), as millions of adults are migrating from developing countries in search of job opportunities and a better life. The UNICEF Working Paper claims that children 'left behind' fail to learn basic skills and have inequitable access to education (UNICEF 2018). But even when children are in school, there is a chance that children may not be learning. These challenges are widespread. According to the Education Commission report, only one in ten children in low-income countries is on track to gain basic skills by 2030 (The Education Commission 2016). Moreover, the obstacles to learning disproportionately affect marginalised children – those in poor households, rural conflict-affected areas (especially girls), and children with disabilities. This is considered a serious root cause of the conflict.

All this contextual information leads to the eventual remarks that more and better learning data are required to diagnose where and why learning gaps exist, and what actions need to be taken to close the gaps, in the belief that this will help to identify alternative ways to solve the conflict by addressing the root causes of the conflict and promoting positive peace. Recent education data show that millions of children are 'left behind', thousands are 'out-of-school', and that even though many children are going to school, their achievements in grade-level assessments are very poor (UNICEF 2018; Education Review Office 2017; and UNESCO Institute of Statistics 2016). Two alternative models of peace education – citizen-led assessments at community level, and non-violent communication tools in classrooms – tend to address these fundamental problems of the learning crisis.

24.2 Methodology

The purpose of this paper is to examine education governance from the perspective of participatory peace education. This text challenges two aspects of current education governance in Nepal – 'pen and paper' learning assessments and 'violent' classroom environments – and proposes two alternative peace education pedagogical approaches to tackle those problematic approaches. Thus, this chapter relies on two peace education pilot projects.

The first pilot project engaged hundreds of young undergraduate students to assess the reading and numeracy skills of all children aged between five and sixteen years in sampled households (regardless of their school-going or out-of-school status), and in partnership generated new approaches to assess their learning achievements. The second project addressed the problem that teachers sometimes use violent disciplinary techniques in school classrooms. Here the chapter discusses the project that engaged teachers in participatory exercises by training them to use non-violent communication.

Collating the observations of the fourteen-day Citizen-Led Assessment Fieldwork alongside notes from more than thirty-four conversations with local officials, youths, teachers and citizens, this text examines several questions to be considered in the implementation of learning assessments and non-violent communication in decentralised contexts. During the citizen-led assessment fieldwork, this research team visited 119 schools and 3820 households in thirteen neighbourhoods in the Budanilkantha Municipality in Kathmandu. The researcher worked with twenty teachers of Brihaspati Vidya Sadan School in Kathmandu for six months in the pilot project of non-violent communication in schools. Although the researcher has developed this paper based on both qualitative and quantitative data, this is more a reflection-based paper developed from an eclectic qualitative approach.

24.3 Citizen-Led Assessment for Participatory Education Governance

As in other parts of the developing world, primary school enrolment rates have improved in Nepal during the last fifteen years – net enrolment has increased from 72.61% in 2000 to 96.33% in 2016 (UNESCO Institute of Statistics 2016). As a result, Nepal has almost met Goal 2 of the *Millennium Development Goals* (MDGs) set in 2001: "achieve universal primary education". After the introduction of the SDGs in 2015, the fourth goal aims to "ensure inclusive and equitable quality education and promote lifelong learning opportunities". This raises some important questions: how do we track and measure the quality of education? The answer could be 'through the learning assessments'.

"Many different written methods of assessment have been developed, each with its own advantages and disadvantages in terms of reliability, validity, educational impact, costs and acceptance" (Schuwirth/Vleuten 2004). In Nepal, the *Education Review Office* (ERO) has been conducting various forms of educational assessment, including the *National Assessment of Student Achievement* (NASA) at various grades of school education. The Education Review Office provides assessment frameworks, audits the performance of educational institutions, and carries out *Early Grade Reading Assessment* [EGRA] (Education Review Office 2017: 4). The United Nations agencies, specifically UNICEF and UNESCO, have also conducted occasional learning assessments in Nepal. The common features of all these existing learning assessments are: (i) school- and grade-based, (ii) mostly 'pen and paper' written tests, (iii) classroom set-up.

Citizen-Led Assessments (CLAs) emerged in India in 2005 as a way to raise awareness and advocacy around low learning levels and to act as a force for bottom-up accountability and action that would improve education quality and learning (ASER Centre 2017; ASER Pakistan 2017). To understand CLAs well, both terms need to be examined independently. 'Citizen-led' means that all activities are conducted by ordinary citizens. These citizens are usually volunteers who take the opportunity to connect with communities across the country and gather data that could contribute towards improving learning outcomes. "What distinguishes CLAs from other assessment tools is that they combine learning measurement approaches (from the education field) with citizen-monitoring approaches (from the transparency and accountability field) to engage ordinary citizens in the assessment of children's learning" (Plaut/Eberhardt 2015). The second term, 'assessment', means checking the results of something. In this case, we are checking whether young learners can read in English and Nepali, as well as do maths at the second- or third-grade level. One of the major methodological strengths of citizen-led assessments is that they focus on measuring only simple learning outcomes in reading and maths. The citizen-led assessments are not led by the government or any technical specialist. Though technical experts design the methods, the assessments are led and executed by ordinary people.

In Nepal, the citizen-led assessment was piloted in 2017–18 and is called the *Annual Status of Education Report* (ASER). ASER is citizen-led because it is a way for all of us, especially young people, to engage in making the education system better for everyone. After implementing the ASER pilot survey, the author received information about three kinds of actors involved in education: schools; households and parents; and children and students.

The data related to schools include the detailed infrastructures available in the schools, such as classrooms, toilets, computers, educational materials etc.; data on management issues, such as the existence of a *School Management Committee* (SMC), *School Improvement Plans* (SIP), teachers' appointments and students; the enrolment and actual attendance of students; water, sanitation and hygiene facilities in the schools; and internet, electricity and other facilities necessary for a modern education at school.

The data related to households and parents include assets and resources in the house, including ownership of motorcycles, fridges etc.; the number of home-owners versus renters in the area; the number of permanent residents in the municipality versus temporary residents; and the education of parents and its impact on the learning of children.

The data related to the children or students includes: the ability to read and understand the English and Nepali language – the *Annual Status of Education Report* (ASER) in Nepal has tried to measure their ability to recognise alphabets, read words, sentences and paragraphs and to understand what they have read; and the ability to do the basic maths necessary for everyday survival.

Now, here, a question comes. How does the ASER assessment work? In Nepal, we had an ASER Nepal research team which provided technical support including trainers, whereas all the work was done by local volunteers under the leadership of the Mayor and Ward Chairs in the local municipality. The process which was followed in Budanilkantha Municipality in 2018 was as follows:

1. The local government identified 200 volunteers between the ages of eighteen and twenty-six who were interested in doing socially productive work. The most effective volunteers were undergraduate students who had to be able to read and understand daily newspapers such as *Kantipur* and *Kathmandu Post* in both Nepali and English.
2. The ASER research team trained all 200 volunteers for five days. The training was conducted in separate locations based on the Wards that each group needed to cover. Required volunteers were chosen based on a practical and written test at the end of the training sessions.
3. Ten volunteers in each ward (neighbourhood unit in Nepal) were mobilised. These volunteers visited 300 households in each ward.

    - The 300 households in each ward were scientifically selected with the help of expert statisticians who are university professors.
    - All children aged between five and sixteen in the sampled households were tested one-by-one on English reading, Nepali reading and maths.

4. Two volunteers in each ward were mobilised to visit schools in their assigned ward to gather information. All schools in the municipality were visited and the standard information was gathered.
5. Data from the households and schools were compiled by the ASER research team and presented to the Mayor's office and also to the group of the young volunteers.

The research team, the young volunteers and the municipality team together developed the ASER 2018. The report aimed to answer a simple but important question: are the children of Budanilkantha Municipality able to read (in Nepali and English) and do maths at grade 2 level? As a secondary goal it also gathered data about the status of schools and youths so that this evidence could be used for action i.e. to formulate policy that allows the local government to secure the education and thus the future of children under its care.

The results showed cause for concern. About 40% of 8–16-year-old students could not read the Nepali story with fluency, about 25% could not read the paragraph in English, and about 30% could not perform a grade 2 division problem. The subject in which government school students in Budanilkantha Municipality performed significantly lower than average was English. Government school students performed at about average in maths and better than average in Nepali. These learning crises are observed in both government and private school students. The evidence in this report suggests there is a systemic problem in the education sector. Thus, policy efforts to improve learning outcomes should encompass all types of school – government, trust and private for profit – and their teachers.

Besides learning outcomes, this report also gathered data about the households and schools. Access to education looks promising, with 99% of school-age children reported as being enrolled in school. An analysis of the last 1% did not reveal any discernible pattern in out-of-school children. Thus getting this last 1% to school will require special efforts at the ward level. Not many state representatives besides ward chairs/members have the kind of granular local knowledge that is needed to help these children get to school. The last 1% is hard work, but ensuring that every child goes to school is worth working hard for.

There were 102 registered schools in the municipality. Three of those schools were non-operational at the time of the fieldwork. One was a religious school, 10 were non-profit or trust schools, 18 were government-funded community schools and 71 were private for-profit schools. About 14% of school-going children were reported as studying in government-funded schools. Government schools taught a larger percentage of the municipality's population in higher grades. While 13% of children attended government schools in grade 1, that number grew to 20% by 8th grade. Government schools did as well as average on many infrastructural measures, such as the use of chalkboards, the electrification of classrooms, and the presence of libraries. On some infrastructural measures – such as having usable toilets, sufficient classrooms and access to filtered drinking water – government schools performed better than average. This reflects an investment by both the state and civil society in school infrastructure over the past decade.

Yet, government schools underperformed on some data points that measured the management and upkeep of this infrastructure. While government schools had more early-grade-friendly hand-washing basins in restrooms (94.7% in the community versus 83.1% in all deemed researchable by early grade students) they also reported more hand-washing basins in restrooms without soap (53.3% in government schools versus 36.7% in all schools). These challenges require not more infrastructure investment but better management and greater accountability. Another area that could be improved with better management in government schools was student attendance. During the time of data collection, only 73.8% of students attended community schools. Across all schools that number was 91.8%. Since it is not possible to teach students who are absent, it is important a greater effort be made to get students to school. Three areas for additional infrastructural investment in community schools were visible in the data – boundary walls, playgrounds and computer labs.

Shockingly, only about 1% of respondents reported having had access to formal vocational or skill-development training in the municipality. Improving access to relevant vocational and skill-development skills, particularly for youths, would therefore be a sensible strategy for the municipality to adopt.

The municipality was slightly older than the nation on average. Nationally the under-24 population is just over 50% compared with 43.5% in Budanilkantha Municipality (Center Bureau of Statistics 2013; ASER Survey 2018). The municipality was also fairly well connected. 84.3% of the households had an internet-capable mobile phone. Finally, two data points stood out as low-hanging fruits that the Municipal government could impact with a limited resource commitment. At 55%, the birth registration rate for children aged 0–4 was better than the national average of 42% (World Bank 2017) . Nevertheless, birth registrations offer an area where the local government can focus its energy and demonstrate quick progress in one of the SDG indicators. Only one in three youths aged between 17 and 24 reported having a bank account. Access to finance and learning responsible fiscal habits is critical to prosperity. Thus an increase in youth bank accounts may be a shared goal in which the private sector *Bank and Financial Institutions* (BFI) and the municipality can collaborate.

After the submission of the report, the municipality government organised a consultative workshop and identified various intervention methods to address the challenges raised by the citizen-led assessment reports. The research team did not intervene directly after the assessment report. The most important argument from the process could be that it set the foundation for 'evidence to action' practice at the local level. The report indicated the areas where the municipality performed well and the areas which needed improvement. Recently, I noticed that a civil society group called *Collaborative School Network* (CSN) is running a few government schools to support the efforts to improve the learning achievements in the marginalised rural communities of that municipality. Similarly, the findings from the report were discussed at the teachers' meetings in the municipality, which allowed them to develop strategies for improving the learning outcomes as well as managing the organisational issues and resources pointed out during the assessment process. So there are various interventions and policies at the municipality level which are informed by the report from the citizen-led assessment and which are partly or fully trying to respond to the identified gaps.

24.4 Non-violent Communication for a Culture of Peace in Schools

After the learning crisis, the second significant problem in Nepalese schools is related to corporal punishments and verbal abuse. It is also not unusual for the situation to trigger several types of conflict, including physical punishments, bullying and verbal violence from teachers against students, or among students, or between teachers. And at the same time, the school dropout rate is very significant as the children reach higher grades (UNICEF 2014). Mostly, parents and children blame the violent (both physical and verbal) and discriminatory school environment and a poor economic family background for school dropouts (based on conversations with parents and youths in May 2018 in Budanilkantha and Hetauda Municipalities). Thus, school-based violence and conflict are major reasons why children leave school. "Managing conflicts in a constructive way creates the foundation for a good working environment in which democratic values are interwoven with teaching, thus contributing to learning" (Gothlin/Widstrand, 2012: 2–3). The government has made primary education free and compulsory (Constituent Assembly Secretariat 2015), but just making the schooling legally compulsory will not ensure that children will learn what they are supposed to learn. Along with the concern about the learning focus of teaching, it is also very important to make sure the next generation grows up in a dignified and just learning environment.

In such a context, the promotion of the culture of peace in schools is crucial in Nepal to help create a conducive and dignified learning environment. The researchers could particate in a pilot project in collaboration with members of a non-violent communication practice group in Kathmandu that supported a non-violent communication project at Brihaspati Vidya Sadan Kathmandu to promote a culture of peace in schools. Marshall Rosenberg has defined non-violent communication as "communication skills that strengthen our ability to remain human" and "a way of communicating that leads us to give from the heart" (Rosenberg 2003: 3). In other words, a non-violent communication approach supported the researcher to reframe the way people verbally express themselves and receive others' expressions.

Teachers' use of violent disciplinary techniques is frequently reported in Nepali schools. Many parents and youths regard it as normal in school and consider it essential for learning (based on conversations with parents in Budanilkantha and Hetauda Municipalities in May 2018). At the same time, almost all students from grades 6 to 8 who were consulted by the researcher when the project started responded that they felt very uncomfortable, sad, angry and sometime frustrated because of the physical and verbal punishments. The new project was introduced at that school to increase the awareness of students, teachers and parents of a culture of peace in schools. The aim of the non-violent communication project in school was to help make communication easier, avoid misunderstandings, and ensure classrooms are a dignified place for teaching and learning by finding mutual solutions to conflicts.

At Brihaspati Vidya Sadan School, twenty teachers from grades 6–8 were involved in the process. At the beginning, there were three days of in-depth training on the concepts of non-violent communication in school for twenty selected teachers. After that, the team met once a week for two hours for six months for regular practice. During that weekly meeting, the researcher took part in the discussion and practice as well. The group explored how the teaching and learning process can occur without physical or verbal punishments. The teachers were also encouraged to tell the group about conflicting issues or the challenges they faced in the classroom and at school, so that the whole group could discuss possible ways to handle the issue or challenge on the basis of non-violent communication values. Every week the teachers practised various exercises, including understanding themselves and others, handling anger, communicating with children respectfully, and differentiating between two types of language: more power-with language (request) and more power-over language (demand).

During the last month, the researchers and teachers co-designed a lesson plan for the classes to evaluate the impact of the pilot project. A scheme of work was developed to examine the understanding of civic-national values by the children in their class using art-based tools. Playback theatre, graffiti, poetry, songs and music were used as teaching pedagogy on evaluating the understanding of the children rather than the conventional data collection method such as interviews, surveys and group discussions. In September 2018, the researcher evaluated the work in two schools: one in the school where non-violent communication was introduced and one in another school where this intervention was not conducted. Twenty students from grades 6–8 at the Brihaspati Vidya Sadan School, where the project was implemented, were asked "what is community for you?" The researcher received eighteen different responses from participating students. For example, one student responded, "For me, community is a group like a basketball team, where one player passes the ball, and another friend can score." Another student answered, "Community is a group of people made of love and support." Another response was: "Community has diverse people like a farmer, teacher, doctor and musician. The farmer grows food for us, and the doctor cures the farmer if they fall sick. We all are dependent on each other."

In another school where the intervention was not implemented, the researcher followed exactly the same process of evaluation among twenty children from grades 6 to 8. Seventeen students offered exactly the same definition of a community and only three responses were different. The answers of all seventeen students were an extract from their textbook which the students had memorised without knowing the contextual meaning of the definitions. It was interesting to observe that the class teacher marked those three different definitions as wrong and marked those seventeen definitions as correct. He argued that all answers must be the same as in the textbook which he taught them. The teachers at the non-intervention school disliked diverse and creative answers, and they wanted their children to be disciplined and follow instructions exactly, whereas in the intervention school, diverse subjective and creative answers of the students were easily accepted after the project intervention.

24.5 A Participatory Peace Education Approach as Part of Education Governance

This paper is based on two examples where the research team challenged two aspects of current education policy in Nepal. The first is the problematic nature of global and national agendas in learning assessment, which don't match the realities and needs of local communities in Nepal, enforcing traditional 'pen and paper' based assessments. The assessment part in this paper demonstrates some of the techniques developed as a result, offering targeted interventions designed to report the situation of children's individual skills and understanding. The *Annual Status of Education Report* (ASER) as a citizen-led assessment research project helped the local government design a plan to gather the data and disseminate the information to the community, schools and teachers. Despite a growing focus on data collection, there is little indication that decision-makers are systematically using data and analysis to inform policies or their decisions. Furthermore, the focus on producing nationally and globally comparative metrics of learning often mean that local decision-makers do not have access to data with enough granularities to produce useful and actionable snapshots of learning and identification where action needs to be taken.

To answer the question, "What can the local government and its leadership do with the data of the survey?", based on the research experiences, it can be said that the *Annual Status of Education Report* (ASER) has tried to support the local government leadership in using the data to guide their own priorities to:

1. Develop more effective policies for the teaching of English in schools. Effective use of English is an important demand of parents across Nepal.
2. Identify which areas and communities need extra support to ensure equity in society.
3. Get a fair and representative idea of access to computers, internet and modern technologies in schools, and thus formulate policies to increase access to the essential aspects of modern life.
4. Identify the challenges and challenge areas facing the constituency and thus decide which sectors/areas to encourage NGOs to work in.
5. Mobilise local youths to understand the realities of education in their locality. Identify exceptional young people within this pool and train them to act on the problems they see.
6. Mobilise local youths and support in voluntary tuition in high-need communities/areas.

The *Annual Status of Education Report* (ASER) project has tried to advance the dissemination of data through fostering a collaborative relationship between civil society and the local government in deciding how, where, and when to disseminate the data and in what format. This had an advance dissemination by leveraging the opportunity to meet and to discuss the data, creating an environment for collectively understanding the problem and exploring community-led solutions. The first project has tried to work closely with one local government authority in Nepal to jointly implement the innovative, low-cost, citizen-led model of conducting household-based learning assessments, ensuring that the data produced is easy to understand and actionable. Five key audiences have been identified during the research process: local government officials; local community leaders; head teachers and teachers; parents; and volunteer data collectors.

In this process, the research analysis explored the motivations of local governments in undertaking citizen-led assessments. These motivations are multifaceted and extend beyond a narrow focus on learning. Several factors are explored that the local government noted as being of interest to them, from the idea of skilling local youth and gathering comprehensive data to looking for opportunites for visibility. The research project with a local government model also explored ways to leverage assessors as a means of ensuring accountability. Several challenges were noted and observed with this approach, and reflect on additional mechanisms that may be necessary to encourage active citizen assessors to become agents for accountability.

Similarly, the second alternative peace education model focused on establishing a culture of peace in schools by addressing the problem of teachers' use of sometimes violent disciplinary techniques in schools. The teachers of one school along with the researcher tried a participatory exercise that used non-violent communication techniques. Teachers and researchers focused on changing the culture of physical and verbal punishment which prevails in the school culture of Nepal. In a school project aiming at a culture of peace the teachers and the researcher tried to work together to establish a dignified learning environment as non-violent communication practitioners.

24.6 Conclusion

This paper has assessed governance in the education field in Nepal from the perspectives of a participatory peace education approach. The researcher has used two innovative intervention models: i) a citizen-led learning assessment model and ii) non-violent communication tools to promote a culture of peace in schools. The findings from both projects revealed that evidence only leads to action if it is easy to understand, discuss and share. Based on both activities, it is evident that this level of clarity and collaboration is crucial. If outcomes/findings from the research works are not easy to understand and actionable by stakeholders from national to local government, from head teachers to parents, then the research findings are of limited value.

This research project concludes that citizen-led assessment methods of measuring the learning outcomes of children are a very useful tool for new local bodies to practise evidence-based decision-making in Nepal. The adoption of such a model could be a solution to the ongoing quest for new techniques for a peaceful decision-making process at the level of local governments, which indirectly contributes to making local government, local youths and civil societies accountable for addressing the causes of structural violence. Engaged citizens can both hold local governments accountable to ensure a better learning experience and serve as resources that can contribute to improved learning. The citizen-led assessment described in this chapter did not just produce research findings to make decisions through peaceful means but also trained a pool of young people who can be used to improve the quality of education locally. Similarly, the culture of peace in schools helps to make the school environment more conducive to teaching and learning activities and minimises both physical and psychological violence. To retain children in school and to address the problem of school-dropouts, it is very important to ensure schools are child-friendly and peaceful.

References

ASER Center, 2017: *Annual Status of Education Report (ASER) 2017*; at: https://img.asercentre.org/docs/Publications/ASER%20Reports/ASER%202017/aser2017fullreportfinal.pdf (5 August 2018).

ASER Pakistan, 2017: *ASER Report 2017*; at: https://aserpakistan.org/index.php (5 August 2018).

Central Bureau of Statistics, 2013: *Nepal Population Census 2011* (Kathmandu: Central Bureau of Statistics).

Constituent Assembly Secretariat; 2015: *Constitution of Nepal 2015*; at: https://www.ilo.org/dyn/natlex/docs/MONOGRAPH/100061/119815/F-1676948026/NPL100061%20Eng.pdf (15 October 2018).

Education Review Office, 2017: "Editorial" in: *Nepalese Journal of Educational Assessments*, 2,1.

Gothlin, Marianne; Widstrand, Towe, 2012: *Nonviolent Communication: A Way of Inspiring Respectful Dialogue in Schools* (Stockholm: Skolande).

Karki, Budhi, 2014: "State Restructuring and Federalism Discourse in Nepal", in: Budhi, Karki; Edrisinha, Rohan (Eds.), 2014: *The Federalism Debate in Nepal: Post Peace Agreement Constitution Making in Nepal*, vol. 2: 1–22 (Kathmandu: United Nations Development Programme [UNDP]).

National Campaign for Education Nepal, 2018: *Brief Analysis of National Education Budget for the F/Y 2018/19.*

Plaut, Daniel; Eberhardt, Molley Jamieson, 2015: *Bringing Learning to Light: The Role of Citizen-Led Assessments in Shifting the Agenda* (Washington, DC: Results for Development).

Rosenberg, Marshall B., 2003: *Nonviolent Communication: A Language of Life* (Encinitas, CA: PuddleDancer Press).

Schuwirth, Lambert W.T.; van der Vleuten, Cees P.M., 2004: *Different Written Assessment Methods: What can be said about their Strengths and Weaknesses?* (Oxford: Blackwell Publishing Ltd.).

Stockholm International Peace Research Institute, 2019: *SIPRI Yearbook 2019: Armament, Disarmament and International Security* (Oxford: Oxford University Press).

The Education Commission, 2016: "The Learning Generation: Investing in Education for a Changing World"; at: https://report.educationcommission.org/report/?cn-reloaded=1 (15 October 2018).

UNESCO, Institute of Statistics, 2016: "Statistical Tables".

United Nations Children's Fund (UNICEF), 2014: "UNICEF Annual Report 2014".

United Nations Children's Fund (UNICEF), 2018: "UNICEF Working Paper: Children 'Left Behind'"; at: https://www.unicef.org/media/61041/file (19 November 2018).
Uppsala Conflict Data Project, 2019: "Conflict: India-Western South East Asia"; at: https://ucdp.uu.se/conflict/13653 (10 October 2018).
World Bank, 2017: "World Bank Annual Report 2017".

Chapter 25
Can Infrastructure-Based Development Bring Peace to West Papua?

Cahyo Pamungkas

Abstract This chapter discusses the question "To what extent did infrastructure-based development projects in West Papua, especially in road construction, provide positive impacts on peace-building between 2014 and 2018?" Since 2014, the Indonesian government under President Joko Widodo has given more attention than previous governments to coping with separatism in West Papua. The separatist movements have already been active since 1965 and have now transformed from armed struggle into political campaigns. One of the national government efforts is promoting many road construction projects to increase connectivity between districts in West Papua. Of the 3259 km of road construction planned in 2014, the government had constructed almost half (48%) by 2018. My argument is that increasing infrastructure-based development has a minor impact on improving livelihoods among the indigenous West Papuans, so this policy is not entirely successful in supporting peacebuilding or coping with separatism. Previous studies mentioned that the Indonesian government has implemented a hybrid peace-building approach but failed to prevent political violence in West Papua. My research suggests that either a liberal or an illiberal peacebuilding approach is necessary but not sufficient to end the political violence. The national government should conduct a mediated political settlement via dialogue with the separatist group. The infrastructure-based development is not enough to promote peacebuilding without a dialogue between Jakarta and West Papua.

Keywords Development · Dialogue · Infrastructure · Peace-building · West Papua

25.1 Introduction

Two mainstream approaches in peace and conflict studies are *illiberal* and *liberal peace*. The former conceptually refers to an approach in which the central government makes use of *illiberal* methods for bringing peace, such as re-establishing existing patrimonial networks, increasing the presence of security forces and providing

Mr. Cahyo Pamungkas, researcher at Indonesia Institute of Sciences;
Email: cahyopamungkas@gmail.com.

Ú. Oswald Spring and H. G. Brauch (eds.), *Decolonising Conflicts, Security, Peace, Gender, Environment and Development in the Anthropocene*,
The Anthropocene: Politik—Economics—Society—Science 30,
https://doi.org/10.1007/978-3-030-62316-6_25

impunity to corrupt local officials in exchange for their loyalty to the central government (Smith 2014: 1511–1512). This approach is often considered inadequate, although necessary, in creating negative peace or the absence of violence (Galtung 1969: 183). In a *liberal peace* approach peace is attained by promoting legal and security reforms, market deregulation, political decentralisation, and empowering democratic institutions. In academic discourses, Bjorkagen (2016: 17) indicates that most studies have pointed to a dominance of the liberal peace approach by focusing on the implementation of the liberal peace model. However, in reality, there has been an illiberal turn in conflict management since the end of the Cold War (Heathershaw 2008; Joshi et al. 2014; Richmond 2006; cf. Lewis et al. 2018: 2). Indonesia is an example of the application of the illiberal peace approach in resolving religious and ethnic communal conflicts in Maluku (Smith 2012). The national government restored the power of local elites and re-established a patronage network with them.

However, in dealing with the secessionist movement of indigenous West Papuans, Indonesia implements a hybrid peace approach which combines both liberal and illiberal approaches. West Papua, at the eastern corner of Indonesia, has uninterruptedly undergone political violence despite having experienced a democratisation process during the last decade. Limited international interest in the cause, immense national pressure on the endeavour of maintaining national integrity, and a high risk of conflict have given rise to illiberal peacebuilding, namely a security approach, with necessary liberal adaptations (Smith 2012). Various examples of liberal policies implemented by the Indonesian government following the 1998 political reforms include the Special Autonomy concession, the provision of Special Autonomy Funds, which have been distributed since 2002, and politically affirmative action for indigenous West Papuans.

The administration of President *Susilo Bambang Yudhoyono* (SBY) attempted to accelerate the economic development in this region by establishing an Office for Development Acceleration. The unit had a military general – who commanded the military operations in Aceh (2003–2004) – as the person in charge. Between 2012 and 2014, the office only succeeded in promoting affirmative policies in the field of education. Under the office's programme, the government granted many scholarships for West Papuan students to study in secondary schools and universities outside West Papua. The government also provided a quota for West Papuan youths to join the military and police forces. In 2015, President Joko Widodo, the successor of SBY, dissolved this office due to the consideration that it had failed to accelerate development in other sectors, such as basic health services, the local economy, infrastructure, and political development. Many indigenous West Papuans saw the President, who gained the majority vote in West Papua, as a new hope for peaceful conflict resolution. He then visited West Papua more than three times every year to ensure the progress of Papuan infrastructure development, in particular the construction of the trans-Papuan highway.

Based on the above description, this chapter aims to answer: (i) to what extent does infrastructure development, especially the Trans-Papua Highway construction, during the President Joko Widodo period, improve the life and livelihood of indigenous West Papuans and their ecological sustainability? (ii) to what extent can such a development support the peacebuilding process? and (iii) which peace-building concepts are suitable for analysing the West Papuan secession conflict? The thesis of this text is that infrastructure development is a necessary but not a sufficient condition to end the political violence as well as the secessionist movement in West Papua. In addition, we need to formulate a new concept of a peace approach that is more appropriate for the West Papuan conflict.

25.2 Methods

There are more contemporary conflict studies about West Papua, e.g. Hernawan (2013), Rutherford (2012), and Kirksey (2012), than peace studies. Widjojo et al. (2008) have filled the literature gap in peace studies regarding the resolution of the West Papuan conflict. Their study proposes several ideas for the Papuan road map to peace, i.e. a peace dialogue, a reconciliation to resolve political violence, a recognition of the basic rights of indigenous West Papuans, and a culture-based socio-economic development. According to Smith (2012), these four ideas can be found in the boundary between liberal and illiberal approaches. Dialogue, for example, is supposedly found between two diametric perspectives, i.e. the state security approach on one side, and political decentralisation or more significant development allocation on the other. The idea of reconciliation lies between the demand for a judicial process for human rights abuses and the implementation of security measures in this region.

This research differs from that of Widjojo et al. (2008) and focuses on the effectiveness of development programmes as a strategy of peacebuilding in West Papua. Additionally, it examines the mainstream perspective in conflict studies that violence and peace are formed by economic dimensions. Massive economic development in conflicted areas is thus supposed to produce more peaceful conditions compared with poor development. In short, progressive economic development, conceptually, is conducive to peace, resulting in the outcomes of improvements in the society's livelihood, capital accumulation by the society, an increase in public services provided by the state, and ecological and social sustainability (International Alert 2015: 8–9).

Economic as well as social development is needed to reduce any kind of conflict, including secessionism, if such conflicts are realistically rooted in competition over scarce resources (Coser 1956: 49). The economic and political domination of a majority over minority groups often produces marginalisation of the latter, prompting them to aspire to the right to self-determination. The chief characteristics of independence can be interpreted as being free from repression, marginalisation, discrimination, and poverty. If a broader concept of poverty is used, a bad economy, lack of political freedom and civil rights, and poor development contribute as drivers to either direct or structural violence (Sen 2000: 3–4; Merino/Condejas 2017: 921).

Reconstruction and social economic development tend to support prolonged sustainable peace in post-conflict areas. The reconstruction of infrastructure and the provision of social services are necessary to improve economic activities and get people involved in peace processes (Panic 2005: 8–11). Due to a rise in per capita income and economic growth, the possibility of violent conflict will decline (McIntosh/Buckley 2015: 4). In general, it can be argued that increasing wealth is likely to make people support peace and reduce war (Humphreys 2003: 2).

Peace does not merely mean the absence of violence, or negative peace, but also includes positive peace, the absence of structural violence (Galtung 1969). The Peace in Papua Network, referring to Galtung's definition, translated the indicators of Papua Peace Land, i.e. the right to freedom of movement, recognition of the fundamental rights of West Papuan indigenous peoples (including rights over land ownership and cultural values), resources exploitation that maintains natural sustainability and local wisdom, health services, and the provision of basic education (PPN 2014: 8–48). In this study, peace is explained by the absence of political violence, improved livelihoods, the preservation of social and cultural life, and sustainable ecology.

This peace concept also supports the idea of Brauch (2009a: 69) for a new peace policy that combines both sustainable peace and sustainable development to overcome the essential problems of humankind. As Brauch (2009b: 21) mentions, the dimensions of security in the post-Cold-War era should include economic, societal, and environmental dimensions, and also sectionalisation of the application of security to various fields, such as livelihood and health. Therefore, every peace policy is not only intended to cope with violent conflict, but also to keep harmony between social groups and between a society and its environment through consensus (Oswald Spring 2000: 235). My argument here is that the infrastructure project in West Papua under President Jokowi is likely to increase positive impacts on indigenous groups and reduce the political violence in this region.

This article is based on my fieldwork supported by the Indonesian Institute of Sciences and the Asia Foundation. The data collection methods involved observations and interviews at the research sites in May and June 2018, in the regencies of Jayawijaya and Yalimo of West Papua. The informants of this research were people who make active use of the Trans-Papua Highway, both indigenous and migrant. They included people of various professional backgrounds, such as village chiefs, tribal leaders, religious leaders, district heads, local marketplace traders, public transport drivers, and plantation labourers. The data collection method used a multi-site approach (Marcus 1995: 97), following the movements of human beings and commodities along the highway. The approach assumes that the reality of a place always has a relational nature to realities in other sites. The dimensions to be studied here are the histories of the highways, improvements to the livelihoods of local societies, the fulfilment of basic services, and the creation of a feeling of security among indigenous West Papuans between 2014 and 2018.

25.3 Findings

25.3.1 The History of the West Papuan Conflict

This text uses the term 'West Papua' to refer to the western part of the island of New Guinea, which has belonged to Indonesia since 1963. West Papua shares cultural traditions and ethnicities with the eastern part of New Guinea, which is the independent state of Papua New Guinea. In 2003 the Indonesian government divided West Papua into two provinces, namely Papua and Papua Barat.

Before 1824, West Papua was under the cultural and political influence of the Sultan of Tidore of the Moluccas. However, the 1824 London Treaty between the Netherlands and the United Kingdom stipulated that the Netherlands would control the western part and the United Kingdom the eastern part of the island. The border between the two territories is at the 141st latitude as their common boundary, according to a treaty between the two colonial powers in 1865 (Drooglever 2009: 8–11). When the Netherlands recognised the Republic of Indonesia's sovereignty in 1949, they were still in dispute over the political status of West Papua. Indonesia insisted that the Dutch had to cede all territories of the former Netherlands East Indies, including West Papua (Kahin 1953: 165–166; Elisabeth et al. 2004: 22–23). However, the Dutch said that the population of West Papua was culturally not part of Indonesia and has the right to self-determination (Saltford 2000: 71–72).

The dispute turned into a military confrontation in 1961, when the Netherlands established the *Nieuw Guinea Raad*, recognised political symbols of Papua, and formed the Papuan Volunteers Force (*Papoea Vrijwilligers Korps*, PVK). President Soekarno, with military support from the Soviet Union and Eastern Bloc countries, declared the *Trikora* (Three People's Commands). The United States then mediated the dispute and military confrontation through the 1962 New York Agreement between Indonesia and the Netherlands. The latter stipulated the power transfer from the Netherlands to Indonesia facilitated by the *United Nations Temporary Executive Authority* (UNTEA) and the implementation of the Act of Free Choice to realise the aspirations of the West Papuans.

The Act of Free Choice was implemented in 1969, after President Soekarno was ousted by General Suharto, who had a closer political relationship with Western countries during the Cold War. The results of the Act of Free Choice showed that 99% of West Papuan people representatives agreed to integrate Papua into Indonesia. However, West Papuan independence activists argued that the Act disrespected the 1962 New York Agreement that required a 'one man one vote' plebiscite. The so-called Act of Free Choice only involved 1024 persons appointed by the Indonesian government to represent the whole West Papuan population.

After their integration into Indonesia, some indigenous West Papuans continued fighting for independence via armed resistance. Between 1970 and 1996, the Indonesian government responded with military operations, whereby civilian casualties reached about 10,000 (Amnesty International 2018: 17–18). Despite implementing political and democratic reforms in 1998, the Indonesian government has not ceased using a security approach, complemented with the Special Autonomy for West Papua, economic development, and local government reform via the division of regencies, districts, and villages.

After the fall of the New Order Regime in 1998, West Papuan leaders openly voiced their aspirations for independence. They conducted a Papuan People's Congress in 2000 and formed the Presidium of the Papua Council to fight for its independence. However, the leader of this movement, Theys H. Eluay, was later assassinated by Indonesian security forces in 2002, so the movement became weaker (Ipenburg 2002; Pekey 2018: 24). The Indonesian government then imposed the Special Autonomy of Papua in 2002, which was followed by the splitting of West Papua into Papua and Papua Barat in 2003. Since then the government has intensively increased social and economic development, including infrastructure in both provinces, using Special Autonomy Funds. Between 2002 and 2016, the national government transferred a Special Autonomy Fund of as many as 77 trillion Indonesian Rupiah (IDR) to the government of Papua and Papua Barat (Catharina 2019: 8). However, the *Human Development Index* (HDI) of Papua and Papua Barat provinces was, respectively, 63 and 64 below the national level of 71. In regencies or cities dominated by West Papua natives, the HDI is generally lower compared with regencies or cities dominated by migrant populations (LIPI/TAF 2018; Afriansyah et al. 2019: 4–5) (Fig. 25.1 and Table 25.1).

The present-day West Papua consists of Papua and Papua Barat (literally translated as 'west Papua') provinces, with an area of 316,533 km^2 and 99,671 km^2 respectively. The combined sum of their territories is 416,204 km^2, equalling roughly 22% of the Indonesian territory, which is 1,913,578 km^2. The population of Papua province was 3,207,444 and of Papua Barat 893,362 in 2016 (BPS Papua 2017: 5; BPS Papua Barat 2017: 37). After fifty-five years under Indonesian control, indigenous Papuans have

Fig. 25.1 Map of the Indonesian State. *Source* https://www.nationsonline.org/oneworld/map/indonesia_map.htm (2 April 2020) (Copyright: nationsonline.org 1998–2020)

Table 25.1 Special autonomy fund for Papua and Papua Barat Provinces (2002–2016) (in thousands of IDR)

Year	Papua		Papua Barat	
	Special autonomy fund	Infrastructure fund	Special Autonomy Fund	Infrastructure fund
2002	1,382,300,000			
2003	1,539,560,000			
2004	1,642,617,943			
2005	1,775,312,000			
2006	2,913,284,000			
2007	3,395,748,000	1,000,000,000		
2008	3,590,142,897	330,000,000		680,000,000
2009	2,609,796,089	880,000,000	1,118,484,600	600,000,000
2010	2,694,864,788	800,000,000	1,154,942,052	600,000,000
2011	3,157,459,547	800,000,000	1,353,196,948	600,000,000
2012	3,833,402,135	571,428,572	1,642,886,635	428,571,428
2013	4,355,950.048	571,428,572	1,866,835,735	428,571,428
2014	4,777,070,975	2,000,000,000	2,047,315,954	500,000,000
2015	4,940,429,880	2,250,000,000	2,117,327,000	750,000,000
2016	5,395,051,859	1,200,000,000	2,329,517,820	1,113,750,000
Total	47,902,990,170	10,402,857,144	13,630,506,744	5,700,892,856

Source Katharina (2019: 8)

not gained many benefits from development programmes yet, because national and international companies have immensely exploited their natural resources. In 2017 both provinces were recorded as the poorest provinces in Indonesia, with 27.74% of the population of Papua and 23.01% of Papua Barat living below the poverty line, while the national average is 9.82% (BPS Papua Barat 2017: 375–377). The Human Development Index of both provinces was also the lowest, namely 58.05 (Papua) and 62.21 (Papua Barat), far below the national average of 70.18 (BPS Papua 2017: 158; BPS Papua Barat 2017: 111).

An implication of Papua's integration into Indonesia is the migration of many non-West Papuan ethnic groups into West Papua, mainly consisting of Buginese, Butonese, Makassarese, Javanese, and various Sumatran ethnicities. The recent studies show that 50% of Papua Barat's population come from migrant ethnicities, based on the 2010 census (LIPI 2019). The migrant population of the province of Papua is estimated at around 30%. Those ethnic groups become dominant in the commercial agriculture and business sectors of the economy, while indigenous peoples only find work in the hunting, forestry, fishery and subsistence sectors. The economic gap is noticeable when we compare the quality of housing of non-indigenous peoples, who tend to live in urban areas or transmigration areas, with the settlement of indigenous peoples. McGibbon (2006: 25) argued that economic marginalisation and political violence during the New Order period (1967–1998) had strengthened the political identity of West Papua, in addition to the policy of Papuanisation begun by the Dutch colonial government in the 1960s, as explained by Chauvel (2005: 2).

The indigenous peoples of West Papua warmly welcomed the victory of Joko Widodo (popularly known as Jokowi) in the 2014 presidential election. Jokowi gained the majority of votes in West Papua, as the Papuans had learned about his track record when he was Mayor of Surakarta and Governor of Jakarta, i.e. they knew he had experience of solving conflicts through peaceful negotiations instead of violence. In his administration, President Jokowi considers the Papuan problem to be one of his priorities, especially with regard to economic and social problems. President Jokowi is attempting to complete the development plans not fully implemented by the previous SBY administration – for example, the construction of modern marketplaces for indigenous Papuans, ports and airports, the Trans-Papua Highway, power plants, health service improvements, and a single-price fuel policy. The primary peace policy of President Joko Widodo in West Papua is to build the Trans-Papua Highway to improve connectivity between regencies and between mountainous and coastal areas.

Among those programmes, this chapter focuses only on one policy, namely infrastructure development, specifically the regional economic corridor associated with the Trans-Papua Highway (Sloan 2019: 98). The road network development is essential, for such a network will inevitably create better connectivity between communities of Papuan indigenous peoples that have long been isolated from each other. The project is addressed to build a connection between mountainous and coastal areas, in order to facilitate the distribution of industrial goods. The expected impact is that the prices of goods, including fuel, in mountainous areas will become equal to those in Jayapura. Inter-area connectivity would smooth the movement of peoples, commodities, and capital, improving the lives and livelihoods of indigenous West Papuans. Life improvement would consequently make them more inclined to reconcile with the Indonesian government, and dissolve their collective memory of suffering and their past hard times (Fig. 25.2).

Under Indonesian rule, the construction of the Trans-Papua Highway was started around 1979 and 1980, mainly to connect urban coastal cities and transmigration areas. In the mid-1990s, many vans, jeeps, and trucks, generally owned by migrant populations, transported people and commodities in coastal areas (Kirksey/Bilsen 2002: 842). However, road construction was much more limited in the central mountains, occurring only in Wamena, the capital of Jayawijaya regency. The development of the Trans-Papua Highway then continued during the administration of President Habibie in 1999. It was also continued by President Yudhoyono through the Papua and West Papua Development Acceleration Unit in 2012–2014. However, it was during the Jokowi administration that the construction is significantly pushed. The highway has a total length of 4365 – 3295 km in Papua and 1070 km in Papua Barat. By the end of 2017, the national government has constructed 4158.29 km of roads. In Papua province, the highway consists of the following lines: (i) Merauke-Tanah Merah-Waropko, (ii) Waropko-Oksibil, (iii) Dekai-Oksibil, (iv) Kenyam-Dekai; (v) Wamena-Habema-Kenyam-Mumugu,

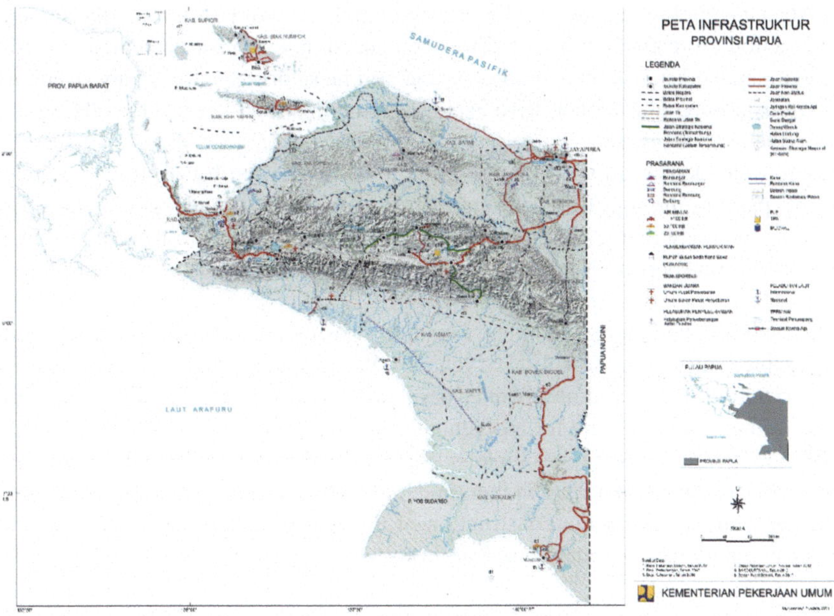

Fig. 25.2 The construction of the Trans-Papua Corridor in the Province of Papua. *Source* Pusat Data and Informasi Kementrian Pekerjaan Umum dan Perumahan Rakyat [The centre of data and information's technology, Ministry of Public Works and People Housing]

(vi) Wamena-Elelim-Jayapura; (vii) Wamena-Mulia-Ilaga-Enarotali, (viii) Wagete-Timika, and (ix) Enarotali-Wagete-Nabire (BBPJN Papua 2017). This chapter focuses its discussion on the Wamena-Elelim-Jayapura line, which had been almost completely connected at the time of writing.

25.3.2 Economic and Socio-cultural Impacts of the Trans-Papua Highway on Papuan Indigenous Peoples

25.3.2.1 History of the Road Construction

In the Wamena-Elelim-Jayapura highway, one of the observed sites was the Kurulu-Abenaho segment, which can be traversed by four-wheel-drive cars. This segment is characteristically mountainous, having steep ascents and descents, and bridges over large rivers. This road segment had been paved in 2016 as part of the construction of the Trans-Papua connectivity project. In the last two years, the entire segment from Kurulu to Abenaho has been asphalted. Meanwhile, at the time of writing, the Abenaho-Elelim segment had not been completed, with 18 km of roads under construction, from km 68 to km 86. This segment has not yet been asphalted. It is still prone to landslides, has many potholes, and still lacks bridges over many rivers. At numerous points, four-wheel-drive cars traversing this road have to ford the rivers, due to the lack of bridges.

Before the construction of the road, the local population had to travel in small groups, passing through tropical jungles from Abenaho to Wadangku district and spending the night there. At dawn, they continued their journey to Kurulu, and then rode public transport to Wamena. Similarly, when returning, they travelled on foot from Kurulu, starting at daybreak and reaching Abenaho at nightfall.

In 1966, the national government began to construct a soil road, but it was still impassable even by four-wheel-drive vehicles. West Papuan indigenous peoples used the road to trade their handmade woven bags (*noken*) in exchange for piglets. In those days, Wamena was a centre for indigenous Papuan pig farming. Most West Papuan indigenous peoples raise piglets for later use as a dowry or for traditional ceremonies. Due to the frequent tribal wars in that time, only men travelling in groups were able to reach Wamena. After the construction of the Trans-Papua road, both male and female populations of Wamena and Kurulu could visit each other in four-wheel-drive vehicles.

The population of Abenaho generally belongs to the Yali tribe of Yalimo, and the Hubla tribe, a branch of the Dani. Migrants of other than West Papuan ethnicities are not allowed to live in the district, except those who work as teachers or health service workers, and live there only temporarily. The local leaders have not even allowed military or police forces to open offices in the district. The reason is that the local populations are still traumatised by the political violence that took place in 1977. Budiarjo and Liong (1988: 78–81) explain that in that year around 15,000 indigenous West Papuans in the Central Mountain area, including Yalimo, mounted resistance against Indonesian armed forces, which cost the lives of many indigenous West Papuans.

Many informants claimed that their area is safer without migrant people from outside West Papua. In the case of crimes, heads of villages invite police officers from Elelim to investigate, and even then, such cases occur very rarely. Such a situation can be expressed by the local population in another way: the less the police and military are present in an area, the safer the area is. Interviews and observations show that there are no business activities run by migrant populations in the district. The only outsiders are exclusively teachers and nurses.

Regarding the history of connectivity between regencies in West Papua, the road was asphalted most recently in 2016. Previously, the compacted gravel and sand road could be traversed by four-wheel-drive vehicles from Wamena to Abenaho. There was also public transport between Abenaho and Elelim, but the number of vehicles was insufficient, and their frequency was limited. The government did not harden and asphalt the road from Abenaho to Elelim until 2016. The Abenaho District has 108 villages spread in the hills. However, the roads between villages and between the villages to the district capital are not yet asphalted, but usually in the form of soil roads, and some of them are compacted with gravel and sand.

Another observation site is the Wamena-Kurulu segment of the Trans-Papua Highway, linking Wamena to Elelim of Yalimo Regency and Karubaga of Tolikara Regency. The Wamena-Kurulu road is the smoothest and most used, and was paved several times in the period between 1997 and 2016. Kurulu was selected as a research site because the government had upgraded the road much earlier, in the 1990s. Thus, our assumption is that this road had a positive impact and contributed towards various changes in public services and the livelihoods of indigenous peoples. The government has built some public facilities, such as primary schools, secondary schools, and a senior high school, a community health centre, a district military base, and a district police station. Public transport in the form of minibuses has connected Wamena to Kurulu and Wossi since 2004. This means that the level of connectivity between the area and Wamena has existed for a long time, compared to that of other districts outside Wamena.

Almost all indigenous peoples living in Kurulu district are members of the Dani tribe. There are also outsiders employed as security workers, military and police officers, teachers and health workers. The observation also indicates that a female migrant woman married a West Papuan man and opened a small kiosk in the district. As in Jayawijaya and Yalimo, the number of the migrant population is limited. Migrant populations usually live in the regency's capitals – towns like Wamena, Elelim, and Karubaga. Indigenous West Papuans in the Central Mountains firmly and actively maintain their respective social structures. In every tribe, the smallest unit is a clan. Sometimes two or more clans make up a tribal alliance. Each alliance has a war chief, a welfare manager, and a medicine expert. The highest rank is the chief of war; the chief leading the group to declare or wage wars against other groups. The welfare manager is in charge of coordinating agricultural and husbandry activities so that the members of the alliance can live optimally. The medicine expert is tasked with providing traditional health services, including traditional rituals. In short, ethnic-based communities, as well as church organisations, are active in West Papua. They are even closer to the population at grass-roots level and use government institutions as an avenue to obtain access to resources and public services.

With regard to connectivity, since 2004 the Wamena-Kurulu road has been vital to the daily life of the indigenous West Papuans. As mentioned above, there is relatively inexpensive public transport for those travelling from Wamena to Wossi. However, four-wheel-drive vehicles have to be used to travel from Wamena to Abenaho and Elelim of Yalimo Regency and from Wamena to Karubaga of Tolikara Regency. Larger trucks carry consumption goods and construction materials to those places, even reaching Ilaga of Puncak Jaya regency. The most extensively used stretch of the Trans-Papua Highway line is between Wamena and Karubaga. Other groups of travellers use the Wamena-Karubaga-Wossi highway, carrying fruit and vegetables from their hometown to sell in Wamena. On their way home, they buy daily necessities to sell in their hometown.

25.3.2.2 Impact on Livelihood

The road is indeed recognised by local populations as bringing changes to the livelihoods of indigenous West Papuans. Kirksey and Bilsen (2002: 846) stated that the indigenous West Papuans benefited from the road construction mainly from being employed in the lowest paying jobs, such as guides, cooks, porters, and manual workers. Other new jobs, such as car drivers, motor taxi drivers and timber workers, flourished among the population. Another impact of the road is the building of small kiosks. The most visible impact of the road is the existence of several kiosks in the Abenaho district selling daily necessities such as dairy products, vegetable oil, instant noodles and bottled water, as well as cigarettes and soft drinks. Other shops provide gasoline, areca nuts, and betel. However, restaurants or eateries are nowhere to be found – such businesses are almost exclusively run by migrants, but tribal groups and local government do not allow migrants to run businesses in this area. The absence of migrants operating restaurants in this district makes it hard to obtain ready-cooked meals.

In terms of food resilience, indigenous West Papuans in Abenaho generally consume sweet potatoes (*hipere*) to fulfil their nutritional needs, as well as taro, rice, and vegetables such as beans, long beans, spinach and scallions. They do not consume meat unless they go hunting in the forest to catch *kuskus* (Australian possum). Consumption of other types of meat is low. However, currently they consume a lot of instant foods, such as noodles, canned fish and instant spices, sold by shops in Wamena.

The indigenous population in the district go to Wamena to buy construction materials and daily necessities; they take money when going and carry industrial products when returning to the district. They use local marketplaces to sell vegetables, horticultural produce and pigs. Some markets are located at certain points along the Abenaho road, opening every Friday. There are also vegetable-sellers who sell their commodities from Abenaho to Elelim and buy manufactured goods such as rice, vegetable oil, and soap. The road makes it easier for indigenous West Papuans to sell their horticultural commodities and buy daily necessities in Wamena.

The road is therefore perceived to have a positive impact on the livelihoods of indigenous populations, including those in the Abenaho district. One of the impacts is a large-scale local contractor who worked on the construction of the Kurulu-Abenaho road segment. The owner of the business is a Christian man from Toraja, who married an indigenous woman, and obtained cultural recognition as an indigenous West Papuan. Other business persons include the owners of construction companies co-operating with the local government who are also indigenous Papuans. Their workers are predominantly indigenous West Papuans.

The road has similarly had a positive impact on the livelihoods of indigenous West Papuans living in the Kurulu district. Many informants there admitted that the road has enabled West Papuans to transport their vegetables and fruit – such as beans, long beans, carrots, cabbages and bananas – to market. In addition, in the district there are two roadside convenience stores that sell community needs, such as cooking oil, various drinks, instant noodles, cigarettes, and other items. There are also several stalls along the Trans-Papua Highway owned by Papuan women who sell areca nuts and betel. Indigenous West Papuans living in this district sell their agricultural commodities to migrants in Wamena and buy their daily necessities from those people.

However, the road construction in Kurulu also has negative impacts, especially for youths and the adult male population. The men frequently travel to Wamena by public transport to sightsee for no particular reason. In the town, they purchase alcoholic drinks and bring them home to drink together. Drinking alcohol results in excessive drunkenness and leads to road accidents and crime.

Another impact is that the indigenous population finds it easier to obtain good quality rice. The ease of access to Wamena permits the people of Kurulu to obtain greater quantities of rice than before. The youth increasingly prefer rice to the traditional staples, taro and sweet potatoes. From an economic perspective, a bunch of sweet potatoes, less than ten pieces, can be exchanged for 3–5 kg of rice. Rice is less expensive than sweet potatoes, and therefore, in the course of time, the former gradually becomes the staple food. However, it makes the indigenous population depend on rice and other industrial foods instead of their traditional foods. The road enables instant food products, such as instant noodles and soft drinks, to be consumed by indigenous West Papuans. This is potentially worrisome because of the possibility that the case of other West Papuan ethnicities such as Asmat tribal groups will recur: the Asmat suffer from malnutrition because they now only consume industrially processed foods and have abandoned their traditional staples. However, at the time of writing there has not been any indication of that trend in Kurulu.

In Kurulu, the road also threatens the socio-economic life of indigenous West Papuans with instability, in particular because of the large number of pigs imported from outside Wamena. Pigs are the most precious assets for the indigenous population, for they have numerous uses, especially for paying customary fines and dowries, and slaughtering in customary ceremonies, such as wedding parties, funerals, and village or district events. In the 1990s, the pigs in Wamena were native to the region: they were fat, large and never underfed. Pork lard can be used for frying, while the pigs were symbols of their owners' prosperity. Persons who own a lot of pigs are more respected by the community. Today, pigs from outside the region overwhelm the district, indigenous pigs become more expensive, and people buy more livestock for they are less costly. As a result, the indigenous pigs become more difficult to sell, resulting in decreased income for their owners, who are indigenous West Papuans. The people of Wamena are not used to large-scale pig farming; they let their pigs roam around, and their number is limited. The Trans-Papua road from Jayapura to Wamena is worrisome, as it will increase the number of pigs from outside the region.

25.3.2.3 Sustainable Environment and Basic Services

Another unintended consequence of the Trans-Papua Highway is that indigenous West Papuan populations living around the road are encouraged to conduct extensive illegal logging. The road makes it easier for the indigenous population to cut down trees on a massive scale and sell the timber to migrant businesses in Wamena and Elelim. The demand for timber in the city is relatively high due to the rapid economic development. Timber is transported from forest areas to the road, and timber traders from Wamena, who have already made appointments with the loggers, take over. The chainsaw operators and timber transporters are generally indigenous.

A similar story is found in Kurulu of Jayawijaya. After the completion of the road construction, trees are more easily cut down and transported by the indigenous population from forest areas to the nearest towns. There are two kinds of timber frequently extracted in Kurulu. The first is firewood for local cooking. The second one is logs sawn into boards and sold to timber traders. Up to the present day, illegal logging by the indigenous population is still conducted, with the timber being sold to migrant traders living in the nearest towns. Such a phenomenon takes place in almost all districts in Central Mountains that still have forests. Logging is considered the easiest way to obtain large sums of money, rather than waiting for harvests. This information is corroborated by another research study undertaken by Global Forest Watch (2018) that observed that between 2001 and 2017, the tree cover in Eastern and Central parts of Papua has lost 567,000 ha, a 1.9% decrease, since 2000. In the Western part, it has lost 235,000 ha, a 2.5% decrease, since 2000. Palm oil plantation and Trans-Papua Highway construction contribute to the loss of tree cover.[1]

Another impact of Trans-Papua iHghway construction in the Abenaho district on the educational sector was identified in the researcher's field observation. In the district, there are already a senior high school and a junior high school, and ten primary schools. The teaching activities in the primary and secondary schools still need more attention due to teachers mostly living in Wamena; they go to the school once a week or once a month. In this district, the government has provided houses for all teachers. Except in the district capital, most teachers are less active; only two or three of the teachers keep teaching on a daily basis, while the rest stay in Wamena. Some informants said that such a condition is common in the Central Mountain of West Papua.

A similar case can be found in Kurulu, even though the road had been paved since 2004, meaning that connectivity is much better there; its educational situation is not much different from that of other districts. The feeling of insecurity among non-Papuan teachers is high. During the observation, a female teacher of a Catholic Foundation kindergarten was stabbed by a drunken young man at night. Even the presence of police and military posts in Kurulu cannot provide security to the teachers. Another case took place in 2017; a female teacher of the Catholic primary school was injured by a stone thrown by her student. Both teachers were indigenous West Papuans but came from other districts.

In addition to security matters, the availability and presence of teachers in the classes are still a problem in Kurulu, except for the Catholic primary school. State-funded schools face the problem of low attendance of teachers in the class. In general, only two assistant teachers and a civil servant teacher are active in the school, while other civil servant teachers stay at home in town and do not always come to teach.

[1] See at: https://www.globalforestwatch.org/dashboards/country/IDN/23 (28 January 2019).

In the Catholic primary school, the situation is relatively better, for all teachers live in a nearby boarding house and they attend the classes every day.

The availability of the good road has not had a particularly positive impact on the provision of health services in Abenaho; the story is similar to that of the education situation. The community health centre, founded in 1995, nominally has one doctor and two trainee doctors. Unfortunately, they are rarely present in the office. The civil servant doctor only visits the health centre a few days a year; at most, he stays in the district for a one-month period per year. In the last five years, the indigenous West Papuans in Abenaho have suffered from the same kind of diseases. The most frequently reported illness is acute respiratory tract infection. A mobile ambulance service had been provided in 2008–2009, and the facility was renewed in 2015. However, the ambulance became unusable in 2017, as it was not well maintained.

A different story of the road's impact on health services can be noticed in Kurulu. From year to year, the community health centre of Kurulu provides better services to the indigenous population. Although one or two nurses return to their homes in Wamena by public transport before the end of service hours, the health services underwent an improvement in 2018, according to local people. However, it is not due to the presence of the good road, but because of a new doctor. The previous doctor stayed in the city, but his successor stays in the house provided for medical workers, and he never leaves his post during service hours.

Those findings support Anderson's research results (2015: 53) that the state is absent in providing health services in villages. Its presence is solely symbolic. He rejected the thesis that the state's presence is too dominant in West Papua, arguing that it was not valid (2015: 18). The real situation is that the state neglects the provision of basic services for indigenous West Papuans. Except in the security dimension, the presence of the security apparatus in a significant number, with the rationale of dealing with separatism, can be questioned critically by intellectuals, considering that the essential problem is the lack of public services.

25.3.2.4 Socio-cultural Values

The development of the Trans-Papua Highway in Abenaho has led to changes in the lifestyle of indigenous populations, including an increase in the consumption of alcoholic drinks. Most informants indicated that there were no shops in Abenaho that sold alcoholic drinks. The district chief and the Protestant Church forbid people to sell alcoholic drinks in all villages. However, the young people can obtain alcohol easily in Wamena, by riding a motorbike. One example of the negative impact of the excessive consumption of alcoholic drinks is a secondary school pupil who committed violent acts at his school. The Trans-Papua Highway, from Jayapura to Wamena, becomes a threat as alcohol and also drugs become more easily obtainable in the future.

The road construction in Kurulu has impacts on the disturbance of social relations between groups. Most informants in this field research, especially public transport drivers with a migrant background, stated that some armed indigenous people often stopped public transport coming into Wamena. They were drunken and forced the drivers to give certain amounts of money. Those drivers said that the highway could be said to be safe but it was not the case; or it could be seen as unsafe but looked as if it were safe.

This indicates that the provision of the road has to be followed by improvements to cultural resources, so indigenous West Papuans will not have to suffer from the negative impacts of modernisation. Indigenous peoples need to be made aware that alcoholic drinks provide no benefit and tend to harm, or alcohol consumption needs to be replaced with more productive activities. For that purpose, one of the methods to be used is to revitalise the cultural values of indigenous West Papuans, such as cultural bonds and communal co-operation.

Today, co-operation has almost vanished, and the spirit of individualism replaces communalism. Young West Papuans have abandoned their customary houses, and they are no longer familiar with the Baliem valley customs. Papuan adult males generally prefer to live in Wamena than in their traditional houses in the villages, because the city provides more pleasures that were not available in the past. The World Bank (2009: 28) has already predicted these phenomena, that the massive infrastructure development in West Papua can destroy local culture and weaken the traditional economic system due to the large inflow of migrants originating outside West Papua. These migrants work mainly as traders, small businessman, and transport services. The negative impact of the Trans-Papua Highway, according to Kusumaryati (2017: 144), is that its construction has generally facilitated and accelerated the exploitation of abundant natural resources, so it is not a road to freedom but a road to disaster.

25.4 Security and Peace Issues

This hybrid peace-building model combining security, reform and patronage approaches in West Papua was fruitful in appeasing some opposition groups among the local political elites between 2002 and 2008. Smith (2012) points out that policy has been successful in accommodating local political elites' interests in autonomy and in weakening the separatist movement. It means that some West Papuan elites, particularly those working for the local government and the Presidium of Papuan Council, have reduced their demand for independence and accepted Special Autonomy. The national government also transferred the Special Autonomy Fund, some of which is distributed to villages in cash.

Political actors changed in 2008 when some West Papua free armed groups set up the *West Papua National Coalition Liberation Front* (WPNCL), based in *Papua New Guinea* (PNG) and Vanuatu. Then, some tribal affiliated groups conducted the second Papuan People's Congress in 2011 and declared the existence of the Federal Republic State of West Papua (Federal Group). One year after the declaration, the youth political movement set up the National Parliament of West Papua (NPWP) and the National Committee of West Papua. The rise of the resistance movement in this region is related to complicated causes. One of them, as shown previously in this chapter, is that the Special Autonomy and its social and economic development programme only benefited the migrant population instead of West Papuan natives. The Special Autonomy Fund allocation is not in accordance with the provisions requiring 30% to be spent on basic education and 15% on basic health services. There are indications that some funds have been misused by government elites for personal or group interests.

Between 2002 and 2016 the national government transferred the Special Autonomy Fund to the tune of IDR 77 trillion to the governments of Papua and Papua Barat (Catharina 2019: 8). However, the *Human Development Index* (HDI) scores of Papua and Papua Barat provinces are respectively 63 and 64, i.e. lower than the national score of 71. In regencies or cities dominated by indigenous West Papuans, the the HDI score is generally lower than that of regencies or cities dominated by migrant populations (LIPI/TAF 2018).

The hybrid peace model has failed to disrupt the continuing cycles of violence conducted by Indonesian police and armed forces in West Papua. The *International Coalition for West Papua* (ICP) (2017: 9) and Amnesty International (2018: 10) have reported a series of extrajudicial killings in Papua even after the 1998 political reforms. The situation in Papua continues to lack negative peace since its integration into Indonesia in 1963.

Present-day West Papua is different from the conditions before 2014. In 2014, all secessionist organisations, including WPNCL, the federal group, and NPWP, were united under the ULMWP (*United Liberation Movement of West Papua*). This group was recognised as an observer in the *Melanesian Spearhead Group* (MSG) in 2015. The ULMWP has actively campaigned for free West Papua via the MSG, the *Pacific Islands Forum* (PIF) and the United Nations.

It seems that progress in infrastructure development has not been followed by an improvement in the sense of security among indigenous West Papuans, particularly those who live in the Central Mountains. Even though the national government's construction of the Trans-Papua Highway in the Central Mountains has facilitated trade, the road has been viewed as a threat to the indigenous population. They are still worried that the road could be used to deploy Indonesian armed forces for military campaigns in this region.

Another issue is the conflict between the Indonesian security forces and the *West Papuan National Liberation Army* (WPNLA), such as the skirmish that occurred in June 2018 at Nduga district; the skirmish discouraged people from using the road and resulted in a temporary halt to the road construction. A similar case was reported in *Lani* magazine (XXIII/September 2014: 1–3): it was not safe to travel the Wamena-Tiom road in July or August after a skirmish occurred between military personnel and the separatist West Papuan National Liberation Army. The high numbers of police and armed forces posts are frequently problematic; on the one hand their presence enhances the security of migrants, but on the other hand they are a cause of insecurity for the indigenous peoples.

A recent issue is the massacre of seventeen civilians who worked on the Trans-Papua Highway in Nduga regency on 2 December 2018. The WPNLA stated that they were responsible for the incident, as the victims were part of the Indonesian military who allegedly spied on them (Darmono 2018). Responding to the attack,

Indonesian armed forces and police sent troops to eliminate the secessionist group in mid-December. This military operation resulted in the death of fourteen civilians, all of whom were indigenous Papuans, and 22,000 people from twelve districts took refuge in jungles around Nduga (Haluk 2019: 6). This phenomenon proves that the Trans-Papua Highway project in the Papua highlands is not accepted by indigenous West Papuans, as they feel threatened by the excessive Indonesian military presence and the influx of migrants from areas outside West Papua. During the rule of President Yudhoyono, the national government used army units to build the road to fulfil the strategic purpose of connecting isolated army units in West Papua (Weatherbee 2016: 27).

Security and peace for indigenous West Papuans are still a significant problem, and the presence of security forces in their villages is often seen as an indicator of lack of safety. Despite the completion of the construction of roads and airports, peace is not automatically created in the areas. Although the intensity has decreased significantly, political violence has not completely disappeared in the Jokowi era. The Amnesty International report (2018) indicated that in 2010–2018, sixty-nine extrajudicial killings have occurred in Papua with ninety-five casualties. Between 2014 and 2018 in the Jokowi era there have been thirty-eight incidents resulting in fifty-one deaths. These extrajudicial killings include those linked with political activism (twenty-eight cases, thirty-nine victims): police arrests following protest actions voicing political demands, allegedly planned murders of pro-independence activists, and fatalities within the course of security operations. Extrajudicial killings not linked to political activism include forty-one cases with fifty-six victims: police arrests in the case of non-political mass gatherings and public order disturbances, criminal acts with the intent of revenge, and killings in the process of arresting crime perpetrators and misbehaving individuals. Those committing the killings are the police (thirty-nine victims), military personnel (twenty-seven victims), joint police-military operations (twenty-eight victims), and civil service police (one victim). Most victims were indigenous Papuans (eighty-five victims), while only ten non-Papuan persons were victimised (Amnesty International 2018: 9).

Other data issued by the International Coalition for West Papua show that some human rights violation cases took place in the Jokowi era, especially against civil rights. Illegal torture cases increased in 2014 but decreased in 2015 and 2016. The number of torture victims rose above the 2012 figure in 2014 and 2015 but decreased in 2016. Extrajudicial killing cases had a similar pattern to that of torture. The data permit us to infer that 2014–2015 were the peak years of the political violence in West Papua, and it was in 2016 that the number of incidents declined, due to more intense efforts by the government to prevent violence because of international pressures (Table 25.2).

In contrast to the patterns above, the number of political arrests increased in 2016, i.e. 5361 people were arrested. Such a great number of arrests occurred when the government forcefully dismissed a peaceful action commemorating the annexation of Papua by Indonesia on 1 May 2016. Similar cases happened to peaceful actions on Papuan issues which were dismissed by security forces; the numbers increased significantly in 2016. In contrast to the decreasing number of political violence cases

Table 25.2 Political violence against West Papuan Indigenous Peoples. *Source* International Coalition of West Papua (2017: 9)

Civil and political rights data	2012	2013	2014	2015	2016
Cases of violence/arrests/threats against local journalists	12	20	18	5	8
Cases of violence/deportation/intimidation/obstruction of foreign journalists and observers providing coverage on Papua-related issues in Indonesia	2	0	2	4	0
Number of reported torture/maltreatment cases	28	30	76	19	20
Number of reported torture/maltreatment victims	99	67	135	244	64
Number of reported cases of extrajudicial killings	11	6	8	10	6
Number of reported victims of extra-judicial killings	18	8	12	14	6
Prosecution of perpetrators (police and military)	0	0	0	1	2
Number of political arrests	210	548	370	1083	5361
Number of political prisoners/detainees	40	70	55	57	27
Number of public peaceful assemblies/gatherings on Papua-related issues, forcefully intervened by security forces	8	29	14	24	35
Violence/threats against human rights defenders	3	1	5	5	2
Cases prosecuted with treason charges (Articles 106 and 110 of the Criminal Code)	27	25	31	14	15

and political detainees in the Jokowi era, freedom of expression has deteriorated, with a significant increase in the forced dismissal of peaceful actions and of arrested activists involved in those actions.

Citing Smith (2012), the Indonesian government faced two difficult options in creating solutions to the West Papuan issue. The first option, conceding more significant political and economic autonomy and formal recognition of Papuan people's political aspirations, would threaten the Jakarta elites' interests in exploiting Papuan natural resources. The second option, exercise of violence, such as during the New Order era, is not a strategic option, for such violence will increase international pressure. The best option lies between those two, i.e. continuing the limited security approach used in limiting freedom of expression, while increasing efforts to improve economic development to win support from a significant number of West Papuans. Such a step is a form of the illiberal peacebuilding approach, accompanied by specific structural adaptations. However, the question is whether such a hybrid concept in peacebuilding is an actual solution for the complicated Papuan conflict.

During the field observation in Yalimo, the chief of the Yali tribe, and an employee in Elelim district office, said that the construction of the Trans-Papua Highway, ports, and airports is an obligation of the government, but self-determination remains the right of the indigenous West Papuans in their deepest hearts. It means that however well the government implements development policies in West Papua, such efforts will not necessarily erase the West Papuans' memory of being oppressed and their hope of independence; such efforts only mitigate the situation at the elite level. West

Papuan distrust in the Indonesian government is so high that certain intellectuals, such as Elmslie/Webb-Gannon (2013: 159), have accused Indonesia of committing a slow genocide of indigenous Papuans. Therefore, peacebuilding efforts via infrastructure and security approaches are not good options.

Dialogue between the national government and representatives of the West Papuan people is the most appropriate option. The Indonesian government may invite all significant West Papuan leaders to talks in an effort to resolve the problem in each particular sector, such as politics and law, human rights, socio-economic and socio-cultural factors. Sectorial dialogues are not formal international negotiations, such as those conducted between the Indonesian government and the Free Aceh Movement in 2005. However, such dialogues are inclusive, involving all civil society elements, including the West Papua freedom political movement. Together with the Indonesian government, they may formulate which national and development strategies are appropriate for West Papuan indigenous people. Such a strategy can be put within the framework already initiated by President Jokowi in 2017 by appointing a *Person in Charge* (PIC) of the West Papuan sectorial dialogue.[2]

25.5 Conclusions

Unlike the arguments in previous studies, the findings of this research indicate that massive infrastructure development in West Papua does not support peacebuilding. This is because such development does not benefit the indigenous population which suffered from the violent conflict. Additionally, the development does not automatically increase their access to public services, but it does encourage them to exploit natural resources excessively. The infrastructure development, referring to Sen (2000) and Merino/Condejas (2017), has not succeeded in making conflict-affected populations free from poverty, discrimination, marginalisation or repression, and in fact worsens these conditions.

There is no negative peace in West Papua. The province is even further from Brauch's (2009a: 69) idea of a peace policy that combines both sustainable peace and sustainable development. The Indonesian government is likely to prolong sustainable political violence and gradually destroy the environment in West Papua through excessive development programmes. This also differs from the peace policy objective referred to by Oswald Spring (2000: 235). In reality, the peace policy in West Papua has widened the gap between the migrant and the West Papuan native populations instead of keeping harmony between social groups. The recent conflict of racism and riots in this region in 2019 showed that Indonesian peace policy in West Papua should be reformulated.

The Trans-Papua Highway has not yet significantly improved the livelihoods of the indigenous population. Job diversification does occur, but it is not capable of

[2]See at: https://www.voaindonesia.com/a/pemerintah-akan-lakukan-dialog-sektoral-untuk-papua/4081405.html (25 December 2018).

shifting them from their subsistence agriculture. On the other hand, the road causes deforestation and dependence on imported foods, which were not originally staple foods in West Papua. Road connectivity provides greater access for the indigenous populations to health and educational services. However, conversely, the road is seen as an immense threat because it facilitates the entry of drugs and alcohol, and also of imported farmed pigs that threaten indigenous pig-raising activities.

Also, infrastructure development has not played a wholly significant role in nurturing security and peace among indigenous West Papuans. The massive development of roads in the Jokowi era has not yet been balanced with a decrease in cases of violence, especially extrajudicial killings and restraints on freedom of expression. Such an effort from the government side is in line with the illiberal peacebuilding approach with specific structural adaptations, i.e. a combination of the security approach and infrastructure development to improve interconnectivity between different territorial areas. Problems arise when indigenous West Papuans are not capable of taking optimal advantage of the facilities provided. In contrast, the migrant population, both transmigrants and spontaneous migrants, whose population is increasing, is more capable of gaining benefits from the development of these facilities.

Based on these findings, the conceptual implication of illiberal and liberal peacebuilding in the case of West Papua is relevant when analysing some government efforts in peacebuilding. This research confirms the study by Smith (2014) that the Indonesian government uses a hybrid peacebuilding approach in West Papua by combining both liberal and illiberal peace. Due to the internal colonialism context between Jakarta and West Papua, such an approach is not able to resolve the secessionist conflict and continuous political violence. A relevant peacebuilding concept for resolving the West Papuan conflict problem that involves all the parties cannot merely refer exclusively to liberal or illiberal peacebuilding approaches, and it is necessary for the problems to be solved through a thorough understanding of the feeling of West Papuans who see themselves as objects of internal colonialism. Peace study is supposed to understand that peace, according to the West Papuan perspective, is rooted in a society's cultural values, and is commonly shaped by state apparatus, businesses, the church, and customary communities. The developmentalist ideology, complemented with the security approach, cannot guarantee peace, not even negative peace.

References

Amnesty International (AI), 1975: *Don't Bother, Just Let Him Die: Killing with Impunity in Papua* (Jakarta: Amnesty International Indonesia).

Anderson, Bobby, 2015: "Papua Insecurity: State Failure in Indonesian Territory", in: *Policy Studies*, 73 (Honolulu: East West Centre).

Afriansyah, Anggi; Kusumaningrum, Dini D.; Baskoro, Andhika A.; Prasojo, Ari P.S, 2019: *Policy Paper: Alternatif Layanan Pendidikan Bagi Orang Asli Papua di Provinsi Papua Barat* [Policy

Paper: Alternative Education Services for Papuans in the Province of West Papua] (Jakarta: Lembaga Ilmu Pengetahuan Indonesia).

Bjorkagen, Martin, 2016: "Understanding Illiberal Peace-building: An Analysis of Conflict, Peace and Reconciliation in North Maluku Province, Indonesia" (MSc thesis, Lund University, Sweden, Department of Political Science).

BPS (*Biro Pusat Statistik*, Statistic Bureau) Papua, 2017: *Papua in Figures 2017* (Jayapura: BPS Papua).

BPS Papua Barat, 2017: *Papua Barat in Figures 2017* (Manokwari: BPS Papua Barat).

Budiardjo, Carmel dan Liong; Lim, S., 1988: *West Papua: The Obliteration of a People*. (Thornton Heath: Tapol).

Brauch, Hans Günter, 2009a: "Securitizing Global Environmental Change", in: Brauch, Hans Günter et al. (Eds.): *Facing Global Environmental Change* (Berlin/Heidelberg: Springer): 65–102.

Brauch, Hans Günter, 2009b: "Introduction: Facing Global Environmental Change and Sectorialization of Security", in: Brauch, Hans Günter et al. (Eds.): *Facing Global Environmental Change* (Berlin/Heidelberg: Springer): 21–42.

Chatarina, Riris, 2019: *Menakar Capaian Otonomi Khusus Papua* [Accounting the Result of Special Autonomy of West Papua] (Jakarta: Buku Obor).

Chauvel, Richard, 2005: "Constructing Papuan Nationalism: History, Ethnicity, and Adaptation", in: *Policy Studies*, 14 (Honolulu: East-West Center).

Coser, Lewis, 1956: *The functions of social conflict* (Glencoe, Illinois: The Free Press).

Drooglever, Pieter, J., 2009: *An Act of Free Choice: Decolonisation and the Right to Self-Determination in West Papua* (Oxford: Oneworld Publications).

Elisabeth, Adriana; Eidjojo, Muridan S.; Cahyadi, Rusli; Blegur, Sinnal, 2004: *Pemetaan Peran dan Kepentingan Para Aktor dalam Konflik Papua* [Mapping the Roles and Interests of Actors in the Papuan Conflict] (Jakarta: Lembaga Ilmu Pengetahuan Indonesia).

Elmslie, Jim; Webb-Gannon, Camellia, 2013: "A Slow-Motion Genocide: Indonesian Rule In West Papua", in: *Griffith Journal of Law and Human Dignity*, 1,2: 142–166.

Galtung, Johan, 1969: "Violence, Peace, and Peace Research: in *Journal of Peace Research*, 6,3: 167–191.

Haluk, Marcus, 2019: *Peristiwa Kali Yigi Mbua di Kabupaten Nduga dan Perjuangan Pengakuan Kedaulatan Politik bagi Bangsa Papua* [The Incident of Kali Yigi, Mbua, in Nduga Regency and the Struggle for Recognition of Political Sovereignty for the Papua Nation] (Jayapura: The United Liberation Movement for West Papua).

Hernawan, Yohanes, B, 2013: "From the Theatre of Torture to the Theatre of Peace: the Politics of Torture and Re-imagining Peacebuilding in Papua, Indonesia" (PhD dissertation, Australia National University, Pacific Institute).

Humphreys, Macartan, 2003: "Economics and violent conflict" (Cambridge, MA); at: *Economics and Violent Conflict* (unicef.org) (1 April 2020).

International Alert, 2015: *Peace Through Prosperity: Integrating Peacebuilding into Economic Development* (London, UK: International Alert).

International Coalition for Papua (ICP), 2017: *Human Rights in West Papua 2017* (Wuppertal & Geneva: ICP Secretariat.

Ipenburg, At, 2002: "The Life and Death of Theys Eluay", in: *Inside Indonesia*, 70 (April–June), at: https://www.insideindonesia.org/the-life-and-death-of-theys-eluay (1 April 2020).

Jaringan Damai Papua [JDP, Papua Peace Network], 2014: *Indikator Papua Tanah Damai* [The Indicators of Papua Land of Peace] (Jayapura: JDP).

Kahin, George M., 1953: *Nationalism and Revolution* (Ithaca: Cornell University Press).

Kirksey, Eben, 2012: *Freedom in Entangled Worlds* (Durham: Duke University Press).

Kirksey, Eben; Van Bilsen, Kiki, 2002: "A Road to Freedom; Mee Articulations and the Trans-Papua Highway", in: *Bijdragen tot de Taal-, Land- en Volkenkunde*, 158,4: 837–854.

Kusumaryati, Veronica, 2017: "The Great Colonial Road", in: High Education Press, 5,2: 137–145.

Lewis, David; Heathershaw, John; Megoran, Nick, 2018: "Illiberal Peace? Authoritarian Modes of Conflict Management", in: *Cooperation and Conflict*, 53,4: 486–506.

Marcus, George, E., 1995: "Ethnography in/of the World System: The Emergence of Multi-Sited Ethnography", in: *Annual Review of Anthropology*, 24: 95–117.
McGibbon, Rodd, 2006: "Pitfalls of Papua: Understanding the Conflict and its Place in Australia-Indonesia Relations" (Sydney: Lowi Institute).
McIntosh, Katie; Buckley, Joanna, 2015: *Economic Development in Fragile and Conflict-Affected States: Topic Guide* (Birmingham, UK: GSDRC, University of Birmingham).
Merino, Leticia; Cendejas, Josefina, 2017: "Peacebuilding from a Commons Perspective", in: *International Journal of the Commons*, 11,2: 907–927.
Oswald Spring, Úrsula, 2000: "The Future of Humanity: Human, Gender And Ecological Security", in: *Peace and Conflict: Journal of Peace Psychology*, 6,3: 229–236.
Panic, Milivoj, 2005: "Reconstruction, development and sustainable peace; A unified programme for post-conflict countries", CDP Background Paper No 8, at : https://www.un.org/en/development/desa/policy/cdp/cdp_background_papers/bp2005_8.pdf (1 April 2020).
Pekey, Frans, 2018: *Otonomi Khusus Papua: Dinamika Formulasi Kebijakan yang Semu* [Papua's Special Autonomy: The Dynamics of Artificial Policy Formulation] (Jakarta: Penerbit Buku Kompas).
Rutherford, Danilyn, 2012: *Laughing at Leviathan: Sovereignty and Audience in West Papua* (Chicago: University of Chicago Press).
Saltford, John, 2000: "United Nations Involvement with the Act of Self-Determination in West Irian (Indonesian West New Guinea) 1968 to 1969", in: *Indonesia*, 69: 71–92.
Sen, Amartya, 2000: "Development as freedom", in: *Development in Practice*, 10, 2: 258–258.
Sloan, Sean et al., 2019: "Hidden Challenges for Conservation and Development along the Trans-Papuan Economic Corridor", in: *Environmental Science & Policy*, 92: 98–106.
Smith, Claire, 2012: "Understanding Illiberal Peacebuilding Models: Government Responses to Ending Mass Violence in Indonesia" (Medford, Massachusetts: World Peace Foundation).
Smith, Claire, 2014: "Illiberal Peace-Building in Hybrid Political Orders: Managing Violence During Indonesia's Contested Political Transition", in: *Third World Quarterly*, 35,8: 1,509–1,528.
The Asia Foundation (TAF); Lembaga Ilmu Pengetahuan Indonesia (LIPI), 2018: *Jalan Untuk Komunitas: Membangun Infrastruktur Konektivitas Jalan Untuk Penghidupan Orang Asli Papua Dan Lingkungan Hidup* [Roads For Community: Building Road Connectivity Infrastructure For Living Original Papua and Environment] (Jakarta: TAF-LIPI).
Weatherbee, Donald E., 2016: "Trends in Southeast Asia: Understanding Jokowi's Foreign Policy" (Singapore: Yusof Ishak Institute [ISEAS]).
World Bank, 2009: *Investing in the Future of Papua and West Papua: Infrastructure for Sustainable Development* (Washington DC/Jakarta: World Bank, 5 June).

Other Literature

Darmono, Bambang, 2018: "Massacre in Nduga: Who Should Protect the Nation?", in: Jakarta Post.com, 14 December; at: https://www.thejakartapost.com/academia/2018/12/14/massacre-in-nduga-who-should-protect-the-nation.html (25 January 2018).
Global Forest Watch; at: https://www.globalforestwatch.org/dashboards/country/IDN/22
Global Forest Watch; at: https://www.globalforestwatch.org/dashboards/country/IDN/23.
"Indonesian Political Map", at: https://www.nationsonline.org/oneworld/map/indonesia_map.htm (2 April 2020).
"*Pastikan Jalur Wamena-Tiom Aman*" [Make Sure Wamena-Tiom Highway is Safe], in: *Majalah Lani*, XXIII, September, 2014: 1.Nationsonline.org, 1998
Pusat Data & Informasi Kementrian Pekerjaan Umum dan Perumahan Rakyat [The Centre of Data and Information Technology, Ministry of Public Works and People Housing], 2016: "Peta Infrastruktur Provinsi Papua".

About the International Peace Research Association (IPRA)

Founded in 1964, the International Peace Research Association (IPRA) developed from a conference organised by the Quaker International Conferences and Seminars in Clarens, Switzerland, 16–20 August 1963. The participants decided to hold international Conferences on Research on International Peace and Security (COROIPAS), which would be organised by a Continuing Committee similar to the Pugwash Conferences. Under the leadership of John W. Burton, the Continuing Committee met in London, 1–3 December 1964. At that time, they took steps to broaden the original concept of holding research conferences. The decision was made to form a professional association with the principal aim of increasing the quantity of research focused on world peace and ensuring its scientific quality.

An Executive Committee including Bert V. A. Röling, Secretary General (The Netherlands), John W. Burton (United Kingdom), Ljubivoje Acimovic (Yugoslavia), Jerzy Sawicki (Poland), and Johan Galtung (Norway) was appointed. This group was also designated as Nominating Committee for a 15-person Advisory Council to be elected at the first general conference of IPRA, to represent various regions, disciplines, and research interests in developing the work of the Association.

Since then, IPRA has held 25 biennial general conferences, the venues of which were chosen with a view to reflecting the association's global scope. IPRA, the global network of peace researchers, held its 25th General Conference on the occasion of its 50th anniversary in Istanbul, Turkey in August 2014 where peace researchers from all parts of the world had the opportunity to exchange actionable knowledge on the conference broad theme of 'Uniting for sustainable peace and universal values'.

Ú. Oswald Spring and H. G. Brauch (eds.), *Decolonising Conflicts, Security, Peace, Gender, Environment and Development in the Anthropocene*,
The Anthropocene: Politik—Economics—Society—Science 30,
https://doi.org/10.1007/978-3-030-62316-6

The 26th IPRA General Conference took place from 27 November to 1 December 2016 in Freetown, Sierra Leone on the theme: *Agenda for Peace and Development: Conflict prevention, Post-conflict transformation, and the Conflict, Disaster and Development Debate.* The 27th IPRA General Conference took place in Ahmedabad, India, 24–27 November, 2018 on the theme *Innovation for Sustainable Global Peace.* The next General Conference is scheduled to be in Nairobi, Kenya during January 2021 (https://www.ipra2021.org/).

On IPRA: http://www.iprapeace.org/
On the IPRA Foundation: http://iprafoundation.org/

IPRA Conferences, Secretary Generals and Presidents 1964–2018

IPRA general conferences	IPRA Secretary Generals/Presidents
1. Groningen, The Netherlands (1965)	1964–1971 Bert V. A. Röling (The Netherlands)
2. Tallberg, Sweden (1967)	1971–1975 Asbjorn Eide (Norway)
3. Karlovy Vary, Czechoslovakia (1969)	1975–1979 Raimo Väyrynen (Finland)
4. Bled, Yugoslavia (1971)	1979–1983 Yoshikazu Sakamoto (Japan)
5. Varanasi, India (1974)	1983–1987 Chadwick Alger (USA)
6. Turku, Finland (1975)	1987–1989 Clovis Brigagão (Brazil)
7. Oaxtepec, Mexico (1977)	1989–1991 Elise Bouding (USA)
8. Königstein, FRG (1979)	1991–1994 Paul Smoker (USA)
9. Orillia, Canada (1981)	1995–1997 Karlheinz Koppe (Germany)
10. Győr, Hungary (1983)	1997–2000 Bjørn Møller (Denmark)
11. Sussex, England (1986)	2000–2005 Katsuya Kodama (Japan)
12. Rio de Janeiro, Brazil (1988)	2005–2009 Luc Reychler (Belgium)
13. Groningen, the Netherlands (1990)	2009–2012 Jake Lynch (UK/Australia)
14. Kyoto, Japan (1992)	Katsuya Kodama (Japan)
15. Valletta, Malta (1994)	2012–2016 Nesrin Kenar (Turkey)
16. Brisbane, Australia (1996)	Ibrahim Shaw (Sierra Leone/UK)
17. Durban, South Africa (1998)	2016–2018 *Úrsula Oswald Spring* (Mexico)
18. Tampere, Finland (2000)	Katsuya Kodama (Japan)
19. Suwon, Korea (2002)	2019–2020 Christina Atieno (Kenya)
20. Sopron, Hungary (2004)	Matt Meyer (USA)
21. Calgary, Canada (2006)	**Presidents**
22. Leuven, Belgium (2008)	The first IPRA President was Kevin Clements (New Zealand/USA, 1994–98)
23. Sydney, Australia (2010)	His successor was *Úrsula Oswald Spring* (Mexico, 1998–2000)
24. Mie, Japan (2012)	
25. Istanbul, Turkey (2014)	
26. Freetown, Sierra Leone (2016)	
27. Ahmedebad, India (2018)	
28. Nairobi, Kenya (2021)	

Ú. Oswald Spring and H. G. Brauch (eds.), *Decolonising Conflicts, Security, Peace, Gender, Environment and Development in the Anthropocene*, The Anthropocene: Politik—Economics—Society—Science 30,
https://doi.org/10.1007/978-3-030-62316-6

About the Editors

Úrsula Oswald Spring (Mexico) is a full-time Professor/Researcher at the National University of Mexico (UNAM) in the Regional Multidisciplinary Research Center (CRIM). She was the national coordinator of water research for the National Council of Science and Technology (RETAC-CONACYT), the first Chair on Social Vulnerability at the United National University Institute for Environment and Human Security (UNU-EHS); founding Secretary-General of El Colegio de Tlaxcala; General Attorney of Ecology in the State of Morelos (1992–1994), National Delegate of the Federal General Attorney of Environment (1994–1995); Minister of Ecological Development in the State of Morelos (1994–1998). She was President of the International Peace Research Association (IPRA, 1998–2000), General Secretary of the Latin-American Council for Peace Research (2002–2006), and Co-Secretary General of IPRA (2016–2018).

She has studied medicine, clinical psychology, anthropology, ecology, and classical and modern languages. She obtained her Ph.D. from the University of Zürich (1978). For her scientific work she received the Sor Juana Inés de la Cruz Prize

Ú. Oswald Spring and H. G. Brauch (eds.), *Decolonising Conflicts, Security, Peace, Gender, Environment and Development in the Anthropocene*, The Anthropocene: Politik—Economics—Society—Science 30,
https://doi.org/10.1007/978-3-030-62316-6

(2005), the Environmental Merit in Tlaxcala, Mexico (2005, 2006), and the UN Development Prize. She was recognized as Woman Academic in UNAM (1990 and 2000); and Woman of the Year (2000). She works on non-violence and sustainable agriculture with groups of peasants and women and is President of the Advisory Council of the Peasant University.

She has written 46 books and more than 328 scientific articles and book chapters on sustainability, water, gender, development, poverty, drug consumption, brain damage due to under-nourishment, peasantry, social vulnerability, genetically modified organisms, bioethics, human, gender, and environmental security, peace and conflict resolution, democracy, and conflict negotiation. Among her recent co-edited Englisch books are: (with Brauch, Mesjasz, Grin, Dunay, Chadha Behera, Chourou, Kameri-Mbote, Liotta): *Globalization and Environmental Challenges: Reconceptualizing Security in the 21st Century*, 2008; (co-edited with Brauch, Grin, Mesjasz, Kameri-Mbote, Chadha Behera, Chourou, Krummenacher): *Facing Global Environmental Change: Environmental, Human, Energy, Food, Health and Water Security Concepts* (2009); (co-edited with Brauch): *Reconceptualizar la Seguridad en el Siglo XXI* (2009); (co-edited with Brauch, Mesjasz, Grin, Kameri-Mbote, Chourou, Dunay, Birkmann): *Coping with Global Environmental Change, Disasters and Security – Threats, Challenges, Vulnerabilities and Risks* (2011); (ed., 2011): *Water Resources in Mexico. Mexico – Scarcity, Degradation, Stress, Conflicts, Management and Policy*; (co-edited with Brauch, Grin, Scheffran, 2016): *Handbook on Sustainability Transition and Sustainable Peace*; (co-edited with Brauch, Bennett and Serrano Oswald, 2016): *Addressing Global Environmental Challenges from a Peace Ecology Perspective*; (co-edited with Brauch, Bennett and Serrano Oswald, 2016): *Regional Ecological Challenges for Peace in Africa, the Middle East, Latin America and Asia Pacific*; (co-edited with Brauch, Collins and Serrano Oswald, 2018): *Climate Change, Disasters, Sustainability Transition and Peace in the Anthropocene.*

Address: Prof. Dr. Úrsula Oswald Spring, CRIM-UNAM, Av. Universidad s/n, Circuito 2, Col. Chamilpa, Cuernavaca, CP 62210, Mor., Mexico.
Email: uoswald@gmail.com
Website: http://www.afes-press.de/html/download_oswald.html

Hans Günter Brauch (Germany), Dr., was, until 2012, Adj. Prof. (Privatdozent) at the Faculty of Political and Social Sciences, Free University of Berlin; since 1987 he has been chairman of *Peace Research and European Security Studies* (AFES-PRESS). He is editor of: *Hexagon Book Series on Human and Environmental Security and Peace*; *SpringerBriefs on Environment, Security, Development and Peace* (ESDP); *SpringerBriefs on Pioneers in Science and Practice* (PSP); *Pioneers in Arts, Humanities, Science, Engineering, Practice* (PAHSEP); and of *The Anthropocene: Politik – Economics – Society – Science* (APESS) with Springer International Publishing (Cham, Switzerland). He was a guest professor of international relations at the universities of Frankfurt am Main, Leipzig, Greifswald, and Erfurt, a research associate at Heidelberg and Stuttgart, and a research fellow at Harvard and Stanford Universities. In the autumn and winter of 2013/2014 he was a guest professor at Chulalongkorn University in Bangkok. He has been a member of the IPRA Council (1996–2000 and 2016–2020) and of its Executive Committee (2016–2018), Chairman of IPRA's Subcommission on Statutes (2016–2018), Co-convenor of IPRA's *Ecology and Peace Commission* (EPC) from 2012 to 2018 and a coeditor of its five peer-reviewed IPRA books (see below).

He has published on security, armaments, climate, energy, migration, Mediterranean issues, new concepts of security, global environmental change and climate change, the Anthropocene, and peace ecology in English and German, and his works have been translated into Spanish, Chinese, Greek, French, Danish, Finnish, Russian, Japanese, Portuguese, Serbo-Croatian, and Turkish. Recent books in English: (co-edited with Liotta, Marquina, Rogers, Selim): *Security and Environment in the Mediterranean: Conceptualising Security and Environmental Conflicts*, 2003; (co-edited with Oswald Spring, Mesjasz, Grin, Dunay, Chadha Behera, Chourou, Kameri-Mbote, Liotta): *Globalization and Environmental Challenges: Reconceptualizing Security in the 21st Century*, 2008; (co-edited with Oswald Spring, Grin, Mesjasz, Kameri-Mbote, Chadha Behera, Chourou, Krummenacher): *Facing Global Environmental Change: Environmental, Human, Energy, Food, Health and Water Security Concepts* (2009); (co-edited with Oswald Spring): *Reconceptualizar la Seguridad en el Siglo XXI* (2009); (co-edited with Oswald Spring, Mesjasz, Grin, Kameri-Mbote, Chourou, Dunay, Birkmann): *Coping with Global Environmental Change,*

Disasters and Security – Threats, Challenges, Vulnerabilities and Risks (2011); (co-edited with Scheffran, Brzoska, Link, Schilling, 2012): *Climate Change, Human Security and Violent Conflict: Challenges for Societal Stability*; (co-edited with Oswald Spring, Grin, Scheffran, 2016): *Handbook on Sustainability Transition and Sustainable Peace*; (co-edited with Oswald Spring, Bennett and Serrano Oswald, 2016): *Addressing Global Environ-mental Challenges from a Peace Ecology Perspective*; (co-edited with Oswald Spring, Bennett and Serrano Oswald, 2016): *Regional Ecological Challenges for Peace in Africa, the Middle East, Latin America and Asia Pacific*; co-edited with Oswald Spring, Collins and Serrano Oswald, 2018: *Climate Change, Disasters, Sustainability Transition and Peace in the Anthropocene.*

On behalf of the Federal President of Germany, Frank-Walter Steinmeier, on 13 July 2020 he was awarded the Federal Cross of Merit on Ribbon (*Bundesverdienstkreuz am Bande*). On the 75th anniversary of the liberation of Germany from Fascism on 8 May 2020 he established a foundation for "Peace and Ecology in the Anthropocene" that will award both local prizes for high school students in his county and an international scientific prize for peace and ecology in the Anthropocene.

Address: PD Dr. Hans Günter Brauch, Alte Bergsteige 47, 74821 Mosbach, Germany.
Email: brauch@afes-press.de
Website: http://www.afes-press.de and http://www.afes-press-books.de/

About the Contributors

Saima Sabit Ali (Pakistan), M.Sc., Urban Management and Development, PGD Environmental Design, B. Architecture, is a multi-disciplinary development professional, researcher and educator, working at present with MAES Pvt. Limited (Engineering and Planning Consultants) in the area of study and policy-planning for water resources development and management in Pakistan. Her current area of interest and work is water security in the Indus Basin. This chapter covers the first part of her (preparatory Ph.D.) research. She has worked for both private and public sector organisations, including regulatory bodies, in the areas of architecture, environmental design, urban management, development and policy-planning, and advised on academic and professional excellence in architecture and town planning. She has served as a thesis supervisor, juror, accreditation team member and inspector for Higher Education Institutions, Visiting Faculty at the University of Karachi, Pakistan. Her other roles include guest lecturer for INTBAU (Pakistan Chapter) and the Heritage Foundation; member of the Expert Working Group of Supreme Court of Pakistan for national policy recommendations for informal settlements in Pakistan; consultant, Fresh Water Action Network South Asia; urban development specialist in Capital Development Authority, Islamabad.

Address: Ms. Saima Sabit Ali, M.Sc., Head Office, MAES Pvt. Ltd. 71-K Block Gulberg III- Lahore-54660, Pakistan.
Email: ar.saimaali@gmail.com

Pietro Ameglio Patella (Mexico), Professor of "Culture of Peace and Nonviolence" at the Faculty of Philosophy and Letters (UNAM), is responsible for the UNAM Project "Education Programme for Peace and Nonviolence". He has specialised in the areas of epistemology, education, culture for peace; non-violent civil resistance; construction of violence and reconstruction of the social fabric; autonomous education; social research and databases. He is the author of the book *Gandhi and*

Ú. Oswald Spring and H. G. Brauch (eds.), *Decolonising Conflicts, Security, Peace, Gender, Environment and Development in the Anthropocene*, The Anthropocene: Politik—Economics—Society—Science 30, https://doi.org/10.1007/978-3-030-62316-6

Civil Disobedience: Mexico Today (2002); the co-editor of the anthology *How to Build Peace in Mexico Today? Texts, Authors and Questions about Construction, Education and Culture for Peace* (2016); and the co-Editor of the *Anthology of the Thought and Work of Juan Carlos Marín.* He wrote the article "Civil Resistance and Non-Violence" in the UNESCO encyclopaedia *International Security, Peace, Development and Environment* (vol. II). In 2014 he received the International Prize for Peace Education – El Hibri Foundation in Washington.

Address: Mr. Pietro Ameglio Patella, Privada Enriqueta, 15, Colonia Lotes Alegría, 62469-Cuernavaca, México.
Email: serpajc@laneta.apc.org
Website: www.serpajmx.org

Aminu Bakari Buba (Nigeria) is an emerging scholar and lecturer in the Department of History and Diplomatic Studies, Federal University of Kashere, Gombe State, where he has been based since 2012. In 2008 he became a member of the *Historical Society of Nigeria* (HSN), and he is a lifetime member of the Eurasia Research Group. Bakari has attended both local and international conferences, where he has presented papers on aspects of Environmental History in Nigeria, and published in journals. Bakari is currently undertaking his Ph.D. in the Department of History, Bayero University, Kano, Kano State, Nigeria. The author's area of research interest is Economic and Environmental History.

Address: Aminu Bakari Buba, Department of History and Diplomatic Studies, Federal University of Kashere, PMB 0182, Gombe State, Nigeria.
Email: bakariamin@gmail.com
Website: http://www.fukashere.edu.ng

Mansee Bal Bhargava (India), Ph.D., entrepreneur, researcher, and educator, is Consulting Partner at Environmental Design Consultants in Ahmedabad, India. She has recently joined Amity University Chhattisgarh in Raipur, India as a professor. She is interested in sustainability and the liveability aspects of the built environment, and makes recommendations on the architecture, design, development, planning, management, and governance of large-scale residential, recreational, institutional, and industrial developments. She teaches architecture and planning and provides guidance on Bachelor dissertations. Her research forte lies in water governance, particularly with regard to urban lakes in India.

Address: Prof. Dr. Mansee Bal Bhargava, 1102-Safal Prelude, Corporate Road, Prahladnaagr, Ahmedabad-380015, Gujarat, India.
Email: mansee.bal.bhargava@gmail.com
Website: www.mansee.in

Hans Günter Brauch (Germany), Dr., Adjunct Professor (Privatdozent) at the Faculty of Political and Social Sciences, Free University of Berlin (retired); since 1987 chairman of *Peace Research and European Security Studies* (AFES-PRESS); see above under Editors.

Address: PD Dr. Hans Günter Brauch, Alte Bergsteige 47, 74821 Mosbach, Germany.
Email: brauch@afes-press.de
Website: http://www.afes-press.de and http://www.afes-press-books.de/

Jude Cocodia (Nigeria), Senior Lecturer and Acting Head of the Department of Political Science, Niger Delta University, Nigeria, has an M.Sc. in international relations from the University of Benin, Nigeria (1999), an M.A. in philosophy from Erasmus University Netherlands (2004), and a Ph.D. from the University of Nottingham, UK (2016). He is a recipient of the International Peace Research Association Foundation award (2014) and an Associate Fellow of the Higher Education Academy, UK (2016). He is the Secretary of the Nigeria Political Science Association South-South Zone, a reviewer for *Armed Forces and Society* and the Editor of the *Wilberforce Journal of the Social Sciences*. His research interests revolve around peace, conflict, security and democracy in Africa. He has published with Routledge and contributed chapters to books published by Palgrave Macmillan and I. B. Tauris.

Address: Jude Cocodia, PO Box 1378, Yenagoa, Bayelsa State, Nigeria.
Email: jcocodia41@gmail.com
Website: https://www.linkedin.com/in/jude-cocodia-02021338/?originalSubdomain=uk; https://scholar.google.co.uk/citations?user=ebqqw_sAAAAJ&hl=en; https://iprafoundation.org/jude-cocodia/

Safiye Ateş Burç (Turkey), Dr., is a research assistant at the Department of Political Science and International Relations, Mardin Artuklu University, Turkey. She gained her Ph.D. in peace studies and conflict resolution at Ankara University and her Master's degree in political science at Hacettepe University. In order to complete her Master's thesis she worked on the political relations of the Metinan tribe in Mardin with the central authority after conducting a three-month field study. During that process (2006–2009) she was supported by the Scientific and Technological Research Council of Turkey (TÜBİTAK) with a political science scholarship. She continues to research on the subjects she focused on in her doctoral and Master's theses, and is especially interested in peace studies, traditional peacemaking and conflict resolution approaches and practices, Kurdish studies, politic anthropology and political philosophy.

Address: Dr. Safiye Ateş Burç, Mardin Artuklu Üniversitesi, Diyarbakır Yolu Üzeri Kampüs Alanı, İİBF, Kat: 1, Ofis: 12, 47100, Artuklu/Mardin, Turkey.
Email: safiyeattes@gmail.com
Website: https://artuklu.academia.edu/SafiyeAte%C5%9F, http://www.artuklu.edu.tr/tr/akademik/dr-safiye-ates-durc, https://scholar.google.com/citations?user=NnsQ4UsAAAAJ&hl=tr, https://www.researchgate.net/profile/Safiye_Ates_Durc

Robert Senath Esuruku (Uganda), Dr., is Head of the Department of Development Studies at Makerere University and a Visiting Professor of Development Studies at the University of Namibia. He is currently evaluating and redesigning Uganda's government-led affirmative framework for the recovery of Northern Uganda, *Peace Recovery and Development Plan* (PRDP). He was the Dean of the Faculty of Humanities and Social Sciences, a member of the University Governing Council and the Director of Research at Uganda Martyrs University. He has published widely in the areas of gender, governance, post-war recovery and development.

Address: Dr. Robert Senath Esuruku, Department of Development Studies, College of Humanities and Social Sciences, Makerere University, PO Box 7062, Kampala, Uganda.
Email: esuruku@chuss.mak.ac.ug or robert.esuruku@gmail.com
Website: https://www.researchgate.net/profile/Robert_Esuruku; https://www.linkedin.com/in/robert-esuruku-50b44842/?originalSubdomain=ug

David Evans Dr., OAM, is a retired medical practitioner who specialised in pathology and sexual health medicine while working in Tasmania, Australia. He was Chairman of the HIV/AIDS Committee, President of the Tasmanian Branch of the Australian Medical Association, and undertook voluntary pathology work in Nepal and East Timor. In retirement, Dr. Evans studied humanities at the University of New England (UNE), Armidale, Australia, completing a Ph.D. in 2013 entitled *Romance Tragedy of 'Romeo and Juliet Style': A Study in Qualitative Research with Special Focus on India and Nepal in the Time Period 1993-2013* – a social emergence study of domestic violence towards eloping couples. He has presented papers at the APPRA Conference in Kathmandu, the IPRA Conference in Ahmedabad, and the UNE Rethinking Peace, Conflict and Governance Conference in Sydney. His peripatetic interests include Real (Royal) Tennis, the Oriental Game of GO, Barbershop Singing and the Quaker Way of Life.

Address: Dr. David Denton Edward Evans, Apt 104/123 Fisher Street, Fullarton, South Australia, Australia, 5063.
Email: ddeevans1@gmail.com

Marcos Alan S. V. Ferreira (Brazil) is Associate Professor in the Department of International Relations at the Federal University of Paraiba – UFPB (João Pessoa, Brazil). He was Visiting Professor or Researcher at the University of Manchester (United Kingdom), Uppsala University (Sweden), Universidad Nur (Bolivia) and the University of Vechta (Germany). He is currently conducting the research "Violent Non-State Actors as a Challenge to Peace in South America", funded by the Brazilian National Council for Scientific and Technological Development (CNPq).

Address: Prof. Dr. Marcos Alan S. V. Ferreira, Avenida Monteiro da Franca, 913, Apto. 1001, Manaíra, João Pessoa/PB CEP 58038-320, Brazil.
Email: marcosalan@gmail.com
Website: https://ufpb.academia.edu/MarcosAlanSVFerreira
ORCID: http://orcid.org/0000-0002-3196-6508

Alfonso-Poncho-Hernandez Gomez (Mexico) has a Bachelor of Arts degree in philosophy from UNAM and a Master's degree in anthropology from UAM-I, where he is currently studying for a Ph.D. in anthropological sciences. He has attended international courses at the Gujarat Vidyapith in Gujarat, India (2012) and Tufts University in Boston, USA (2015), and participated in a research exchange programme with the University of Texas in Austin (2018). He has been a scholarship holder from FONCA in Mexico, CONACYT, and he is a fellow of the International Center on Nonviolent Conflict (ICNC), and a trainer for the United States Institute of Peace (USIP). He has given courses and conferences at universities in the United States, India, Colombia, Ecuador, Honduras, Costa Rica, Brazil, and various cities in Mexico.

Address: Alfonso-Poncho-Hernandez Gomez, 2da cerrada de Emilio Carranza, San Andres Tetepilco, Iztapalapa, 09440, Mexico City.
Email: ahimsamusik@gmail.com
Website: www.globalunityac.org

Egbert Jahn (Germany), Prof. Dr., Emeritus Professor of Political Science and Contemporary History at the University of Mannheim (1993–2005), was born in 1941 in Berlin and studied history, geography, political science and education in Marburg/Lahn, Berlin, and Bratislava (1961–1969), obtaining a State examination (1968) and a Ph.D. (1969) in Marburg/Lahn, where he was Assistant Professor (1969–1970). He was a researcher (1970–1974) and research director (1975–1993) at the Peace Research Institute in Frankfurt and simultaneously Professor of Political Science and Political Sociology at the University of Frankfurt (1975–1993) before becoming Professor of Political Science and Contemporary History at the University of Mannheim (1993–2005) and research area and project director at the Mannheim Center for Social Research (1993–2009). Since his retirement (2005) he has continued to lecture at the University of Frankfurt. He has also been a Visiting Professor at the University of Copenhagen (1986–1987), the University of California, Irvine (1988) and the University of Vilnius (1993).

Address: Prof. Dr. Egbert Jahn, Gagernring 96, 65779 Kelkheim, Germany.
Email: egbjahn@web.de
Website: https://en.wikipedia.org/wiki/Egbert_Jahn; https://www.mzes.uni-mannheim.de/d7/de/profiles/egbert-jahn; http://fkks.uni-mannheim.de/mitarbeiter/egbert_jahn/; https://www.uni-frankfurt.de/76982883/Egbert_Jahn

Putri Ariza Kristimanta (Indonesia) is a researcher from the Research Center for Politics at the Indonesian Institute of Sciences (LIPI), a government-funded research institution. Her research broadly focuses on peace, conflict, and security studies. She specialises in the study of Indonesia's security sector reform and governance. She is also interested in the women's peacebuilding movement and political violence. Currently, she is leading a multi-year research project on the governance of non-traditional security in Indonesia while finishing research about the democratic oversight of intelligence bodies. Before joining LIPI, she completed her Master's degree in peace and conflict resolution studies at the Indonesia Defence University in 2017. Her contribution to this book is a shortened and refined version of her thesis manuscript. She has been actively involved in the International and Asia-Pacific Peace Researcher Association (IPRA and APPRA) since 2018.

Address: Ms. Putri Ariza Kristimanta, Pusat Penelitian Politik LIPI, Widya Graha Lt. XI Jl. Jenderal Gatot Subroto, Kav. 10, DKI Jakarta 12710, Indonesia.
Email: putri.ariza@gmail.com, putr018@lipi.go.id

Katarina Marej (Germany), M.A., is a Ph.D. candidate, and a cultural anthropologist, sociologist and historian currently working on issues of citizenship education in hyper-diverse societies and the promotion of a culture of peace and sustainability. She conducts research on functions and constructions of collective identities and memories and their transformation modes and conditions in micro-macro processes. The focus is on the manifold connections between culture, politics, and education and the role of academia for socio-cultural change. She is orientated towards qualitative approaches, including interdisciplinary transformative research, semiotics, and deep-hermeneutics as well as (inter)textuality, (de-)construction, and postcolonial analyses.

Address: Ms. Katarina Marej, Institute of Political Sciences, University of Münster, Scharnhorststr. 100, 48151 Münster, Germany.
Email: katarina.marej@uni-muenster.de
Website: https://www.uni-muenster.de/IfPol/personen/marej.html

Md Nurul Momen (Bangladesh) is a professor in the Department of Public Administration at the University of Rajshahi (Bangladesh). Dr. Momen offers courses for different semesters, conducts academic research, and provides supervision and consultation on the academic activities of students and researchers. Dr. Momen has published many articles in international journals and chapters in various books dealing with a range of debates on public policy and law, governance, and public sector reform in South Asia. He has participated in many national and international seminars, workshops and conferences, and has been a member of several professional networks, journal editorial boards, research institutes, and civil society organisations. He has also supervised the research degree students in every academic year.

Address: Prof. Dr. Md Nurul Momen, Department of Public Administration, University of Rajshahi, Rajshahi-6205, Bangladesh.
Email: nurulmomen78@gmail.com (personal); nurulmomen@ru.ac.bd (work)
Website: http://103.79.117.242/ru_profile/public/teacher/21601645/profile

Amar K. J. R. Nayak (India) is a tenured full time Professor of Strategic Management at the Xavier Institute of Management Bhubaneswar, Xavier University Bhubaneswar. He engages in research, teaching, training and policy advice to various organisations and institutions, including United Nations Agencies, the Government of India, state governments, corporations, non-profit organisations and others on Transition Strategies for rebuilding Sustainable Ecological Systems. He has published six books, two monographs, and over a hundred case studies and research articles. He teaches strategic management, research methodologies, and transition strategies for sustainability. He is an alumnus of the leading engineering, technology and management institutions in India – e.g. the National Institute of Technology in Rourkela, the Indian Institute of Management in Bangalore and the Indian Institute of Technology in Kharagpur – and of Kobe University, Japan. He has travelled widely for studies, lectures, seminars and conferences, and has lived in different parts of the world, including Asia, Europe, North America and South America.

Address: Professor (Dr.) Amar K. J. R. Nayak, Xavier Institute of Management, Xavier Square, Chandrasekharpur, Bhubaneswar-751013, District-Khurda, Odisha.
Email: amar@ximb.ac.in
Website: https://xub.edu.in/cses/objective.html, http://hib.xub.edu.in/Hibiscus/Pub/faccvDet.php?client=xu&facid=XF241, https://www.researchgate.net/profile/Amar_Nayak5

Janet Monisola Oluwaleye (Nigeria) is a lecturer in the Department of Political Science at Ekiti State University, Ado Ekiti, Nigeria, where she received her Ph.D. in political science in 2014. Her research interests include development, governance, public policy, gender, political behaviour and public administration. Her research has been published in local and international journals. She has attended both local and international conferences and workshops. She is a member of IPRA and has also participated in the *American Political Science Association* (APSA) African Workshop programmes. She won an APSA African Alumni research grant in 2017.

Address: Janet Monisola Oluwaleye, Ekiti State University P.M.B. 5363, Ado Ekiti, Ekiti State, Nigeria.
Email: jmoluwaleye@gmail.com; monisola.oluwaleye@eksu.edu.ng.
Website: www.eksu.edu.ng

Mokua Ombati (Kenya) is a hands-on research anthropologist doctoral fellow affiliated to the Anthropology and Human Ecology Department at Moi University, Kenya. For his Master's degree he researched women's status and participation in Kenya's legislative leadership. His work has been published by renowned international publishers, including *Africa Development*, the longest standing Africa-based social science journal, published by CODESRIA, the apex pioneer independent Pan-African social sciences research organisation in the African continent, and Springer Nature, the second-largest global scientific publisher. His other published works are by Cambridge Scholars Publishing. Mokua has facilitated and presented at scientific gatherings locally, regionally and across the globe. He has been involved in local and international research and training, partnerships and collaborations. He maintains a concentrated focus on African Indigenous Knowledge Systems, gender, peace and development, security and non-violence, children and youth, and climate change adaptation from a socio-cultural epistemology.

Address: Mokua Ombati, Anthropology and Human Ecology Department, School of Arts and Social Sciences, Moi University Main Campus. PO Box 3900-30100, Eldoret, Kenya.

Email: keombe@gmail.com
Websites: https://mu-ke.academia.edu/MokuaOmbati; https://www.linkedin.com/in/mokua-ombati-72b71932/; https://scholar.google.co.za/citations?user=JgJsAioAAAAJ&hl=en
Phone: (+504)97862636
ORCID ID: https://orcid.org/0000-0002-4458-6731

Úrsula Oswald Spring (Mexico), full-time Professor/Researcher at the National University of Mexico (UNAM) in the Regional Multidisciplinary Research Center (CRIM), Co-Secretary General of IPRA (2016–2018); see above under 'On the Editors'.

Address: Prof. Dr. Úrsula Oswald Spring, CRIM-UNAM, Av. Universidad s/n, Circuito 2, Col. Chamilpa, Cuernavaca, CP 62210, Mor., Mexico.
Email: uoswald@gmail.com
Website: http://www.afes-press.de/html/download_oswald.html

Cahyo Pamungkas (Indonesia) has been a researcher of conflict studies at the Centre of Areas Studies, the Indonesia Institute of Sciences since 2003. At the Institute, he has conducted studies on the relationship between ethno-religious groups in South-East Asia and Eastern Indonesia, mainly Maluku and Papua. In 2005 he received the Asia Public Intellectual (API) fellowship from the Nippon Foundation to study "The Effectiveness of the Autonomous Region in Muslim Mindanao (ARMM) in Coping with Separatism and the Role of the National Reconciliation Commission of Thailand in Peace Building". In 2008 he graduated with a Master of Science degree in sociology from the University of Indonesia. Between 2005 and 2008 he was assigned by the Institute to study separatist conflict in Papua and became co-author of a model for conflict resolution called "Papua Road Map". In 2015 Cahyo finished his Ph.D. in social sciences at Radboud University in Nijmegen, the Netherlands, under the Ethno-Religious Conflict in Indonesia and the Philippines Programme.

Address: Mr. Cahyo Pamungkas, Indonesia Institute of Sciences (Lembaga Ilmu Pengetahuan Indonesia), Widyagraha Building 8th Floor, General Gatot Subroto Street No. 10, Jakarta Selatan, Jakarta, Indonesia, 12710.
Email: cahyopamungkas@gmail.com

Esteban A. Ramos Muslera (Honduras/Argentina/Spain) Ph.D. in Political Science from the University of Valladolid (U. Va), Master's degree in Participatory Research for Local Development from the Complutense University of Madrid (UCM) and Political Science graduate from the same university. He is currently Coordinator of the Peace Area of the University Institute for Democracy, Peace and Security (IUDPAS) at the National Autonomous University of Honduras (UNAH), Council member of the International Peace Research Association (IPRA), Co-Secretary General of the Latin American Council for Peace Research (CLAIP), and Director of the Latin American Journal of Peace and Conflict Studies.

His latest scientific publications include: (2020) “Investigación y formación nacional e internacional para la paz transformadora en la Universidad Nacional Autónoma de Honduras”, *Revista Utopía y Paxis Latinoamericana, año 25, 90, 264–281*, http://doi.org/10.5281/zenodo.3872576; “Práxis participativa de la paz transformadora: teoría y método”, *in: Pax Crítica. Aportes teóricos a las perspectivas de paz posliberal (Madrid: Tecnos, 2019)*: 425–471; *Guía de Capacitación en Paz y Violencia.* Tegucigalpa: IUDPAS-UNAH; and “La participación del estudiantado en el diseño curricular de la Educación para la Paz Transformadora” (2019), *OBETS. Revista de Ciencias Sociales*, 14 (2), 437–470, http://doi.org/10.14198/OBETS2019.14.2.06.

Address: Prof. Dr. Esteban A. Ramos Muslera, Edificio IUDPAS, Ciudad Universitaria, Tegucigalpa MDC, 11101, Francisco Morazán, Honduras.
Email: esteban.ramos@unah.edu.hn, esteban.ramos@gmail.com
Website: https://unah.academia.edu/EstebanRamos; https://scholar.google.es/citations?user=A093qm8AAAAJ%26hl=es;https://iudpas.unah.edu.hn/area-de-paz/sobre-el-area-de-paz; https://iudpas.unah.edu.hn/area-de-paz/revista-latinoamericana-estudios-de-la-paz-y-el-conflicto

Maria Subert (USA), Ph.D., is Doctor of Communication Studies (rhetoric and public culture; related area: media) and Assistant Professor at the City University of New York, Hostos Community College. She has studied at Eotvos Lorant University, Hungary; Goethe University, Germany; and Scripps College of Communication, Ohio University, United States. Her research focus is moral and ethical communication, social justice, peace and conflict, communication among cultures, and communicating with marginalised communities.

Address: Prof. Dr. Maria Subert, City University of New York, Hostos Community College, 450 Grand Concourse, C-417, Bronx, NY 10451, USA.
Email: subert.ma@gmail.com
Website: https://cuny.academia.edu/MariaSubertPhD
ORCID iD: https://orcid.org/0000-0002-2836-0564

Rajib Timalsina (Nepal) is assistant professor at the Department of Conflict, Peace and Development Studies (CPDS), Tribhuvan University. He is a governing council member of the International Peace Research Association (IPRA) for the 2016–2020 term. Mr. Timalsina is a non-violent communication practitioner and co-director of a research project on learning assessment in Nepal. He is a co-investigator on the Changing the Story Phase 2 Project and the Mobile Arts for Peace Research Project supported by AHRC-UK.

Address: Ass. Prof. Rajib Timalsina, Department of Conflict, Peace and Development Studies (CPDS), Tribhuvan University, General Post Box No. 9434, Gyaneshwor, Kathmandu, 44600, Nepal.
Email: rajib.timalsina@gmail.com
Website: https://www.linkedin.com/in/rajib-timalsina-34ba8460

Professor Helen Ware (Australia/UK) is currently the foundation Professor of Peace building at the University of New England in Armidale, New South Wales, and leader of the Peace Studies Centre, with more than twenty international Ph.D. students studying peace, and environmental and gender issues across Africa and Asia. She has studied history, economics and sociology, gaining a Ph.D. from the University of London in 1970. She ran a social statistics/teaching change project across twelve African countries whilst conducting research at the Australian National University, Canberra until 1981. She directed research and projects for the Australian Human Rights Commission from 1981 to 1985 before moving to AusAID, the Government's aid agency. From 1986 to 1989 she was Australian High Commissioner to Zambia and Malawi, and Ambassador to Angola, liaising with the liberation movements in Southern Africa before returning to AusAID in charge of evaluations. In 2002 she returned to academia, drawing upon her experience for peacebuilding research with an African focus, chairing the University's Ethics Committee and teaching/mentoring research students. She now specialises in training Ph.D. students from countries in transition in peace studies, covering aspects ranging from gender issues to terrorism and refugee studies.

Address: Professor Helen Ware, Peace Studies Centre, University of New England, Armidale, New South Wales, Australia, 2351.
Email: hware@une.edu.au
Website: http://une.edu.au/staff-profiles/hass/hware

Charles Ndalu Wasike (Kenya), Ph.D., is the current head of the *African Peace Research and Education Association* (AFPREA), an Africa chapter of the International Peace Research Association. He specialises in matters concerning peacebuilding. He has widely written on the subject and presented on it at international conferences. Wasike is a lecturer at Moi University in Kenya and at the adjunct faculty at the National Aviation College in Addis Ababa, Ethiopia. He lectures and consults on matters of finance, risk management, peace and development.

Address: Dr. Charles Ndalu Wasike, Moi University, PO Box 24404, GPO00100 Nairobi. Kenya.
Email: ndaluwasike2006@yahoo.com
Website: www.mu.ac.ke

Index

Ú. Oswald Spring and H. G. Brauch (eds.), *Decolonising Conflicts, Security, Peace, Gender, Environment and Development in the Anthropocene*,
The Anthropocene: Politik—Economics—Society—Science 30,
https://doi.org/10.1007/978-3-030-62316-6

H

J

K

The manufacturer's authorised representative in the EU is Springer Nature Customer Service Centre GmbH, Europaplatz 3, 69115 Heidelberg, Germany. If you have any concerns regarding our products, please contact ProductSafety@springernature.com

Printed and bound by CPI Group (UK) Ltd, Croydon, CR0 4YY
22/09/2026
02220979-0017